Discovering the Deep

A Photographic Atlas of the Seafloor and Ocean Crust

The deep oceans and global seafloor are truly Earth's last frontier. They remain largely unexplored, yet are critical to our survival on this planet. This magnificent volume transports you to volcanoes, boiling hot springs, and fault zones in mountains a mile beneath the ocean surface, where bizarre landscapes host exotic life forms that rival the most imaginative science fiction.

Illustrated throughout with spectacular color photographs, the book starts with a historical summary of seafloor exploration and the developing technologies used to study this extreme environment. It then presents a series of chapters that describe the distinctive geological components of the Earth's ocean floor and the myriad environments found along mid-ocean ridges where new crust is created. It is here that discoveries of chemosynthetic biological communities revolutionized our ideas about the origins of life and the potential for life's existence elsewhere in the Universe. The authors provide elegant explanations of how the various geological and hydrothermal features of the seafloor are formed through physical, chemical, and biological processes, and present comparative examples from places where the seafloor is exposed on land.

Discovering the Deep is an indispensable reference for researchers, teachers, and students of marine science, and a visually stunning resource that will enlighten and intrigue oceanographers and enthusiasts alike. A suite of online resources, including photographs and video clips, combine with the book to provide fascinating insights into the hidden world of seafloor geology and biology using the latest deep-sea imaging and geological concepts.

Jeffrey A. Karson is Professor of Geology in the Department of Earth Sciences at Syracuse University and a Fellow of the Geological Society of America. He investigates the interplay of magmatic construction and mechanical deformation of extensional and transform fault environments from ophiolites, contemporary oceanic crust, Iceland, continental rifts, and continental margins. Professor Karson has participated in numerous field projects and seafloor research cruises worldwide. He is the author or co-author of over 150 publications on seafloor spreading and related phenomena.

Deborah S. Kelley is a Professor in the School of Oceanography at the University of Washington and also Associate Director of Science for the first US underwater high-power, high-bandwidth cabled observatory, funded by the NSF. She is the former Chair of the UNOLS Deep Submergence Science Committee. Her research examines the linkages between submarine volcanoes, underwater hot springs and how they support life in the absence of sunlight. Dr. Kelley has participated in over 35 blue water cruises routinely using robotic vehicles, and has been on over 50 *Alvin* deep-sea submersible dives, to depths as great as 4000 meters.

Daniel J. Fornari is a marine geologist and Senior Scientist at the Woods Hole Oceanographic Institution (WHOI). He was the first Chief Scientist of Deep Submergence at WHOI, a past Director of WHOI's Deep Ocean Exploration Institute (DOEI), and was a Chair of the Ridge 2000 Program, a NSF-funded national program studying mid-ocean ridges. Dr. Fornari is the author or co-author of over 100 research publications and has participated in over 100 research cruises and as many dives in deep-sea submersibles. His research focuses on volcanic and hydrothermal processes along the global mid-ocean ridge, using a variety of surface and deep-towed sonar systems, submersibles, ROVs, and AUVs.

Michael R. Perfit is Professor of Geological Sciences at the University of Florida, where he has received several awards for excellence in teaching. He has co-authored more than 100 scientific publications on volcanism and magma genesis at oceanic spreading centers and seamounts. Dr. Perfit has participated in 30 oceanographic research cruises and taken more than 35 dives in *Alvin*. He has served as the Chair of the Deep Submergence Science Committee, and is also a Fellow of the Geological Society of America and the American Geophysical Union, and a Trustee of the Consortium for Ocean Leadership.

Timothy M. Shank is an Associate Scientist in the Biology Department of the Woods Hole Oceanographic Institution and Director of WHOI's Ocean Exploration Institute. His research focusses on ecological and evolutionary processes that affect ecosystem structure, faunal evolution, and biodiversity in the deep sea. Dr. Shank has conducted more than 57 scientific expeditions, using a wide variety of submersibles, ROV and AUV systems (including more than 50 *Alvin* dives). He is currently the Director of the Hadal Ecosystems Studies Program, pioneering the first research use of the full-ocean depth hybrid ROV (*Nereus*) in the deepest trenches.

Praise for
Discovering the Deep

"Though it's fashionable to say we know little about the mysteries of the deep, this book shows the vast wealth of understanding that pioneering researchers have already gleaned, with their probing sound waves, persistent robots and courageous submariners. This is the book I wish I'd had on my eight deep ocean expeditions, to better understand the wonders I was gazing upon. A must-own for anyone in the ocean sciences, and for those simply curious about what lies down there in the most remote realm on our planet."

James Cameron, explorer and film-maker

"*Discovering the Deep* will open your eyes to the largest and most unexplored region on Earth – the global Mid-Ocean Ridge. My own introduction to the Mid-Ocean Ridge began with the FAMOUS Expedition to the Mid-Atlantic Ridge in 1974, and this beautifully illustrated and comprehensive account shows how far we have come over the last 40 years in our understanding of this fundamental tectonic feature of the Earth and the technology required to investigate it. Not since Bruce Heezen and Charlie Hollister's classic book, *The Face of the Deep*, first published in 1971, have we seen such a comprehensive photographic atlas of what lies in the hidden depths of the sea."

Dr. Robert D. Ballard, President of the Ocean Exploration Trust

"*Discovering the Deep* is an exquisite synthesis of the complexity of natural processes, the beauty of oft unseen environments, and the critical scientific discoveries of our recent past. To read this volume is to become an enlightened traveler into the fascinating realm of the seafloor and ocean crust. No student of the natural world should be without this magnificent atlas of our planet's richly structured deep ocean environment."

Dr. Kathryn D. Sullivan, Under Secretary of Commerce for Oceans and Atmosphere and NOAA Administrator; former NASA astronaut

Discovering the Deep

A Photographic Atlas of the Seafloor and Ocean Crust

JEFFREY A. KARSON
Syracuse University

DEBORAH S. KELLEY
University of Washington

DANIEL J. FORNARI
Woods Hole Oceanographic Institution

MICHAEL R. PERFIT
University of Florida

TIMOTHY M. SHANK
Woods Hole Oceanographic Institution

CAMBRIDGE
UNIVERSITY PRESS

CAMBRIDGE
UNIVERSITY PRESS

University Printing House, Cambridge CB2 8BS, United Kingdom

Cambridge University Press is part of the University of Cambridge.

It furthers the University's mission by disseminating knowledge in the pursuit of education, learning and research at the highest international levels of excellence.

www.cambridge.org
Information on this title: www.cambridge.org/9780521857185

First published 2015
Reprinted 2015

Printed in the United Kingdom by Bell and Bain Ltd, Glasgow

A catalog record for this publication is available from the British Library

Library of Congress Cataloging in Publication data

Karson, Jeffrey A.
Discovering the deep a photographic atlas of the seafloor and ocean crust / Jeffrey A. Karson, Syracuse University, New York, Deborah S. Kelley, University of Washington, Daniel J. Fornari, Woods Hole Oceanographic Institution, Massachusetts, Michael R. Perfit, University of Florida, Timothy M. Shank, Woods Hole Oceanographic Institution, Massachusetts.
pages cm
1. Ocean bottom–Pictorial works. 2. Earth (Planet)–Crust–Pictorial works. I. Title.
GC87.K37 2015
551.46′800222–dc23 2014028680
ISBN 978-0-521-85718-5 Hardback

Additional resources for this publication at www.cambridge.org/discoveringthedeep

Contents

Foreword

An alien approaching Earth for the first time would observe a watery planet but would have no clue as to what lies beneath the blue-gray liquid surface. This book unveils and illuminates, in ways never before possible, that hidden world and the astounding processes that occur at the seafloor and within the oceanic crust. Drawing from unparalleled imagery and data from twenty-first century technology, which includes human-occupied and robotic vehicles, sophisticated sensors and analytical techniques, the authors have succeeded in "pulling the plug" on the global ocean so that the awe-inspiring terrane of the ocean floor and processes that form it are revealed.

It was 44 years ago that Bruce Heezen and Charles Hollister published their seminal book, *The Face of The Deep*, an engaging and informative text that used analog seafloor photographs to highlight and interpret the myriad processes operating in the global abyss. *Discovering the Deep* sets a new benchmark for a comprehensive treatise about visualizing the seafloor, how ocean crust forms, and the amazing life forms that thrive in one of the most extreme environments on Earth – submarine hot springs. It presents up to date, integrated multidisciplinary concepts about the mid-ocean ridge and its role in plate tectonics and developing the architecture of the oceanic crust. The authors also synthesize published data from mid-ocean ridge analogs on land – ophiolites – to better understand how oceanic crust is generated and the far-reaching implications of these processes for interpreting Earth and ocean history.

As a student of Heezen, a contemporary of Hollister, and someone who has spent most of his career studying the mid-ocean ridge using ships and the submersible *Alvin*, I was drawn into all aspects of this compelling and masterful synthesis, especially the high-resolution images of mid-ocean ridge volcanic and hydrothermal terranes and biota. The more than 500 stunning images of the seafloor, many never before published, were taken using the latest in deep-sea digital camera and lighting technology. The remarkable images provide glimpses of underwater eruptions where molten rock turns to glass almost instantly, and black-smoker vents where the most exotic life forms known on Earth live. Spanning environments from the Arctic, Atlantic, Pacific, and Indian Oceans, the authors provide captivating images of these seafloor oases, which host novel microbes that use toxic gases rising from Earth's interior as their energy source.

The authors have succeeded in presenting a holistic view of the oceanic crust from top to bottom. They present the current state of knowledge on the complex and beautiful forms of volcanic flows that pave the mid-ocean ridge crest to the massive basal layers of intrusive, plutonic, and ultramafic rocks that until recently have been difficult to sample and place in their proper geological context. The variations in style and structural complexity of these primary units are now known to be critical to the construction of all mid-ocean ridges, from ultra-slow to fast, and the hydrothermal systems and life that they support. This work ends with an important look towards the future and new technologies that will be required to address upcoming opportunities and challenges in understanding the vast oceans that make Earth habitable.

The five authors of this atlas are all recognized as individuals who have made significant contributions to our understanding of the mid-ocean ridge system. They have succeeded in synthesizing the many different, but complementary data sets – weaving together a text that elegantly takes the reader on a journey from hydrothermal vents down through the oceanic crust. They tell a story that is entertaining, illuminating, and scholarly. For a layperson interested in learning more about fundamental planetary process, a student of marine geology just embarking on a career, or a seasoned investigator of oceanic crustal processes, this atlas of the seafloor and oceanic crust is sure to expand horizons and enrich understanding of how our planet works for all readers.

PAUL J. FOX
Professor Emeritus
Department of Geology & Geophysics
Texas A&M University
McLeod, Montana

Preface

New Views of Earth from Below the Oceans

How inappropriate to call this planet Earth, when it is quite clearly Ocean.
Arthur C. Clarke (1957)

Covering 70% of our planet and reaching depths of nearly 11,000 meters (36,000 feet), the ocean is truly the last frontier on Earth. In this dark and largely unexplored environment the potential exists for profound scientific discoveries. Despite the opaque blue mask of the oceans as depicted on many maps, the Earth's crust extends continuously beneath the ocean surface. Here, the Earth beneath the sea is marked by spectacular landforms created by distinct geological phenomena and inhabited by exotic life forms (Figs. P.1 and P.2).

Since light does not penetrate the ocean depths, imaging and exploring the vast seafloor environment presents special challenges. The global seafloor is now coming into detailed focus as a result of the application of new remote-sensing technologies (Fig. P.3). In this book, we explore this expansive, hidden, and largely inaccessible part of our planet – the oceanic crust and the mid-ocean ridge system where magmatic, tectonic, and hydrothermal processes continuously generate the most extensive volcanic terrane on Earth. This work examines the links between submarine volcanoes and the life that thrives in this extreme environment. At the mid-ocean ridge crest, chemosynthetic ecosystems provide the closest analogs to life on early Earth and offer insights into how life began on our planet. This has implications for the search for life elsewhere in the solar system and beyond.

Ask anyone to name Earth's most prominent natural features, and you will get some familiar answers: the Grand Canyon, Mt. Everest, the Amazon River. Ask an oceanographer or marine geologist, and you will get a single response: the mid-ocean ridge. Discovered and initially

Figure P.1 The deep ocean hosts astonishing life forms that thrive in some of the harshest conditions on Earth – the deep seafloor. At 1500 m depth (4920 ft) beneath the ocean surface a 0.5 m diameter "Dumbo" octopus traverses a lobate lava flow on Axial Seamount, Juan de Fuca Ridge.

mapped just 60 years ago, this ~60,000 km long chain of active volcanoes, rift valleys, and fault zones is the planet's most dynamic and continuous tectonic feature, and it is here where most of the volcanism on Earth occurs (Figs. P.4 and P.5). Recognition of the mid-ocean ridge paved the way for placing the historical concept of "continental drift" into a global tectonic context. This led to the development of the plate tectonics paradigm, which provided the foundation for understanding Earth's evolution and the effective exploration of its mineral, hydrocarbon, and biological resources.

The mid-ocean ridge is an essential research environment for the twenty-first century, where investigations will continue to change our concepts of biological and biogeochemical processes. Less than 40 years ago, submarine hot springs (hydrothermal vents) were found along the mid-ocean ridge at the Galápagos Spreading Center in the eastern equatorial Pacific. That discovery revolutionized biological and oceanographic sciences, revealing for the first time the bizarre assemblages of animals and symbiotic microbes that thrive in the dark, high-pressure abyss in an exotic chemical system. It is within this alien environment that chemosynthesis was first discovered, a remarkable biochemical process that allows organisms to obtain their energy not from sunlight but from volcanic gases. Its occurrence in the deep ocean has led to important new research, some of which has significant implications for biomedical and industrial applications. There is no doubt that the next 50 years of oceanographic research will yield new discoveries that have the potential to fundamentally alter the understanding of evolution of life on Earth and how that knowledge can be used to benefit society.

Most of the seafloor lies at depths greater than several thousand meters (more than a mile) beneath the ocean's surface. This environment contrasts starkly with familiar features on land. It is a place with no sunlight; light is absorbed by the upper few hundred meters of ocean water, effectively casting the deep seafloor in perpetual darkness. The temperature in the deep ocean is typically only a few degrees above the freezing point of water. Unlike our airy atmosphere, the ambient environment in the deep ocean is defined by denser, liquid water with crushing pressures of hundreds to over a thousand atmospheres, which equates to tons per square inch. The ocean basins contain incredibly flat abyssal plains hundreds of kilometers wide, interrupted by rugged mountainous features and deep canyons with vertical relief of thousands of meters. Much of the deep seafloor is buried or partially covered by a nearly continuous layer of tiny carbonate or silicic shells of organisms that once lived in warm, sunlit surface waters and then slowly rained down on to the seafloor below. This muddy blanket remains nearly undisturbed because there is no rainfall, rivers, or glaciers to disrupt it (Fig. P.6).

From this quiescent sediment-covered abyss emerge the jagged mountainous features of the most volcanically and seismically active part of our planet: the mid-ocean ridges.

Figure P.2 Frozen lava columns mark the drain out and subsequent cataclysmic collapse of lava lake ceilings formed during ponding of 1200 °C (2192 °F) melt at the MOR. At a mile or more beneath the ocean surface, the near freezing seawater rapidly quenches the molten lava, forming glass-covered flows that reflect the dynamic nature of submarine eruptions.

These broad and rugged volcanic mountain ranges snake their way around the Earth in a complex, interconnected network of fault zones, rift valleys, and volcanic ridges. This mountain range is unlike those on the continents. Mountain belts on land are mostly formed by violent collisions that occur as continental masses impact upon one another as a consequence of the movement of tectonic plates. This involves the titanic "bulldozing" of continental basement rocks and overlying sedimentary rocks into complexly faulted and folded belts of metamorphic rocks that are uplifted, even as erosion works to cut them down. In stark contrast, mid-ocean ridges are formed by processes related to the divergence and pulling apart of the rigid outer shell of the Earth: the lithosphere. The magmatism and faulting at these ridges create new swaths of the Earth's surface in a process referred to as "seafloor spreading." The ridges are elongate features that are not so much uplifted as they are left as high-standing regions flanked by older, cooler seafloor that has subsided as it ages and is transported away from either side of the spreading axis. Through much of Earth history, seafloor spreading has created about 60% of Earth's crust, with older lithosphere recycled, or "subducted," back into the mantle through plate tectonics at active margins where deep-sea trenches occur. Thus, seafloor spreading is a primary process on a planetary scale.

Across most of the seafloor, life forms familiar to humans do not exist; there are no forests, savannahs, or tundra with their characteristic ecosystems. The absence of light penetration means organisms that rely on photosynthesis cannot thrive on the deep-ocean floor. Without light to support the base of the food chain, larger organisms such as fish and

Figure P.3 The remotely operated vehicle *ROPOS* surfaces in the Pacific Ocean after a dive to 2500 m (8200 ft) on the Juan de Fuca Ridge. Robotic vehicles similar to this one explore the depths using high-bandwidth fiber-optic cables, thereby permitting detailed exploration of over 98% of the global seafloor down to 6500 m (21,320 ft) depth.

Figure P.4 For more than 4 billion years the oceans and the volcanic processes within them have co-evolved. The plate boundaries and locations of MORs today are vastly different from where they were located in early Earth history. This present-day view shows the age of the ocean basins based on magnetic data. It highlights the spreading centers that wrap around the globe like the seams on a baseball (thin black lines within the red colors that denote the youngest crust). This dynamic system forms the largest volcanic mountain chain on Earth and hosts the youngest seafloor.

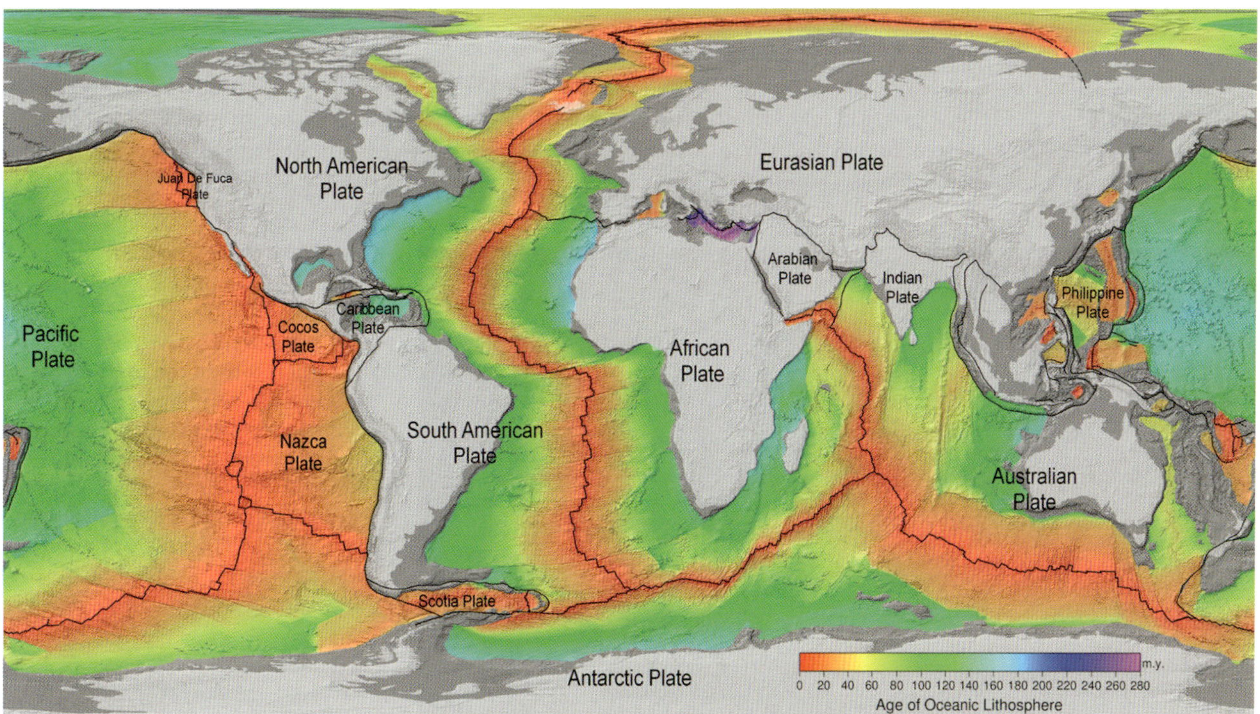

Figure P.5 Mid-ocean ridges are spectacular features exemplified by seafloor spreading processes that create major rift zones or valleys along the spreading axes. Largely hidden beneath the darkness of the sea, these areas were historically difficult to image. In places such as Iceland, however, where volcanism has been ongoing for millions of years, the volcanoes that make up the island are considered to be the subaerial continuation of the Mid-Atlantic Ridge. The rift shown here is part of a spreading system that separates the North American Plate from the Eurasian Plate.

marine mammals find slim pickings at depth. But within this unlikely place, significant biomass exists, and it is where Earth's life probably first appeared. Most of the life forms inhabiting this remote ecosystem are believed to be micrometer-sized microbes living in cracks and crevices in the rocks beneath the seafloor.

Nowhere is the connection between volcanism and life more evident than along mid-ocean ridges, where hot springs at over 300 °C (572 °F) host some of the most amazing life forms on the planet (Fig. P.7). These include single-cell organisms that thrive at temperatures up to 122 °C (251 °F), and vibrant red tubeworms, 3 m long, which grow at rates of 50 cm/yr and contain hemoglobin in their blood – the same compound found in human blood. Hydrothermal vents are sites where superheated seawater, modified by chemical reactions with hot subsurface rocks, gushes out from the seafloor. Entrainment of volcanic gases and metals leached from the surrounding rocks into the

Figure P.6 Sediment-covered abyssal plains make up more than 50% of the Earth's surface. These low-relief areas preserve a remarkable historical record of the evolution of the ocean basins, climate change, and mountain formation and demise. In this nutrient-poor environment, animals such as brittle stars forage on organic particles that rain down from above, as well as on plankton and bacteria.

Figure P.7 The discovery of high-temperature hot spring systems in 1979 was made at a MOR spreading center, the East Pacific Rise at 21° N. These high-temperature vents issue jets of metal-rich superheated fluids from the seafloor and are one of the most profound scientific discoveries ever made. Within these extreme environments, characterized by darkness and hundreds of atmospheres of pressure, life was found to not only survive, but flourish. The lush and vibrant communities of tubeworms, sprouting bright red plumes, in areas typically devoid of much color was a startling discovery. Equally surprising was the presence of microorganisms thriving, in the absence of sunlight, by metabolizing volcanic gases at temperatures in excess of 90 °C (194 °F). With the advent of genomic sequencing, we are now gaining insight into the vast diversity of microbes that live on and within the seafloor.

Figure P.8 One of the most spectacular events in crustal formation is the release of billions of microbes from the seafloor during and following seafloor volcanic eruptions. The microbes form whitish sulfide-rich particles that are ejected in dense, billowing streams from the seafloor – called snowblowers, such as the one imaged here following the 2011 eruption on Axial Seamount.

hydrothermal fluids provides nutrients that sustain the microbial base of a "chemosynthetic" food chain. Larger organisms feeding off these microbes are forced to live close to the hydrothermal vents where their food source is concentrated. Perhaps even more spectacular are dramatic outpourings of massive white microbial biomass from the crust during and following seafloor volcanic eruptions. These are reminiscent of blizzards of snow (Fig. P.8). Many of these microorganisms are archaea that optimally grow at temperatures in excess of 80 °C (176 °F). They are thought to be some of the most ancient of extant organisms. Understanding the processes occurring in the dynamic environments along mid-ocean ridges is key to comprehending not only the origins of the seafloor, but also life on our planet.

Highlighted by hundreds of high-resolution seafloor images, this book explores the formation of the oceanic crust at mid-ocean ridge spreading centers and the diverse processes occurring there. They include: hydrothermal phenomena, volcanism that creates the uppermost layer of ocean crust, crystallization and flow of magma to form the lowermost layers of the ocean crust, and tectonic processes that create the distinctive geology and physical properties of oceanic lithosphere. The compelling stories of how new ideas about mid-ocean ridges developed provide the background to understand the importance of this branch of science and an appreciation of the complexity of the diverse processes occurring in the deep ocean and beneath the seafloor.

The future of Earth's inhabitants unquestionably relies on the oceans and the Earth beneath them. Marine scientists using twenty-first-century oceanographic techniques are poised to explore and develop a deeper understanding of these scientific frontiers. Through these studies new insights into oceanic crustal evolution will emerge; from creation at mid-ocean ridges to destruction beneath subduction zones and the life that thrives, survives, and expires within these environments. Join us in our exploration to discover the deep.

Acknowledgments

We conceived this book with the goal of engaging a wide readership and exposing them to the mysteries and fantastic visual perspectives we now have of deep-sea processes and features at the mid-ocean ridge. Our objective was to unveil the deep seafloor and the details of how oceanic crust is generated so that its beauty and complexity could be revealed and appreciated, for both its aesthetic and scientific value. This book represents the compilation and synthesis of a vast amount of information collected over the past 60 years by our mentors, colleagues, and scientific associates here in the USA and internationally, and by each of us during the nearly 150 years, collectively, we have been involved in studying the seafloor and ocean crust.

Each of us owes a debt of gratitude to numerous individuals who facilitated access to imagery, as well as field and geochemical data that are crucial to the stories we present about how the oceanic crust forms and the many interrelated processes associated with seafloor spreading. The images presented in this book were taken from numerous research submersibles, HOVs, ROVs, AUVs, and towed digital camera systems. We have utilized imagery from a variety of sources. They include 35 mm images acquired by these vehicles throughout the 1990s and, over the past decade, high-definition images taken with more advanced imaging systems, which used the latest advances in deep-sea metal halide and LED lighting employed by scientists, engineers, and filmmakers. We have identified the source of the imagery and at the end of the book provide detailed credit information, and, where possible, we have included the vehicle type, facility operator, engineering or film group that provided the images, and, where appropriate, the funding agency that supported the research. The images we present were taken primarily with the US submersible *Alvin* (WHOI), the French *Nautile*, and Russian *MIRs*; US ROVs *Jason* (WHOI), *Tiburon* (Monterey Bay Aquarium Research Institute, MBARI), *Hercules* (University of Rhode Island), the Canadian *ROPOS* (Canadian Scientific Submersible Facility), German *Quest* (MARUM, University of Bremen), Norwegian *Bathysaurus* (University of Bergen), and the WHOI *TowCam*, a towed digital camera system. We are indebted to our colleagues, the vehicle pilots, and vessel and deep-submergence system operators for their collaboration and generosity in making these extraordinary images available to us.

We also acknowledge the numerous colleagues who helped us with key aspects of this work, reading and commenting on early drafts of chapters, and encouraging us to finish this volume despite the nearly 10-year gestation period it took to complete. Steven Ross provided important editorial comments on initial drafts of all chapters and we are very grateful for his input and encouragement as we finished the book. Lonny Lippsett also provided editorial advice on early drafts of various chapters, and we are grateful for his insights and guidance. Janyne Ste. Marie expertly indexed the entire book. The following individuals deserve our personal and professional appreciation: Jean-Marie Auzende, Robert Ballard, Rodey Batiza, Francoise Boudier, Christian Broski, Dave Butterfield, James Cameron, Suzanne Carbotte, David Caress, John Chadwick, William W. Chadwick, Jr., Walter Cho, Anna Cipriani, David Clague, Lawrence Coogan, James Cowen, John Delaney, Alden Denny, Cornel de Ronde, Colin Devey, John Dewey, Henry Dick, Dolly Dieter, Margo Edwards, Mitch Elend, Don Elthon, Bob Embley, David Epp, Javier Escartín, P. Jeff Fox, Gretchen Früh-Green, Allison Fundis, Allison Gale, David Garrison, Denny Geist, Chris German, Kathy Gillis, Marguerite Godard, Hunter Hadaway, Karen Harpp, Nick Hayman, Rachel Haymon, Taylor Heyl, Jim Holik, Susan Humphris, Steve Hurst, David Jousselin, Emily Klein, Ken Kostel, Marv Lilley, Lonny Lippsett, Amanda Loman, Richard Lutz, Ken Macdonald, Brian Midson, D. Jay Miller, John Mutter, Adolphe Nicolas, Mark Olsson, Rolf Pederson, Sven Petersen, Meagan Pollock, Barbara Ransom, Volker Ratmeyer, Ian Ridley, Ken Rubin, Bill Ryan, Keith Shepherd, Alexander Shor, George Shor, John Sinton, Debbie Smith, Matt Smith, S. Adam Soule, Mark Stoermer, Philip Taylor, Maurice Tivey, Meg Tivey, Maya Tolstoy, Doug Toomey, Robert Varga, Karen Von Damm, Rachel Walters, Dorsey Wanless, William Wilcock and Maria Wilhelm.

The archives at both Woods Hole Oceanographic Institution and the Scripps Institution of Oceanography were instrumental in providing historical images, and many other individuals provided us with the excellent illustrations from their publications.

Partial funding to help support the collaboration that produced this book was provided by Cambridge University Press, The National Oceanic and Atmospheric Administration's Ocean Exploration Program, The National Science Foundation, Syracuse University, The University of Florida – Geological Sciences, The University of Rhode Island – Institute for Exploration, The University of Washington – School of Oceanography, and The Woods Hole Oceanographic Institution and its Deep Ocean Exploration Institute.

Finally, we are grateful for the patience and commitment by our publisher Cambridge University Press and Susan Francis, who has encouraged us throughout this too-long process, and her colleagues Laura Clark, Zoë Pruce and Jessica Murphy, who helped edit and produce the book.

JAK thanks the collaborators, post-docs, and graduate students who shared so much time in the field and at sea investigating black rocks in cold, wet places. Thanks also to those mentors who provided introductions and encouragement in areas that contributed so much to this book: John Dewey and Bill Kidd (ophiolites and kinematics), Jeff Fox (seafloor geology), Bruce Rosendahl (East African Rift), Kent Brooks (East Greenland), and Kristján Sæmundsson and Bryndís Brandsóttir (Iceland). My undying gratitude goes to Eileen and my family for their love, encouragement, and infusions of energy in support of my efforts.

DSK offers special thanks to her family, Steve, Ann, and John, and especially to her mother, Hazel, who supported her during her long absences at sea and during her long onshore hours finishing this book. DSK also offers a special thank you to long-time mentor, collaborator, and friend Dr. John Delaney, for his scientific and educational vision that led to the collection of many of the images used in this book and the science behind them.

DJF is grateful for the love and support of his family, CL, Sasha, Simon, and Riley, and their understanding of the long absences at sea and in the lab needed to accomplish his work, and his brother Jim, and his parents who always were proud of his research accomplishments.

MRP thanks his colleagues, post-docs, and students at the University of Florida for their collaboration and support. He particularly wants to thank his family, Renee, Carol, Ryan, and Melissa, for their love and support, and understanding the many dinners that were missed. He is also grateful to the outstanding professors at St. Lawrence University who first got him hooked on geology and gave him the encouragement to attend graduate school. The support and guidance provided by Doc Ewing, Bruce Heezen, Bob Kay and Ian Ridley were invaluable during his graduate career. Lastly, he is grateful to his late grandfather, Sol, who always lived by the ocean and thought it was a good idea for him to attend college.

TMS extends his deep appreciation to his mother; for her strength, unconditional love and support in all aspects of his career and life, and to his father, who showed him the gift of following life's passion and the levity to persevere through challenging times. He expresses immense gratitude and love for his two brothers Michael and Mark, and for the love and understanding of his two daughters, Emma and Callie, and wife Leslie, who supported him during his combined four years of research and discovery at sea. Thanks also goes to the many mentors, colleagues, and members of his laboratory who supported and enriched his research and life over the years and not only made this book possible but a living entity.

1 Entering the abyss: oceanographic technology

The sea, once it casts its spell, holds one in its net of wonder forever.

J.-Y. Cousteau (1953)

1.1 First encounters

The fantasy of every marine geologist is to "pull the plug" on the oceans – to drain the water from the 70% of Earth's surface hidden beneath the waves and have a direct view of the seafloor's myriad shapes and structures. Over the last 75 years, scientists and engineers have taken advantage of innovative technologies (Figs. 1.1 and 1.2) and analytical methods to produce detailed images and geophysical

Figure 1.1 The deep-diving submersible *Alvin* preparing to dive to the axis of the Juan de Fuca Ridge. *Alvin* carries one pilot and two observers and has made more than 4700 dives to the seafloor since it was built in the 1960s. *Alvin* currently descends to a maximum depth of 4500 m; however, its titanium personnel sphere was replaced in 2013 with one that is certified to dive to 6500 m. The support ship for *Alvin* is the research vessel (R/V) *Atlantis*, operated by the Woods Hole Oceanographic Institution.

Figure 1.2 Two generations of deep-submergence vehicles meet nose to nose. *Alvin* has been central to the development of numerous deep-ocean technologies and, along with the French IFREMER submarine *Cyana*, provided some of the first access to the abyss for routine scientific observations. The yellow vehicle in the foreground is the *Sentry* autonomous underwater vehicle (AUV). It represents the latest generation of robotic underwater vehicle systems that is now being used to map the seafloor at very high resolution, and to take measurements of the chemical, biological, and physical properties the deep ocean.

representations of the deep ocean floor and ocean crust, spanning resolutions from kilometers to sub-millimeter.

Oceanographers cover disciplinary spectrums including chemistry, biology, physics, geology, and engineering; oceanography as a science is, by its nature, multidisciplinary. The iconographic oceanographer today is just as likely to be a professor soaked by spray on deck in a screaming gale, as an engineer who develops a new multi-sensor vehicle system that is innovative, progressively less inexpensive, and both small and light enough to be launched from almost any ship or submersible system in the world.

Since the early twentieth century, oceanography has relied on an increasingly sophisticated fleet of research ships that have been used as floating laboratories to study, sample, and access the oceans. These ships, and the dedicated, skilled personnel that operate them, provide the capacity to conduct complex experiments in the harsh conditions often encountered at sea. Traditionally, the technology used to study and sample the oceans has progressed from simple instruments deployed from wires and conducting cables to more intricate devices that collect better samples more efficiently. Modern sensors often provide the capability to better resolve the chemical, biological, and physical characteristics of processes or features in and at the bottom of the ocean. In the past few decades, oceanography, like many other fields of science, has benefitted from the explosion in digital electronics, new battery technologies, and robotics engineering.

New technology is rapidly changing oceanography, and the kinds of questions oceanographers can investigate, both in terms of the physical scale of various phemonena and how they change over time. Today, robots are deployed that slice through the ocean on their own for weeks or months at a time. Seafloor instrument arrays, collecting physical, chemical, and biological data, have also been installed and planned at numerous sites. During the past decade, scientists and engineers have conjured up and placed in operation a vast array of sophisticated buoys, floats, gliders, moorings, and seabed sensors that can report back 24–7 via satellite, providing monitoring in real time of the oceans as never before.

The thread of human history has been fundamentally influenced by the sea. The oceans have been sources of unknown terror – both real and imagined – the means by which empires expanded and how the outcome of global conflicts was determined, and a source of vital nutrition throughout human history. Because of these intricate and innate links between humans and the ocean there has always been a strong imperative to explore it.

Leonardo da Vinci made the first drawings of a submarine more than 500 years ago. William Bourne, a British mathematician, drew plans for a submarine in 1578, and in 1620 Cornelius van Drebbel, a Dutch inventor, was the first to actually build a submarine. Nathaniel Bowditch was an American mathematician whose *New American Practical Navigator*, published in 1802, was the first comprehensive book on maritime navigation and observations on oceanic meteorology, currents, and winds. Four centuries after da Vinci, Jules Verne still fantasized about exploring the deep in his novel *Twenty Thousand Leagues under the Sea*, published in 1875. In the 1860s, Ernest Bazin began to take photographs from a diving bell. Fellow Frenchman Louis Boutan developed one of the first underwater cameras in 1893. Around the same time, oceanography as a scientific field began in the United Kingdom with the *Challenger* Expedition (1872–1876). It was the first expedition organized specifically to gather data on a wide range of oceanic features and processes, including ocean temperatures, seawater chemistry, currents, marine life, and the geology of the seafloor.

The expedition employed HMS *Challenger*, a British Navy corvette (Fig. 1.3). This small warship was converted into the first dedicated oceanographic ship. It carried laboratories, microscopes, and other scientific equipment onboard. During this 127,580 km voyage circumnavigating the globe,

Figure 1.3 HMS *Challenger* in St. Thomas harbor during the southbound leg of its global voyage from 1872 to 1876, during which the first routine oceanographic observations were made and samples of the seafloor recovered.

Figure 1.4 Illustration of the dredging apparatus on HMS *Challenger*. New technologies of the time included specialized ropes and mechanical winches that allowed scientists to collect some of the first samples of rocks, sediment, biology, and water from the seafloor on this first global oceanographic expedition.

new methods for dredging the seafloor and sampling ocean water were developed. New technologies, such as deep-sea winches and specialized lines for lowering equipment and measuring the depth to the ocean bottom, were tested (Fig. 1.4). One of the seminal discoveries of this expedition, among the many it produced, was that there appeared to be a shallow region in the middle of the North Atlantic basin that later proved to be the Mid-Atlantic Ridge.

German, Swedish, and other global oceanographic expeditions followed in the late nineteenth and early twentieth centuries. These paved the way for modern oceanographic research, encouraging development of new technologies and the discovery of new paradigms that fundamentally changed how we understand Earth history and ocean processes.

In 1932, William Beebe and Otis Barton descended in their bathysphere to about 1000 m off Bermuda, ushering in the era of modern deep submergence and humans' direct observations of deep-ocean and seafloor life (Fig. 1.5). Less than 30 years later, in 1960, Don Walsh and Jacques Piccard dove to 10,915 m to set the deep diving record in the US Navy bathyscape *Trieste*, touching down on the bottom of the Challenger Deep in the Mariana Trench in the western

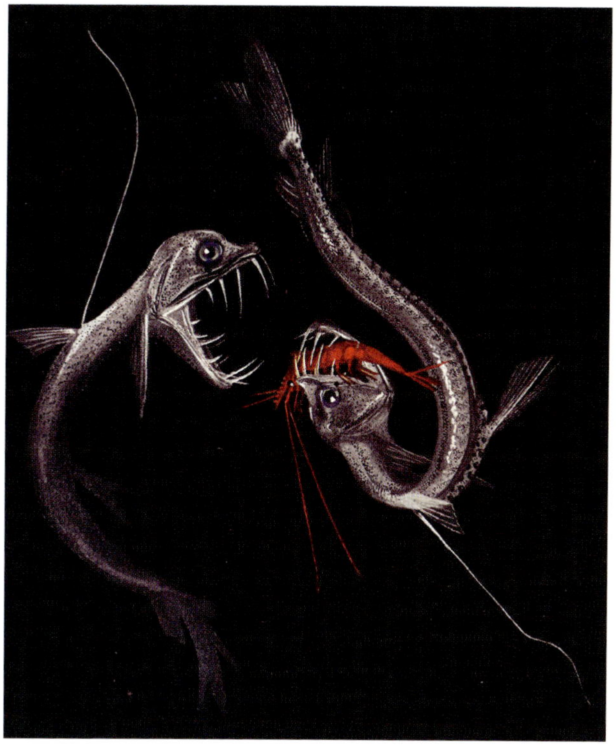

Figure 1.5 Drawing by Estelle Bostelmann of two young viper fish as observed by Beebe and Barton from their bathysphere in the depths off Bermuda. These drawings were published in a series of *National Geographic* magazine articles detailing the revolutionary descent of the explorers in 1932. Detailed drawings such as these were the common way to convey the physical characteristics of marine life and nature prior to the routine use of photography in the ocean.

Pacific. In the intervening years, basic and applied oceanographic research played a significant role in the outcome of World War II. This, in turn, led to the recognition of the strategic and societal importance in understanding the physics, geology, biology, and chemistry of the ocean, and the subsequent investment in basic oceanographic sciences by government funding agencies worldwide.

In the 140 years since the *Challenger* Expedition, oceanographic science and technologies have advanced significantly, allowing scientists to develop and pursue new hypotheses for the underlying, interacting processes that govern how the oceans and Earth work. Scientists now use sophisticated vehicles and imaging systems to reveal the wonders of the abyss, especially the realm below 400 m where sunlight does not penetrate. These discoveries have provided keys to understanding Earth history and planetary processes and the fundamental importance of the mid-ocean ridge (MOR), ocean trenches, transform faults, seamounts, and the great variations in terrain making up the continental margins throughout the global ocean (Fig. 1.6). We have realized da Vinci's and Verne's dreams of exploring the ocean depths and routinely taking humans to the deep seafloor.

The methods used to "reveal" the seafloor can be roughly grouped as either acoustic or optical, with each having vastly different resolution capabilities. To better appreciate the significant technological advances that have been applied to exploration of the deep ocean and seafloor, we provide some historical context for acoustic and optical imaging systems employed to map and image seafloor features.

The following historical discussion illustrates how far oceanography has advanced in a short period of time. The chronology emphasizes the important impacts that technological advances have made toward understanding oceanic crustal architecture, hydrothermal and volcanic features present on the ocean floor, and the novel biological communities that thrive at depth in the absence of sunlight.

1.2 Acoustic imaging

1.2.1 Echo sounding

In the early twentieth century, in response to the *Titanic* disaster, the German physicist Alexander Behm developed echo sounding – a method of transmitting sound waves through water to reflect off objects. Echo sounding proved to be incapable of locating icebergs on the sea surface, but scientists found it was an excellent tool to measure the depth of the ocean. Sound waves, traditionally ranging in frequency from 3.5 kilohertz (kHz) to 12 kHz, were transmitted into the ocean, reflected off the seafloor, and recorded at the surface by microphones or hydrophones. The deeper the seafloor beneath the ocean's surface, the longer the sound waves took to echo back.

Strategic requirements during World War II, particularly how seafloor topography provided hiding places for submarines to avoid detection, as well as the influence of seafloor features on acoustic propagation, provided a major impetus for developing sonar, originally an acronym for "**so**und **na**vigation and **r**anging." From the post-World War II era to the 1970s, a higher-resolution echo sounder, called a precision depth recorder (PDR), provided the standard means for academic scientists and the military to survey the ocean floor and measure its depth (Fig. 1.7). The sound

Figure 1.6 R/V *Vema* of the Lamont-Doherty Geological Observatory of Columbia University (now Lamont-Doherty Earth Observatory) operated from 1953 to 1981. *Vema* collected a wide range of oceanographic data during its many voyages, which spanned over a million miles and paved the way for a revolution in our understanding of the Earth and oceans.

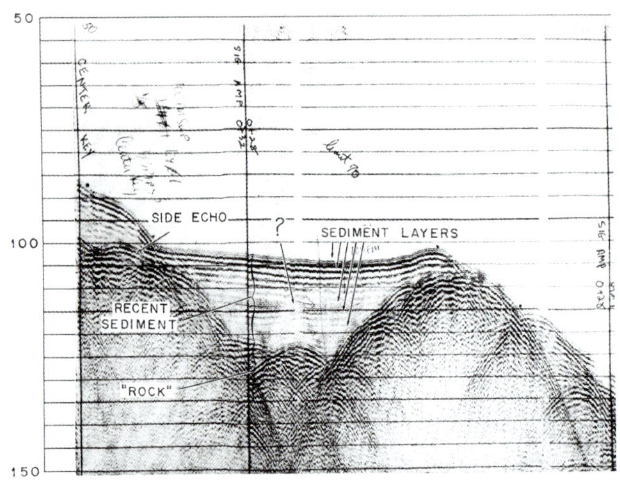

Figure 1.7 Precision depth recorder (PDR) record made using a 3.5 kHz echo sounder. Depth in fathoms (a fathom is 6 feet, or almost 2 meters) is on vertical axis, time (which was linked to the ship's location) is on horizontal axis. Complex reflection patterns generally indicate that there is rugged seafloor topography on either side of the ship.

beam of PDR systems is transmitted from the ship through the water and reflected off the bottom. The depth is calculated using the round-trip time through the water, as measured on the PDR and divided in half, multiplied by the speed of sound in seawater (average ~1500 m/s). Because of the wide beam of sound energy transmitted from the surface transducer mounted on the ship's hull, this type of early bathymetric mapping system had spatial resolution of the order of hundreds of meters at depths below 3000 m, so the measured depths were really an average of the depth over the ensonified area.

In the 1950s, Bruce Heezen and Marie Tharp from the Lamont Geological Observatory of Columbia University in New York used thousands of US Navy soundings and other bathymetric data to compile their now classic physiographic maps of the ocean basins (see Fig. 2.5). These maps revolutionized the way scientists viewed the seafloor and led to the discovery of the global MOR system and its role in seafloor spreading and plate tectonics. That, in turn, provided new frameworks for explaining how the continents and ocean basins evolved. William "Bill" Menard at the Scripps Institution of Oceanography in California used similar approaches to map large areas of the central and eastern Pacific, recognizing the importance of the MOR and seafloor spreading on the geologic evolution of the western margins of South, Central, and North America (Menard, 1964, 1967; Menard and Atwater, 1969).

1.2.2 Satellite oceanography

As in many fields, oceanographic science has instigated and benefitted from revolutionary advances in technology over the past half-century. A notable example, at the most basic level, is resolving the topography of the ocean basins, which is essential to understanding their evolution. The application of new satellite and ship-based mapping technologies in the early 1970s, developed largely in response to the Cold War, allowed scientists to map bathymetric features at unprecedented resolution throughout the global ocean.

Satellite-based geophysical data, especially gravity and magnetic studies, complemented sonar data to achieve a holistic understanding of the evolution of ocean basins. Gross topography of the global seafloor was produced at a resolution of about 1 km per pixel (Fig. 1.8). Even with

Figure 1.8 The digital bathymetric map of the global ocean seafloor compiled from satellite altimetry data and available depth sounding and gravity data produced by Walter Smith and David Sandwell (Smith and Sandwell, 1997). The spatial resolution of features mapped using these data ranges from 1 to 12 km. These images provide the most accurate and holistic view of the topography of ocean basins "with the water drained out." Compare this figure with the Heezen and Tharp compiled bathymetry shown in Fig. 2.5.

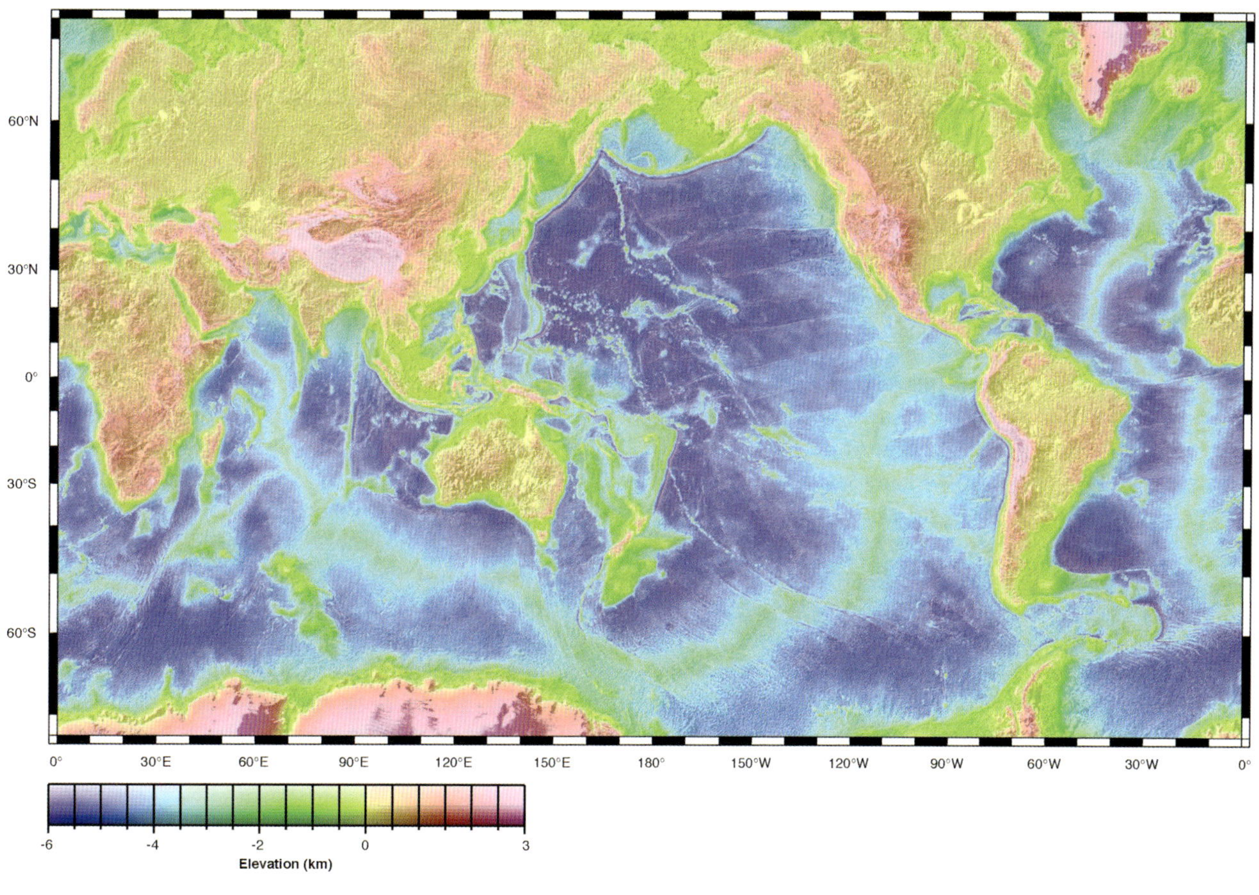

these coarse regional data, it is possible to resolve detailed topography of the global *and* local seafloor as well as the gravity signature of the crust and upper mantle with remarkable clarity. These perspectives permit the study of areas that span hundreds of thousands of square kilometers using satellite geophysical data, and areas of thousands of square kilometers using ship-based surveys.

Early oceanographic surveys, prior to 1970, did not have the benefit of routine, continuous satellite navigation. Thus, it was often difficult to return to exactly the same location to conduct time-series or repeat surveys or sampling. Features mapped on one expedition might not be relocated for sampling on subsequent cruises, or at times features might be repeatedly mapped or misidentified because of navigational errors. A critical technological development that helped to advance seafloor mapping and imaging was the deployment of the Global Positioning System (GPS) satellites in constellations that yielded global coverage of all oceans. GPS now provides critical location information that allows the position of ships to be determined to within 10 m on the sea surface. This accuracy in geodetic positioning meant that an entire range of oceanographic survey data could now be accurately placed (within a few meters) and compared to higher-resolution surveys and detailed observations using newly developed deep-submergence vehicles, effectively revolutionizing the ability to map and image the seafloor.

An additional benefit of GPS technology and advances in ship-based sensors and control systems is that ship positions could now be very accurately controlled in real time, allowing them to maintain geodetic position in a range of oceanographic conditions throughout the globe. Called dynamic positioning, or DP for short, it provides for autonomous vessel positioning though a combination of computer-controlled systems. These are tied to both GPS navigation input at high data rates and simultaneous autonomous control of ship propulsion systems in response to the magnitude of current, wind, and wave forces acting on the vessel. Dynamic positioning is now a routine capability used in most oil and gas deep-water exploration and drilling activities and is essential for many remotely operated vehicle operations, where vessel position must be maintained in order to carry out seafloor surveys and sampling. The benefits of GPS and DP navigation are many and they translate directly to our ability to navigate deep-submergence vehicles (which cannot use GPS below the ocean surface) with a high degree of both precision and accuracy in geodetic space.

More recently, because of advances in sonar technology and attitude sensing of vehicle motions, ultra-short baseline navigation for the spectrum of deep-sea vehicles has revolutionized the ease with which vehicles can be navigated without sacrificing the precision and accuracy needed to produce quantitative map-based imagery of the seafloor.

1.2.3 Multibeam sonar

Post World War II technological advances in both computing and acoustics led to the development of multiple beam bathymetric sonars by the US Navy. These systems relied on not just one acoustic pulse, but multiple pulses sent and received by acoustic transducers and receivers configured as arrays on ships' hulls, gathering a swath of bathymetric information as the ship moved through the water at 8–10 knots (kts). The Navy system, called SASS (Sonar Array Bathymetry Survey System), transformed how scientists mapped the global ocean seafloor. SASS data played a major role in mapping the Mid-Atlantic Ridge at the site of the FAMOUS (French–American Mid-Ocean Undersea Study) expedition in 1974 (Ballard and Van Andel, 1977a), in the first major science expedition for the human-occupied submersible *Alvin*, and in the Galápagos Rift, where the first low-temperature hydrothermal vent communities were discovered in 1977 (Corliss *et al.*, 1979) (see Section 4.14).

Academic scientists have had access to multibeam sonar systems since the early 1980s, when they were first used to map the topography of the northern East Pacific Rise crest over a distance of more than 1000 km (Macdonald *et al.*, 1992; Macdonald, 1998). Since that time, the early hull-mounted multibeam systems, which had 12 beams in an array oriented across the keel of research ships, have evolved into arrays of more than 200 beams using frequencies up to 30 kHz (Fig. 1.9).

Shallow-water multibeam systems operating at higher frequencies, generally above 100 kHz, for use at depths shallower than 1000 m, provide even greater precision and accuracy for coastal seafloor mapping and locating water column features such as rising methane bubble plumes. Advances in sonar processing and acoustic array technologies have also increased the swath that multibeam sonar systems can cover in a single pass. Older systems were able to yield swaths of integrated bathymetric data that were approximately as wide as the mean water depth over which the survey was being carried out. Modern systems with more than 200 beams cover swaths that are three to four times the water depth.

Multibeam bathymetry systems afford stunning improvements in the resolution of seafloor features (Fig. 1.10). While the old PDR echo sounders could resolve the seafloor depth to within a few tens of meters, that depth was an average over the footprint of the beam on the seafloor. At depths greater than a few thousand meters, the diameter of the older sonar footprint is several hundred meters, resulting in greatly diminished spatial resolution. In steep terrains, such as the escarpments bounding transform faults or fracture zones, even multibeam bathymetry cannot resolve small (10 to 50 m) steps in the relief.

By comparison, a modern 30 kHz multibeam hull-mounted bathymetry system can resolve the topography of the seafloor down to depths of about 4000 m with a vertical resolution of about 10 m and spatial precision of

Figure 1.9 An illustration showing the swath coverage of multibeam sonar used on many modern oceanographic vessels. Multibeam sonar can resolve seafloor bathymetric variations of 10 m vertically over horizontal distances that range from 25 to 100 m, depending on the frequency of the system. Continuous bathymetric maps with spatial resolution of 50 to 80 m and vertical resolution of 10 to 15 m covering a thousand square kilometers can be acquired in only 10 to 12 days. The yellow signals from the ship to shore depict the real-time satellite transmission of data and images that are used for "telepresence" (see Section 1.7).

about 25 m. As an example of how important accurate bathymetric mapping is, HMS *Challenger* made the first lead-line sounding to the bottom of the Mariana Trench in 1875 and recorded a depth of 8,184 m. In 2010, a multibeam system was used to map the Challenger Deep, yielding the deepest part to be 10,984 m, with an accuracy of ±25 m (Gardner *et al.*, 2014).

Today, multibeam sonar mapping is a prerequisite for almost any type of seafloor research. In some cases, investigators choose to map relatively small areas – as small as 100 square kilometers – specifically targeting structural or volcanic features they plan to explore later in more detail using various types of deep-submergence vehicle systems. In other cases, efforts to map larger areas of seafloor with multibeam sonar have produced exceptionally detailed maps revealing large seafloor features and how changes in morphology reflect a host of ocean crustal formation processes (see, for example, the MOR segments mapped in specific areas during the RIDGE and Ridge 2000 Programs from the early 1990s to 2012; www.ridge2000.org).

1.2.4 Sidescan sonar

While bathymetric sonar provides quantitative measures of the vertical relief of seafloor features, sidescan sonar – where the sound is projected laterally to either side of a ship, towed vehicle, or robotic vehicle – can be used to provide a fine-scale textural perspective of the seafloor. This type of imaging of the seafloor is often called "backscatter" data because the spatial distribution of dark to light patterns in the recorded image depends on the type of ocean floor being surveyed as well as the slopes in the terrain. The scattering of the sound waves from irregularities on the seafloor and whether the bottom is composed of rock or sediment, and the distribution of those materials, creates characteristic acoustic reflectivity patterns in sidescan imagery. Faults – large and small – as well as constructional volcanic features and lava flows, and variations in seabed "roughness" all reflect sound energy differently. Therefore, sidescan sonar images commonly resolve more subtle variations in seafloor morphology than bathymetric sonars, and as such are valuable for interpreting both tectonic processes and field relationships on the ocean floor.

Sidescan sonar, developed in the 1960s, added a new dimension to precision depth recording and multibeam sonar surveying, providing quantitative data on the texture, spatial dimensions, and geomorphology of seafloor features. The Scripps Institution of Oceanography at the University of California, San Diego, had assembled an exceptional group of scientists and engineers interested in developing new seafloor mapping technologies. Their leader was a marine geophysicist, regarded by many as the father of academic deep-submergence vehicle systems: Dr. Fred Spiess. Among his many accomplishments, Dr. Spiess developed the *Deep Tow* system, a deep-sea vehicle hosting an array of down-looking and side-looking sonars (Fig. 1.11). The sonars operated at different frequencies, optimizing the acoustic backscatter of the seafloor to gain a lateral perspective of the terrain. In addition, because this sonar used higher frequency sound at about 100 to 200 kHz, it also revealed the fine-scale textural and structural fabric of the terrain, which greatly enhanced the quality of images of geological features on the seafloor.

Over the years, *Deep Tow* went through numerous system upgrades as new technologies were developed. It hosted sensitive temperature sensors with a camera and lights, so that it could collect photographic images and use them to confirm the nature of specific features mapped using various sonar systems. *Deep Tow* is commonly credited with providing the roadmap to discovering hydrothermal vents at the Galápagos Rift. During an expedition in May 1976 to the Galápagos Rift, cameras on *Deep Tow* imaged fields of what were subsequently interpreted to be large clams, and

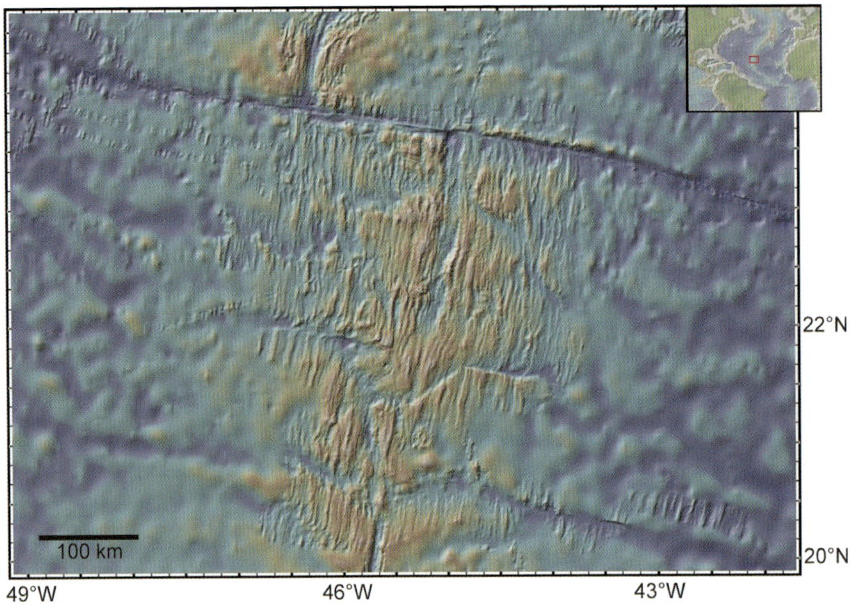

Figure 1.10 A bathymetric map of the Mid-Atlantic Ridge in the 20–24° N region showing the prominent rift valley, rough topography of the ridge flanks, and discontinuities along-axis. The map is a combination of data sets including coarser 1 km gridded data derived from satellite altimetry along the east and west edges of the map and 100 m gridded multibeam data in the center along the ridge axis. Lighter colored areas are shallower depths of the MOR crest and upper flanks.

its sensors registered bottom-water temperature anomalies of 0.25 °C, both of which were completely unexpected on the deep seafloor. The images and data it collected were a crucial antecedent to the follow-on discoveries of hydrothermal vents at the Galápagos Rift in 1977 that were made by geologists diving in the research submersible *Alvin* (Lonsdale, 1977; Corliss *et al.*, 1979) (see Section 4.14).

Over the last 40 years, *Deep Tow*'s success encouraged the development of an array of sidescan sonar vehicles capable of mapping both large areas and specific features of the seafloor, providing key information on its structure and geological history and the discovery of seafloor biological communities.

Figure 1.11 The *Deep Tow* vehicle, as configured in the mid 1970s, was developed by Dr. Fred N. Spiess and colleagues at the Scripps Institution of Oceanography's Marine Physical Laboratory.

Different types of seafloor, comprising sediments, rocks, lava surfaces (see Chapter 5), fissures and faults, reflect sound differently, and variable slopes also impact patterns of reflectivity in sidescan sonar data. The imagery produced by sidescan sonar helps achieve another important way of "draining the water" out of the oceans, thereby revealing variations in seafloor textures that are important for studying the evolution of different areas and the processes affecting them.

More recent sidescan sonars, using frequencies from 12 to 200 kHz, reveal the texture of the ocean floor and MOR crests and flanks in both stark and subtle ways that multibeam bathymetry does not portray (Fig. 1.12). Some sidescan sonar systems (e.g., GLORIA, SeaMARC-2, MR1–2) (Figs. 1.13 and 1.14) are towed a few hundred meters below the ocean's thermocline, the boundary between the shallow mixed layer and the deep ocean, where there is a steep temperature gradient from the surface down to several hundred meters that impacts sound velocity and acoustic raypaths. The swath width of these systems can be 10–20 km. These systems ensonify large areas of seafloor with lateral resolution varying between about 8 and 50 m per pixel. Other sidescan sonar systems (e.g., TOBI, SeaMARC-1, DSL-120A, IMI-30) are towed within a few hundred meters of the seafloor, and have correspondingly greater resolution that ranges between 1 and 5 m per pixel, but with narrower swaths between about 1 and 5 km wide (Blondel, 2009) (Fig. 1.15).

Sidescan systems have also been designed to operate on remotely operated vehicles (ROV) and particularly on autonomous underwater vehicles (AUV), whose stability provide an excellent platform for collecting these types of acoustic data. Recent examples of how important near-bottom sidescan data are to the study of seafloor volcanic eruptions include data from the East Pacific Rise near

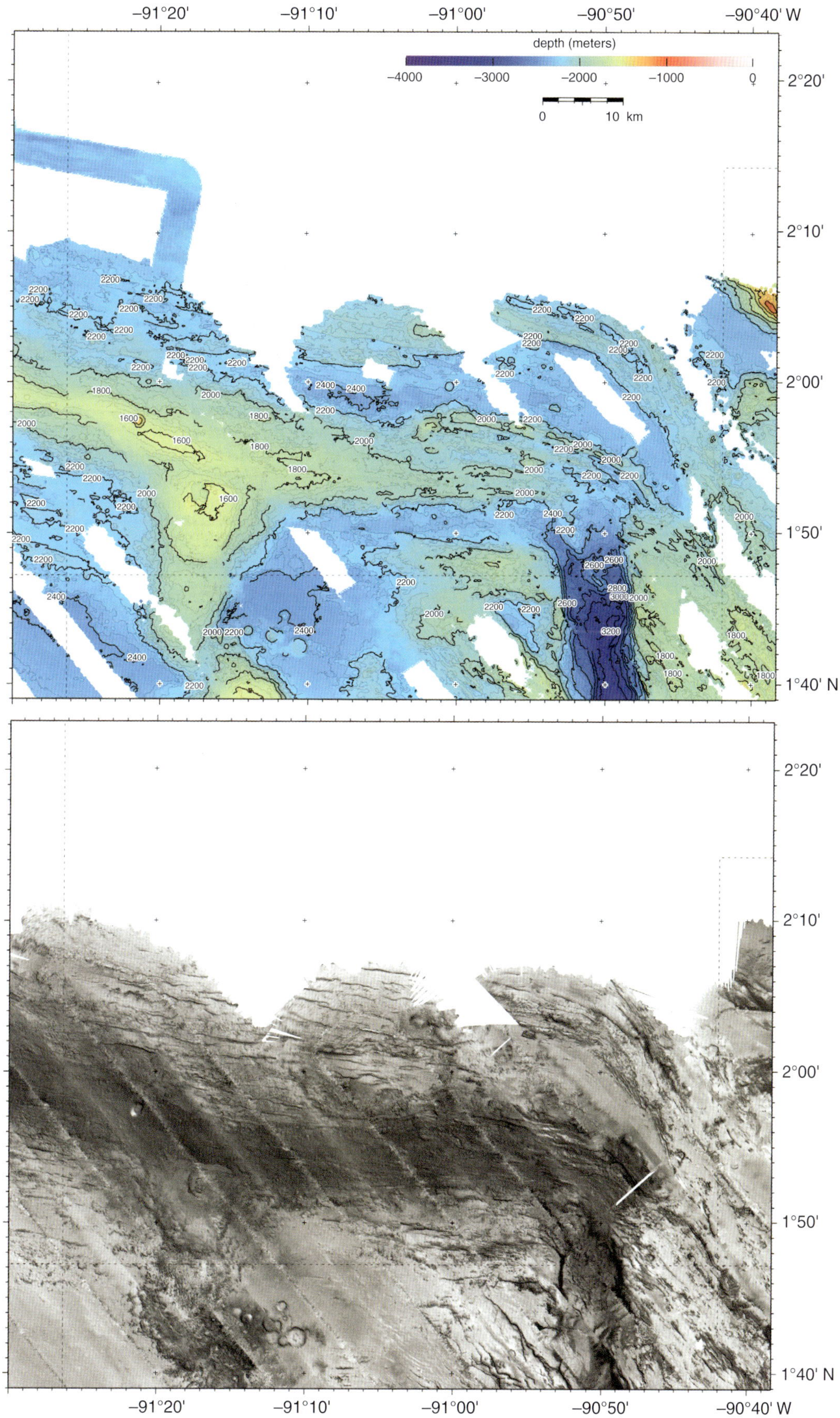

Figure 1.12 Sidescan sonar data provide important backscatter maps that depict both large- and small-scale seafloor structures, which are important for proper interpretation of the geologic history of seafloor terranes. MR1 sidescan sonar (bottom) and compiled multibeam bathymetry maps of the Galápagos Spreading Center (GSC) (top). The E-W swath of high backscatter (dark gray) in the bottom sidescan image is the GSC axis and upper flanks where lava flows are exposed at the seafloor and are thus more reflective to the sound energy. The linear features parallel to the GSC in the sidescan swath are abyssal hills and low-relief faults that are not well resolved in the bathymetry. Light gray areas are sediment covered, so have low-reflectivity. Small seamounts and volcanic-flow fronts occur south of the GSC near 91° 20′W and are part of the seamount province in this region.

Figure 1.13 The Hawaii Mapping Research Group, MR1 sidescan sonar system during deployment. The system operates at 12–13 kHz and collects large-area sidescan imagery swaths (10–20 km width) which have a resolution of 2–5 m per pixel.

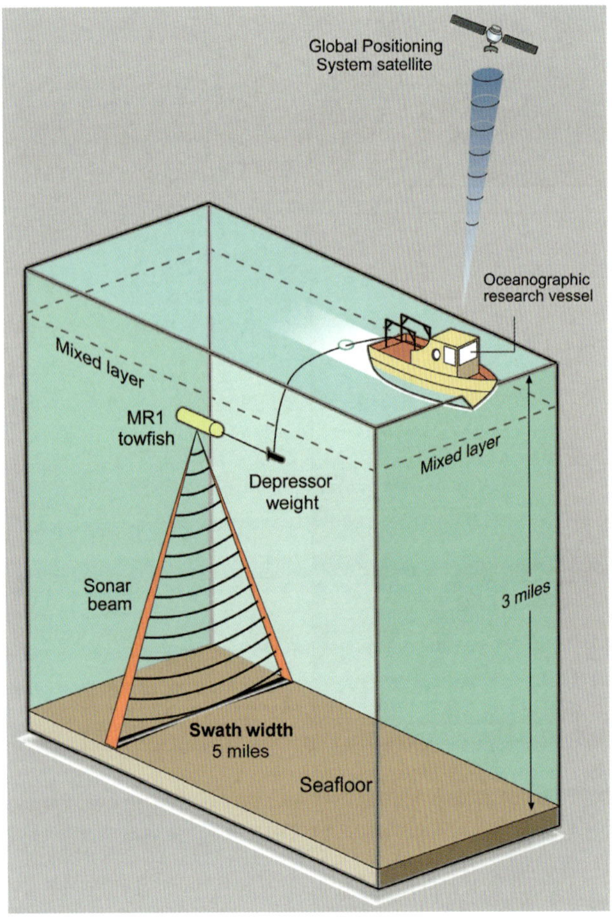

Figure 1.14 Illustration depicting the geometry of sidescan sonar acquisition using the near-surface towed MR1 system operated by the Hawaii Mapping Research Group at the University of Hawaii. The sonar towfish is towed below the thermocline to avoid acoustic ray-path problems. The towfish is also towed behind a depressor weight to minimize the effects of surface sea state and motion that are transmitted mechanically down the tow cable.

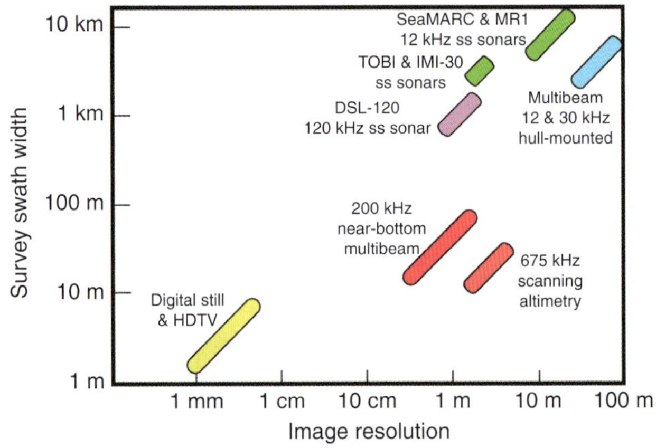

Figure 1.15 Plot showing the variations in image resolution as a function of the survey method and swath width of modern sonar systems used for oceanographic research compared to the resolution obtained from digital cameras and high-definition TV (HDTV).

9°50′ N where "before and after" DSL-120 sidescan sonar revealed the significance of channelized lava flows (Fig. 1.16) in distributing lava at a fast-spreading MOR during the 2005–2006 eruptions (e.g., Soule *et al.*, 2007, 2012) (see Chapter 5, Figs. 5.20, 5.31, 5.33 and 5.40).

Ultra-high-resolution sidescan data acquired by AUVs revealed the detailed distribution of new seafloor created by a submarine volcanic eruption on Axial Seamount in 2011 on the Juan de Fuca Ridge (Caress *et al.*, 2012). Results from near-bottom AUV-based sidescan sonar surveys that have raised public awareness of deep-sea mapping techniques and new technologies include the discovery of the wreckage of an Air France aircraft found in about 4 km of water on the east flank of the equatorial Mid-Atlantic Ridge by the REMUS AUVs in April 2011 (see Section 1.4.3).

1.3 Twentieth-century developments in deep-sea optical imaging technology

In the early 1950s, Harold Edgerton, an engineering professor at MIT, developed the deep-sea strobe light, providing the "sunlight" required to take photographs of the deep ocean and seafloor for the first time. That development, coupled with the engineering efforts of Maurice "Doc" Ewing and Lamar Worzel at the Lamont Geological Observatory and J. Brackett Hersey, Allyn Vine, and David Owen at Woods Hole Oceanographic Institution, led to the first generation of deep-sea cameras (Ewing *et al.*, 1946; Hersey, 1967). These systems (Fig. 1.17) were simple by today's standards, but they provided key photographic evidence of animals and seafloor features over small areas in the deep ocean. They were also indispensible in carrying out crucial deep-sea search missions, such as those required after the tragic accident that sank the Navy submarine *Thresher* in 1963 and the loss of an atomic bomb off the coast of Spain in 1966 (Fig. 1.18).

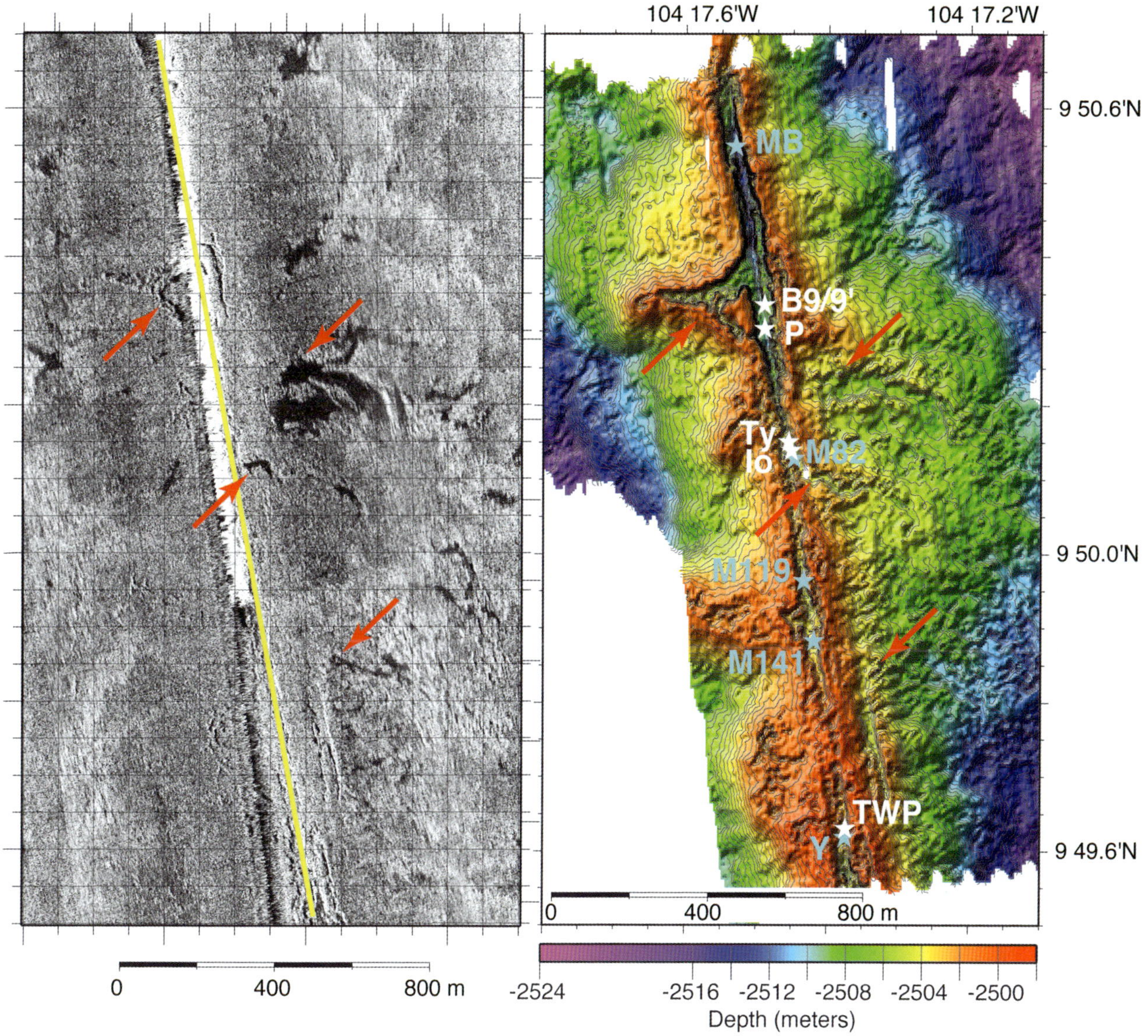

Figure 1.16 DSL-120 sidescan sonar image (left) and near-bottom bathymetric map made using the AUV *ABE* (right) of an approximately 2 km^2 section of the fast-spreading East Pacific Rise axis near 9°50′ N. The gray, white, and dark patterns in the sonar imagery result from variable backscatter (reflection) of sound energy generated from the volcanic seafloor by the near-bottom (100 m altitude) towed sonar instrument. The red arrows on the sidescan image point to lava channels that are floored by smooth sheet flows. The same arrows on the bathymetric map show the features resolved by the near-bottom mapping. High-temperature (white) and low-temperature (blue) vent locations are noted as mapped prior to the 2005–2006 eruptions. The trace of the spreading axis is shown as a yellow line in the sonar image. For comparison see Fig. 5.40.

Photographs of the seabed and fauna have provided important causal links between geological and biological phenomena on the deep-ocean floor, and photomosaics have facilitated seafloor mapping and spatial analysis of seabed features. From the 1960s to the 1980s, marine geological and biological investigations of deep-ocean terrains have relied heavily on traditional optical cameras and deep-sea strobe lighting to photograph the abyssal seafloor accurately and clearly (Heezen and Hollister, 1971; Spiess and Tyce, 1973; Ballard and Moore, 1977; Grassle *et al.*, 1979; Lonsdale and Spiess, 1980).

Photographs of the seafloor provided some of the first clues about the abundance and diversity of life in the deep ocean, morphology of lava flows on the mid-ocean ridge crest, and evidence for hydrothermal venting. Today, photographic and direct observations of the ocean floor are intimately tied to understanding the dynamic and interactive physical, chemical, and biological processes occurring there.

Bruce Heezen and Charles Hollister, in their seminal 1972 book *Face of the Deep*, provided the first grand compilation of deep-sea imaging taken with the newly developed underwater cameras and strobes, and they discussed the scientific implications and insights that those images provided. The mysteries of the deep were no longer hidden from view; a "plug" had been found to drain small areas of the ocean basins.

Figure 1.17 Maurice "Doc" Ewing (top, left) and Allyn Vine (top, right) on the R/V *Atlantis* holding one of the first deep-sea cameras in the late 1950s. David Owen deploying a deep-sea camera from R/V *Vema* in the late 1950s (bottom).

Deep-sea photographic imaging systems continued to be developed through the 1960s to the 1980s. Camera systems such as LIBEC (Light BEhind the Camera) and MITZAR towed cameras were developed by the US Navy in the late 1960s to image areas several tens of meters on a side. This made creating manual photomosaics of the images easier and provided expanded areal coverage; nevertheless, the images were still largely analog in nature and relied on traditional 35–70 mm film technology (Fig. 1.18). It was clear, however, that much broader spatial coverage was needed to comprehend the shapes and structures of seafloor topography over hundreds to thousands of kilometers. Light attenuation generally limited deep-ocean photography to objects less than 20 m from the camera, continuing to create a significant roadblock to large-area imaging of the seafloor (Benthos, Inc., 1984).

In the 1980s to 1990s, new towed cameras with higher-capacity strobe lighting and large film magazines were deployed for both basic seafloor geological and biological mapping along the MOR crest and other terrains, as well as for archeological and search missions, such as the one that discovered the wreck of the *Titanic* (Ballard, 1985). These systems (e.g., *ANGUS* and *ARGO* operated by WHOI's Deep Submergence Laboratory; Harris and Ballard, 1986) relied on rudimentary electronic controls to trigger the cameras and strobes at specific intervals and also provided some primitive acoustic telemetry to transmit information such as bottom water temperature, depth, and altitude along the survey track.

The next generation of deep-sea cameras began to be developed in the mid to late 1990s as professional and consumer digital cameras became available and the size of digital electronics and controls were miniaturized. Equally important was the development of high-precision optics that could be manufactured to withstand the extreme pressures at seafloor depths down to ~6000 m. Once such system developed at WHOI was the *TowCam* system (Fornari, 2003). The WHOI *TowCam* (Fig. 1.19) is an internally recording digital deep-sea camera system that uses a 16 megapixel color camera and 600 watt/second strobes. The system is designed to also permit acquisition of volcanic glass samples using small rock-core winches, and water-sampling bottles can also be triggered based on CTD water properties data that are collected in real time. *TowCam* is towed on a standard coaxial CTD sea cable, so it can be operated from many ships in the worldwide oceanographic fleet, thereby permitting real time acquisition of digital depth, altitude data, and low-resolution images from the camera. The high-resolution images are stored within the camera and are downloaded after the photographic transect is completed.

Recent advances in digital imaging technology as applied to deep-sea photography have transformed the oceanographic sciences. Digital imaging permits a wide disciplinary range of investigations in many seafloor terrains without the cumbersome and time-intensive requirements needed to develop long rolls of 35 mm film, particularly at sea. Digital imaging is required for near-bottom scientific studies on many levels and in different disciplines. From a geological perspective, the ability to digitally image a large area and merge the constituent images together in a photomosaic provides a powerful tool for mapping and understanding the geological relationships between features of various dimensions – from centimeter scale to tens or hundreds of meters in size or area (Fig. 1.20). Digital photography permits integration of images with other digital data sets including near-bottom sidescan sonar, near-bottom scanning/multibeam sonar bathymetry, and geophysical data. Biological features on the seafloor, and the distribution of biota in different environments, also lend themselves to precision study using digital images and mosaics (e.g., Sarrazin *et al.*, 1997; Delaney *et al.*, 1998; Shank

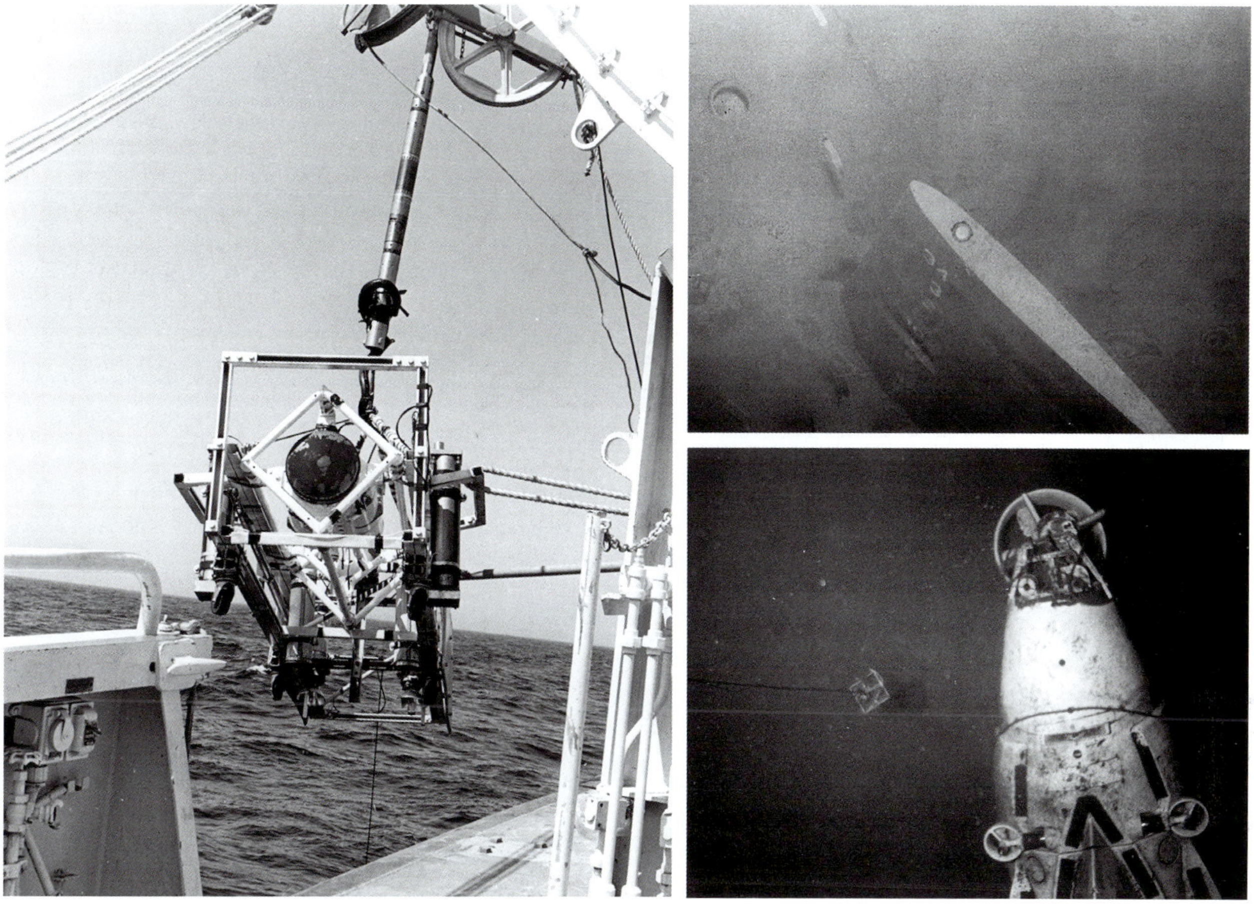

Figure 1.18 The US Navy's MITZAR deep-sea camera system (left) used in the 1960s to search for the wreckage of the US Navy submarine *Thresher*. Photograph (right, top) taken by MITZAR in 1964 of the upper rudder of the sunken *Thresher*. Photograph (right, bottom) taken in 1969 by the MITZAR of the hull of the *Alvin* submersible after it sank in rough seas during a recovery in 1968 and landed on the seafloor at a depth of 1500 m.

Figure 1.19 The WHOI *TowCam* is a 6000 m rated deep-sea towed digital camera system that has multiple rock and water sampling capabilities. It has been used on over 40 expeditions throughout the global ocean (see http://www.whoi.edu/page.do?pid=17630).

et al., 1998; Sarrazin and Juniper, 1999; Trask and Van Dover, 1999; Powdoski *et al.*, 2009; Waller *et al.*, 2011). Revisiting various deep-sea study areas for geological, geotechnical, biological, and geochemical studies is now common, and has been an important research theme within the Ridge 2000 Program. These time-series studies and other field investigations rely heavily on accurate maps of relatively small tracts of seafloor (hundreds to a few thousands of square meters) and successive re-mapping and imaging of the same areas over time to deduce the biological, geochemical, and geological changes that have occurred there (e.g., Sarrazin *et al.*, 1997; Shank *et al.*, 1998; Ferrini *et al.*, 2007, 2008; Yoerger *et al.*, 2007; Powdoski *et al.*, 2009; Waller *et al.*, 2011).

The past decade has seen a revolution in both computing and imaging technology, and both have had major impacts on all branches of science. Many features on the seafloor have dimensions that are larger than the field of view of single images taken by currently available cameras. For marine scientists, the ability to digitally image large areas of seafloor to characterize biological, geological, geochemical, or archeological features and processes represents a new approach to both planning and implementing deep-ocean fieldwork. In

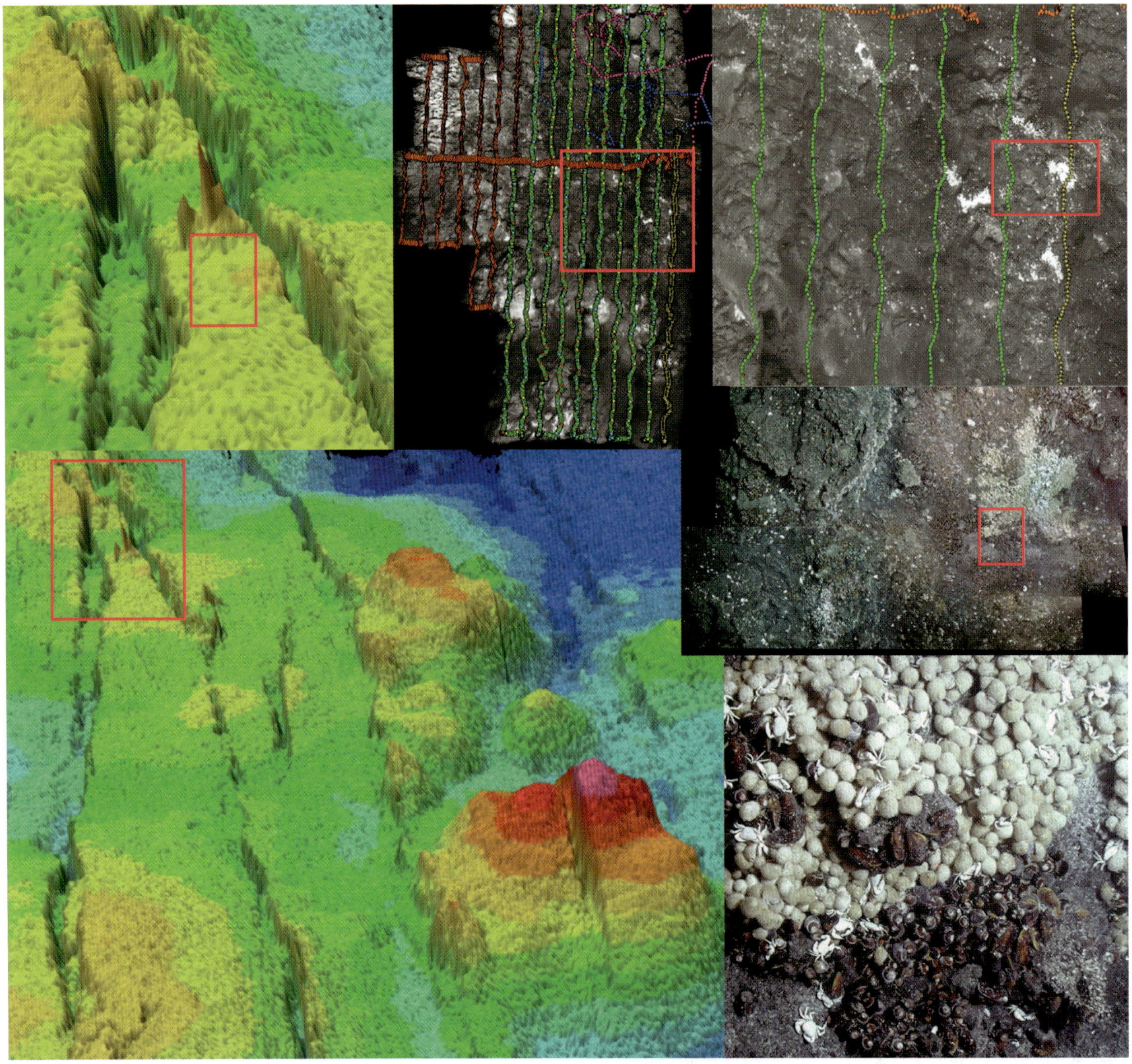

addition, digital imaging technology provides the potential for immediate or near-real time review of high-quality photographs and the capacity for quantitative analysis and integration with other high-resolution data sets. Digital imaging also provides an important capability for conducting time-series studies and monitoring of seafloor sites where near-real time image analysis is necessary to discern long-term and short-term changes in terrain characteristics (e.g., Fornari *et al.*, 1996; Humphris and Kleinrock, 1996; Kleinrock and Humphris, 1996a; Humphris *et al.*, 1996; Ferrini *et al.*, 2007, 2008; Delaney *et al.*, 1998; Hart *et al.*, 2000), faunal community structure, or activity at seafloor hydrothermal vents or other dynamic phenomena (e.g., Sarrazin *et al.*, 1997; Delaney *et al.*, 1998; Shank *et al.*, 1998; Sarrazin and Juniper, 1999; Van Dover and Trask, 2001).

For example, hydrothermal vents and surrounding volcanic seafloor have been digitally imaged on a scale of tens to hundreds of meters at several MOR vent sites. These include: the TAG and Lucky Strike Vent Fields on the

Figure 1.20 Composite illustration showing 2 m resolution near-bottom multibeam bathymetry acquired using the AUV ABE over the Kilo Moana Vent Field in the Lau Basin Spreading Center (in the two left side images, the upper left is a close-up of the vent chimney structure shown in the red box). Scale across lower left bathymetry image is ~100 m. Scale across upper left bathymetry image is ~25 m. Red-pink is 2500 m depth range, green-blue is >2600 m depth. The three ABE photomosaics in the upper right corner of the illustration show progressively zoomed-in views of the mosaic at top middle. The red box on those three mosaics shows the area covered by each successive enlargement. The vent animal distributions are shown by individual images as color-coded squares (middle top mosaic is overall view and upper right mosaic is close-up of the middle portion of the larger mosaic). Middle right mosaic shows the gastropod community imaged in the single image at lower right (outlined by red box in mosaic above it).

Figure 1.21 Composite image of support vessels used to launch and recover the submersible *Alvin* from the mid 1960s to the present – R/V *Lulu* is in the background center and was used from 1970 to 1984; R/V *Atlantis II* at right was used as the *Alvin* support ship from 1984 to 1996; and the R/V *Atlantis*, the current support ship for *Alvin*, is on the left.

Figure 1.22 The French research submersible *Nautile* can dive as deep as 6 km and is operated by the French Institute for Exploitation of the Seas (IFREMER). *Nautile* has been involved in many research programs throughout the Pacific, Atlantic, and Indian Oceans, and is also used by French government agencies for applied engineering and oceanographic studies. An array of cameras and lights is on the front of the submarine above the basket and manipulator arrangement. *Nautile* dives with two pilots and one observer to make observations (http://flotte.ifremer.fr/fleet/Presentation-of-the-fleet/Underwater-systems/Nautile).

Mid-Atlantic Ridge (Fornari *et al.*, 1996; Humphris and Kleinrock, 1996; Van Dover and Trask, 2001; Escartin *et al.*, 2008), on the Juan de Fuca Ridge (Delaney *et al.*, 1998; Yoerger *et al.*, 1999; Embley *et al.*, 1999), in the Guaymas Basin (Lizarralde *et al.*, 2011), on the northern and southern East Pacific Rise (EPR) (Fornari *et al.*, 1998, 2004, 1998, 2012; Kurras *et al.*, 2000; Sinton *et al.*, 2002; Shank *et al.*, 2003; Hey *et al.*, 2004; Soule *et al.*, 2005, 2006, 2007; Garry *et al.*, 2006; Rzhnanov *et al.*, 2006; Cowen *et al.*, 2007; Ferrini *et al.*, 2007; Fundis *et al.*, 2010), and Eastern Lau Spreading Center (e.g., Ferrini *et al.*, 2008; Podowski *et al.*, 2009) (see Chapter 4). These studies have been instrumental in providing synoptic views of individual vent edifices, vent fields, lava flow morphology and their context within the surrounding volcanic and structurally modified seafloor.

New technologies have enabled discoveries, such as those made at the novel Lost City Hydrothermal Vent Field in 2000 (Kelley *et al.*, 2001b,c), the 2005–2006 eruptions at the East Pacific Rise (EPR) (Soule *et al.*, 2007), the 2011 Axial Seamount eruption (Caress *et al.*, 2012; Chadwick *et al.*, 2012), and other study sites over the last ~10 years (see Chapters 4 and 5). Digital images can be used to quickly verify interpretations made using sonar data, and can enable rapid modifications in reconnaissance plans during the same expedition (Shank *et al.*, 2003), thereby greatly improving the efficiency of seafloor mapping studies and improving the chances for serendipitous discoveries. The widespread availability of digital seafloor imagery today, and the connectivity to ships at sea via the constellation of global communications satellites, has also greatly facilitated the dispersal of this information via the World Wide Web so that shore-based scientists, students, and the general public can participate in the research experience.

Figure 1.23 The *Shinkai 6500* research submersible operated by the Japan Marine Science and Technology Center (JAMSTEC) has carried out extensive deep-sea research expeditions in the Pacific and Indian Oceans since 1991 and has accomplished over 1300 dives. *Shinkai* also dives with two pilots and one observer (http://www.jamstec.go.jp/e/about/equipment/ships/shinkai6500.html).

Figure 1.24 The Russian research submersible *MIR II*, capable of diving to depths of 6000 m, is operated by the Institute of Oceanography. There are two *MIR* submarines and both have been used for research and cinematographic deep-sea filming involved in numerous Hollywood movies, as well as exploration of the wreck of the *Titanic*. *MIR*s dive with two pilots and one observer.

1.4 Submergence vehicles

The ability for scientists to make direct observations in seafloor environments has yielded significant discoveries, including the discovery of hydrothermal vents. The

Figure 1.25 The newest deep-diving human-occupied submersible – *Alvin*, of WHOI – was placed into service and certified in early 2014. It is currently certified to ~4500 m, but will eventually be able to dive to 6500 m with two observers and one pilot, as it has done throughout its operational career (http://www.whoi.edu/main/hov-alvin/).

emergence of vehicles that could transport researchers directly to the seafloor to conduct fieldwork began in the mid 1960s, with initial engineering efforts driven by the strategic needs of the Cold War (Sontag and Drew, 1998). The confluence of corroborating data to support seafloor spreading, and the need to conduct fieldwork to map and sample the seafloor, led to a groundswell of support to allow small rescue submarines to be used for basic oceanographic research.

1.4.1 Human-occupied vehicles

For the past 40 years, submersible human-occupied vehicles (HOVs) such as the Deep-Submergence Vehicle (DSV) *Alvin* developed at the Woods Hole Oceanographic Institution on Cape Cod, Massachusetts, have given scientists direct access to the seafloor and the ability to explore it from a firsthand and up-close perspective – one they could only fantasize about from the decks of ships. Since *Alvin* was first put into service in 1964, it has served as a model for subsequent research submersible technological advancements both in the USA and abroad (Figs. 1.21–1.25).

Figure 1.26 The remotely operated vehicle (ROV) *Jason*, operated by WHOI's US National Deep Submergence Facility, is a fly-away ROV capable of operating at depths of 6500 m throughout the world's oceans, as it can be operated from many research vessels and can be shipped globally via sea freight. It has a removable sled on its underbelly where instruments can be mounted and samples are stored. It can also accommodate a horizontally oriented drill to take well-located core samples of the seafloor, including the walls of active black smokers, thus providing a 3D perspective into the chimney (http://www.whoi.edu/ndsfVehicles/Jason/).

Figure 1.27 (Top) The ROV *Tiburon*, from the Monterey Bay Aquarium Research Institute (MBARI), is operated from the support ship R/V *Western Flyer* and lowered through a moon pool in the center portion of the ship. *Tiburon* operated from 1996 to 2008, and has been replaced by the ROV *Doc Ricketts* (bottom) (http://www.mbari.org/dmo/vessels_vehicles/Doc_Ricketts/Doc_Ricketts.html).

Today, there are deep-diving (greater than 4 km depth capability) HOVs routinely operated by government agencies and organizations in France (*Nautile*) and Japan (*Shinkai* class) (Figs. 1.22 and 1.23). The USSR had developed two *MIR* submarines (Fig. 1.24), which operated routinely for science missions in the 1980s and1990s, but are now primarily used for underwater cinematography. China has recently entered the community of research submarine operators with the development of the *Jiaolong* submersible that can dive to a depth of 7 km.

For over three decades (1965–1995), *Alvin* played a significant role both as a test bed for technical development of deep-diving research and rescue submarines, and as a routine operational system for conducting basic research in the water column and on the deep seafloor. By July 2014, *Alvin* had made more than 4700 dives. Over that time period, much of what has been learned about the detailed structure and composition of the oceanic crust from site-specific surveys, mapping and sampling, as well as relationships between deep-sea life and hydrothermal processes, by US scientists was accomplished through research expeditions utilizing *Alvin*.

Beginning in early 2011, *Alvin* underwent a major rebuild that includes a new 6500-meter-rated titanium pressure sphere and vast improvements in pilot and observer observational capabilities (Fig. 1.25). There are five viewports, with the three larger diameter viewports providing a 245° field of view across the front of the submarine, giving greater than 40% overlap in fields of view between the pilot and adjacent observers.

Figure 1.28 The ROV *ROPOS* is operated by the Canadian Scientific Submersible Facility. It is a state-of-the-art vehicle for installation of underwater, cabled observatories with the capability of installing high-power, high-bandwidth extension cables on the seafloor using the CSSF ROCLS (Remotely Operated Cable Laying System). On its underbelly, *ROPOS* hosts a latch that allows it to connect into ROCLS (shown here), and packages such as underwater junction boxes weighing almost 2000 kg. ROCLS can be removed from *ROPOS* when operations require it to be in more standard scientific configurations for instrument deployments, sampling, and mapping of the seafloor (http://www.ropos.com).

Figure 1.29 The ROV *Victor*, operated by IFREMER, was developed in the mid to late 1990s and placed into service in 1999 (http://flotte.ifremer.fr/fleet/Presentation-of-the-fleet/Underwater-systems/VICTOR-6000). It has worked extensively throughout the North and South Atlantic Ocean, extending up to the Arctic, conducting multidisciplinary research and observatory science, primarily at the MOMAR observatory at Lucky Strike Seamount on the Mid-Atlantic Ridge near 37° N (Michel *et al.*, 2003; Colac *et al.*, 2011; http://www.ipgp.fr/rech/lgm/MOMAR/).

1.4.2 Tethered robotic vehicles

Although human, cognitive presence has proven extremely valuable and is still essential for certain facets of deep-submergence science, a new class of vehicles was needed that would permit routine access to the deep ocean for repeated measurements over long periods of time and to greater depths. Over the past decade, the development of increasingly sophisticated ROVs by research and engineering groups, principally in the USA, United Kingdom, Canada, France, and Japan, has greatly expanded our ability to observe, map, and sample the ocean and seafloor (Figs. 1.26–1.29). Most recently Germany, China, and Korea have also become major operators of deep-submergence vehicle systems of all types. ROVs and AUVs now offer significant advantages for multidisciplinary studies of oceanographic and seafloor processes (UNOLS, 1994).

ROVs utilize innovative steel-armored, fiber-optic cables with copper conductors that connect the deep-sea vehicle to the support ship. The fiber-optic cable that tethers the ROV to the surface ship supplies power to operate the vehicle's propulsion, navigation, data acquisition, and lighting systems, and transmits gigabytes of data from the depths through glass optical fibers back up to the laboratory on the ship for scientists to analyze in real time. Using ROVs, dozens of scientists can now simultaneously participate in seafloor observations and sampling for days at a time, around the clock.

1.4.3 Autonomous underwater vehicles

AUVs, the latest in a generation of innovative deep-submergence systems, have greatly expanded the capabilities of Earth and ocean scientists to investigate the far

Figure 1.30 The AUV *ABE* (*Autonomous Benthic Explorer*) developed and operated by WHOI (left) being launched in 2001. The WHOI AUV *Sentry* (right) has a dive capability of 6 km and includes multibeam, sidescan, and CHIRP sub-bottom sonar, and a variety of water properties sensors, as well as a digital still camera and strobes that enable it to take continuous photographic images of the seafloor. *Sentry* has now replaced *ABE* as the AUV system in the National Deep Submergence Facility at WHOI (http://www.whoi.edu/main/sentry).

reaches and depths of the global ocean with high efficiency in both the spatial dimension and time. Current oceanographic capabilities include ships with large deck areas and over-boarding infrastructure, generally large berthing capacity for multidisciplinary science parties, as well as improved robotic vehicle capabilities providing the capacity to operate several vehicles in sequential operations on a research cruise. In the case of AUV operations coupled with either HOV or ROV operations, this offers the potential to carry out detailed AUV surveys of specific areas in advance of HOV or ROV diving. This means that detailed maps, as well as reconnaissance images of areas of interest, can be acquired before determining the scope and locations of specific sampling objectives. This type of "nested survey" science (Fornari *et al.*, 1997; Shank *et al.*, 2003) was carried out in a sulfide recovery and mapping project on the Juan de Fuca Mothra Vent Field (Delaney *et al.*, 1998), on the southern East Pacific Rise to interpret DSL-120 sonar data and explore for fresh lava flows and hydrothermal vents (Baker *et al.*, 2002), for the documentation of the Lost City vents (Kelley *et al.*, 2001c, 2005; Karson *et al.*, 2006), to locate new vent sites at 86° W at the Galápagos Rift (Shank *et al.*, 2003), and on the summit of Lucky Strike Seamount on the Mid-Atlantic Ridge near 37° N to map the distribution of hydrothermal and volcanic facies (Scheirer *et al.*, 2000; Colaco *et al.*, 2002; Humphris *et al.*, 2002; Escartín *et al.*, 2008; Barreyre *et al.*, 2012).

AUVs, because of their stability and hovering capabilities, have produced unprecedented nested observations, including bathymetric maps at the sub-meter scale, three-dimensional imaging of hydrothermal plumes, and time-series data of water properties at 1–20 Hz (Figs. 1.30–1.36). These capabilities result from the significant payload of sophisticated sensors carried by modern AUVs, which includes cameras; multibeam, sidescan and CHIRP sub-bottom sonars; chemical sensors including reducing potential (Eh), pH, dissolved O_2 and mass-spectrometers; and precision navigation systems. In the USA, the *ABE* AUV, rated to 4.5 km operating depth, was responsible for

Figure 1.31 Illustration showing the AUV *ABE* acquiring near-bottom bathymetric data and co-registered water properties data as it flies over a seafloor terrain where a hydrothermal vent is present. The thin colored trail behind the vehicle represents bottom-water temperature anomalies – the red color shows the presence of the vent plume. The blue-green swath is the near-bottom multibeam bathymetry.

Figure 1.32 The MBARI AUV *D. Allan B.* is a torpedo-shaped vehicle rated for operations to 6 km depths. It is equipped with multiple sonar systems that operate simultaneously during a mission, providing exceptional resolution of seafloor features on and below the bottom. The vehicle acquires multibeam, sidescan, and sub-bottom sonar data that can resolve topography and features smaller than 1 m, equivalent to high-resolution mapping done on land that is used to assemble digital elevation models of various terrains. The image also shows the Launch and Recovery System (LARS) for the vehicle (http://www.mbari.org/auv/MappingAUV/Default.htm).

collecting some of the first very high-resolution (1 to 5 m) bathymetric maps of seafloor along MORs at the Juan de Fuca Ridge and the northern and southern East Pacific Rise in the late 1990s and early 2000s. Unfortunately, in 2011, *ABE* was lost during a mission to the Chile Ridge in the Southeast Pacific. This was likely due to a failure in one of its glass pressure spheres that housed the electronics, causing an implosion that destroyed the vehicle.

Recent developments in high-capacity lithium-ion batteries and vast improvements in robotics software and command/control capabilities, coupled with miniaturization of in-hull computer systems, have resulted in much longer and more far-reaching AUV surveys in a wide range of seafloor environments. These capabilities have greatly expanded the potential for adaptive autonomous mission profiles, where the vehicle can respond to environmental conditions to change mission profiles and zero-in on important features or processes.

Throughout the community of deep-submergence vehicle operators, data acquisition capabilities of research submarines, ROVs, and AUVs now support each other – spatially, temporally, and in their relative payloads – making the integrated data sets more comprehensive, geospatially registered, and thus more information-rich. This synergy among vehicle systems (Shank *et al.*, 2003) also results in significant improvements in the ability to direct and modify data collection in real time during a field program, thus allowing scientists to study multi-scale problems (centimeters to kilometers) and efficiently react to at-sea discoveries.

One of the most important capabilities afforded by modern AUV systems is their ability to acquire near-bottom sonar data of various types to produce high-resolution maps

Figure 1.33 Near-bottom multibeam bathymetry maps acquired by the MBARI AUV *D. Allan B.* during surveys at the summit of Axial Seamount prior to 2011 (left) and in 2012 (center), showing the changes in seafloor topography created by the 2011 submarine eruption. The right map shows the difference in seafloor depth caused by the latest volcanic eruption (Caress *et al.*, 2012; Clague *et al.*, 2013).

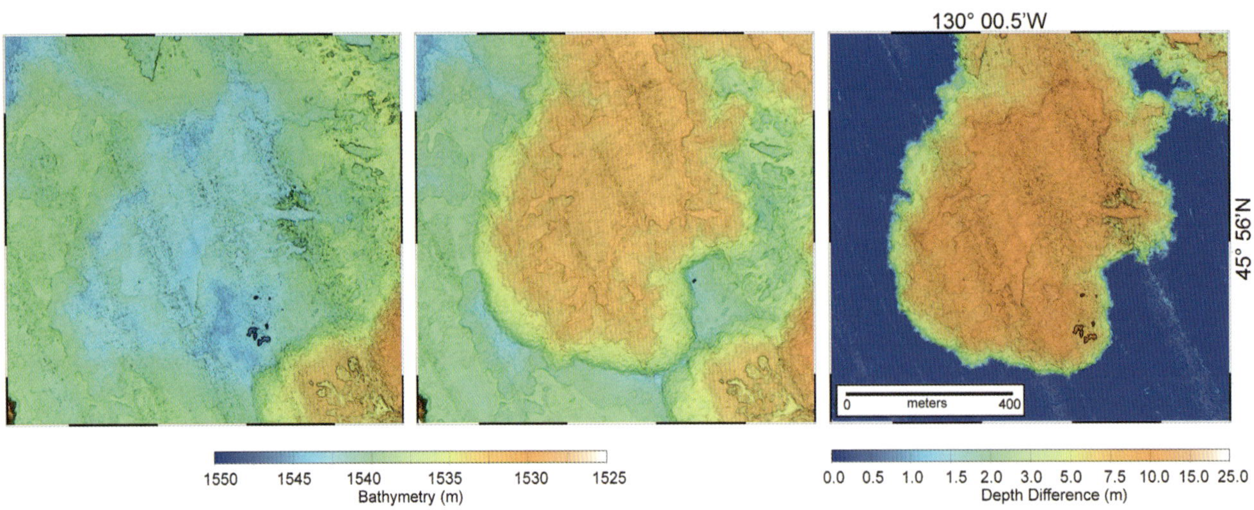

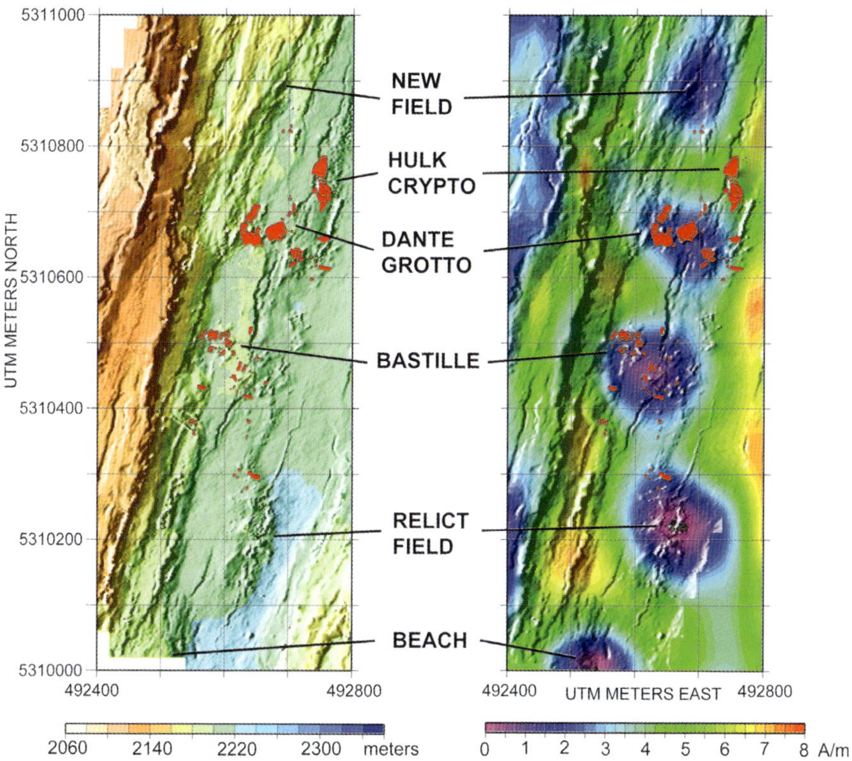

Figure 1.34 Near-bottom bathymetry (left) collected during AUV and ROV dives over the Main Endeavour Vent Field and corresponding near-bottom magnetic intensity map showing low magnetization "bull's eyes" over areas affected by long-term hydrothermal circulation in the crust. The corrosive hydrothermal fluids alter the magnetic minerals in the host basaltic rocks (Tivey and Johnson, 2002).

Figure 1.35 REMUS (Remote Environmental Monitoring UnitS) are a class of torpedo-shaped AUVs (left) operated by the Ocean Systems Laboratory (OSL) at WHOI that provide autonomous survey capability over a depth range from 100 to 6000 m. The vehicles have been used for measurement of coastal water properties and currents, high-resolution sidescan and multibeam mapping, and for military applications associated with mine clearing operations in coastal waters. A REMUS 6000 AUV being recovered (right). The OSL Group, operating three REMUS 6000 vehicles simultaneously during several cruises, was responsible for the search and discovery of the black box of Air France Flight 447 (Fig. 1.36), which crashed into the mid-Atlantic near 2° N and 32° W in June 2009, on the east flank of the Mid-Atlantic Ridge in water depths of approximately 4 km.

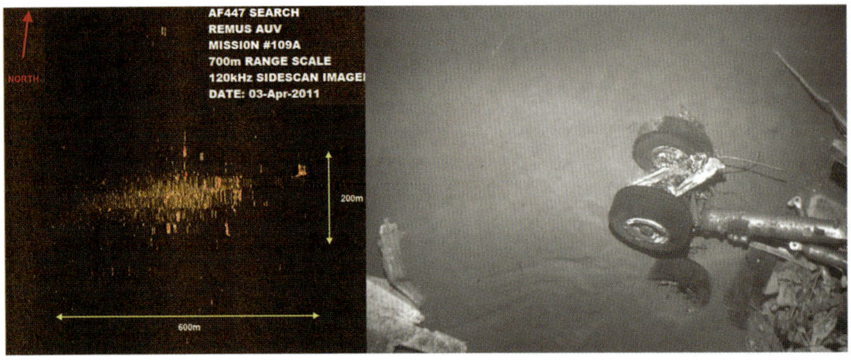

Figure 1.36 Sidescan sonar image (left) acquired by the REMUS 6000 AUV during the search for the Air France 447 crash site, showing acoustically bright targets in an elongate pattern made by the crash debris in a sediment pond on the east flank of the Mid-Atlantic Ridge near 2° N. REMUS 6000 photograph (right) of one of the landing struts of AF447 partially buried in sediment.

of the seafloor. The higher frequencies of the multibeam, sidescan, and CHIRP sub-bottom data used on AUVs (ranging between 200 and 500 kHz) require the vehicle to be able to traverse the seafloor along precise track lines, separated by ~10–100 m at altitudes that range from about 5 to 50 m, to produce sub-meter resolution bathymetry (Fig. 1.33). The sophisticated navigation and attitude sensing systems on AUVs use combinations of inertial navigation, fiber-optic gyroscopes, and bottom-lock acoustic Doppler current profiler (ADCP) dead reckoning at rates higher than 20 Hz to achieve highly precise vehicle motion records and geodetically referenced navigation. The integrated navigation and vehicle attitude systems are critical to accurate processing of high-frequency sonar data. This permits the generation of digital bathymetric maps with sub-meter spatial and vertical resolution and produce near-bottom geophysical images that provide new perspectives on subseafloor processes (Figs. 1.33–1.36).

1.4.4 Hybrid remotely operated vehicles

In May and June 2009, in the Challenger Deep of the Mariana Trench, the WHOI hybrid remotely operated vehicle (HROV) *Nereus* successfully performed scientific observation and sampling at 10,903 m. The *Nereus* HROV was designed to perform scientific survey and sampling to the full ocean depth – significantly deeper than the dive capability of all vehicles operating today. *Nereus* operated in two different modes (Fig. 1.37). For broad-area survey, the vehicle could operate untethered as an AUV capable of exploring and mapping the seafloor with sonars and cameras. *Nereus* could be converted at sea to become an ROV for close-up imaging and sampling. The ROV configuration incorporated a lightweight fiber-optic tether for high-bandwidth, real-time video and data telemetry to the surface. This enabled high-quality tele-operation. A manipulator, lightweight hydraulic power unit, and sampling instruments could be added to provide sampling capabilities. This technology provided a practical solution to achieving access to the deepest ocean floor. While science questions for accessing hadal depths (below 6 km) led to the development of an HROV, the successful development of *Nereus* (Bowen *et al.*, 2009) prompted scientists from around the world to come together to identify and investigate the foremost questions in deep-ocean science using this novel vehicle technology. Unfortunately, the HROV *Nereus* was lost while diving at 10 km depth during a science mission to the Kermadec Trench in May 2014 on a dive to 10,000 m depth, after performing numerous dives from 6000 m to 10,000 m for the HADES international research program (http://www.whoi.edu/hades/).

In March 2012, a seminal event occurred in ocean exploration with the successful descent of James Cameron's *DEEPSEA CHALLENGER* deep-diving vehicle to the deepest depth in the ocean – the Mariana Trench's Challenger Deep (Fig. 1.38). That experience, and certainly others that will follow in the twenty-first century, will continue to foster awareness about the oceans, how little we know about its inhabitants and the many terranes it contains, its beauty, and the importance of oceanographic science to global society's future.

1.5 Seismic monitoring and imaging

Scientists use seismometers to decipher the internal structure of Earth. Seismic waves generated by earthquakes are distorted, accelerated, or slowed down as they pass through different types of rocks inside the Earth. Seismometers equipped with precise clocks record the shape and speed of these waves when they arrive at the recording sensor. After an earthquake, data from many seismometers distributed around the globe help seismologists locate the earthquake and calculate (using triangulation methods) the depth at which it was generated (the epicenter). Using the seismic data geophysicists can calculate the velocity structure of Earth's crust and mantle, which are used to estimate the composition of those parts of Earth we cannot see or sample.

1.5.1 Revealing the details of crustal and upper mantle structure at MORs using multichannel seismic techniques

Multichannel seismic (MCS) methods use multiple streamers containing hundreds of hydrophone receivers, and tuned airgun arrays that provide a sound pulse of controlled-source

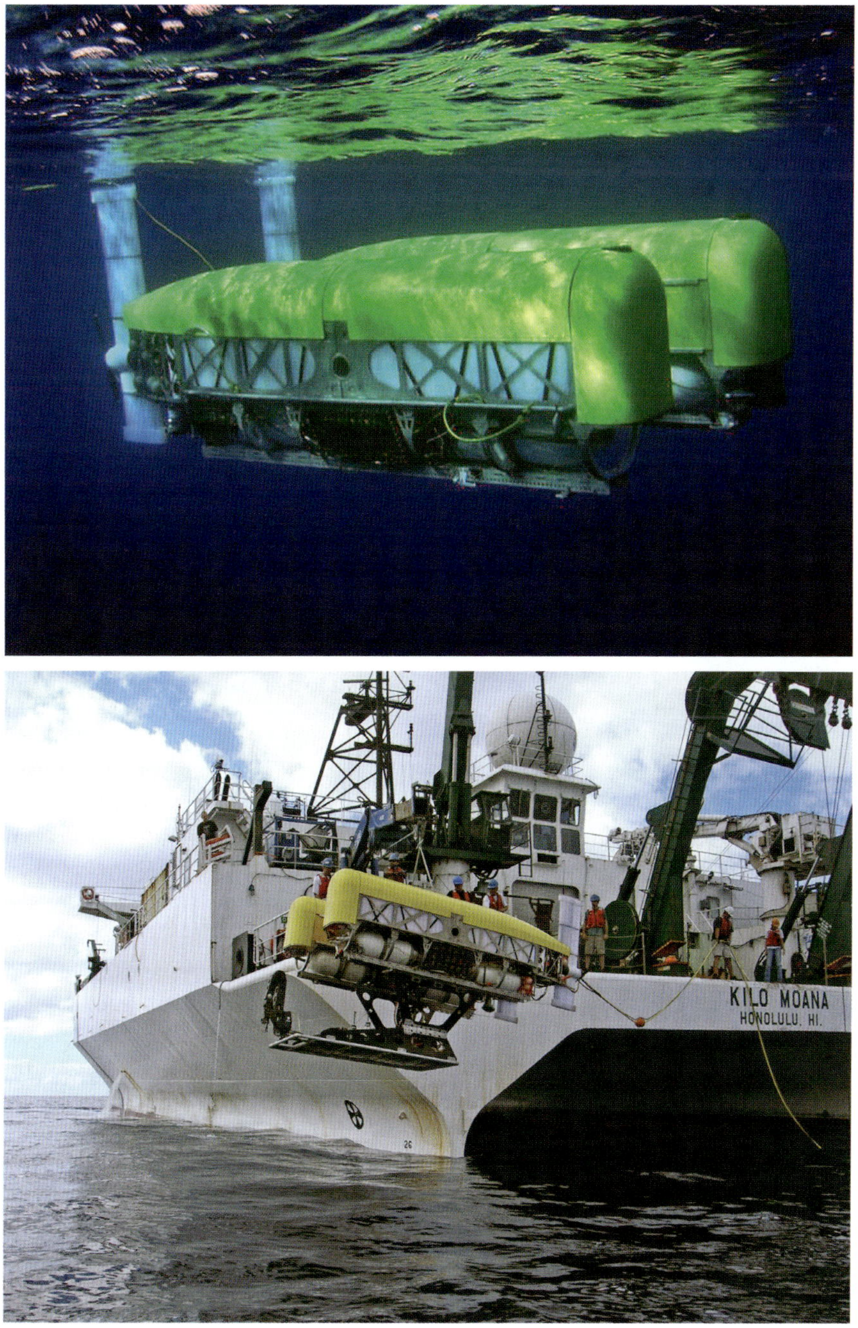

Figure 1.37 The WHOI hybrid remotely operated vehicle (HROV) *Nereus* (top) shown in AUV mode. It can be configured at sea and transformed into an ROV with manipulative capability and real-time video and command/control functions using novel micro-fiber technology. It is rated for operations to full ocean depth (11,000 m). *Nereus*, configured (bottom) in its hybrid ROV mode with a tool sled and manipulator, shown during sea trials in 2008 using the University of Hawaii research ship R/V *Kilo Moana* offshore Oahu, Hawaii. Unfortunately, Nereus was lost in 2014 during diving in the Kermadec Trench.

compressional energy (Mutter *et al.*, 2009). The streamer arrays (Fig. 1.39) are towed behind survey vessels to acquire detailed subseafloor scans of sections of the oceanic crust and upper mantle. These advances in imaging the oceanic crust have been made possible by researchers' access to state-of-the-art MCS acquisition systems now commonly used in deep-water oil and gas exploration, and by development of data processing and modeling techniques that specifically deal with the particularities of oceanic crustal structure and the challenges of subseafloor imaging in the abyss (Canales *et al.*, 2012b; Carbotte *et al.*, 2012).

A strength of 3D MCS reflection imaging lies in its ability to "sense" the seismic wave field in three dimensions, resulting in improved accuracy and resolution of structure, particularly in rugged terrains such as at MORs, where energy scattered by rough seafloor contaminates and degrades the recorded signal. Magmatism along the MOR crest drives a host of processes related to crustal generation. Thus, an ability to image magma bodies beneath the MOR axis is crucial to understanding accretionary processes. Similarly, faulting modifies the newly created crust. Understanding the internal structure of MORs along- and across-axis aids in identifying the role of tectonics in modifying and aging the crust as it spreads away from the axis.

Because MOR magmatic and structural processes are very dynamic in time, the size of the axial magma body can vary

Figure 1.38 James Cameron's one-person submersible, *DEEPSEA CHALLENGER*, which he co-designed, transported him to the deepest region of the global ocean, the Mariana Trench, to sample, observe, and film in 3-D.

over decades due to partial draining during eruptions and replenishment with magma in between eruptions (Perfit and Chadwick, 1998; Rubin *et al.*, 2012; Chadwick *et al.*, 2013). Time-lapse techniques using repeat 3D surveys ("4D seismics") are well established in the hydrocarbon industry for monitoring reservoir depletion, and can be readily applied to MOR settings. Axial melt bodies beneath MORs at a range of spreading rates and with variable segment structure have been investigated with 3D MCS over the past decade, including the EPR 9° N OSC and the EPR between 9°30′ N and 9°55′ N (Kent *et al.*, 2000; Carbotte *et al.*, 2012, 2013), the melt lens beneath Lucky Strike seamount at the MAR 37° N (Singh *et al.*, 2006), the Cleft and Endeavour segments of the Juan de Fuca Ridge (Canales *et al.*, 2009; Carbotte *et al.*, 2012) (Fig. 1.40). At the northern EPR near 9°50′ N, a time series of MCS data has provided the first glimpse of temporal variations over decadal scales in subcrestal melt distribution at a fast-spreading MOR (Mutter *et al.*, 2009; Carbotte *et al.*, 2012, 2013). In addition, because of the higher resolution of the 2008 MCS acquisition at the northern EPR, off-axis magma bodies have been recognized there for the first time, east of the axis between 9°53′ N and 9°55′ N using 3D MCS techniques (Canales *et al.*, 2012).

1.5.2 Ocean-bottom seismometers

About 90% of all natural earthquakes occur beneath the ocean surface, where high pressure and near-freezing temperatures make measurements difficult. The ocean-bottom seismometer (OBS) was developed to measure earthquakes that occur throughout the ocean basins and to extend our understanding of crustal and upper mantle structure across the margins of continents into the ocean basins. Earthquakes along the MOR occur at generally shallow depths, less than 10 km below the seafloor, and tend to be weak, less than magnitude 2 on the Richter scale; while earthquakes above subduction zones associated with deep-ocean trenches typically occur at greater depths (>20 km) and can have magnitudes above 9, with over 200 times the energy release (Satake and Atwater, 2007; Wilcock *et al.*, 2009; Weekly *et al.*, 2013).

OBSs are also used as seafloor receivers during active source MCS 3D seismic experiments; however, they generally are autonomous, recording data internally for downloading and processing after recovery. Some instruments have been wired to shore in areas that frequently experience seismic hazards, such as offshore eastern Japan, where the Japan Trench lies above a subduction zone and large deep earthquakes occur often (https://www.jamstec.go.jp/donet/e/).

Seismometers work using the principle of inertia. The seismometer body rests securely on the seafloor. Inside, a heavy mass hangs on a spring between two magnets. When the ground moves, the seismometer and its magnets move too, but the mass briefly stays where it is. As the mass oscillates through the magnetic field, it produces an electrical current, which the instrument measures.

The seismometer itself (Fig. 1.41) is a small metal cylinder; the rest of the OBS instrument consists of a data logger and batteries to run the seismometer, often a weight to sink it to the seafloor, a remote-controlled acoustic release, and flotation to bring the instrument back to the surface. Numerous countries are now installing cabled versions of OBS for early warning systems for offshore earthquakes that may generate tsunamis (see Sections 4.10 and 4.11).

Engineers have designed two basic kinds of seismometers that measure ground motion caused by earthquakes, both small motion (less than 1 mm) and large motion (typically more than a meter). Short-period OBSs record high-frequency motions (up to hundreds of times per second). They can record small, short-period earthquakes and are also

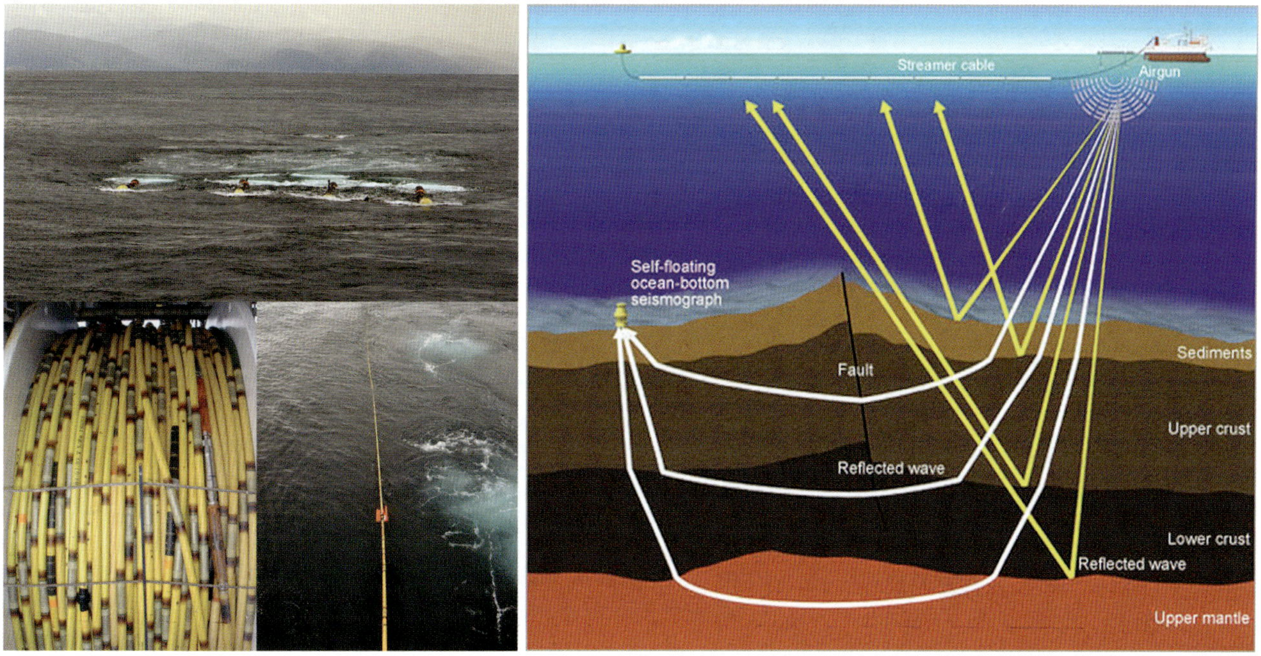

Figure 1.39 Composite image (left) showing multichannel seismic (MCS) acquisition system deployed from the R/V *Langseth* off the coast of Oregon during one of the CASCADIA experiment legs. Top left image shows the airgun array. Lower left image shows the hydrophone streamer reel. Lower middle image shows the streamer being deployed. Illustration (right) shows the methodology for acquiring seismic data using reflection techniques.

useful for studying the crust and upper mantle beneath MORs (see Fig. 4.119). Long-period OBSs (also known as "broadband" instruments) record a much broader range of motions, with frequencies of about 10 per second to once or twice a minute. They are used for recording mid-sized earthquakes and seismic activity far from the instrument.

1.5.3 Recording seismicity along the global MOR axis using autonomous hydrophones and the SOFAR channel

During World War II, Maurice Ewing and J. Lamar Worzel, two US scientists working at the Lamont-Doherty Geological Observatory of Columbia University, embarked on

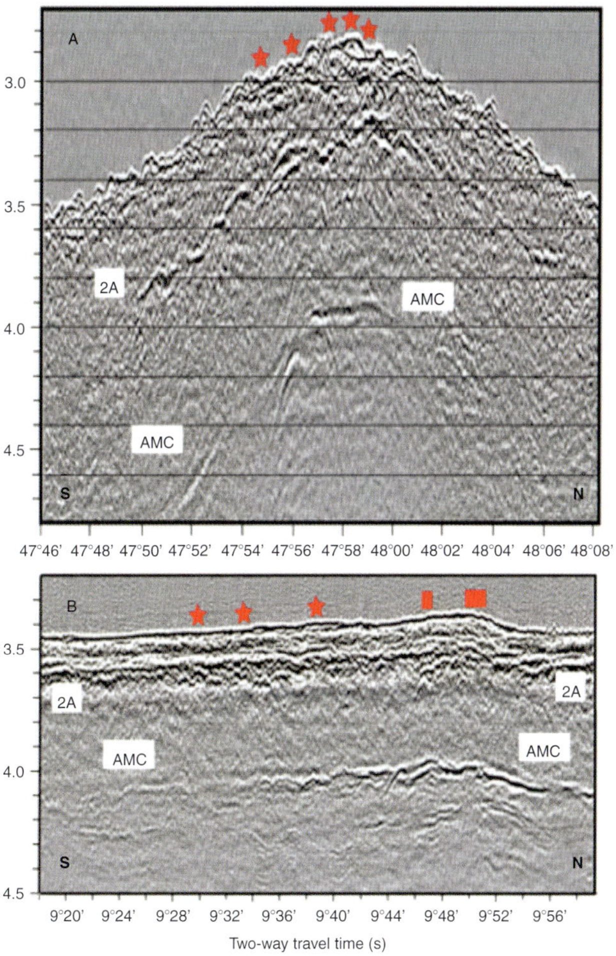

Figure 1.40 Multichannel seismic reflection profiles along the axis of the Juan de Fuca Ridge over the Main Endeavour Hydrothermal Vent Field (top) and East Pacific Rise (above) commonly show a strong subsurface reflector interpreted as the top of an axial magma chamber (AMC). Highly reflective upper crust interpreted to be high-porosity lavas is commonly referred to as Layer 2A (base of this layer is noted). These seismic profiles show that the AMC and Layer 2A are nearly continuous along the axis with disruptions at spreading segment boundaries, and that AMCs are present above areas where hydrothermal vents are present. The vertical scale is in two-way travel time of seismic waves and corresponds to a depth for Layer 2A of about 300 m and for the AMC of about 1200 m. The red stars and boxes show locations of high-temperature hydrothermal vents (Carbotte *et al.*, 2012).

Figure 1.41 Ocean-bottom seismometers (OBSs) on the back deck of the R/V *Wecoma* of Oregon State University (top) in preparation for deployment for the CASCADIA experiment, an onshore and offshore seismic experiment designed to study large magnitude (>9) megathrust earthquakes along the Washington and Oregon coast (http://cascadia.uoregon.edu/CIET/). An OBS being deployed from the R/V *Oceanus* (bottom left) and the electronics of an OBS housed in a glass pressure sphere (bottom right).

a basic research program to study acoustics in the ocean. Their immediate goal was to develop strategies that would help the US Navy detect enemy submarines or mines, or help US submarines avoid detection.

The key experiment they conducted involved detonating 1 pound (0.45 kg) of TNT explosive under water near the Bahamas islands, and detecting the sound thousands of kilometers to the east near West Africa. The test confirmed Ewing's theory that low-frequency sound waves were less easily scattered or absorbed by water and could travel long distances without losing energy. The key, however, was the discovery of a layer of water in the ocean that acted like a pipeline to channel low-frequency sound and transmit it over great distances. This sound pipeline, called the SOFAR (sound fixing and ranging) channel, was also independently discovered by Russian scientist Leonid Brekhovshkikh, working with underwater explosions in the Sea of Japan.

The key to the SOFAR channel (Fig. 1.42) is that sound energy, traveling in waves, speeds up in waters where temperatures are warmer (near the surface) or where

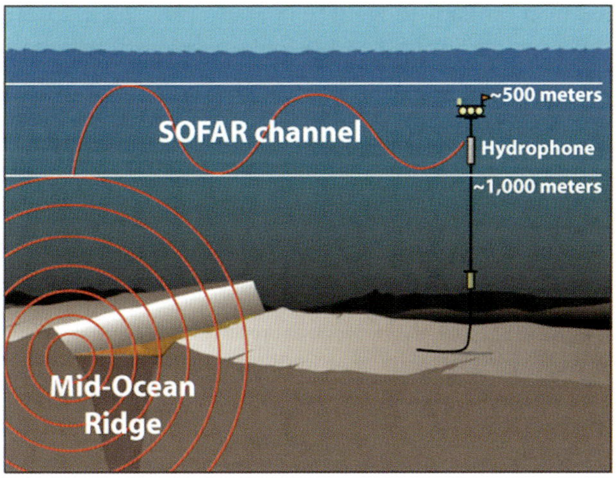

Figure 1.42 Hydrophones can be deployed in the middle of ocean basins, in arrays that encompass large segments of the MOR crest and upper flanks. The instruments record autonomously and have been developed to detect and monitor previously unheard/undetected low-magnitude (less than about magnitude 2) volcanic and earthquake activity in the oceans that cannot be detected using land-based seismic arrays. They are deployed in the SOFAR channel, a layer of water in the ocean that channels sound waves generated by seismic events, allowing them to be transmitted and detected thousands of miles away.

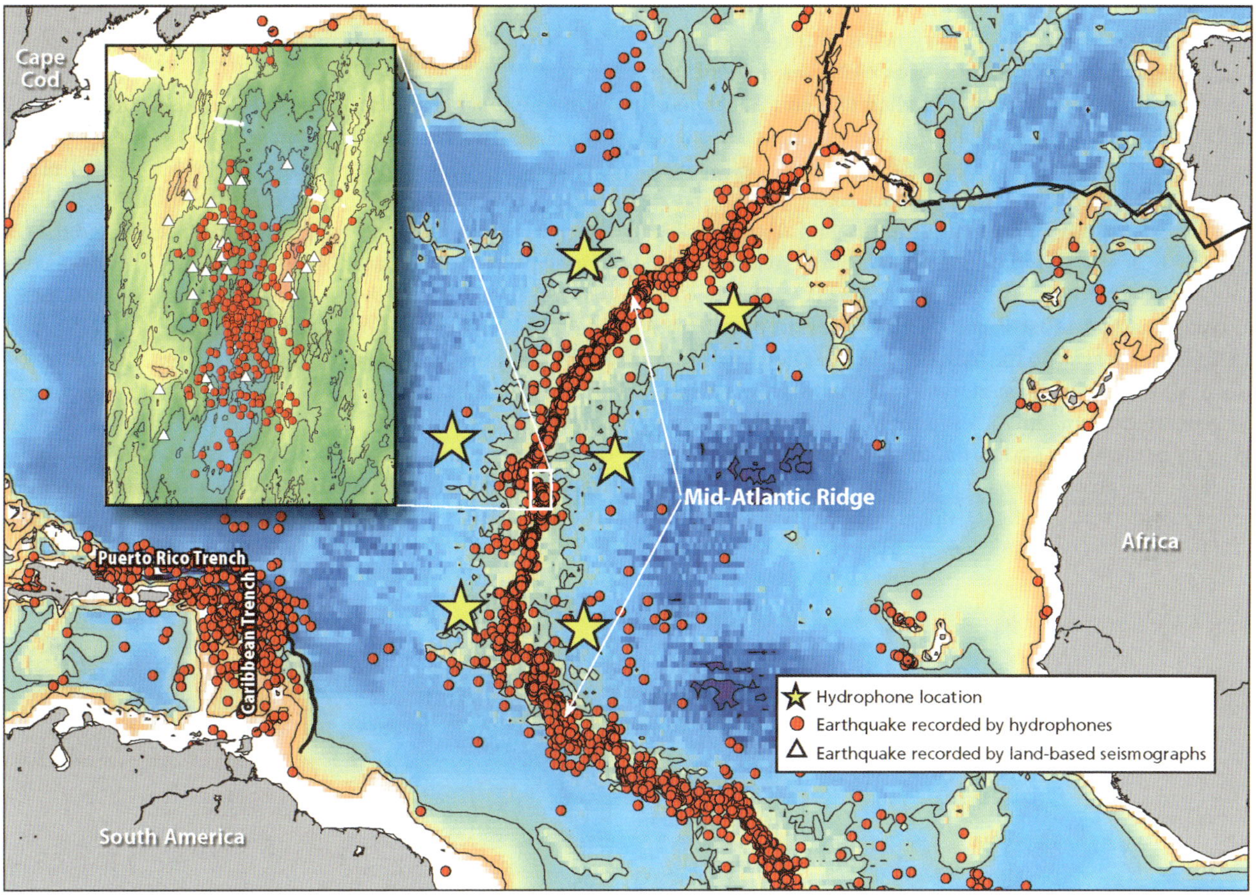

Figure 1.43 Deployed in 1999, an array of six hydrophones that use the SOFAR channel to register small earthquakes generated at the Mid-Atlantic Ridge axis and flanks, as well as elsewhere in the North Atlantic basin, recorded over 7000 earthquakes between 1999 and 2002. These data provide a new perspective on MOR seismicity patterns along the ridge axis that may help to identify areas of thicker versus thinner crust along the axis, as well as areas that may be more prone to magmatic events such as submarine eruptions or dike intrusions. The inset map shows the location of the many earthquakes recorded by the autonomous array (red dots) and the much smaller number of land-based detected earthquakes over the same time period (white dots).

pressure is higher (at depths). However, in between those bodies of water the SOFAR channel is bounded on the top and bottom by water layers where sound velocities increase. The boundaries act like a ceiling and floor. When sound energy enters the channel from below it slows down when it hits the top layer (the "ceiling"). It does not keep going, but rather it is refracted back downward. Then it hits the "floor" or bottom layer, and it is refracted back upward. In this way, sound is efficiently channeled horizontally with minimal loss of signal over thousands of kilometers (Smith, 2004).

The US Navy immediately saw the value of this research and the SOFAR channel. The research program eventually developed into the Navy's Sound Surveillance System, or SOSUS. In the mid to late twentieth century, the Navy deployed a network of underwater microphones, called hydrophones, connected by undersea cables to onshore facilities at strategic points around the globe. The hydrophones were installed in locations to optimally exploit the SOFAR channel. The Navy could detect radiated acoustic energy of less than a watt at ranges of several hundred kilometers – a sensitivity that can distinguish the number of propellers on a submarine. Clearly, basic oceanographic geophysical research made a substantial contribution to US strategic interests, primarily in monitoring Russian and other nations' submarine movements during the Cold War. While the strategic value of SOSUS was evident, the scientific application of the system had to wait until the end of the Cold War in 1991, when the Navy declassified some of the SOSUS capability and allowed academic research to be carried out using it (Fox *et al.*, 1995, 2001).

Scientists at the National Oceanic and Atmospheric Administration's (NOAA) Pacific Marine Environmental Laboratory (PMEL) in Seattle, led by Chris Fox, had the idea of using the Northeast Pacific SOSUS array to detect low-level seismic waves on the Juan de Fuca Ridge off the Pacific Northwest, with the assumption that sound waves would be generated by magmatic eruptions or earthquakes along the ridge-transform system in the Northeast Pacific. That strategy was groundbreaking; it led to important discoveries of detected magmatic events at various locations along the Juan de Fuca Ridge (Dziak *et al.*, 1999, 2011, 2012a). However, the SOSUS network in the Northeast Pacific only covers a small segment of the global MOR, so PMEL developed moored, autonomous hydrophones that could be deployed for as long as two years in the SOFAR channel to monitor seismicity in more remote ocean regions of the MOR. In 1996, PMEL deployed long-term hydrophones in the eastern equatorial Pacific between 20° N and 20° S and began monitoring them successfully (Fox *et al.*, 2001). In 1999, the network was expanded into the North Atlantic (Smith *et al.*, 2003) (Fig. 1.43), and in 2004 an autonomous array was deployed in the Southwest Pacific, east of Fiji in the back-arc spreading centers in the Lau Basin (Dziak *et al.*, 2012b).

1.6 Ocean drilling: probing into the seafloor

Oceanographers have been able to collect sediment and rock samples from the ocean bottom ever since the *Challenger* Expedition, but they did not have the technology to probe very far beneath the seafloor until the late 1960s. In 1968, an international group of oceanographic institutions and the US National Science Foundation created a program for ocean drilling. For 25 years, the Deep Sea Drilling Project (DSDP) operated the *Glomar Challenger*, a research ship 122 m in length equipped with a drilling platform and scientific laboratories. The DSDP was so successful that a new Ocean Drilling Program (ODP) was created in 1985. Ocean-floor drilling continues today with a larger and more technologically advanced US drill ship, the *JOIDES Resolution*, a 143 m long vessel that can drill in water over 8 km deep (Fig. 1.44),

Figure 1.44 The ocean drilling ship *JOIDES Resolution* (left) has been drilling academic boreholes into the ocean crust for the past 2.5 decades. The drilling ship *Chikyu* (right), operated by the Japan Agency for Marine-Earth Science and Technology (JAMSTEC), has the capability to drill holes with riser technology to depths of ~4000 m.

Figure 1.45 An IODP scientist studying a core of basalt approximately 5 cm in diameter onboard the *JOIDES Resolution*.

as well as several other academic drill ships and platforms operated by other countries, notably Japan, the United Kingdom, and France (http://www.iodp.org) (Smith *et al.*, 2010; Bickle *et al.*, 2011; Humphris *et al.*, 2011).

Initially, DSDP cores collected on either side of MORs provided crucial evidence to confirm seafloor spreading and plate tectonics by determining the age of the sediments just above the volcanic basement rocks at known distances from the modern spreading axis. The drill cores were also instrumental in determining and refining the geomagnetic polarity time scale, an essential reference frame that allows scientists to determine rates of Earth and ocean processes from the present to more than 150 million years ago (Tominaga and Sager, 2010; Malinverno *et al.*, 2012).

The long sections of sediments that accumulate layer-by-layer on the seafloor and remain undisturbed for millions of years before being sampled by ocean drilling also provide a record of how Earth's climate has changed during its history. Drilling provides the only practical way to access continuous marine geological records of ocean temperature, chemistry, and circulation for at least the last 65 million years. As we now know, Earth's climate has not been constant over the past few hundreds of thousands of years, as well as further into the past hundreds of millions of years. Drilling studies show that Earth temperatures were much warmer than at present for millions of years, despite similar levels of heating by the Sun. Chemical information in the cores is also being used to evaluate the past levels of atmospheric carbon dioxide (CO_2) and to understand the sensitivity of global temperatures to greenhouse gases. All of these studies require the ability to drill long cores in sediments and rocks deep into the Earth's crust at many geographic locations and in many tectonic environments (Fig. 1.45).

Scientific ocean drilling provides an ability to conduct experiments with novel sensors and recording systems within the Earth's crust, to determine chemical and biogeochemical reactions between rocks and fluids within the crust and how they are influenced by temperature, rock type, and tectonic environment. Drilling studies have determined that hydrothermal fluids can travel tens of kilometers within the crust, but little is known about the rates of transport, the geometry of the plumbing system, and the influence these processes have on how the Earth's lithosphere, ocean, and biosphere have evolved (Davies and Elderfield, 2004). Some of the boreholes drilled by scientific ocean drilling experiments are used as observatories to monitor the temperature and pressure of crustal fluids, and to collect samples of these fluids for analyses both *in situ* as well as in shore-based laboratories. By conducting these types of experiments in adjacent boreholes simultaneously, scientists can identify how these properties vary between locations and in time. Boreholes are also being used as sites where seismometers can be installed and wired into ocean observatory networks to help monitor earthquakes and provide early-warning capability for large earthquakes that may generate tsunamis. Ocean drilling has and will continue to be an important technological capability for ocean and Earth scientists well into the twenty-first century and beyond. It provides the definitive linkage between remote sensing acquired using seismic and other geophysical techniques in a variety of seafloor environments and the actual sedimentary, metamorphic, or igneous rocks present in the oceanic crust.

1.7 Real time access to the ocean realm: ocean observatories

For generations, research ships have been the primary way for oceanographers to go to sea for weeks or months at a time, to carry out their expeditions, and to collect

Figure 1.46 Locations of installed (green dots) and planned cabled seafloor observatories (red dots). MOMAR, NEPTUNE and the Regional Scale Node (RSN) cabled observatories are dedicated to real-time monitoring of MOR processes.

data. The need for ships will never go away, because there is no substitute for being at sea, immersed in the environment when deploying and recovering instruments and samples, with the synergy that exists among scientists and engineers in the field. However, newly available sensor and submersible technologies have significantly increased the scope, efficiency, and accuracy of observations and field experiments that are not necessarily shipboard-monitored. These are collectively called ocean observatories (Figs. 1.46–1.48).

As investigators have learned more and more about oceanographic processes, they have come to realize that they are not often in the right place at the right time, or making observations for long enough to capture the full time scale of an event or process. Ocean observatories provide the capability to investigate and measure some of the major processes that result in significant heat, chemical, and biological fluxes in the oceans – large storms, underwater eruptions, landslides, harmful algal blooms.

With this recognition, John Delaney from the University of Washington and Alan Chave from WHOI, nearly 20 years ago (Delaney *et al.*, 1987), began formulating plans to bring power and the Internet directly into the oceans for a sustained, continuous presence, through the use of high-power, high-bandwidth cabled observatories hosting hundreds of sensors distributed on the seafloor and throughout the water column on moorings.

This dream is now realized, with the installation of two cabled observatories off Vancouver Island, Canada, and Washington and Oregon, USA, that include an array of 1700 km of fiber-optic cables (Fig. 1.47) (Kelley *et al.*, 2014). These two systems, NEPTUNE Canada (now called Ocean Networks Canada – ONC) and the Regional Scale Nodes (RSN) observatory (http://www.interactiveoceans.washington.edu/) funded by the US National Science Foundation as part of the Ocean Observatories Initiative (OOI), represent the largest infrastructure investment in the world's oceans for oceanographic studies in the twenty-first century.

Providing up to 8 kW power and 240 Gb/s communications, these new ocean observing systems host hundreds of seafloor and water column sensors. They include high-definition (HD) cameras imaging hydrothermal vents (Fig. 1.48); mass spectrometers for measuring gas concentrations in methane seeps and vents; seismometers that allow detection of magma movement beneath the seafloor and large subduction zone earthquakes; and tilt meters to measure the inflation and deflation of underwater volcanoes prior to and following eruptions.

Instrumented robots, such as the deep-sea crawler called *Wally* on NEPTUNE Canada, are investigating methane seeps, and providing video and chemical data directly to shore. In addition, water column moorings spanning 3 km vertically in the ocean and hosting instrumented wire crawlers are now providing information on changing acidity of the oceans, climate change, and periodic anoxia events resulting in large die-offs of benthic animal communities. These distributed sensor arrays will provide continuous communications and transmission of video and data for many years in real time over the Internet to the global community (Fig. 1.49).

Numerous countries are now investing in this new technology because of its importance to science and also to society. For instance, Japan is continuing to invest millions of dollars into its offshore, cabled observatories to provide early warning for earthquakes and tsunamis. Installation of

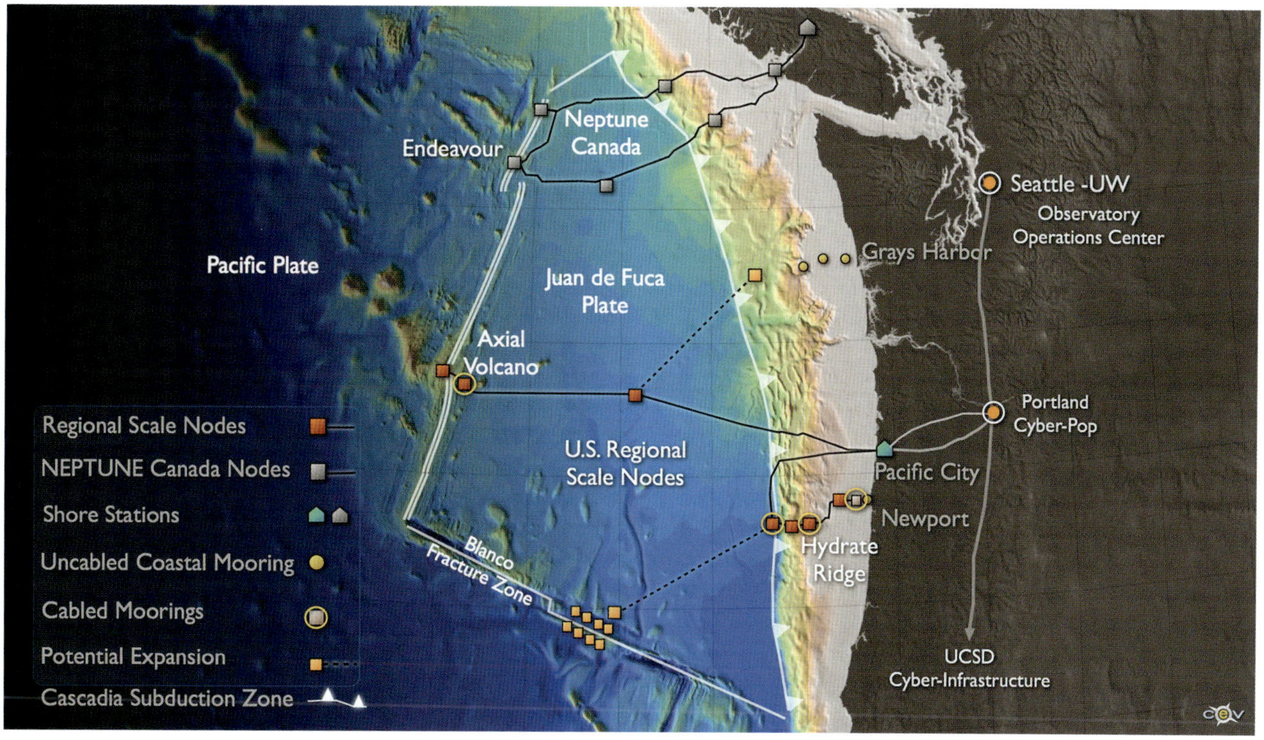

Figure 1.47 The Regional Scale Nodes (US) and Ocean Networks Canada submarine cabled observatories, located off the Pacific Northwest coast, include 1700 km of high-power, high-bandwidth telecommunications cables, 13 large seafloor nodes that distribute power and communications to smaller junction boxes, and sensors at key scientific sites. Extension cables provide power and communications to hundreds of instruments on the seafloor and throughout the water column on instrumented moorings that rise 3 km above the seafloor.

Figure 1.48 Illustration showing the ROV *Jason2* and a deep-sea light system used for illuminating the seafloor for high-definition video imaging during a University of Washington experiment on the Juan de Fuca Ridge in 2005, in preparation for the installation of NSF's Regional Scale Nodes seafloor observing system.

these systems has required the development of new deep ROVs, such as the Canadian system *ROPOS*, that can install kilometers of seafloor extension cables. AUVs will be able to plug into docking stations that will allow transfer of data and *in situ* recharging of the vehicles' batteries for a sustained presence beneath the ocean's surface.

The most vital processes at work in the deep ocean and within the Earth are happening continuously, over great

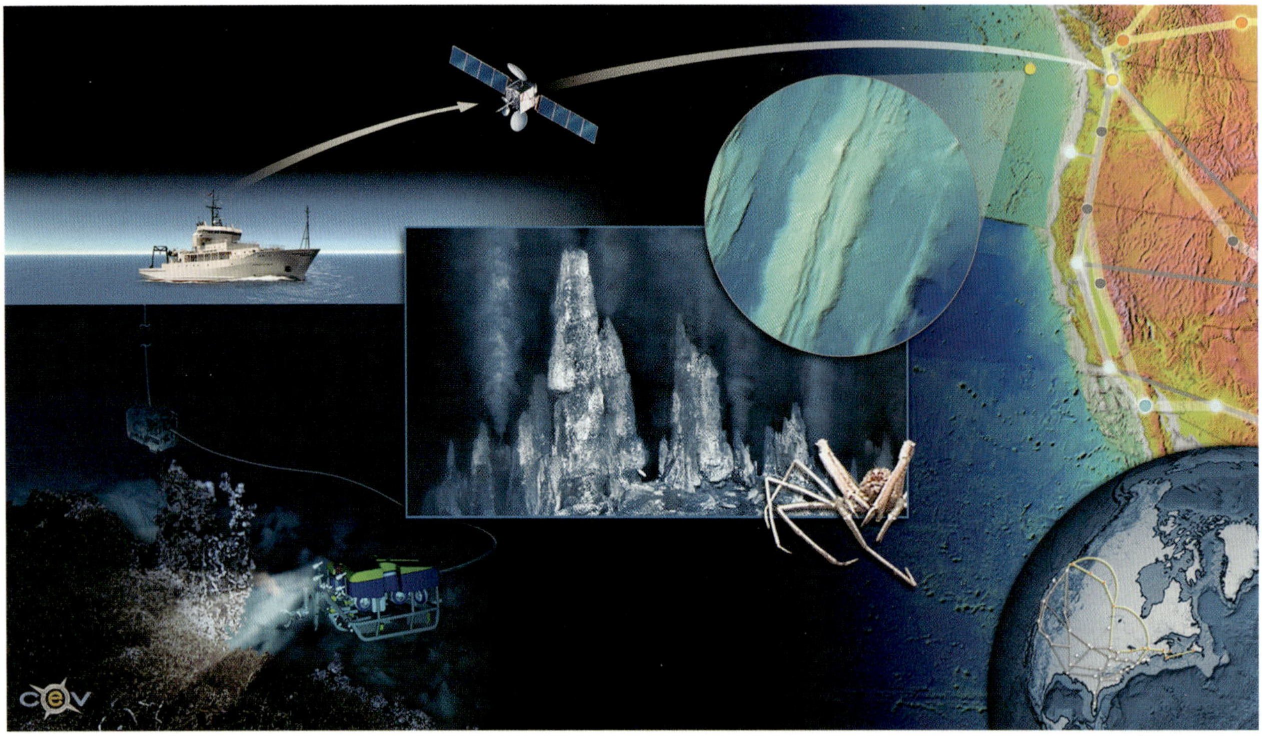

Figure 1.49 Illustration depicting the main components of a telepresence system – the ROV at the seafloor making observations (lower left and center), and the fiber-optic cable transmitting data and imagery from the ROV up the cable to the support ship and the scientists onboard (upper right). Via the Internet, these data and imagery are now streamed live to shore-based laboratories, educators and to the general public, who experience real-time interactions with scientists and engineers while getting information from the ROV as it explores the ocean depths (right).

sweeps of time and space – from seconds to millennia, and from millimeters to thousands of kilometers. Today, we are on the threshold of being able to measure and record the biological, physical, geological, and chemical pulses of the oceans and the Earth as easily as reading the news (or checking email) over a cup of coffee. These chemical, physical, geological, and biological processes are intertwined. The only way we can truly understand these processes is to be able to measure them, all of them, simultaneously *in situ* – where they are happening, in real time, over as wide, or as small an area as necessary.

This effort will lead to a new understanding of how our ocean and our planet works from the colonization of the smallest microbes to variations in global ocean chemistry and the forces that control Earth's climate over various time scales. Ocean observatories will provide the capability to understand Earth–ocean interactions at a level of detail previously not possible, but which are essential for us to explore in this century and beyond.

1.8 Telepresence: seagoing participation and research through the Internet

There have been notable contributions to the increasingly important task of fostering public awareness and scientific collaborations for understanding the ocean's role in both Earth history and its profound influence on controlling climate and weather.

In the 1940s and 1950s, Rachel Carson's books *Under the Sea-Wind* and *The Sea Around Us* were perhaps the first eloquent presentations of both the beauty and natural processes occurring in the world's oceans and their scientific importance in the field of oceanography. At the same time, Jacques Cousteau began his pioneering development of underwater breathing apparatus. His engineering accomplishment also led to his quest to understand and protect both the oceans and the life within them and to bring that beauty to the public's eye through stunning photographic and moving imagery.

The advent of high-definition video on tethered deep-sea vehicles, coupled with ship-based satellite communications and ever-increasing bandwidth capabilities result in the ability to stream live, high-definition video and other data from sea onto the Internet. One of the first times that this was done was in 2005 on a University of Washington cruise led by John Delaney to the Endeavour Segment of the Juan de Fuca Ridge. From a depth of 2200 m and 200 miles offshore Vancouver Island, high-definition video was streamed from an ROV in real time across North America and to Europe, Australia, and Japan using a 15 Mbps satellite link. During a two-week period, over a million viewers logged on to experience this expedition live.

Figure 1.50 A high-definition still camera (foreground) deployed in front of "Mushroom Vent" on Axial Seamount on the Juan de Fuca Ridge as part of the Ocean Observatories Initiative, and a 3D thermistor array (blue rods in background) positioned over a diffuse flow vent. The orange coil of cable is the power and data "extension" cord, a fiber-optic cable that transmits the imagery and thermistor data from the instruments back to shore-based laboratories for analysis. Field of view ~6 m.

The streaming of high-resolution video from sea to the Internet has become routine (Fig. 1.50). This capability has led to the concept of "telepresence," allowing teams of scientists, educators, and the world public to directly observe deep-sea research activities in real time and communicate with at-sea and shore-side scientists through blogs, social media platforms, and Voice over Internet Protocol (VoIP) phones.

More recently, the visionary work by Robert Ballard has fostered robotic technology to explore and study the oceans and seafloor. His public outreach efforts (Jason Project, http://www.jason.org/), funded in part by the NOAA's Ocean Exploration and Research Program, as well as private funding, present the excitement of oceanography and marine archeology to the pubic using satellite technology and the Internet. Exploration of the seafloor is broadcast to the land-based public and researchers, delivering a new and exciting vision of the oceans and oceanographic science to world audiences (Fig. 1.49).

Simultaneously, various groups, including notably NOAA's Ocean Exploration and Research Program and the National Science Foundation's Ocean Sciences Directorate, have been fostering telepresence-capable expeditions to expand the capacity of the science teams at sea with collaborators on shore. The shore-based team participates in the expedition in real time as full partners with the seagoing scientists to investigate oceanographic phenomena using diverse robotic technologies. Telepresence (Fig. 1.49) is an important part of the cabled component of the Ocean Observatories Initiative, where land-based scientists talk directly to engineers and scientists onboard the ship during installation of infrastructure on the seafloor and throughout the water column. These programs and future ones like them will expand the number of students who can experience oceanographic field research and data acquisition, thereby providing a new way to engage future oceanographers and engineers.

These new technologies have allowed millions of viewers to participate in sea-going research through programs offered by the University of Washington, NOAA Ocean Exploration, the Institute for Exploration, the Ocean Exploration Trust, and the Jason Foundation. In addition to video, other data also are now streamed to shore live, such as multibeam bathymetry and AUV-based remote-sensing data, allowing shore-based scientists to process the data and beam them back to the ship in rapid turn-around. Telepresence will continue to broaden the participation of scientists, students, educators, and the public in the discovery process of sea-going oceanography.

Over the past few years, telepresence technologies have become a valuable and dynamic tool for not only creating collaborations among scientists and stakeholders for understanding Earth systems and human interaction, but also educating scientists with specific expertise outside of their discipline to address fundamental issues, including biodiversity and habitat baselines, in anticipation of future change in ocean conditions.

In 140 years, we have witnessed oceanographic science mature from a largely unknown and unconnected group of disciplines to an integrated field of scientific study that has revealed the enormous complexities of Earth, ocean, and atmosphere phenomena. From imagining the global ocean abyss as a "featureless biological desert" at the start of the nineteenth century, we now realize that the oceans and ocean floor are what make Earth so unique in our solar system, and that the origin and survival of life on this planet is intimately tied to oceanographic phenomena. Technology has revealed the myriad physical, chemical, and biological processes within the oceans and seafloor, and in so doing is helping us to better understand Earth history and plan for the future. The twenty-first century undoubtedly holds new discoveries yet to be made using traditional sea-going methods that employ new technologies, as well as ocean observing systems and telepresence – all of which will help to "pull the plug" in different and novel ways to reveal the inner-workings of the global ocean and seafloor (Fig. 1.50).

2 Earth beneath the sea

The birth of the oceans is a matter of conjecture, the subsequent history is obscure, and the present structure is just beginning to be understood. Fascinating speculation on these subjects has been plentiful, but not much of it predating the last decade holds water.

J. H. F. Umbgrove (1947)

2.1 A new understanding of Earth and ocean history

For those born in the middle of the twentieth century, and having embarked on the professional journey to understand the seafloor and oceanographic processes, it continues to boggle the imagination that the guiding theories about how the Earth "works" are only a few decades old. Ideas of how the ocean formed, mountains are built, and the relationships between earthquakes and volcanoes have been debated for thousands of years. However, it was only in the middle of the last century that the confluence of new

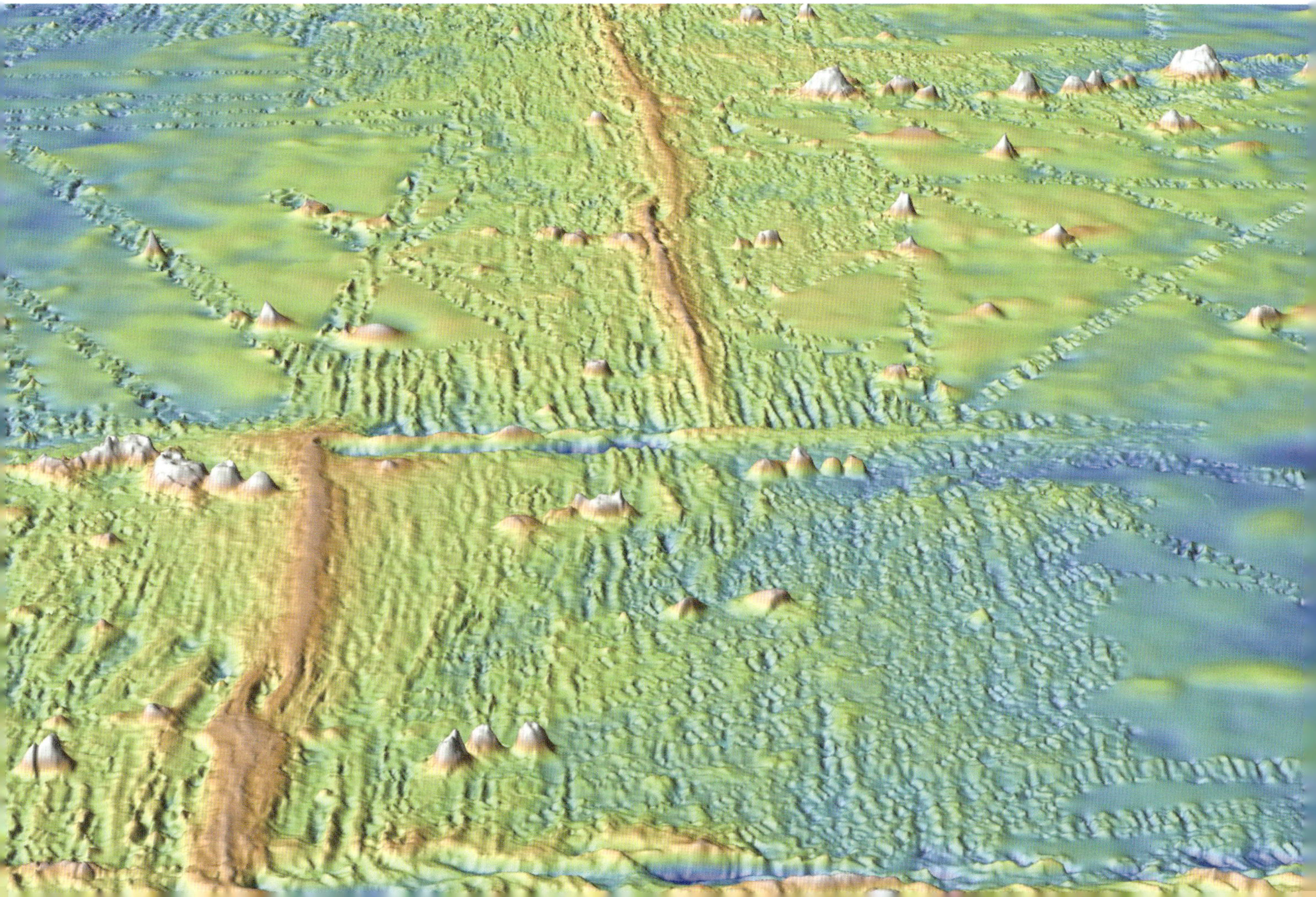

Figure 2.1 The topography of more than two-thirds of Earth lies hidden beneath the global ocean. This perspective view of the eastern equatorial Pacific seafloor, viewed to the northeast, shows part of the East Pacific Rise, transform faults, seamounts, and other features. The terrains are as complex and awe inspiring as any on land, but the nature of this vast expanse and how it contrasts with continental crust has only emerged in the past 60 years.

technology for measuring the properties of Earth both on land and under the oceans (e.g., topography, gravity, magnetics, and seismicity), and new concepts for how to interpret those data led to the guiding theories of seafloor spreading and plate tectonics. These ideas nucleated and gained momentum in the 1950s and 1960s at a handful of universities and newly minted oceanographic institutions around the world. The history of this scientific field of study is elegantly presented in Naomi Oreskes' (2001) compendium of essays authored by ocean and Earth scientists who led the revolution in developing plate tectonics as a holistic concept to explain Earth and ocean evolution. What is clear from those essays is that the way in which we understand the processes that shape the Earth and ocean floor (Fig. 2.1) results from open collaborations across many disciplines of science and engineering. Without the ability to do basic research we would not have achieved the knowledge and progress we enjoy today, and which are fundamental to our progress as a global society.

The story of how the new concept of seafloor spreading developed is a highly instructive tour of a rapidly evolving branch of Earth sciences that both stimulated and, in turn, fed off the momentum of the plate tectonic revolution. The definition and description of seafloor spreading is an essential part of how lithospheric plates interact and indeed how planet Earth works. Progress in this area did not develop in a linear fashion. Instead, the research trajectories of geophysics, seafloor geology, and terrestrial geology surged ahead individually, sometimes out of sync, but frequently intersecting and stimulating significant advances in other areas. These intertwined research paths have led to the current, but still incomplete, understanding of the composition and structure of the oceanic crust.

The practical challenges of doing research on the seafloor are compounded by the nature of geological processes that create the oceanic crust and lithosphere. As new oceanic crust is produced by seafloor spreading, deeper levels of the ~7 km thick crust are buried beneath layers of lava and sediment and thus are not directly accessible. Thus, deep-sea exploration is pushed to extremes that are comparable to robotic missions to other planetary bodies. Methods and practices developed for use on land must be adapted to investigate physical, chemical, and biological features that are distinctly different in the deep-sea environment (see Chapter 1). Looking backward from our current tenets in Earth sciences, it is possible to trace the progress through a series of discoveries. The recognition of seafloor spreading as a first-order planetary process in the early 1960s played a key role in the acceptance of plate tectonics as we know it today (Fig. 2.2).

2.2 Seafloor spreading: a consequence of plate tectonics

Plate tectonics is a conceptual model that describes the first-order mechanical behavior of Earth's rigid outer shell, the lithosphere, which includes the crust and upper mantle. A basic precept of plate tectonics is that this outermost layer is composed of a mosaic of relatively cool, rigid, "lithospheric plates," commonly hundreds to thousands of kilometers across but typically less than 200 km thick. These giant slabs slide over the hotter, weaker, underlying mantle "asthenosphere," moving across the spherical face of the planet at rates that vary from less than 10 mm per year to about 150 mm per year (Fig. 2.2). The plates diverge, converge, or slide past one another in an ever-shifting pattern of interactions. The continents are embedded within the lithospheric plates and slowly migrate across the surface of the planet as a consequence of plate movements. Thus, the long-recognized notion of continental drift (e.g., Wegener, 1912) is simply a corollary of plate tectonics.

Divergent plate boundaries, referred to as "spreading centers," are places where new lithosphere is created to fill the widening gap between plates that are moving apart (Fig. 2.3). As plates diverge, roughly equal amounts of new lithosphere are "accreted" to the receding edges of both plates as spreading proceeds. Some spreading centers are located near the middle of ocean basins – hence the term "mid-ocean ridges" (MORs) or simply "ridges." In some areas of the ocean, spreading centers are located nearer the margins of ocean basins as a result of plate movements over time and the consumption of oceanic lithosphere at subduction zones (a prime example is the Juan de Fuca Plate off the coasts of Washington and Oregon). Regardless of location, these terms are commonly used synonymously to refer to the narrow seams along which new oceanic lithosphere is produced. The creation of new tracts on the Earth's surface by seafloor spreading is compensated on a global scale by the recycling or "subduction" of oceanic lithosphere at deep-sea trenches and volcanic arcs, such as those found around the periphery of the Pacific basin. At the third type of boundary, "transform faults," plates slide past one another with no significant growth or destruction of lithosphere (Fig. 2.3).

As lithospheric plates diverge at spreading centers, hot, weak, but still solid mantle asthenosphere rises to fill the gap. As the asthenosphere rises and decompresses it partially melts, producing magma. The magma separates from the flowing mantle material and rises toward the surface, where it solidifies to form the distinctive assemblage of igneous rocks that defines the oceanic crust. It is within the relatively narrow zone defined by spreading centers that 85% of the Earths magmatic budget is concentrated and where the vast majority of Earth's crust beneath the deep seafloor has been created and modified by seafloor spreading.

As the mantle lithosphere moves laterally away from a spreading center, it cools, increases in density, and systematically subsides with age. In this way, rising and laterally flowing asthenosphere is converted to cooler, rigid lithosphere with a cap of igneous crustal rocks in a slow, but continuous process that ultimately forms the foundation for the great expanses of ocean floor.

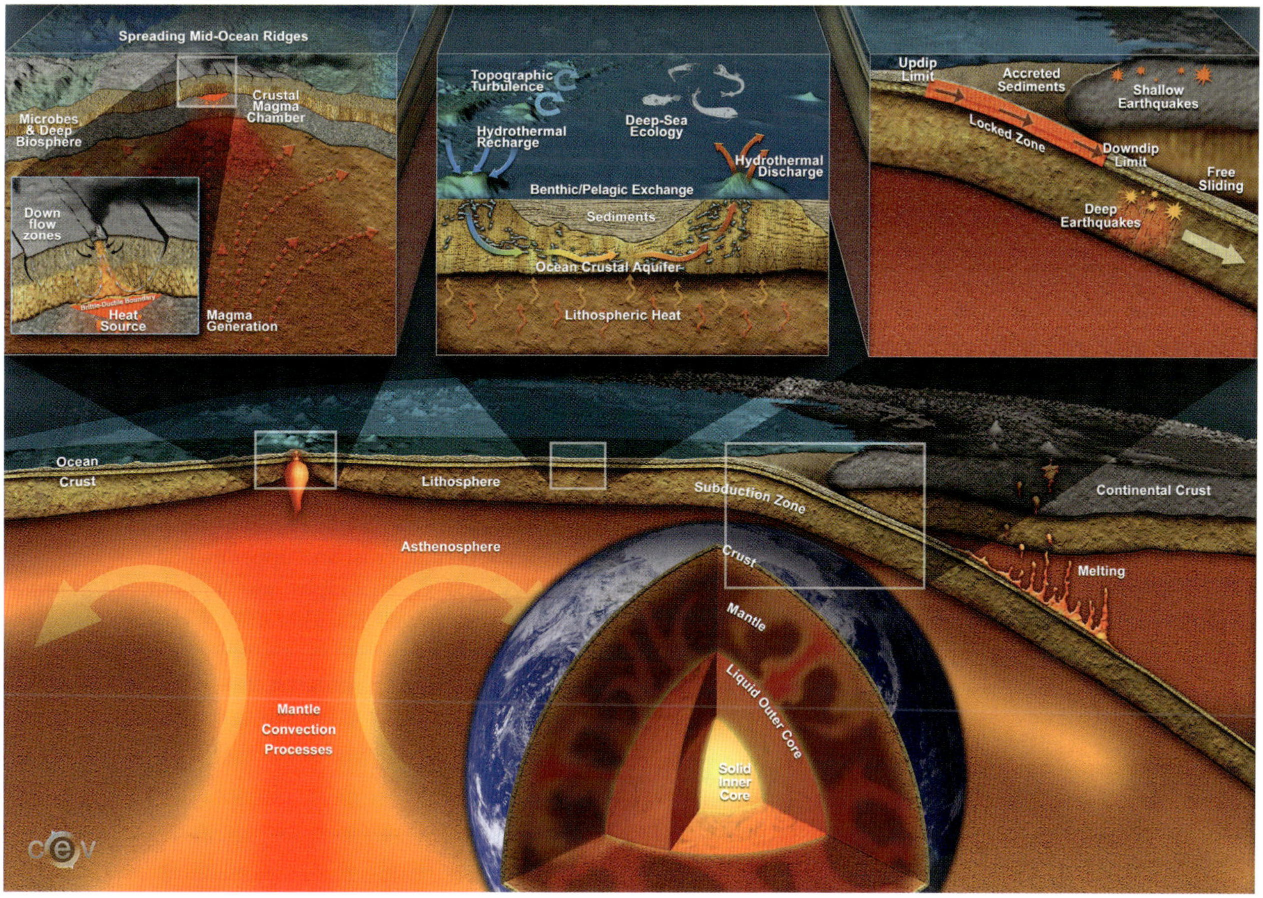

Figure 2.2 Rigid lithospheric plates move over the hotter, weaker asthenosphere (lower image), constantly creating new oceanic crust and lithosphere at spreading ridges (upper left) and recycling oceanic lithosphere back into the mantle as it is subducted at deep-sea trenches above volcanic island arcs (upper right). This "life cycle" of oceanic lithosphere dictates the shapes of the ocean basins, the geology of the seafloor, and the ecology of the marine environment (upper middle) (from Delaney and Barga, 2009).

With age the seafloor is slowly buried by the accumulation of pelagic sediments. Some of these sediments are scraped off adjacent continental margins as the oceanic plates are subducted. These oceanic sediments, coupled with those eroded from the continents, create the thickest sedimentary volumes on Earth. The sedimentary layers with their abundant microfossils – accumulating from bottom to top – provide a nearly continuous record of Earth's climate for millions of years. Although most of the oceanic crust created over Earth's history has been subducted back into the mantle, some rare fragments of it ("ophiolites") have survived as they were thrust upon the edges of continents and incorporated into mountain chains such as the Appalachians, Urals, and the Alpine–Himalayan Belts.

Seafloor spreading has probably been generating new oceanic crust and upper mantle for at least 3 billion years, more than half of Earth's history, and possibly longer. Over geologic time, oceanic plates comparable in size to those of today's oceans have been created and destroyed many times over. The oldest tracts of seafloor on Earth today are less than 200 million years old, in contrast to the much older continents. This cycling of Earth's lithosphere, driven by a combination of gravity and convection in the Earth's mantle, is the most important means of transferring heat, mass, and chemicals between the Earth's interior and the surface.

2.3 Continental drift, "geopoetry," and seafloor spreading

More than half of the twentieth century passed before the rapidly accumulating marine geophysical data collected in the wake of World War II gave birth to the concept of plate tectonics as a comprehensive description of large-scale Earth processes. At that time, careful acquisition of geological and geophysical data, led by such luminaries as Maurice Ewing, Bruce Heezen, and Bill Menard, provided the grist needed for the Earth science community to come to grips with

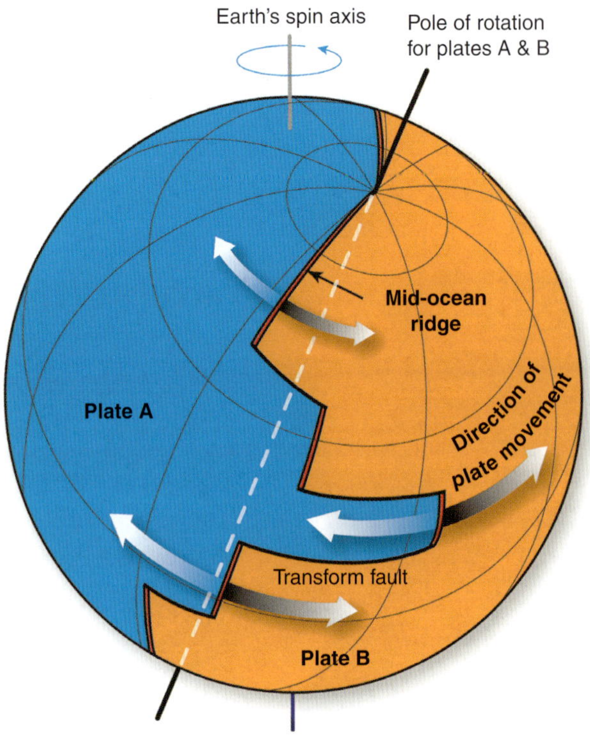

Figure 2.3 Pairs of curved lithospheric plates diverge by movements related to a pole of rotation. Spreading centers (red) lie along great circles that pass through the pole of rotation (similar to lines of longitude). Transform fault offsets (black) lie along small circles (lines of latitude) centered on the pole. Note that the spreading rate increases with distance away from the pole.

understanding the evolution of ocean basins and the implications of those data sets for revealing Earth history. The newly acquired data provided insights into ideas that had been in play since the early 1900s. These included Alfred Wegner's controversial theory of continental drift based on the matching of continental coastlines and geological features on either side of the North and South Atlantic basin and similarities in fossils found in West Africa and eastern South America (Wegener, 1912). Harry Hess's novel idea – "geopoetry" in his words – was that thermal convection in plastically yielding mantle rocks beneath Earth's crust provided the driving force to split apart the MOR and raft away the continents to either side of it (Hess, 1962). Bob Dietz (1961) described the "spreading sea-floor theory" and suggested that melts rising from below the crust are intruded into fissures on the seafloor, forcing the crust beneath the MORs to spread apart and feeding basaltic volcanic eruptions on the seafloor. Thus the concept of seafloor spreading was born. This chapter outlines the historical development of ideas that have led to the current understanding of seafloor spreading phenomena and the nature of oceanic crust. They provide a context for more detailed discussions in topic-specific chapters in this book.

2.4 Unveiling the oceanic lithosphere

Investigation of the crust and deeper lithosphere beneath the seafloor has involved a complex interplay of data and concepts from both oceanic and land-based investigations. Fundamental data on the composition and internal structure of oceanic lithosphere and the processes that create it come from three different, but mutually influential perspectives. The three key areas of investigation include (1) marine geophysical studies using seismic, magnetic, gravity, and remote-sensing data (see Chapter 1); (2) direct studies and sampling of the seafloor; and (3) investigations of "ophiolite complexes," on-land exposures of ancient oceanic lithosphere (crust and upper mantle) (Fig. 2.4). Individually, these perspectives provide critical information that has advanced our understanding of the oceanic lithosphere. Collectively, they have shaped and continue to refine the perception of the nature of the oceanic crust and upper mantle.

Geology and oceanography are both highly observational sciences. However, within the last century they have shared increasingly quantitative geophysical and geochemical data sets resulting in progressively more detailed definition of processes and models based on them. The current view of the oceanic lithosphere and how it is generated is largely the product of the creative interpretation of available data in an effort to accommodate information from diverse perspectives and to infer what is not known in a logical way. As is commonly the case in assembling a puzzle, the addition of new pieces results in an evolving perception of the final form and the context within which each piece fits. Seeing the pieces from a different perspective, for example,

Figure 2.4 Ophiolite complexes are fragments of ancient oceanic crust and mantle that are exposed on land. They provide direct access to extensive exposures of materials that ordinarily exist on the seafloor or deep in the oceanic crust and mantle. Here geologists examine ophiolite outcrops on a mountainside of upper mantle rocks in the northern Oman Mountains.

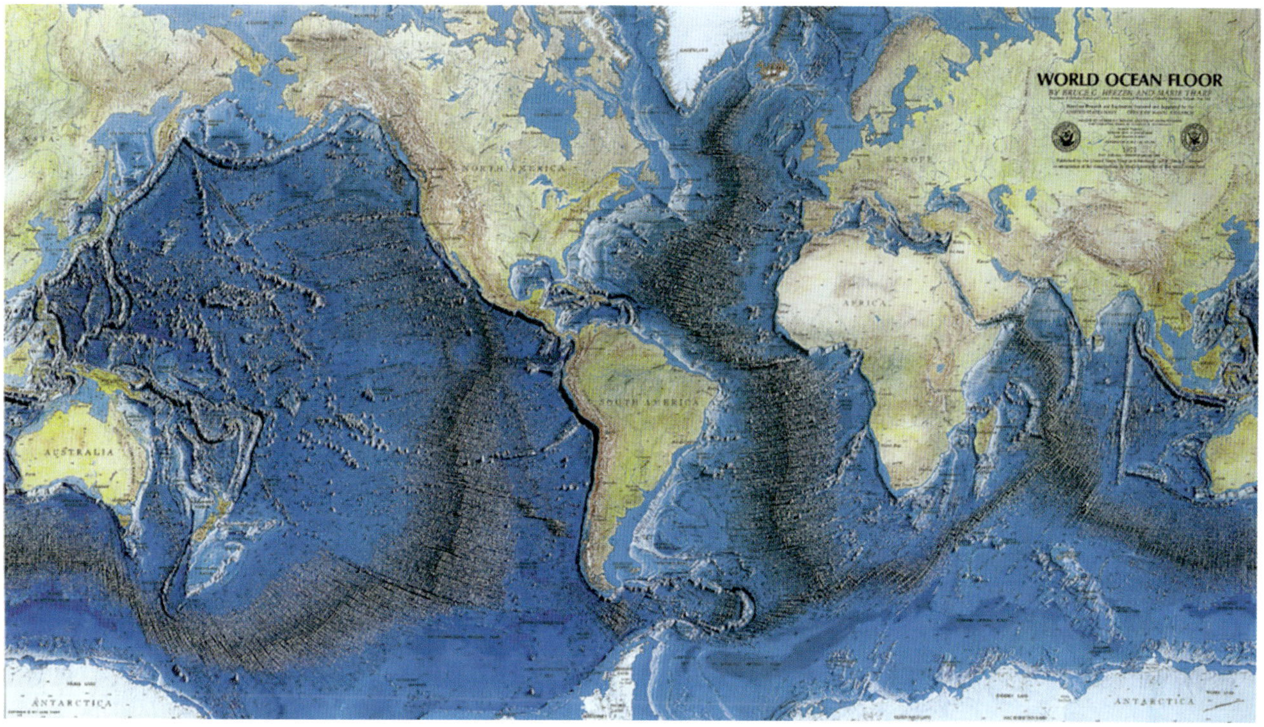

can lead to important breakthroughs and new solutions. Because our current perceptions rely so heavily on indirect information, exceptional new visual and remote-sensing images of the seafloor and oceanic lithosphere, afforded by rapidly evolving technologies (see Chapter 1), provide a new level of understanding of these important, but still mostly hidden and largely inaccessible parts of Earth.

Figure 2.5 Maps of the seafloor created by Bruce Heezen and Marie Tharp (Heezen *et al.*, 1959) revealed the MOR system as a series of linked volcanic mountain ranges encircling the globe, the longest continuous physiographic feature on Earth.

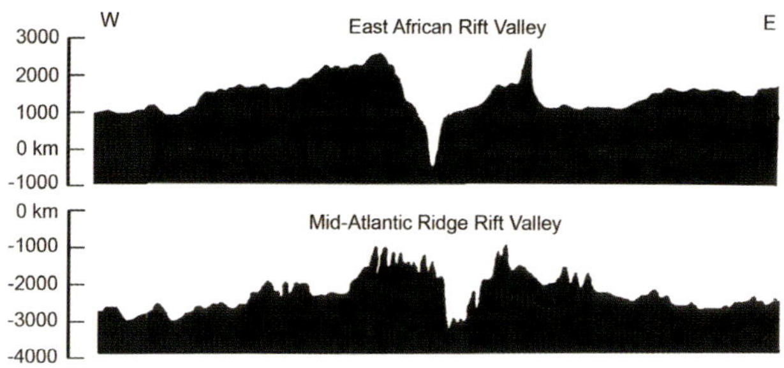

Figure 2.6 Bruce Heezen (1960) made the important conceptual leap of comparing the shape and scale of the Mid-Atlantic Ridge to the East African Rift, proposing that the former is also a young volcanic rift valley where Earth's crust is actively being pulled apart.

2.5 Geophysical framework of seafloor spreading

In the late 1950s and early 1960s, the shape of the seafloor in the North Atlantic and Northeast Pacific was revealed by the systematic integration of echo-sounder data from densely spaced survey lines (Fig. 1.7). The compilations and interpretations of Bruce Heezen and Marie Tharp (Heezen *et al.*, 1959; Heezen and Tharp, 1977) (Fig. 2.5) clearly delineated the Mid-Atlantic Ridge bisecting the seafloor of the North Atlantic basin with a distinct rift valley along its crest (Fig. 2.6) and major east–west trending fracture zones forming scars hundreds of kilometers in length across the seafloor.

From the rapidly growing body of data on marine magnetic anomalies collected by Fred Vine and Drummond Matthews and correlative studies of magnetic reversal stratigraphy in terrestrial lavas and sedimentary rocks by Alan Cox and others, a convincing case was assembled for the bilaterally symmetrical accretion of oceanic lithosphere at divergent plate boundaries (Fig. 2.7), thus confirming the concept of seafloor spreading (Vine and Matthews, 1963; Cox, 1969).

Mapping of the magnetic anomaly "stripes" on the seafloor in many localities provided a global view of the age of the oceans (Fig. 2.8). The nearly bilaterally symmetric pattern of anomalies across spreading centers reveals how spreading rates vary along the MOR system and how they have changed with time. Reconstructions of seafloor areas guided by matching magnetic anomalies on either side of spreading centers is a fundamental technique employed in plate tectonic kinematic models. By comparing the widths of the stripes with the time intervals that they represent, average rates of spreading can be determined. Spreading rates

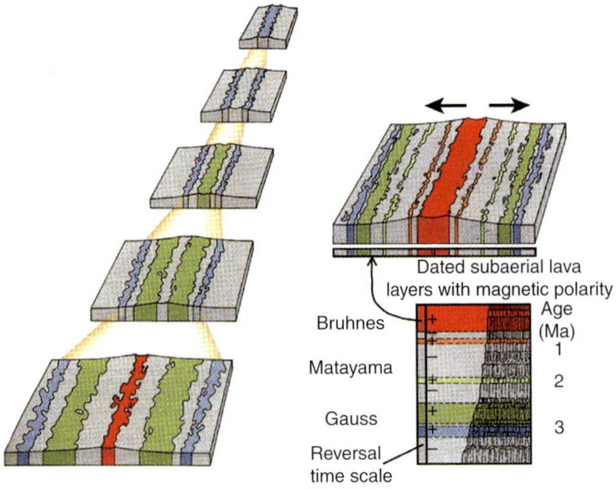

Figure 2.7 Magnetic minerals in lava flows record the polarity of Earth's magnetic field at the time they erupt and cool. Each new strip of oceanic crust created at a spreading center splits the previously formed crust resulting in matching strips of crust of the same age on either side of the spreading center (left). Lavas with normal (+) or reversed (−) polarity are correlated with dated polarity reversals from lava flows on land (right). Identification of patterns of swaths of seafloor with different polarities and widths provides a means of determining the age of large tracts of the seafloor and measuring spreading rates at different segments of the MOR.

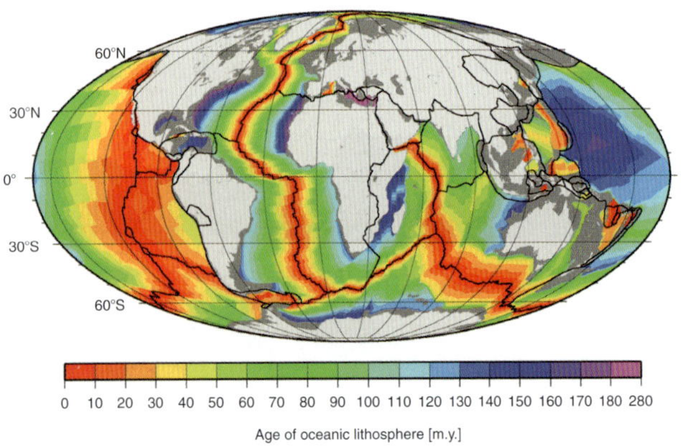

Figure 2.8 The age of the seafloor is primarily determined by the identification of magnetic anomaly patterns on the seafloor and by the ages of microfossils in sedimentary rocks directly overlying the oceanic basement (Müller *et al.*, 2008). The youngest seafloor is symmetrically centered on MORs (red color). Note that there is very little surviving seafloor older than about 180 million years as a consequence of subduction at deep-sea trenches.

on MORs range from as low as about 10 mm/yr to as fast as 150 mm/yr (see Chapter 3), with evidence for substantially higher rates in the past. Although these rates seem exceedingly slow (comparable to the rate of growth of human hair or fingernails), over millions of years they are sufficient to account for the broad expanses of the oceanic basins on Earth today.

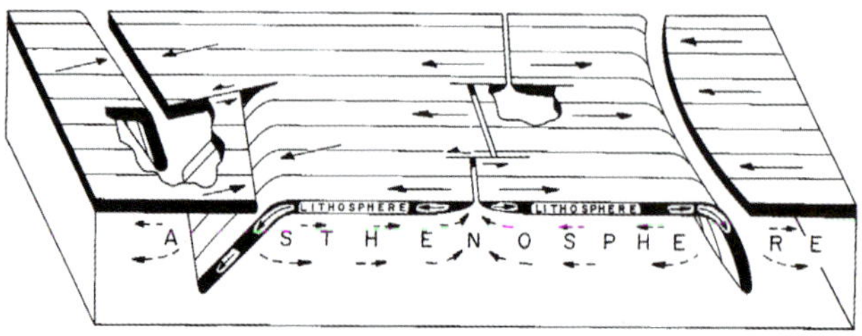

Figure 2.9 An early representation of the lithosphere and asthenosphere and the relationships among different types of plate boundaries (Isaaks *et al.*, 1968).

Improvements in seismic techniques and instrumentation and global integration of seismic data with tectonic and bathymetric information by Isaaks, Oliver, and Sykes (Isaaks *et al.*, 1968) firmly established the locations of major plate boundaries. These data also constrained the relative motions of plates, and the present-day interactions of lithosphere plates across different types of plate boundaries (Fig. 2.9).

The new global synthesis of seismicity was especially important in delineating the MOR system. The relative motions of lithospheric plates were confirmed by earthquake studies that showed divergence at oceanic ridges, strike-slip movements at transform faults, and convergence at deep-sea trenches linked to Benioff seismic zones (Isaaks *et al.*, 1968). By analyzing earthquake data, seismologists are able to determine the location, depth, energy released, and direction of crustal movements associated with earthquakes along major fault zones beneath the seafloor (Sykes, 1967; Isaaks *et al.*, 1968; Baranzangi and Dorman, 1969). Prior to publication of these data, the mechanism linking earthquake activity along spreading centers and connecting transform faults was unknown.

Results recorded over several decades reveal that earthquakes along the MOR system occur mostly at depths of 1–8 km, consistent with extensional faulting (Sykes, 1967, 1969). Larger and more frequent earthquakes occur along transform faults (Fig. 2.10) and indicate strike-slip (lateral) displacements in the sense predicted by geophysicist J. Tuzo Wilson (Wilson, 1965). Transform earthquakes are in some cases "slow earthquakes," with displacements taking place over hours to months (Okal and Stewart, 1982). Earthquake data compilations show concentrations of activity in areas of intense tectonism especially near transform faults and other spreading-center discontinuities (see Chapter 3).

Compilations of seismic data show that the frequency and size of earthquakes are much greater on slow-spreading ridges than fast-spreading ridges (Bergman and Solomon, 1984). This relationship is a direct reflection of the cooler, thicker axial lithosphere along the plate boundary zone in these environments. On some parts of slow-spreading ridges, seismic activity occurs in the mantle beneath spreading axes, indicating temperatures less than 750 °C, probably corresponding to relatively cool

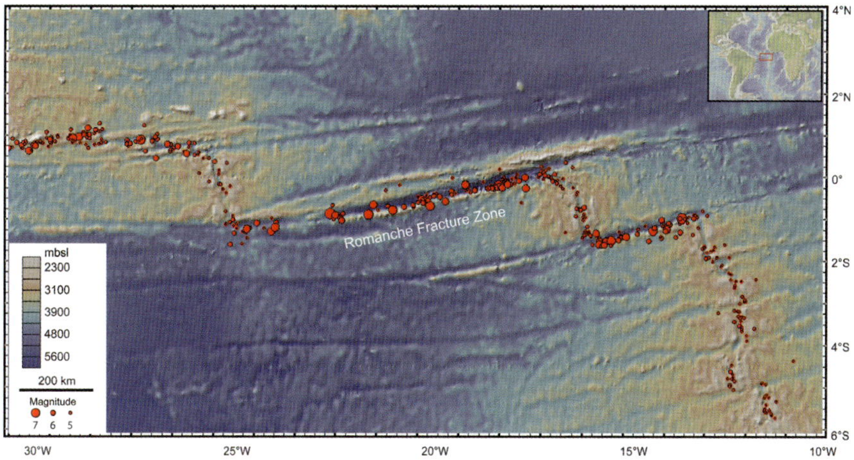

Figure 2.10 Major transform faults such as the Romanche Fracture Zone in the Equatorial Atlantic are clearly defined by their deep bathymetry (>7 km) and seismicity (red dots indicate earthquakes epicenters with magnitude >5, 1973–2013; from USGS-NEIC).

areas that have not experienced recent magmatic activity (Toomey *et al.*, 1985).

Microearthquake activity (less than magnitude 1) detected by local arrays, typically deployed for weeks to months at a time at a specific seafloor study site, help resolve specific faulting processes and locations. On slow-spreading ridges, microearthquakes are focused along major rift valley faults (Toomey *et al.*, 1988; Kong *et al.*, 1992; deMartin *et al.*, 2007). On intermediate- to fast-spreading ridges, microseismicity is commonly linked to hydrothermal cooling and fracturing in the crust, defining the subsurface extent of hydrothermal circulation systems (Wilcock *et al.*, 2002; Tolstoy *et al.*, 2006).

The energy released by faulting during earthquakes can account for only part of the deformation associated with plate movements along MORs, indicating that aseismic slip ("creep") must play a significant role in accommodating spreading (Bergman and Solomon, 1984; Solomon *et al.*, 1988). This behavior may be facilitated by slow movement along major fault zones, distributed deformation over broader areas, or slip on zones of very weak materials.

2.6 Composition of the oceanic crust

The recognition of plate tectonics and the significance of oceanic lithosphere naturally led to questions about composition and processes associated with the formation of oceanic crust and underlying mantle. Even as late as the mid 1960s, there were very few samples of basement rocks recovered by dredging of major seafloor escarpments. Dredged samples from the 1940s and 1950s included mostly variably altered basalts and serpentinites (altered upper mantle rocks) (Table 2.1; Fig. 2.11) (e.g., Shand, 1949). The

Table 2.1 Major rock types from ophiolites and the seafloor and compressional wave velocities

Rock type	Significance	V_P (km/s)[a]
Basalt	Black to dark gray, fine-grained rock from lava flows or shallow intrusions, glassy to crystalline	4.5–6.5
Gabbro	Gray, coarse-grained plutonic rock formed in magma chambers, massive to layered	6.5–7.6
Peridotite	Dark-colored, coarse-grained, olivine-rich upper mantle rock, high-temperature metamorphic fabric	>8.1
Serpentinized peridotite	Black to dark green, peridotite partially altered at <500 °C, massive to sheared	2.8–8.0

[a] Typical range of compressional wave velocities measured in samples

Figure 2.11 Rock samples recovered from the seafloor include serpentinite and basalt. Serpentinites (left) are samples of hydrated upper mantle rock. This sample from the Southwest Indian Ridge has large light-colored pyroxene crystals surrounded by black serpentine minerals. White serpentine veins highlight the deformation fabric. Basalt samples (right) are commonly fragments of pillow lava with glassy rinds (shiny black areas).

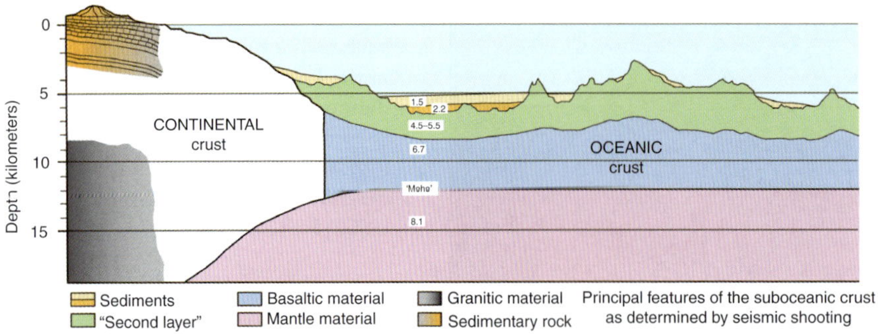

Figure 2.12 Early seismic refraction surveys showed that the oceanic crust is much thinner than that of the continents and that it has a consistent layered structure (Ewing and Engel, 1962).

abundance of serpentinites inspired Princeton professor Harry Hess to propose that the oceanic crust was composed primarily of serpentinites with a thin veneer of basaltic lava (Hess, 1954, 1962).

It was, however, marine seismic refraction studies that provided critical constraints on the nature of the oceanic crust and lithosphere. In these investigations, the velocity of compressional sound waves passing through the crust and upper mantle were determined as a function of depth. The velocities corresponding to different depths were compared to laboratory measurements of the transmission of compressional waves passing through different rock types sampled from the seafloor. Correlating rock velocities to those determined in refraction experiments provided an indication of the geology of the oceanic crust and upper mantle.

Three essential pieces of evidence emerged in the early 1960s that strongly influenced the perception of the nature of the oceanic lithosphere and, by inference, how it formed.

First, it was discovered that oceanic crust has a consistently layered, horizontal seismic velocity structure with seismic compressional wave velocities of the layers increasing downward (Fig. 2.12). Analysis of compressional wave arrivals indicated that these layers have relatively sharply defined interfaces, within the resolution of the technique (tens to hundreds of meters). The oceanic crust was determined to be significantly thinner and denser (higher compressional wave velocities) than continental crust.

In more recent investigations, the seismic structure of the oceanic crust has more typically been described in terms of discrete layers or intervals with velocity gradients. Compressional wave velocities increase downward, reflecting increasing density and lithostatic stress and closure of pore spaces (Christeson *et al.*, 1992; White *et al.*, 1992).

Second, subsequent investigations in many different parts of the seafloor found that the seismically determined thickness of oceanic crust (excluding sedimentary accumulations) was remarkably constant, averaging about 7 km (Fig. 2.13). Historically, this seismic crustal thickness has commonly been equated with the thickness of igneous (magmatic) rocks that compose the oceanic crust. This led to the general correlation of variations of seismic crustal thickness with magma production and, by inference, mantle temperature. However, this correlation has a

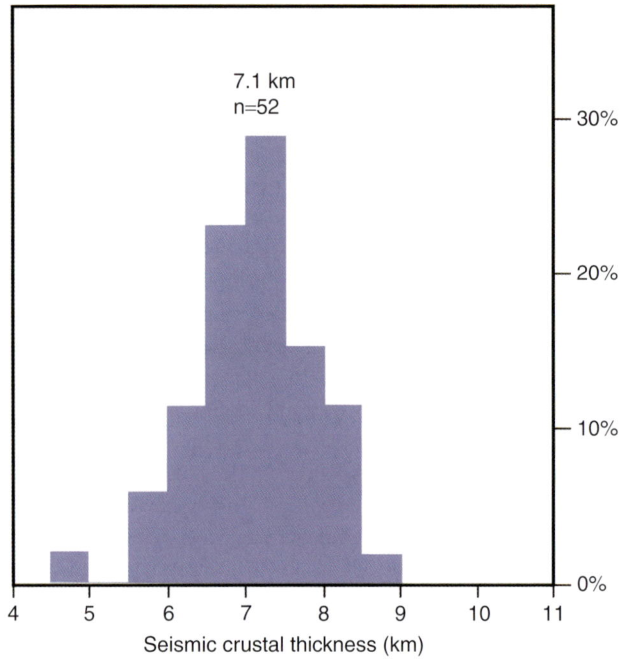

Figure 2.13 A compilation of seismic refraction data from the seafloor (after White *et al.*, 1992) shows that, regardless of spreading rate, the seismic thickness of the crust is ~7.1 km. Vertical axis shows the frequency of interpretations of thickness measurements.

number of important uncertainties. For example, elevated temperatures or hydration ("serpentinization") of mantle peridotites can lower seismic velocities in mantle rocks. Likewise, fracturing of the upper crustal layers can decrease seismic velocities. Sealing of fractures by hydrothermal minerals can potentially increase velocities. These complications preclude any simple one-to-one correlation of velocities to specific rock types. Thus, without direct sampling of the deep oceanic crust, there are significant uncertainties about its detailed internal structure and composition.

Oceanic crust with a seismic thickness that is greater than normal occurs near spreading ridges proximal to hot spots associated with rising mantle plumes, where elevated mantle temperatures and enhanced magma generation occur. Iceland is an extreme example, with a seismic crustal

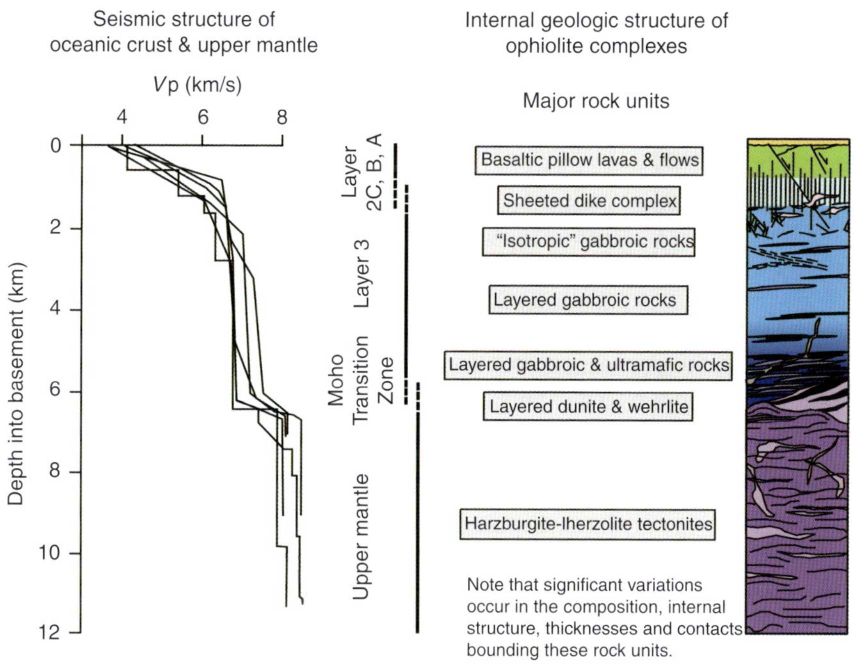

Figure 2.14 A step-like pattern of increasing compressional wave velocity with depth is typical of the layered structure of the oceanic crust (left). Contacts between layers and the internal structure of layers are best modeled as velocity gradients corresponding to gradational compositional changes (White *et al.*, 1992). Laboratory measurements of compressional wave velocities of oceanic and ophiolite rocks link the internal structure of ophiolite complexes to the velocity structure of oceanic crust (right).

thickness of more than 25 km near this very active hot spot (White *et al.*, 1995). Thinner crust occurs in lithosphere formed in relatively cold environments, for example near oceanic fracture zones (Detrick *et al.*, 1993). Much of the variation in seismic thickness occurs in the mid-crustal layers thought to represent gabbroic rocks that crystallize from subsurface magma bodies (Mutter and Mutter, 1993).

Third, the compressional wave velocities of the oceanic crustal layers were documented and correlated with specific rock types (Fox *et al.*, 1973; Christensen, 1978; Christensen and Mooney, 1995). Laboratory measurements of compressional wave velocities of rocks from the seafloor and ophiolites (determined at appropriate confining pressures) provided a link between the seismic structure and geology of the oceanic lithosphere (Fig. 2.14). Reconstructions of the velocity structure of ophiolite complexes based on rock velocity data compared very well with those of the oceanic crust. This allowed a direct correlation of the composition and structure of ophiolite rock units to specific velocity intervals in the oceanic crust (Fox *et al.*, 1972; Peterson *et al.*, 1974; Christensen, 1978; Christensen and Salisbury, 1982).

2.7 Direct access to the oceanic crust

Marine geologists have known that there are volcanic rocks on the seafloor since at least the 1870s, when submarine basalts were recovered from the deep sea by trans-Atlantic cable repair ships (Hall, 1876) and the *Challenger* Expedition (Murray and Renard, 1891). Systematic sampling of the oceanic crust first began with dredging of steep seafloor escarpments to recover rock samples (Fig. 2.15) and coring to collect sedimentary deposits (see Chapter 1). In particular, major efforts by scientists at Scripps Institution of

Figure 2.15 The first samples obtained from the seafloor were recovered by dredging – dragging a strong, steel mesh bag across steep, rugged areas of the seafloor. The dredge is connected to the ship by ~1–2 cm thick steel wire rope able to withstand 10–25 tons of tension. Dredging continues to be an important means of collecting large volumes of oceanic rocks and for reconnaissance petrological and geochemical studies of the crust.

Figure 2.16 For over 40 years, *Alvin* has transported thousands of scientists to the deep seafloor to conduct experiments and investigate geological, geochemical, and biological phenomena there.

Oceanography and Lamont-Doherty Geological Observatory (now the Lamont-Doherty Earth Observatory) resulted in collections of samples that began to define the distinction between continental and oceanic crustal geology (Heezen *et al.*, 1959; Menard, 1964; Engel *et al.*, 1965; Heezen and Hollister, 1971; Engel and Fisher, 1975).

In many cases, however, precise locations of the samples were poorly constrained and they were widely spaced, making it impossible to relate them to one another. Only in the 1960s, after many years of routine collection of seafloor rocks, did the first awareness of the general distribution of rocks and sediment types beneath the global ocean begin to emerge.

Since that time, a rapid progression of technological developments has provided increasingly sophisticated and higher-resolution imaging and sampling capabilities (see Chapter 1). In particular, the development of deep-submergence human-occupied vehicles (HOVs) specially equipped for scientific purposes, such as DSV *Alvin* in 1964, permitted the first opportunities for scientists to directly examine and sample the geology of the seafloor (Fig. 2.16).

The development of modern navigation using satellites of the Global Positioning System (GPS) to locate ships, and sophisticated acoustic methods to track deep-sea vehicles on the seafloor, provided the ability to pinpoint seafloor samples and seafloor measurements with an accuracy of a few meters or better. The rapid evolution of computing technology and invention of fiber-optic communications over the past few decades has enabled the development of sophisticated remotely operated vehicles (ROVs) and autonomous underwater vehicles (AUVs). These systems have greatly expanded the access to the global seafloor over the full range of ocean depths from 1 to 11 km (see Chapter 1).

These technologies have made it possible to map and sample the seafloor and oceanic crust with resolutions comparable to those used in subaerial investigations. Still missing, however, is the ability to view seascapes on scales of tens to hundreds of meters (Fig. 2.17), which commonly provide important spatial perspectives of geological features and processes. The limited reach of current lighting systems and the attenuation of light in the deep oceans has continuously improved through the application of visual and acoustic imaging including high-resolution, near-bottom sidescan sonar mapping, and the ability to assemble large-area mosaics (see Chapter 1).

Seafloor investigations involving direct observations using deep-submergence vehicle systems have focused on two different but related perspectives on the oceanic lithosphere (Fig. 2.18). One approach is directed at the axial regions of spreading centers where new oceanic lithosphere is actively being created. These investigations concentrate on the young (years to tens of thousands of years old), uppermost surface of the seafloor and provide information on surficial and shallow-crustal processes such as volcanic eruptions and hydrothermal venting and their products (see Chapters 4 and 5).

The second approach takes advantage of exposures of the crust and upper mantle along major escarpments produced by faulting. These cliffs can be thousands of meters high and many tens of kilometers in length. They produce "tectonic windows" into the oceanic lithosphere because they expose natural cross sections of the upper few kilometers of the oceanic crust and in some cases upper mantle rocks (Karson, 1998). These escarpments expose the final products of seafloor spreading generated over tens of thousands to millions of years. From the structure and composition of the rocks that make up these terranes, the

Figure 2.17 Investigations on land benefit from broad vistas that provide a context for samples and direct measurements of various features. Glaciated escarpments provide cross sections of the basaltic lavas in Iceland (left). Layered gabbroic rocks of the lower oceanic crust are widely exposed in the Oman Ophiolite (right). Comparable optical views are not possible on the seafloor with current technology.

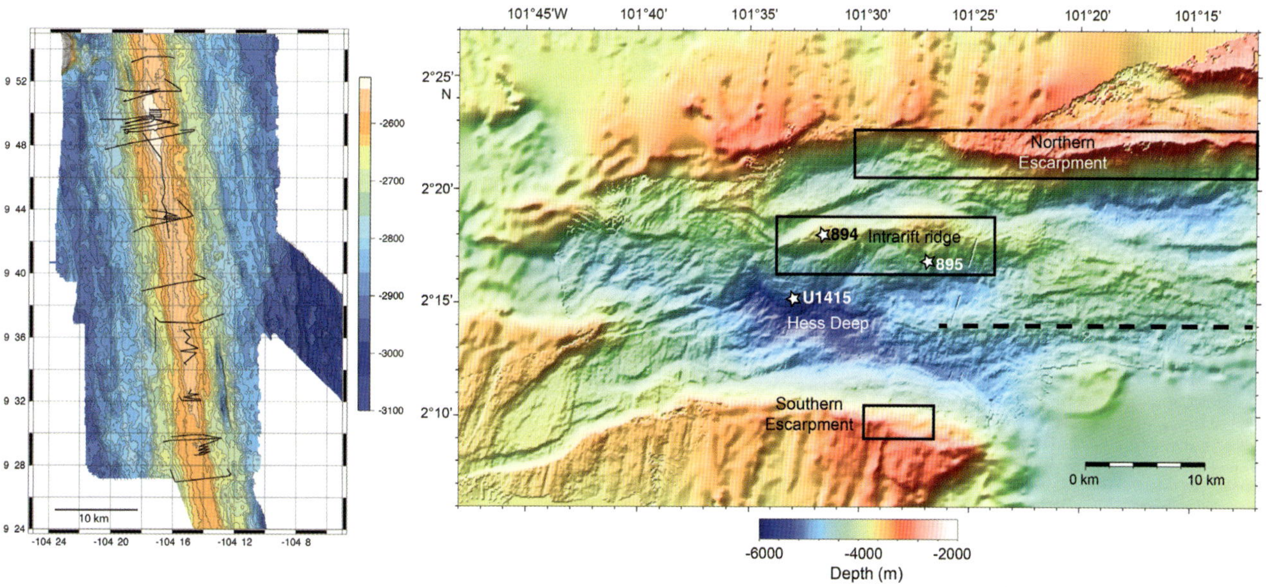

Figure 2.18 Investigations of the axial regions of active spreading centers (left, EPR 9° 50′N) and major escarpments "tectonic windows" (right, Hess Deep Rift) provide different but complementary perspectives on the oceanic crust and seafloor spreading processes. Numbered black lines show locations of near-bottom studies – both submersible dives and camera tows. White stars show locations of ODP/IODP drill holes.

mode of construction of the oceanic crust can be inferred by providing indirect information about the igneous and metamorphic processes that occur beneath spreading centers.

Investigations in these areas have the advantage of an extensive three-dimensional perspective that can be exploited with a variety of seafloor investigative techniques. The tectonic processes responsible for their exposure have affected rock units exposed in tectonic windows, which may be rearranged or physically and chemically altered compared to typical oceanic lithosphere distant from these locations. These effects must be taken into account in the interpretation of crustal features found in these settings.

2.8 Unraveling geology on the seafloor

While studies of ophiolites and marine geophysics began to build a view of seafloor spreading and the oceanic lithosphere, direct investigations of the deep seafloor provided additional new perspectives. In 1974, the FAMOUS (French–American Mid-Ocean Undersea) project carried out the first comprehensive geological and geophysical investigation of a MOR spreading center (Heirtzler and LePichon, 1974). In this study, investigators used HOVs and an array of deep-towed vehicles for the first time to study the geology of an actively spreading ridge. The results of this investigation generally confirmed the shallow crustal part of the ophiolite model by documenting the presence of abundant basaltic pillow lavas and interpreting volcanic and tectonic processes along the ridge axis. Similar results followed from investigations of an adjacent part of the Mid-Atlantic Ridge (MAR) in the Azores MAR (AMAR) area (Stakes *et al.*, 1984) and soon after on the East Pacific Rise (EPR) (Francheteau and Ballard, 1983).

During early geological and geophysical investigations of the Galápagos Spreading Center in 1977, actively venting hydrothermal systems with exotic chemosynthetic biological systems were discovered (Corliss *et al.*, 1979; Anderson *et al.*, 1982). Realization of the interrelated aspects of seafloor volcanic, hydrothermal, and biological systems had an immense impact on science, ultimately inspiring the creation of the US RIDGE (Ridge Interdisciplinary Global Experiments) Program by the National Science Foundation (Delaney *et al.*, 1987b). This program for the first time fostered a broad integrated approach to the study of MORs by geologists, geophysicists, chemists, and biologists. Major contributions of this program were the numerous surveys and sampling programs of large, previously poorly known parts of the global MOR system (Macdonald, 1998). The effort also gave birth to the new field of astrobiology by providing potential analogs for extraterrestrial systems that could support life (see Chapter 4).

In the 1990s, the international InterRIDGE Program and successor Ridge 2000 Program extended these investigations (Fornari *et al.*, 2012). The Ridge 2000 Program focused investigations at three Integrated Study Sites: 8–11° N on the EPR, the Endeavour Segment of the Juan de Fuca Ridge, and the Lau Basin, a back-arc spreading system. These programs ushered in a new era of MOR investigations, by revealing the causal and feedback relationships among diverse physical, chemical, and biological processes across different styles of spreading centers.

In parallel with studies of active MOR spreading centers, investigations of major tectonic escarpments along rifted spreading centers and transform faults (Fig. 2.19) provided

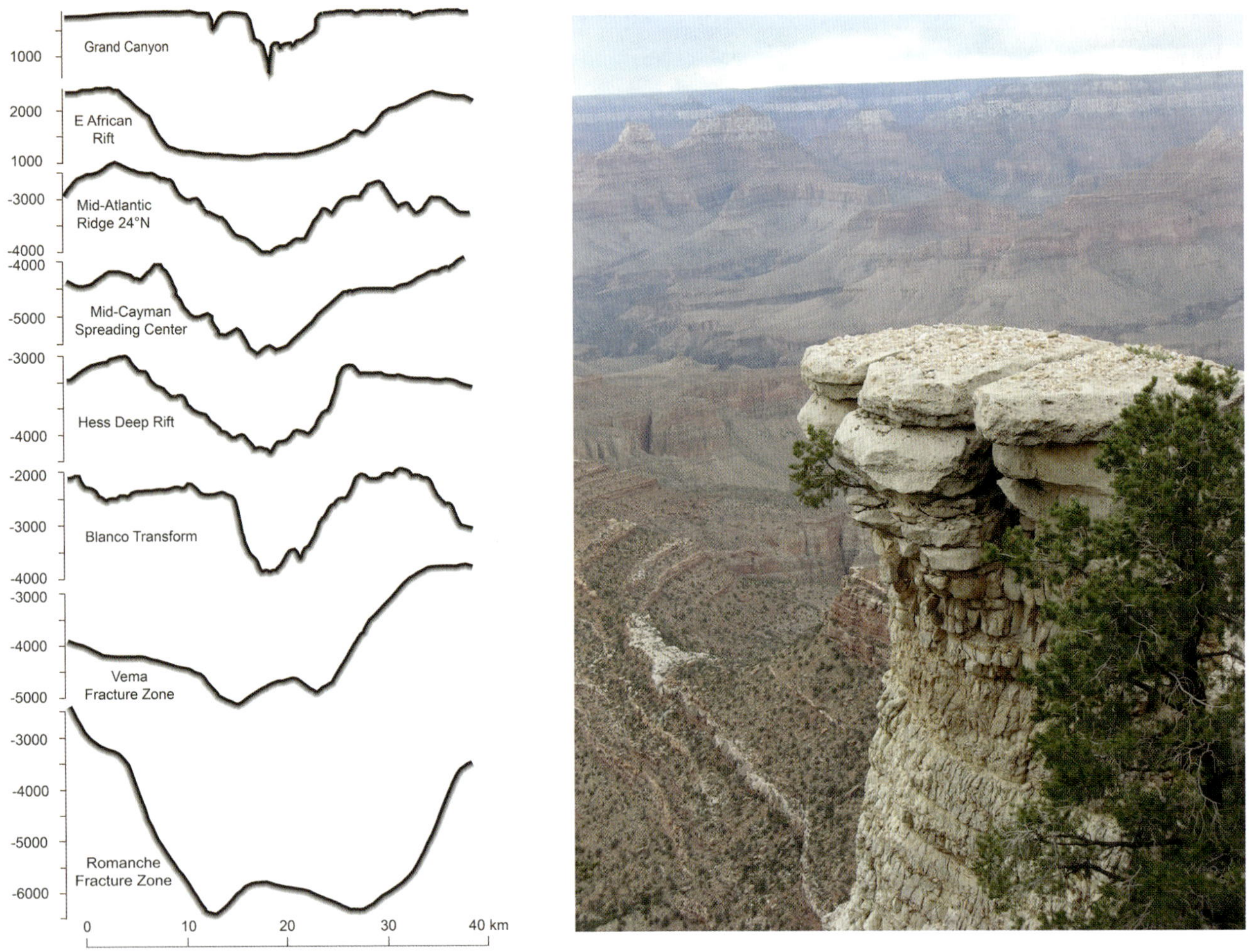

Figure 2.19 Huge faulted escarpments in the seafloor (left) provide "windows" into the structure and composition of the upper oceanic crust. Impressive features of the continents such as the Grand Canyon (right) and the East African Rift have relatively small vertical relief compared to many seafloor escarpments at oceanic rifts and fracture zones.

access to a diverse range of rock types analogous to the deeper crustal levels of ophiolites (see Chapters 7 and 8). For many years, dredging at these kinds of locations yielded many different types of variably deformed and metamorphosed gabbroic rocks and peridotites. Beginning in the late 1970s, HOVs were used to study and sample these materials, especially from the many escarpments along slow-spreading ridges. The geology of these exposures proved to be much more complex and varied than the middle to lower crustal parts of ophiolites, forcing a reconsideration of the prevailing perceptions of seafloor spreading processes.

2.9 Deep crustal drilling

Another major approach to determining the physical and chemical composition of oceanic crust is through crustal drilling using specially designed ships (Fig. 1.44). These vessels have the capability to drill holes as deep as 2 km through the sedimentary cover and into the oceanic basement rocks below (see Chapter 1). Deep drilling is an important approach because the precise location of the ship and, therefore, the drill, allows access to sites in a wide variety of tectonic environments including those far from areas of tectonic disruption (Figs. 2.20 and 2.21). Unlike surface sampling, drilling can potentially collect nearly continuous vertical cores. Geophysical investigations of the walls of boreholes (borehole logging) and experiments done within the holes provide additional information about the composition of the crust and the physical structures within the rock formations drilled. The main limitations to drilling are the difficulties of recovering material in fractured rock units, the limited depth of penetration, and the very limited lateral dimensions of each drill core (typically 7 cm in diameter) relative to broader regional geological relations and the resolution of most geophysical investigations (tens to hundreds of meters in the shallow crust and greater at depth).

In the 1970s a major engineering effort by the Deep Sea Drilling Project (DSDP) resulted in a sustained effort to drill deeply into the oceanic crust generated at the Costa Rica Rift in the eastern Pacific. After seven legs of drilling over nearly two decades, borehole logging, and engineering operations

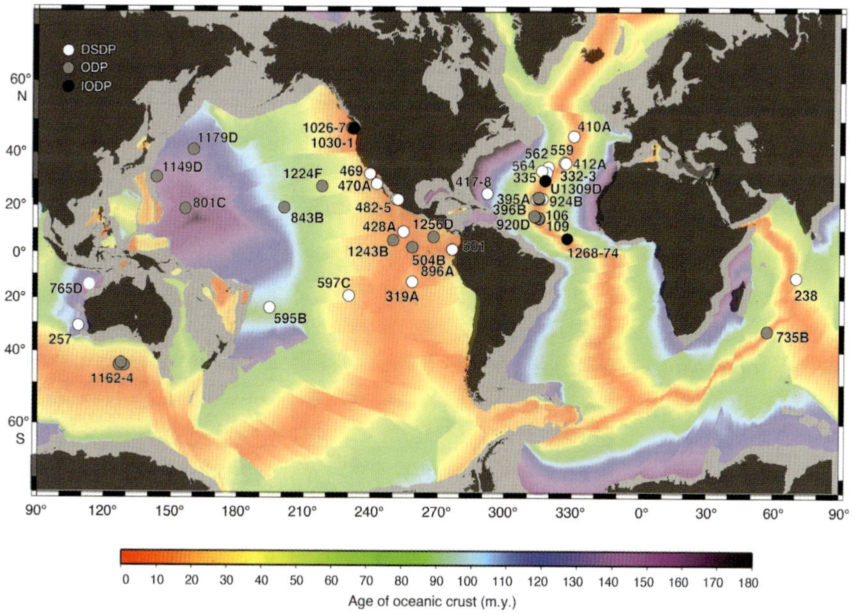

Figure 2.20 Locations of drill holes with greater than 100 m of penetration into the oceanic crust accomplished by DSDP (white), ODP (gray), and IODP (black). Note that this is a small fraction of the total number of holes drilled by these projects in other parts of the seafloor over the past four decades, including sedimented areas throughout the global ocean.

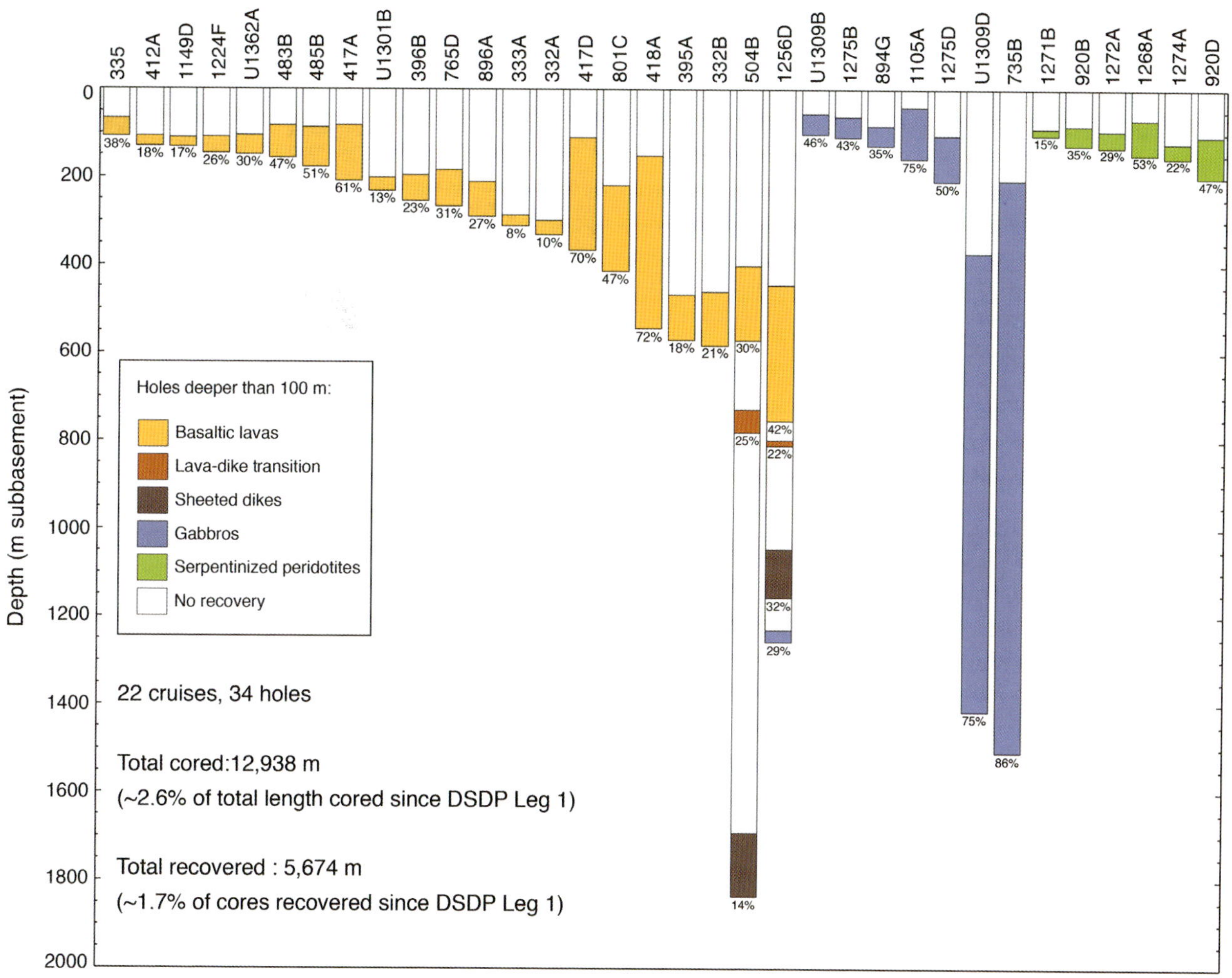

Figure 2.21 Summary of crustal drilling results for holes that penetrated greater than 100 m depth into the oceanic crust from DSDP, ODP, and IODP drilling legs. Colors show the rock units penetrated and approximate depths. Numbers at base of columns indicate the percentage of core recovered (also shown schematically by color bars). Note that drilling only at sites 504B and 1256D has penetrated intact crust of fast-spreading ridges to depths greater than a kilometer. Holes 735B and U1309D are both in faulted and uplifted crust of slow- to ultraslow-spreading ridges. See Fig. 2.20 for locations.

at Hole 504B reached a depth of almost 2000 m into the oceanic basement. Drilling penetrated the entire thickness of basaltic lavas, a transition zone, and about 1000 m of sheeted dikes (see Figure 2.14). This landmark achievement provided the first reference hole into the upper oceanic crust (Anderson *et al.*, 1982). Importantly, this hole, despite rather limited core recovery, provided a confirmation of the applicability of the ophiolite model to understanding the structure and composition of the upper layers of the oceanic crust. It was not until 2005 that Hole 1256D drilled through lavas and sheeted dikes and into about 200 m of gabbroic rocks (see Chapters 6 and 7), the next deeper layer of the oceanic crust. This hole further confirmed the validity of the ophiolites as useful analogs for the oceanic crust.

These investigations have been substantially augmented by systematic mapping and sampling of tectonic windows into oceanic crust generated at relatively fast spreading rates (Karson *et al.*, 2005). Studies at major transform faults and near propagating rifts confirm the expected layered sequence of lavas, sheeted dikes, and gabbroic rocks in the upper oceanic crust, but also show that there are substantial variations in the thickness, internal structure, and composition of these rock units. Research that integrates drilling results with studies of tectonic windows has begun to provide a three-dimensional perspective on the geology of the upper oceanic crust.

Beginning in the late 1980s, drilling at tectonic escarpments directly into exposures of plutonic rocks (Figs. 2.21 and 2.22) was successful in recovering important vertical sections of variably deformed and metamorphosed gabbroic material and peridotites (Dick *et al.*, 1991; Gillis *et al.*, 1993; Cannat *et al.*, 1995a). Placing these crustal sections in the context of the generalized ophiolite model of oceanic crust has, however, proven to be difficult. Combined results of drilling and near-bottom studies of tectonic escarpments have shown that, at relatively slow spreading rates, extreme faulting and limited magmatic construction commonly result in the exposure of complexly deformed and metamorphosed gabbros and mantle peridotites at exceptionally shallow crustal levels (Karson, 1998). These exposures appear to be analogous to crustal sections exhumed by major low-angle normal (detachment) faults in continental metamorphic core complexes (Karson and Dick, 1983; Karson *et al.*, 1987; Karson, 1990; Tucholke *et al.*, 1998; Smith *et al.*, 2006).

Progress in understanding the complex processes associated with tectonic extension and strike-slip faulting in the oceanic lithosphere has gone hand-in-hand with that on the continents. Although studies of continental rifts and major fault zones have taken the lead in advancing our knowledge of tectonic processes in these environments (for example, along the San Andreas Fault; Mount and Suppe, 1987; Holdsworth *et al.*, 2011), investigations of major tectonic features of the oceanic lithosphere have also made significant contributions.

Most active plate boundaries, including spreading centers and transform faults, are situated in oceanic lithosphere.

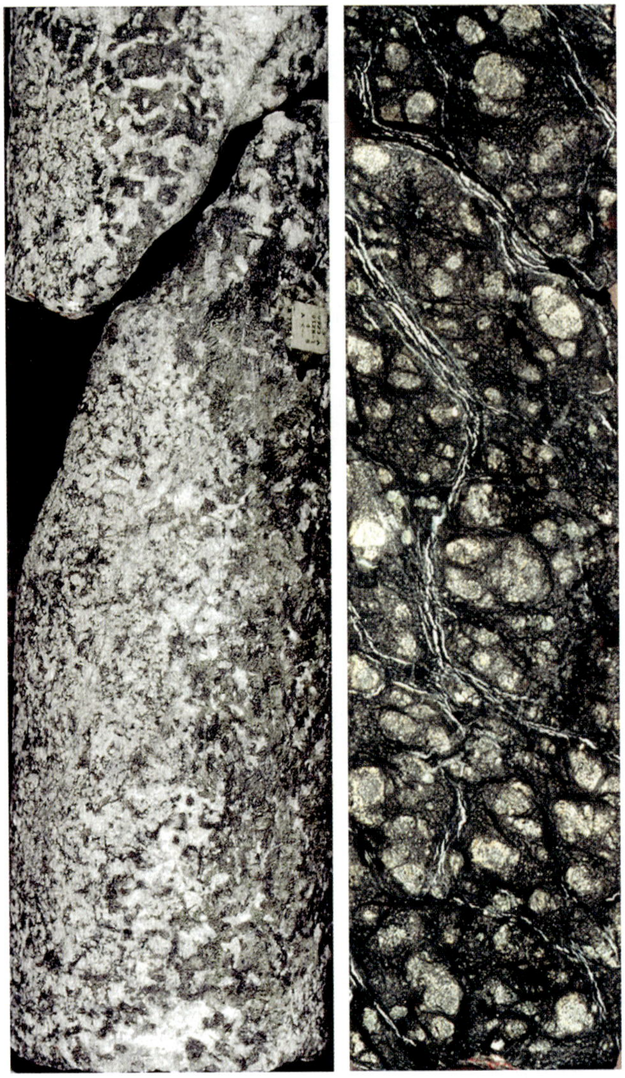

Figure 2.22 Drilling into exposures of plutonic rocks on the seafloor has recovered extensive cores of deep crustal gabbroic rocks with diverse textures and mineralogies (left) and serpentinized upper mantle peridotites (right). ODP Leg 153, Holes 924 and 920. Cores are about 5 cm wide.

Therefore, detailed information on the composition and physical properties of the oceanic lithosphere and its components is essential to understanding earthquakes and major crustal movements. Plate boundaries on the seafloor can be investigated in ways that are not possible for plate boundaries on land. Seafloor plate boundaries can be surveyed from surface ships or the seafloor with continuous crossings without obstructions or other limitations encountered on land and especially near population centers. For example, marine seismic refraction and reflection surveys provide detailed images of plate boundaries that cannot easily be acquired on land. Data available from these settings are likely to help elucidate deformation mechanisms and other fundamental tectonic processes relevant to all plate boundaries.

Today marine science is on the threshold of a new era of drilling exploration where, in addition to targeted shallow drilling (hundreds of meters or less) of key areas of the oceanic lithosphere, there are plans for continuous drilling through the entire thickness of the oceanic crust (typically 7 km) and into the underlying lithospheric mantle. The "Mohole to the Mantle" or M2M program (Umino *et al.*, 2013) renews the dream of drilling through the crust/mantle boundary (the seismic Mohorovičić discontinuity) initiated in 1959 (Bascom, 1959; Greenberg, 1964). This project promises to provide a continuous reference hole through the oceanic crust and to provide a context for material recovered in previous drill cores using the unique capabilities of the drilling vessel *Chikyu* (see Chapter 1).

2.10 Oceanic crust and mantle on land

Mountain-building processes, which result from the collision of lithospheric plates, commonly result in uplift and exposure of volumes of intact to highly fragmented pieces of oceanic crust and upper mantle that collectively make up an ophiolite. In some cases, large (tens to hundreds of kilometers across), continuous, layered ophiolite complexes that have been tilted and eroded provide direct access to nearly complete sections of the oceanic crust and upper mantle (Fig. 2.23). Some of the best-preserved and intensely studied ophiolite complexes include the ~90-million-year-old (90 Ma) Oman Ophiolite in the northern Oman mountains (Glennie *et al.*, 1973; Coleman, 1981; Nicolas, 1989), the ~90 Ma Troodos Ophiolite of Cyprus (Gass, 1968; Moores and Vine, 1971), and the ~500 Ma Bay of Islands Ophiolite in Newfoundland, Canada (Casey *et al.*, 1981, 1983; Girardeau and Nicolas, 1981).

Figure 2.23 Tilting, faulting, glaciation, and erosion dissect ancient oceanic lithosphere in ophiolite complexes. The contact between gray lower crustal rocks to the left and brown upper mantle rocks to the right in the Bay of Islands Ophiolite Complex represents the oceanic Moho.

Because similar rock types and structures can form in a variety of extensional environments where basaltic magmatism occurs (for example, different types of oceanic spreading centers, island arcs, back-arc basins, volcanic rifted margins, and ocean islands), there are uncertainties surrounding the setting in which individual ophiolites have formed. Even if an ophiolite complex were known unequivocally to have formed at an oceanic spreading center, there are no indepentent criteria to determine the spreading rate, spreading direction (direction to the ridge axis), or details of its tectonic setting when it formed because features of rock units indicative of those aspects of spreading are not yet clearly defined in the oceans or recognized in ophiolites. Still, ophiolites provide extremely important information about the products and processes of seafloor spreading and a framework for interpreting geological and geophysical data collected on the seafloor (Chapters 6–8). The following historical perspective on ophiolite studies provides a back-story and corollary to the evolution of MOR investigations and how geologists have worked to piece together the puzzle of oceanic crustal formation and evolution.

The earliest inklings of the composition of the oceanic crust came from geologists in the Swiss Alps in the early 1800s who were studying dark green to black rocks – distinct from more common lithologies of the continents (Table 2.1). These included basalt (a black to gray volcanic rock typical of rifts and other large igneous provinces), gabbro (the coarsely crystalline chemical equivalent of basalt formed by relatively slow subsurface cooling), and peridotite (a highly magnesium-rich rock derived from the mantle). Metamorphosed basalts, gabbros, and peridotites ("serpentinite") (Fig. 2.24) were discovered to be spatially related to fine-grained, deep-water marine sedimentary rocks such as cherts and limestones.

Figure 2.24 Serpentinite forms by hydration of mantle peridotites. Sheared, phaccoidal structures are common in serpentinite shear zones (left). Field of view is ~3 m across. Olivine and orthopyroxene, the main constituents of peridotites, are converted to a dark, fine-grained aggregate of serpentine minerals cut by numerous cross-cutting veins, which create a snake-skin-like surface on this hand sample (right).

Thus, these dark colored (mafic) rocks were logically inferred to have formed in deep-water environments and to represent fragments of the oceanic "basement," the hard-rock foundation of the seafloor. Collectively, early workers, for example Brongniart (1813), referred to these rock associations as "ophiolites" (*ophis* meaning "snake" in Greek) because of their reptile-skin-like weathered surfaces (Fig. 2.24). Later workers, notably Steinmann (1905), recognized the recurrent, systematic association of cherts, basaltic lavas, and serpentinites (altered peridotites), subsequently dubbed the "Steinmann Trinity." It is important to note that although this association was well documented and its deep-marine provenance was established, there was no conceptual framework that linked these rocks to the vast expanses of crustal material beneath the oceans. Instead, they were thought to be related to ephemeral "geosynclines" – giant marine depressions in Earth's crust from which mountain belts were born (e.g., Kay, 1951; Aubouin, 1965).

Further, there were very few constraints on the petrogenesis of these peculiar rock types, so their specific modes of formation were very vague at best. At the dawn of the development of plate tectonic theory, in the early to mid 1960s, the connection between masses of ultramafic (highly magnesian, olivine-rich) mantle rocks and basalts incorporated into mountain belts (collectively referred to as "Alpine Peridotites" by Benson) and oceanic material was not entirely clear. The writings of Robert Dietz (1961), Harry Hess (1955, 1962, 1965), and Peter Wyllie (1967) during this period paved the way toward a deeper understanding of the nature of oceanic rocks and how they formed. At that time, ophiolites were known to be common components of mountain belts, but their origin and mode of formation were poorly understood (Fig. 2.25).

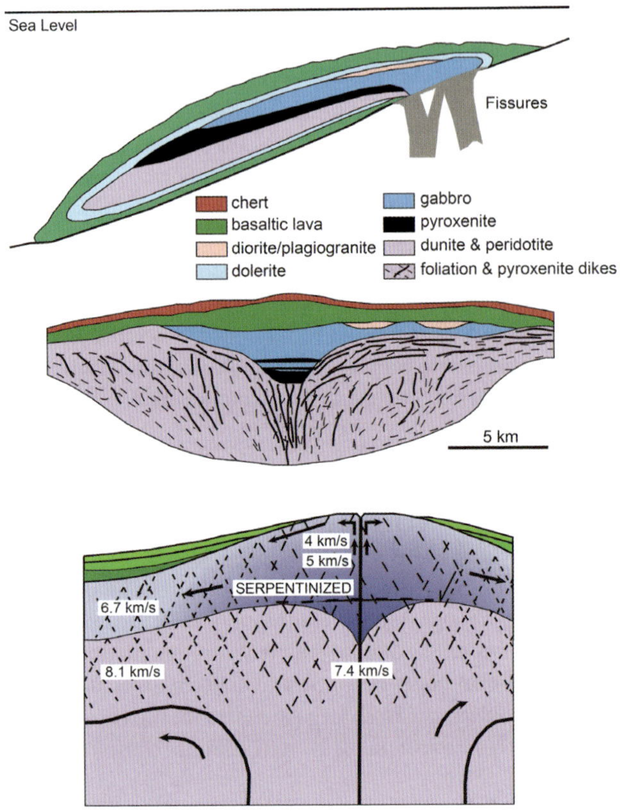

Figure 2.25 Prior to the discovery of seafloor spreading, ophiolite complexes were interpreted as massive, submarine, mafic to ultramafic magmatic outpourings in "geosynclines" (top). Later investigations, strongly influenced by modern experimental petrology, inspired more complex models of deforming, intrusive bodies (middle) (Moores, 1969). The abundance of serpentinites in ophiolites prompted the suggestion that these altered upper mantle rocks constitute the bulk of oceanic crust (bottom) (Hess, 1962). Once considered unlikely, this idea now appears to have been ahead of its time.

2.11 Linking ophiolites to oceanic lithosphere

In the late 1960s and early 1970s several workers, notably Ian Gass (Gass, 1968), John Dewey and Jack Bird (Dewey and Bird, 1971), as well as Eldridge Moores and Fred Vine (Moores and Vine, 1971), were quick to note the general similarity between the overall thickness, layered geological structure, and reconstructed seismic velocity structure of some major ophiolite complexes. The correlation between the rock units of ophiolite complexes and seismically defined layers of the oceanic lithosphere provided an indirect yet important window into the composition and structure of the oceanic crust (Fig. 2.26). The structure and composition of rock units in ophiolites facilitated the investigation of processes associated with seafloor spreading without direct access to the seafloor.

A landmark in the rapidly developing ideas regarding the constitution of the oceanic lithosphere was established at the 1971 Penrose Conference, sponsored by the Geological Society of America. At this meeting, a new definition of the term "ophiolite" was proposed and strongly linked to oceanic lithosphere. This definition was built upon roughly 150 years of terrestrial geology and preceded even the earliest direct, near-bottom investigations of seafloor geology. Although it could not draw on direct information from the seafloor, clearly it was strongly influenced by the prevailing notions of a layered oceanic crust and upper mantle inferred from seismic refraction studies and the prevailing concepts of seafloor spreading. The "Penrose" definition (Anonymous, 1972) outlined in Table 2.2 was modified later to include overlying volcanic and sedimentary sequences that hold important clues to the tectonic setting in which an ophiolite was formed (Moores, 1982). Importantly, it recognized that hydrothermal deposits produced by high-temperature seafloor hydrothermal vents and even fossilized vent fauna were also important components of ophiolites (Haymon *et al.*, 1984, 1989; Zierenberg *et al.*, 1988).

In the early 1970s, Akiho Miyashiro (Miyashiro, 1973, 1975a) pointed out that the geochemistry of lavas from the Troodos Ophiolite, on the island of Cyprus, was not typical

Table 2.2 Penrose Conference definition of "ophiolite" (Anonymous, 1972)

Ophiolite refers to a distinctive assemblage of mafic to ultramafic rocks. It should not be used as a rock name or as a lithologic unit in mapping. In a completely developed ophiolite, the rock types occur in the following sequence, starting from the bottom and working up:

- *Ultramafic complex*, consisting of variable proportions of harzburgite, lherzolite and dunite, usually with a metamorphic tectonite fabric (more or less serpentinized)
- *Gabbroic complex*, ordinarily with cumulus textures commonly containing cumulus peridotites and pyroxenites and usually less deformed than the ultramafic complex
- *Mafic sheeted dike complex*
- *Mafic volcanic complex*, commonly pillowed

Other rock types commonly associated include

(1) an overlying sedimentary section typically including ribbon cherts, thin shale interbeds, and minor limestones
(2) podiform bodies of chromite generally associated with dunite
(3) sodic felsic intrusive and extrusive rocks

Faulted contacts between mappable units are common. Whole sections may be missing. An ophiolite may be incomplete, dismembered, or metamorphosed ophiolite. Although ophiolite generally is interpreted to be oceanic crust and upper mantle, "the use of the term should be independent of its supposed origin."

of MOR basalts and, in fact, was much more typical of lavas from island arcs above subduction zones. At this time, the Troodos Ophiolite was the "gold standard" of an on-land exposure of oceanic crust. Miyashiro's comments drew considerable fire from several investigators (Hynes, 1975; Miyashiro, 1975b,c; Moores, 1975) and these exchanges stimulated an effort to develop objective statistical geochemical criteria for the tectonic provenance of ophiolites. This was done by comparing the geochemical characteristics of ophiolite lavas with the growing body of data from the oceanic crust of the deep seafloor and lavas generated above subduction zones, particularly in back-arc basins, and especially by Joe Cann and Julian Pearce (Pearce and Cann, 1971, 1973). In this evolving controversy, different ophiolites were classified and reclassified as having a provenance at MORs or in "supra-subduction zone" settings as new geochemical data were obtained and new criteria were developed (Shervais, 2001; Pearce, 2003). Interestingly, much of this discussion occurred with little reference to regional tectonic and geologic constraints on the ophiolites (Dewey and Bird, 1971). This geochemical "fingerprinting" took precedence over a discussion of fundamental igneous processes and how these could be reconciled with all the data at hand. Opinions on the origins of various ophiolites inevitably became somewhat polarized (Moores *et al.*, 2000).

In the mid 1980s, detailed geochemical work on the least-altered Troodos lavas and glasses showed that these rocks are indeed more similar to volcanic rocks of subduction zone settings than MORs. These results confirmed Miyashiro's earlier suspicions and called into question the provenance of Troodos and other ophiolites (Robinson *et al.*, 1983; Rautenschlein *et al.*, 1985). Meanwhile, new data from the western Pacific, in particular the Woodlark and Lau Basins, showed lavas with the chemical affinities of both MOR and subduction zone lavas formed in spreading centers associated with back-arc basins above subduction zones (Leitch, 1984; Pearce *et al.*, 1984; Perfit *et al.*, 1987; Hawkins, 2003; Pearce, 2003). These lavas from the western Pacific were described as having formed in a "supra-subduction zone" environment, that is, as a result of processes above a down-going slab of lithosphere (Fig. 2.2). The notion that ophiolite formation by seafloor spreading occurs in these environments has become a unifying concept that has received wide acceptance.

The rapid accumulation of geochemical data on crust formed in subduction zone environments helped to provide a means of discriminating between subduction zone and MOR crust in ophiolites. However, the waters have been muddied by the discovery of lavas with subduction-zone geochemical characteristics erupted along the Chile Rise, a *bona fide* MOR, that is not in a supra-subduction zone setting (Sturm *et al.*, 2000). Adding to the complications are ophiolites on the adjacent subducting Andean Margin, which might be fragments of crust produced at the Chile Ridge or material generated near its present position in a supra-subduction setting (Lagabrielle *et al.*, 1994).

Today, many ophiolites are thought to have formed in supra-subduction zone settings. However, most workers also support the idea that the magmatic and tectonic processes associated with crustal spreading and accretion should be essentially the same for these settings as those occurring at MOR spreading centers. Ophiolites continue to be one of the most important geologic windows into the oceanic crust and underlying mantle rocks and, by inference, the processes by which they are created.

2.12 Cross-pollination of data and concepts

The confluence of information from marine geology and geophysics with contributions from ophiolites has resulted in an almost inevitable drift toward a highly generalized

Figure 2.26 Summary of the layered structure seen in many ophiolite complexes correlated with the oceanic crust and upper mantle. The recurring sequence of rock units in ophiolite complexes inspired the views adopted by the 1972 Geological Society of America Penrose Conference (Table 2.2).

view of seafloor spreading and the oceanic lithosphere (Table 2.3). Direct investigations of the oceanic crust formed at relatively fast spreading rates, where a nearly continuous magma supply is inferred, have generally confirmed the upper ~2 km (lavas and sheeted dikes) of the expected layered structure of the ophiolite model. At slower spreading rates, major faulting and variable magma supply result in a more complicated and widely disrupted crustal structure that commonly lacks a simple layered structure. In extreme cases, spreading in these environments results in the widespread exposure of oceanic mantle rocks (serpentinized peridotites) with only limited exposures of more typical oceanic crustal rocks (basalts and gabbros). The internal composition and structure of the oceanic crust in these environments, and how it may relate to the layered seismic structure, remains open to question. Available data come from different techniques and different locations along MORs rather than a comprehensive body of data from a single spreading environment. The resulting view is a patchwork of loosely related parts that do not all necessarily provide an unambiguous picture of seafloor spreading. This generalized view represents an evolving work-in-progress that is destined to change as new and higher-resolution information is acquired.

Results of investigations outlined above highlight the extreme diversity of seafloor spreading processes as well as the nature of oceanic crust and lithosphere created by those processes (see Chapter 3). Visualizing seafloor spreading as a simple process that can be represented in a single schematic diagram (Fig. 2.27) does not reflect the extreme diversity of processes documented along the global MOR and in the oceanic crust. Just as continental rifts and rifted margins show extreme variations in faulting and magmatic features, MORs and the crust produced along them reflect a very broad spectrum of phenomena collectively included under the current view of seafloor spreading.

As discussed above, the perception of the oceanic crust and upper mantle and the processes by which they are accreted at MOR spreading centers has historically leaned heavily on studies of ophiolites. The rapid development of marine geophysical techniques and the capability of detailed near-bottom geological investigations have led to increasingly higher resolution data sets over the past few decades. The results of these studies are discussed in the

Table 2.3 Time line of key developments in perspectives of the oceanic lithosphere

	Ophiolite geology	Marine geology	Marine geophysics
			OOI begins
		2nd Axial eruption	EPR 3D AMC survey
	Oman data synthesis		3D seismic surveys on Langseth
	Zr-dating in Oman		NEPTUNE Canada on-line
2010			Mismatch of Layer 2A and lavas
		IODP Hole 1256D reaches gabbro	EPR and JdF AMC surveys
		Live net broadcasts from sea	Deep seismic zone imaged at TAG
		2nd EPR 9° 50′N eruption	MAR OCC seismic studies
			ARAD 3D seismic study of OSC
	Macquarie Island studies		Large AMC on MAR at 34° N
		Subaxial subsidence of upper crust	MAR segment-scale seismic studies
2000		Ridge 2000 Program	
	Penrose Conference re-definition	Lost City Vent Site discovered	Seafloor geodesy begins
		Tectonic windows in Pacific	Magma chamber compliance data
		1st Axial eruption	"Megamullions" mapped on MAR
		Pet-B database available	EPR and JdF dike intrusions
	Gabbro sills in Oman Ophiolite	"Arc" rocks on Chile Rise	MELT Experiment on SEPR
		EPR axial summit trough	
	Gabbro glacier model	Segmentation of MORs	Global satellite gravity
		Co-axial eruption	T-phase seismicity studies
		East Pacific Rise Mapping	No AMC at robust MARK Area
		1st EPR 9° 50′N eruption	Gravity bullseyes and ribbons at MAR
1990	Ophiolite core complexes	RIDGE Program begins	
			SOSUS array in academic studies
			Discovery of AMC on EPR
	Mantle diapirs in Oman Ophiolite	Reykjanes Ridge study	Seismic Layer 2A mapping
		MAR studies show spreading variations	Expanding spread experiments in Atlantic
		Detachment faults and OCCs identified	First MCS studies in Atlantic oceanic crust
	Supra-subduction zone ophiolites defined	TAG site discovered	MAR microseismic studies
		OSCs identified	
		AMAR investigation	
1980	Oman and Bay of Islands magma chamber models	DSDP Hole 5048 started	
	Cumulus theory challenged	Rift propagation studies	Refined seismic structure of oceanic crust

Table 2.3 (*cont.*)

	Ophiolite geology	Marine geology	Marine geophysics
	Geochemical "fingerprinting" of ophiolites	Mid-Cayman Spreading Center study	Thin crust found at fracture zones
	Transform faults in ophiolites	Hydrothermal vents discovered	
	Early magma chamber models	Fracture zone studies	
		East Pacific Rise studies	
	Troodos Ophiolite debated as an arc		
	Intensive studies in Oman, Troodos, Bay of Islands Ophiolites	Seafloor depth/age relationship	
1970	Penrose Conference definition	Project FAMOUS	Rock velocities matched to seismic layers
	Ophiolites accepted as oceanic lithosphere		First academic multibeam echo sounding
		DSDP begins	
			Plate tectonics linked to global seismology
		"Seafloor spreading" coined	Seismic structure of oceanic crust defined
	Studies of layered igneous rocks		
		MORB defined	
		Studies of dredged rocks	
		MAR rift valley recognized	Magnetic anomaly stripes found
	Experimental petrology of mafic and ultramafic rocks	Early deep-sea photography	
1960		Heezen and Tharp maps published	

following chapters. While marine geological and geophysical investigations once helped validate the ophiolite model of oceanic crust and upper mantle, they have also shown that both the oceanic lithosphere and the processes that create it are much more complex and diverse than originally thought. New results have stimulated a re-examination of ophiolites to try to obtain more detailed information on processes that were not evident in earlier studies. They have also encouraged a reconsideration of highly faulted ophiolite complexes once thought to have been dismembered by post-spreading tectonic processes. These areas may provide details of processes at spreading centers now known to include complex faulting and intermittent magmatism, which are prevalent especially at slow- to ultraslow-spreading ridges.

In the following chapters, the focus is on images of the seafloor and rock units formed at spreading centers and the processes they reflect. These include hydrothermal vents and the exotic fauna supported by chemosynthetic systems (see Chapter 4). In most places along the global MOR system, hydrothermal vents and submarine lava flows (see Chapter 5) can be imaged and investigated in detail using a variety of novel deep-submergence vehicle systems and sensors (see Chapter 1). The corresponding chapters that follow reflect the intense investigations of these features conducted over the past few decades.

Unlike the upper part of the oceanic crust (hydrothermal vents and lava flows), deeper crustal and upper mantle rock units are exposed only where faulting has removed overlying crustal material. Accordingly, exposures of these rocks are more limited, so less is known about them than shallower crustal units. For the deeper crustal units (dikes, gabbros, and peridotites) details of geological relationships in ophiolites provide an important context for the interpretation of their less accessible counterparts on the seafloor (see Chapter 6–8).

The collection of data and the development of concepts surrounding seafloor spreading have been extremely rapid over the past few decades and show no signs of slowing down. To the contrary, the pace of discovery continues to

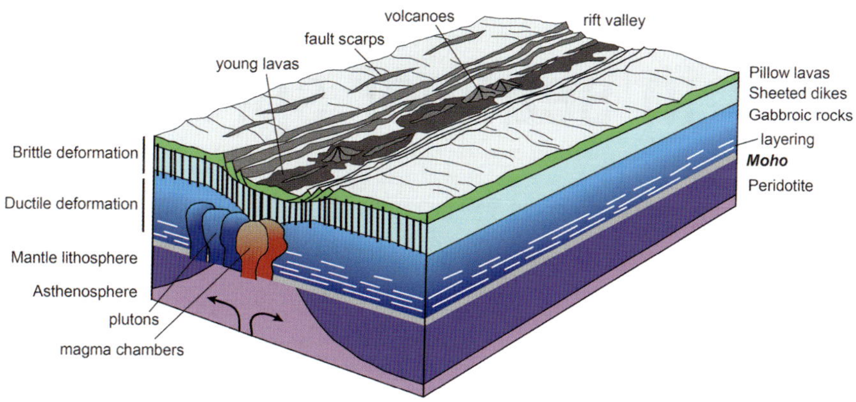

Figure 2.27 Highly generalized representations of spreading centers and oceanic crust suggest a relatively simple, uniform process of accretion and crustal structure (after Twiss *et al.*, 1992). Although useful in concept, much more complex structures and processes occur along the global MOR system, primarily as a function of magma budget and spreading rate, with the primary controlling factor being magma supply rate.

accelerate along with improved techniques for data collection and analysis, and more extensive surveying of the global MOR system. The remarkable images in the chapters that follow and the inferences they make possible set the stage for the next generation of multidisciplinary investigations of MORs. Future studies will certainly refine concepts regarding the creation and evolution of oceanic lithosphere and the varied chemical and biological processes occurring within the vast ecosystem of the Earth beneath the sea.

3 Diversity in seafloor spreading

In this seasonless abyssal world of total and eternal night, "scenery" acquires special meaning. Mountains, plains, valleys, and canyons of the deep sea are neither smaller nor less impressive than those on land, but we simply cannot see them with our own eyes nor are they seen by the inhabitants of the sea.

B. Heezen and C. Hollister (1971)

3.1 Seafloor spreading: a first-order global process

The largest and most continuous mountain ranges on Earth are the extensive systems of ridges, rifts, fault zones, and volcanic features that make up the globe-encircling mid-ocean ridge (MOR) system. Unlike the great mountain ranges of the continents that form by major tectonic collisions, the 60,000 km long MOR is created as lithospheric plates diverge, generating new oceanic crust and mantle by the process known as seafloor spreading. The interactions of

Figure 3.1 The surface of a Hawaiian lava lake on Kilauea Volcano shows cracks of glowing lava that outline plates of darker, cooled lava as they diverge and slide past one another. Lava rises to fill cracks where the plates diverge creating a pattern analogous to the interconnected spreading centers and transform faults of the global MOR system.

solid plates of newly formed lava pulling apart over the surface of active lava lakes provide a stunning analog for spreading centers and transform faults (Fig. 3.1).

As lithospheric plates diverge, a wide range of geological features form at MORs. These features are determined by the complex interactions between magmatic construction (the creation of the igneous crust) and mechanical deformation processes (brittle faulting to ductile flow in the crust and mantle). Although generally not exposed above sea level (Iceland and the Afar region of East Africa are exceptions), the MOR system is a first-order feature of Earth (Fig. 3.2). It is the most continuous volcanic and tectonic feature of our planet, and the locus of the most intense flux of mass and energy from Earth's interior to its surface.

For over half of Earth's ~4.6 billion year history, about 60% of the surface of the planet has been oceanic lithosphere (Burke and Kidd, 1978; Abbot and Hoffman, 1984). The linked processes of seafloor spreading and subduction have resulted in the continuous "repaving" of the seafloor over geologic time. Through Earth's evolution, seafloor spreading has been the most important mechanism of planetary heat loss. It is generally thought that, globally, spreading was much faster on the hotter, rapidly cooling, early Earth than it is today (Burke and Kidd, 1978; Abbot and Hoffman, 1984).

Since early investigations of the MOR system, there has been a tendency to view all spreading centers as a single feature where relatively simple, layered crust is created (see Chapter 2). However, marine geological and geophysical research have demonstrated that both the processes at spreading axes and the crust and lithosphere produced at them can vary significantly as a function of spreading rates and magma budgets along the global MOR system. In general, fast-spreading ridges have high magma budgets and appear to conform well to generalized models of spreading. Here the crust produced appears to have a relatively uniform, layered structure that may be analogous to that of some ophiolite complexes. In contrast, slow- and ultraslow-spreading ridges generally have low magma budgets and intermittent magma supplies. Plate separation is accommodated largely by mechanical extension (faulting and plastic deformation), resulting in an array of tectonic features similar to those found in continental extensional regimes. The combined influences of thermal and tectonic processes on magmatic construction result in crust and lithosphere that are heterogeneous (both structurally and petrologically) with large variations in seismic and geologic structures both along and across the axis. Simple, layered structures, as expected in crust formed at faster-spreading ridges, may be rare and limited to areas with robust, persistent magma supplies.

Along the MOR system, there are numerous tectonic features related to spreading centers, including many types of ridge-axis discontinuities, transform faults, propagating rifts, and microplates. Spreading ridges similar to those of the MORs also occur in back-arc basins above subduction zones. In these settings, spreading processes are strongly influenced by subduction processes and they are not discussed further here. Variations in the style of seafloor spreading in different settings reveal important details of the processes that build and modify the oceanic lithosphere and can impact hydrothermal systems and the life that they support (see Chapter 4). In this chapter, the processes associated

Figure 3.2 The global MOR system is a continuous network of volcanic and tectonic features marking divergent lithosphere plate boundaries. This global perspective highlights the MOR (relatively shallow, light-green area on the seafloor) as it encircles the globe. Differences in morphology of these different parts of the MOR reflect different structural, petrological, and morphological characteristics as a function of spreading rate and magma budget. The East Pacific Rise (left globe) is a fast- to superfast-spreading ridge, while the Mid-Atlantic Ridge (middle) is a slow-spreading ridge. MORs in the floor of the Indian Ocean (right) include the Southeast Indian Ridge and Central Indian Ridge spreading at intermediate rates, and the ultraslow-spreading Southwest Indian Ridge.

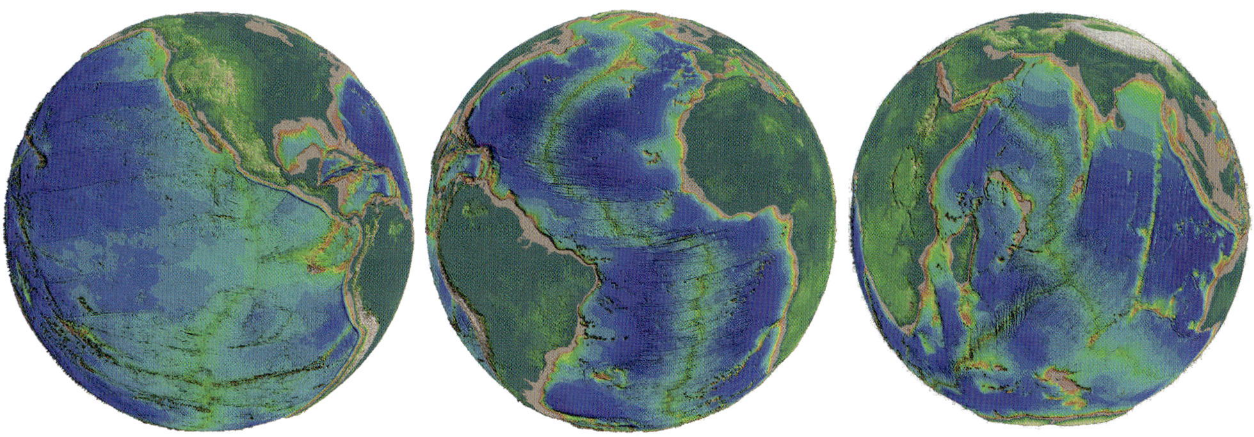

with different types of MOR spreading centers are contrasted to highlight these variations.

In general, processes that determine the morphology of spreading centers and geology of the oceanic crust are a function of spreading rate, that is, the rate at which the plates on either side of the spreading center are diverging. Globally, spreading centers are commonly grouped in terms of their full spreading rates: ultraslow, slow, intermediate, fast, and superfast (Table 3.1; Fig. 3.3). Spreading is bilaterally symmetrical in most cases, with some notable exceptions such as the Australia–Antarctic Ridge, south of Australia, and the Mohns Ridge in the Arctic. Spreading is also mostly orthogonal to the trend of ridges, but there are numerous instances of oblique spreading, where the plate motion is at an angle of less than 90° to the ridge axis, notably the Reykjanes Ridge, south of Iceland, and the Southwest Indian Ridge (Fig. 3.4).

Table 3.1 Spreading rates along the mid-ocean ridge system (see Fig. 3.3 for locations)

Major ridge segment	Full spreading rate (mm/yr)
Ultraslow (<20 mm/yr)	
Southwest Indian Ridge	14
Gakkel Ridge	12–16
Mid-Cayman Spreading Center	16
South America–Antarctic Ridge	18
Slow (20–50 mm/yr)	
Red Sea Spreading Center	22
Northern Mid-Atlantic Ridge	22
Central Mid-Atlantic Ridge	24
Southern Mid-Atlantic Ridge	34
Gulf of Aden Spreading Center	33
Central Indian Ridge	42
Intermediate (50–80 mm/yr)	
Chile Ridge	59
Galápagos Spreading Center	59
Juan de Fuca Ridge	60
Southeast Indian Ridge	68
Pacific–Antarctic Ridge	73
Fast (80–120 mm/yr)	
Northern East Pacific Rise	110
Superfast (>120 mm/yr)	
Southern East Pacific Rise	149

Note that these are average rates based on the width of magnetic anomaly stripes (which can form over tens of thousands to more than 1 million years). Actual rates may be significantly faster or slower on shorter time scales.

Spreading rates are reflected in the width of seafloor formed in a given time interval (Fig. 2.8). Seafloor ages based on magnetic anomalies show that spreading rates are not necessarily constant for spreading centers over millions of years and that they may speed up or slow down as dictated by global plate motions. Spreading rates averaged over thousands to millions of years (the width of magnetic anomaly stripes) vary from less than 10 to about 150 mm per year (full spreading rates), comparable to the rate of growth of human fingernails. Actual rates may be significantly faster or slower on shorter time scales. Although most of the length of the MOR system today is spreading at relatively slow rates, most of the seafloor created during the past 200 million years formed at relatively fast rates, in some cases much faster than the fastest spreading rates of today (Müller *et al.*, 2008).

Over the last four decades of the twentieth century, there has been a natural tendency to generalize seafloor-spreading processes, creating the impression that all parts of the MOR system are similar and that the oceanic crust has a uniform composition and structure worldwide. This is reflected in the schematic diagrams featured in many textbooks and research articles showing a classic "layer cake" structure in illustrations (Fig. 2.27). However, a closer look reveals that spreading processes and the oceanic crust and mantle generated at different spreading centers are highly variable and geologically complex.

3.2 Magma budget: key to variability along the MOR system

Major differences in the morphology and structure of spreading centers are a function of the interplay of two major processes: magmatic construction and tectonic deformation. Both processes are strongly focused along these plate boundaries. The relationship between these processes is reflected in the morphology of the axial region of the spreading center (Fig. 3.5). At faster rates, decompression melting, melt segregation, and the transport of melt to the surface are more efficient, and magmatic construction can keep pace with or even overwhelm the effects of faulting. At slower rates, magmatic construction does not fully accommodate plate separation and tectonic deformation. The geophysical properties and geology of oceanic crust formed at different parts of the MOR system reflect this diversity of processes.

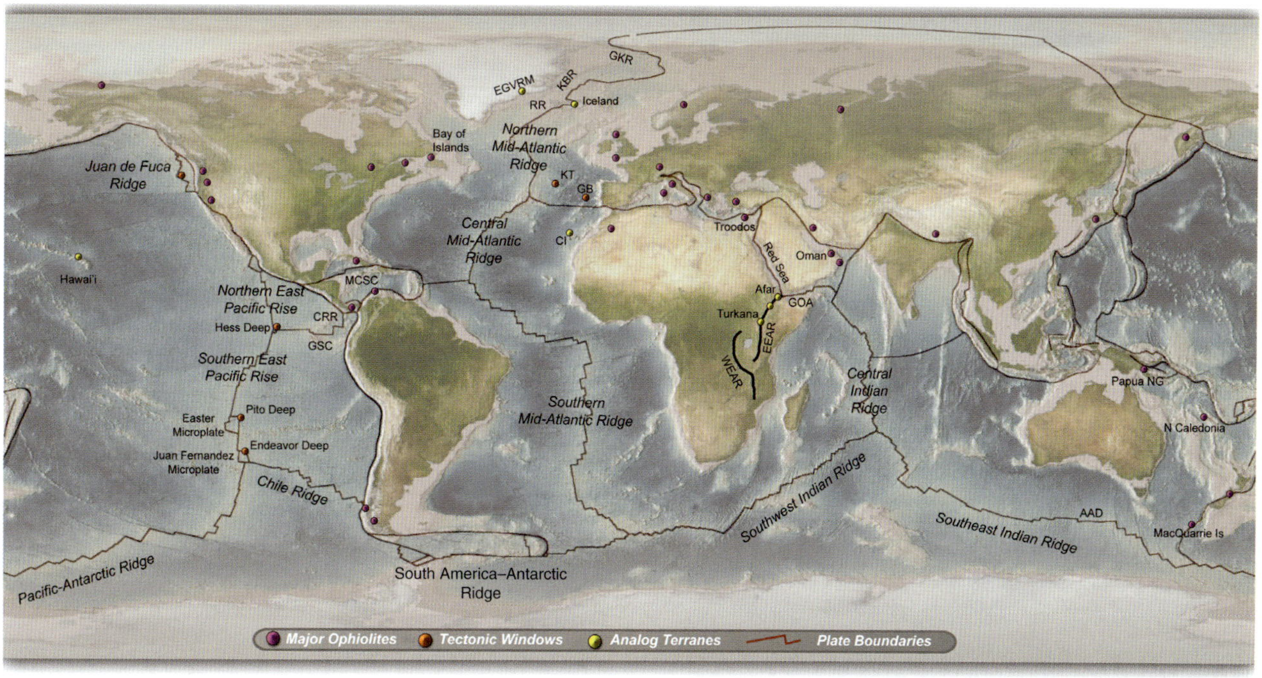

Isolated areas of significant mantle upwelling referred to as hot spots are likely related to plumes of mantle material rising from Earth's deep mantle or core–mantle boundary. Hot spots are linked to the rupture of continental lithosphere and large igneous provinces (LIPs) (Coffin *et al.*, 2006), including mid-plate oceanic plateaus where extensive ocean island chains originate (e.g., the Hawaiian–Emperor Seamount Chain). Some hot spots occur on or very near to MOR spreading centers. In these settings, enhanced melting results in a higher magma supply, which can produce V-shaped ridges as plates spread laterally away from these centers. Major hot spots on or near MORs have elevated topography regardless of spreading rate for hundreds of kilometers along the ridge axis. For example, on the Reykjanes Ridge, near the Iceland hot spot, voluminous magmatism results in spreading processes that resemble those of a much faster spreading ridge (Fig. 3.6). Thus, "magma budget," defined as the volume of magma delivered to a spreading segment over a given length of time, may be a more useful concept in comparing MORs than spreading rate alone (Karson, 1998).

3.3 Diverse morphologies of spreading centers

Despite the importance of MOR magmatic systems, the default response to plate separation is mechanical stretching of the lithosphere. This manifests itself as faulting near the seafloor and plastic flow in the lower crust and mantle. Without magmatic construction, plate separation would result in thinning and rupture of the lithosphere similar to what is seen in some continental rifts. Magmatic activity in the form of volcanism, dike injection, and intrusion of magma bodies interrupts lithospheric deformation and accommodates at least part of the tectonic extension. One of the most obvious indications of differences in spreading processes at different parts of the MOR is the range of morphologies of spreading centers. This is readily apparent in cross-sectional profiles (Fig. 3.5) and map views (Figs. 3.6 and 3.7).

Figure 3.3 Locations of key study areas referred to in this volume, including MOR spreading centers, major ophiolite complexes, tectonic windows at major escarpments, and relevant analog terranes. Note that ophiolites (purple dots) occur along all major continental mountain belts. Other localities: AAD, Australia–Antarctic Discordance; CI, Canary Islands; CRR, Costa Rica Rift; GSC, Galápagos Spreading Center; GB, Gorringe Bank; GOA, Gulf of Aden; GR, Gakkel Ridge; KR, Kolbeinsey Ridge; KT, King's Trough; MCSC, Mid-Cayman Spreading Center; RR, Reykjanes Ridge; EGVRM, East Greenland Volcanic Rifted Margin; EEAR, Eastern Branch East African Rift; WEAR, Western Branch East African Rift.

3.3.1 Fast-spreading MORs

At superfast- and fast-spreading ridges, the morphology of the axial region is a smooth, elevated swell covered with young basaltic lava flows (e.g., Macdonald, 1982) that some have likened to a highly elongated shield volcano (Lonsdale, 1985). The voluminous lava production results in little, if any, evidence of faulting until the crust has moved a few kilometers away from the axis where linear abyssal hills develop (Fig. 3.7) (Haymon *et al.*, 1991; Fornari *et al.*, 1998; Escartín *et al.*, 2007).

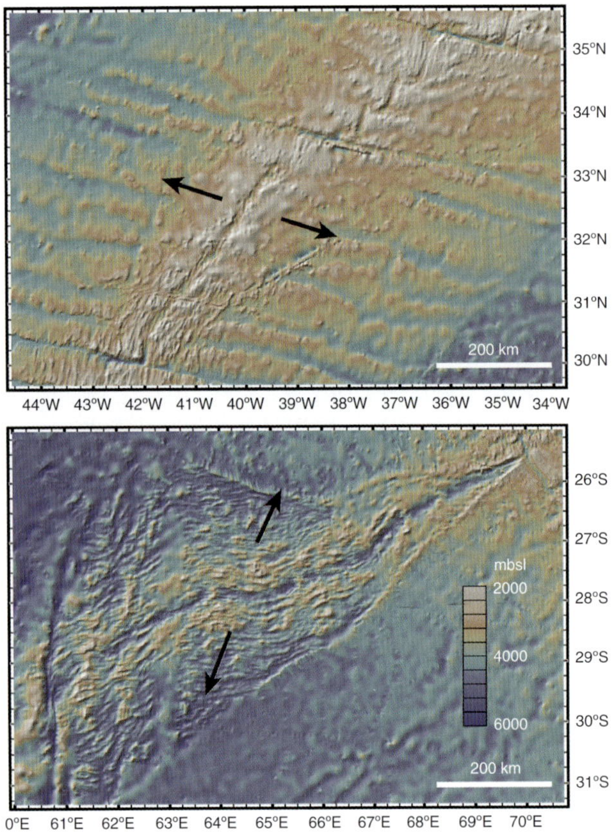

Figure 3.4 Most sections of the MOR are oriented with spreading segments orthogonal and transform faults parallel to the spreading direction, as in the case of the central MAR (top). In some cases, especially along some slow-spreading ridges, the axis is oriented at an angle to the spreading direction, as in the Southwest Indian Ridge (bottom). Both areas are at the same scale and have the same color-coded depth scale. Black arrows show the spreading directions.

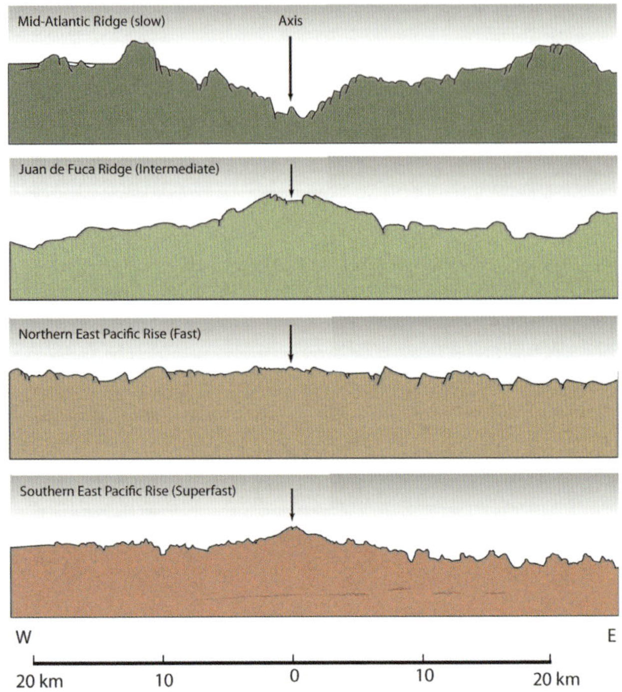

Figure 3.5 Bathymetric profiles across MORs with different spreading rates reflect different geological processes, specifically the relationship between magmatic construction and mechanical deformation. At slower-spreading ridges (with a generally lower magma budget), faulting dominates in a sputtering magmatic system. At fast to superfast rates, magmatic construction overwhelms faulting at the axis. Intermediate-rate ridges show a range of morphologies suggesting that magmatism and extension are nearly balanced in that range (Macdonald, 1982; Perfit, 2001).

3.3.2 Slow-spreading MORs

At slower spreading MORs, major faults create a broad rift valley at least several kilometers wide and commonly more than a kilometer deep (Figs. 3.5 and 3.7). In these settings of more restricted magma budgets, faulting and ductile deformation of the axial lithosphere are major processes and these types of spreading centers are shaped primarily by tectonic processes that can be highly variable in time and space (Stakes *et al.*, 1984; Karson *et al.*, 1987; Cannat *et al.*, 1997; Smith *et al.*, 2006; Escartín *et al.*, 2008).

At slow and ultraslow spreading rates (<40 mm per year, Table 3.1), the morphology of the spreading centers resembles those of continental extensional terranes with broad rift valleys and locally highly extended areas (Figs. 3.5 and 3.7). In some places, extreme stretching and faulting of the axial lithosphere result in exposures of deep crustal and even upper-mantle rocks in dome-like massifs referred to as oceanic core complexes (OCCs), which are analogous to similar structures found in some continental rifts (Buck, 1991).

3.3.3 Intermediate-spreading MORs

Spreading centers with intermediate rates (40–80 mm/yr, Table 3.1) exhibit transitional morphologies and structures between those at faster and slower rate MORs (Figs. 3.5 and 3.7). Temporal and spatial variations in magma supply appear to control variations between rifted, magma-poor and smooth, magma-rich modes of spreading (Kappel and Ryan, 1986; Smith *et al.*, 1994; Stakes *et al.*, 2006). Lithosphere formed in these environments appears to be relatively continuous, similar to that of fast-spreading environments, but with superimposed faulting that can extend more than 1–2 km beneath the seafloor.

3.3.4 MOR segmentation

Although MORs are remarkably continuous globally at scales of tens to hundreds of kilometers, on smaller scales they are made up of discrete, morphologic "spreading segments" which are separated by a range of discontinuities along the ridge axis (Fig. 3.8). Ridge segments are typically classified in terms of their length scales and morphologies.

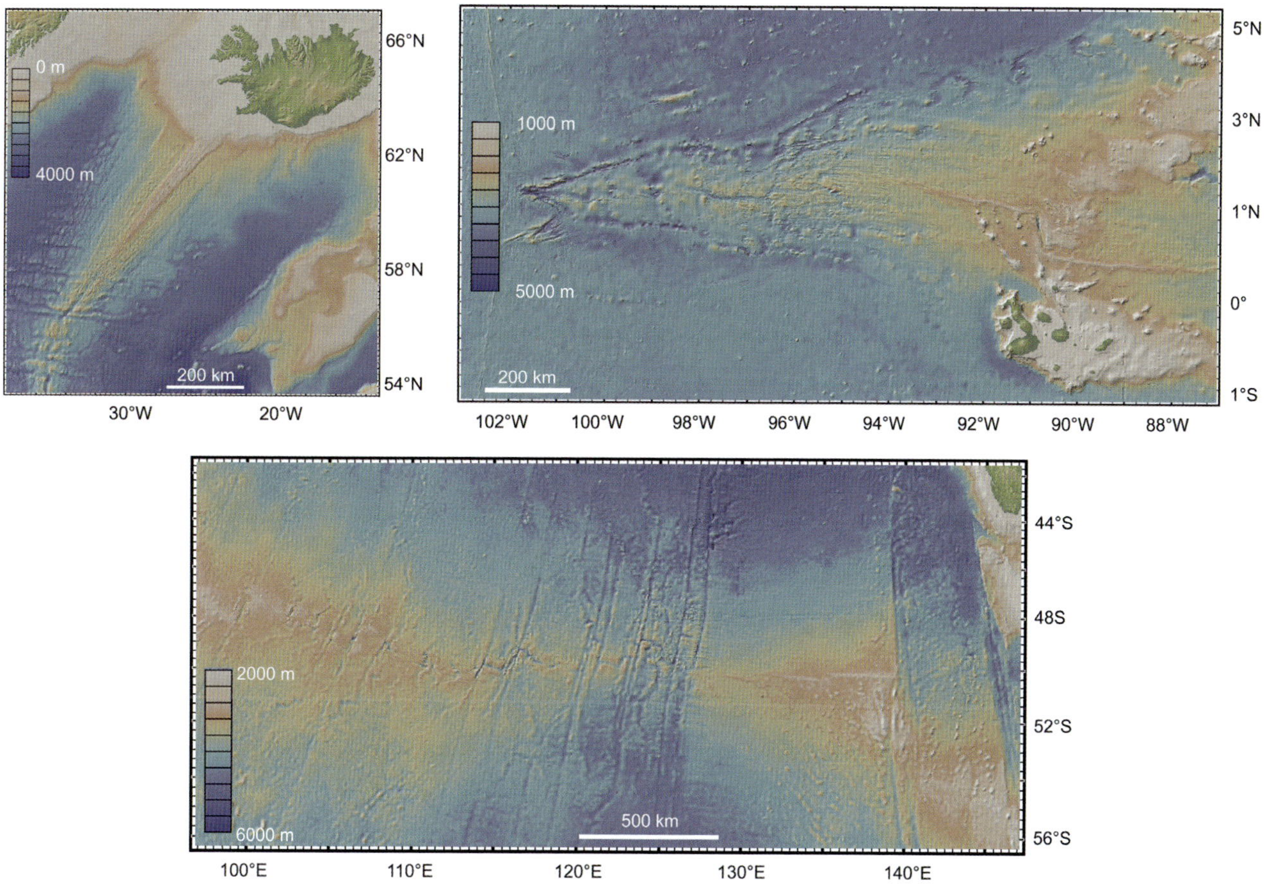

Figure 3.6 Spreading rates vary little along the slow-spreading northern MAR (top left) and intermediate rate Galápagos Spreading Center (top right), but rift valleys transition into smooth axial highs and areas of anomalously thick oceanic crust as the magma budget increases proximal to the hot spots located beneath Iceland and the Galápagos Islands, respectively. Along the Southeast Indian Ridge (bottom), the Australia–Antarctic Discordance (AAD) is a relatively deep interval of the MOR with a well-developed rift valley interpreted as an area of relatively cool mantle temperatures and hence low magma budget.

First-order segments are on the order of tens to hundreds of kilometers in length and separated by major transform faults. First-order segments are subdivided into progressively smaller, second- to fourth-order segments, which are separated by a variety of discontinuities specific to spreading rate and magma supply.

Within individual segments, geologic structures and dominant processes vary systematically from their centers to their endpoints (Macdonald, 1998a,b). However, spreading processes and ridge morphology can vary substantially between ridge segments separated by only a few kilometers. Faulting, magmatic construction, basalt geochemistry, and hydrothermal processes are commonly interrupted at segment boundaries (Ballard *et al.*, 1981; Langmuir *et al.*, 1986; Haymon *et al.*, 1991; Macdonald *et al.*, 1991; Hooft *et al.*, 1997; Carbotte *et al.*, 2000, 2013; Smith *et al.*, 2001).

Discontinuities that define segment boundaries range from major transform faults, up to hundreds of kilometers in length, to much more subtle offsets and morphologic changes along the axis of the spreading center (Lonsdale, 1977; Macdonald and Fox, 1983; Macdonald *et al.*, 1991). Segments and segment boundaries along relatively fast-spreading ridges reflect the dominant magmatic processes, for example dike intrusion in the shallow crust (Figs. 3.8 and 3.9). At slower spreading rates, the axial lithosphere is cooler and thicker and both segments and segment boundaries are defined by a combination of tectonic deformation features and local volcanic structures (Figs. 3.8 and 3.10).

Discrete segments of the MOR, which are tens of kilometers in length, host features indicative of different amounts of magmatic activity. Segments with significant magmatic activity are referred to as magmatically "robust" or, conversely, with diminished magma budgets as "starved." Robust segments have relatively shallow bathymetry, with a well-developed neovolcanic zone. Robust segments on slowly spreading ridges are characterized by axial volcanic ridges (AVRs) or other discrete volcanic centers and relatively narrow, shallow rift valleys, for example the classic FAMOUS Area on the MAR (Macdonald *et al.*, 1975). Along fast-spreading ridges, robust segments feature pillow mounds and broad (few kilometers) axial swells typified by large ("inflated") cross-sectional areas

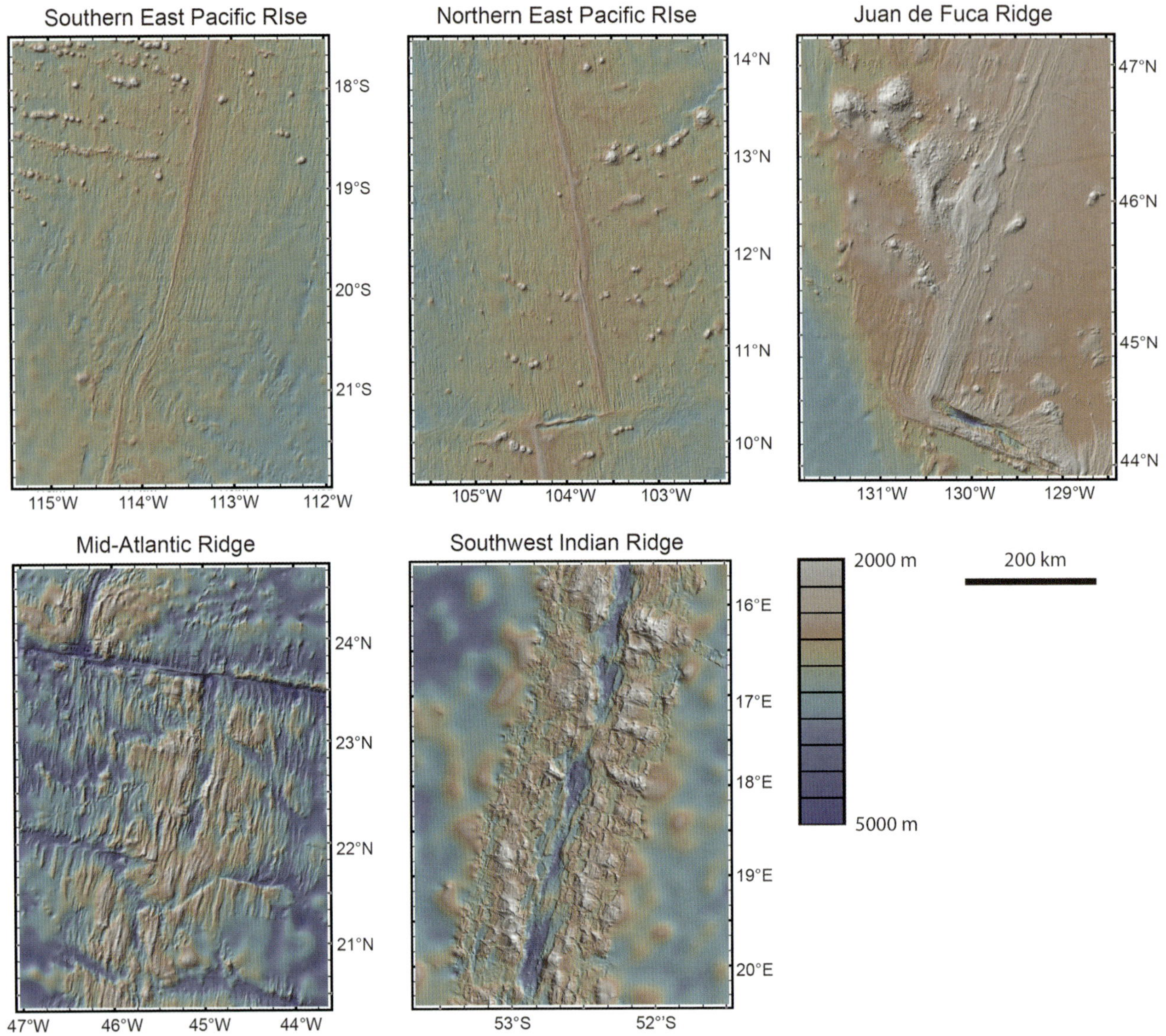

Figure 3.7 The morphology of spreading centers and adjacent abyssal hills varies dramatically with spreading rate. These images are all at the same scale and have the same color-coded depth scale (lower right). Smooth axial highs with small axial summit depressions occur at high spreading rates, such as the East Pacific Rise. Both axial highs and rift valleys occur at ridges with intermediate rates, such as the Juan de Fuca Ridge. The slow-spreading MAR has a broad rift valley and rugged terrain including dome-like oceanic core complexes. Ultraslow-spreading ridges, such as the Southwest Indian Ridge, have segmented rift valleys and complex, rugged terrain with abyssal hills that are much more irregular and varied than the others shown here.

(Macdonald and Fox, 1988). On parts of fast-spreading ridges with a relatively low magma supply, the axial swell has a relatively small cross-sectional area and may have prominent faults (Macdonald and Fox, 1988). On slow-spreading ridges, starved segments have a deep, broad rift valley with only sparse young volcanic features. Lava flows and volcanic cones look old and can have a substantial sedimentary cover.

Regional surveys of bathymetry, magnetics, and gravity show that the distinct, segmented character of the MAR is related to corridors of oceanic lithosphere with similar depth and crustal thickness that extend for hundreds of kilometers to either side of the ridge axis (Fig. 3.11). Shallow swaths of thick crust that formed near segment centers are bounded by deeper strips of thin crust that formed near segment ends. These strips lie along the traces of oceanic fracture zones. The observed relationships imply that the particular segmented character of a ridge axis can persist for millions of years of spreading history (Tucholke *et al.*, 1997; Wang *et al.*, 2014). If crustal thickness is controlled by mantle temperature, this also implies that mantle upwelling patterns are stable for long periods of time. Transform faults and fracture zones, which result from plate kinematics, may impose lithosphere-scale boundaries on mantle upwelling.

3.4 Components of spreading systems

In order to discuss the diversity of processes at MOR spreading centers, it is useful to compare some of the major volcanic and tectonic features that are common across a range of spreading regimes. Direct comparisons are not possible in all cases because the same types of features are not present in all spreading systems and because the same types of studies have not been carried out at all types of MORs. In addition, information is commonly available only in limited areas that may not be representative of the broad, regional character of the oceanic crust. As noted above, seafloor geology along the axial regions of MORs correlates well with the morphology of spreading segments. Geophysical data provide constraints on deeper level processes discussed in a later section, resulting in a more three-dimensional view of crustal accretion.

Major disruptions of spreading processes and oceanic lithosphere occur at large-scale discontinuities in the global MOR system. These include transform faults and their fracture zone extensions, propagating rifts, and oceanic microplates. These features affect the full thickness of the oceanic lithosphere and therefore modify the types of spreading-related structures described above. In cases where spreading centers intersect these features, thermal and mechanical effects alter spreading processes for tens of kilometers along the MOR axis. In others, fully formed oceanic lithosphere is modified by thermal and mechanical processes specific to these settings. Although these features are well documented from bathymetry, seismicity, and magnetic anomaly patterns, relatively little is known of the internal structure of the crust and lithosphere in these settings.

3.4.1 Neovolcanic zone

The axial region of the MOR, within which recent volcanic eruptions and high-temperature hydrothermal activity are concentrated, is called the neovolcanic zone (Ballard and

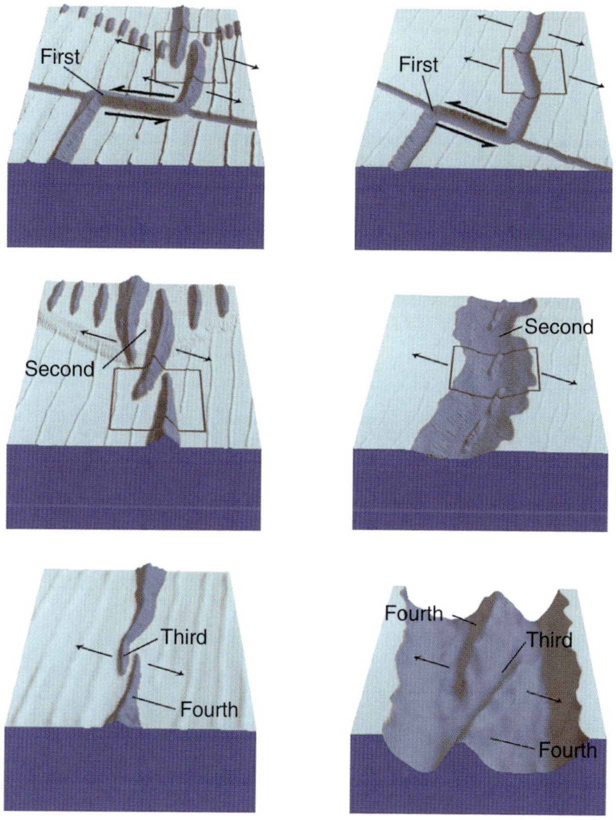

Figure 3.8 The morphology of MORs shows a hierarchy of first- through fourth-order segmentation patterns that vary with spreading rate or magma budget (Macdonald, 1998a). Left, fast-spreading; right, slow-spreading.

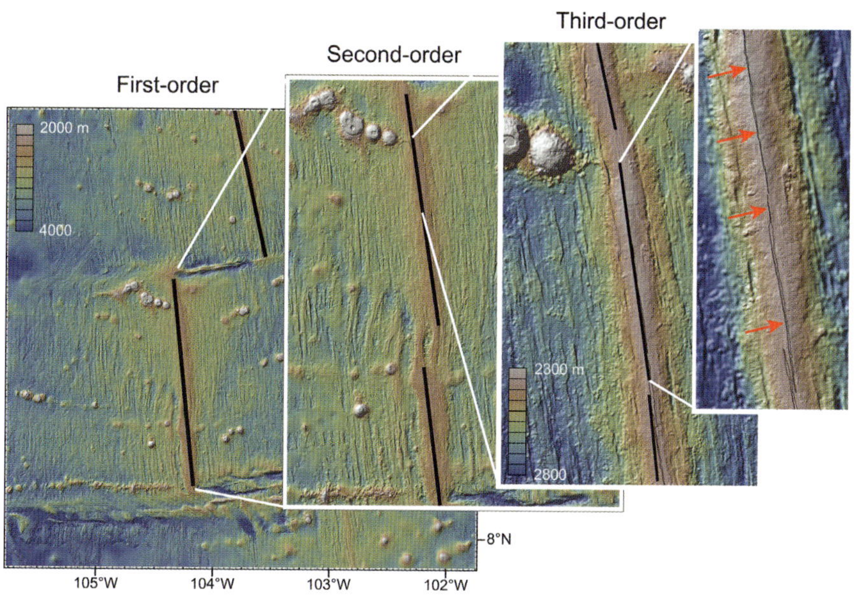

Figure 3.9 MOR spreading centers can be subdivided on the basis of their morphology as a hierarchy of features across a range of length scales. The East Pacific Rise at 8°–11° N illustrates first- through fourth-order segmentation of the spreading center. Different order segments have been found to correlate with along-axis variations in hydrothermal, geological, and geophysical characteristics.

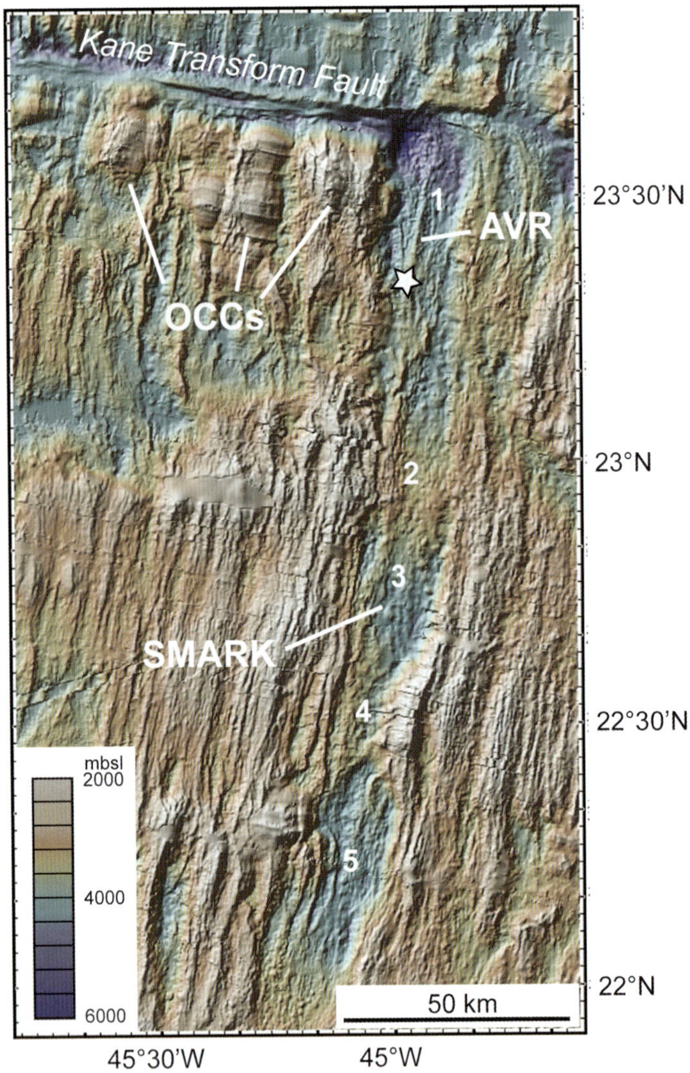

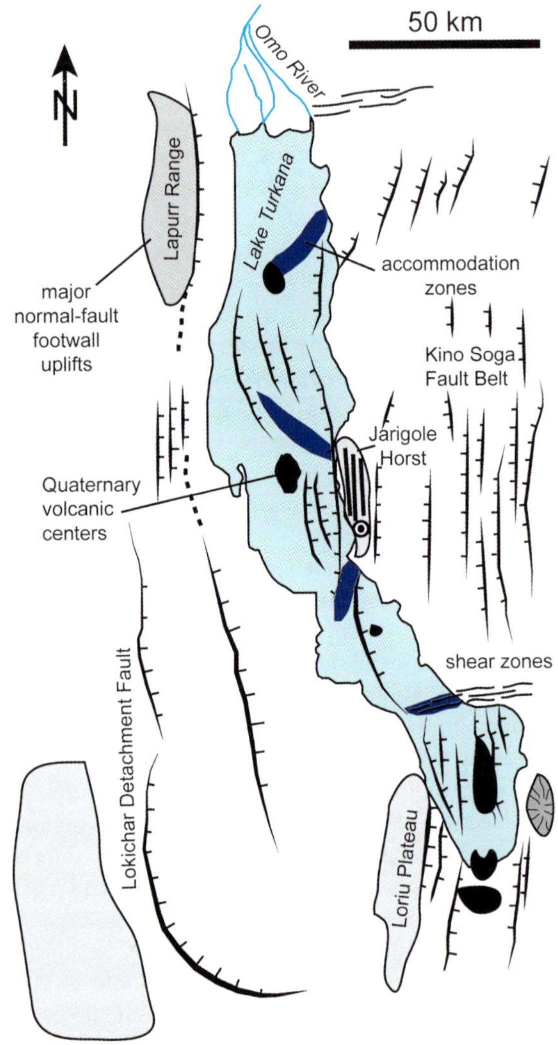

Van Andel, 1977b; Macdonald, 1982; Fornari *et al.*, 1998; Perfit and Chadwick, 1998). Regardless of spreading rate, the neovolcanic zone is a hummocky, relatively low-relief (tens to hundreds of meters) terrain. It comprises variably fissured basaltic lava flows with low- to modest-relief (typically <100 m) faults, especially at distances greater than about 500 m to 1 km from the spreading axis. At 23° S on the superfast-spreading southern EPR there is a fault-bounded axial valley a few hundred meters deep and less than a kilometer wide (Sinton *et al.*, 2002). Eruptive fissures and relatively low-relief volcanic edifices are typical. The neovolcanic zone varies dramatically with spreading rate or magma budget in its dimensions, volcanic morphology, and age of lavas (see Chapter 5). Although young lavas occur along most parts of the axial region of the MOR axis, the age of those lavas may range from zero to tens of thousands of years.

Active hydrothermal vents are important features of the neovolcanic zones of MORs and are known to exist along all spreading centers (Fig. 3.12). Their form, setting, chemistry, and biology vary substantially with spreading rates and

Figure 3.10 The bathymetry of the MARK Area (MAR 22°–24° N) (left) shows several morphologic spreading segments that correlate well with the geology and geophysics of the area. From north to south, spreading segments include (1) a northern segment with a prominent axial volcanic ridge (AVR) and a major detachment fault on the western median valley wall that exposes lower crustal and mantle rocks in highly extended OCCs; (2) a shallow symmetrical segment (saddle) with relatively young faulted pillow lavas; (3) a deep half-graben with seismicity in the axial mantle and no young axial volcanic rocks (SMARK Area); (4) a shallow, asymmetrical saddle; and (5) a deep, symmetrical rift valley. Each segment appears to have its own distinct blend of tectonic and magmatic processes. They are separated by poorly defined boundaries similar to accommodation zones in continental rifts. Sketch map of the Turkana Rift, Kenya (right), showing structures comparable to those of the MARK Area, including discontinuous Quaternary volcanic centers (black), variable rift structures (graben and half-graben) defined by major normal faults and associated footwall uplifts (gray), and a major detachment fault (Lokichar Fault). Rift segments are linked by accommodation zones (dark blue) (Dunkelman *et al.*, 1988; Mutter and Karson, 1992).

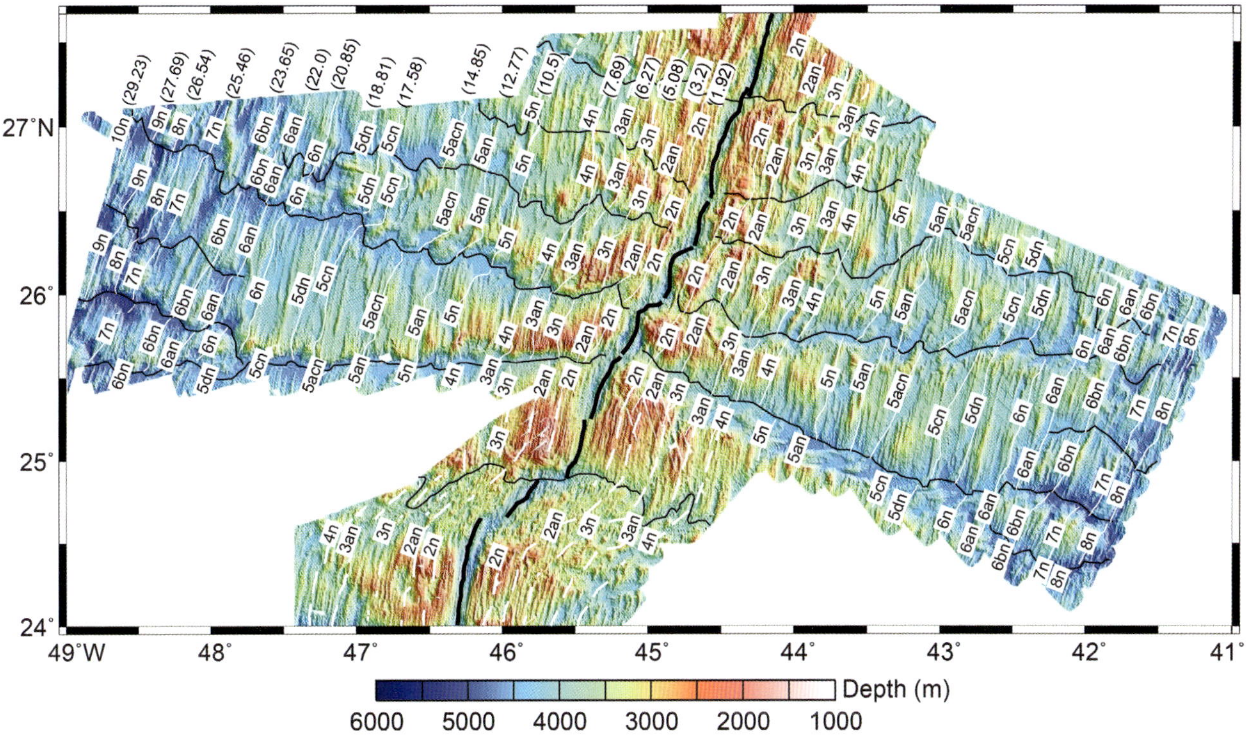

Figure 3.11 Corridors or strips of relatively thick crust (shallower) formed at segment centers are bounded by thin crust (deeper) formed at segment ends and can be traced for hundreds of kilometers off-axis. This indicates that spreading center segments are stable for millions of years. Widening or narrowing of crustal strips indicates lengthening or shortening of axial segments over time, respectively. This map of the central Atlantic shows depth variations for ~30 million years of spreading history. Black lines highlight the centers of deep, thin crust along the traces of fracture zones and segment boundaries (from Wang *et al.*, 2014).

globally (see Chapter 4). Hydrothermal processes are primarily controlled by heat from subsurface magma bodies and/or hot crustal rocks and faulting. Thus, their seafloor distribution, size, and the character of mineralized chimneys and chemistry of their fluids are all functions of primary spreading processes (e.g., Kelley *et al.*, 2002; Kelley and Shank, 2013).

High-temperature (~350 °C) "black smoker" hydrothermal vents are strongly focused in the axial rift of fast-spreading ridges and in the axial valley of intermediate-rate ridges (Haymon *et al.*, 1991; Delaney *et al.*, 1992; Baker *et al.*, 1994, 2001; Kelley *et al.*, 2012, 2014; Jamieson *et al.*, 2013). The high rates of fluid flow (meters per second) emanating from these vents imply that they are located along high-permeability fault zones and fissures. Studies of major escarpments in the upper crust, deep crustal drill cores, and ophiolites show that hydrothermal metamorphism in the subsurface is highly heterogeneous and strongly controlled by fracture networks and fault zones (Kelley *et al.*, 1993; Alt *et al.*, 1996; Gillis *et al.*, 2001, 2005).

At fast spreading rates, the neovolcanic zone is continuous along-axis and commonly defines a broad (few kilometers), low-relief (100–200 m) swell (axial plateau) that in many places has a narrow (~50–500 m wide), central axial summit graben or collapsed trough produced by subsurface dike intrusion and collapse related to magma withdrawal (Fig. 3.13) (Chadwick and Embley, 1998; Fornari *et al.*, 1998).

Lavas emanating from the axial region flow over the flanks of the spreading axis, resulting in a swath of young lavas ~2 to 4 km wide (Fornari *et al.*, 2004; Soule *et al.*, 2005) (see Chapter 5). Lavas of the neovolcanic zone on fast-spreading ridges are a few years to thousands of years old based on U-series dates and observations of fresh lava related to recent eruptions (Sims *et al.*, 2003; Rubin *et al.*, 2012). At intermediate rates, neovolcanic zones are similar to those of fast-spreading ridges but they may be enclosed within a rift valley, as at the Juan de Fuca Ridge (e.g., Stakes *et al.*, 2006) (Fig. 3.14).

At slower spreading rates, the neovolcanic zone is up to 4 km wide and contained within the central part of the axial rift valley (Fig. 3.15). Young volcanic terrane is typically discontinuous, with isolated volcanic cones and larger axial volcanic ridges (AVRs) up to tens of kilometers long, about 4–8 kilometers wide, and a few hundred meters high. Some slow-spreading segments do not have young-looking lavas; the rift valley floor is covered only with older, Mn-encrusted and visibly altered lavas partially buried by pelagic sediment. There are relatively few near-bottom surveys of slow-spreading axes, so it is not clear how common this situation may be. Available

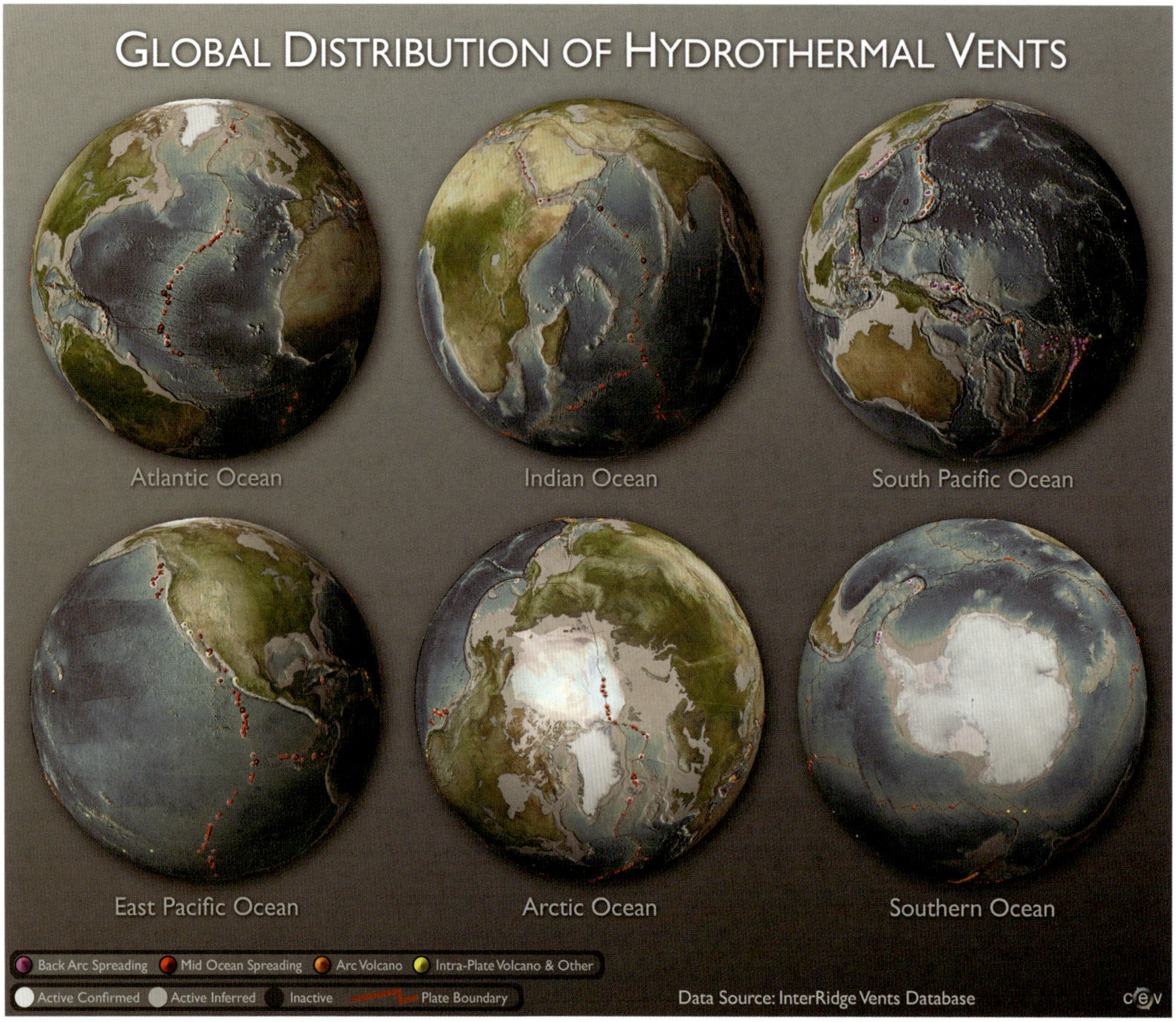

Figure 3.12 Global views of the MOR axis and known hydrothermal vent sites.

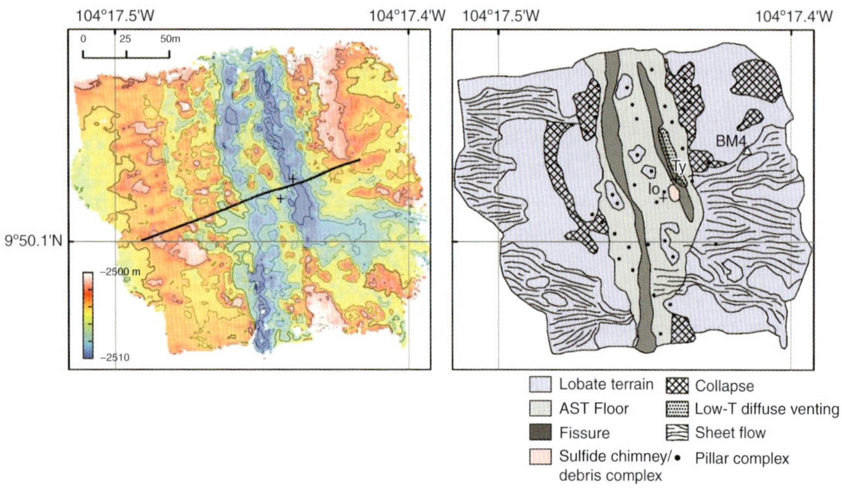

Figure 3.13 Near-bottom investigations using both visual observations and high-resolution sonar permit this type of detailed geologic mapping of the neovolcanic zone of the East Pacific Rise, showing the axial summit trough (AST), young lavas, hydrothermal vents, and fissures along the axis of the spreading center (Soule *et al.*, 2005).

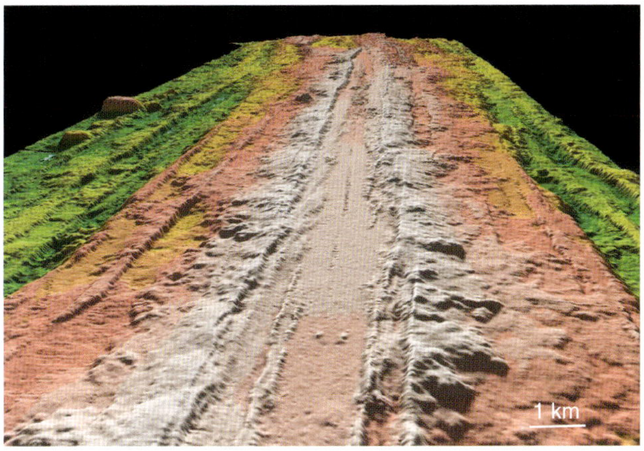

Figure 3.14 Shaded relief image of the Cleft Segment of the Juan de Fuca Ridge, one of the most intensively studied intermediate-rate ridges, looking from the south (~ 44°33′ N) to the north (~ 45°05′ N). The elevated summit area (gray to pink) is cut by a distinct ~2 km wide rift valley, bounded by fault scarps, which bisects the linear ridges to either side. Young lavas occur in the axial valley (neovolcanic zone) but also occur as isolated flows on the flanks of the bounding crestal ridges. Note the AST or "cleft" is apparent as a thin linear fissure that bisects the axis in the middle of the axial valley. Light gray color is depth range from 2100 to 2200 m; dark green >2600 m (modified from Stakes *et al.*, 2006).

U-series dates show that even some very young-looking, glassy, sediment-free lavas on the MAR are 10,000–20,000 years old (Sturm *et al.*, 2000). Only rarely have recent neovolcanic zone eruptions been documented on slow-spreading ridges, for example at the Turtle Pits (Haase *et al.*, 2007; German *et al.*, 2008a) site at 5° S on the MAR (see Chapter 4).

3.4.2 MOR faults

On fast-spreading ridges, linear fault scarps only rarely occur along the broad volcanic swell that dominates the spreading axis. Instead, fault scarps a few tens of meters high emerge outside of the neovolcanic zone on the lower flanks of the axial summit plateau region a few kilometers from the axis (Fornari *et al.*, 1998, 2012). These faults and volcanic constructions create linear abyssal hills, which are spaced at intervals of a few kilometers and can be tens of kilometers in length and parallel to the spreading axis (Lonsdale, 1977; Macdonald *et al.*, 1996). Thus, it appears that faulting near the axis of fast-spreading ridges may be masked by frequent axial volcanism and only become obvious beyond the ~1–4 km extent of lava flows emanating from the neovolcanic zone. Abyssal hills are rafted laterally away from the axes as spreading proceeds and thus they provide a record of the local ridge-axis geometry as the oceanic crust is generated over geologic time (Malinverno and Cowie, 1993; Macdonald *et al.*, 1991, 1992).

The shapes and dimensions of abyssal hills vary substantially with spreading rate (Fig. 3.7) and may be produced

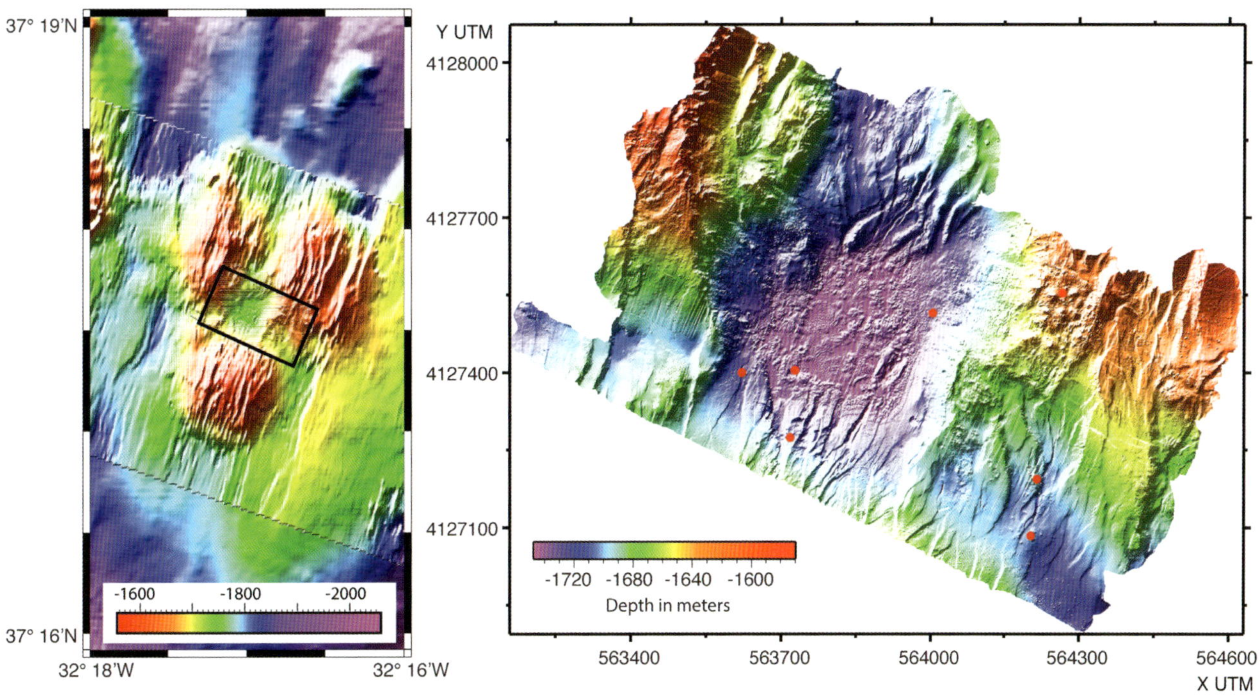

Figure 3.15 The Lucky Strike segment near 37° N on the MAR is a particularly magmatically robust ridge segment with the composite Lucky Strike Seamount situated in a well-developed rift valley (left). Multibeam and AUV bathymetric surveys provide detailed maps of the area. Black box on left map shows the area of detailed AUV bathymetry (right) at the summit of the volcano. Red dots on right map show high-temperature hydrothermal vent locations (Cannat *et al.*, 1999; Escartín *et al.*, 2001; Combier, 2007; Ondreas *et al.*, 2009; Barreyre *et al.*, 2012).

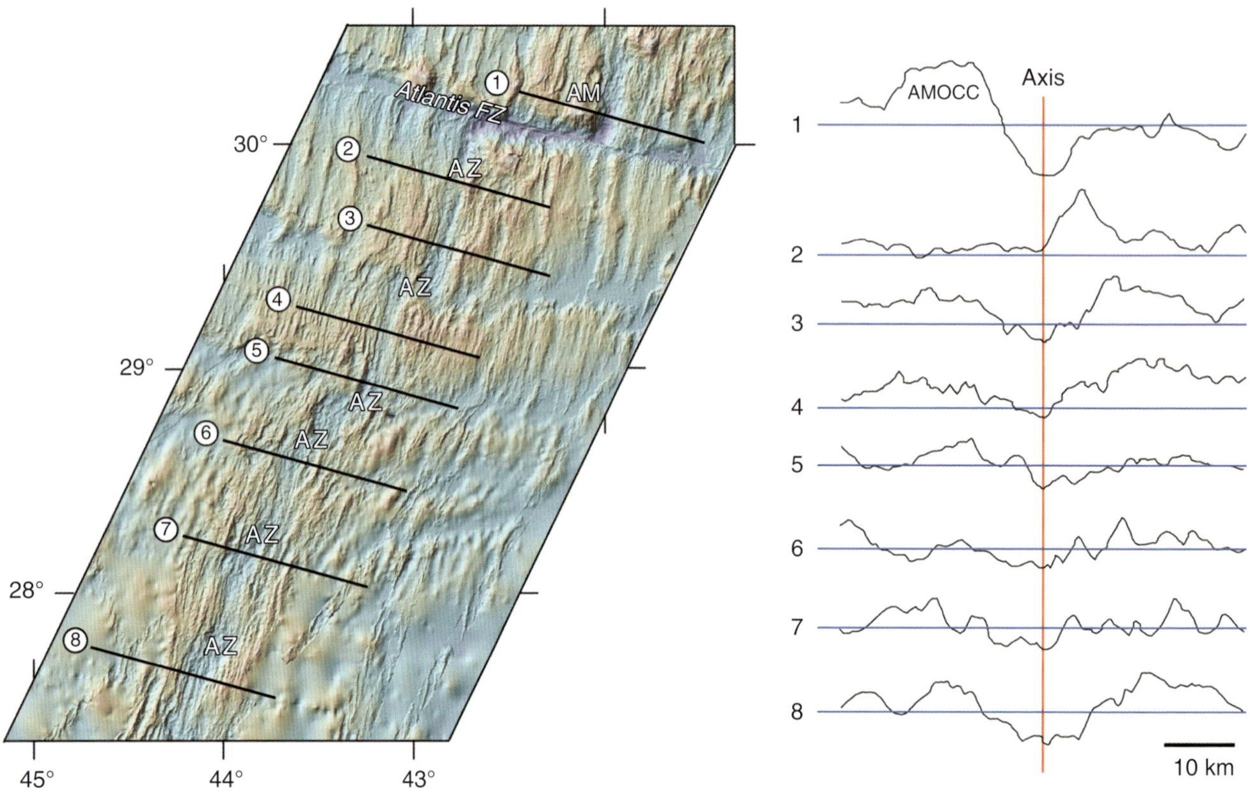

Figure 3.16 The morphology of slow- to ultraslow-spreading ridges is highly variable along the MOR axis and appears to depend strongly on the regional or local magma budget. In general, faulting at slow-spreading MORs creates a well-developed rift valley with many features similar to those of continental rifts. This section of the MAR rift valley consists mostly of segments with the form of symmetrical grabens or asymmetrical half-grabens linked end-to-end by accommodation zones (AZ) or "non-transform offsets." Bathymetric profiles 1–8 on the right illustrate the large variations in rift valley morphology. AM, Atlantis Massif VE = 5×.

by somewhat different processes in different settings (Macdonald and Luyendyk, 1977, 1985; Edwards *et al.*, 1991; Macdonald *et al.*, 1996; Buck *et al.*, 2005). Without the overlying ocean and sedimentary cover, the seafloor would have a strongly lineated pattern of parallel elongated ridges and valleys on either side of the MOR axis over most of the planet. Abyssal hills formed at fast-spreading centers have relatively low relief (typically 100–200 m), forming continuous (tens of kilometers), linear ridges spaced at intervals of a few kilometers. Abyssal hills in these environments are likely to define the trace of either normal faults or linear volcanic ridges.

At slow-spreading ridges, faulting is generally much more obvious, but highly variable. Major fault zones bound the rift valley, which is typically a few to ten kilometers across and more than a kilometer deep (Fig. 3.16) (Shih, 1980; Dick *et al.*, 2003). The relief and shape of MOR rift valleys globally and regionally vary significantly. Faulting at slow-spreading ridges clearly accommodates part of the strain related to plate separation and accounts for the relatively large-magnitude earthquakes ($M > 5$) that occur along slow-spreading MOR rift valleys (Bergman and Solomon, 1984; Solomon *et al.*, 1988; Smith *et al.*, 2003). Cross-sectional profiles have a wide range of expressions, including symmetrical graben and asymmetrical half-graben with variable widths, as well as areas where the axial region is a rugged faulted terrain with no clear axial depression. Spreading segments with different forms are linked by structures including transform faults and poorly defined "non-transform offsets" which are analogous to accommodation zones of continental rifts.

Major normal faults, tens to hundreds of meters high, bound the axial valley floor at intermediate- to slow-spreading MORs (Macdonald and Luyendyk, 1977; Kappel and Ryan, 1986; Kappel and Normark, 1987; Karson *et al.*, 1987; Zonenshain *et al.*, 1989; Karson and Rona, 1990; Delaney *et al.*, 1992; Glickson *et al.*, 2007). Slip on faults at slow- and ultraslow-spreading centers, which can result in hundreds of meters of displacement, causes uplift of the axial crust. These rocks are incorporated into the nearby rift mountains, thereby exposing the upper levels of the oceanic crust (Fig. 3.17). Major escarpments bounding the rift walls typically have a stair-step form defined by alternating steep faults and intervening terraces. Talus and debris slide deposits are extensive at the base of the rift walls (Figs. 3.18

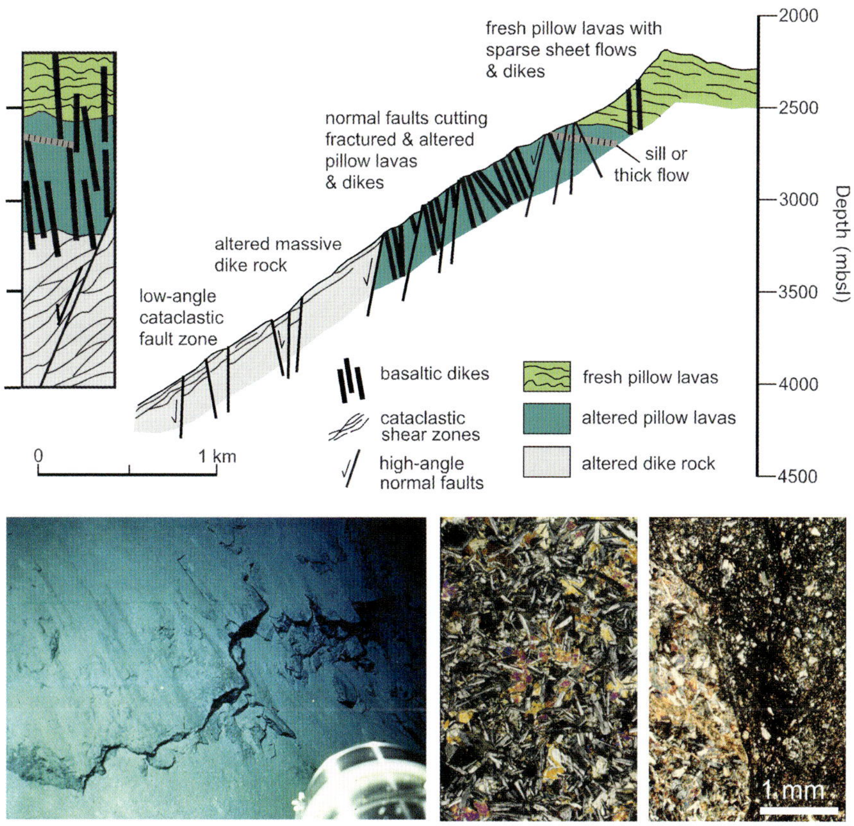

Figure 3.17 A near-bottom study, which used the submersible *Alvin* at a major fault zone defining the eastern median valley wall of the MAR at 23° N (SMARK Area), shows a cross section of the upper crust (above) exposed by faulting (Lawrence *et al.*, 1998). The crustal section is represented as a columnar section at the same depth scale to the left. Outcrops (lower left) have a shingled structure created by closely spaced low-angle (~30°) fault surfaces in basaltic rocks. Field of view ~2 m. Photomicrographs (lower right) taken in cross-polarized light show typical undeformed igneous diabasic textures (left) and cataclastic textures (right) with intense grain-size reduction from brittle deformation along the fault zones.

Figure 3.18 Poorly sorted, angular, blocky talus accumulates at the base of active fault zones on the seafloor. This example is from the Pito Deep Rift. Field of view ~2 m.

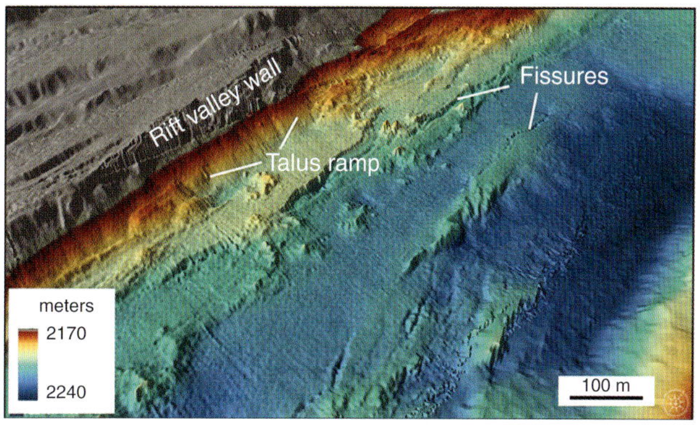

Figure 3.19 Bathymetry and sidescan sonar mapping of the seafloor along the axis of the Endeavour Segment of the Juan de Fuca Ridge show a continuous talus ramp at the base of the rift valley wall (smooth, orange-yellow area). This is typical of major fault scarps along spreading centers and transform faults.

and 3.19) where they create continuous aprons of variably sized rock fragments.

In crust formed at slow rates, abyssal hills are much more variable in form than faster-spreading crust. They can have vertical relief higher than 600 m, but tend to be less than about 30 km in length and spaced at intervals of about 10 km. In addition to the linear abyssal hills that form at many spreading center segments, more irregular to equant, dome-shaped features form as oceanic core complexes.

3.4.3 Oceanic core complexes

In areas subject to extreme extensional faulting, unroofing of the crust results in isostatic uplift and arching of corrugated (detachment) fault surfaces. This process creates broad,

Figure 3.20 Dome-like OCCs along the Kane Transform Fault (right side of image) at ~24° N on the MAR have vertical relief of over 4 km. They are marked by spreading-parallel (~E–W) corrugations, across arched, low-angle detachment faults. Variably deformed and metamorphosed lower crustal and upper mantle rocks are exposed on major escarpments including the rift and transform valley walls as well as those separating the massifs. A young (<2 million years old) core complex defines the western median valley wall adjacent to the ridge–transform intersection (>5000 m depth). Successively older OCCs to the west were formed over several million years of spreading. North is to the right and the image is about 30 km wide in that direction.

Figure 3.21 Large corrugated fault surfaces occur along the median valley wall of the MAR – in some places providing a snapshot of the early stages of development of an oceanic core complex. This example from 13°19′ N was imaged by high-resolution sidescan sonar surveys and is ~10 km from the top to the bottom of the slope along the corrugated surface. (MacLeod *et al.*, 2009).

dome-like massifs with spreading parallel lineaments (Cann *et al.*, 1997; Blackman *et al.*, 1998; Tucholke *et al.*, 1998; Smith *et al.*, 2006). These oceanic core complexes (OCCs – also referred to as "megamullions") represent very distinct morphologic features of crust formed at ultraslow- and slow-spreading MORs and rarely on intermediate-spreading MORs (Karson, 1990; Tucholke *et al.*, 1998; Karson, 1999; Escartín *et al.*, 2003; Smith *et al.*, 2006; Escartín *et al.*, 2008; MacLeod *et al.*, 2009; Pedersen *et al.*, 2010a).

The OCCs contrast in morphology with the more typical linear abyssal hills on the flanks of many MORs (Fig. 3.20). In general, hydroacoustic and teleseismic activity along the MAR is concentrated along segments with asymmetrical rift valleys or OCCs (Smith *et al.*, 2006). Marine magnetic anomalies in some cases transect OCCs without deflections, indicating that horizontal spreading rates must be comparable along-strike with or without OCCs (Hosford *et al.*, 2003). Oceanic core complexes appear to form mostly in areas of highly asymmetrical spreading with extreme deformation of the lithosphere on one side of the median valley and essentially normal processes on the other (Karson and Dick, 1983; Tucholke *et al.*, 1998; Karson, 1999; Escartín *et al.*, 2008; MacLeod *et al.*, 2009).

In some cases, the massifs that make up OCCs appear to be composed primarily of serpentinized mantle peridotite, suggesting a very low magma supply (Karson and Lawrence, 1997; Karson, 1999; Boschi *et al.*, 2006; Karson *et al.*, 2006a; Kelemen *et al.*, 2007; Dick *et al.*, 2008). Other OCCs have thick (more than a few kilometers) sections of gabbroic rocks, implying significant magmatic activity during crustal accretion (Karson and Dick, 1983; Dick *et al.*, 2000; Godard *et al.*, 2009). More generally, OCCs have geologic structures that are somewhere between these end-members with serpentinized peridotites intruded by a collage of relatively small (tens to hundreds of meters across, and meters to tens of meters thick) gabbroic bodies (Cannat, 1996; Dick *et al.*, 2008; Escartín *et al.*, 2008; Godard *et al.*, 2009; MacLeod *et al.*, 2009).

The tops of OCCs are marked by corrugated and striated surfaces interpreted as major fault zones (detachment faults). These faults arch over the tops of the domes and plunge beneath the adjacent median valley (Figs. 3.20 and 3.21), or are abruptly truncated by major, spreading-related structures (Tucholke *et al.*, 1998; Smith *et al.*, 2012). The geology of the lower plate or footwall of an OCC is extremely complex, with major ductile to brittle shear zones cutting a heterogeneous mix of gabbroic and ultramafic rocks. Unraveling the complex history of these terranes is the key to understanding the distinct style of spreading that creates them.

Paleomagnetic studies of drill cores from the footwalls of OCCs demonstrate that large-scale horizontal-axis rotations have occurred (Garces and Gee, 2007; Morris *et al.*, 2009).

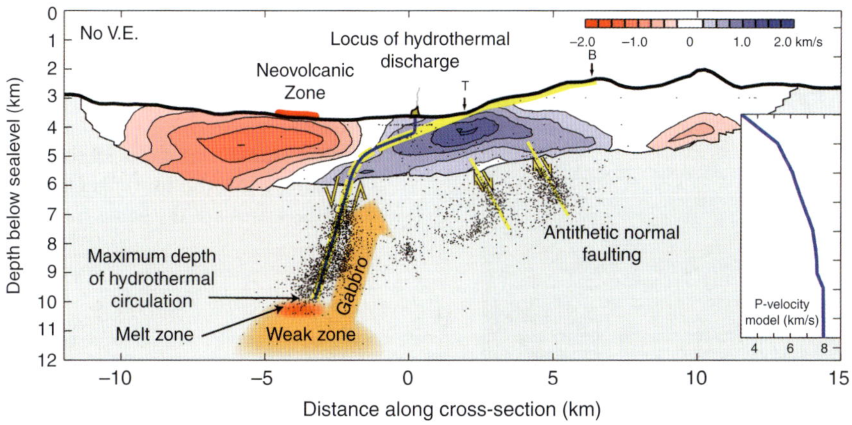

Figure 3.22 Detailed earthquake locations in the TAG area (MAR at 26° N) show strong clusters of small earthquakes (small black dots) that define a number of steep, seismically active fault zones beneath the eastern median valley wall (deMartin *et al.*, 2007).

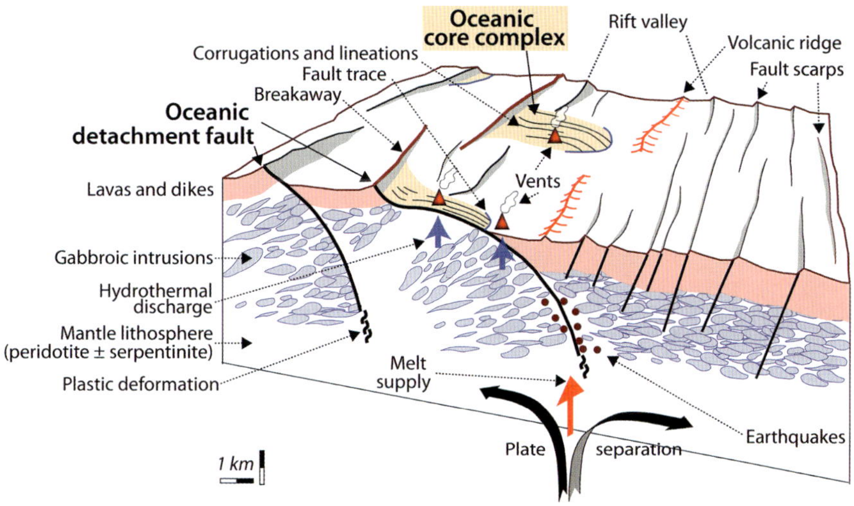

Figure 3.23 Oceanic core complexes appear to develop from steeply dipping shear zones in highly heterogeneous oceanic crust at slow-spreading MORs. Magmatic intrusions occur along the shear zone that rolls over from an originally steep orientation into an arched, dome-like structure capped by a low-angle (rotated) detachment fault. The result is a highly asymmetrical spreading system with an OCC forming on one side of a ridge segment and more normal oceanic crust on the opposing side (Escartín and Canales, 2010).

This supports the interpretation that detachment fault zones that cap OCCs slipped while in steeper orientations before rolling over into their present arched forms.

A glimpse of an early stage in the development of an OCC may be provided by seismic data that define a steep fault zone at the MAR axis in the TAG area (Fig. 3.22) (deMartin *et al.*, 2007). Although no low-angle fault zone has been documented in this area, and the seismicity cannot be directly linked to slip on low-angle surfaces at depth, it is possible that the steep (~70°) seismically defined fault could roll over to a low-angle orientation as spreading proceeds.

Numerical models for the formation of OCCs employ constraints from the length scales of surface features and rheological models for the lithosphere (Lavier *et al.*, 1999; Tucholke *et al.*, 2008). These models conclude that asymmetrical spreading leading to OCC formation relies on magmatic construction in which intrusion at shallow levels is sufficient to account for about half of the spreading. This allows the footwall of an OCC to move laterally away from the axis as normal spreading persists on the opposite side of the spreading center. These models provide excellent hypotheses to be tested by future seafloor investigations.

Recently, detailed studies have begun to quantify the processes by which OCCs form (Grimes *et al.*, 2008; Morris *et al.*, 2009; Cheadle and Grimes, 2010; Escartín and Canales, 2010). New investigative approaches, in part building on experience from continental core complexes, have begun to clarify the mode of formation of these features and frame the next generation of questions that will contribute to a better understanding of this distinctive style of seafloor spreading (Fig. 3.23).

3.4.4 Transform faults

Ridge–ridge transform faults are the active strike-slip segments of oceanic fracture zones that link MOR spreading centers (Figs. 3.24 and 3.25). As such they are an integral part of the MOR system and a fundamental type of plate boundary.

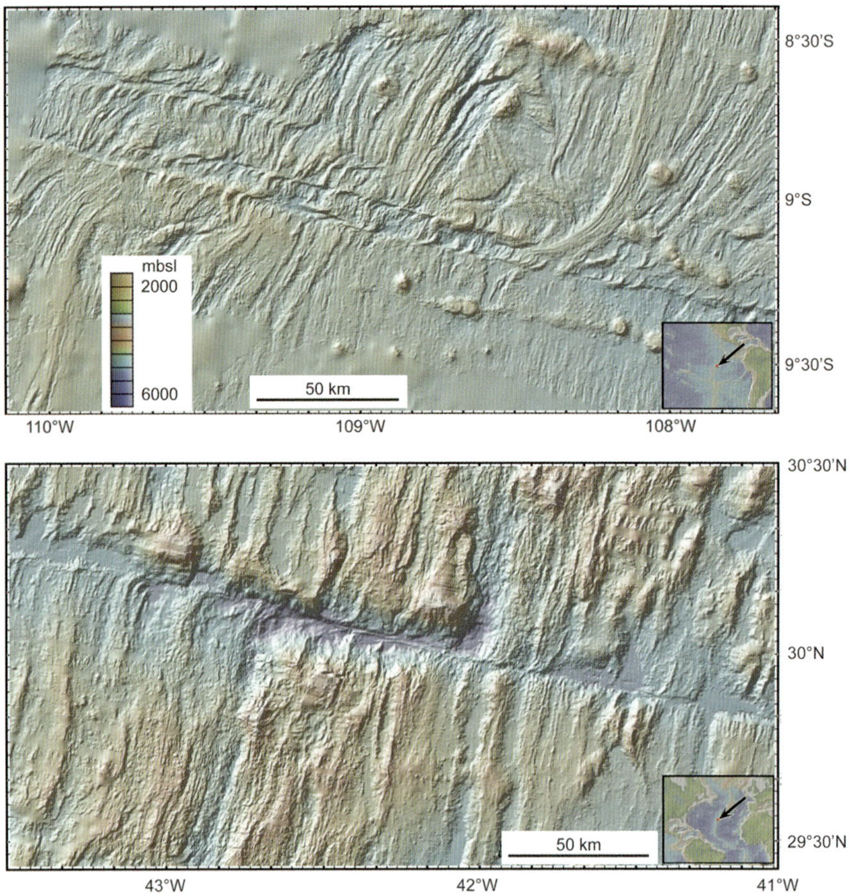

Figure 3.24 Examples of major transform faults on the MOR in fast- and slow-spreading environments: Wilkes Transform on the fast-spreading southern EPR (top) and Atlantis Transform on the slow-spreading MAR (bottom). Depth and distance scales are the same for both images.

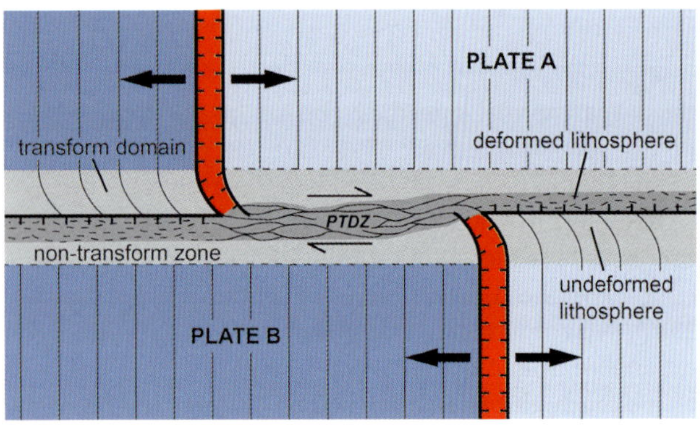

Figure 3.25 Oceanic fracture zones include the active transform fault zone that links spreading centers as well as the non-transform zones that extend away from the plate boundary. The shearing stresses associated with the transform zone affect the transform domain that may be tens of kilometers wide (light gray). A narrower area affected by strike-slip deformation (dark gray) includes the active principal transform displacement zone (PTDZ). Note that the intraplate non-transform zones are contacts between somewhat older sheared lithosphere (dark gray) and younger lithosphere formed at the offset ridge axes.

As pointed out by J. Tuzo Wilson (1965) ridge–ridge transform faults have a complex geometry (Fig. 3.25). The only active strike-slip part of a fracture zone is the section linking the spreading centers. Beyond the offset ridge axes, no strike-slip displacement occurs. These parts of fracture zones are contacts between lithosphere of different ages with different geologic histories. The age difference is a function of the length of the transform offset and the spreading rate. The older side will have an edge that has passed through the active transform section. As spreading proceeds, this material passes the offset ridge axis as younger lithosphere (which will not experience strike-slip faulting) is accreted against it. Together these two sections of lithosphere move laterally away from the spreading axis as parts of the same plate. Isostatic adjustments related to the age and thermal differences across the fracture zone may result in vertical displacements. The geologic and kinematic complications of fracture zones make them extremely interesting and complex components of the MOR system (DeLong *et al.*, 1977, 1979; Karson and Dewey, 1978).

Seafloor studies show that the deep valleys along transform faults are largely blanketed by turbidites, pelagic sediments, and talus produced by debris slides from the

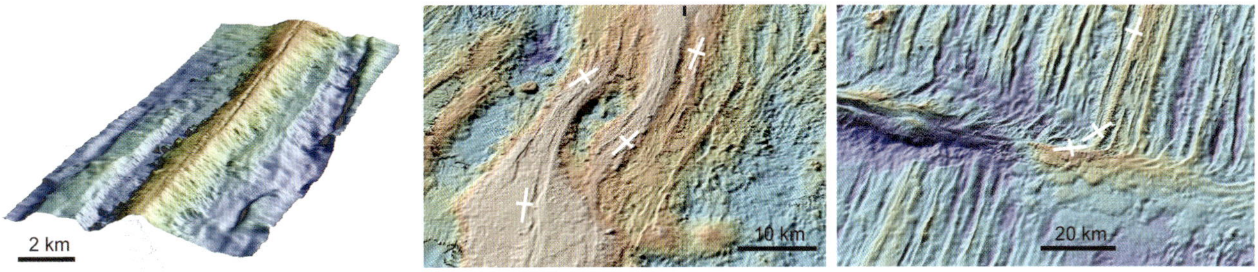

Figure 3.26 Seafloor lineaments such as the axial summit trough of the EPR (left) and abyssal hills that form parallel to spreading centers are related to dike intrusion in the subsurface. Dike trends are an indication of the stress field (white crosses show orientations of maximum and minimum horizontal compressive stresses). Stresses and inferred dike trends change around overlapping spreading centers (middle) and near transform faults (right). At transform faults, dikes are expected to be deflected toward the offset ridge segment in the transform domain, the opposite of the shape expected from shearing.

adjacent valley walls (Fox and Gallo, 1986, 1989; Gallo *et al.*, 1986). Some transform faults show the effects of compression or extension related to regional plate motions or more localized releasing or restraining bends (Pockalny *et al.*, 1997). Some transform faults, such as the Siqueiros and Wilkes Transforms (Fig. 3.24), have small, intra-transform spreading centers that manifest as pull-apart structures linking strike-slip transform strands (Fornari *et al.*, 1989; Cochran *et al.*, 1993; Perfit *et al.*, 1996).

Floors of transform valleys typically lack exposures of material that have been affected by transform fault deformation. Rare examples include sheared chalks at Transform B on the MAR (Goud and Karson, 1985) and fault-gouge-bearing sedimentary rocks at the Clipperton Transform (Barany and Karson, 1989). Consequently, the nature of transform fault deformation and metamorphism is not directly known.

Deformed and metamorphosed oceanic basement rocks have long been recovered from the walls of fracture zones (Fox and Stroup, 1981) and interpreted as examples of transform fault processes (Miyashiro, 1972, 1973a,b). These include a diverse suite of metagabbros and variably serpentinized peridotites. More recent near-bottom investigations now suggest that most, if not all, of these materials were deformed and metamorphosed in extensional regimes related to OCCs and exposed by relatively small offset (tens to hundreds of meters) normal faults along transform valley walls. Although direct evidence of transform fault deformation and metamorphism on the seafloor is lacking, some ophiolites are cut by major strike-slip shear zones with deformed and metamorphosed rocks that are interpreted in terms of transform fault processes.

Near transform faults, processes along MORs are modified over distances of tens of kilometers. Decreasing seismic crustal thickness and extensive outcrops of deep crustal and upper mantle rocks are interpreted as indications of decreased magma supply (Fox *et al.*, 1980). As fracture zones are approached, the trends of abyssal hill lineaments curve, reflecting changes in the stress field near the transform fault or other perturbations (Figs. 3.25 and 3.26). More specifically, seafloor lineaments and abyssal hills, near fracture zones, are thought to be related to faults and dike-fed volcanic ridges that are both strongly controlled by regional stresses along ridge axes (Lonsdale, 1977; Macdonald and Luyendyk, 1985; Macdonald *et al.*, 1991).

Near ridge–transform intersections, lineaments are deflected toward the offset ridge axis in so-called "J-shaped ridges" (Pockalny *et al.*, 1988, 1997; Tucholke and Schouten, 1988; Croon *et al.*, 2010). These occur along the non-transform parts of fracture zones, indicating that they formed near ridge–transform intersections under the influence of transform shearing stresses. The effects of these deflections can be seen more than 10 km away from transform faults (Fig. 3.26), indicating that the area is affected by shearing stresses.

Along the active transform, the curved lineaments are cut by faults and in some cases rotated to bend away from the offset ridge axis, as would be expected for deformation by transform shearing. This creates the opposite sense of curvature produced by the stress-controlled lineaments described above. At the surface, the bending is probably the result of lateral offsets on small faults, but at depth, where ductile deformation occurs, more continuous bending of dikes and related structures can take place, probably producing very different dike geometries at different crustal depths (Fig. 3.27). It is possible that an individual dike intruded near a ridge–transform intersection could initially curve inward toward the offset ridge (J-shape). Then, as the dike moves through the transform domain, it can be progressively cut by faults and rotated in the opposite sense. Rotations of this sense should be locked into the crust on the older side of the non-transform parts of fracture zones.

3.4.5 Propagating rifts

Propagating rifts occur where spreading centers lengthen into older oceanic lithosphere, and create a distinctive family of structures and deformation features (Hey, 1977;

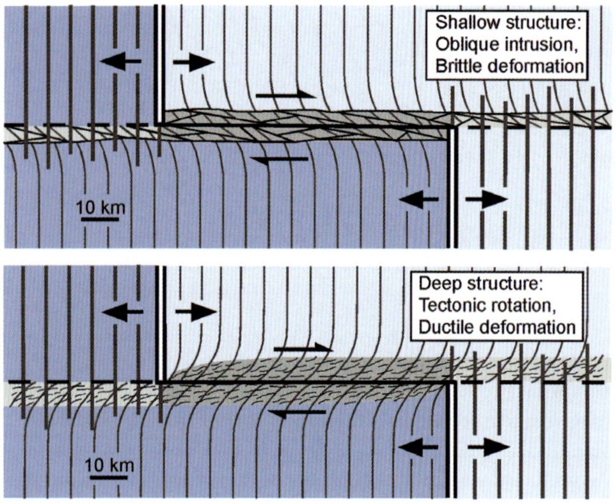

Figure 3.27 Dike trends near transform faults are expected to be different at shallow and deeper structural levels. In the upper crust (top), dike trends (black lines) are deflected toward the offset ridge axis within about 10 km of the ridge–transform intersection. Within a few kilometers of the transform fault, strike-slip fault strands may cut and displace these dikes. At deeper levels (below) dikes are progressively rotated as they pass through a ductile shear zone (Karson, 1987).

Hey *et al.*, 1986, 1989) (Fig. 3.28). Rifts propagate much like giant cracks in the lithosphere. Near the crack tip, rifting with no volcanism takes place. This process has created some of the major tectonic windows into the oceanic crust of the Pacific. Spreading-related topography more typical of MORs trails behind the rifted tip of the propagator by tens of kilometers. Lavas erupted near propagator tips, where the lithosphere is often colder, have distinctive geochemistry that can be related to this particular tectonic setting (Christie and Sinton, 1981, 1986; Perfit *et al.*, 1983; Sinton *et al.*, 1983).

Some of the most recognizable types of features of propagating rifts are major V-shaped discontinuities on the seafloor, referred to as pseudofaults (Fig. 3.28). These major structures bound the wedge-shaped area of new lithosphere created along the propagating spreading center, separating it from older lithosphere to either side (Hey, 1977; Perfit *et al.*, 1983) (see Chapter 5). Transform offsets at propagating rift tips create additional complications, causing a wave of deformation in lithosphere in the wake of migrating transform zones that are incorporated into pseudofaults (Kleinrock and Hey, 1989).

The kinematics of rift propagation and associated transform fault migration result in shearing at a high angle to existing spreading structures. This leads to reactivation of spreading structures in a pattern known as "bookshelf faulting," where deformation is accomplished by small amounts of horizontal slip on a family of parallel fault planes. This type of deformation is well documented

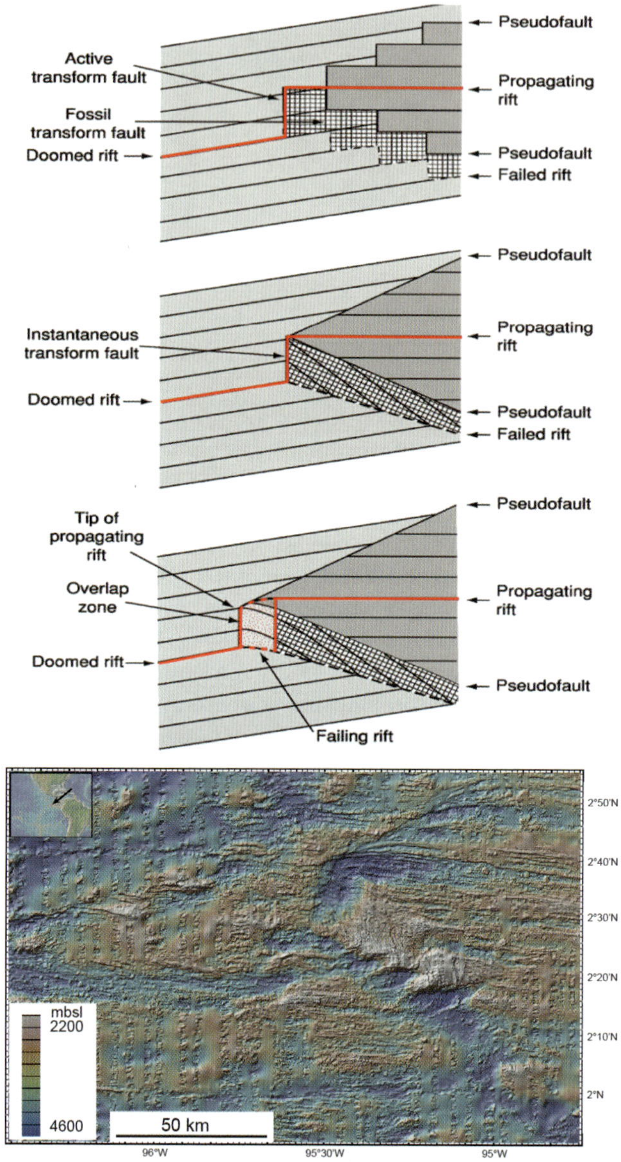

Figure 3.28 Propagating rifts produce a distinctive pattern of features on the seafloor. Diagrams (above) show features resulting from propagation of a rift with a transform offset with incremental (top) and continuous (middle and bottom) propagation and transform migration (Hey *et al.*, 2004). Bathymetry of a similar propagating rift system on the Cocos–Nazca (Galápagos) Spreading Center shows shallow spreading centers and pseudofaults corresponding to features in the upper diagrams.

in transform faults in the thick oceanic crust of Iceland (Einarsson and Ericksson, 1982; Young *et al.*, 1985). A similar deformation occurs in microplates near migrating transform zones related to propagating rifts (Acton *et al.*, 1991; Wetzel *et al.*, 1993). This type of deformation is probably widespread in oceanic crust with its strong mechanical anisotropy created by pervasive parallel dikes, faults, and fissures.

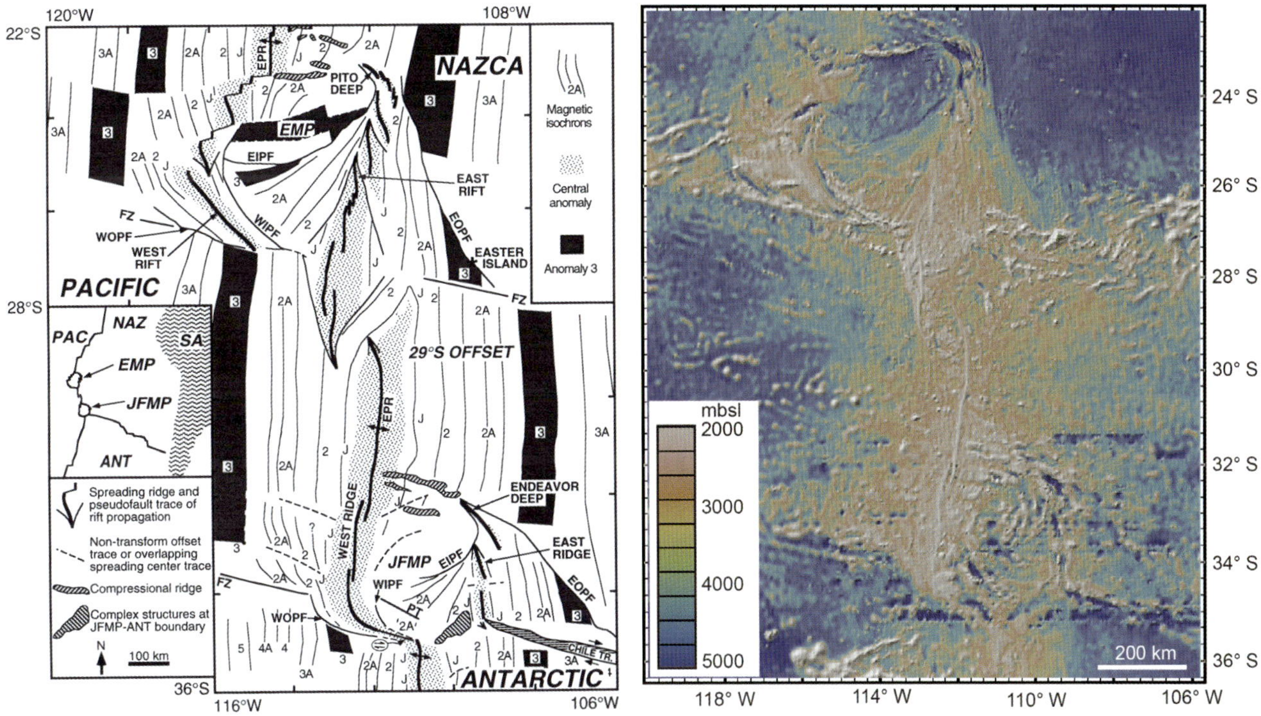

Figure 3.29 The Easter and Juan Fernandez Microplates occur on the southern EPR. The diagram on the left shows how the microplates are bounded by propagating rifts and transform faults (Hey *et al.*, 2004). The interiors of the microplates deform internally and rotate relative to the major plates that surround them. Bathymetry of the same area of the seafloor (right) shows shallow spreading centers and deeper areas of older lithosphere. The deepest areas are along transform faults and at propagating rift tips.

3.4.6 Microplates

Oceanic microplates form along the superfast-spreading southern East Pacific Rise where they are surrounded by linked transform faults and spreading centers with propagating rift tips (Hey, 2004). They occur where spreading centers are laterally offset by hundreds of kilometers. Along the spreading axes that bound microplates, the amount of spreading increases away from propagating rift tips. This results in the rotation of the older oceanic lithosphere trapped in the middle of the microplate. Complex patterns of lineaments and magnetic anomalies form in the microplates (Fig. 3.29) as a result of rift propagation, rotation, and internal deformation (Hey *et al.*, 1985; Searle and Francheteau, 1986; Francheteau *et al.*, 1987, 1988; Searle *et al.*, 1989; Hey, 2004). Microplates have complex kinematic relations to the larger oceanic plates that bound them and can rotate about nearby axes rapidly relative to other plates (Schouten *et al.*, 1993).

3.5 Internal structure of the oceanic crust

The bathymetric and seafloor features described above are the result of magmatic and tectonic processes that occur and extend into the crust and upper mantle beneath MORs. To get a more accurate and detailed knowledge of the processes at MORs we need a three-dimensional view. This necessarily involves using exposures at OCCs described above, as well as information from drilling, tectonic windows, and a variety of geophysical techniques to image subsurface structures.

Compared to the extensive bathymetric mapping of the MOR system, what is known of the underlying oceanic crust from direct observations of exposed rock units is much more fragmentary and spatially isolated. Large exposures of at least the upper few kilometers of the oceanic crust are available at faulted escarpments (tectonic windows), but these commonly occur near spreading segment boundaries. Drilling into the oceanic crust has been focused near some tectonic windows (Fig. 2.18, see Chapter 2) and in areas considered to have formed near segment centers. In fast-spreading crust, two drill holes have provided important reference sections of the upper ~2000 m of the crustal structure (see Chapters 6 and 7). In slow-spreading crust, drilling has been much less successful, except in a few areas where lower crustal and upper mantle rocks have been uplifted and exposed on the seafloor. Collectively, near-bottom studies along tectonic windows and the results of drilling provide details on the composition and structure of the upper part of the oceanic crust at the scale of hundreds of meters to millimeters, but not necessarily in places that are optimal for directly verifying interpretations of geophysical data.

Thus, correlating the geology of the oceanic crust with its broader scale geophysical properties involves extrapolations across scales of observations that do not overlap substantially and that must be made over great distances along the MOR. Consequently, current correlations are subject to some significant uncertainties. Future investigations will be required to help close the gaps between these investigative approaches.

3.5.1 Upper crust

The upper oceanic crust formed at MORs is generally considered to be composed of basaltic lavas (see Chapter 5) and dikes (see Chapter 6), which are the solid remnants of conduits that supplied magma to seafloor eruption sites. Based on crustal drilling, seafloor observations, and studies of ophiolite complexes, the upper 1–2 km of the oceanic crust is thought to be composed of more or less equal amounts of pillow lavas and more massive flows and dikes (see Chapter 2). Dikes cut the lavas and increase proportionally downward in the crust. A continuous sheeted dike complex composed of more than 90% dikes intruded side-by-side forms the lowest part of the upper crust. In crust formed at fast to intermediate rates, crustal drilling and tectonic windows provide a glimpse of the upper crust and hints to how it is constructed in these types of MORs. At slow- to ultraslow-spreading ridges, lavas and rare sheeted dikes occur discontinuously with large areas of deeper-level rocks exposed directly on the seafloor. In these settings, the lavas and sheeted dikes have either been removed by faulting or were never part of the accreted crust. Drilling has penetrated the upper part of lava sequences in some places but not reached deeper level rocks. Because of the limited direct access to the upper oceanic crust, geophysical investigations are used to try to map this part of the crust. This requires interpretation of seismic data in the context of the available geology.

With the advent of multichannel seismic reflection studies in the oceanic crust in the 1980s, improved vertical (tens of meters) and horizontal (hundred meters) resolution of the upper crustal structure became possible. Surveys along and across many segments of the MOR resulted in a systematic view of the seismic structure of the uppermost crust, traditionally identified as seismic Layer 2 (see Chapter 2).

The seismic structure of the uppermost part of the oceanic crust, referred to as Layer 2A (Fig. 3.30), is now generally interpreted as a low-velocity zone that has a low vertical gradient underlain by a higher velocity interval with a steep velocity gradient that creates a mappable reflection (Houtz, 1976; Christeson *et al.*, 1992; Harding *et al.*, 1993; Kent *et al.*, 1994; Vera and Diebold, 1994; Hussenoeder *et al.*, 2002a). The seismic structure of Layer 2A is interpreted as a result of high porosity near the surface that rapidly decreases downward. Seismic Layer 2A near fast-spreading ridge axes has a typical thickness of only 150–250 m with

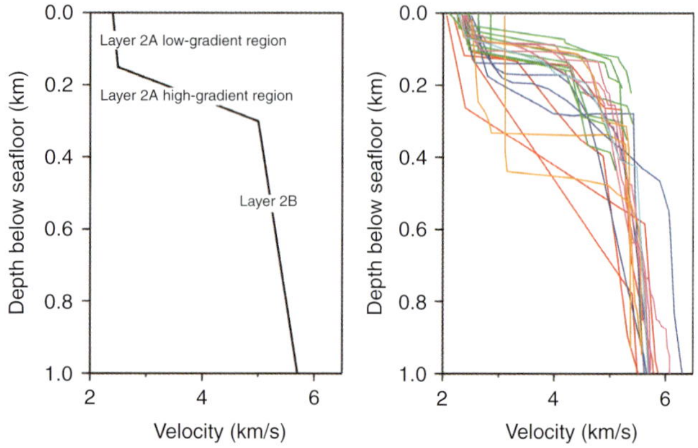

Figure 3.30 Summary of results of determinations of seismic Layer 2A in young crust from the superfast- to fast-spreading EPR. Velocity–depth plot on left shows the general character of the seismic structure of the upper oceanic crust with Layers 2A and 2B. Plot on right shows data from several different experiments. The rapid downward increase in velocity is the result of a decrease in porosity from a transition to more massive basaltic material or sealing of cracks by hydrothermal mineralization (Christeson *et al.*, 2012).

compressional wave velocities less than 3 km/s (Carbotte *et al.*, 2012; Christeson *et al.*, 2012). Within 1–2 km of the axis, the thickness of Layer 2A increases by a factor of two or three (300–700 m thick) and its velocity increases to greater than 4 km/s (Kent *et al.*, 1994). In mature crust (older than 10 million years), Layer 2A ranges from less than 200 m to nearly 800 m, with an average thickness of about 500 m (Christeson *et al.*, 1996, 2012).

Where present, Layer 2A is more variable and somewhat thicker on-axis at intermediate to slower rate ridges, but off-axis, thicknesses and velocities are similar at all spreading rates. The velocity of Layer 2A increases rapidly at its base to 5–6 km/s into the underlying Layer 2B, which does not appear to change in thickness with age (Carbotte *et al.*, 1997, 2012; Van Ark *et al.*, 2007).

Layer 2A thickens across the axial region of spreading centers where lava flows accumulate, suggesting that it may correspond to high porosities in fractured and loosely constructed lavas related to extrusive processes (see Chapter 5). The observed velocity gradient at the base of Layer 2 could correspond to the top of underlying sheeted dikes or could result from the interplay of other processes that can modify the porosity of the upper crust – for instance, faulting (increasing porosity), interacting with compaction, alteration and deposition of vein materials (decreasing porosity). Interpretation of the geophysical data relies on correlations with the rock record in tectonic windows and drill cores.

Multichannel seismic surveys within a few kilometers of major tectonic escarpments, where the thickness of lava has been mapped, show that in some places, especially where lavas are relatively thick (for example, >1000 m at the Blanco Transform), the Layer 2A/2B boundary lies within the lava unit. Elsewhere, this boundary lies close to the base

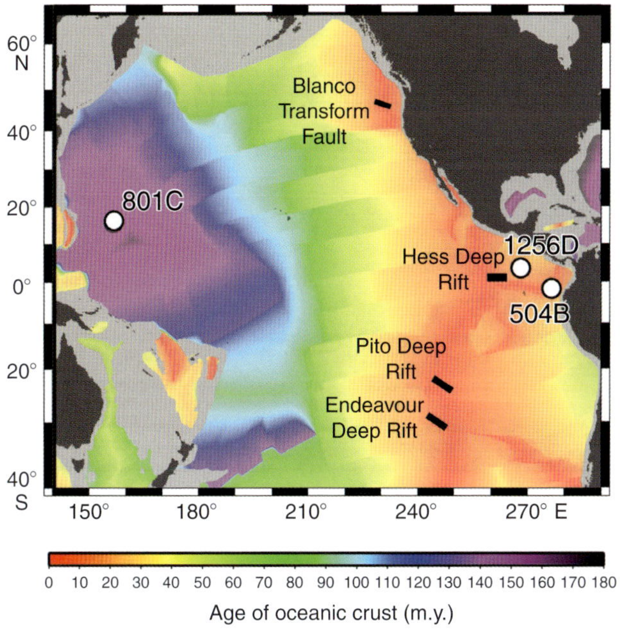

Figure 3.31 Unlike crust formed at slow spreading rates, relatively fast-spreading crust has very few large tectonic escarpments that may provide "tectonic windows" into the internal structure of the oceanic crust. Major escarpments that have been investigated (black boxes) include the Blanco Transform Fault and propagating rift escarpments at the Hess Deep, Pito Deep, and Endeavour Deep Rifts. Although there are many crustal drill holes that have reached the oceanic basement in the relatively fast spreading crust of the Pacific, only a few have penetrated more than about 100 m (white dots): Holes 801C, 504B, 1256D.

of the lavas near the lava–dike transition zone (Christeson *et al.*, 2007). The relatively constant depth of the Layer 2A/2B boundary compared to the variable thickness of lavas in measured sections suggests that this boundary may be controlled by depth(pressure)-dependent processes such as alteration and/or crack closure (Christeson *et al.*, 2007, 2010).

Deep crustal drilling and near-bottom studies of tectonic windows in the Pacific (Fig. 3.31) also provide important information about the upper 2 km of the oceanic crust formed at superfast to intermediate rates. In general, the rock types and structures found in these areas match those anticipated from studies of the upper part of ophiolite complexes. These include basaltic lavas underlain by a sheeted dike complex, which in turn passes downward into coarse-grained gabbroic rocks (Figs. 3.32 and 3.33; Table 3.2). The available results suggest possible systematic variations in thickness and internal structure of the rock units with spreading rate or magma supply. These results are summarized briefly below. More details are provided in later chapters.

The lavas are dominantly pillow lavas with lesser lobate and sheet flow morphologies. These are often difficult to discern in fractured, vertical sections of the crust. Downward through the sections, the dip of lava flows increases from typically less than 20° to more than 30° and as much as 50° (Karson, 1998, 2002). A transition zone between the lavas and dikes contains a mix of fractured and altered lavas and dikes. The sheeted dike complex is variably fractured and altered with the metamorphic grade continuing to increase downward (see Chapter 6). The dikes are not vertical and dip less than 50° in some places. Paleomagnetic studies confirm that the lavas and dikes have been rotated from initially subhorizontal and subvertical orientations, respectively. Fracturing and metamorphic grade both increase downward. Material recovered from the deepest drill holes and structurally deepest seafloor exposures only represents the upper part of the gabbroic unit, which appears to be much less altered and fractured.

Unfortunately, much less information is available for the upper crust generated at slow-spreading MORs. This is in part because upper crustal lavas and dikes are commonly removed by faulting in many tectonic windows. These results may be biased because of the highly variable nature of slow-spreading crust, where spreading processes appear to vary systematically from magma-rich segment centers to magma-poor segment ends, at which faulting creates the crustal sections that have been investigated. Deep drilling has had only limited success in drilling into the lava units of slow-spreading crust, so data comparable to those from faster spreading environments are not yet available.

In general, it appears that the thickness of lavas at slow-spreading ridges is highly variable, echoing the along-strike variations in the apparent thickness and character of the neovolcanic zone in different spreading segments. In some places, lava sequences a few hundred meters in thickness have been documented. In contrast, other areas appear to have experienced stretching and faulting of the axial crust with little or no lava erupted as part of the process. These types of variations may take place along the axis of the MOR on the scale of spreading segments (tens of kilometers). The same general result is suggested in seismic data showing that seismic Layer 2A is highly variable in thickness in slow-spreading crust and that, where present, it can be much thicker than that of faster-spreading MORs (Hussenoeder *et al.*, 2002b; Seher *et al.*, 2010; Arnulf *et al.*, 2014).

3.5.2 Lower crust and Moho transition zone

Indirect geophysical techniques and ophiolite analogs are used to constrain the composition, structure, and processes of the lower oceanic crust and upper mantle (see Chapters 7 and 8). Widely spaced and in some cases very fragmented exposures of deep level rocks aid the interpretation of geophysical data. Most of these exposures are in slow-spreading crust and they are all near the ends of spreading segments or in settings where major faulting has caused significant crustal disruption. It is not clear how representative these areas are of the regional composition and

Figure 3.32 Tectonic windows into intermediate- to fast-spreading crust of the Pacific provide extensive exposures of upper crustal rock units. Exposures of pillow lavas (upper left), sheeted dikes (upper right), faulted lavas and dikes (lower left), and gabbroic rocks (lower right) occur in a regular layered sequence as found in many ophiolite complexes. Field of view ~2 m for all images.

structure of relatively nearby slow-spreading crust or the crust formed in faster-spreading environments.

One of the most exciting results in recent decades has been the discovery of magma bodies beneath MOR spreading centers (Detrick *et al.*, 1987; Sinton and Detrick, 1992; Kent *et al.*, 1993). Seismic reflection imaging of the subsurface beneath spreading axes identifies the tops of lens-shaped magma bodies located beneath the axis of superfast- to intermediate-spreading ridges (Detrick *et al.*, 1987; Kent, 1998; Van Ark *et al.*, 2007; Carbotte *et al.*, 2012, 2013). The seismic reflection signature of this feature is referred to as the axial magma chamber (AMC) reflector. These melt lenses are typically 1–2 km wide and 50–100 m thick, and are situated at depths of 1–2 km beneath the seafloor (Fig. 3.34).

Seismic properties of these melt lenses indicate that they are likely crystal mushes with as much as 60% melt (Sinton and Detrick, 1992; Crawford *et al.*, 1999; Dunn *et al.*, 2000; Canales *et al.*, 2006, 2009, 2012b). Shallow axial melt lenses are perched above a much larger volume of material containing 10–20% melt. The latter have been characterized in seismic records as a "low-velocity zone" (LVZ), reflecting the fact that sound waves move more slowly through hot, melt-rich material than solid rock. It is common for the shallow melt lens and LVZ together to be referred to as an axial magma chamber, and this term is used throughout this book.

Data from drilling, tectonic windows, and ophiolites all indicate that at deeper than 1 km the top of the AMC would be below the lavas and sheeted dikes of the upper crust, and near the top of the gabbroic rocks (Fig. 3.33). Coarse-grained gabbroic rocks from ophiolites and the seafloor imply crystallization in magma bodies and seismically defined AMCs provide further evidence for these features.

The tops of AMCs are clearly imaged by a strong reflector that is nearly continuous beneath MORs that are spreading at superfast to intermediate rates (Carbotte *et al.*, 2012) (see Chapter 1). Discontinuities between kilometer-scale magma lenses and changes in the depth and form of AMCs correlate well with the morphology of the ridge axis (Van Ark *et al.*, 2007; Carbotte *et al.*, 2012, 2013). In some places, AMCs

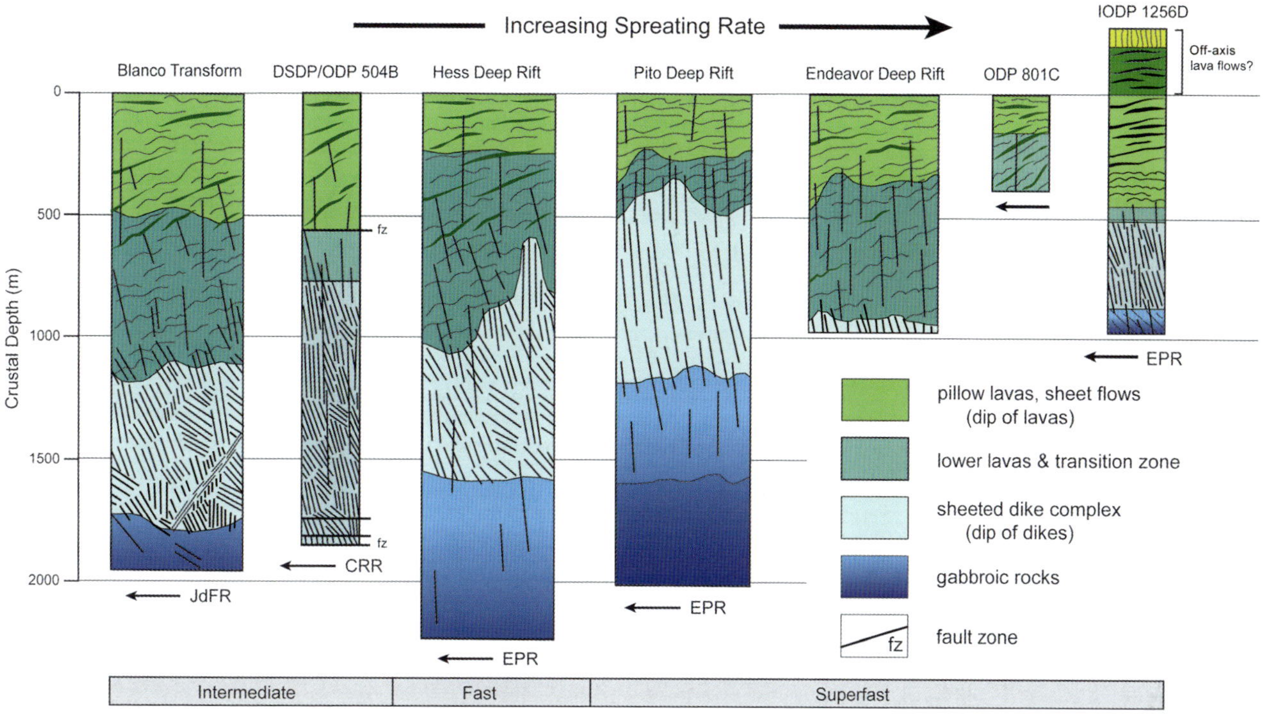

Figure 3.33 Columnar sections of upper crustal structures found in crust formed at intermediate to superfast spreading rates (left to right). Wider columns summarize relationships at tectonic windows; narrower ones are for DSDP/ODP/IODP drill holes. Data from Blanco Transform (Karson *et al.*, 2002b), DSDP/ODP Hole 504B (Pariso and Johnson, 1989; Alt *et al.*, 1993), north wall of the Hess Deep Rift (Karson *et al.*, 2002a), Pito Deep Rift (Karson *et al.*, 2005; Chutas, 2007), Endeavour Deep (Larson *et al.*, 2005), ODP Hole 801C (Pockalny and Larson, 2003), and IODP Hole 1256D (Wilson *et al.*, 2006; Umino *et al.*, 2008). Arrows at base of columns show the direction to spreading centers relative to the structures: EPR, East Pacific Rise; CRR, Costa Rica Rift; JdFR, Juan de Fuca Ridge. (See Fig. 3.31 for locations.)

become less distinct and even discontinuous near spreading segment ends. However, in other places, seismically imaged magma bodies are anomalously large beneath ridge-axis discontinuities such as overlapping spreading centers (Kent *et al.*, 2000). Where detailed studies have been conducted, the geometry of AMCs correlates well with surface features such as active hydrothermal vents (Carbotte *et al.*, 2000, 2012; Van Ark *et al.*, 2007).

Initial studies of AMCs along fast and intermediate rate ridges suggested that they were located in the area immediately beneath the MOR axis and only at relatively shallow depths. More recently, detailed seismic investigations have documented reflectors similar to the top of the AMC deeper beneath the axis of spreading centers at various depths down to the Moho (Canales *et al.*, 2009). Melt lenses have also been imaged at various crustal depths at distances as much as 7 km off-axis in somewhat older crust (Figs. 3.35 and 3.36) (Toomey and Hooft, 2008; Canales *et al.*, 2009; Carbotte *et al.*, 2012; Han *et al.*, 2014). Thus, the view of crustal construction has changed from a strongly focused mode of intrusion directly beneath a MOR axis to a broader and more diffuse process that can occur over a 2–8 km swath centered on the ridge crest.

Unlike the nearly continuous AMCs beneath faster-spreading MORs, investigations of slow-spreading ridges have rarely identified magma chambers. In some places, focused seismic investigations identify areas of low seismic velocity that might be crystal mush or areas of very young, hot rock (Magde *et al.*, 2000). In only two locations, the Reykjanes Ridge and the Lucky Strike segment of the MAR (37° N), have well-characterized subaxial magma bodies been identified (Sinha *et al.*, 1997; Singh *et al.*, 2006). Both of these ridge segments are influenced by thermal anomalies associated with hot spots. The magma bodies in these areas are larger and deeper than those of fast-spreading ridges and appear to be confined to discrete spreading center segments (Fig. 3.37).

Across the spectrum of spreading rates, the depth to the top of magma bodies has a systematic relationship (Fig. 3.38). At slower rates, where MORs tend to have well-developed rift valleys, magma bodies are deeper and occur near the base of the crust. At higher spreading rates, magma bodies tend to be shallower, located in the upper crust beneath the spreading axis. However, as noted above, magma lenses also occur at various depths down to the base of the crust and can be located several kilometers away from the spreading axis. Exceptions occur in areas of anomalously hot or cool mantle temperatures that tend to cause deviations from this systematic relationship. Overall, these relationships suggest a non-linear relationship between

Table 3.2 Geology of oceanic crust: intermediate to superfast rates (see Fig. 3.33 for references)

Rate	**Intermediate**		**Fast**	**Superfast**			
Location	Blanco FZ	ODP 504B	Hess Deep	Pito Deep	Endeavor Deep	ODP 801C	IODP 1256D
Tectonic setting							
Full rate (mm/yr)	54	72	110	144	151	160	220
Growth rate (kyr/km)	37	28	18	14	13	12.5	10
Age of crust (Ma)	0.6–1.6	5.9	1–1.5	3	2.6–3.6	165	15
Geological characteristics							
Upper lavas (m)	200–500	~200	100–300	100–400	~200	188	184
Upper lava dip	*<30°*	*20–30°*	*20°*	*<20°*	*0–30°*	*<25°*	*5–20°*
Lower lavas (m)	300–900	~400	500–600	100–300	~200	>218	466
Lower lava dip	*<50°*	*50°*	*40°*	*<40°*	*<30°*	*35–43°*	*<20°*
Transition Zone 1 (m)	200–900	~180	20–200	80–275	~500		61
(typical)	*(450)*			*(160)*			
TOTAL lavas	700–1800	780	220–840	100–400	~900	>406	710
(typical)	*(1100)*		*(450)*	*(230)*			
Sheeted dike unit (m)	200–550	>1056	300–1025	180–595	>100?		346
(typical)	*(450)*		*(650)*	*(400)*			
TZ and SDU dike dip	*>40°*	*68°*	*70°*	*>70°*	*>65°*		*Ave70°*
TOTAL upper crust (m)	1100–2000	>1836	810–1365	760–1430	>1000		1057
(typical)	*(1500)*		*(1100)*	*(1100)*			
Transition Zone 2 (m)	<100?			200			>50
Gabbroic rocks (m)	>200		>1000	>900			>100

spreading rate (and mantle temperature) and magma delivery to the crust: highly focused and deeper magmatic construction at slower rates and a shallower and broader region of melt intrusion at faster rates. Differences in occurrences at fast- and slow-spreading ridges provide direct evidence of more persistent, continuous melt delivery at faster rates and an inconsistent, sputtering magmatic input at slower rates.

Globally, the seismic thickness of oceanic crust away from MORs is broadly consistent; however, large variations in thickness can occur more locally, especially at slow-spreading ridges. In general, crust produced at relatively fast spreading rates is 6–7 km thick (Fig. 3.39) with minor variations near ridge-axis discontinuities. In contrast, at slow-spreading ridges, the seismic thickness of the crust can vary between 2 and 8 km over distances as short as a few kilometers. In addition, its internal seismic reflection character can also be highly variable (Detrick *et al.*, 1993; Canales *et al.*, 2000; Hooft *et al.*, 2000) (Fig. 3.39). Most of the thickness changes appear to be in the deeper levels of the crust (Mutter and Mutter, 1993).

The uniform seismic structure of the oceanic crust formed at superfast to intermediate spreading rates suggests a relatively simple, layered geologic structure that correlates with that of some ophiolite complexes. It is tempting to correlate velocity intervals and transitions with specific rock types and geologic contacts. However, in the cases where drilling or tectonic windows have permitted direct comparisons of geology with seismic data, the geology has not matched the predictions (Detrick *et al.*, 1994; Collins *et al.*, 2001; Christeson *et al.*, 2007).

The oceanic crust–mantle transition, or Moho, is an important reference horizon. It is usually interpreted as the boundary between magmatic crustal rocks and underlying residual upper mantle peridotite (see Chapter 2). Variations in seismic crustal thickness are commonly taken as indications of variations in magma supply either along the MOR or with time.

Multichannel seismic reflection surveys commonly show a distinct reflector with high impedance contrast appropriate to a gabbro/peridotite contact (Harding *et al.*, 1989; Mutter and Karson, 1992; Mutter and Carton, 2013). Despite a clear Moho seismic refraction detected on a large scale (spanning several kilometers), indications of Moho reflections are developed along only about 50% of areas surveyed in both fast- and slow-spreading crust (Mutter and Carton, 2013), suggesting shorter wavelength (few kilometers) variability.

Near the EPR, Moho reflections have been mapped within a few kilometers of the spreading axis. In detail, the character of the Moho reflection varies from sharp to gradational and in some places has a "shingled" structure, with seismic reflectors intersecting the upper mantle at a low angle. Different expressions of the reflection Moho are related to crustal thickness variations, third-order morphologic segmentation of the ridge axis, and mantle melt supply (Aghaei *et al.*, 2014).

Seismic reflection profiles of the deep oceanic crust reveal major structures in the crust that can be interpreted in terms of their geometry (Fig. 3.40). Within both fast- and slow-spreading crust continuous reflectors cut through the middle to lower crust and have been interpreted as faults or shear zones or as igneous contacts. Whereas major faults are well documented in slow-spreading crust, such large-scale fault structures are very unlikely in fast-spreading crust. Thus, major intracrustal reflectors may correspond to different types of geological features in crust formed at different rates (Mutter and Carton, 2013). Deep crustal drilling has not penetrated sections of the crust that would produce such reflections and possible analogous features have not been identified in ophiolites.

The internal structure of slow-spreading crust provides some hints to the origin of OCCs. Seismic reflection images of these areas commonly show low-angle reflections interpreted as major shear zones (McCarthy *et al.*, 1988; Mutter and Karson, 1992; Morris *et al.*, 1993; Ranero and Reston, 1999; Reston and Ranero, 2011). Overall, the structures have braided geometry and commonly flatten downward near the top of the mantle. Similar complex structures occur in areas of highly extended continental crust, such as the Basin and Range of the western USA, suggesting that crust of slow-spreading MORs may have been stretched internally

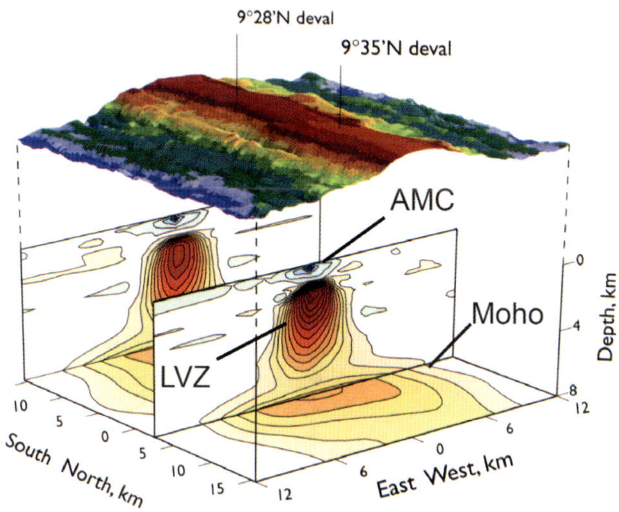

Figure 3.34 Seismic data provide this 3D view of the region beneath the axial topographic swell of the East Pacific Rise and show the underlying axial magma chamber comprising a shallow, melt-rich, magma lens (blue) and a deeper, broader, low-velocity zone (red) with a lower percentage of melt (Dunn *et al.*, 2000).

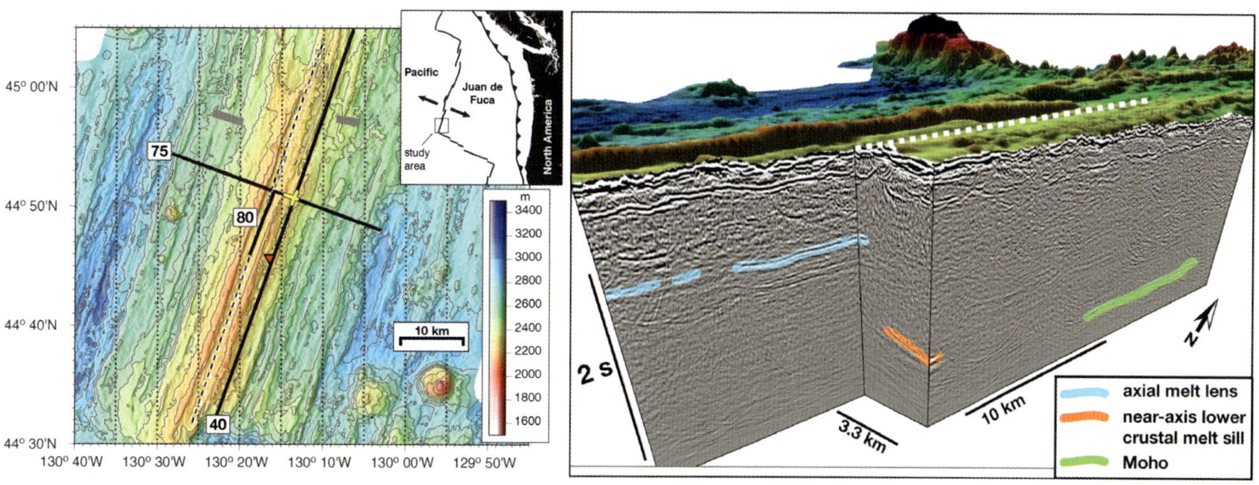

Figure 3.35 A three-dimensional seismic volume (right) showing the seismic reflection structure of crust that makes up the southern Juan de Fuca Ridge (left). Right image shows a well-developed axial magma chamber (blue) and also magma bodies several kilometers off-axis (orange) near the base of the crust (Moho, green) (Canales *et al.*, 2009).

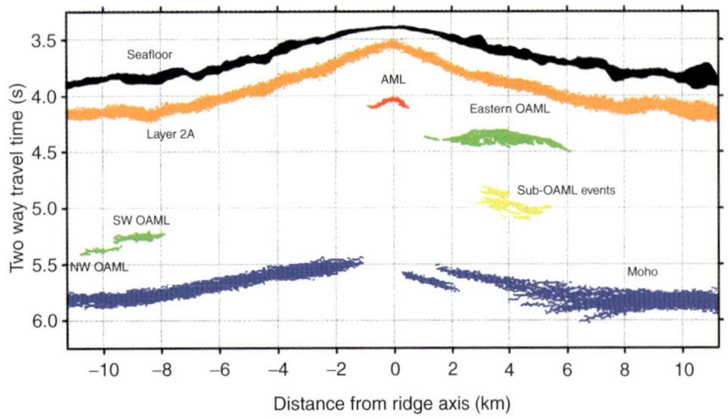

Figure 3.36 Three-dimensional seismic data reveal the internal structure of the oceanic crust near the East Pacific Rise. Here the Layer 2A reflector (orange) is at a depth of about 600 m beneath the seafloor (black). The crust–mantle boundary (Moho, in blue) is at a depth of about 6 km. In addition to a small AMC reflection (red), this image shows deeper reflectors (green and yellow) interpreted as off-axis melt lenses. Two-way travel time (vertical scale) is related to depth as determined by generally downward-increasing crustal velocities (Han *et al.*, 2014).

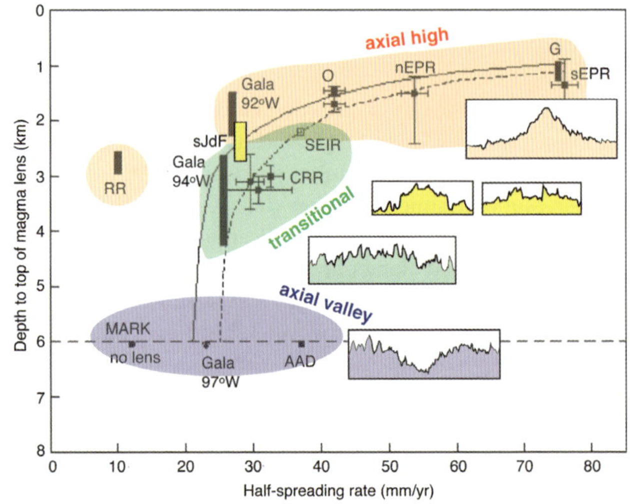

Figure 3.38 Summary of seismic reflection results from MORs with different spreading rates show that AMC reflections at faster spreading rates are systematically shallower than at slower rates (Canales *et al.*, 2005).

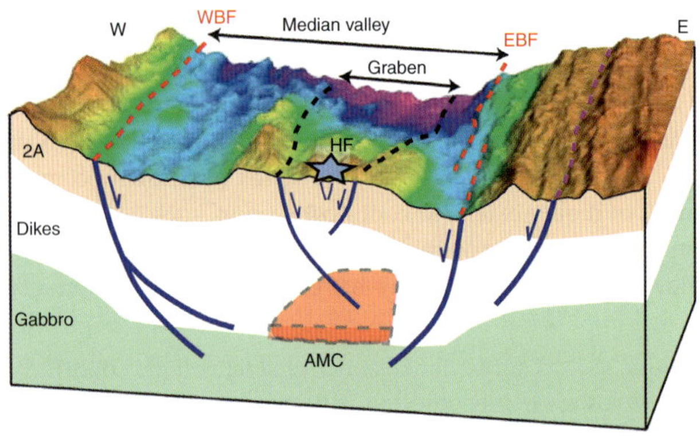

Figure 3.37 Schematic diagram showing the size and location of a large axial magma lens (AML) that is 4 km across and at a depth of 4 km beneath the Lucky Strike ridge segment on the MAR. HF, hydrothermal field; WBF and EBF, western and eastern boundary faults, respectively (from Singh *et al.*, 2006).

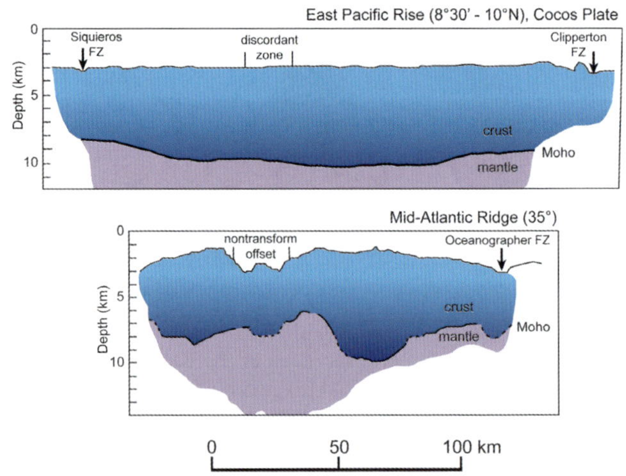

Figure 3.39 Line drawings of the seismic structure of oceanic crust highlight major differences between crust formed at fast- and slow-spreading ridges (crust, blue; mantle, purple). Fast-spreading crust (above) has a relatively uniform, continuous internal structure even at ridge-axis discontinuities. Slow-spreading crust (below) can have large variations in thickness and average compressional wave velocity, especially at spreading segment boundaries (Canales *et al.*, 2003).

in ways that are not necessarily apparent from surface structures. The OCCs may be examples of places where focused slip occurs on individual shear zones, resulting in localized crustal thinning and uplift (Mutter and Karson, 1992; Ranero and Reston, 1999; Reston and Ranero, 2011). Deep crustal drilling and studies of tectonic windows provide direct samples of plutonic rocks and deformation structures that may be correlated with seismic features and hence shed light on deep crustal and upper mantle spreading processes.

Samples of relatively deep-level rocks from beneath the volcanic carapace in relatively fast-spreading crust are rare. Tectonic windows (Fig. 3.41) provide exposures of this material in a few localities, which have been exploited by dredging, near-bottom studies, and deep drilling (see Chapter 7). The largest vertical sections are at the tectonic window at the Hess Deep Rift, which extends downward more than 1500 m beneath the base of the sheeted dike complex, to about halfway to the Moho from the top of the scarp. Exposures also occur at the Pito Deep Rift (Fig. 3.33).

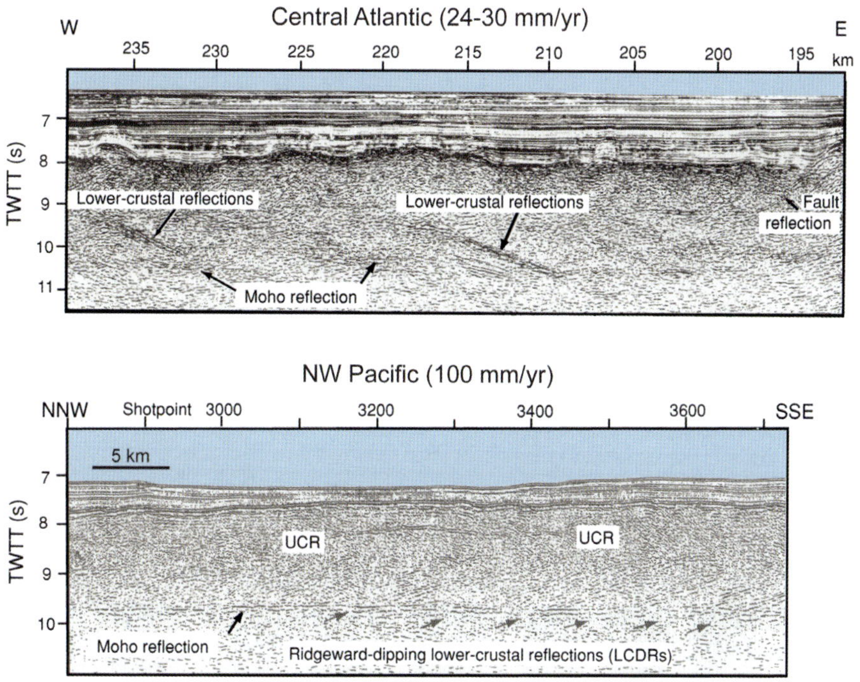

Figure 3.40 Examples of seismic reflection profiles in crust formed at slow-spreading MORs (above) (Banda *et al.*, 1992) and fast-spreading MORs (below) (Reston *et al.*, 1999). The reflection character of slow-spreading crust is more complex than that of faster-spreading crust. Fast-spreading crust has a uniform reflectivity with few upper crustal reflectors (UCR) and a continuous Moho reflection. Slow-spreading crust commonly has strong reflectors, dipping toward the spreading axis where it formed, that transect the entire thickness of the crust, in some places displacing the seafloor and/or the Moho. These reflectors are interpreted as major fault zones. In some places the reflectors merge and flatten near the Moho, suggesting a major lithospheric detachment system.

Ophiolite complexes, assumed to be representative of oceanic lithosphere, also provide access to deep-level rocks that formed in analogous environments. These perspectives help fill the gap between indirect geophysical data and the geology of the oceanic lithosphere.

3.5.3 Upper mantle

Beneath MORs, hot, buoyant mantle material rises toward the surface and rolls over as it moves laterally away from spreading centers. As the basaltic material rises and decompresses, it partially melts and is consequently less dense than the surrounding mantle material. The melts move rapidly toward the surface where they supply magma bodies in the crust and upper mantle. The flow of this material has been likened to a conveyor belt upon which the oceanic crust is built.

In general, models of mantle melting and transport suggest that, beneath fast-spreading ridges, the geometry of melt delivery from the mantle is likely to be in the form of a continuous vertical, sheet-like channel (Lin and Phipps Morgan, 1992). However, at slow-spreading ridges, a more complex three-dimensional mantle melt supply is expected (Forsyth, 1992). These differences may help account for the differences in crustal thickness observed in crust formed at different rates.

Along the southern EPR, the region in which melting takes place has been imaged with different techniques and at different scales. At the largest scale, an area of relatively low seismic velocity in the mantle beneath the southern EPR near 17° S, a superfast MOR segment, was the subject of a major geophysical investigation in the late 1990s: the MELT Experiment (Melt Seismic Team, 1998). The results of that work have been interpreted in terms of a broad, triangular-shaped region of partial melting with its apex beneath the EPR axis. This region is nearly 800 km wide and extends to a depth of more than 150 km. Remarkably, rising melts from this large area of melt generation are focused into an area only a few kilometers wide at the axis, possibly by flow constrained by the overlying rigid lithosphere (Fig. 3.42). At sites along both the southern and the northern EPR, the upper ~100 km of the melting regime has been imaged by magnetotelluric methods, which are sensitive to the presence of fluids in rocks so they capture the high electrical conductivity of the melt-rich region beneath the axis (Baba *et al.*, 2006; Key *et al.*, 2013).

At a still shallower level, three-dimensional (tomographic) seismic methods (see Chapter 1) have been used to map the locus of melt concentrations at the top of the mantle beneath several segments of the MOR. Despite the along-axis continuity of the southern EPR and the AMC in the shallow crust beneath the ridge crest, discontinuous, circular low-velocity areas are imaged beneath the axis (Fig. 3.43). These are interpreted as the tops of melt-rich zones extending upward from the melting regime below. These relationships suggest that melts from the mantle are distributed along the MOR axis from regions of localized mantle supply. Most of the low-velocity areas are centered near the ridge axis, but some can be as much as 10 km off-axis (Toomey *et al.*, 1990; Dunn *et al.*, 2000).

The results from the EPR help outline the general form of the melting regime beneath this fast-spreading MOR. To

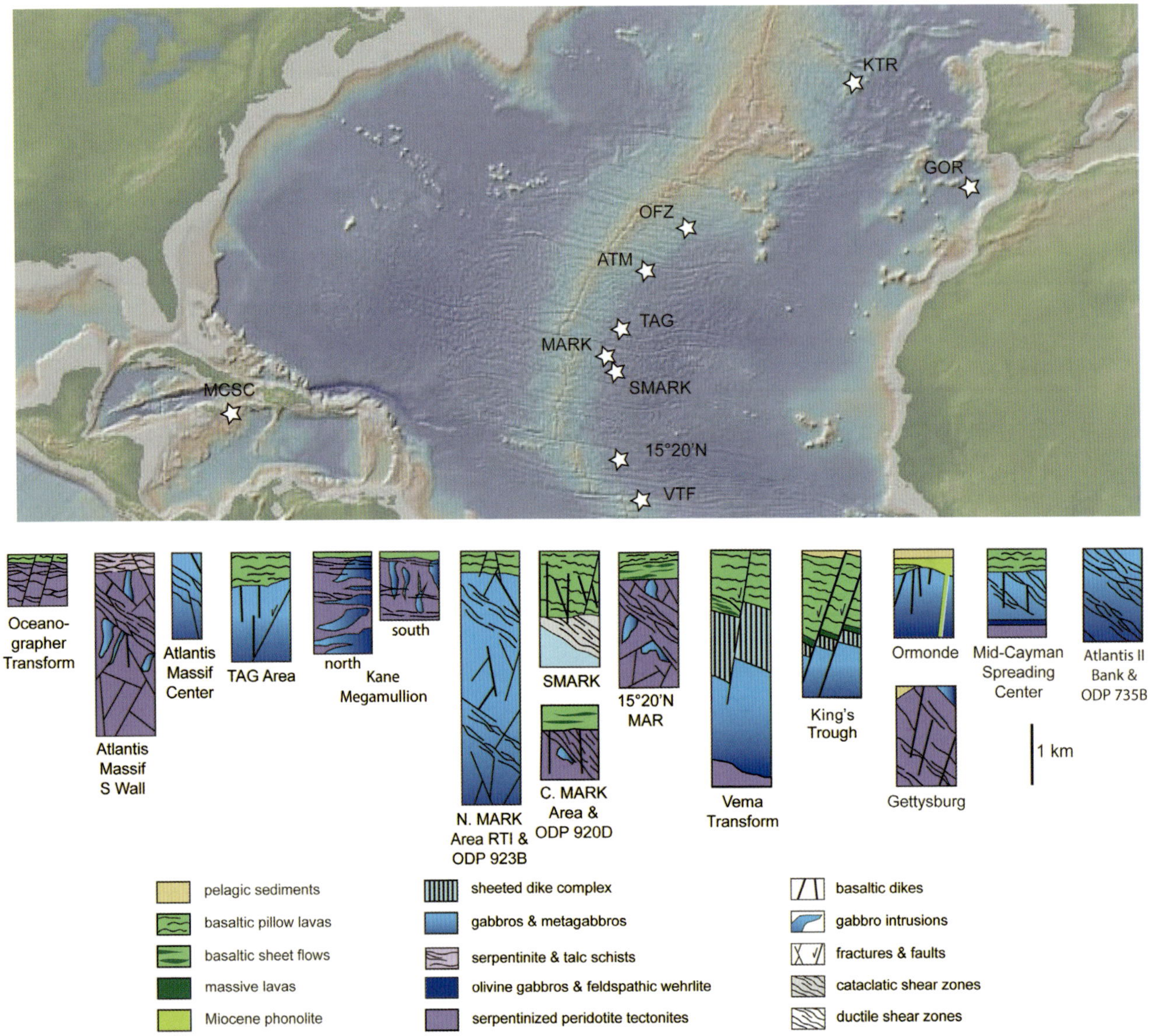

Figure 3.41 Locations of tectonic windows (escarpments) and deep crustal drill holes in slow- and ultraslow-spreading crust (above): ATM, Atlantis Massif; GOR, Gorringe Bank (Gettysburg and Ormonde Seamounts); KTR, King's Trough; MARK, MAR at Kane Transform; MCSC, Mid-Cayman Spreading Center; OFZ, Oceanographer Transform Fault; SMARK, Southern MARK Area; TAG, TransAtlantic Geotraverse Area; VTF, Vema Transform Fault. Note that the Atlantis II Transform is on the Southwest Indian Ridge. Columnar sections (below) summarizing geology in tectonic windows and deep crustal drill holes in ultraslow- to slow-spreading crust reveal complex geological structures. Only at the Vema Transform does there appear to be a "normal" crustal structure with the rock units anticipated from ophiolites. Other areas show lava lying directly over variably deformed and metamorphosed gabbroic and serpentinized ultramafic rocks. (Data from OTTER, 1984; Karson, 1998; Blackman *et al.*, 2002; Godard *et al.*, 2003; Dick *et al.*, 2008.)

date, there are no comparable studies from MORs with intermediate or slow spreading rates for comparison.

3.6 Seafloor spreading: fast and slow

The past few decades of research on MORs have shown that seafloor spreading involves a highly diverse family of processes that generate a variety of crustal structures. At one end of the spectrum are fast-spreading ridges with robust, continuous magma supplies that produce a relatively uniform type of oceanic crust with an internal structure similar to that anticipated from ophiolite studies. A high magma budget results in magmatic construction dominating over plate stretching at all levels (Fig. 3.44). At fast-spreading MORs, brittle deformation in the upper crust resembles a continuously collapsing caldera-like structure in the lavas and sheeted dikes. At middle to lower crustal levels, magmatic flow is likely to be an

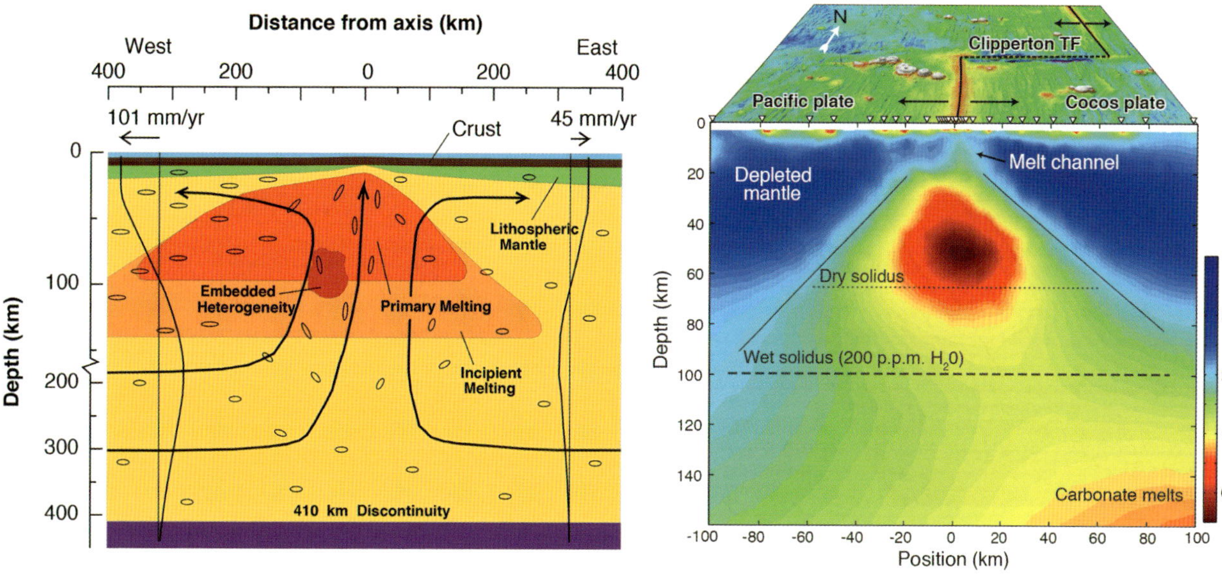

Figure 3.42 The results of the large-scale MELT Experiment on the southern EPR (left) imaged the melting regime (red and orange) beneath the spreading center. Melts from the broad (~400 km) and deep (>150 km) region of melting are channeled upward and mainly erupt at the surface in the very narrow (<1 km) axial region of the spreading center (MELT Seismic Team, 1998). Magnetotelluric imaging beneath the northern EPR (right) shows areas of high electrical conductivity (red) corresponding to somewhat shallower concentrations of melt in the upper few tens of kilometers of lithosphere (Key *et al.*, 2013).

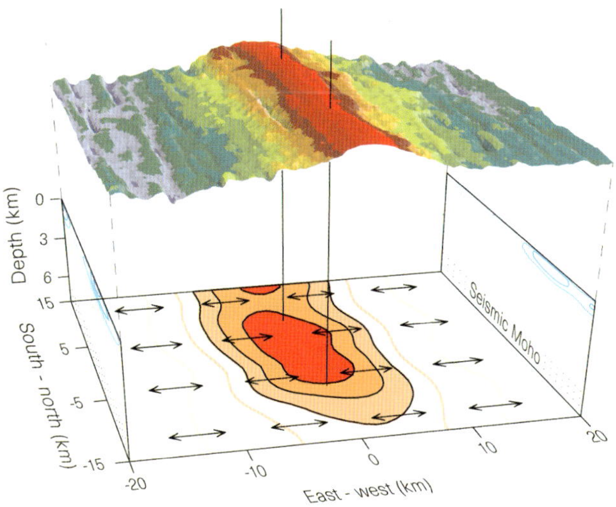

Figure 3.43 Low-velocity areas (red in bottom of diagram) beneath the northern EPR crest at 9°30′ N are interpreted as the tops of melt-rich conduits at the top of the mantle melting regime (Dunn and Toomey, 1997).

important process of mass redistribution as relatively small magma bodies incrementally build this part of the crust. There is little direct information about the composition and structure of the lower crust in these environments. Information on this region of the crust will be required to understand the mode of crustal accretion in the middle to lower crust and how it is related to the petrogenesis of the lavas and dikes of the upper crust at fast-spreading MORs. Deep crustal drilling in bare rock environments at various sites will be necessary to address these questions in the future.

At slow-spreading ridges, spreading is much more complex and variable, as is the crust generated at these types of MORs (Fig. 3.45). The overall form of spreading centers and details of exposed crustal materials can be related to an inconsistent, sputtering magma supply and a generally low magma budget. Major variations in the composition and structure of the crust occur over length scales of tens of kilometers. Thanks to the uplift and unroofing of the crust in areas of extreme tectonic disruption, more direct information regarding crustal construction is available for these complex areas than from fast-spreading ridges with more robust magmatism. The understanding of OCCs is still incomplete because of the diversity and complexity of these features. But the nature of the crust in areas of more "normal" slow spreading are also poorly constrained because they tend not to be as well exposed.

The combination of limited exposures of deep crustal rocks in specific tectonic settings and the highly variable thickness of crustal seismic units in slow-spreading crust raise important questions regarding the interpretation of these data. Specifically, it is not clear how the geological structure of tectonic windows (exclusively at segment ends) might relate to the geophysical data from areas generated near segment centers where faulting generally does not provide exposures of deep-level rocks.

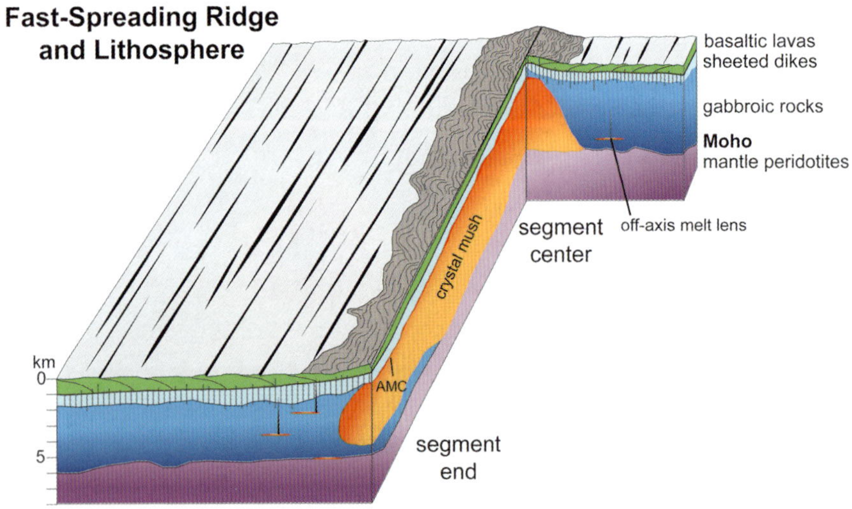

Figure 3.44 Schematic diagram of a fast-spreading ridge and crustal accretion processes creating a uniform crustal structure from a continuous axial magma chamber.

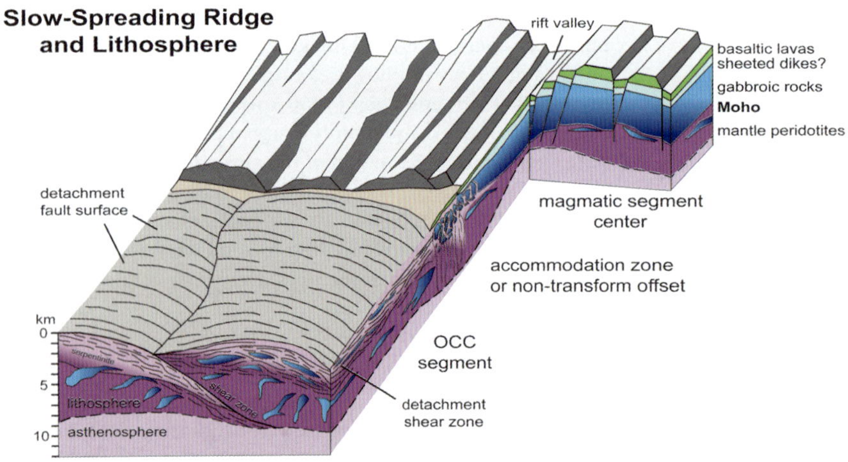

Figure 3.45 Schematic diagram of a slow-spreading ridge and crustal accretion processes creating heterogeneous and discontinuous crustal structure.

This overview of the morphological and structural details of MORs provides the basic information for an understanding of seafloor spreading processes and how oceanic crust is built. We devote separate chapters to the specific phenomena involved in emplacing oceanic crust and modifying it as it ages. Even as ocean crust forms at the axis through magmatic and volcanic processes it is impacted by hydrothermal processes. The following Chapter 4 provides details about diversity in hydrothermal venting throughout the global MOR. Chapter 5 focuses on seafloor volcanism and the surficial expression of magmatic processes that are central to creating oceanic crust. Chapters 6 through 8 focus on successively deeper regions of the oceanic crust from the dikes below the surficial lava flows to the intrusive rocks that form the lower crust and upper mantle. Many of the topics introduced in this chapter are explored in greater detail in subsequent chapters to provide details of the diverse processes associated with seafloor spreading throughout the global MOR system.

4 Hydrothermal vents

And, when he came near, the water grew boiling-hot, . . . he was on the edge of a vast hole in the bottom of the sea, up which was rushing and roaring clear steam, enough to work all the engines of the world at once. . ."

Charles Kingsley (1863)

4.1 Oases of the deep

In 1977, the scientific community was astonished by the discovery of underwater hot springs on the seafloor, thousands of meters below the ocean surface (Corliss *et al.*, 1979; Spiess *et al.*, 1980). Warm fluids issuing through cracks and fissures in the basaltic seafloor supported rich biological communities (Figs. 4.1–4.3) that thrived in the absence of sunlight. Equally surprising was the finding that these unusual animal colonies were sustained by microorganisms fueled by the products of magmatic degassing and alteration of the oceanic crust through deep-seated convection of hot seawater (Lister, 1974; Lilley *et al.*, 1983; Wilcock *et al.*, 2004).

Figure 4.1 This 1 m tall "chimlet" at the top of the black smoker called Sully, is at a depth of 2200 m on the Juan de Fuca Ridge. It billows 360 °C, metal-rich fluids and is built on a ~5 m tall mound of sulfide debris. It hosts a vibrant colony of tubeworms with bright red plumes.

These vent ecosystems were initially viewed as isolated, widely spaced oases on an otherwise barren seafloor. In the past 35 years, however, the discovery rate of vent sites has dramatically increased in response to maturation of search strategies and to improved detection techniques (see Chapter 1). More than 300 of these sites hosting superheated hydrothermal fluids that result in the formation of "black smokers" (>300 °C, Figs. 4.1–4.5) have now been documented. There is evidence of many more along the global mid-ocean ridge (MOR).

Ongoing discoveries about these remarkable environments continue to astound us, from the chemical and physical nature of black smokers and hydrothermally active fissure systems, to the multitude of highly adapted life forms thriving in these extreme ecosystems. For example, in 1995 the black smoker "Godzilla" on the Endeavour Segment of the Juan de Fuca Ridge stood as high as a 15-story building (Fig. 4.4) (Robigou *et al.*, 1993). In 2000, a completely different type of hydrothermal system called the Lost City Hydrothermal Field (Kelley *et al.*, 2001c, 2005) was serendipitously discovered on the Mid-Atlantic Ridge (MAR). Unlike black smoker deposits, which are dominantly composed of metal sulfides, chimneys at Lost City are composed of limestone (calcium carbonate) rising 60 m above the seafloor. Equally surprising, in 2004 and 2010, was the discovery of shallow volcanic calderas in the Mariana arc and volcanoes in Lau Basin hosting pools of molten sulfur (Embley *et al.*, 2006; Lupton *et al.*, 2006), explosive

Figure 4.2 (opposite) Sully, in 2006, hosted multiple black smoker orifices and extinct oxidized sulfide debris. Areas of robust, low- to moderate-temperature diffuse flow support "fat" healthy tubeworms (*Ridgeia piscesae*) reaching ~ 0.75 m in length.

Figure 4.3 First published image of a black smoker taken in 1979 by *Alvin* during the RISE (Rivera Submersible Experiment) expedition on the East Pacific Rise (Spiess *et al.*, 1980).

Figure 4.4 This drawn-to-scale sketch shows the massive black smoker called Godzilla. This 50 m tall edifice toppled over in 1996: one year later it had regrown 10 m. Note *Alvin* for scale.

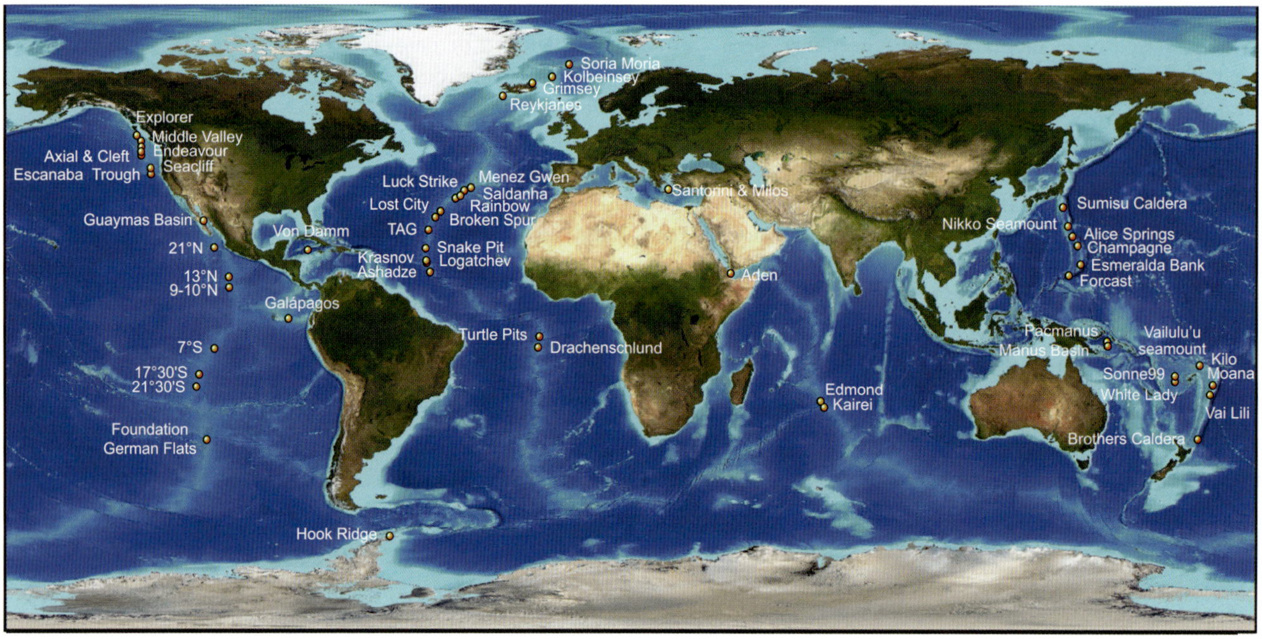

Figure 4.5 Over 300 submarine hydrothermal vent systems have been discovered, with many more hinted at by detection of temperature anomalies that suggest venting of fluids from the seafloor. In the coming decade, exploration of the high-latitude regions, and other portions of the MORs and arc systems, will likely lead to the discovery of many more vents.

lava bubbles, and 1200 °C lava flows (Resing *et al.*, 2011). Perhaps the most far-reaching consequence of these discoveries, however, is the realization that life itself may have originated within these dynamic planetary systems in which geological, chemical, and biological processes are intimately linked (Baross and Hoffman, 1985; Martin *et al.*, 2008; Langmuir and Broecker, 2012).

This chapter provides a broad overview of seafloor hydrothermal systems and examples of the life that they support. A brief discussion of the driving forces operative in the mantle and plutonic sections of the oceanic crust (see Chapters 3, 7, and 8) is also provided, for it is from low- to high-temperature fluid–mineral reactions in mantle and/or shallower crustal rocks that hydrothermal fluids obtain their unique chemical signatures, life-sustaining nutrients, and heat (Fig. 4.6). These nutrients are supplied to novel and diverse microorganisms that use inorganic compounds as a chemosynthetic energy source within the vent environments (Karl, 1995). Recent discoveries such as submarine diking–eruptive events provide not only new insights into the Earth's geologic engine, but also unequivocal evidence for the presence of an active subseafloor biosphere, including the initial colonization of nascent venting systems by biological communities (Haymon *et al.*, 1993; Juniper *et al.*, 1995; Delaney *et al.*, 1998; Holden *et al.*, 1998; Shank *et al.*, 1998; Marcus *et al.*, 2009).

The heart of this chapter is a journey through the diverse spectrum of some of the most well-studied hydrothermal systems currently known on MORs. A series of selected images illustrate the striking geological and biological diversity of vents along: (1) ultraslow- and slow-spreading regions of the Mid-Atlantic and Central Indian Ridges in the Atlantic and Indian Oceans; (2) intermediate-spreading systems such as those on the Juan de Fuca Ridge and Galápagos Rift in the Pacific; and (3) fast- and ultrafast-spreading ridges such as the East Pacific Rise off the west coast of Central America and Mexico (Fig. 4.5).

The coming decades will surely see profound new discoveries about our oceans and the life they host. The finding of diverse submarine hot-spring systems within myriad seafloor environments may have important implications for our search for life on other tectonically active planetary bodies as well. The certainty that water exists, and has existed on Mars (where there are giant shield volcanoes and evidence for ultramafic rocks; Head *et al.*, 1999; Titus *et al.*, 2003; Wilson and Head, 2007; Ehlmann *et al.*, 2010), and the presence of a liquid ocean on Europa (Fig. 4.7) (Kivelson *et al.*, 2000) raises fundamental questions as to whether geobiological systems similar to submarine hot springs are present, or were once present, elsewhere in the solar system.

4.1.1 Fluid overturn in the oceanic crust

Vigorously venting black smokers on the ocean floor, which reach temperatures above 400 °C, are fueled by heat from hot rocks and circulation of seawater at depths of about 1 km to perhaps more than 8 km below the seafloor (based on earthquake data and axial magma chamber depths) (Fig. 4.6) (Sinton and Detrick, 1992; Wolfe *et al.*, 1995; Kelley *et al.*, 2002; Davis and Elderfield, 2004; Carbotte *et al.*, 2012). The same processes also support

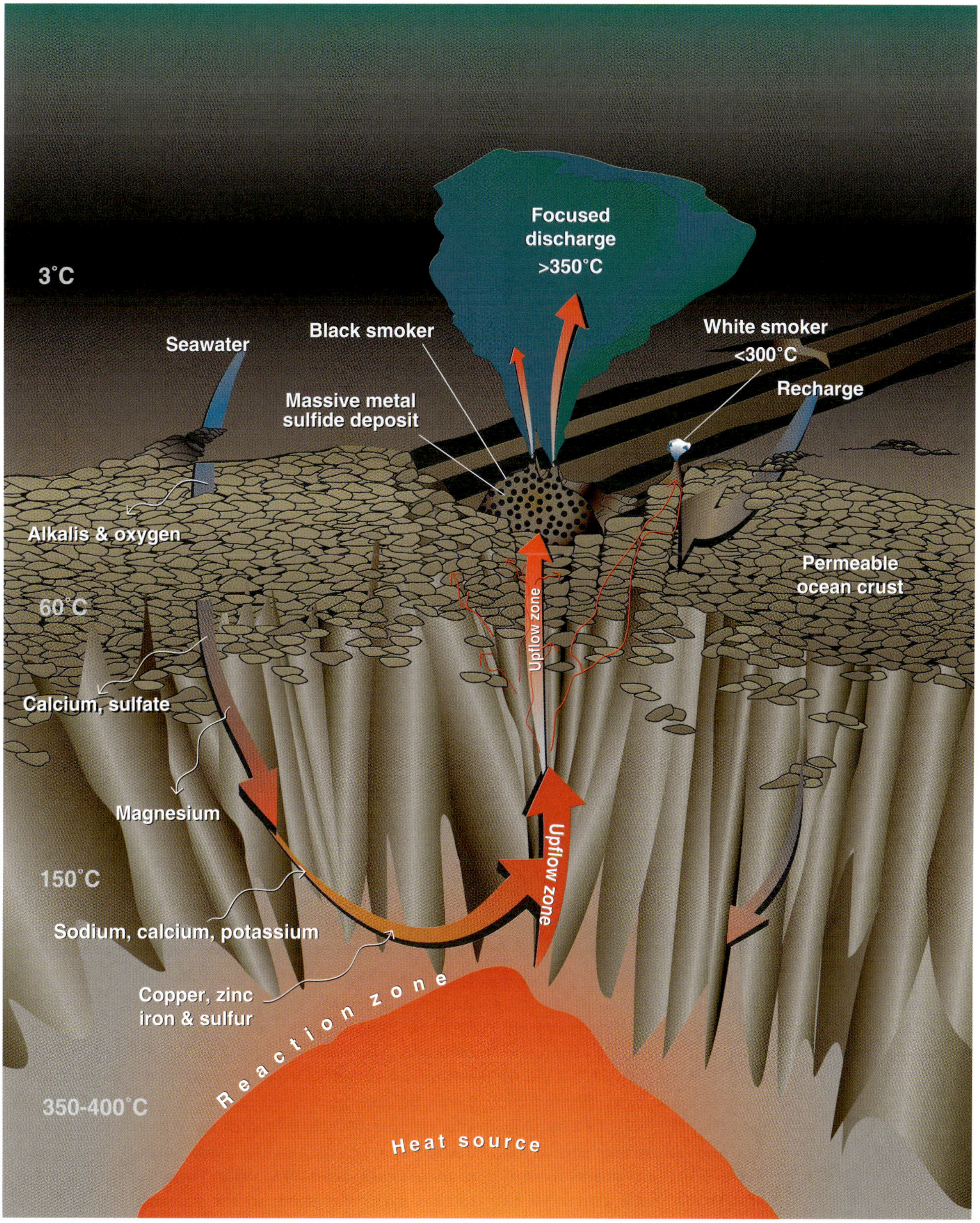

Figure 4.6 Down-welling of cold seawater and progressive heating with depth causes profound changes in resultant hydrothermal fluid compositions, properties, and temperature. For example, interaction with basaltic host rocks causes the fluids to lose all magnesium, oxygen, and sulfate. At their deepest penetration, the resultant fluids are enriched in metals and magmatic gases, and they are extremely hot: >400 °C. The fluids are also buoyant, causing them to rush toward the seafloor through narrow, pipe-like conduits. Upon mixing of the superheated fluids with cold seawater, metals precipitate out – forming the black smoker edifices.

Figure 4.7 Europa, an ice-covered moon of Jupiter, is believed to host a liquid ocean, perhaps 100 km deep. Its nearest neighbor Io is the most volcanically active body in our solar system.

Figure 4.8 Small (~1 km across) intrusion (cupola) of compositionally evolved gabbro overlain by a dike complex in the Troodos Ophiolite, Cyprus. This body is a 91-million-year-old fossilized magma chamber that once was beneath the seafloor. Extensive alteration by circulating fluids occurs at the contact.

more subdued, lower-temperature, diffusely venting systems that typically emit fluids at 5–50 °C (Butterfield *et al.*, 2004; Von Damm and Lilley, 2004). In volcanically active areas such as the fast-spreading East Pacific Rise (EPR), convection of heated seawater, or hydrothermal fluid, is driven by heating from magma chambers at depths as shallow as 1 km below the seafloor (Kent *et al.*, 1993). Circulation is thought to occur through axial faults aligned parallel and perpendicular to the ridge axis, and through the highly porous and fractured extrusive layer, where permeabilities may reach 10^{-13} m^2 (Fisher, 1998; Wilcock and Fisher, 2004). Hydrothermal circulation has been demonstrated to be of fundamental importance to the cooling of Earth and the regulation of the ocean's geochemical composition (Davis and Elderfield, 2004).

As seawater-derived fluids are heated from below by mantle and plutonic rocks (Fig. 4.8), large-scale convection cells are set up with limbs that typically exceed 1–2 km in height (Fig. 4.9). It is the convergence of these cells and the discharge of high- and low-temperature fluids that result in the formation of vent deposits and diffuse flow sites at the crust–hydrosphere interface (Fig. 4.9). With establishment of hydrothermal upflow, cold seawater must downwell to recharge the system. Although black smokers and diffuse flow fields, densely colonized with animal and microbial communities, are conspicuous hallmarks of the surface expression of the upflow zone, the geographic distribution and nature of the downflow zones are difficult to observe directly. However, conductive heat flow measurements along the bare-rock Endeavour hydrothermal system on the Juan de Fuca Ridge are consistent with downflow occurring as cross-axis flow, and as annular haloes surrounding high-temperature vent fields (Johnson *et al.*, 2010).

At deeper crustal levels, the exact nature of fluid pathways is unclear, but many models support a fine fracture network called a cracking front (Lister, 1974, 1983). In this model, the transition from the magma chamber roof to the base of the hydrothermal cell is bounded by a narrow highly conductive boundary layer, below which the hot rocks are too ductile to sustain significant stress, which thus prevents fracturing. This thin conductive boundary layer, perhaps only tens of meters thick, is required to maintain a high heat flux for extensive periods of time (Lister, 1983; Lowell *et al.*, 1995; Wilcock and Delaney, 1996). During fluid circulation within the narrow crack network, heat is mined from the surrounding wall rock, setting up a series of complex feedback loops (Norton, 1984). Because heat is withdrawn, the host rocks contract and fracture, which promotes fluid access to fresh rock, additional deepening of the crack network, and removal of more heat (Lister 1974, 1983). Thousands of microearthquakes each year occur in the subsurface below active hydrothermal fields (see Chapter 2), and are believed to be signatures of this activity (Wilcock *et al.*, 2002, 2009; Wilcock 2004; Tolstoy *et al.*, 2008).

The highest temperature and deepest portion of the hydrothermal cell is commonly called the reaction zone (Alt, 1995), although reactions occur along all paths of hydrothermal flow. Geothermometric and isotopic

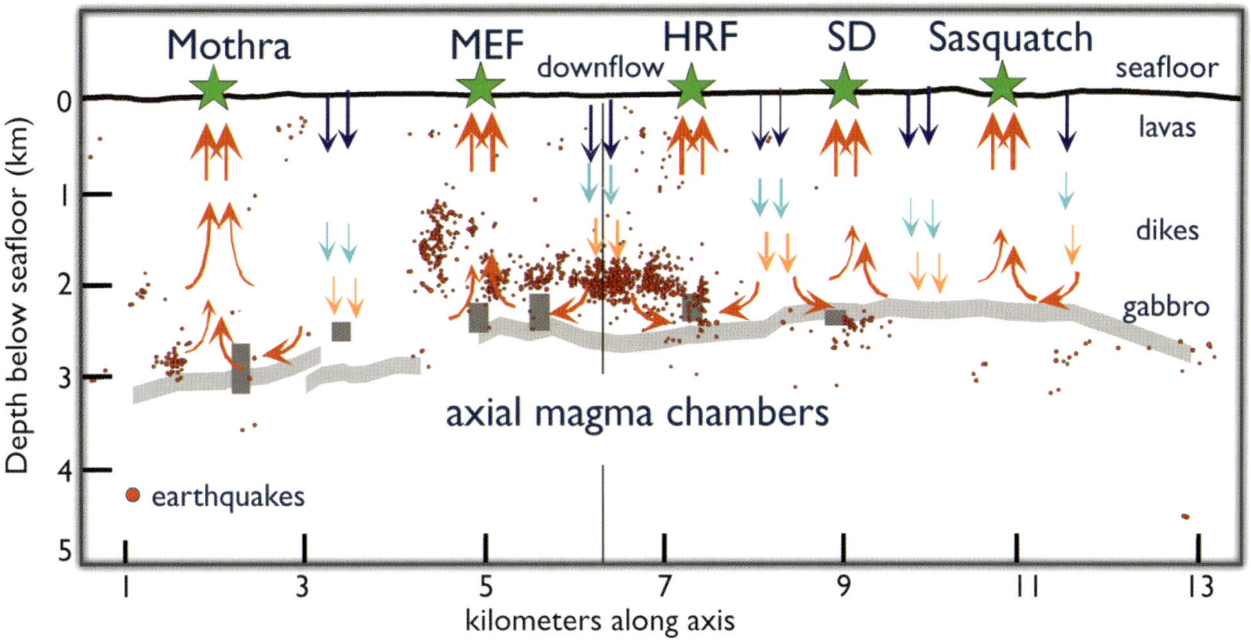

Figure 4.9 Along the Endeavour Segment of the Juan de Fuca Ridge, a strong seismic reflector (thick gray line) indicates that magma chambers underlie all of the five known hydrothermal fields [from south to north: Mothra, Main Endeavour Field (MEF), High Rise (HRF), Salty Dawg (SD), and Sasquatch] (after Wilcock *et al.*, 2009). Down-welling cold seawater (blue arrows) reaches depths of ~2–3 km, as indicated by the location of small magnitude earthquakes (red dots). The fluids progressively heat up as they near the tops of the magma chambers. Here, the hydrothermal fluids are very hot (>400 °C) and buoyant, causing them to rise rapidly to the seafloor along upflow zones (red arrows). Black smoker fields are formed where two upflow limbs of the resulting convection cells merge and high-temperature fluids are emitted into the overlying water column.

analyses of deep-seated plutonic rocks indicate that penetration of fluids may occur at temperatures from 500 °C to as high as about 750 °C (Gillis and Roberts, 1999; Manning *et al.*, 2000; Coogan *et al.*, 2006). The exact structure, nature of flow, and the evolution of this high-temperature zone are poorly understood. But field studies of submarine scarps, ophiolites, and drilled rock cores recovered by the Ocean Drilling Program suggest that this zone may be characterized by a downward-propagating set of fluid-filled fractures that are eventually filled with high-temperature minerals such as amphibole, with greater or lesser epidote, and quartz (Fig. 4.10) (Nehlig and Juteau, 1988; Kelley *et al.*, 1993; Nehlig *et al.*, 1994; Gillis, 2002; Bosch *et al.*, 2004).

4.1.2 The metamorphosis of seawater to hydrothermal fluids in underwater volcanoes

Along the downward journey from the ocean floor to depths of 2–8 km beneath it, seawater-derived fluids undergo significant changes in temperature, composition, and volatile concentration as the heat source is approached. As cool, dense seawater migrates deep within the crust along networks of cracks, the increasingly hotter fluids interact and exchange elements with the surrounding host rock (Fig. 4.6). Both rock and fluid compositions are profoundly altered (Figs. 4.10–4.12) (Kelley *et al.*, 2002). In the shallow, cooler portions of fluid pathways, away from hydrothermal discharge sites, low-temperature oxidation and alkali fixation dominates (Gillis and Robinson, 1990; Alt and Teagle, 2000).

In down-welling zones, extensive replacement of the glassy rinds of pillow basalts and sheet flows occurs by minerals such as iron oxyhydroxides, zeolites, and clays, all of which are common at temperatures below about 100 °C (Gillis and Robinson, 1990). These minerals are also common as veins that fill radial fractures in the basalts. Alkalis such as potassium, rubidium, and cesium are fixed

Figure 4.10 Amphibole veins, formed at above ~350 °C, are thought to represent deposition of minerals as part of a fossilized cracking front.

Figure 4.11 Epidosites in the sheeted dikes of Cyprus mark zones of hydrothermal upflow. They are commonly depleted in metals, but rich in quartz and chlorite. Typically, the glass-rich chilled margins alter more readily than the inner cores of the dikes. Dikes are typically ~1 m across. These rocks are rarely recovered from the seafloor.

Figure 4.12 A fossilized upflow zone (or stockwork system) in the Skuriotissa copper mine in Cyprus. The gray-white rocks in the funnel-shaped zone were once basalts, but they are now pervasively altered through extensive exchange with upwelling, hot, acidic, hydrothermal fluids. Mineralized tubeworm casings and massive sulfide near the top of the deposit mark a ~91-million-year-old fossilized black smoker system. Each bench is ~3 m tall.

within the replacement minerals, as are rare earth elements and water. Uptake of these elements is balanced in the rocks by loss of magnesium, silicon, sulfur, and in some cases calcium (Gillis and Robinson, 1990; Alt, 1995). While rock-alteration processes start at the ridge axis, they continue for millions of years as the crust is transported away from the spreading center (Davis and Elderfield, 2004)

As fluids migrate deeper into the crustal section, elements such as copper, zinc, iron, lead, and sulfur, as well as silica, are leached out of the rocks at temperatures of 350–550 °C, and are incorporated into the hydrothermal fluid (Seyfried *et al.*, 1988; Von Damm, 1995). At temperatures above about 150 °C, uptake of magnesium in the rocks is accompanied by loss of alkalis (potassium, rubidium, lithium) from the basaltic material (Seyfried, 1987; Gillis and Robinson, 1990; Alt, 1995) and sulfate (SO_4^{2-}) is removed from seawater during the precipitation of anhydrite (Barker *et al.*, 2010). Sodium, magnesium, and calcium are added to the rock to form greenschist facies mineral assemblages (a metamorphic term for basaltic composition rocks that have been altered at around 300–450 °C) that include minerals such as chlorite, amphibole, and sodium-rich plagioclase (see Chapters 2 and 3).

The base of the reaction zone, where hydrothermal fluids reach their highest temperature end-member composition, is thought to occur either near the sheeted dike–gabbro interface or within the underlying gabbroic rocks (see Chapters 6 and 7). Here the altered rocks contain amphibolite facies mineral assemblages that include amphibole and plagioclase. In most basaltic environments, fluids that eventually reach the seafloor at temperatures of 350–400 °C or more are reduced, acidic (pH 2–6) and highly enriched with respect to seawater in lithium, potassium, rubidium, cesium, copper, iron, and silica (Von Damm, 1995). Rocks recovered from the seafloor and from ophiolites that are rich in epidote, chlorite, and quartz are thought to represent fossilized upflow zones (Fig. 4.11).

4.1.3 Formation of black smoker chimneys

Once a high-temperature venting system is established, the precipitation of minerals and metals at the interface of high-temperature fluids and cold seawater forms black smoker chimneys (Fig. 4.13). They grow and evolve as venting continues at the seafloor. Examination of stockwork deposits (fossilized plumbing networks) in ophiolites (Fig. 4.12), the large size of some deposits, and rare dating studies of black smokers and associated debris indicate that, in some environments, flow channels supporting black smoker venting may switch on and off for several thousands of years (Richards *et al.*, 1989; Hannington *et al.*, 1995; Lalou *et al.*, 1998; Humphris and Tivey, 2000; Jamieson *et al.*, 2013). Resulting sulfide deposits in slow- and intermediate-spreading environments are typically larger than those in fast-spreading systems: faster-spreading environments are more frequently impacted by diking–eruptive events that reorganize flow channels or cover deposits with new lava. The upflow zones develop intense alteration haloes analogous to those in terrestrial porphyry copper ore bodies, where high-temperature mineral assemblages are focused near the core of the upflow zone, and lower-temperature minerals occur in the outer and shallower portions of the system. In ophiolites such as those in Oman and Cyprus, exposed upflow zones in quarries show alteration so intense and pervasive that it is difficult to recognize the original rock types or minerals (Fig. 4.12). During the lifetime of a hydrothermal system, individual conduits may close and reform, creating veins of a highly complex stockwork array that records multiple hydrothermal pulses and events (Hannington *et al.*, 2005).

The metal-rich, low pH, reduced, 350–400 °C (or hotter) fluids are initially "clear" within the vent orifice and at its outer edge (Fig. 4.13). However, as the fluids rapidly and turbulently mix with oxygen-rich, cold seawater (~2 °C), the drastic temperature and chemical changes cause minerals such as anhydrite, barite, pyrite, chalcopyrite, and sphalerite to precipitate (Figs. 4.13 and 4.14) (Tivey and Delaney, 1986; Hannington *et al.*, 1995; Tivey, 1995; Tivey *et al.*, 1999; Kristall *et al.*, 2006).

Many of the fine-grained sulfide particles are carried upward into the buoyant, jet-like plumes that spew out of the vent at more than a meter per second (Fig. 4.13); the

Figure 4.13 Small chimlets vent clear, ~350 °C hydrothermal fluids at the summit of the 18 m tall black smoker called El Guapo (the handsome one) in the International District Hydrothermal Field, Axial Seamount, Juan de Fuca Ridge. At a water depth of ~1500 m, these fluids are boiling. Sulfide worms and tubeworms thrive in cooler diffuse fluids flowing from the porous chimney walls.

Figure 4.14 Cut cross sections of three chimneys (Finn, Roane, and Gwenen) that emitted fluids ranging from 168 °C to 302 °C, recovered from the Mothra Hydrothermal Field (Delaney *et al.*, 2001). Finn exhibited a chalcopyrite-lined conduit, several tens of centimeters across, while Roane and Gwenen contained a network of branching conduits and rare, very large open voids. Finn was a vigorous black smoker chimney, while Roane and Gwenen were dominated by diffuse flow.

particles are carried 100 m to as high as 200 m up into the overlying ocean water, forming broad, extensive hydrothermal plumes (Lupton, 1995; Baker, 2009; Baker *et al.*, 2012). Some of these particles sink back to the ocean floor, oxidize, and become hydrothermal sediment, others are scavenged by microbial communities within the plumes (Winn *et al.*, 1986; Lilley *et al.*, 1995, Cowen *et al.*, 1998, 2001; McCollom, 2000). Sulfide and sulfate minerals not carried away in the hydrothermal plumes are deposited at the vent orifice, causing the chimneys to grow upward and outward (Fig. 4.13) (Haymon, 1983; Tivey and Delaney, 1986; Hannington *et al.*, 1995; Tivey, 1995; Kristall *et al.*, 2006).

Analyses of chimney deposits in a variety of environments have allowed general models of their evolution to be developed (Haymon, 1983; Tivey and Delaney, 1986; Hannington *et al.*, 1995; Tivey, 1995, Tivey *et al.*, 1999; Kristall *et al.*, 2006; Kelley *et al.*, 2012). In many chimneys, anhydrite ($CaSO_4$) and fine sulfide particles form the initial chimney structure. Anhydrite precipitates from seawater at temperatures of ~150 °C, and during seawater mixing with hot hydrothermal fluids, sulfides and anhydrite co-precipitate. With continued growth, the consolidation of this fine-grained, friable structure allows the inner fluids to become insulated to surrounding seawater. As a result, turbulent mixing is reduced within the structure, temperatures rise, and higher-temperature sulfide minerals precipitate. At temperatures above ~250 °C, chalcopyrite begins to form (Fig. 4.14).

Smaller chimneys commonly exhibit strong mineralogical zonations with annular rings of chalcopyrite lining the inner open channels, followed by variable mixtures of sphalerite, marcasite, and pyrite in the outer walls. In chimney deposits that are several meters across and high, the inner structures can be exceedingly complex, with anastomosing channels, highly varying porosity and mineralogical zones, through-going fractures, and progressive incorporation/fossilization of animal habitat structures (for example, tubeworm casings) that once occupied the exterior chimney walls (Fig. 4.14) (Tivey *et al.*, 1999; Kelley *et al.*, 2002; Kristall *et al.*, 2006, 2011; Zhu *et al.*, 2007). Intense recrystallization of sulfide minerals is common in many structures, making them analogous to metamorphic rocks.

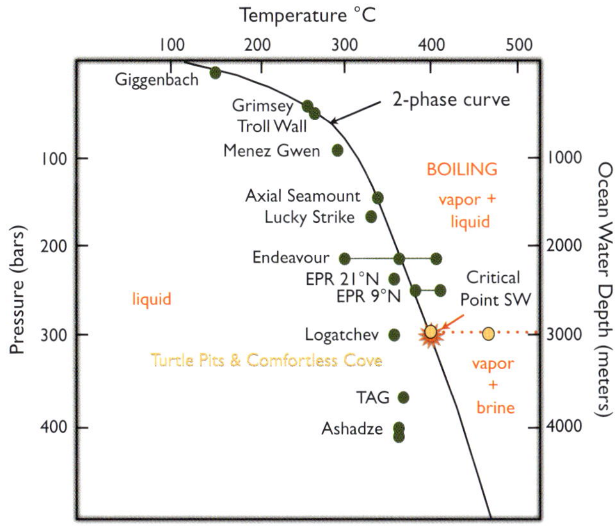

Figure 4.15 Two-phase curve for seawater (SW) (3.2 wt% NaCl) and depths and temperatures of black smoker fluids along the MOR. Boiling is particularly common in shallow vent environments.

Figure 4.16 In 1998, this 405 °C black smoker chimney called Brandon, on the Southern East Pacific Rise, was venting fluids of less than seawater chlorinity from two orifices and fluids with more than seawater chlorinity from three orifices – indicating phase segregation of vapors and brines. At a depth of 2870 m (287 bars pressure) these fluids are very close to the critical point of seawater (Von Damm *et al.*, 2003).

In the outer zones, minerals such as anhydrite, barite, and clay commonly fill fine pore spaces, indicating infiltration of seawater into the structure. With aging of the structure, conductive cooling of high-temperature fluids that circulate within the pores of the chimney promotes precipitation of amorphous silica. This gives the chimney strength (Tivey, 1995; Tivey *et al.*, 1999; Kristall *et al.*, 2006). No data exist on modification of hydrothermal fluid chemistry within the pores of the chimneys over time, but it is likely that microbial communities within the chimney walls impact both fluid and mineral chemistry at highly localized spatial scales.

Numerous studies of black smoker systems show that they exhibit highly variable morphology and vary dramatically in size. Their ecology varies not only within different oceans, but also at the regional scale of a ridge segment and at the local scale of structures within a vent field (Juniper and Tunnicliffe, 1997; Sarrazin *et al.*, 1997; Van Dover, 2000; Kelley and Shank, 2010). There is much that remains to be discovered about these systems. For example, the interactive relationships among microbial, geochemical, and geophysical conditions over time and space, which could reveal controls on microbial growth and their survival, remain uncharted.

4.1.4 Boiling fluids in and at the seafloor

In addition to water–rock reactions within high-temperature reaction zones, there are numerous other processes that can impact the composition of vent effluent during ascent to the seafloor (Figs. 4.15 and 4.16). Two of the most important of these processes are condensation of brines and boiling. Because seawater contains 3.2 wt% (weight percent) sodium chloride (NaCl), it behaves differently than pure water as it is heated (Bischoff and Rosenbauer, 1989). For example, as seawater migrates to depth, at temperatures above 407 °C and pressures above 298 bars (2980 m at hydrostatic pressures), the fluids separate into droplets of brine in a vapor (Figs. 4.15–4.18). As the buoyant fluids rise adiabatically (that is, with no transfer of heat between the wall rock and rising fluid) along the two-phase curve, they pass through the critical point and then undergo boiling.

The boiling–convective processes are highly similar to what happens when a pot of cold water is heated on a stove. Cool, dense fluids that sink to the bottom of the pan are heated and begin to rise, and cooler fluids are entrained downward. Eventually the fluids boil and vapor bubbles form. In the ocean depths, these processes, in part, govern the gas concentration of the vent fluids [e.g., carbon dioxide (CO_2), hydrogen sulfide (H_2S), and hydrogen (H_2)] and are of particular importance in partitioning of chloride and metals (Figs. 4.19 and 4.20). There is in fact growing interest in mining polymetallic sulfides from shallow submarine volcanoes where boiling has increased the concentration of gold, silver, and copper (Hannington *et al.*, 2005; Davies *et al.*, 2007; Clark *et al.*, 2010).

In 2006, one of the most extreme examples of phase separation and supercritical fluids was documented at the Turtle

Pits and Comfortless Cove Hydrothermal Fields in the South Atlantic (see Section 4.9). There, two small mounds capped by active, 407 °C black smoker chimneys were emitting gas bubbles (Haase *et al.*, 2007; Koschinsky *et al.*, 2008). At a depth of 3000 m, these fluids were venting at pressure–temperature conditions around the critical point for fluids of seawater salinity (Fig. 4.15) (Koschinsky *et al.*, 2008). A short burst of fluid (20 s) venting in the Comfortless Field was measured at 464 °C, clearly above the critical point and far above previously measured vent fluid temperatures (Fig. 4.15) (Koschinsky *et al.*, 2008). These low-salinity fluids were enriched in iron, copper, cobalt, and molybdenum, reflecting the extremely high temperature of the fluids and the importance of phase separation and chloride complexing in metal transport. Building on models of vent–fluid evolution associated with magmatic–volcanic events (Butterfield *et al.*, 1997; Delaney *et al.*, 1998; Von Damm, 2000), Haase *et al.* (2007) suggest that the high-temperature fluids and their vapor-dominated nature are influenced by recent volcanic activity at 5° S on the MAR. This interpretation is consistent with the presence of fresh, highly glassy lava flows that partially cover sulfide debris at the Comfortless Field. The eruption of these lavas may have occurred in 2002, as evidenced by seismic events detected in this area (Haase *et al.*, 2007).

Such high fluid temperatures have only rarely been measured at MORs. A temperature of 405 °C was measured in 1998 at the Brandon Vent on the southern EPR (Fig. 4.16). This chimney was venting a vapor and brine at two different orifices on the structure and the fluids were critical (Von Damm *et al.*, 2003). Previous to this, 402 °C fluids were measured in 1984 in the Main Endeavour Field on the Juan de Fuca Ridge (Fig. 4.18) (Delaney *et al.*, 1984). The development of low-salinity vapors (less than the 3.2 wt% NaCl in normal seawater) and brines (up to 10 wt% NaCl) by phase separation has been inferred to be an active process in

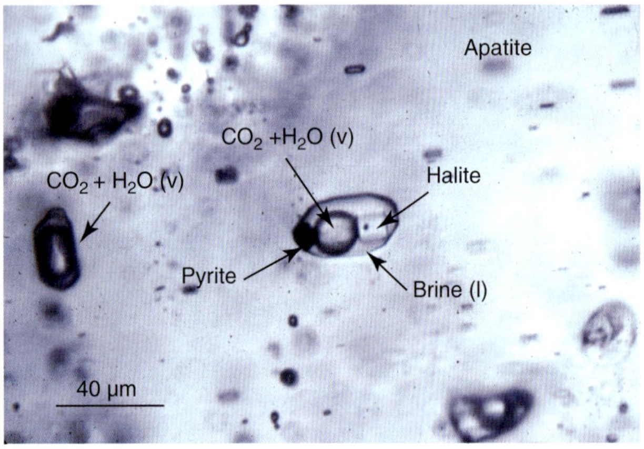

Figure 4.17 Silica-enriched submarine plutonic rocks commonly host small capsules of entrapped magmatic fluids called inclusions. The fluid inclusions shown here were trapped at above 700 °C in apatite – a late-forming mineral in MOR magma chambers. The fluid has a salinity of 40 wt% NaCl, and is rich in metals (e.g., pyrite). The CO_2-bearing, brine-rich inclusions are associated with 1–2 wt% NaCl vapor-rich inclusions that formed by condensation of magmatic fluids at pressures of ~1.2 kbar (Kelley *et al.*, 1987).

Figure 4.18 In the Main Endeavour Field (MEF) on the Juan de Fuca Ridge, vents are at a water depth of ~2200 m (220 bar) and seismic constraints place a magma chamber at ~2.3 km beneath the seafloor. As seawater-derived fluids migrate to depth within the reaction zone, they begin to rise buoyantly and intersect the two-phase curve. At above 298 bar and 407 °C (the critical point of seawater – CP), small droplets of brine in a low-salinity vapor are condensed out. During adiabatic rise to shallower depths, the fluids pass through the critical point and undergo boiling. This process results in the formation of vapor bubbles in a continuous liquid. Both condensation and boiling occur beneath numerous submarine hydrothermal fields and these processes have been well documented at Endeavour and the East Pacific Rise.

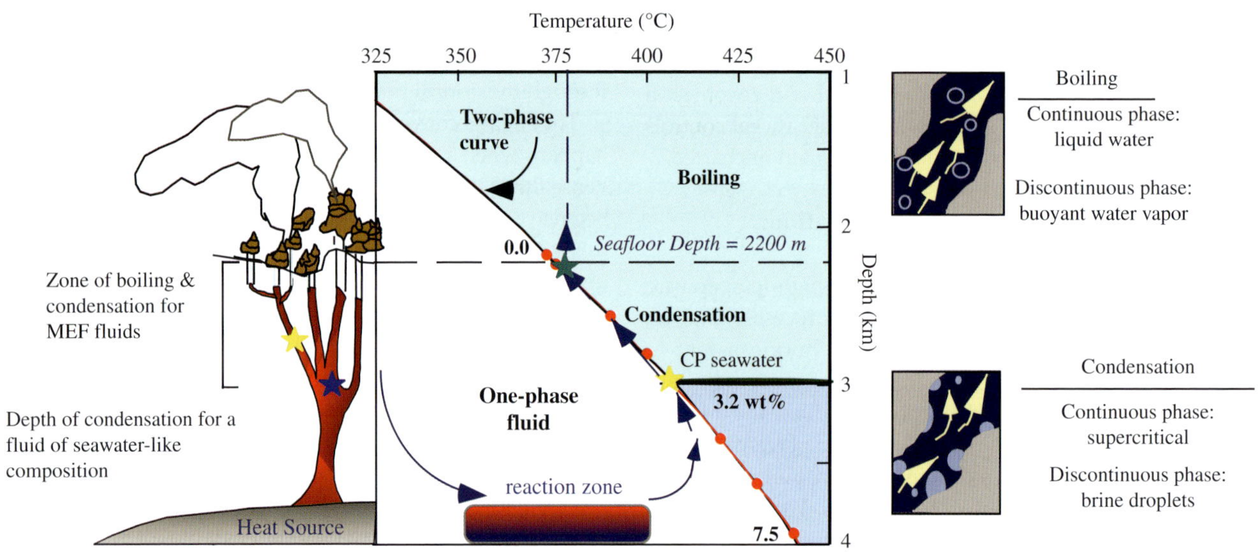

numerous high-temperature systems operative beneath hydrothermal vents in both the Pacific and Atlantic Oceans (Fox, 1990; Delaney *et al.*, 1992; Butterfield *et al.*, 1994; Von Damm *et al.*, 1997, 2003; Von Damm, 2000; Kelley *et al.*, 2002).

Microscopic inclusions of fluid with salinities of 5 wt% to 20 wt% NaCl preserved in oceanic plutonic rocks indicate that segregation of vapors and brines occurs deep in these systems as well (Delaney *et al.*, 1987a; Fournier, 1987; Kelley and Früh-Green, 2001). Brine condensation is an extremely common process in submarine magma chambers, where NaCl concentrations can reach beyond 70 wt% (Fig. 4.17) (Kelley *et al.*, 1987, 2004; Vanko, 1988). In continental systems, fluids with compositions similar to these are responsible for the development of large mineral deposits mined for copper, zinc, and gold (Reynolds *et al.*, 1985; Hannington *et al.*, 2005).

Modeling (Fontaine and Wilcock, 2006) and high chlorine concentrations measured in young, unaltered lavas (e.g., Michael and Cornell, 1998; le Roux *et al.*, 2006) supports the notion that seawater-derived brines may be stored in the shallow crust. As hydrothermal systems wane and cool, the brines may be entrained into upwelling fluids. This release may produce the high-salinity fluids observed at systems such as Endeavour, where fluids reach twice normal seawater salinities.

4.1.5 Gases from the deep

The large volume of melt generated at MORs is a key contributor of volatiles to the hydrosphere. Gases such as CO_2 and helium (He) in magma chambers originate from the mantle and are incorporated during melting processes. Exsolution of magmatic gases out of rising magmas helps drive submarine eruptions. Indeed, although rare, degassing at high concentrations causes highly explosive events in the deep sea (Head and Wilson, 2003; Clague *et al.*, 2013). Some of the highest measured concentrations of CO_2 (9160 ppm) occur at Axial Seamount on the Juan de Fuca Ridge (Helo *et al.*, 2011) (see also Sections 4.11 and 5.4).

Volatiles are routinely incorporated into hydrothermal fluids through direct degassing of magma chambers, by high-temperature leaching from the enclosing host rock, and by inorganic synthesis at moderate to high temperature. The conditions under which carbon-, hydrogen-, and sulfur-rich volatiles form, and their flux to the overlying ocean are of great interest because many of these gases are nutrients, which are key to microbial proliferation and nutrient cycling (Wilcock *et al.*, 2004).

The concentrations of CO_2 in end-member fluids venting from submarine volcanic systems are typically in the range of 2–20 mmol/kg; however, in active areas where volcanic gases are directly incorporated into the high-temperature limb of the convection cell, or where boiling occurs, values of up ~300 mmol/kg have been measured (Butterfield *et al.*, 1990; Lilley *et al.*, 2003; Kelley *et al.*, 2004). Indeed, significant increases in CO_2 concentrations in the vents may precede volcanic eruptions (Lilley *et al.*, 2003). In shallow systems, such as those in the Mariana Arc volcanoes, CO_2 concentrations are so high that liquid CO_2 bubbles form and hot vent fluids contain 2.7 moles/kg – the highest concentrations ever reported for submarine systems (Figs. 4.19 and 4.20) (Lupton *et al.*, 2006).

In addition to CO_2, vent fluids are also commonly enriched in hydrogen (H_2) above background ocean waters, and can reach values up to 15 mmol/kg (Proskurowski *et al.*, 2006). It is believed that the primary source of H_2 is water–rock reactions involving oxidation of iron-bearing mineral phases at moderate to high temperatures (Neal and Stranger, 1983; Horita and Berndt, 1999; McCollom and Seewald, 2001; Lilley *et al.*, 2003; Seyfried *et al.*, 2010). Dissolved H_2, as well as methane (CH_4), are particularly important in venting systems impacted by the presence of ultramafic rocks in the subsurface (Charlou *et al.*, 2002; Früh-Green *et al.*, 2004; Proskurowski *et al.*, 2006; McCollom and Bach, 2009).

There are two principal sources of H_2S in hydrothermal systems, seawater sulfate and the sulfur bound in the rock. Isotopic studies indicate that a quarter to a third of hydrothermal H_2S in bare-rock environments is derived from reduced seawater sulfate (Shanks *et al.*, 1995). Sulfur concentrations in fresh mid-ocean ridge basaltic glasses range from ~700 to ~1600 ppm with about 85% of this as sulfide and the remainder as sulfate (Sakai *et al.*, 1984; Jambon, 1994). Alteration of basaltic glasses and of sulfide phases in the rock contribute sulfur within the hydrothermal fluids resulting in H_2S concentrations of ~150–300 ppm (Von Damm *et al.*, 1995). In magmatically robust systems, up to a third of the sulfur in MOR basalts may be transferred directly into seawater during crystallization and eruption (Moore and Fabbi, 1971). Recent discoveries at Rota Volcano in the Mariana Islands indicate that these arc magmatic systems are also enriched in SO_2, which leads to sulfuric acid and elemental sulfur (Embley *et al.*, 2006). A surprising discovery at Rota was molten pools of sulfur and volcaniclastic ejecta that include sulfur globules on the seafloor at the volcano's summit.

Sources for CH_4 in submarine environments may include re-equilibration of magmatic fluids during degassing and cooling, water–rock reactions, thermogenic decomposition of organic matter, and microbial processes (Welhan, 1988; Lilley *et al.*, 1993; Kelley and Früh-Green, 1999, 2001; Proskurowski *et al.*, 2008). The proportional distribution among these various sources varies among sites. In low-temperature diffuse-flow environments, the CH_4 concentration commonly exceeds that in nearby high-temperature vents. It is likely that this increase is due to robust communities of methanogens that thrive in the lower-temperature fluids in the subseafloor environment (Von Damm and Lilley, 2004; Love *et al.*, 2008).

4.1.6 Impact of volcanic eruptions

Perhaps one of the most important discoveries made in the last decade is the realization of the profound impacts that diking–eruptive events have on submarine hydrothermal activity and on chemical and biological fluxes from below the seafloor to the overlying ocean (Lupton, 1995; Delaney *et al.*, 1998; Embley and Lupton, 2004) (Figs. 4.21–4.24). Eruptions not only provide a "time-zero" with which to construct manipulative temporal experiments (Fig. 4.22), they also provide a contextual framework within which the effects of eruptions can be assessed. These include the disturbance they have on ecological interactions in different biogeographic regions of the world, and the time scales over which they vary. Since 1986, more than 36 seafloor eruptions or tectonic events have been inferred or directly observed (Baker *et al.*, 2012; Rubin *et al.*, 2012; Kelley *et al.*, 2014). Even more numerous are diverse geological disturbances (for example, diking and cracking events) that have resulted in a broad spectrum of impacts on seafloor phenomena.

These events serve as key agents for creating new hydrothermal habitats because they provide bursts of nutrients that support microbial blooms and the development of macrobiological communities (Figs. 4.21–4.24) (Haymon *et al.*, 1993,

Figure 4.19 Close-up of $CO_2 \pm H_2S$ bubbles exiting from Eifuku Volcano in the Mariana Arc. The bubbles are rimmed by clathrate (a solid mixture composed of water and CO_2) (Lupton *et al.*, 2006).

Figure 4.20 Chimneys at the Champagne Vent site, Eifuku Volcano in the Mariana Arc. The chimneys are ~50 cm tall, venting fluids at 103 °C. Note bubbles in the upper left portion of the image. They contain CO_2 concentrations of ~2 moles/kg, values an order of magnitude higher than any previously reported for submarine hydrothermal vents (Lupton *et al.*, 2006).

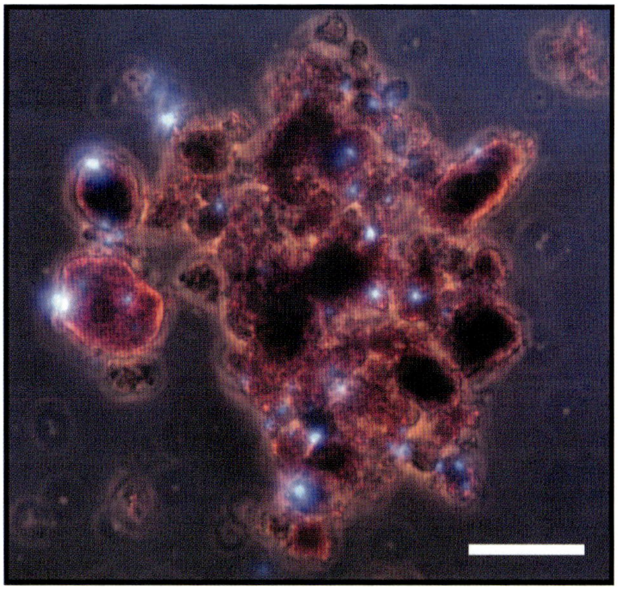

Figure 4.21 Novel, iron-utilizing microorganism grown from megaplume fluids recovered from the 1996 Gorda Ridge eruption (see Fig. 4.23). These organisms grow optimally at up to 88 °C in the presence of CO_2 and are dependent on iron reduction. Cells in this image are blue, forming on sodium sulfide aggregates. They cause precipitation of iron oxide. Scale bar is equal to 10 μm.

Delaney *et al.*, 1998; Shank *et al.*, 1998; Kelley *et al.*, 2014). They also impact magmatic/hydrothermal fluxes, control fluid geochemical composition, and alter the successional development of faunal communities (Fig. 4.22). The disturbances, in part, govern the temporal and spatial scales of local extinction and recolonization for biological communities and direct the evolution of physiological adaptations (Shank *et al.*, 1998).

Eruptions have been documented on the EPR, Southern MAR, Galápagos Rift, CoAxial, Axial Seamount, Northwest Rota Volcano, West Mata Volcano, and Loihi Seamount, as well as numerous other ridge and arc environments (Baker *et al.*, 2012). The eruptions display diverse characteristics

Figure 4.22 Time-series images of 9°50′ N vent communities following the eruption in 1991, showing profound changes in microbial and macrofaunal communities. (Top) Microbially produced flocculant material and cloudy shimmering water vent from an area near the central fissure in 1991. (Second from top) By March 1992, a small number of vestimentiferan tubeworms *Tevnia jerichonana* colonized the area that was still venting cloudy water. (Third from top) Between December 1991 and 1994, tubeworm communities dramatically increased. (Bottom) By 1995, the vestimentiferan *Riftia pachyptila* colony increased by ~50 individuals, while small patches containing dead *T. jerichonana* were evident. Hydrogen sulfide in the venting fluids had decreased by half two years following the eruption (Shank *et al.*, 1998).

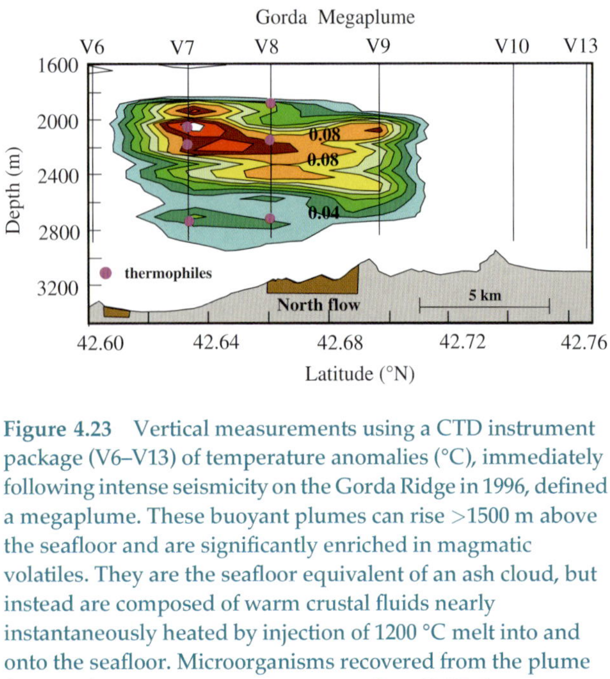

Figure 4.23 Vertical measurements using a CTD instrument package (V6–V13) of temperature anomalies (°C), immediately following intense seismicity on the Gorda Ridge in 1996, defined a megaplume. These buoyant plumes can rise >1500 m above the seafloor and are significantly enriched in magmatic volatiles. They are the seafloor equivalent of an ash cloud, but instead are composed of warm crustal fluids nearly instantaneously heated by injection of 1200 °C melt into and onto the seafloor. Microorganisms recovered from the plume (Fig. 4.24) require temperatures more than 80 °C above background seawater values – this and similar discoveries were pivotal in providing unequivocal evidence for a hot, subseafloor biosphere (e.g., Holden *et al.*, 1998; Summit and Baross, 1998).

Figure 4.24 Dense colonies of microbes cover a three-month-old lava flow in the caldera of Axial Seamount in 2011. In areas of more intense diffuse flow, white bacterial mats line fractures and cavities in the glassy lava flows. Orange-brown, iron-oxidizing and sulfur-utilizing bacteria rapidly covered the new lava and were thickest at the flow fronts where pillow basalts dominated.

and styles from explosive pyroclastic deposits, to thin lava flows, to lava flows attaining more than 10 m in thickness (Caress *et al.*, 2012) (see Section 5.10).

A key feature of submarine diking–eruptive events appears to be the voluminous release of volatile-enriched, warm buoyant hydrothermal plumes that rise up to 800 m or more above the seafloor. These "megaplumes" can extend for up to 70 km along the axis and achieve a thickness of several hundred meters (Fig. 4.23) (Baker *et al.*, 1987; Baker and Lupton, 1990; Lupton, 1995; Delaney *et al.*, 1998; Lupton *et al.*, 1999, Murton *et al.*, 2006). Their importance lies not only in their thermal and chemical properties, but also in what they may reveal about the subseafloor biosphere. Extensive degassing of magma/lava during eruptions, coupled with melt intrusion in the subsurface and ponding of lava beneath the seafloor, can produce temporary "spikes" in the volatile concentrations in stable hydrothermal vent systems (Butterfield *et al.*, 1997; Delaney *et al.*, 1998; Lilley *et al.*, 2003). It is thought that the release of volatiles, coupled with an increase in subseafloor fluid circulation due to melt intrusion, results in extensive subseafloor microbial blooms (Fig. 4.24; see Section 4.11) (Haymon *et al.*, 1993; Delaney *et al.*, 1998; Holden *et al.*, 1998; Summit and Baross, 1998; Love *et al.*, 2008; Kelley *et al.*, 2014).

In general, CO_2, H_2, H_2S, He, and CH_4 are found in high concentrations immediately following intrusion and eruption of basaltic material onto the seafloor (Butterfield *et al.*, 1997; Von Damm, 2000; Lilley *et al.*, 2003; Von Damm and Lilley, 2004; Love *et al.*, 2008). The elevated concentration of

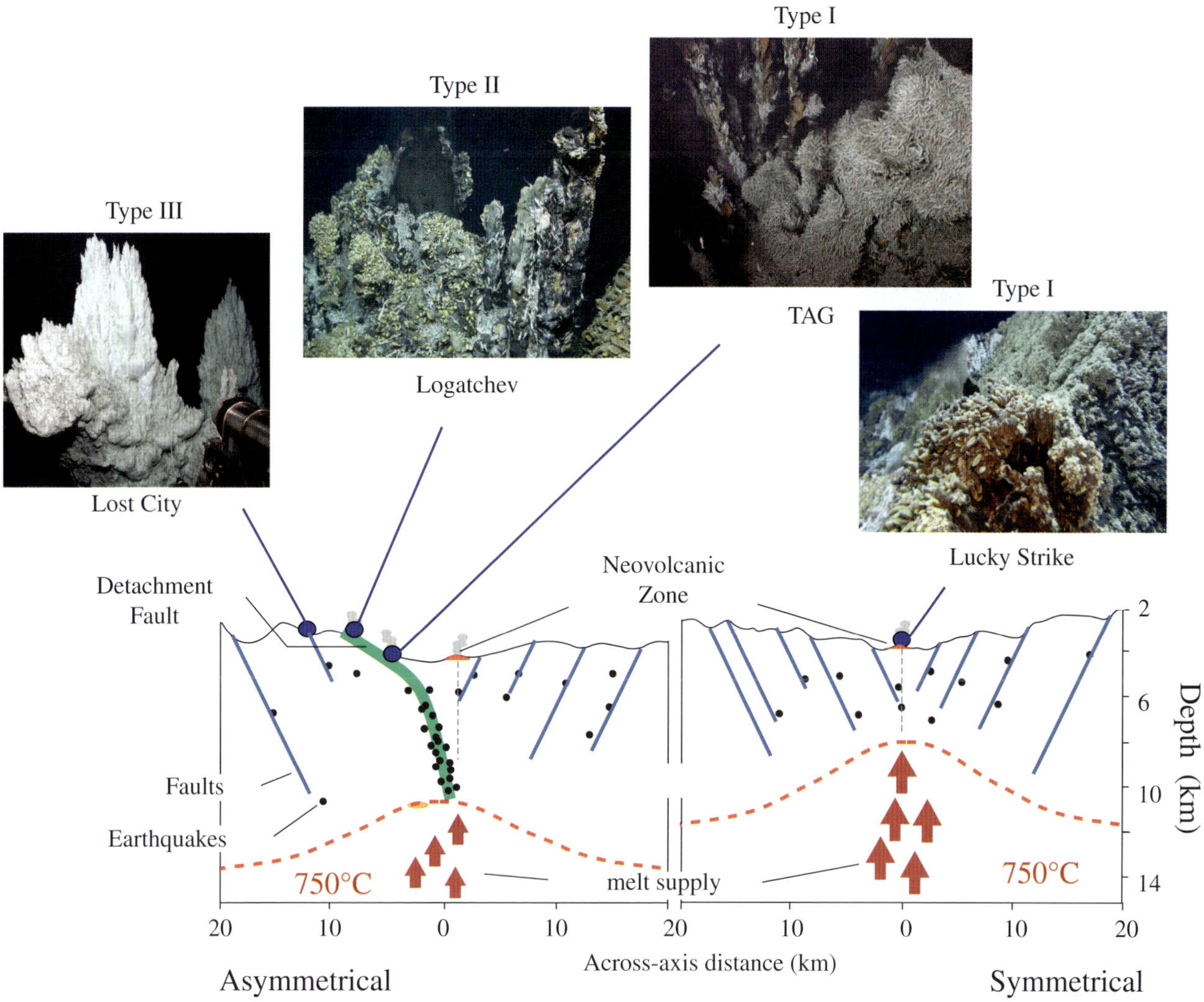

CH$_4$ in the event plume fluids, its isotopically light nature, and geochemical measurements from near-shore eruptions on Kilauea indicate that CH$_4$ is not produced during transient high-temperature alteration episodes associated with new surface eruptions (Sansone *et al.*, 1991). Instead, the data indicate that much of the source fluid for event plumes is from within the deeper crust and not from the surface lava flows (Kelley *et al.*, 1998; Lupton *et al.*, 2000).

4.1.7 Oceanic crust, tectonics, and vent diversity

The prior sections of this chapter have focused on hydrothermal systems driven by cooling of gabbroic bodies and overlying crust beneath submarine volcanoes (Fig. 4.25). For over three decades, these types of systems, located along the narrow MOR axial corridor, were thought to be the norm. However, in 2000, the Lost City Hydrothermal Field was found 15 km west of the MAR (Fig. 4.26). The site was not on a volcano, but instead on a tectonically uplifted section of upper mantle material (see Chapter 8). This discovery, once again, fundamentally changed our views of where and how life can exist in the oceans and about the nature of submarine hydrothermal systems (Kelley *et al.*, 2001b; Martin *et al.*, 2008).

Figure 4.25 Cartoon of Types I, II, and III hydrothermal systems with respect to magmatic centers and faults [after Escartín *et al.* (2008) and Kelley and Shank (2010)]. Black smoker fields similar to Lucky Strike are favored in magmatically robust areas along spreading centers. In zones of extreme crustal attenuation, variable mixtures of altered and deformed peridotitic and gabbroic rocks form extensive seafloor exposures. Here, cooling of mantle and gabbroic material results in Type II, high-temperature black smoker systems with fluid chemistries enriched in magmatic gases, and dissolved volatiles derived from fluid–rock interactions in the peridotites. Logatchev image was taken with the ROV *Quest 4000* (MARUM, University of Bremen). High concentrations of CH$_4$, H$_2$, CO$_2$, and low molecular weight hydrocarbons are hallmarks of these systems. Type III systems, driven by cooling of mantle material, host carbonate chimneys venting fluids rich in CH$_4$ and H$_2$, and abiogenically produced low molecular weight hydrocarbons. The high-pH, moderate- to low-temperature fluids are extremely metal-poor.

Lost City reveals a combination of extreme conditions never before seen in the marine environment (see Section 4.6). These

Figure 4.26 A wreckfish (*Polyprion americanus*) swims by a 30 m tall carbonate tower in the Lost City Field at a water depth of 750 m. The wreckfish is ~1 m in length. The shingled, bright white carbonate deposit growing on the side of the main tower is venting clear fluids at 80 °C. Here, vent-endemic organisms are typically <1 cm in size, and the thick swarms of shrimp that typify many MAR black smokers are absent. Venting at Lost City has been active for over 120,000 years.

conditions include venting of basic, 40–91 °C, metal-poor hydrothermal fluids with high concentrations of dissolved H_2, CH_4, and other low molecular weight hydrocarbons from carbonate chimneys rising 60 m above the seafloor (Kelley *et al.*, 2005; Ludwig *et al.*, 2006; Proskurowski *et al.*, 2008). The fluid chemistry is driven by fluid–rock reactions in the underlying ultramafic rocks (see Chapters 2 and 8) at temperatures up to 200 °C (Früh-Green *et al.*, 2003; Proskurowski *et al.*, 2006). Hydrothermal activity has been ongoing for more than 120,000 years (Ludwig *et al.*, 2011), making this field the most long-lived hydrothermal system known in the deep sea.

Hints that systems such as Lost City could exist began with investigation of exposed oceanic mantle rocks in ophiolite deposits (Barnes and O'Neil, 1969) and in slow- and ultraslow-spreading environments where variably altered ultramafic rocks are particularly abundant (Rona *et al.*, 1987). These spreading systems are characterized by low magma supply, and physical offsets and discontinuities in linearity of the spreading axis. They commonly host exceptionally long-lived (1–2 million years) detachment faults (see Chapter 3). Additional focused investigations of these tectonically complex environments reveal that there may be a progression of vent field characteristics away from the spreading axis that results in three general classes (Fig. 4.25) (Kelley and Shank, 2010):

Type I vent fields are predominantly influenced by basaltic crust, as described in Section 4.1.3. Here, fluid circulation is driven by heat extracted either from a cooling magma chamber in which melt is present or from proximal gabbroic crust, or both. These systems host high-temperature, black smoker chimneys and fluid chemistries are commonly impacted by phase separation at depth. The fluids are low in pH, and are enriched in magmatically derived CO_2, with variable concentrations of CH_4, H_2, and H_2S. The vents at the Lucky Strike Field, at Axial Seamount, and along the EPR are type examples of basaltic-driven systems (see Sections 4.4, 4.11, and 4.13, respectively).

Type II high-temperature black smoker environments are driven by heat extracted from variable mixtures of gabbro and ultramafic material (harzburgite – see Chapter 8)

(Fig. 4.25). They are common near the base or part way up the exposed footwall of detachment faults (see Chapter 3), or in areas where there are discontinuities in the axis of spreading centers (Charlou *et al.*, 2002; Escartín *et al.*, 2003, 2008; Petersen *et al.*, 2009). Similar to volcanically driven systems, boiling is also common where there is a high-temperature gabbroic heat source at depth. Dissolved volatiles in the vent fluids are enriched in CO_2, reflecting the magmatic–gabbroic influence, but the fluids also contain highly elevated concentrations of CH_4 and H_2 that are believed to be derived primarily from abiogenic sources involving reduced conditions and alteration of mantle material (Charlou *et al.*, 1988, 2002; Holm and Charlou, 2001; Schmidt *et al.*, 2007; Konn *et al.*, 2009; Petersen *et al.*, 2009). In addition, the fluids contain variable concentrations of low molecular weight hydrocarbons (Konn *et al.*, 2009). The Logatchev Hydrothermal Field is a type example of a gabbroic–ultramafic-hosted system (see Section 4.8).

Type III systems are low-temperature, peridotite-hosted vent environments where fluid circulation is driven by a combination of cooling of predominantly mantle material and lesser exothermic serpentinization reactions (Figs. 4.25 and 4.26) (Kelley *et al.*, 2001b, 2005; Früh-Green *et al.*, 2003). In stark contrast to Type I and Type II vent systems, the fluids in Type III environments are low to moderate in temperature (up to 90 °C), high in pH (9–11), and lack a CO_2 component (Proskurowski *et al.*, 2008). Similar to Type II systems, however, the fluids are enriched in H_2 and CH_4 (Kelley *et al.*, 2005; Proskurowski *et al.*, 2006, 2008). They also contain high concentrations (relative to background seawater) of low molecular weight hydrocarbons that include acetate and formate (Proskurowski *et al.*, 2006, 2008; Konn *et al.*, 2009; Lang *et al.*, 2010, 2012). Here, the chimneys are carbonate: Lost City is the type example for an ultramafic-dominated system (see Section 4.6).

4.1.8 Life in the extreme: microbes of the deep

Of all the impressive natural phenomena that researchers have witnessed on the seafloor, the discovery of massive effusions of biogenic material being emitted from fissures, cracks, and pits following eruptive events is one of the most remarkable and important findings since deep-ocean vents were discovered (Fig. 4.24) (Haymon *et al.*, 1993; Embley *et al.*, 1995; Delaney *et al.*, 1998; Shank *et al.*, 1998; Chadwick *et al.*, 2013; Kelley *et al.*, 2014).

The potential significance of volcanically supported microbial biomes to the origin of life and the possibility of life on other planets is a major driving force for continuing exploration of these dynamic systems on Earth. The greater integration of microbiology with earth sciences at hydrothermal systems is already having profound implications not only for understanding early life on Earth and the co-evolution of microbes with their physical, chemical, and biological environments, but also offers an approach and an Earth-based laboratory that will aid in the hunt for extant or extinct life on other solar system bodies.

The discovery of abundant chemosynthetic life instantly swept away old notions that had bounded our understanding of where and how life could exist. The cornerstone sources of life on Earth – sunlight and photosynthesis – are absent at the bottom of the ocean. Instead, the planet itself seemingly provided all the heat and chemical materials necessary to support life (Fig. 4.27). Life not only survives, it thrives under extreme ocean-bottom pressure and amid fluids emitted from seafloor hydrothermal vents that contain toxic (at least to familiar life forms) chemicals.

Studies of submarine eruptions reveal unequivocal evidence that beneath the seafloor, there are diverse populations of single-celled, heat-loving organisms that thrive in the absence of sunlight and oxygen (Summit and Baross, 1998; Summit, 2000) (Fig. 4.21). Much is yet to be learned about this biotope, and the biological responses to submarine eruptions, but it is clear that microbes are intimately linked to the mantle, for it is from the mantle that the ocean crust, associated gases, and nutrient sources are ultimately derived.

Contemporaneous with changes in the temperature, chemistry, and distribution of vent fluids during volcanic events, there is a rapid and substantial effusion of microbial-generated particles, called floc, entrained in diffusely venting fluids emanating from the seafloor (Figs. 4.22 and 4.24). This intense, particle-laden hydrothermal flow creates "snowstorms" reaching 50 m above the seafloor (Nelson *et al.*, 1991; Haymon *et al.*, 1993). In addition, mats of filamentous bacteria, 1–10 cm thick, cover areas as large as 50 m^2 of fresh lava flows, and floc can accumulate as deep drifts. This flocculant material is composed primarily of inorganic, sulfur-rich filaments with a low organic carbon content (Nelson *et al.*, 1991).

Laboratory work indicates that the flocculant material may be rapidly produced by microscopic organisms growing at the interface between sulfide and oxygen (Taylor and Wirsen, 1997). Recent genetic analyses of flocculant material ejected into the water column following the April 2011 eruption at Axial Seamount show that the floc was dominated by sulfur-oxidizing bacteria, and that the dominant Archaea were CH_4-producing organisms (thermophilic *Methanococcales*) requiring warm temperature and anoxic conditions (Fig. 4.24) (Meyer *et al.*, 2013).

Although investigation of these systems is relatively young, data collected during rapid-response cruises, weeks to months following an eruption, support a model in which subseafloor biotopes are periodically perturbed by diking–eruption events that temporarily destroy or enrich thermophilic zones within the subsurface (Delaney *et al.*, 1998). This model is supported by the correspondence between the massive effusion of microbial material from the seafloor, the bloom of microorganisms for several weeks following an eruption, and the enriched volatile concentrations in megaplume fluids and in hydrothermal vents perturbed by eruptions (Haymon *et al.*, 1993; Embley *et al.*, 1995; Delaney *et al.*, 1998; Holden *et al.*, 1998;

Figure 4.27 (Top) Wispy white strands of filamentous bacteria colonize tubeworms *(Ridgeia piscesae)* in a diffuse flow site on the flanks of a massive black smoker chimney called Hulk in the Main Endeavour Field, Endeavour Segment of the Juan de Fuca Ridge. Here "skinny" tubeworms are less than a centimeter across. The brown tubes' lack of red plumes indicate a potentially stressed community.

Figure 4.28 (Bottom) Dense white and orange microbial mats characterize the sediment-hosted hydrothermal mounds in the Guaymus Basin. Heating of organic material in the sediments produces hydrocarbon-rich fluids, which in places contain oil droplets. Active microbial communities include anaerobic methane oxidizers (*Methanosarcinales*).

Kelley *et al.*, 1998, 2014; Shank *et al.*, 1998; Baker *et al.*, 1999; Love *et al.*, 2008; Chadwick *et al.*, 2013).

Much of the supporting energy for these microbes likely comes from the degassing of volatiles associated with the eruption and from volatiles released during high-temperature interaction of seawater and dike–extrusive material. The massive injection of magmatically derived gases such as CO_2 prior to and following an eruptive event, and an associated increase in circulation of warm fluids through the porous oceanic crust, is believed to cause a microbial "bloom" in the subsurface that may last for several months (Kelley *et al.*, 2014). This is consistent with recent findings of thermophilic *Methanococcales* at Axial Volcano (Meyer *et al.*, 2013).

4.1.9 Chemosynthesis and the metabolic menu

Within the seafloor and black smoker chimney walls, microbial communities occupy both aerobic and anaerobic environments and thrive at *in situ* temperatures that range from ambient seawater temperature (2 °C) to temperatures greater than 121 °C (Baross and Deming, 1995; Karl, 1995; Kashefi and Lovely, 2001; Schrenk *et al.*, 2003; Takai *et al.*, 2008). In general, increases in temperature are accompanied by decreases in oxygen and increases in hydrothermally derived chemical compounds and elements. This is also reflected in the microbial communities with strictly anaerobic organisms dominating environments exceeding 50 °C. The concentration of microbes in vent fluids ranges from 10^7 to 10^9 per liter.

Within vents, the predominant energy and carbon sources used by microbes originate *de novo* as a result of hydrothermal processes. Thus, chemical energy replaces light energy as the fuel that supports the primary producers; the gases derived from the volcanoes form the thread of life on and within the seafloor. Only O_2, the key electron acceptor for all metazoan animals and aerobic microorganisms, and possibly some organic compounds, originates from pelagic photosynthesis.

The primary producers use CO_2 as their main carbon source, although there are organisms at vents that use CH_4 and organic compounds as well (Fig. 4.28). The recognition of the importance of chemosynthesis in the vent environment was one of the most profound discoveries in marine science. It has fundamentally changed the view that the base of the food chain was driven by photosynthetic production of biomass with sunlight (Jannasch, 1995). A general reaction for anaerobic chemosynthesis can be expressed as:

$$2CO_2 + 6H_2 \rightarrow [CH_2O] + CH_4 + 3H_2O$$

It has become increasingly clear that there is an extraordinarily high diversity of microorganisms associated with submarine hydrothermal vent environments. In part, this is because of the highly diverse biotopes associated with vent microhabitats, which span so many extremes in physical and chemical conditions. However, little is known about most of these organisms other than their phylogeny based on small subunit ribosomal RNA sequences, because a relatively few species have been cultured in laboratory settings.

The important primary producers in vent food webs are the bacteria that oxidize sulfur, iron, CH_4, and H_2. The sulfur-oxidizing bacteria represent the dominant CO_2-using microbes in vent environments (Jannasch, 1995). They are the key symbionts associated with tubeworms, shrimps, and mollusks and are a significant fraction of the free-living organisms in diffuse-flow fluids and making up microbial mats (Figs. 4.29 and 4.30). They are also common in sediments where consortiums of sulfate-oxidizing and CH_4 producers are common.

The aerobic microorganisms use oxygen or NO^{3-} as the electron acceptor and derive their energy from the oxidation of reduced sulfur compounds, CH_4, H_2, NH_4^+, or organic compounds. The oxidation of reduced sulfur compounds is the most important of these processes, as it is the basis of most of the dominant symbiotic associations with vent-endemic fauna.

Methane-oxidizers are the only other group of bacteria that enters into symbiotic associations with vent animals such as mussels (Cavanaugh *et al.*, 1987; Streams *et al.*, 1997; Cary *et al.*, 1998a; Dechaine *et al.*, 2006), although, like H_2-oxidizing and ammonia-oxidizing bacteria, they are present on the surfaces of animals and rocks, and in hydrothermal plumes (Mevel *et al.*, 1996).

Isolated metal-oxidizing microbes usually require organic compounds for growth and metal oxidation. There is also a diverse group of aerobic heterotrophic microorganisms in vents that use a wide variety of organic compounds including organic acids, carbohydrates, and fatty acids (Karl, 1995).

Methanogens have been isolated or detected using molecular methods from most high-temperature anaerobic vent habitats. The best-studied methanogens are hyperthermophiles capable of growing at greater than 80 °C. These organisms are extraordinarily fast growing; some species have a doubling time of less than 30 minutes at their optimum growth temperature.

In addition to energy sources and electron acceptors and donors, organisms require other chemical species including a source of nitrogen and phosphorus. All organisms have requirements for a variety of trace metals. These metals are incorporated as metal clusters in enzymes and other proteins.

Hyperthermophiles are defined as organisms that have a maximum growth temperature of 90 °C or higher and grow optimally at 80 °C or higher. They have received considerable attention for their ability to survive and grow at these extreme high temperatures and low pH, their frequent phylogenetic position as the most ancestral of extant organisms, and their high potential for biotechnological and biomedical applications (Wilcock *et al.*, 2004). Most of the hyperthermophiles isolated from hydrothermal vent environments are Archaea and are represented by anaerobic heterotrophs and methanogens that cannot survive in the presence of oxygen (e.g., Ver Eecke *et al.*, 2013). Most hyperthermophiles are strict anaerobes; only three species use oxygen. The use of sulfur (as S°) as an electron acceptor is a hallmark of heterotrophic hyperthermophiles; for most species, S° is a requirement for life.

The past decade has seen a fundamental change in our ideas concerning the linkages among geological, chemical, and biological processes at MOR spreading centers. This change has been brought about in part by the growing realization that many processes intrinsic to the dynamics of the Earth (e.g., volcanism, earthquakes, fluid flow, sediment maturation) provide partial or complete energy/nutrient fluxes necessary to support substantial, diverse microbial communities. Also, one of the predictions from global phylogenetic models based on 16S rRNA sequence analyses is that the most ancient of extant organisms were anaerobic hyperthermophiles that inhabited volcanic environments (Ciccarelli *et al.*, 2006).

4.1.10 Life in the extreme: vent fauna

Perhaps the most striking discoveries associated with hydrothermal vents are the unique chemosynthetic fauna they host (Fig. 4.31). The tight coupling between geological processes and living organisms at hydrothermal vents provides a singular opportunity to study how fundamental planetary processes shape the evolution of life. Scientists now look to understand how evolutionary processes shape observed faunal patterns in different geological settings along the global MOR system (Juniper and Sarrazin, 1995).

Deep-sea hydrothermal vents provide ephemeral and disjunct habitats for highly specialized groups of animals. These transient habitats and the animals they support are subjected to the volcanic dynamics of active submarine systems around the globe, and thus species have adapted to both the "harsh" geochemical milieu of venting systems and the vents' stochastic and ephemeral time scales – on the order of months to decades.

To date, more than 750 new species have been discovered, although less than 50 active vent sites have been biologically investigated in great detail (Baker and German, 2004; Desbruyères *et al.*, 2006; Bachraty *et al.*, 2009). While, on average, a new species has been described every 10 days over the past 30 years (Shank, 2004), well over half of the

Figure 4.29 A community of tubeworms (*Riftia pachyptila*), brachyuran crabs, and vent fish (*Thermarces cerberus*) thrive in nutrient-rich, warm fluids issuing from Tubeworm Pillar on the EPR at 9° N. Crabs are ~5 cm across. In 2004, these animals perished at this site due to a rapid cessation of venting, and the 11 m tall sulfide pillar was subsequently destroyed during eruptive events in 2005–2006.

Figure 4.30 The TAG Hydrothermal Field supports dense aggregations of shrimp (*Rimicaris exoculata*). These alvinocarid shrimp are of interest because, even though they have no eyestalks, they have a dorsal organ hosting a visual pigment that absorbs maximally near 500 nm. High-temperature chimneys emit light in the non-visible spectrum, supporting the idea that shrimp detect dim light with their non-image-forming "eyes."

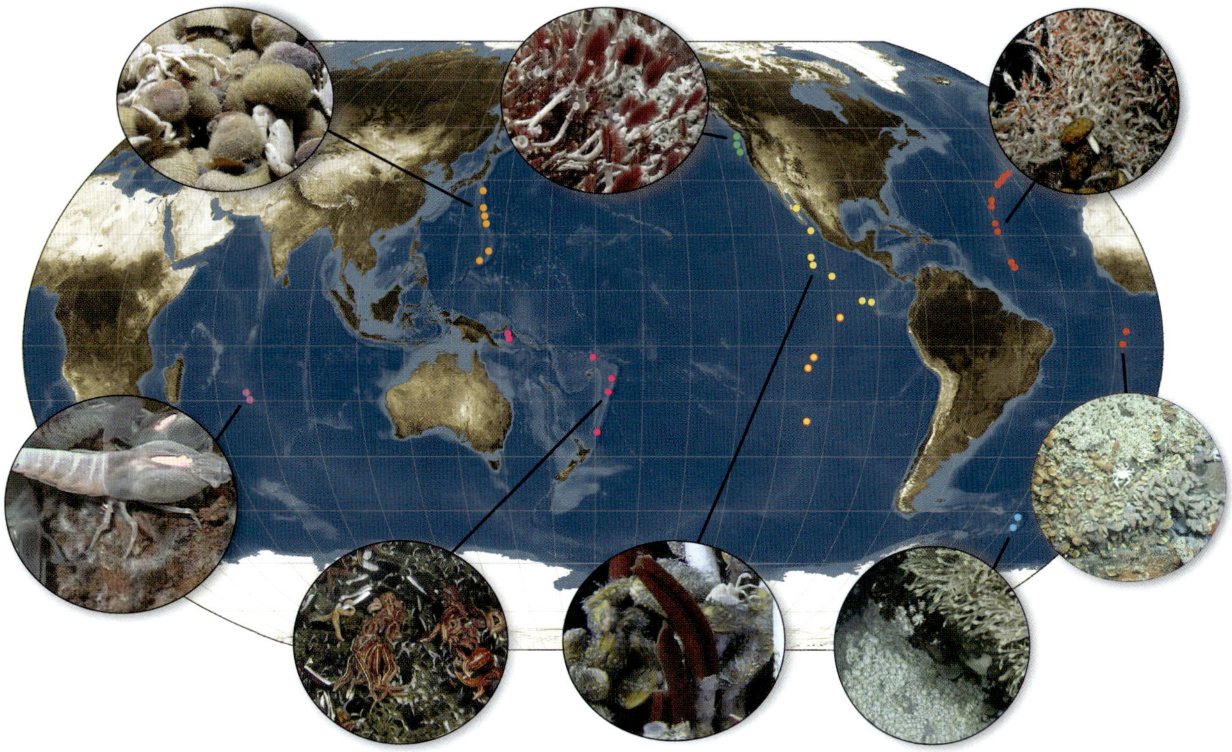

Figure 4.31 Global distribution of distinct biogeographic regions currently recognized for MOR hydrothermal vent fauna [including some non-ridge sites; after Rogers *et al.* (2012)]. The regional suites of species characteristic of hydrothermal vents can be considerably different in various parts of the world ocean as a consequence of the evolutionary history of the different ocean basins. Vicariant events and physical oceanographic processes, species-specific variations in the life histories of individual species, rates of seafloor spreading, and the frequency and spacing of active venting along the MOR system impact this faunal distribution.

~60,000 km long global ridge crest remains completely unexplored for hydrothermal venting, with the southern hemisphere being particularly neglected. This is of notable importance because not only does the discovery of each new chemosynthetic ecosystem or site invariably lead to discovery of new species and novel ecologies worthy of further study, but hypothesis-driven investigations of ecosystems in different regions have provided immediate insights into the geological and geochemical processes controlling regional biogeographic patterns and the evolution of chemosynthetic fauna (Van Dover *et al.*, 2002; Shank, 2004).

Distinct faunal assemblages are associated with different MOR systems within and between ocean basins (Figs. 4.25–31). The evolutionary history of vent-endemic organisms inhabiting these deep-sea hydrothermal provinces, and their subsequent biogeographic diversification, have been considered to be tightly linked to the history of regional seafloor spreading and barrier-forming events, such as the linking of North and South America in the Pliocene epoch about 4.5 million years ago (Tunnicliffe and Fowler, 1996; Tunnicliffe *et al.*, 1998). Comparison of these biogeographical provinces and reconstructions of historical MOR spreading events have been used to infer migratory pathways through the MOR system and explain the mechanisms driving the evolution of chemosynthetic fauna along contiguous ridge segments (Tunnicliffe and Fowler, 1996; Van Dover and Doerries, 2005).

Exploring new ridge segments and studying their associated fauna is essential to understanding the processes that create and maintain the global biodiversity of vent fauna (Fig. 4.31). Global biogeographic patterns suggested by the recent discovery of distinct Southern Ocean vent fauna (Rogers *et al.*, 2012) raise fundamental questions about the marked "disconnect" between Pacific, Atlantic, and Indian Ocean fauna. The past decade of research on global biogeographic patterns resulted in the discovery of the Central Indian Ridge (Gamo *et al.*, 2001; Van Dover *et al.*, 2001, 2002) and East Scotia Rise (Rogers *et al.*, 2012) faunal provinces, where fauna are distinct from vent ecosystems of the MAR and Eastern Pacific. Understanding the phylogeographic relationships of vent fauna with other chemosynthetic fauna at cold seeps and at wood and whale falls is crucial for determining the origin, evolution, and biogeographic patterns of chemosynthetically driven species. The recent discovery of vent fauna in the Southern Ocean, along the East Scotia Rise and Bransfield Strait, is shedding new light on the global evolutionary pathways and apparent isolation that have been controlled by the Southern Ocean Ridges (Fig. 4.31).

In the following sections we present a "tour" of hydrothermal systems along the global MOR and highlight geological features and life forms that make these systems some of the most remarkable environments on Earth.

Arctic ridges

4.2 Fire and ice

The character of hydrothermal venting is extremely diverse on ridges within the Arctic Circle, reflecting the numerous magmatic, tectonic, and bare rock to sedimented environments that it encompasses. Until recently, Arctic ridges were some of the least-explored places in the ocean basins. However, there was, and is, intense interest in the Arctic ridge system because: (1) it is an ultraslow-spreading environment where ultramafic rocks are exposed at the seafloor (see Chapter 8), with possible implications for the presence of Lost City-type environments; (2) it is impacted by thick sediments along some hydrothermally active segments with resulting hydrocarbon production potential; (3) many of the vent systems are shallow (>500 m) (Figs. 4.32–4.34), and the faunal communities may have evolved to form a distinct biogeographic province in the presence of potential barriers to the migratory exchange of fauna (e.g., Iceland and the Fram Strait); and (4) major portions of the ridge network north of the Fram Strait are ice-covered during substantial portions of the year.

Systematic exploration of MORs in the Arctic Circle began a decade ago with recent expeditions documenting

Figure 4.32 A black smoker in the Loki's Castle Vent Field. Temperatures reach 317 °C with a pH of 5.5. The Arctic ridges are some of the slowest spreading on Earth, and the hydrothermal vent systems host a distinct vent fauna. Surprisingly, some of the animals may have developed by migration of fauna from cold seeps and from the Pacific (Pedersen *et al.*, 2010a).

Figure 4.33 A complex of weakly venting, multi-flanged sulfide structures in the Soria Maria Hydrothermal Field at a water depth of 700 m. White filamentous bacteria coat many of the structures. These low-temperature edifices reach 8–9 m tall and are composed of barite and silica with minor pyrite, sphalerite, and galena (Pedersen *et al.*, 2010b).

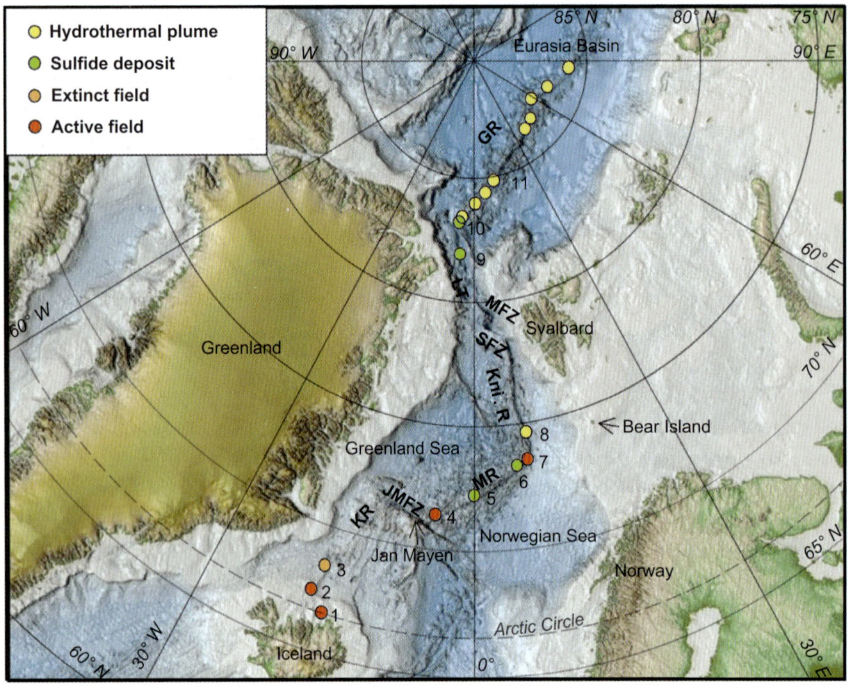

Figure 4.34 The Arctic MORs showing the locations of active and extinct vent fields (Pedersen *et al.*, 2010a). (1) Grimsey (Hannington *et al.*, 2001); (2) Kolbeinsey (Olafosson *et al.*, 1990); (3) Squid Forest (extinct; Nygaard *et al.*, 2003); (4) Jan Mayen, Soria Moria, and Troll Wall (Pedersen *et al.*, 2005, 2010b); (5) Copper Hill sulfide mineralized breccias (Nygaard *et al.*, 2003); (6) Mohns Treasure sulfide deposit; (7) Loki's Castle (Pedersen *et al.*, 2010b); (8) hydrothermal plume (Connelly *et al.*, 2007); (9) sulfide deposit (Snow *et al.*, 2001); (10) sulfide deposit and shimmering water (Edmonds *et al.*, 2003); and (11) hydrothermal plumes (Edmonds *et al.*, 2003).

about 20 active and inactive venting sites from north of Iceland to the Eurasia Basin (Fig. 4.34) (Pedersen *et al.*, 2010a,b). A significant portion of this 4000 km long ridge system is in Norwegian waters. It includes six major segments, which from south to north include the Kolbeinsey Ridge, Mohns Ridge, Knipovich Ridge, Molloy Ridge, Lena Trough, and Gakkel Ridge. The southern portion of the ridge system is strongly impacted by the Iceland hot spot, with elevated topography and significant melt generation in the subsurface. In contrast, the more northern segments lack major volcanic edifices, spreading centers are both orthogonal and oblique, and rift valleys become deeper and more pronounced (Pedersen *et al.*, 2010b).

4.2.1 Grimsey Field, north of Iceland

One of the most well-studied fields is the Grimsey Field, discovered in 1997 north of Iceland (Fig. 4.34). The field is only about 30 km offshore of the island of Grimsey, in a sedimented graben in the Tjornes Fracture Zone. A large area, 1 km^2, is actively venting with shallow aquifers undergoing boiling, reflecting the shallow nature of this system – the water depth is ~400 m (Hannington *et al.*, 2001; Riedel *et al.*, 2001). Intense gas bubble plumes reach more than 300 m above the seafloor (Hannington *et al.*, 2001). Venting is fed by a fault system that intersects a deep normal fault with a reaction zone perhaps 1–2 km beneath the seafloor (Hannington *et al.*, 2001; Riedel *et al.*, 2001; Kuhn *et al.*, 2003). The field hosts mounds 20 m across and rising 10 m above the seafloor, with 1–3 m tall anhydrite and talc chimneys at the mound summits (Hannington *et al.*, 2001). Fluids reach temperatures up to 250 °C. It is believed that sulfide precipitation occurs at depth, resulting in the emission of clear, metal-depleted fluids on the seafloor. Hydrocarbons in the fluids form during thermogenic breakdown of organic materials in the subsurface (Riedel *et al.*, 2001).

4.2.2 Trollveggen (Troll Wall) Field, Mohns Ridge

The Troll Wall Vent Field is located on the Mohns Ridge, along a normal fault on the eastern wall of the rift valley at a water depth of ~550 m (Figs. 4.35 and 4.36) (Pedersen *et al.*, 2010a,b). The field hosts at least 10 major vent sites, each consisting of multiple 5–10 m tall chimneys. The chimneys are white smokers that emit particle-free, gas-rich (CO_2) fluids at temperatures up to 270 °C (Pedersen *et al.*, 2010a, b). Bubble-rich plumes reach high into the water column (Fig. 4.36). Proximal diffuse flow through talus ramparts supports white bacterial mats of sulfur-oxidizing bacteria.

A distinctive feature of this field is extensive areas of low-temperature diffuse flow that support large deposits of yellow-brown iron hydroxide extending for several kilometers (Pedersen *et al.*, 2010a). This area is marked by mounds and chimney-like structures composed of millimeter- to centimeter-thick layers with microtextures reminiscent of branching, twisted filaments similar to iron-oxidizing bacteria (e.g., *Gallionella ferruginea* and *Mariprofundus ferrooxidans*) (Thorseth *et al.*, 2007).

4.2.3 Soria Moria Field, Mohns Ridge

The Soria Moria Vent Field is ~5 km south of Troll Wall on top of a volcanic ridge at a water depth of ~700 m (Figs. 4.35 and 4.37) (Pedersen *et al.*, 2010a). There are at least two

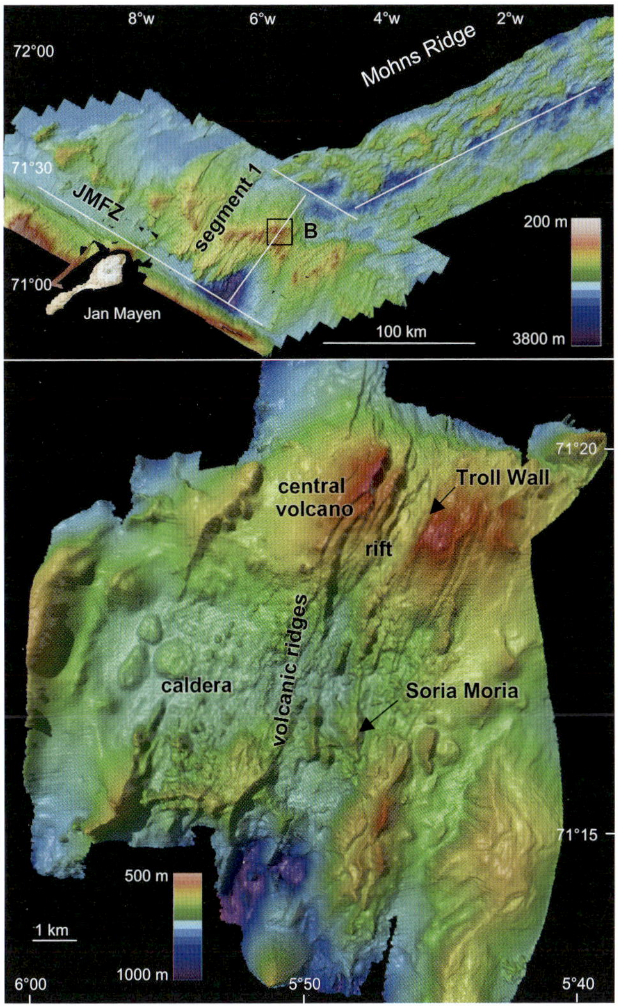

Figure 4.35 (Top) Map of the southwestern Mohns Ridge and parts of the Jan Mayen fracture zone (JMFZ). (Bottom) Bathymetry of the central, shallow part of the Mohns Ridge showing the location of the Jan Mayen vent fields (Troll Wall and Soria Moria). The Troll Wall Field is located along a normal fault that defines the eastern margin of a small rift transecting the shallowest part of the ridge segment. The Soria Moria Field is located on a volcanic ridge ~5 km south of Troll Wall in slightly deeper water. Bathymetric data were acquired using a SIMRAD 300 echo sounder system and gridded at a 7 m cell size. Images are from Pedersen *et al.* (2010a).

venting areas 100–200 m or more across, which are underlain by lava flows forming the volcanic high. The field hosts 8–9 m tall sulfide chimneys and irregularly shaped multi-flanged edifices, which can be 15–20 m across and up to 10 m in height, and issue clear fluids (Fig. 4.37). These edifices are composed of barite, silica and minor pyrite, sphalerite, and the lead-bearing mineral galena (Figs. 4.33 and 4.37). Vent fluids in the Soria Moria and Troll Wall hydrothermal vent fields reach temperatures of ~270 °C, with elevated CO_2 concentrations that are up to 91.7 mmol/kg (Pedersen *et al.*, 2010a).

4.2.4 Lokeslottet (Loki's Castle) Field

The Loki's Castle Vent Field, discovered in July 2008 (73°30′ N, 8° E), was the first black smoker field found in the Arctic Circle (Pedersen *et al.*, 2010b). The Loki's Castle Field is located where the Mohns Ridge transitions into the Knipovich Ridge and where there is an 80° change in the ridge-axis orientation to a more northerly trend (Figs. 4.34 and 4.38) (Pedersen *et al.*, 2010a,b). Venting occurs at a water depth of about 2400 m at the summit of a volcanic ridge, which is bounded 5 km to the east by the extensive Bear Island sedimentary fan (Fig. 4.38) (Pedersen *et al.*, 2010b). To the west, the ridge flank is marked by steep normal faults and truncated oceanic core complexes that expose lower crustal and mantle rocks. The field is at least 3 km long, terminating in an area of well-defined normal faults. Venting is marked by an intense plume, with anomalies in oxidation/reduction potential (Eh), CH_4, and H_2 that rise 300–400 m above the seafloor (Pedersen *et al.*, 2010b).

Although the vent field is clearly underlain by volcanic rocks, it is likely that sediments are comingled with subsurface lavas, resulting in this anomalous vent-fluid chemistry for a bare-rock hosted hydrothermal system (Pedersen *et al.*, 2010a). Black smoker chimneys stand atop two sulfide mounds that reach 150–200 m across and stand 20 to 30 m tall; black smoker edifices reach 13 m in height (Fig. 4.36) (Pedersen *et al.*, 2010a,b). Black smoker fluids are at 310–320 °C and have very high NH_4^+ concentrations (1.6 mmol/kg), as well as elevated values for CO_2 (23.8 mmol/kg) and H_2 (4.9 mmol/kg). Methane values are among the highest reported for a volcanic-hosted vent field (13.5 mmol/kg) (Pedersen *et al.*, 2010b). The high CO_2 values seem to reflect magmatic inputs, while the high H_2 values may reflect fluid interaction with mantle material. Similar to the Endeavour hydrothermal system (see Section 4.10), the high NH_4^+ and CH_4 values point to a sedimentary influence (Lilley *et al.*, 1993; Proskurowski *et al.*, 2006; Pedersen *et al.*, 2010b).

4.2.5 Arctic vent fauna

Within the Mohns Ridge system, 180 taxa have been identified in the vent fields, but only a few are potentially vent-endemic (Schander *et al.*, 2010). Three types of sponges (*Asbestopluma pennatula*, *Cladorhiza* sp. 1 and *Cladorhiza* sp. 2) occur in high abundances directly associated with the smokers and in areas covered by dense bacterial mats. The sponges feed on small crustaceans. This is the first area where calcareous sponges have been reported associated with a black smoker system (Schander *et al.*, 2010). The venting sites also host numerous species (53) of annelids (segmented worms) that are strongly similar to polychaetes from the Jan Mayen systems to the south, with most annelids occurring in non-vent environments in the Nordic seas (Schander *et al.*, 2010).

Several species of mollusks have been identified in the Arctic Circle vent areas. One type occurs on smokers and

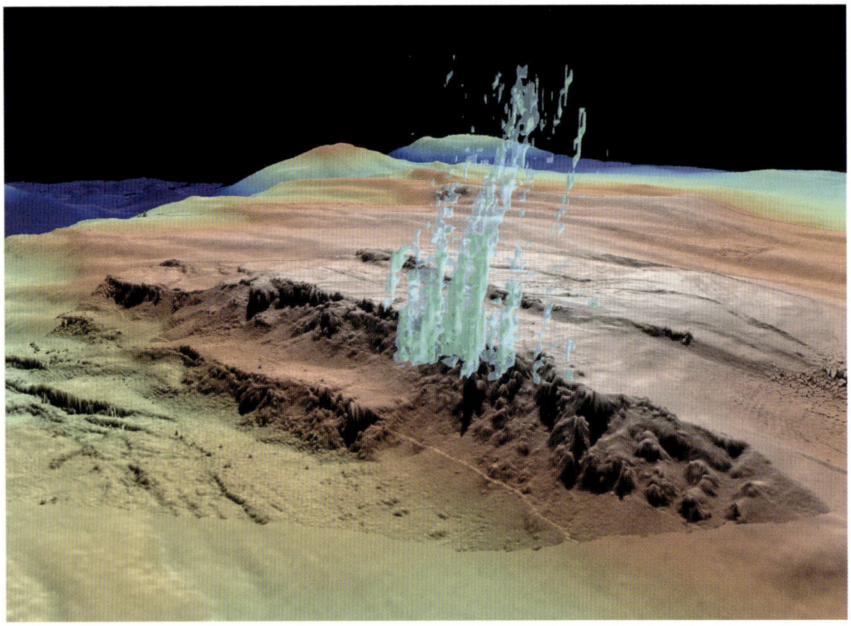

Figure 4.36 Gas-rich bubble plumes, imaged with an echo sounder on the Norwegian research vessel the R/V *G. O. Sars*, rise from the Troll Wall Vent Field. The field is located on a talus deposit along the foot of a 100 m high normal fault that dips beneath the Mohns Ridge (Pedersen *et al.*, 2010a). The fault defines the eastern margin of the rift. Chimneys reach 5–10 m in height and vent 270 °C fluids; at a water depth of 550 m, these fluids are close to the boiling curve (see Fig. 4.15).

Figure 4.37 Complex multi-flanged chimneys in the Soria Moria Vent Field. The small chimlet in the foreground is venting clear hydrothermal fluids (Pedersen *et al.*, 2010a).

iron-hydroxide-dominated mounds covered in filamentous bacteria. The bacteria are reminiscent of *Beggiatoa* (Figs. 4.39 and 4.40) (Schander *et al.*, 2010). In contrast to vent fields south of Iceland, the Mohns Ridge Hydrothermal Fields do not support a high biomass of vent-endemic organisms, and both mussels and alvinocarid shrimp that typify fields such as Menez Gwen and TAG are absent (see Sections 4.3–4.8). There is little difference between the vent faunal assemblages and those in "non-vent" distal waters (Schander *et al.*, 2010).

It may be that the lack of specialized vent fauna represents an early evolutionary stage of these systems to a

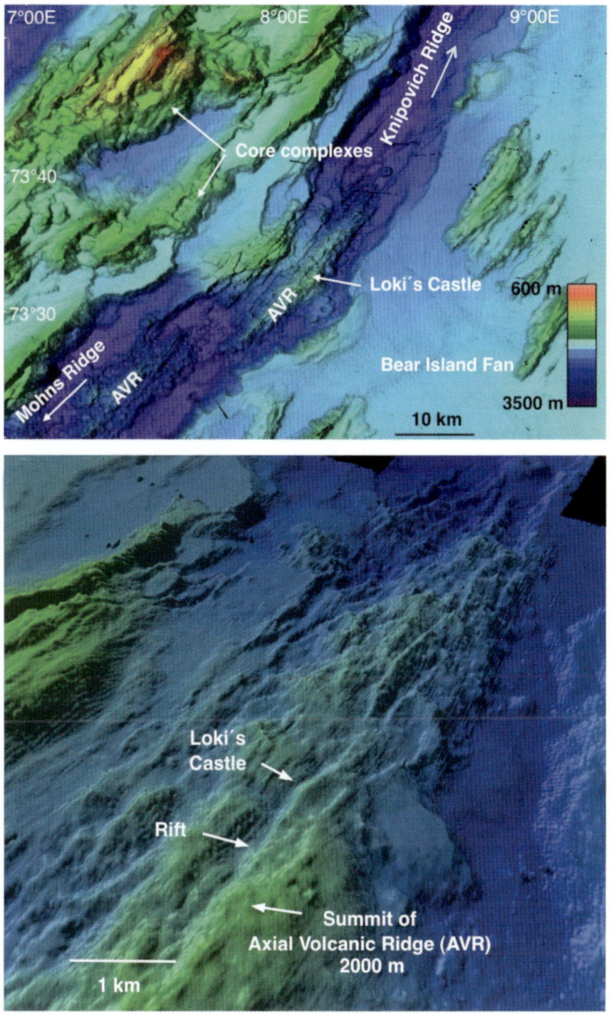

Figure 4.38 (Top) Bathymetric map of the 30 km long northern portion of the Mohns Ridge as it transitions into the Knipovich Ridge (Pedersen *et al.*, 2010b). The Loki's Castle Vent Field is located on a small volcanic high at a water depth of ~2400 m. Two oceanic core complexes bound the ridge to the west, and the sedimentary Bear Island Fan bounds the ridge to the east. The fan developed during repeated Arctic glaciations during the last 3 million years (Pedersen *et al.*, 2010b). (Bottom) Close-up of the axial volcanic ridge hosting the Loki Vent Field, which is located near the summit of the ridge. Normal faults cutting the crest of the ridge may host the conduits for hydrothermal flow (Pedersen *et al.*, 2010a,b).

more mature vent-endemic community, as proposed for the Kolbeinsey vents (Fricke *et al.*, 1989). It may also be that the fauna are impacted to some extent by the shallow depths of the vent fields (Schander *et al.*, 2010). Investigations are still underway to address these controlling factors.

In contrast to the Jan Mayen system, preliminary studies of vent fauna at Loki's Castle show that it harbors amphipods, polychaetes, and gastropods with symbiotic microorganisms that support a strong endemic link to MAR vent systems (Kongsrud and Rapp, 2012). At Loki's Castle, a new genus and species of amphipod was discovered, *Exitomelita sigynae* gen. et sp. nov., living in cracks and crevices on hydrothermal vents (Tandberg *et al.*, 2012). This melitid amphipod is of interest because it may host ectosymbionts embedded in a mucus layer covering its gills. Another important discovery was made at this site of dense

Figure 4.39 A zoarcid-like fish settles amongst dense tufts of thick gray filamentous mats of *Beggiatoa*-like bacteria that thrive on hydrothermal mounds within Arctic Circle vent sites.

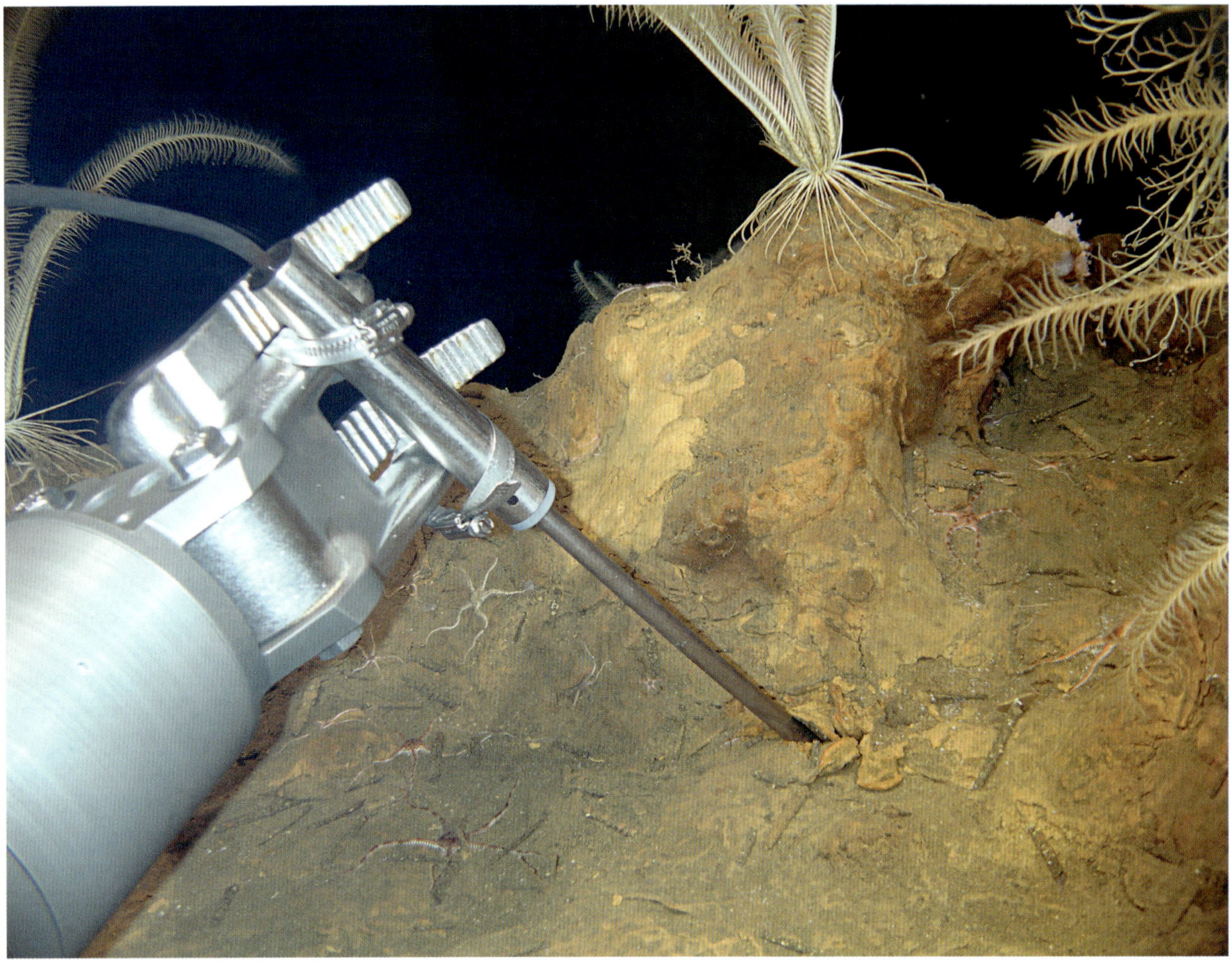

Figure 4.40 A temperature sensor probes the iron oxyhydroxide surface of a hydrothermal mound colonized by small ophiuroid brittle stars and larger attached crinoid feather stars (upper right). This crinoid, most likely *Heliometra glacialis*, dominates large areas surrounding the vent fields.

tubeworm colonies (*Sclerolinum contortum*) typically associated with seeps in the sediments surrounding hydrothermal deposits and within areas of low-temperature venting (~20 °C) (Kongsrud and Rapp, 2012).

A newly proposed polychaete species [*Nicomache (Loxochona) lokii* sp. nov.] discovered at Loki's Castle grows on black smoker walls and within the low-temperature sediment deposits (Kongsrud and Rapp, 2012). Affinities of this worm and other vent fauna within Loki's Castle suggest that the fauna have evolved through local speciation, as well as by faunal migration from cold seeps (Pedersen *et al.*, 2010b; Kongsrud and Rapp, 2012).

Menez Gwen

4.3 A shallow vent system on the MAR

Menez Gwen (Figs. 4.41–4.44) is the northernmost known hydrothermal field in a series of fields that stretches north from the Oceanographer Fracture Zone to the Azores Archipelago. The Azores Plateau is a region of elevated bathymetry influenced by ridge–hot spot interactions for the past 85 million years (Lourenço *et al.*, 1998; Gente *et al.*, 2003) (Fig. 4.43).

Menez Gwen was discovered at 37°50′ N on a French-led cruise in 1994. The field is named after the Menez Gwen segment of the MAR, which extends 55 km in a S-SW to N-NW orientation. The hydrothermal field resides at water depths of 840–870 m on a circular, on-axis volcano that is about 15 km in diameter (Charlou *et al.*, 2000). The volcano is dissected by a 9 km long axial graben floored by fresh lava and volcanic breccia (Fouquet *et al.*, 1995; Ondreas *et al.*, 1997; Marques *et al.*, 2009).

Five major sites of hydrothermal activity have been documented at Menez Gwen: Atos 10, Cage Site, Marker 4, White Flames, and Woody (Marcon *et al.*, 2013). This region is dissected by several N-NE-striking fractures that are thought to serve as conduits for hydrothermal flow (Marcon *et al.*, 2013). Most other hydrothermal sites on the MAR are substantially deeper than Menez Gwen, making it unique in terms of the possible impacts of lower pressures both on boiling processes in the subsurface and on faunal community composition.

Figure 4.41 Small, isolated chimneys, <5 m in height, characterize structures in the Menez Gwen Field. They are predominantly composed of anhydrite, a sulfate mineral that forms when seawater is heated to more than 150 °C. This vent site and the Lucky Strike and Rainbow Fields are marine protected areas for conservation and scientific inquiry. They are managed by Portugal with reference to the Convention for the Protection of the North Mid-Atlantic Network of Marine Protected Areas.

Figure 4.42 A weakly venting anhydrite chimney sits atop a small circular mound composed of barite-rich sediments. In 1994, when the field was discovered, a temperature of 281 °C was the highest temperature measured using a probe on the French HOV *Nautile* (Charlou *et al.*, 2000). Field of view ~1 m.

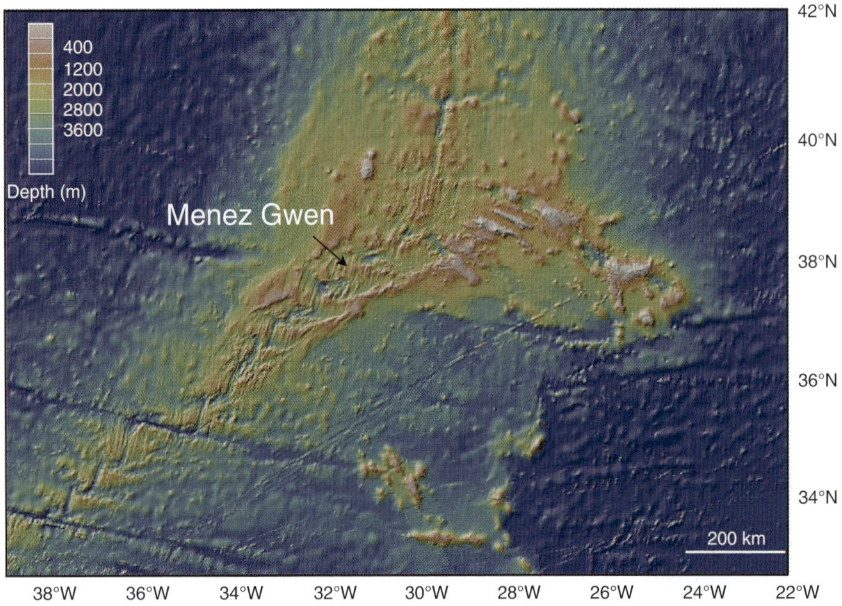

Figure 4.43 The Azores Plateau is formed from excess melt production associated with ridge–hot spot interactions that have been ongoing for ~85 million years (Gente *et al.*, 2003). The Menez Gwen hydrothermal site is located on the western side of this plateau along the S-SW–N-NW-oriented Menez Gwen segment at a water depth of 840–870 m.

Figure 4.44 Menez Gwen hydrothermal vents are at the summit of a young volcano. The vents are typically <5 m in height and are underlain by fresh pillow basalts. Their small size is believed to reflect their youth. Diffuse fluids vent from the small mounds. Field of view ~4 m.

The chimneys and zones of diffuse flow occur within the axial graben. In 2000, chimneys at this site appeared young, as evidenced by their small size and because they are predominantly composed of anhydrite, giving them a gray-white appearance (Figs. 4.42, 4.44 and 4.45). Clear vent fluids exiting from small chimneys had temperatures up to 281 °C (Charlou *et al.*, 2000). The anhydrite forms during mixing of iron-poor hydrothermal fluids and seawater at temperatures above 150 °C. Oxidation of sediments forms slag-like deposits along the mounds (Fig. 4.46).

Vent fluids are depleted in chlorinity (66–70% of normal seawater values), indicating that boiling, seawater mixing, and perhaps conductive cooling are occurring in the

Figure 4.45 Hydrothermal areas extend from 59 to 200 m^2 at Menez Gwen, with most areas covered by poorly lithified sediments. The small chimneys are friable, and are likely composed of sulfide minerals and anhydrite. Fluids exiting small fractures in some localities support mussel patches that prefer hard, basaltic substrates. Field of view ~2 m.

Figure 4.46 Oxidation of iron-bearing sulfide and biological material forms fine, several-centimeter-long stalagtite-like deposits that "drip" from the sides of variably altered metalliferous hydrothermal sediment deposits. Field of view ~1 m.

subsurface (Charlou *et al.*, 2000). The fluids are also highly depleted in iron (under 30 μmol/kg) compared to other vent sites along the MAR. These characteristics are believed to reflect low temperatures in the reaction zone and interaction with basaltic rocks that have already undergone extensive hydrothermal alteration.

The relatively shallow depth (850 m) of Menez Gwen compared to other sites along the MAR may play a large role in the faunal biodiversity and composition of this vent field. Of the more than 225 faunal species now recognized as inhabiting vent sites along the MAR, Menez Gwen hosts more than 80 species, the upper end of the biodiversity range for MAR vents (Kelley and Shank, 2010). Mid-Atlantic vent sites typically host between about 35 (for example Rainbow and Lost City – see Sections 4.5 and 4.6) and 80 species (Menez Gwen and Snake Pit). With the exception of crabs and shrimp (Fig. 4.47), which show a relatively even species diversity throughout Atlantic vent sites, copepod and amphipod species diversity is greatest at the shallower Menez Gwen and Lucky Strike vent sites. Sponge and anemone species are also more diverse at Menez Gwen than at other MAR hydrothermal fields.

Comparisons of the geological settings, habitat longevity, vent fluid chemistry, and faunal composition at well-known vent fields between the Vema Fracture Zone (11° N) and the Azores Plateau (38° N) reveal that the strongest compositional similarity exists between the two shallowmost sites, Menez Gwen and Lucky Strike (Desbruyères *et al.*, 2000, 2001).

As axial depth increases along the MAR, there is a continuum of community dominance, from *Rimicaris* vent shrimp to *Bathymodiolus* mussel-dominated assemblages. This continuum has an even stronger correlation with the

Figure 4.47 Deep water geryonid crabs at Menez Gwen reach up to 14.5 cm (carapace width) and are commonly observed on pillow lavas, and at active vent sites where they forage on mussel beds (*Bathymodiolus azoricus*). This crab species, *Chaceon affinis*, is found from Iceland to Cape Verde and is common at depths of 600–1000 m. Gastropods (*Protolira valvatoides*) and limpets (*Lepetodrilus atlanticus*) are common on the shells of mussels (Desbruyères *et al.*, 2006).

concentrations of metals than with either geographic or bathymetric depth zonation (Cravo *et al.*, 2008). Mussels prefer rocky substrate for colonization, in contrast to shrimp and gastropods, which prefer being in the vicinity of hydrothermal outflow (Riou *et al.*, 2010; Macron *et al.*, 2013).

It may be that symbiotic sulfur- and CH_4-oxidizing bacteria in the gills of the mussels provide up to 52% of their nutrition, with the rest supplied by particulate organic matter obtained during filter feeding. Their flexible feeding strategies allow them to tolerate a wide range of environmental conditions.

Lastly, despite the proximity of the Azores Triple Junction and an apparent absence of geographical discontinuity, the similarity between Menez Gwen and Lucky Strike can be considered low, suggesting that they are "faunal islands" that have distinct composition and habitat requirements.

Lucky Strike

4.4 Hot-spot influenced vent systems

The Lucky Strike Hydrothermal Field is located ~400 km south of the Azores on the Lucky Strike segment of the MAR. In 1992, the field was discovered serendipitously during the French–American Ridge Atlantic program (FAZAR) when a dredge came on deck containing sulfide material covered with vent fauna – hence the field's name (Figs. 4.48–4.50) (Langmuir *et al.*, 1992).

Hydrothermal activity at Lucky Strike (47°17′30″ N, 32°16′30″ W) is mainly focused in, and along, the sides of a young solidified lava lake about 200 m across, within the summit depression of Lucky Strike Seamount at ~1700 m depth (Figs. 4.49 and 4.50) (Humphris *et al.*, 2002). This is one of the largest axial volcanoes within the rift valley on the MAR. The seamount is underlain by a 3 to 4 km wide magma chamber at ~3.5 km depth beneath the seafloor; it extends nearly 7 km along the axis (Singh *et al.*, 2006). This magma chamber is thought to provide the heat that drives hydrothermal flow at this site, with recharge channels likely focused along steep faults along the axial valley walls.

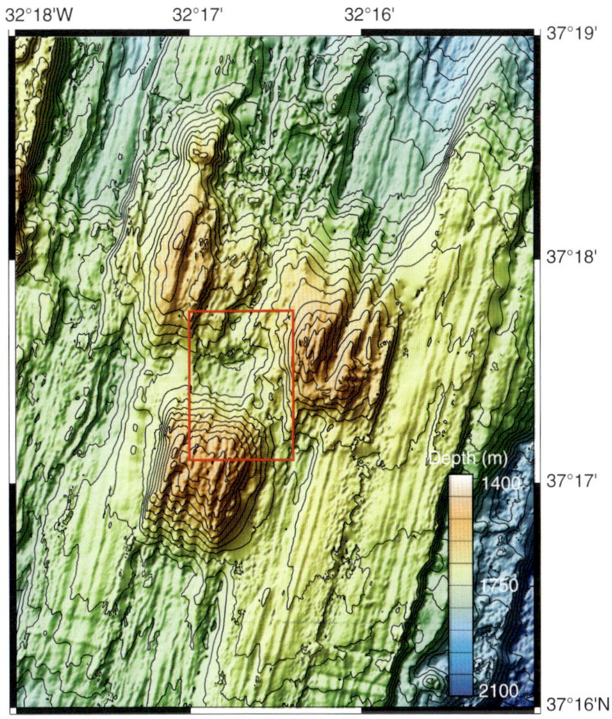

Figure 4.49 Bathymetry for the Lucky Strike Seamount with a contour interval of 20 m. The red box shows the area in Fig. 4.50.

Figure 4.48 Active sulfide mounds support dense mussel communities (up to 12 cm in length) and grazing shrimp, limpets, and crabs.

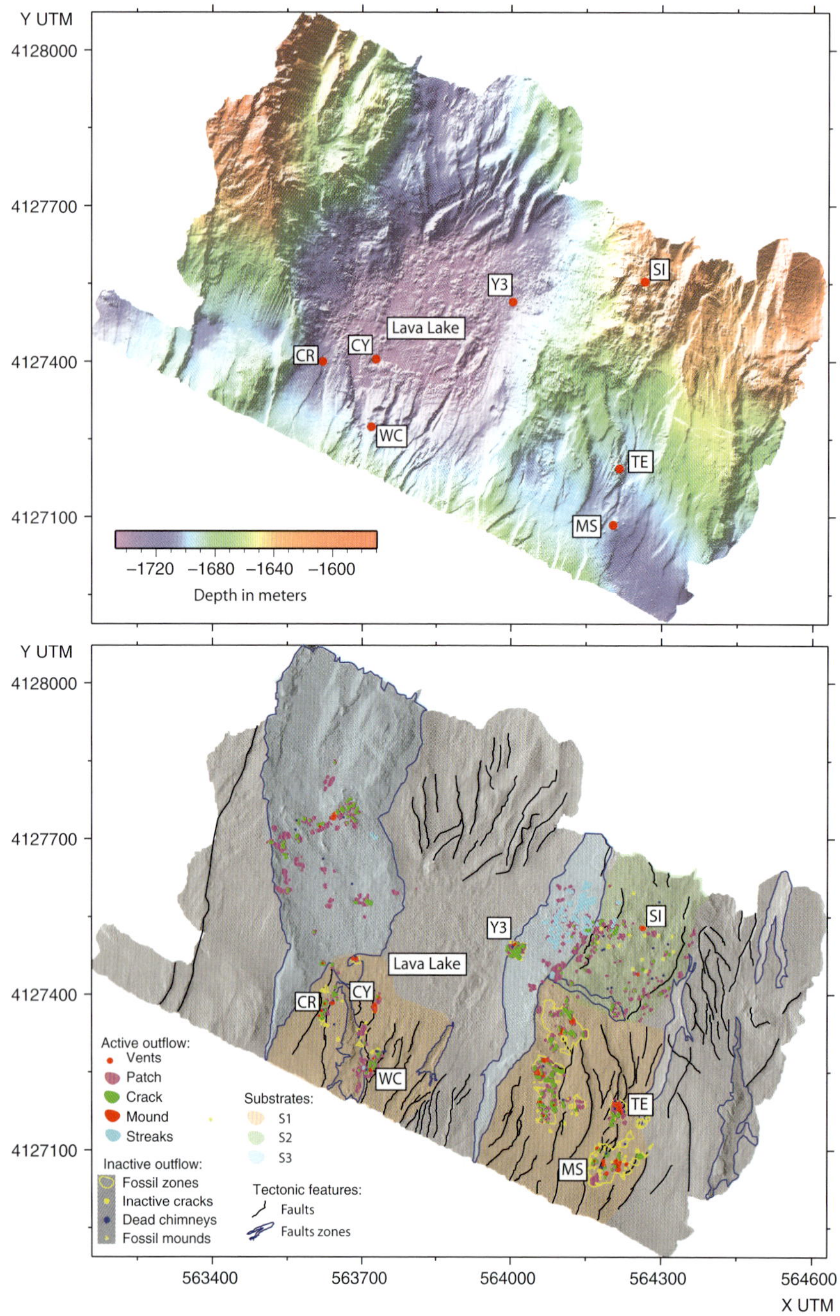

Figure 4.50 High-resolution bathymetric and geological map from Barreyre *et al.* (2012). The bathymetry is gridded at 0.5 m using a RESON Seabat 7125 multibeam echo sounder (Ondreas *et al.*, 2009). (Bottom) Locations of active and inactive hydrothermal deposits, as well as fault zones and underlying substrates (Barreyre *et al.*, 2012). S1, cap of hydrothermally cemented breccia; S2, laterally continuous, massive hydrothermal deposits forming rugged terrain; and S3, slopes of the major axial graben-bounding fault scarps showing a complex morphology formed by talus deposits. Also shown are hydrothermal sites: CR, Crystal; CY, Cypres; IS, Isabel; MS, Montsegur; SI, Sintra; TE, Tour Eiffel (a 19 m tall chimney); WC, White Castle; and Y3.

4.4.1 Hydrothermal venting

Active and inactive hydrothermal deposits are distributed over an area of ~1 km^2, where there are about 30 vent sites (Fig. 4.50). Numerous inward-dipping normal faults cut the area. These faults serve as outflow conduits for fluids (Ondreas *et al.*, 2009). Hydrothermal venting is located on extensive deposits of sulfide rubble and hydrothermally cemented hyaloclastite deposits both east and west of the lava lake (Humphris *et al.*, 2002; Ondreas *et al.*, 2009; Barreure *et al.*, 2012; Mittelstaedt *et al.*, 2012).

In the eastern zone, black smoker chimneys can be up to 20 m in height, and deposits have complex morphologies that include spires, flanges hosting underlying reflecting pools of hydrothermal fluid, and small structures atop diffusely venting mounds (Figs. 4.51–4.54). The surfaces of some active chimneys and mounds are covered in dense mussel communities (Figs. 4.48 and 4.52–4.56). The mounds can be up to 20 m in diameter and rise several meters above the surrounding seafloor (Barreyre *et al.*, 2012).

The most vigorously venting, high-temperature (<300 °C) chimneys are dominated by chalcopyrite and anhydrite, with minor iron and zinc sulfide minerals, as well as amorphous silica (Langmuir *et al.*, 1997; Rouxel *et al.*, 2004). Diffusely venting structures with metal-depleted fluids are predominantly composed of iron sulfide minerals and barite, while 180–220 °C fluids near the bases of spires are barite and zinc sulfide rich (Rouxel *et al.*, 2004). The basalts, chimneys, and fluids are 10–30 times enriched in elements such as barium, potassium, and strontium, reflecting the presence of the Azores hot spot 600 km to the north (Von Damm *et al.*, 1998; Charlou *et al.*, 2000). Carbon dioxide concentrations are elevated, reaching 39.9 mmol/kg, and isotopic compositions reflect a magmatic source (Charlou *et al.*, 2000).

Vent fluid compositions ostensibly reflect supercritical phase separation at depths greater than 1300 m below the seafloor (Von Damm *et al.*, 1998). Spatially, there may be different plumbing systems for the vents in the northeast and southeast portions of the field. In contrast, there seems

Figure 4.51 The top of a small mound is capped by a <0.5 m tall, fragile sulfide chimney venting particle-laden fluids. Typically, adjacent to such chimneys are mussel beds and bacterial mats, which can be seen in the background. Sulfide mounds provide niches for grazing limpets, shrimp, carnivorous crabs, and several species of deep-sea fish.

Figure 4.53 A ~20 cm tall, fragile chimney vents particle-poor fluids.

Figure 4.52 Mussel communities on small sulfide mounds flanked by sulfide debris and hydrothermal sediments at Lucky Strike. As shown on the mound in the background, *Bathymodiolus*, mussels completely obscure the outer surfaces of the deposits. Field of view ~6 m.

Figure 4.54 Mounds at Lucky Strike reach 20 m across and several meters in height. Diffuse flow supports filamentous white bacterial mats interspersed with mussels. Estimates indicate that 75–90% of the heat flux is due to diffuse venting at Lucky Strike, resulting in the generation of ~200–1000 MW (Barreyre *et al.*, 2012). Field of view ~3 m.

Figure 4.55 Mussel assemblages thrive on the edge of a linear sulfide mound in the southern part of the field. Tufts of filamentous bacteria (mid-ground) are interspersed amongst mussels, potentially indicating higher levels of diffuse fluid flux. Field of view ~1 m.

to be just a single source for vents located around the summit depression and lava lake, which are separated by a distance of up to ~450 m (Humphris *et al.*, 2002).

In March 2002, the area was thought to have been perturbed by a diking event, as evidenced by a swarm of 147 earthquakes during a two-day period (Dziak *et al.*, 2004).

Figure 4.56 Wispy thick growths of filamentous bacteria cover oxidized sulfide debris and mussels in zones of diffuse flow. Bythograeid crabs are abundant within mussels and on sulfides deposits (Desbruyères *et al.*, 1994). Mussels may reach >50 mm in size.

Tremor-like characteristics for the swarm, which occurred simultaneously over a 50 km length of the segment, appear to be magmatic in origin (Dziak *et al.*, 2004).

4.4.2 Animal communities

Lucky Strike may host the highest species diversity (more than 110 faunal species) of any vent field on slow-spreading ridges. The faunal biomass at Lucky Strike is dominated by thriving mussel communities (*Bathymodiolus azoricus*), with individuals hosting the red commensal polynoid polychaete *Branchipolynoe seepensis* (Figs 4.52–4.57) (Van Dover *et al.*, 1996; Van Dover, 1999; Van Dover and Trask, 2001).

The mussels form assemblages of thick carpets and hummocks on the sides of active black smokers, in diffuse vent sites, and along small cracks in the seafloor. The dense mussel beds on active sulfide deposits are one of the most notable features at Lucky Strike. Larger mussels on the chimneys are closer to zones of outflow, perhaps reflecting their tolerance to harsher conditions than smaller mussels (Cuvelier *et al.*, 2009). This is the only site in the Atlantic where such mussel abundance in relation to actively venting "chimney" habitats occurs. The mussels host both sulfide-oxidizing and CH_4-oxidizing endosymbiotic bacteria. At least four species of alvinocarid shrimp, a variety of gastropod, and several amphipod species also thrive within the warm fluids (Fig. 4.57). The high species diversity can, in part, be attributed to the higher relative sampling effort at this site over the past decade and, in particular, to the large numbers of described copepod, amphipod, and fish species.

Mirocaris shrimp represent the second highest biomass-dominant fauna, scavenging and hunting among the surfaces of the mussels. Alvinocaridid shrimp associated with sulfide chimneys may consume free-living bacteria

Figure 4.57 High abundances of shrimp and mussels thrive near the underside of a sulfide flange hosting a shallow reflecting pool of high-temperature hydrothermal fluid. At least two species of shrimp, *Chorocaris chacei* and *Mirocaris fortunata*, quickly amble over the substrate, feeding on organic "debris" and microbes present on both large mussels (left) and small (perhaps recently colonized) mussels (lower right), as well as on the surrounding sulfide surfaces. Field of view ~ 0.5 m.

in the hydrothermal fluids or ingest other small invertebrates (Cuvelier *et al.*, 2009).

Relatively rare fauna peripheral to the main vent areas include serpulid polychaetes, stalked barnacles, and even scleractinian hard corals. New, when Lucky Strike was discovered, was the presence of *Echinus* urchins living in clusters of up to 15 individuals in above-ambient temperature water at the immediate periphery of individual vents. Echinoderms are relatively rare at hydrothermal vents and no other echinoderms have been observed at this site. Lucky Strike diffuse flow sites also host numerous mats of white filamentous bacteria (Fig. 4.56).

Rainbow

4.5 Hybrid gabbroic- and ultramafic-hosted vents

The Rainbow Hydrothermal Field (36°13.80′ N, 33°54.12′ W) was discovered in 1997 during the French FLORES diving program, based on exploration of CH_4 and manganese anomalies on the AMAR segment of the MAR (Charlou *et al.*, 2002; Douville *et al.*, 2002). The black smokers are located on the western flank of the Rainbow Ridge, on a fault-bounded tilted block of ultramafic rocks and lesser basaltic material (Figs. 4.59–4.61). The field is bounded on the west by a 25 m tall fault scarp hosting sulfide mineralization (Charlou *et al.*, 2010). Here, a stockwork system hosts sulfide-bearing veins, believed to represent a paleo-conduit system for venting.

Ramparts of sulfide debris underlie the inactive and active chimneys. Basalt outcrops occur 2 km to the east, near the summit of Rainbow Ridge, in the axial floor west of the ridge, and as an isolated occurrence in the northern portion of the field. Eleven major vent sites are distributed over a 100 m long, 200 m wide area at 2270 m depth, with many sites supporting dense colonies of mussels and shrimp (Figs. 4.58 and 4.61–4.65) (Charlou *et al.*, 2002; Seyfried *et al.*, 2004, 2011).

Hydrothermal venting at Rainbow produces a significant plume enriched in CH_4, manganese, and particles rising ~200 m above the seafloor (Charlou *et al.*, 2002; German *et al.*, 2010). Heat flux calculations based on measurements of hydrothermal plumes issuing from high-temperature smokers and diffuse flow sites have yielded a value of about 0.5 gigawatts (German *et al.*, 2010).

Figure 4.58 Thousands of *Rimicaris exoculata* shrimp congregate in areas of turbulent mixing between vent fluid and seawater. Field of view ~2 m.

4.5.1 Carbon dioxide-, hydrogen-, and methane-enriched fluids

Vent fluids issuing from beehive-like structures are highly acidic, hosting elevated chlorinities (750 mmol/kg) with respect to seawater (547 mmol/kg) and temperatures up to 370 °C (Fig. 4.64) (Charlou *et al.*, 2002; Douville *et al.*, 2002; Seyfried *et al.*, 2011). These fluids are lower in silica than other high-temperature black smoker systems on the MAR. Brine generation from phase separation processes at depth, high temperatures, and high acidity promote enrichment in metals (Douville *et al.*, 2002). The fluids are also highly enriched in dissolved CO_2 and CH_4, and they have extraordinarily high concentrations of H_2 (13–16 mmol/kg) (Charlou *et al.*, 2002; Seyfried *et al.*, 2011). The high CO_2 concentrations and elevated temperatures likely reflect the presence of a high-temperature gabbroic heat source in the subsurface (Marques *et al.*, 2007), while the high CH_4 and H_2 concentrations are now recognized as hallmarks of venting

Figure 4.59 Small, bulbous sulfide pillars and oxidized extinct structures are sparsely visited by shrimp. Field of view ~2.5 m.

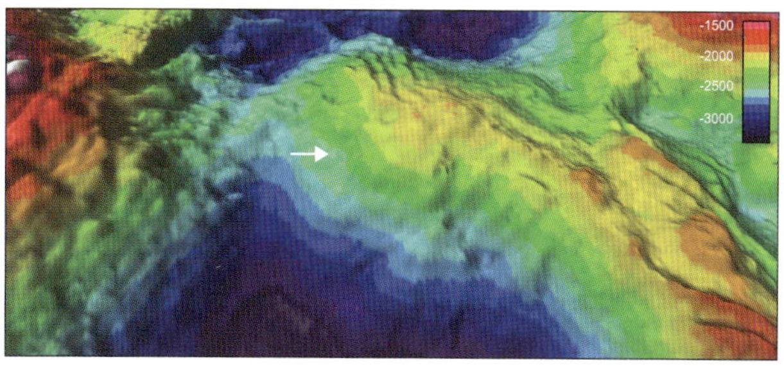

Figure 4.60 Perspective of Rainbow Ridge looking north along the trend of the axial ridge. The ridge separates the two non-transform discontinuities to either side of the ridge. The red arrow indicates the location of the Rainbow Hydrothermal Field. The Rainbow vents are located over a small ~100 × 300 m area. Vertical exaggeration is 2×.

Figure 4.61 Smokers in Rainbow Field reach temperatures of 370 °C (Seyfried *et al.*, 2011) and commonly form ribbed, beehive deposits. High-temperature chimneys are rich in iron-zinc sulfides, chalcopyrite and pyrrhotite. Field of view ~1.5 m.

Figure 4.62 A small forest of active and extinct chimneys provide foraging grounds for shrimp at Rainbow. Temperature measurements show that they forage in a limited range of 5–8 °C, with more elevated temperatures measured in the densest portions of the swarms where the most intense mixing occurs. The shrimp host chemosynthetic microbial epibionts in their branchial cavities. Field of view ~3 m.

systems impacted by fluid interaction with ultramafic rocks (Charlou *et al.*, 2002; Allen and Seyfried, 2004; Kelley *et al.*, 2005; Proskurowski *et al.*, 2006, 2008; Konn *et al.*, 2009).

Rainbow vent fluids have the highest concentrations of iron, copper, cobalt, zinc, manganese, and nickel documented for MAR vent sites (Douville *et al.*, 2002; Marques *et al.*, 2007). This reflects the high concentrations of copper, zinc, and cobalt within the massive sulfide deposits (Marques *et al.*, 2007). Sulfide debris, sediments, and fauna at Rainbow are also impacted by the exceedingly high iron concentrations in the vent fluids (more than 100 times those of other black smoker systems). The high iron results from the production of H^+ in low pH conditions during serpentinization reactions, which causes dissolution of magnetite (Fe_3O_4) in the ultramafic rocks (Douville *et al.*, 2002). Such systems are of particular interest because of the hypothesis that serpentinization reactions in ultramafic rocks may produce the initial building blocks of life (Holm and Charlou, 2001; Martin *et al.*, 2008; Konn *et al.*, 2009).

Figure 4.63 Clusters of shrimp, common in high-energy mixing zones, may derive their energy from iron- and possibly hydrogen-oxidation. Unlike most black smoker systems, end-member vent fluids at Rainbow do not have zero magnesium values (they are as high as 2 mmol/kg Mg). Exiting fluids also contain chlorinities that are higher than seawater (up to 759 mmol/kg, seawater chlorinity is 547 mmol/kg). *In situ* measurements of pH yield values of 4.97 ± 0.7, compared to pH values of 2.5 calculated at 25 °C (Seyfried *et al.*, 2011). Field of view ~0.5 m.

Figure 4.64 Dense aggregates of mussels colonize areas of diffuse flow issuing from a talus rampart of hydrothermal sulfide debris and oxidized sediment. Field of view ~1 m.

4.5.2 Animal communities

More than 32 vent-endemic species inhabit Rainbow, including several that were new to the MAR when the site was discovered. These include the zoarcid fish species *Pachycara sp. n.*, the mussel *Bathymodiolus azoricus* and the polychaete *Bathypolynoe seepensis* that dominate the fauna in active vent areas (Figs. 4.63 and 4.64). These community members, including the shrimp *Mirocaris fortunata* and the polychaete *Amathys lutzi*, commonly reach high abundances (Figs. 4.58 and 4.63). Dense polychaete aggregations are formed by *Spiochaetopterus* sp., mostly in the young eastern sites. The MAR vent-endemic white crabs *Segonzacia mesatlantica* are active on the walls of active smokers.

Comparisons of faunal community structure along the MAR, in particular the mussel beds at Lucky Strike, Snake Pit, and Logatchev (Van Dover and Doerries, 2005), provide growing evidence for a distinct northern, shallow (less than 2500 m) fauna and a southern, deep (greater than 3000 m) fauna on the northern MAR. The likely explanations of biogeographic controls separating potential shallow and deep faunal provinces have focused on depth (Van Dover, 1995), particularly in relation to larval physiological tolerances and life history differences among observed taxa.

Figure 4.65 Mussels (*Bathymodiolus azoricus*) and a bythograeid crab are encrusted with iron oxide-rich sediments. Fluids issuing from the Rainbow vents are highly enriched in both iron and manganese, as well as copper, zinc, and cadmium. Field of view ~0.3 m.

Rainbow is part of a recognized compositional continuum of alvinocarid shrimp and *Bathymodiolus* mussel-dominated assemblages along the MAR. This species distribution cannot be explained by bathymetric zonation or geographic distance alone and may be associated with species-specific tolerances to the metallic content of the end-member fluids. At Rainbow, both *Rimicaris exoculata* and *Microcaris fortunata* shrimp host high metal concentrations, with mercury accumulated in their antennae and digestive glands (Kadar *et al.*, 2006). Thus, differences in faunal composition and distribution at vent sites at different depths – for example Menez Gwen (850 m), Lucky Strike (1710 m), and Rainbow (2300 m) – may reflect differences in vent fluid toxicity and suspended mineral particles. This is particularly true for the ultramafic-hosted Rainbow area in which fluids have high metal concentrations (Desbruyères *et al.*, 2000).

Figure 4.66 Sparse, but dense hummocks of *Bathymodiolus* mussels coated with iron oxide-encrusted filamentous bacteria are striking features within Rainbow. These mussels typically live in areas venting at temperatures from 6 °C to 30 °C. They reach up to 12 cm in length and host chemosynthetic microbes in their gills. Field of view ~0.2 m.

Lost City

4.6 A new kind of vent field

The discovery of the Lost City Hydrothermal Field forever changed our views about submarine hydrothermal systems and where and how life can exist in the oceans (Kelley *et al.*, 2001b,c, 2005; Brazelton *et al.*, 2006, 2010, 2011; Kelley and Shank, 2010). Results from Lost City have led to new models on the origin of life on Earth, and are serving as guides in the search for life on other planets (Martin *et al.*, 2008; Russell *et al.*, 2010).

The field (located at 37°07′ N, 42°07′ W) was discovered serendipitously in 2000 during a cruise to investigate the Atlantis Massif at the intersection of the MAR and the Atlantis Transform Fault (Fig. 4.67). Here, majestic limestone towers rise 60 m above the seafloor (Fig. 4.68). Hydrothermal chemistry is driven by serpentinization reactions in the underlying mantle rocks that produce abiogenic hydrocarbons and 90 °C, pH 11 fluids that support novel microbial communities. Unlike black smoker systems, the field is hosted not within the axial valley, but 15 km west of the rift valley.

4.6.1 Lost City geology

The Atlantis Massif rises nearly 4 km above the surrounding seafloor over a horizontal distance of ~20 km (Fig. 4.67). Although similar in size to Mt. Rainier in the Cascade Range, the Atlantis Massif is not formed by volcanic activity, but instead by extreme crustal extension and long-lived faulting and uplift associated with a major detachment fault (see Chapters 3 and 8). In concert, these processes have resulted in the stripping-off of shallow crustal volcanic material and exposure of ultramafic rocks on the seafloor that were once deep beneath the axis of the MAR (Fig. 4.69) (Boschi *et al.*, 2006; Karson *et al.*, 2006). Based on magnetic surveys, the mantle and shallow crustal rocks that make up this portion of the mountain are roughly 1 to 2 million years old (Blackman *et al.*, 2002).

Bounded to the south by the 150 km long Atlantis Transform Fault, the dissected southern face of the massif provides an unprecedented view into the internal structure of the footwall of this oceanic core complex. Here, the massive, embayed cliffs are composed predominantly of variably deformed serpentinites and lesser, small gabbroic intrusions. Long-lived faulting, extreme crustal extension, and exposure and alteration of shallow mantle material is characteristic of submarine mountains such as the Atlantis Massif, and the processes involved in their construction are probably critical to formation of Lost City-like hydrothermal systems (Kelley *et al.*, 2005).

At 750 m depth, the Lost City Field resides on a fault-bounded promontory on the south face of the massif (Figs. 4.67, 4.69 and 4.70) (Kelley *et al.*, 2005). Lost City is underlain by a 100 m thick, mylonitic, low-angle detachment shear zone composed predominantly of altered ultramafic rocks with lesser gabbroic material (Fig. 4.71) (Kelley *et al.*, 2001b,c, 2005; Früh-Green *et al.*, 2003; Boschi *et al.*, 2006, 2008; Karson *et al.*, 2006).

The highly deformed rocks grade downward into more massive, less deformed serpentinites, peridotites, and gabbros. The shear zone is overlain by a 1–3 m thick pelagic limestone cap rock that is underlain by a 1–3 m thick sequence of sedimentary breccias (Figs. 4.69–4.72). The field is bounded to the east and west by significant embayments formed through faulting and landslides (Fig. 4.70). The exhumed mafic to ultramafic rocks forming the skin of the south face of the massif have undergone intense interaction with circulating hydrothermal fluids (Delacour *et al.*, 2008a, b). This flow has resulted in remobilization of chemicals during focused upflow (Boschi *et al.*, 2006) leading to pervasive (70–100%) alteration and hydration of the mantle rocks and the formation of serpentinites.

Tectonic and hydrothermal processes at Lost City are directly linked. The field is located on a down-dropped

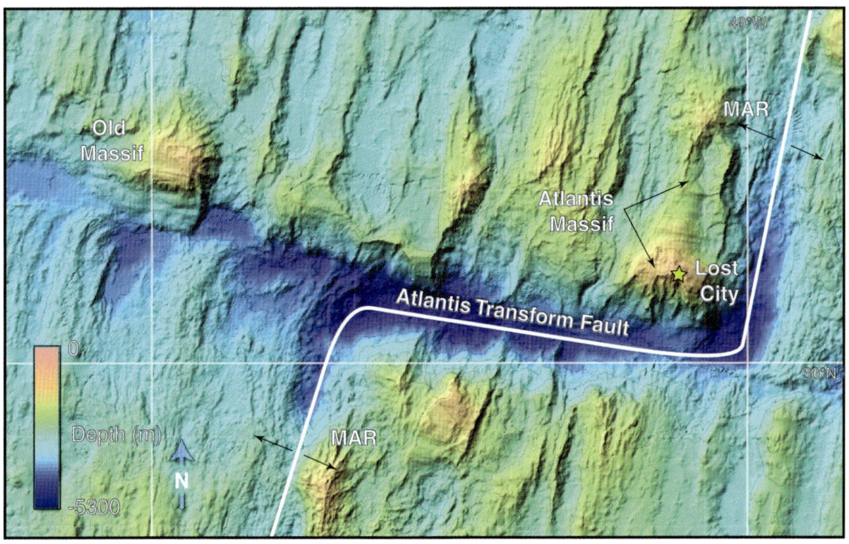

Figure 4.67 The MAR near 30° N hosts numerous massifs, formed through detachment faulting, which may last for >2 million years. Extreme topographic relief is a common feature of these systems. The Atlantis Masssif is bounded by a >5000 m deep nodal basin to the east (dark blue) and the summit is at ~700 m below sea level.

Figure 4.68 (opposite) The actively venting Nature Tower, on the east side of Lost City, rises 30 m above the surrounding seafloor. Significant limestone talus at the base of the structure attests to the collapse of spires and rebuilding as part of the evolution of these complex chimneys. The ROV *Hercules* and support vehicle *Argus* are embedded in the photo for scale. Marker at vent summit is 3 m tall.

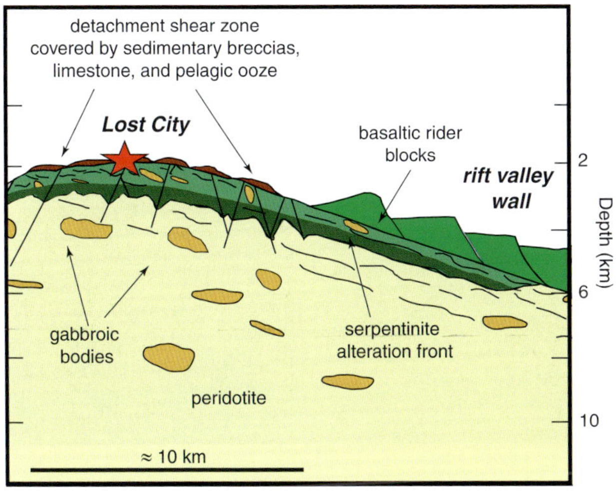

Figure 4.69 The southern central portion of the Atlantis Massif is composed of variably deformed serpentinites with lesser, small gabbroic bodies (after Boschi *et al.*, 2006). The eastward-dipping detachment fault is defined by a ~100 m thick mylonite zone. Long-term focused flow of fluids into the detachment zone and in discrete shear zones has resulted in pervasive alteration of the gabbros and serpentinites (dark and light green). Topographic forcing of fluids, active serpentinization at depth, and normal faulting control the position of the current Lost City Field.

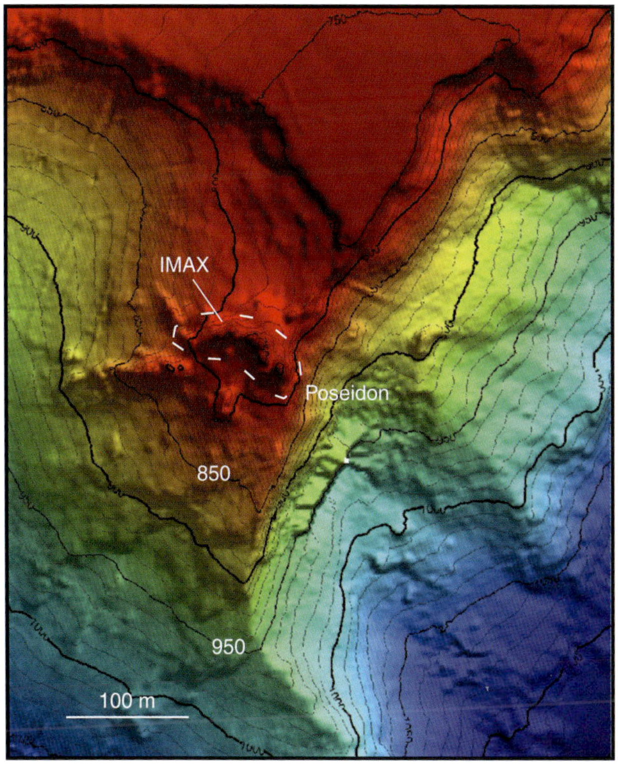

Figure 4.70 The south central summit of the Atlantis Massif is surprisingly flat-topped. The Lost City Field is located on a down-dropped block on a sharp promontory that is bounded by steep normal faults. The faults contribute to formation of large embayments that characterize the south face of the massif. The massive 60 m tall hydrothermal edifice called Poseidon dominates the core of the field.

promontory, about 70 m from the summit of the massif (Fig. 4.72). Venting within the field is focused along discrete normal faults and along anastomosing shear zones that act as conduits for seeps on the east side of the field (Fig. 4.73). An array of fissures less than a meter wide cut the cap rock along the southern edge of the summit of the massif, and in some places fissures are filled with delicate hydrothermal deposits (Fig. 4.74). The heat source required to drive hydrothermal flow is likely derived from cooling of the crustal rocks that make up the massif, with lesser contributions from exothermic reactions at depth in the underlying mantle rocks (Kelley *et al.*, 2001b,c, 2005; Früh-Green *et al.*, 2003).

The largest and most hydrothermally active carbonate towers define an east–west trending lineament (Fig. 4.70). Subvertical faults channel fluids to the largest structures (Kelley *et al.*, 2005). The 60 m tall, actively venting monolith called Poseidon dominates the core of the field (Figs. 4.75–4.79). This composite structure has four large columns that coalesce at their base to form a massive east–west trending edifice more than 100 m across (Figs. 4.70 and 4.75). The youngest and most active areas of venting are bright white and are composed of highly friable and porous carbonate, while the oldest deposits are extremely well lithified, gray-brown in color, and have a much smoother outer surface (making them difficult to sample). Parasitic chimneys, resembling inverted stalactites, are particularly abundant on the northwest face of Poseidon, and form actively venting flanges, or ledges, which protrude several meters from the main trunk. The largest of these structures, called the IMAX flange, stands three stories tall and vents clear 55 °C fluids (Figs. 4.77–4.79). The active chimneys are composed of aragonite and calcite ($CaCO_3$), with lesser amounts of brucite [$Mg(OH)_2$] (Figs. 4.77–4.83) (Kelley *et al.*, 2001b,c, 2005; Ludwig *et al.*, 2006).

4.6.2 An ancient field

Lost City is perhaps the longest-lived venting system currently known in the global oceans. The longevity of venting at Lost City may have important implications for ecological succession of microorganisms within the serpentinite-hosted vent environments and larger vent fauna as well.

Results of age dating on more than 90 carbonate samples, using techniques that were originally developed for dating

Figure 4.71 The summit of the Atlantis Massif above Lost City is capped by a 1–3 m thick, massive white pelagic limestone deposit. This deposit is underlain by a 1–3 m thick sedimentary breccia. Blocks of limestone calve off large fissure systems that cut the limestone cap rock. The capping sequence is thought to have been important during formation of the field, acting as a trap for heat and leading to enhanced hydrothermal flow.

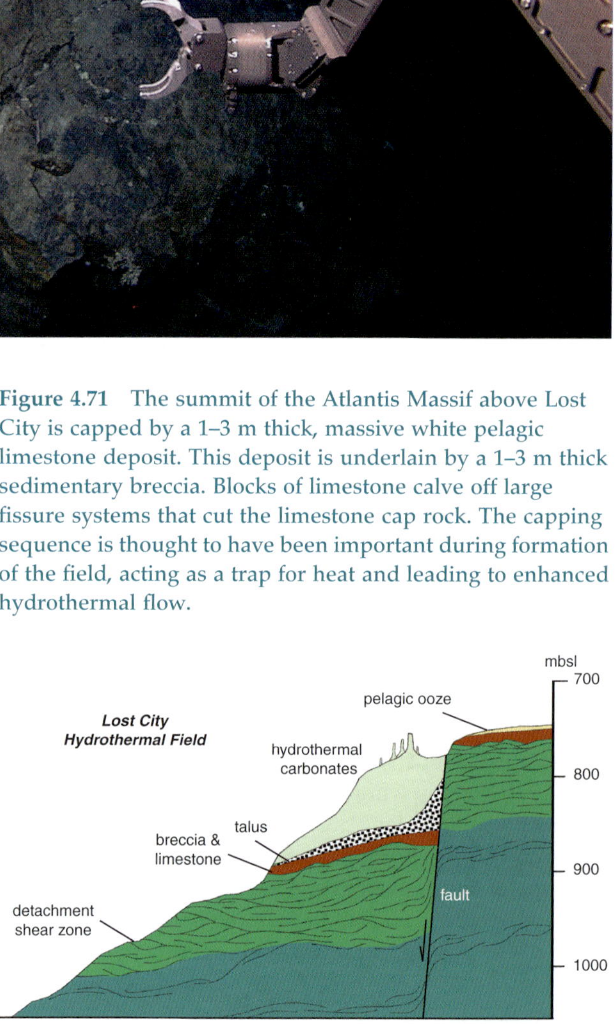

Figure 4.72 The limestone towers and talus deposits that define Lost City are underlain by a carbonate cap rock and sedimentary breccia sequence that abruptly transitions into a 100 m thick shear zone, representing the detachment fault. Hydrothermal venting is focused along a steeply dipping fault system that has resulted in formation of the down-dropped promontory on which Lost City rests.

of corals (U–Th isotopes), reveal that Lost City has been active for more than 150,000 years (Ludwig *et al.*, 2011). This is likely a minimum age, given that sampling has been biased toward the younger samples. Far away from the influence of volcanism, this system is free from perturbations associated with diking–eruptive events, and is most heavily influenced by seismic activity. It is likely that earthquakes and active faulting associated with the Atlantis Massif and Transform Fault, coupled with volumetric expansion of the underlying serpentinized host rocks, play major roles in sustaining permeable pathways through the underlying rocks that feed hydrothermal activity at this site.

4.6.3 Alkaline fluids rich in hydrocarbons

In stark contrast to the high-temperature, acidic fluids in black smoker systems, the carbonate chimneys at Lost City vent 50–90 °C, high-pH (9–11) fluids that are particle free (Figs. 4.81 and 4.83). Beneath the Lost City system, serpentinization reactions in the subsurface produce calcium-enriched, pH 9–11 fluids. Upon mixing with seawater, carbonate precipitates out, driven by the reaction:

$$Ca^{2+} + HCO^{3-} + OH^{-} \rightarrow CaCO_3(\text{solid}) + H_2O$$

(Ludwig *et al.*, 2006).

Petrographic examination of the youngest carbonate material, coupled with direct imaging of filamentous bacteria that are bathed in the warm fluid, suggests that the carbonate precipitates onto the bacteria, progressively "fossilizing" them over time (Figs. 4.92 and 4.93).

Although the concentrations of metals and particulates in the vent fluids are very low (Fig. 4.81), they are significantly enriched in CH_4 (1–2 mmol/kg) and H_2 (0.5–14.4 mmol/kg) and have elevated concentrations of low molecular weight hydrocarbons relative to seawater (Proskurowski *et al.*, 2008; Lang *et al.*, 2010).

Figure 4.73 (Top) The steep eastern walls of Lost City are marked by zones of past and current seepage of hydrothermal fluids leaking out of the anastomosing shear zone. Here, hydrothermal flow results in the formation of myriad deposits that include massive shingle-like formations, towers several meters tall, and delicate structures that are reminiscent of upturned hands. These diffuse seeps are in sharp contrast to black smoker systems dominated by distinct chimney deposits resulting from focused flow.

Figure 4.74 (Bottom) The southern summit of the Atlantis Massif is marked by a network of fractures and fissures less than a meter wide that cut the carbonate cap rock. Some of these fissures are sites of past and active hydrothermal flow. Significant mass wasting along the steep embayments, coupled with this fracture network, indicate that hydrothermal venting is most likely progressing northward at the summit of the massif. Two extinct hydrothermal fields have been documented to the west and at deeper water depths than the current position of the field, respectively.

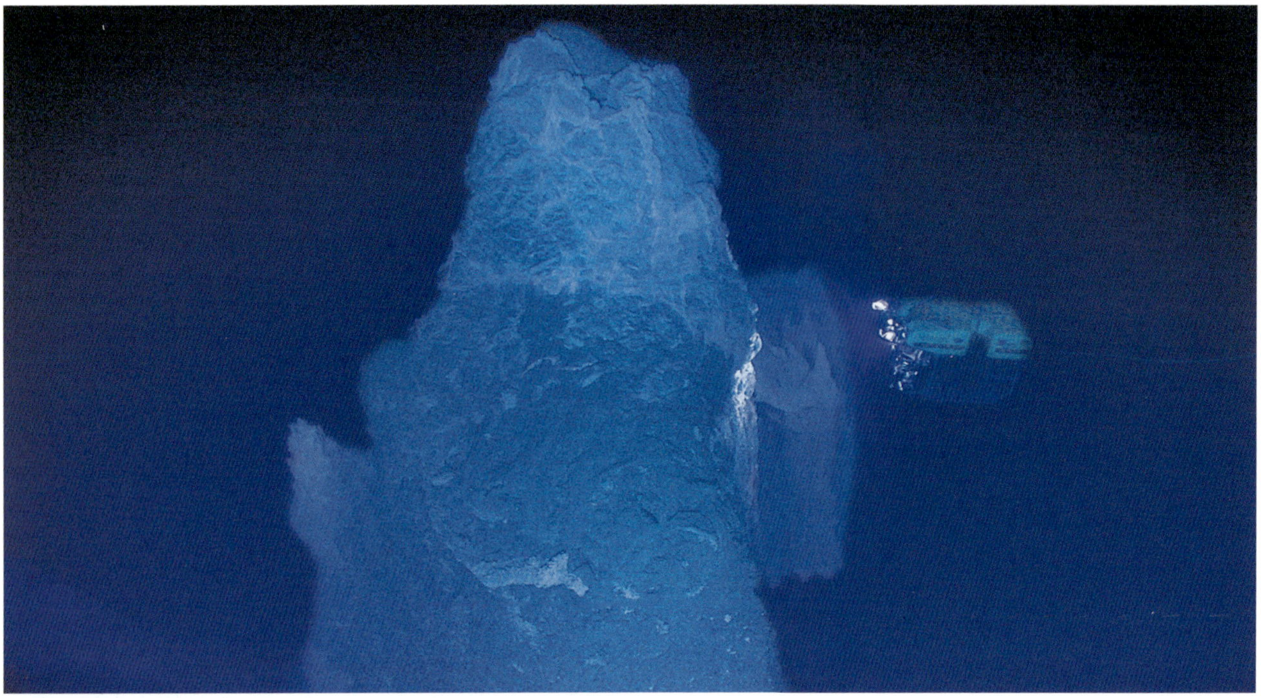

Figure 4.75 (Top) The ROV *Hercules* investigates the summit of the massive edifice "Poseidon," which rises 60 m above the surrounding seafloor. This complex structure is composed of four venting towers that have coalesced over the >150,000 years of hydrothermal activity at this site. The most active sites are bright white, while inactive sites are gray to brown with smoother relief. Venting from the main trunk of Poseidon resulted in growth of parasitic chimneys (e.g., left and below ROV *Hercules*) three stories tall.

Figure 4.76 (Bottom) The highest-temperature vent at Poseidon is a small "christmas tree"-like structure, ~2 m in height. The fluids are enriched in H_2, CH_4, and other low molecular weight hydrocarbons. A complex fracturing history of the walls of Poseidon has resulted in formation of cross-cutting veins that mark multiple hydrothermal flow events out of the trunk of the edifice. Wreckfish are commonly observed within the field, known to follow *Alvin* for hours during seafloor investigations.

Figure 4.77 The three-story-tall IMAX flange protrudes from the north face of the Poseidon edifice. White veins in the massive walls mark sites of active and recently active venting. Sustained fluid flow through fractures in the wall of Poseidon at the IMAX site has resulted in growth of this majestic structure.

Figure 4.78 (Top) The base of the IMAX flange hosts numerous ledges with beautiful, delicate, and fragile limestone deposits, which are home to billions of novel microorganisms. The flanges reach several meters across.

Figure 4.79 (Bottom) Warm (50 °C), buoyant fluids flow out of the walls of Poseidon and are trapped under the concave roofs of the flanges. The resultant reflective pools form "upside-down" waterfalls, with warm fluids flowing up over the lips of the flanges. This flange is located underneath the stalagmite-like structure in the left side of image 4.78.

Figure 4.80 Photomosaic of a 13 m tall carbonate chimney called Ryan. Long-term seepage of fluids from steep cliffs bounding the eastern side of the Lost City Field has resulted in beautiful arrays of narrow pinnacles that reach many tens of meters in height. Collapse and regrowth of chimneys and spires forms extensive carbonate talus deposits beneath the steep scarps.

Figure 4.81 Diffusely venting fluids (90 °C) flow out of a small chimney growing from the side of the Poseidon edifice. This chimney was sampled using the ROV *Hercules* manipulator in 2005. This was the highest temperature chimney in the field. Oxygen isotopic measurements on carbonate veins indicate fluid temperatures of up to ~150 °C in the underlying basement rocks.

Figure 4.82 (Left) Delicate strands of carbonate form the top of a thicker deposit growing from a fissure at the summit of the Atlantis Massif. This type of deposit typifies the very early stages of chimney development at Lost City. The center deposit is ~6 cm in height. (Right) Photomicrograph of a paper-thin wafer (30 μm) of a young carbonate chimney shows a highly porous, anastomosing network of carbonate precipitates reminiscent of filamentous bacteria that grow on the outside chimney walls. The outer walls are awash in hydrothermal fluid–seawater mixtures. The delicate, fine carbonate fingers may represent "fossilization" of filamentous bacteria during nucleation of aragonite crystals on the strands.

The extreme enrichment of H_2 with lesser CH_4, and relative enrichment of hydrocarbons, is a hallmark of fluid chemistries formed through serpentinization reactions at depth, where seawater in contact with olivine oxidizes iron (Neal and Stranger, 1983; McCollom and Seewald, 2001; Proskurowski *et al.*, 2008; McCollom and Bach, 2009).

Detailed chemical and isotopic analyses of hydrothermal fluids at Lost City provide compelling evidence that low molecular weight hydrocarbons are formed abiogenically during water–rock reactions in the underlying basement rocks (Proskurowski *et al.*, 2006; Lang *et al.*, 2010). Significant attention has been given recently to abiogenic versus biogenic

Figure 4.83 The outsides of the active chimneys at Lost City are covered in dense strands of filamentous bacteria that thrive in the mixing zones of high-pH, CH_4- and H_2-rich hydrothermal fluids and seawater. The carbonate interstices shrouded by these filamentous microbes serve as microhabitats for numerous species of small gastropods, polychaetes, and amphipods.

origins for CH_4 and low molecular weight hydrocarbons, in part because of the possible role they have played in development of life within vent systems and also because of implications for the search for life on other planets (e.g., Kelley, 1996; Kelley and Früh-Green, 1999; Holm *et al.*, 2001; Foustoukos and Seyfried, 2004; McCollom and Seewald, 2006; Lang *et al.*, 2010; Holm and Baltscheffsky, 2011). The purest vent fluids contain exceptionally little CO_2. It is thought that the mantle CO_2 has been removed prior to venting, by near-quantitative reduction of mantle CO_2 to hydrocarbons (Proskurowski *et al.*, 2008; Lang *et al.*, 2012; Müntener, 2010).

4.6.4 Life in high-pH systems

Microbial life within the Lost City system is tightly coupled to chimney evolution: it is one of the few vent fields on the MAR and global MOR system where microbiological investigations have been conducted at numerous venting sites that span the entire range of venting characteristics within the field (Schrenk *et al.*, 2004; Brazelton *et al.*, 2006, 2010).

The significant difference in fluid chemistries at Lost City (pH 9–11, trace CO_2 and metals), as compared to black smoker systems (pH below 2–5, high CO_2 and metals), has had a profound impact on the microbiological communities; the organisms have had to develop metabolisms that do not involve chemosynthetic processes necessarily based on CO_2. Here, the active structures are awash in buoyantly rising mixtures of warm, CH_4–H_2-rich vent fluid and cooler seawater, resulting in diverse environmental conditions and formation of a variety of biotopes within the carbonate and serpentinite rocks.

Microbial communities within this ultramafic-influenced environment benefit from having nearly twice the metabolic energy available than those in black smoker systems (McCollom, 2007). The high porosity (Fig. 4.82), warm and nutrient-rich environment within the chimney walls promotes the development of thick mucilaginous biofilms on the carbonate surfaces (Schrenk *et al.*, 2004; Brazelton *et al.*, 2006, 2010; Brazelton and Baross, 2009).

Microscopic analyses of areas within the walls where porosity is greater than 40% show tens of micrometer-thick microbial biofilms composed of irregular coccoid cells with metabolisms that likely involve anaerobic CH_4 cycling (Schrenk *et al.*, 2004; Brazelton *et al.*, 2006). In the highest temperature, oxygen-absent zones within the interiors of the chimneys (above 90 °C), Archaea show a surprisingly low diversity based on their 16S rRNA gene sequences and are dominated by a single group of organisms related to *Methanosarcinales* (Schrenk *et al.*, 2004; Brazelton *et al.*, 2006).

In chimneys where microbial sulfate reduction is active, up to half of the microbial biomass is synthesized from mantle carbon (Lang *et al.*, 2012). Other CH_4-consuming organisms are present in sites that are cool, inactive, or weakly venting (Brazelton *et al.*, 2006). They are also present in cool (<40 °C) carbonate vein environments that cut the serpentinite bedrock. The highest concentrations of microorganisms occur on the outside of diffusely venting chimneys, where they form white to light gray filamentous strands several centimeters long (Fig. 4.83). These venting areas are high-energy environments because of the mixing of oxygen- and sulfate-rich cool seawater, and high-temperature hydrothermal fluids enriched in CH_4 and H_2. In contrast to the results from archaeal 16S rRNA, there is a high diversity of bacterial gene sequences recovered from Lost City (Brazelton *et al.*, 2006).

4.6.5 A rare biosphere at Lost City

Lost City is also the only MOR vent environment where age dating has been coupled with genetic analyses of microbes to examine ecological succession. Using sophisticated genetic analyses techniques on four samples ranging in age from 43 ± 23 years to 1245 ± 257 years, more than 800 distinct taxonomic units (used roughly as a synonym for species) were identified. These are interpreted to represent unique microbial types at the genetic level (Brazelton *et al.*, 2010, 2011). The implication from these results is that there is a much greater diversity of Archaea at Lost City than that detected using historically standard 16S rRNA analyses, which revealed only one genetic species of Archaea in the interior of the carbonate chimneys.

This finding is consistent with the concept of a rare biosphere (Sogin *et al.*, 2006), which suggests the idea that in natural samples microbial abundances are dominated by a relatively small number of different populations. A much greater diversity of organisms and species are considered rare or low in number (and hence difficult to measure because of analytical detection limits). Results at Lost City are not only consistent with the concept of the rare biosphere, but they also show that even over a 1000-year period the community of organisms changed over time (Brazelton *et al.*, 2010). Rare nucleotide (haplotype) sequences in young chimney samples are commonly much more abundant in the older samples, suggesting that organisms can remain rare for long periods until environmental changes allow them to proliferate.

4.6.6 Macrofaunal communities

The biomass at Lost City is markedly smaller than the dense high-biomass macrofaunal assemblages that typify most known high-temperature vent environments (Shank *et al.*, 1998; Van Dover, 2000; Desbruyères *et al.*, 2006; Kelley and Shank, 2010). The animals that live within the small pores and highly sculpted cavities on the outsides of the chimneys and on serpentinite outcrops are typically smaller than 1 cm in size, with transparent-to-translucent shells that make them difficult to observe *in situ*. These animals include a variety of gastropod, polychaetes, and amphipod species (Fig. 4.84) (Kelley *et al.*, 2005; Kelley and Shank, 2010).

Vent-endemic habitats are relegated to the hydrothermally active portions of the nearly vertical and highly porous carbonate edifices, flanges, and spires. Actively venting carbonate habitats (~10–40 °C) are dominated by several species of gastropods and amphipods, including *Bouvierella curtirama* and *Primno evansi* (Kelley and Shank,

Figure 4.84 Brachiopod-like bivalves and urchins colonize steep serpentinite walls beneath the Lost City Hydrothermal Field. These communities were rarely observed on the bedrock beneath the field. Steep cliffs near the summit of the massif host a variety of corals, likely supported by upwelling currents that are funneled up the steep slope of the mountain.

Figure 4.85 Geryonid crabs are common features among the ledges and outcrops in the periphery of the main field. Males of this genus will hold the female underneath for long periods of time (e.g., weeks) prior to mating.

Figure 4.86 A chyrostild crab lives among the branches of an antipatharian (black) coral extending from rock on the periphery of Lost City Field. Growing evidence suggests that certain species of these crabs will spend their entire life on a single coral colony.

Figure 4.87 An *Actinoscyphia* venus flytrap (anemone) is attached to altered mantle rocks on the eastern side of the Lost City Field. Cold water corals are abundant near the top, and at the summit, of the massif where topographic forcing results in strong currents flowing up over the top of the mountain.

Figure 4.88 A plexaurid coral colony is collected from the summit of Lost City. These cold-water gorgonian corals form habitats for at least 2000 invertebrate species around the world. Brown hydroids can be seen growing on several branches.

2010). Present on the friable flanges and spires are species of polychaetes, nematodes, euphausids, foraminifera, ostracods, stomatopods, and demosponges.

Non-venting habitats (such as the sides of inactive solidified carbonate structures, sedimented areas, and breccia cap rock) are dominated by *Lophelia*, gorgonian, and *Desmophyllum* corals, galatheid crabs, turrid gastropods, foraminifera, pteropods, urchins, asteroids, limpets, ophiuroids, and barnacles (e.g., *Poecilasma aurantia* and *Metaverruca inermis*) (Figs. 4.84–4.88).

The most diverse group of invertebrates are polychaetes from the Dorvelleidae family (*Ophiotrocha* n. sp. and two other species), Ceratulidae (*Aphelochaeta* sp.), Glyceridae, Amphinomidae, and the Polynoidae, with a minimum of nine species (at least two of them new species). Mobile megafauna such as the wreckfish *Polyprion americanus* (Fig. 4.76), cut-throat eels, *Synaphobranchus kaupi*, and geryonid crabs (Fig. 4.85) are readily visible surrounding active vents during intensive surveys using submersible imaging systems. A diverse array of corals is particularly abundant at the top of the massif where strong currents are frequent (Figs. 4.86–4.88).

The lack of observable vent-faunal biomass can be attributed to the paucity of reduced chemical nutrients that typically support high biomass in other vent systems (Fustec *et al.*, 1988; Gebruk *et al.*, 2000). The high-pH (9–11), and CH_4–H_2-rich fluids are poor in sulfide species. However, in marked contrast to the observed low biomass, Lost City habitats supports at least as high, if not greater, species diversity than other MAR vent sites (Kelley and Shank, 2010). Vent sites on the MAR typically host 30 to 50 macrofaunal species, with a total of about 100 species residing in the eight major ridge vent sites. Surprisingly, from the 10 active sites sampled at Lost City, over 65 morpho-species in 13 phyla have been documented.

Faunal differences among the active collection sites are likely due to availability of microhabitat (variation in venting intensity) and carbonate porosity, rather than to geographic separation or bathymetric differences spanning the field. Recent assessment at Lost City shows that 58% of the fauna are endemic to this vent environment (Kelley *et al.*, 2005).

TAG

4.7 The largest field known

The largest single known vent deposit along the MOR system occurs on the MAR at 26°08′ N. This hydrothermal field, known as TAG (Trans-Atlantic Geotraverse), hosts low- and high-temperature zones of venting and two major areas of extinct sulfide deposits, which have been mapped in a 5 km by 5 km area (Figs. 4.89–4.91) (Rona *et al.*, 1984, 1993; Thompson *et al.*, 1985; Humphris and Tivey, 2000). It is located at water depths ranging from ~2300 to 4000 m.

Discovered in 1972 during the NOAA Trans-Atlantic Geotraverse expedition (Scott *et al.*, 1974; Rona, 1980), TAG is one of the best studied hydrothermal systems on the MAR and has been one of only a handful of hydrothermal sites drilled by the Ocean Drilling Program (ODP Leg 158) (Rona *et al.*, 1984, 1986; Hannington *et al.*, 1998; Tivey *et al.*, 1998; Humphris and Cann, 2000; Humphris and Tivey, 2000; deMartin *et al.*, 2007; Tivey, 2007). Hydrothermal processes similar to those operating beneath TAG have been documented in ore deposits in Cyprus, which provide three-dimensional views of the subsurface (see Chapter 3) (Figs. 4.92 and 4.93). Humphris and Tivey (2000) provide an excellent review of TAG.

Four main areas of hydrothermal deposits have been mapped that include (1) the active TAG mound hosting numerous high-temperature chimneys (Figs. 4.94 and 4.95); (2) the Alvin Zone composed predominantly of isolated sulfide outcrops, relict sulfide mounds, and an active mound the mound is ~150 m in diameter and 68 m high, venting low-temperature fluids (Humphris and Tivey, 2000); (3) a low-temperature zone that is on the eastern medial valley wall ~4 km east of the active TAG mound

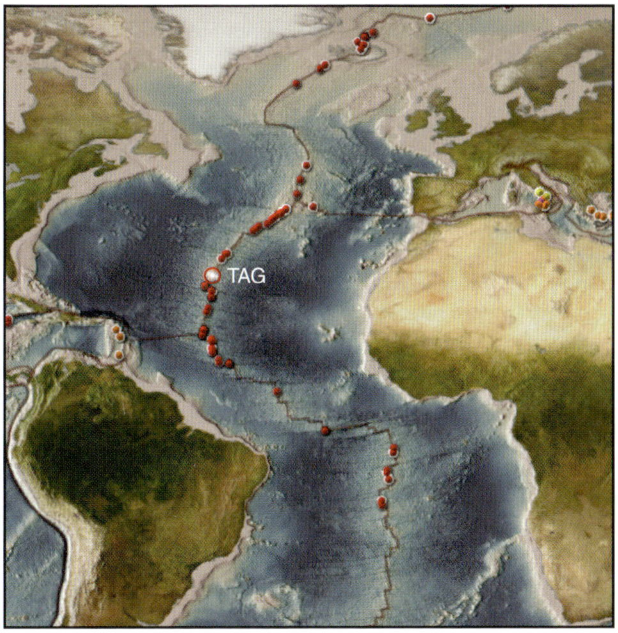

Figure 4.89 Location of the TAG Hydrothermal Field on the MAR.

Figure 4.90 TAG is a ~5 × 5 km^2 mound rising ~50 m above the surrounding seafloor, comparable in size to the Colosseum in Rome.

Figure 4.91 Vigorously venting black smoker chimneys on the summit of the TAG hydrothermal mound reach temperatures of 363 °C. Swarms of alvinocarid shrimp (*Rimicaris exoculata* and *Chorocaris chacei*) completely cover the outside of the chimneys, foraging on bacteria in the vent fluids and on each other's carapaces. Shrimp are typically ~4–5 cm in length. Field of view ~2 m.

(Rona *et al.*, 1993; Humphris and Tivey, 2000); and (4) the MIR zone, located on the lower portion of the eastern valley wall ~2 km E-NE of the active TAG mound (Humphris and Tivey, 2000). The MIR zone includes discontinuous sulfides and low-temperature deposits.

The TAG hydrothermal mound is located ~1 km west of the base of the eastern valley wall in a zone of N-NE-trending active faults and fissures (Kleinrock and Humphris, 1996a,b). The geological processes that have promoted hydrothermal activity in the field for at least 150,000 years (Lalou *et al.*, 1995) have been the focus of speculation. Numerous researchers propose that TAG is located on the hanging wall of a detachment fault (Tivey *et al.*, 2003; deMartin *et al.*, 2007; Canales *et al.*, 2008; Zhao *et al.*, 2012). An ocean-bottom seismometer study at TAG in 2003–2004 documented two distinct zones of seismic activity with events along the eastern valley wall defining a roughly 70°, west-dipping fault extending 7 km beneath the seafloor. These events were interpreted to reflect activity along a detachment fault beneath TAG and tectonic extension along the spreading center (deMartin *et al.*, 2007). The transition from intense seismic activity in the deeper parts of the fault to an aseismic section at shallow depths is interpreted to reflect the rollover to a more domal feature with rotation at depths of 2 to 3 km below the ocean floor (deMartin *et al.*, 2007; Canales *et al.*, 2008).

4.7.1 The TAG mound

The combination of faulting, deep-seated intense seismicity, and perhaps a magma reservoir more than 7 km below the ocean floor has led to the development of the immense TAG sulfide deposit (Figs. 4.90 and 4.93–4.97) (Canales *et al.*, 2007; deMartin *et al.*, 2007). It is 200 m in diameter and 50 m high, and was probably formed by individual venting structures that combined into one deposit over time. Much of the mound is composed of massive sulfide blocks and black-smoker and white-smoker debris that is variably cemented and hydrothermally recrystallized (Fig. 4.93) (Hannington *et al.*, 1998; Humphris and Tivey, 2000). Examination of ~6200 microearthquakes (M_L ~ 1) showed that they clustered in a narrow zone ~50–125 m beneath the mound on the south and west flanks where mineral deposition is occurring (Pontbriand and Sohn, 2014). It is hypothesized that these smaller earthquakes reflect cracking events resulting from thermal contraction of the basement rocks during cooling by hydrothermal fluids (Pontbriand and Sohn, 2014). Seismic refraction data do not support the presence of a mid-crustal magma chamber beneath the valley floor (Canales *et al.*, 2007).

Active venting occurs in a black smoker complex at the mound summit, on a 10–15 m tall conical structure that is about 30 m across at its base (Figs. 4.93–4.95). The edifice hosts multiple, spire-shaped chimneys (Figs. 4.93–4.96) that vent copper-rich fluids at temperatures up to 363 °C. Fluid chlorinities are slightly elevated with respect to seawater values in the black smokers. Within TAG's upflow zones, intense mineral–fluid reactions pervasively alter and mineralize the wall rocks (Fig. 4.93). Carbon dioxide (3.31–3.56 mm/kg), CH_4 (0.15–0.16 mm/kg) and H_2 (0.18–0.23 mm/kg) concentrations (Charlou *et al.*, 1996) are consistent with fluid interaction with basaltic composition rocks in the subsurface.

Significant entrainment of seawater occurs beneath the deposit. Subsurface fluid temperatures reach 380–390 °C (Tivey *et al.*, 1998). These fluids are heated during mixing with high-temperature hydrothermal fluids and by conductive heating during transport through the underlying channel network (Tivey *et al.*, 1998). The entrainment results in generation of acidic fluids and metal remobilization beneath the deposit.

TAG is one of the few hydrothermal sites where there are abundant white smokers. Zinc-rich white smokers occur on the southeast margin of the mound, and discharge fluids that range in temperature from 265 °C to 300 °C. This zone, known as the Kremlin area, hosts 1–2 m tall spires with bulbous tops composed of sphalerite, amorphous silica, marcasite, and pyrite (Tivey *et al.*, 1995). White smoker fluids are about 70% of normal seawater chlorinities (Humphris and Tivey, 2000). In addition to these sites, diffusely venting fluids percolate locally through the top and sides of the mound. Steep-sided ramparts of variably oxidized sulfide blocks and hydrothermal sediment form the sides of the mound (Figs. 4.96 and 4.97).

4.7.2 Long-lived venting

TAG is one of the few sites on the MAR where information is available about the subsurface geometry of the upflow

Figure 4.92 Fossilized upflow zone in the Cyprus Ophiolite showing intense alteration and mineralization caused by similar processes as those operating beneath TAG. In Cyprus, several of these sites have been mined since the Romans inhabited the island.

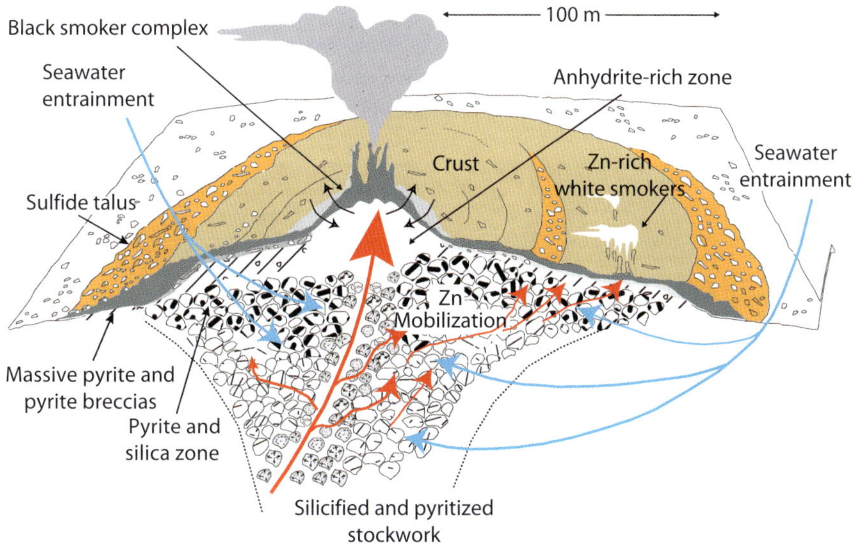

Figure 4.93 The TAG active hydrothermal mound accumulates from vigorous venting black smokers that reside predominantly on the summit of the mound, and variable cementation of older sulfide debris that forms a large apron around the edges of the mound (after Hannington *et al.*, 1998). Subsurface upwelling of high-temperature black smoker fluids, and significant entrainment of seawater into the mound results in deposition of metals and anhydrite in the subsurface (Tivey, 2007). These processes also result in remobilization of zinc and trace metals that are deposited in the outer edges of the mound (Humphris and Tivey, 2000; Tivey, 2007).

Figure 4.94 High-temperature black smoker chimneys at TAG vent copper-rich fluids at temperatures up to 363 °C. They typically host multiple conduits and have high copper concentrations [up to 23 wt% (Humphris and Tivey, 2000)]. The chimney walls are composed of chalcopyrite, anhydrite, pyrite, and trace sphalerite. The outer walls of the chimneys are commonly highly oxidized. The manipulator arm of *Alvin* holds a titanium fluid sampler used to collect vent fluids for follow-on chemical analyses. Shrimp typically range from ~3 to 6 cm in length.

zone. Magnetic surveys, in concert with results from Ocean Drilling Program Leg 158, indicate that the deposit is underlain by a cylindrical pipe or silicified-pyritized stockwork system that is 80 to 100 m in diameter (Fig. 4.90) (Tivey *et al.*, 1996; Hannington *et al.*, 1998). The mound and its underlying network of channels are believed to contain about 3.9 million tons of massive sulfide (Hannington *et al.*, 1998). The large intersecting faults have channeled flow to the hydro-

Figure 4.95 Jets of hot hydrothermal fluid billow from black smoker orifices at the summit of the TAG mound. *Rimicaris exoculata* and *Chorocaris chacei* graze in the plumes. Field of view ~3 m.

Figure 4.96 Bulbous-shaped beehive structures grow through oxidized sulfide debris. The 1–2 m tall spires have high porosity (50%) and are composed predominantly of low-iron sphalerite with minor amorphous silica, marcasite, and pyrite (Tivey *et al.*, 1995).

Figure 4.97 The summit of TAG hosts massive chalcopyrite, pyrite, and/or marcasite blocks with corroding massive anhydrite (Humphris and Tivey, 2000). Collapse of chimneys and variable cementation of the sulfide talus aides in growth of the mound. Friable material is gathered with the arm of *Alvin* and placed in a "biobox" for safe return to the surface.

Figure 4.98 Anemones (*Actiniara*), up to 10 cm in diameter, colonize oxidized sulfide debris and highly oxidized hydrothermal sediment on the outskirts of the TAG deposit.

thermal mound episodically over a 40,000–50,000-year period (Karson and Rona, 1990; Lalou *et al.*, 1995, 1998; Humphris and Kleinrock, 1996). The long history of venting at TAG and the acidic nature of the fluids has led to a mineralogically zoned deposit and remobilization of many of the metals.

4.7.3 Aggregating shrimp

There is a zonation in the biological communities at TAG that is well correlated with hydrothermal activity. The black smoker complex supports aggregations of *Rimicaris exoculata*, with densities estimated at more than 40,000 individuals per cubic meter (Figs. 4.91, 4.95 and 4.96) (Gebruk *et al.*, 1993; Segonzac *et al.*, 1993). The extreme density of these shrimp commonly masks entire chimneys from view. Individuals of *Chorocaris chacei* and *Alvinocaris markensis* shrimp are typically found sparsely at the base and apron of the complex. *Rimicaris exoculata* are of particular interest because they possess a novel photoreceptor considered capable of detecting low light levels (Van Dover *et al.*, 1989; White *et al.*, 2002).

The non-image-forming photoreceptor hosts pigment that absorbs maximally at 500 nm. Detailed analyses of hydrothermal vents along the MAR show that the highest emissions occur in the range of 700–1000 nm due to black body radiation. However, they also emit non-thermal-origin radiation in the visible region of the spectrum (White *et al.*, 2002). Although *Rimicaris exoculata* may detect light, it also possesses high degrees of chemical sensitivity, which it may also use to locate the sides of active sulfide structures and their food sources (Renninger *et al.*, 1995).

Limpets also occur in areas of hydrothermal activity, but are rare within the zones occupied by shrimp. In addition to the high shrimp concentrations at TAG, brachyuran crabs and anemones dominate the periphery of the field, as do larger gastropods (Fig. 4.98) (Van Dover, 2000).

Logatchev

4.8 Ultramafic-influenced vent sites

The Logatchev Hydrothermal Field is one of the best studied ultramafic/gabbroic-hosted sites on the MAR. It provides important insights into ultramafic-hosted hydrothermal systems where active, high-temperature serpentinization occurs at depth (Petersen *et al.*, 2009) (Figs. 4.99–4.103). Hydrothermal plume anomalies rich in CH_4 and manganese were recognized in the early 1980s south of the 15°20′ N fracture zone, indicating active venting (Klinkhammer *et al.*, 1985; Rona *et al.*, 1992; Bougault *et al.*, 1993).

In 1993 and 1994, this area was investigated (with the R/V *Professor Logatchev*, a Russian research ship) using geochemical sensors and a deep-towed camera system, resulting in the discovery of the fields at 14°45′ N at a water depth of ~3000 m (Batuev *et al.*, 1994; Krasnov *et al.*, 1995). Follow-on cruises to this area designated two main hydrothermal sites, Logatchev 1 and Logatchev 2, located ~5.5 km apart (Schmidt *et al.*, 2007) (Fig. 4.103). The fields are on the eastern side of the rift valley on the 15° N segment, in

Figure 4.99 Unusual high-temperature flow occurs out of depressions called Smoking Craters. The craters are 2–5 m deep. Field of view ~3 m.

Figure 4.100 Shrimp (*Rimicaris exoculata*) cluster around a black smoker in the Irinia II mound, with dense mussel colonies (*Bathymodilius puteoserpentis*) populating the lower flanks. Field of view ~1.5 m. This image and Fig. 4.99 were taken with the ROV *Quest 4000* (MARUM, University of Bremen).

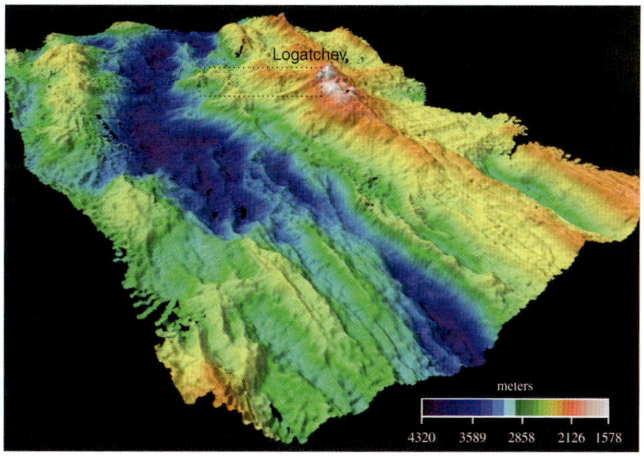

Figure 4.101 Logatchev vents are on the eastern wall of the axial valley south of the 15° 20′ N fracture zone. During Hydromar-IV cruises with the R/V *Meteor* and R/V *Maria S. Merian* between 2004 and 2007, the area was extensively mapped and characterized with direct observations using the ROV *Quest 4000* (MARUM, University of Bremen) (Petersen *et al.*, 2009).

tectonically complicated, highly attenuated areas characterized by abundant faulting and intermixed outcrops of serpentinized harzburgites, and gabbros-gabbronorites. Irregular intrusive bodies, dikes, and dikelets are common in the mantle rocks (Fig. 4.103) (See Chapters 6 and 8) (Augustin *et al.*, 2008; Petersen *et al.*, 2009). In contrast, volcanic edifices and basaltic outcrops intermixed with moderately sedimented areas characterize the rift valley floor (>3900 m water depth).

The Logatchev core complex is proximal to the field. Locations of the vent sites are controlled by a series of NW–SE- and W-NW–E-SE-striking scarps (faults), and slump structures that are believed to be related to uplift of the valley walls (Fig. 4.104) (Kuhn *et al.*, 2004; Augustin *et al.*, 2008; Petersen *et al.*, 2009). Investigations using submersible, robotic vehicles, and by drilling (ODP Leg 209; Site 1270) indicate that serpentinization is a dominant alteration

Figure 4.102 Vigorously venting black smoker chimney in the Candelabra Site in the Logatchev Hydrothermal Field imaged by the ROV *Quest 4000* (MARUM, University of Bremen). The fluids are rich in transition metals (Fe, Mn, Co, Ni, Cu, and Zn) reflecting the impact of ultramafic rock–fluid interaction at high temperatures (Lein *et al.*, 2003). Field of view ~4 m.

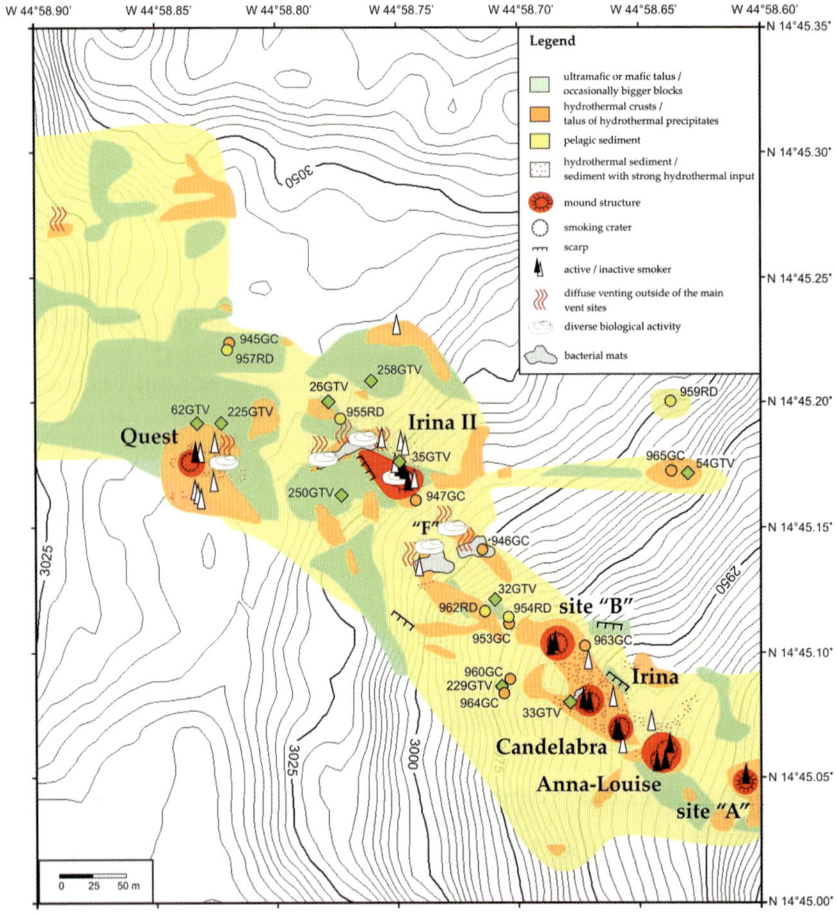

Figure 4.103 Generalized geological map of the Logatchev-1 Hydrothermal Field from Petersen *et al.* (2009). Numbers refer to sites sampled by a grab sampler (GTV), by drilling (RD), and by gravity coring (GC).

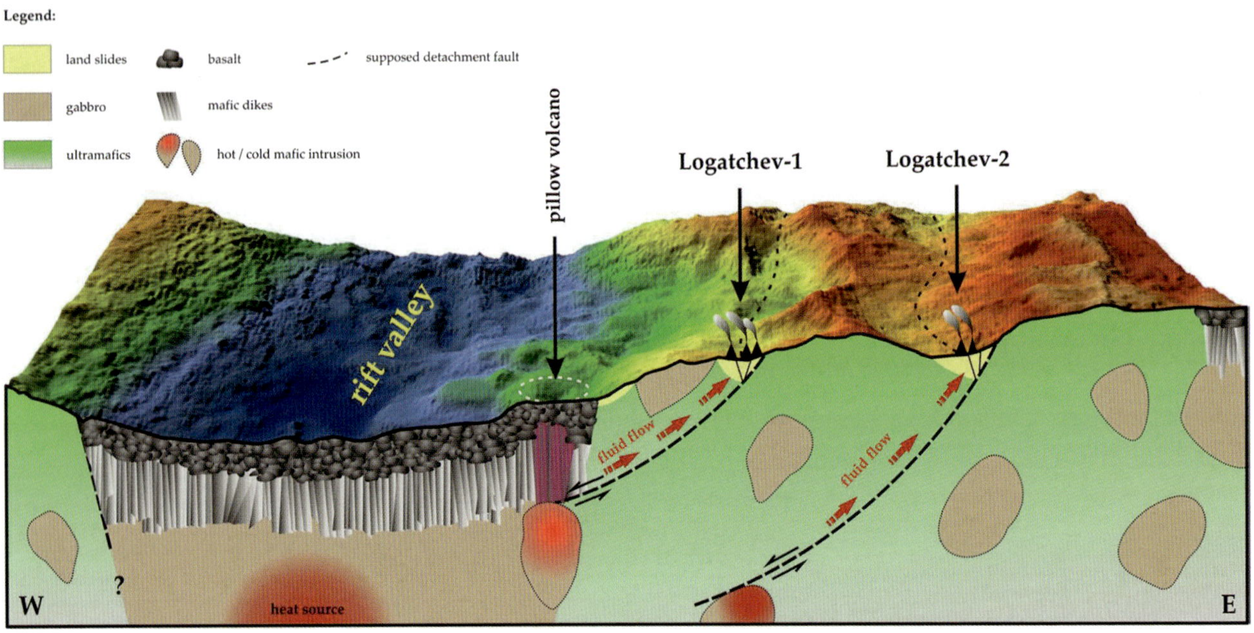

Figure 4.104 Across-axis block diagram from Petersen *et al.* (2009) showing their model for fluid circulation at Logatchev. Detachment faults mine heat from gabbroic heat sources, providing high CO_2 concentrations and temperatures. They act as conduits for high-temperature fluid flow to the Logatchev-1 and Logatchev-2 fields. The position of the gabbroic intrusions is inferred from the presence of a pillow volcano west of the fields. The lithologies are based on direct observations and from sampling.

process in this area. Rocks recovered at ODP Site 1270, less than 4 km away from the field, show that the subseafloor is composed of variably altered and deformed ultramafic rocks hosting abundant, highly deformed gabbroic intrusions with focusing of mafic and evolved melts and hydrothermal fluids into the shear zones (Paulick *et al.*, 2006; Kelemen *et al.*, 2007). Petersen *et al.* (2009) provides an excellent overview of the Logatchev Hydrothermal Field. They propose a model involving low-angle detachment faults serving as fluid flow channels with gabbroic intrusions as the heat sources to drive flow (Fig. 4.99) (Petersen *et al.*, 2009).

4.8.1 Hydrothermal venting

Venting temperatures in the Irina site at Logatchev reach 348 °C, and temperatures of 300–350 °C have been measured in black smoker chimneys at Site B (Schmidt *et al.*, 2007). The highest reported venting temperature measured directly (353 °C) was by Douville *et al.* (2002). A second site of venting is a large, 60 m diameter, steep-sided mound known as Irina II, which rises about 15 m above the seafloor (Fig. 4.103) (Kuhn *et al.*, 2004; Augustin *et al.*, 2008; Petersen *et al.*, 2009). Four chimneys up to 1.5 m high were observed in 2004 at the summit of the mound, as were several beehive structures (Schmidt *et al.*, 2007). Measured vent fluid temperatures at Irina II range from 170 °C to 225 °C.

A third venting site, Quest, is about 130 m W-NW of Irina II in a small depression; temperatures there reach 280 °C (Schmidt *et al.*, 2007). In concert, this area has also been called Smoking Craters, a novel area that hosts 2–5 m deep depressions bounded by sulfide and serpentinite debris (Fig. 4.103). One-meter tall chimneys sprout from the rim of the depressions, while black smoke vents directly from small holes in the craters (Petersen *et al.*, 2009).

A key feature of the black smoker chimneys is that they are exceptionally rich in copper (48 wt%) and commonly include the copper-bearing minerals chalcopyrite and chalcocite (Krasnov *et al.*, 1995) (Figs. 4.100, 4.102, and 4.105). Concentric, zoned aggregates of bornite, chalcopyrite, isocubanite, and pyrite are also present, as well as barite, sphalerite, and amorphous silica (Krasnov *et al.*, 1995; Gablina *et al.*, 2006). Copper-enrichment (up to 28 wt% Cu) in sulfides is also a key feature in the ultramafic-hosted

Figure 4.105 Black smokers host ribbed edifices (foreground) and young "beehive" structures. Shrimp (*Rimicaris exoculata*), 4–5 cm in length, forage in the more vigorously venting areas, while dense aggregations of mytilid mussels (*Bathymodiolus puteoserpentis*) hosting methane- and sulfur-oxidizing bacteria thrive in sites of diffuse flow. Imaged by the ROV *Quest 4000* (MARUM, University of Bremen). Field of view ~2 m.

Figure 4.106 Aggregations of shrimp and mussels near small, vigorously venting chimlets. Field of view ~30 cm.

Rainbow Hydrothermal Field, which is venting 365 °C, low-pH fluids (Marques *et al.*, 2006). In addition, both Logatchev and Rainbow are characterized by high copper/zinc ratios, and high cobalt and nickel (Marques *et al.*, 2006, 2007). Dating of weathered sulfide material shows that venting here has been active for at least 60,000 years (Lalou *et al.*, 1996; Kuznetsov *et al.*, 2006).

The fluids at Logatchev are extremely enriched in CH_4 and H_2 (2.1 mmol/kg and 19 mmol/kg, respectively), which is a hallmark of serpentinization reactions at depth (Charlou *et al.*, 2002; Schmidt *et al.*, 2007); they also contain high CO_2 concentrations (10.1 mmol/kg) (Charlou *et al.*, 2002). The CH_4, H_2, and CO_2 concentrations fall in the range of Rainbow fluids (see Section 4.5), with CH_4 at 0.08–2.5 mmol/kg, H_2 at 5.0–16.0 mmol/kg, and CO_2 at 6.4–16.2 mmol/kg (Charlou *et al.*, 2002). The fluids from both Rainbow and Logatchev also host unusually high iron concentrations, which most likely reflect the leaching of magnetite from the ultramafic and underlying gabbroic rocks by the acidic fluids (Hannington *et al.*, 2005).

4.8.2 Life in high CO_2, CH_4, and H_2 systems

Microbial communities are tightly linked to fluid chemistry at Logatchev. Genetic analyses (16S rRNA sequencing) of material from 300–350 °C fluids at the Smoking Craters sites, Irina I, and Site B, and from a 170 °C smoker in Irinia II, show that a significant component of both the bacterial and archaeal communities is related to known autotrophic H_2-oxidizers (Perner *et al.*, 2007). Methanococcales that use H_2 and CO_2 to produce CH_4 were found at three sites.

Macrofaunal communities at Logatchev differ significantly between the high-temperature black smoker sites, chimney complexes (e.g., Irina II and MAR-2), sulfide mounds, and sedimented areas of diffuse flow (e.g., Anya's Garden) (Gebruk *et al.*, 2000). Sedimented diffuse flow areas are dominated by vesicomyid clams, mussels, gastropods, and numerous species of polychaetes. In contrast, the Irina II area contains abundant mytilid mussels (*Bathymodiolus puteoserpentis*) hosting both methane- and sulfur-oxidizing symbionts in their gills, numerous polynoid polychaetes, conid gastropods, and clusters of *Rimicaris exoculata* shrimp (Shank *et al.*, 1998; Gebruk *et al.*, 2000).

Fauna at Logatchev differ from other sites along the MAR as well. For example, a detailed examination of macrofaunal community structure at Logatchev, Snake Pit, and Lucky Strike revealed that the deeper Logatchev and Snake Pit mussel beds share more than half their associated macrofaunal species. In contrast, these two sites share only 20 to 25% of their macrofaunal species with Lucky Strike (see Section 4.4) (Van Dover and Doerris, 2005). While nearly 50 species of vent fauna have been reported at Logatchev (Gebruk *et al.*, 2000), this site has the distinctive feature of hosting an extremely high abundance (up to 2390 individuals per liter of mussel-volume sampled) of the brittle star *Ophioctenella acies*.

South Atlantic

4.9 Extreme vent temperatures

From 2001 to 2006, the Southern Mid-Atlantic Ridge (SMAR) between about 2° S and 10° S was investigated on several cruises using a nested survey strategy focused on exploring for hydrothermal vents south of the largest offset equatorial fracture zones in the world (Figs. 4.107–4.109). These prominent features, particularly the Romanche Fracture Zone north of Ascension Island, separate the northern and southern MAR. The surveys included multibeam sonar to resolve the large-scale structural and morphological features of the ridge axis, in combination with near-bottom mapping using the *TOBI* sidescan sonar system, CTD surveying and water sampling, and finally AUV near-bottom mapping, water properties sensing, and imaging, which resulted in the discovery of several important hydrothermal sites (Figs. 4.109 and 4.110) (German *et al.*, 2008a). Focused exploration of the 5° S region in 2005 by the AUV *ABE* (see Chapter 1), coupled with CTD surveys, resulted in

Figure 4.107 *Rimicaris exoculata* shrimp (4–5 cm in length) forage on the surface of an active black smoker. Crabs, *Segonzacia sp.* (lower center), feed on microbes and shrimp. This image was taken with the ROV *Quest 4000* (MARUM, University of Bremen).

Figure 4.108 The Turtle Pits vent site consists of two active sulfide mounds (Two Boats and One Boat) and one active chimney – Southern Tower (shown here). This image was taken with the ROV *Quest 4000* (MARUM, University of Bremen).

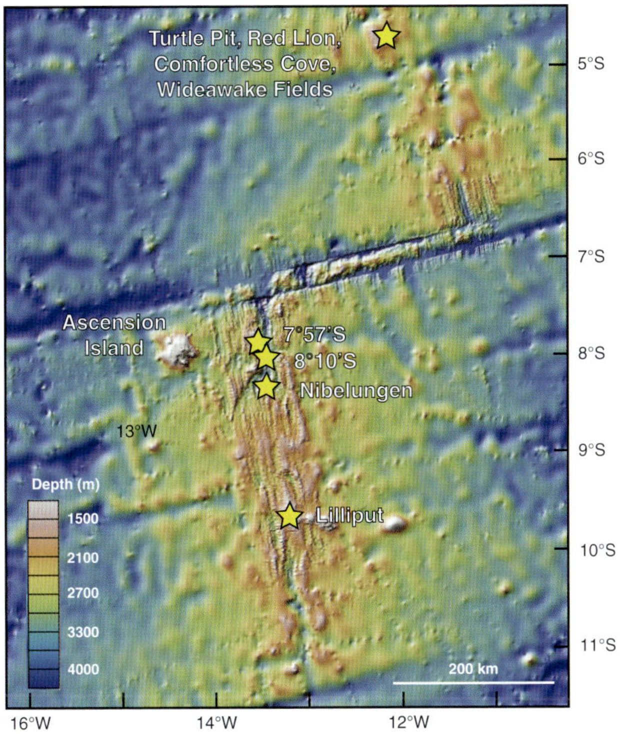

Figure 4.109 Location of hydrothermal vents discovered since 2005 along the southern MAR by the German GEOMAR team using a variety of nested-survey techniques including CTD tow-yo exploration, AUV near-bottom mapping and water property sensing, and ROV direct imaging surveys and sample collections (e.g., Devey *et al.*, 2005; Haase *et al.*, 2007, 2009; German *et al.*, 2008b).

the discovery of the Turtle Pits, Red Lion, and Wideawake Fields (Shank, 2006; German *et al.*, 2008b). Follow-on German-led cruises brought discovery of the Comfortless Cove Field and the Nibelungen and Lilliput Hydrothermal Fields at 8°18′ S and 9°33′ S, respectively (Figs. 4.111–4.114) (Koschinsky *et al.*, 2006; Haase *et al.*, 2009).

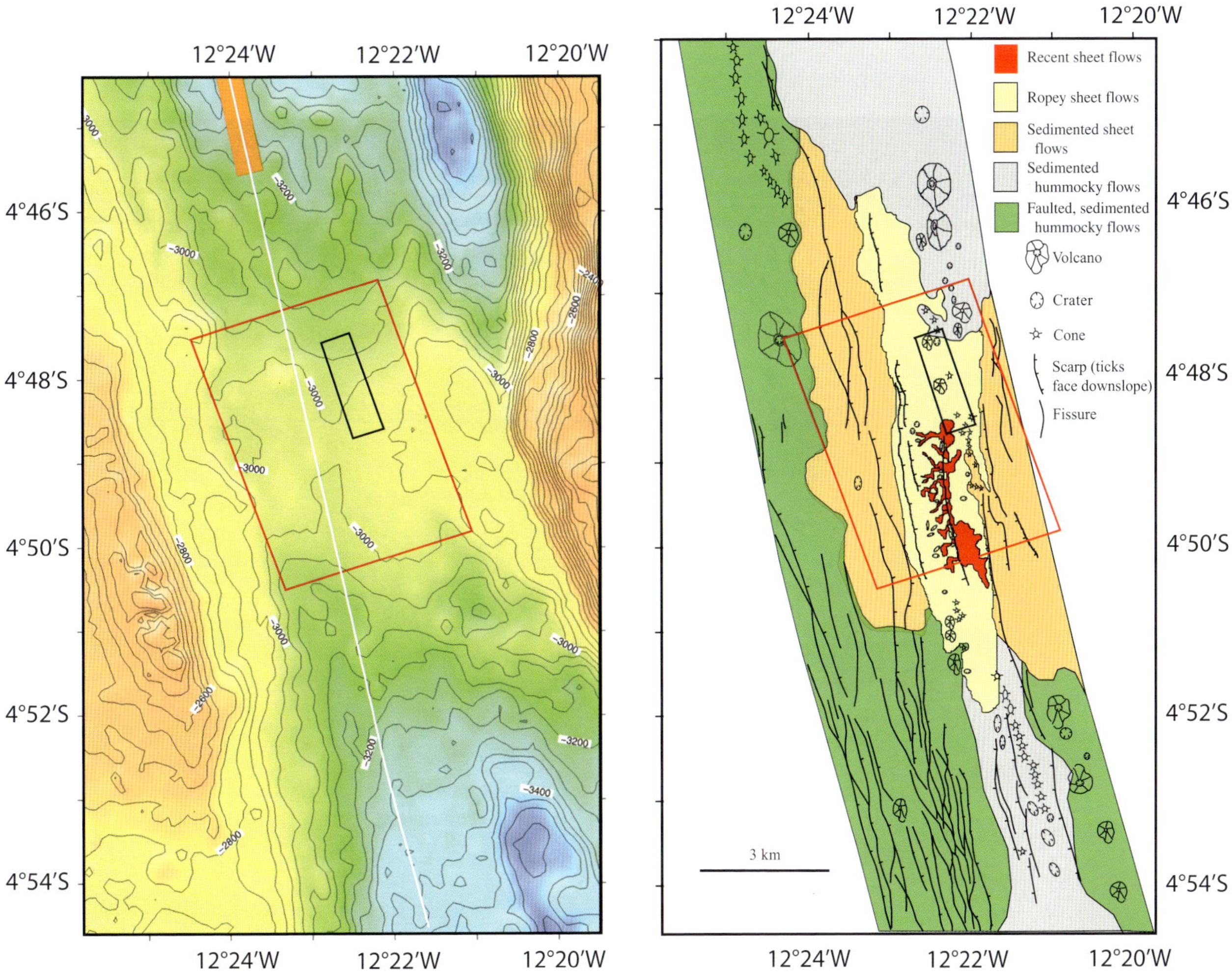

Figure 4.110 bathymetry of the second-order segment of the southern MAR showng tracklines of the sidescan sonar system *TOBI* (white), and nested *ABE* surveys (red and black boxes). (Right) Geological setting of the ~5° S MAR hydrothermal vent fields. Unsedimented sheet flows extend for ~12 km along-axis and ~1.5 km across-axis at the center of this magmatically active second-order ridge segment (from German *et al.*, 2008b).

A major scientific driver for exploring the SMAR for hydrothermal sites has been its possible importance as a barrier for genetic connectivity via larval transport to other vent sites along the MAR and, hence, the distribution of biogeographic provinces globally (Van Dover *et al.*, 2002; Shank, 2004).

The SMAR in this region is a slow-spreading (2–3 cm/yr) center and exhibits variable axial topography including non-transform discontinuities. This results in the likely exposures of ultramafic rocks in core complexes, and, in areas with a shallow rift valley, elevated topography and axial volcanic ridges that can be much shallower (1.5–3 km) than the normal depth of the northern rift valley floor on the MAR, which is typically at a depth of 3.5–4 km.

Direct imaging by the AUV *ABE* and the ROV *Quest 4000* shows that the area hosting the 5° S fields is predominantly bare, fresh basaltic rock. The geology of the high-temperature Turtle Pits Field is reminiscent of the EPR with laminated and jumbled basaltic sheet flows, collapse pits, truncated lava tubes, and hackly sheet flows. Fresh pillow flows border the Wideawake diffuse venting system and in some areas the lavas cover vent fauna (Haase *et al.*, 2007), suggesting very recent submarine eruptions. The Lilliput low-temperature field is also hosted in fresh basaltic rocks and likely reflects hydrothermal circulation associated with a recent eruption and lava flow (3 km^3 volume) from the prominent N–S-trending, 3 km long volcanic ridge that lies at 1.5 km depth just west of the vent site (Haase *et al.*, 2009).

Vent fluid chlorinities range from 546 mmol/kg at Red Lion to 224 mmol/kg at Comfortless Cove (Fig. 4.111). Methane and H_2 concentrations are similar to other basaltic-hosted environments, ranging from 0.013 to 0.045 mmol/kg and 0.12 to 0.664 mmol/kg, respectively (Haase *et al.*, 2007).

Venting in the Turtle Pits Field reaches extreme temperatures. Here, in 2006, a hot smoker, "Two Boats," was venting 407 °C fluids, characterized by phase separation and emanation of a salt-depleted vapor phase with high concentrations of H_2, iron, and copper (Fig. 4.108). The pH was as low as 3.1. The black smoker called Sister Peak in the Comfortless Cove, located between the Turtle Pits and Red Lion sites, has highly similar chemistry. Although boiling was not directly observed, depleted chlorinity and high metal concentrations indicate a common fluid source for the two vents (Fig. 4.111). In contrast, in the Red Lion Field northwest of Comfortless Cove, the four different smokers are cooler (between 193 and 349 °C) and show no phase separation.

Figure 4.111 A large, active black smoker in the Comfortless Cove Field vents gas-rich fluids at temperatures in excess of 400 °C. A short burst (20 s) of fluids at 464 °C was measured at this field (Koschinsky *et al.*, 2008). This image was taken with the ROV *Quest 4000* (MARUM, University of Bremen). Field of view ~1.5 m.

Figure 4.113 Lilliput is named after the numerous, small bathymodioline mussels present between pillow lavas stained with iron oxyhydroxides. The white bacteria-lined collapse cavities are reminiscent of snowblowers (see Section 4.11). This image was taken with the ROV *Quest 4000* (MARUM, University of Bremen). Field of view ~2.5 m.

Figure 4.112 Aggregates of shrimp forage on the surface of a large black smoker edifice in the Comfortless Field, while rare mussels colonize the talus. These shrimp are typically 4–5 cm in length. This image was taken with the ROV *Quest 4000* (MARUM, University of Bremen).

While the fluids at 5° S are clearly a result of water–rock interaction in a basaltic system, the chemistry of the vent crater "Drachenschlund" (Dragon's Throat) at Nibelungen indicates fluid reactions with mantle rocks (serpentinization reactions). Extraordinarily high gas concentrations along with high Fe and Cu concentrations confirm this assumption. The high H_2/CH_4 ratio and the extremely high Cu concentrations indicate an end-member fluid temperature in the range of 360 to 400 °C (Schmidt *et al.*, 2011).

4.9.1 SMAR vent fauna

Chemosynthetic faunal communities at 5–9° S are biomass-dominated by *Rimicaris* shrimp, *Bathymodiolus* mussels, and vesicomyid clams, with high levels of variability in composition and abundance among sites. The abundance of shrimp

Figure 4.114 *Bathymodiolus* mussel beds in a highly oxidized area within the Lilliput Field as imaged in 2005. Most mussels in this area are more than 6.5 cm in length. This image was taken with the ROV *Quest 4000* (MARUM, University of Bremen).

on the sides of active chimneys is relatively small compared to the northern MAR (Figs. 4.107 and 4.112), and clams, while present in mussel beds, are also low in abundance. Notable is the remarkable similarity of the mussel and clam bed bare-basalt habitats in this region (the Wideawake Field) to those on the EPR, including the shared observations of lobate and sheet flow lava over-running these bivalve assemblages (Haase *et al.*, 2007).

In addition to shared species known from the northern MAR (e.g., *Rimicaris exoculata*), new species, including the shrimp *Opaepele susannae*, have been described (Komai *et al.*, 2007). Now more than 17 (morpho) species have been putatively identified from these southern Atlantic sites. Initial genetic comparisons of the dominant and taxonomically similar fauna (shrimp and mussels) from the northern and southern MAR reveal both genetic connectivity and genetic divergence across the equator (Teixeira *et al.*, 2011; Shank, unpublished data).

The diffuse vent sites (Wideawake at 5° S, and Lilliput at 9° S) are dominated by dense thickets of *Bathymodiolus* mussels, highly variable in size. Larger mussels make up 80% of the community and they are more than 8 cm long at Wideawake. In contrast, smaller specimens dominate the Lilliput area, with only 1.5% of the population longer than 6.5 cm: post-settlement stages prevail (Fig. 4.114). Mussels are commonly attached to crevices and vertical smoker walls, and host commensal *Branchipolynoe* polychaetes. Within these mussel assemblages the following are also present: shrimp (*Opaepele*), crabs (*Segonzacia* sp.), abundant detritus feeders (*cf. Amathys*) extending their tentacles out of small tubes, and rare gastropods (*cf. Lirapex*) are also encountered. Mussel aggregations in Wideawake also host the white clam *Calyptogena* sp. Suspension feeders, such as sea anemones or gorgonians are in the periphery of most of the 5–9° S vent fields.

The Nibelungen and Dragon's Throat (8°18′ S) Fields are devoid of the typical Atlantic vent fauna except for numerous parchment-like, flexible tubes of chaetopterid polychaetes of unknown identity. These worms attach to loose, altered rocks and sand at the crater walls. The absence of both vent-related macrofauna and sediment meiofauna might reflect difficulties for larvae to settle on loose sediment, or perhaps high concentrations of toxic heavy metals bound as sulfides that are dispersed through the sediments.

Small gastropods (probably *Lirapex*) are abundant at the southern fields, while several kinds of limpets (lepetodrilids and fissurelids) are known at the northern sites in the 5° S area. The limpets thrive in live mussel clusters grazing on microbial communities. Terebellid and spionid polychaetes (*cf. Amathys*) are found building small tubes of fine particles, feeding in sediment. Both the crab (*cf. Segonzacia*) and the large conid snail *Phymorhynchus* scavenge in mussel fields and on vent chimneys. The sea spider (*Sericosura heteroscela*) also exists within mussel beds, while chaetopterid polychaetes dominate the crater walls of the smoker Drachenschlund. On the periphery of the Nibelungen Vent Fields are suspension feeders, including soft corals (gorgonians), hydrozoan colonies (e.g., *Candelabrum*), and small actiniarian anemones.

Endeavour

4.10 One of the most remarkable places

The Endeavour Segment of the Juan de Fuca Ridge is one of the most remarkable submarine volcanic–hydrothermal systems on Earth. Located 300 km west of Vancouver Island, British Columbia, this intermediate-spreading center is also one of the most well-studied hydrothermal sites, with more than 30 years of interdisciplinary investigations (Fig. 4.115) (Kelley *et al.*, 2014). Within 15 km along the central portion of the ridge, five major hydrothermal fields have been documented; they commonly host large sulfide edifices in addition to numerous smaller fields (Figs. 4.116 and 4.117; Table 4.1).

The fields are some of the most intensely active known, with more than 800 individual active and extinct chimneys documented within the central portion of the ridge (Delaney *et al.*, 1992; Clague *et al.*, 2008; Kelley *et al.*, 2012, 2014). Long-lived chimneys, hosting dense macrofaunal and microbial communities, have grown to 45 m above the surrounding seafloor (Robigou *et al.*, 1993) (Figs. 4.117 and 4.118) and microbes recovered from one of the fields grow at temperatures of 121 °C, 1 °C shy of those currently

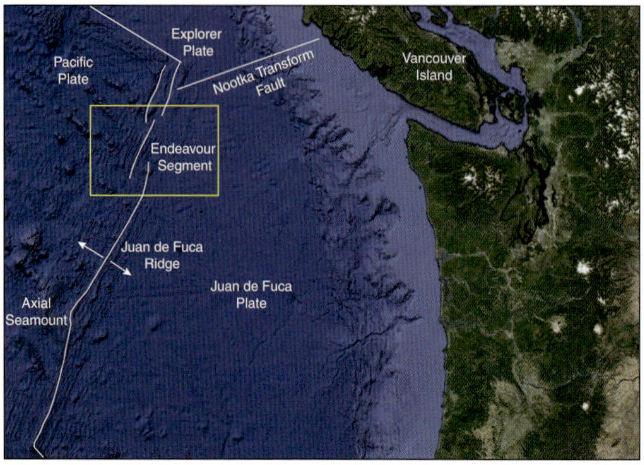

Figure 4.115 The Endeavour Segment of the Juan de Fuca Ridge is located ~300 km off the coast of Vancouver Island. It is bounded to the north by Middle Valley and West Valley, which are part of the Sovanco and Nootka triple junction. To the south, the Endeavour Segment is bounded by the Cobb propagator. It is a Canadian Marine Protected area.

Figure 4.117 (opposite) This black smoker edifice called Bastille, in the Main Endeavour Field, rises more than 15 m above the seafloor and is composed of multiple towers. It hosts dense colonies of tubeworms (*Ridgeia piscesae*), palm worms, and limpets. Scale bar is equal to 1 m.

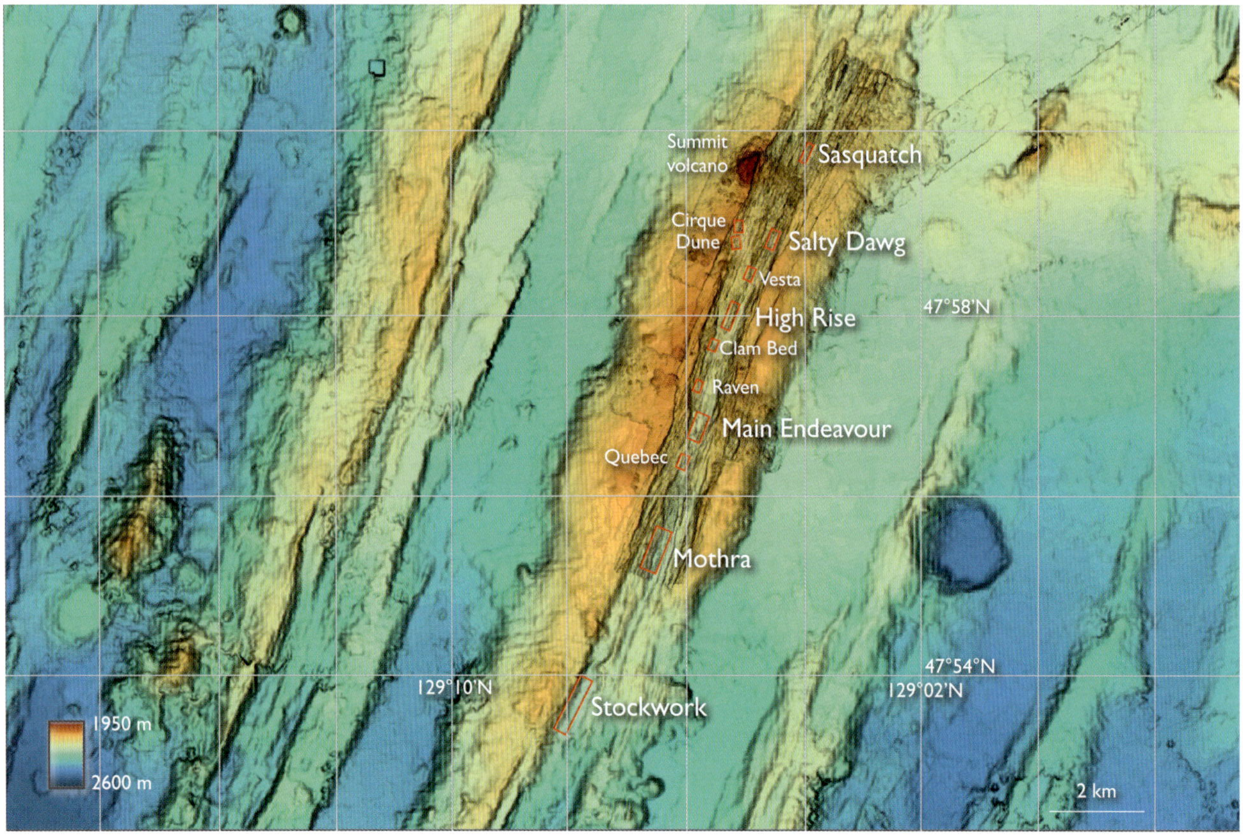

Figure 4.116 The central portion of the Endeavour Segment is one of the most intensely active hydrothermal sites along the global ridge system, hosting five major hydrothermal fields (Sasquatch, Salty Dawg, High Rise, Main Endeavour, and Mothra) and numerous smaller sites, over a distance of ~15 km. Here, the ridge is bisected by a narrow, ~1 km wide valley with 100–200 m tall rift valley walls. All hydrothermal fields are underlain by a magma chamber. Topographic highs and lows flanking the ridge attest to a complex volcanic and tectonic history.

m

Table 4.1 Summary of vent field characteristics: Endeavour

Field	Location	Extent, activity
Sasquatch	47°59.85′ N, 129°4.0′ W, in the center of the axial valley	~10, <1–10 m tall, fragile sulfide chimneys with an areal extent of only ~20 × 20 m; extinct chimneys (up to 20 m tall); fluid temperatures up to 287 °C; rapid growth up to 10 m in 1 year.
Salty Dawg (SD)	47°58.9′ N, 129°04.6′ W, near eastern axial valley wall	>150 m in length, with core of field hosting 35 m long band of massive continuous sulfides; largest structure Grendl reaches 25 m across and 25 m high; field contains at least 25 sulfide structures with flanges common; fluid temperatures up to 305 °C; intense hydrothermal plume encountered in 2005 above field – source unclear; diffuse sites to west on axial valley wall include Cirque (sulfides) and Dune (tubeworms).
High Rise (HRF)	47°58.00′ N, 129°05.50′ W, near center of valley on summit and sides of elongate horst	>10 large structures, currently site of most intense hydrothermal activity; largest structure Godzilla rose 45 m above the seafloor, but collapsed in 1995–1996; similar to SD and MEF, flanges are common; Clam Bed diffuse + smoker site is ~0.5 km to the south.
Main Endeavour (MEF)	47°57′ N, 129°06′ W, near western valley wall along major N–S-trending fissure/fault	>17 large, multi-flanged edifices, which up until ~2005 had >100 black smoker chimneys; new chimneys initiated post 1999, but since then southern complex (Bastille) has been significantly waning; fluid temperatures up to 402 °C; white smoker complex called Cathedral formed post 1999–2000; proximal diffuse flow sites common; significantly perturbed by 1999–2000 events.
Mothra	47°55.2′ N, 129°06.3′ W, near western valley wall	Composed of six clusters of chimneys reaching up to 20 m in height, spaced ~600 m along-strike and ~200 m across-strike; many tens of active and inactive chimneys, typically tall, slender and lacking pronounced flanges; numerous sites of diffuse flow; temperatures up to 320 °C; smokers not perturbed by 1999–2000 events though diffuse flow was impacted.
Stockwork	47°54.00′ N, 129°07.45′ W	Along western axial valley wall with 8 m tall extinct structures at base of wall; continuous exposure of upflow zone 200 m across, 100 m high on wall with massive sulfide, sulfide breccia, well-lithified metalliferous sediments; CTD results indicate weakly venting system immediately to northeast.

marking the upper temperature limit to life (Kashefi and Lovely, 2001).

Endeavour is the only area where seismic intensity has been linked directly to heat flux at the individual vent field scale (Wilcock *et al.*, 2007; Kellogg, 2011). Until about 2001, there was intense phase separation in the subseafloor (boiling, condensation, and brine formation) (see Section 4.1) that had been ongoing for over two decades with fluid temperatures up to 402 °C and fluids with a tenth of normal seawater salinity (Delaney *et al.*, 1984; Lilley *et al.*, 2003).

4.10.1 Endeavour geology

The 90 km long Endeavour Segment is part of a dueling propagator with the Cobb Segment to the south and the Middle/West Valley to the north (Figs. 4.116 and 4.119). It is marked by a central rift valley that hosts the five major hydrothermal vent fields (from north to south): Sasquatch, Salty Dawg, High Rise, Main Endeavour, and Mothra. More recently discovered smaller fields are Raven, Stockwork, and Vesta (Table 4.1). There are also several distal, diffusely venting fields (Cirque, Dune, Clam Bed, Quebec) and very local zones of diffuse flow (Fig. 4.116) (Kelley *et al.*, 2012, 2014).

The central third of the segment is a 25 km long volcanic high, split by a 75 to 200 m deep, 0.5 to 1 km wide, steep-sided axial rift. A thin sill of magma is present beneath the 25 km long central portion of the segment, deepening from 2.2 to 3.3 km beneath the seafloor from the north to the south (Fig. 4.119) (Van Ark *et al.*, 2007). From variations in the depth, width, and amplitude of the magma lens reflection imaged on a suite of across- and along-axis seismic lines, the magma body is inferred to be segmented into about four main sections that underlie all of the major vent fields (Carbotte *et al.*, 2012). High Rise, Salty Dawg, and Sasquatch are all within the central portion of the shallowest lens segment, while the Main Endeavour Field is located above a zone of complex reflectivity at the southern end of this segment. Mothra is underlain by a deeper lens segment to the south, with a strong reflector about 2.5 km beneath the spreading axis (Glickson *et al.*, 2007). In 2005, a fossilized upflow zone was discovered along the western valley wall coincident with a series of extinct chimneys and weak plume anomalies; it is also underlain by a strong seismic reflector suggesting that a magma lens is present (Kelley *et al.*, 2012).

4.10.2 Vent fields

Vent field abundances within the Endeavour Segment increase to the north as the distances between them decrease. Their spacing is closest at the shoalest portion of the segment, where basalt chemistry (MgO chemistry) indicates melts are the hottest and where the axial magma chamber is at its shallowest depth beneath the seafloor (Fig. 4.119) (Kelley *et al.*, 2012). Hydrothermal flow is focused predominantly along normal faults, trending at ~020° within the axial rift and inner walls.

Each of the major high-temperature fields extends over several hundred meters along-axis, and hosts multiple sulfide structures, commonly rimmed by sites of diffuse flow (Figs. 4.120–127). The sulfide edifices are large compared to those at many slow- and fast-spreading ridges (Fig. 4.120 and Table 4.1) (Delaney *et al.*, 1992, 1997; Robigou *et al.*, 1993; Kelley *et al.*, 2001c; Glickson *et al.*, 2007; Kelley and Shank, 2010). Within the Main Endeavour Field, in 1999–2000, there were 19 actively venting sulfide structures, which hosted more than 100 individual black smoker orifices and the only white smoker complex on the segment – Cathedral (Figs. 4.120–4.123). The Main Endeavour and High Rise Fields are the most robustly venting systems, have the largest hydrothermal edifices, and historically have had the highest vent temperatures (Delaney *et al.*, 1992, 1997). Sasquatch is the smallest field with fewer than 10 active chimneys (Fig. 4.125), while Mothra is the largest field, extending more than 600 m along-axis and hosting at least six active sulfide complexes and more than 50 sulfide structures (Fig. 4.126).

Venting has been long-lived at High Rise, Main Endeavour, and Mothra, as evidenced by the massive sizes of the active chimneys and by extinct massive sulfide deposits that reach more than 300 m in length (Fig. 4.120). Some chimneys are 50 m across, and in 1996 the chimney Godzilla was more than 45 m tall (Delaney *et al.*, 1992; Robigou *et al.*, 1993). There are also smaller sites of black smoker activity that include a small venting area south of Salty Dawg called Vesta, and the Raven field (Fig. 4.108) (Kelley *et al.*, 2012). Less commonly, there are also present and past sites of black smoker activity on the western and eastern axial rift walls.

In addition to the myriad sites of black smoker activity, Endeavour also hosts numerous sites of diffuse flow both proximal and distal to the high-temperature fields. The discrete distal sites from north to south include Cirque, Dune, Clam Bed, Beach, and Quebec (Fig. 4.116). In contrast to diffuse sites at the EPR, which are sparse and only about 10 m by 10 m across, diffuse sites at Endeavour are numerous and commonly reach 20 m by 50 m across and sometimes larger (Fig. 4.124) (Table 4.1). The higher frequency and abundance of both sulfide structures and diffuse flow sites at Endeavour versus the EPR likely reflect the relative stability of heat sources and longer periods between eruptive events at Endeavour, which maintain permeability pathways. In contrast, the higher frequency of eruptive events at the EPR leads to disruption of flow paths and/or covering of hydrothermal flow sites by lava (Fornari *et al.*, 2012).

Figure 4.118 The black smoker complex called Sully issued 380 °C fluids in 1999–2000 following a diking event that fundamentally perturbed the hydrothermal plumbing system beneath the Main Endeavour Field. Hydrophone and temperature–resistivity (chlorinity) sensors showed that the vents are acoustically very "loud" (Crone *et al.*, 2006) and that both brines and vapor were generated during the intrusive event (Larson *et al.*, 2009). The pressure housing on the left is about 0.75 m long.

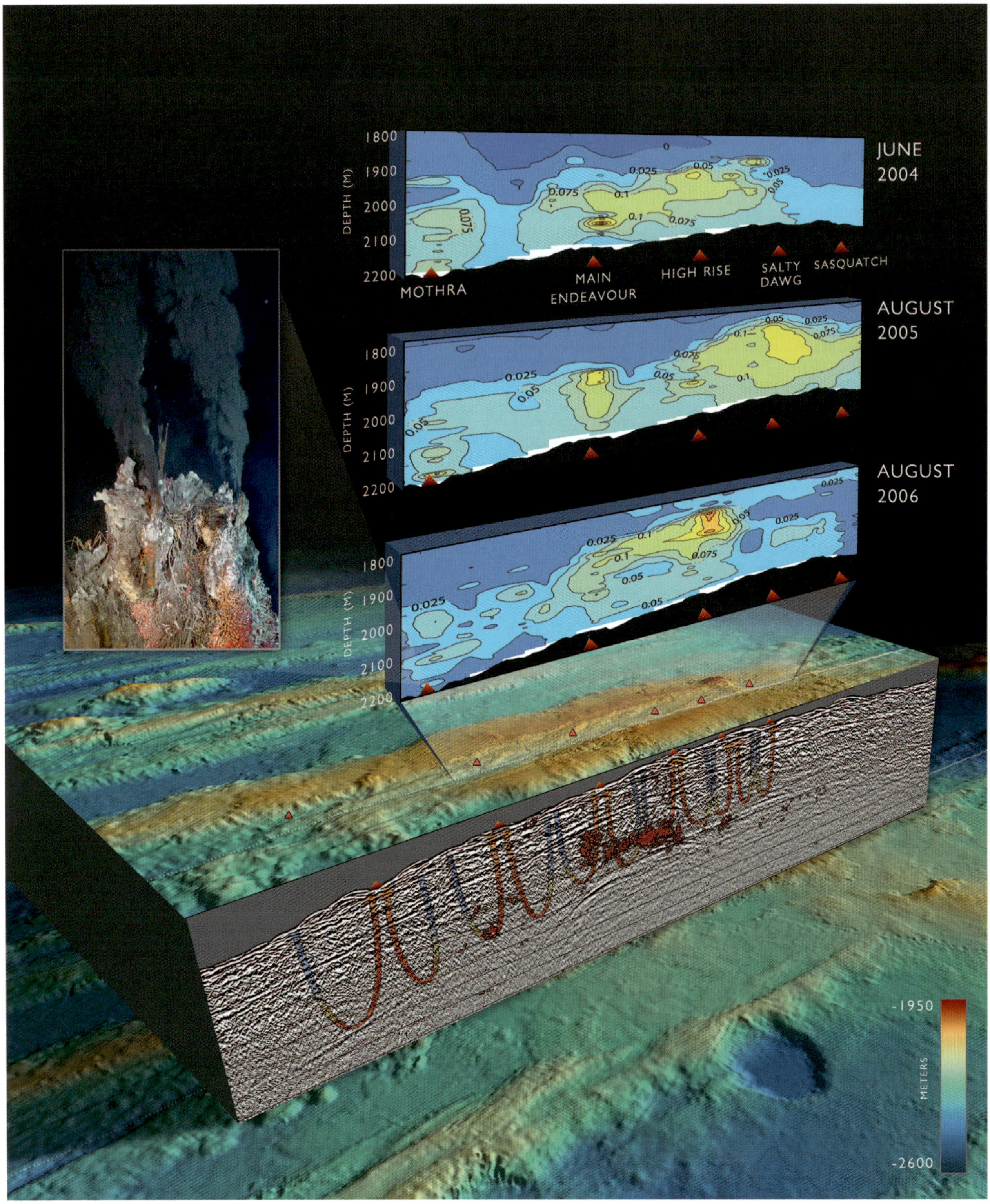

Figure 4.119 Magmatic, seismic, and venting processes at the Endeavour Segment of the Juan de Fuca Ridge are intimately intertwined. The black and white area under the topography of the segment shows a multichannel seismic cross section that defines strong reflections from the tops of magma chambers present beneath all five of the vigorously venting hydrothermal fields – Mothra, Main Endeavour, High Rise, Salty Dawg and Sasquatch (red triangles) (Van Ark *et al.*, 2007; Carbotte *et al.*, 2012, Kelley *et al.*, 2012). Heat extracted from the crust drives hydrothermal flow (gradient colored arrows), yielding black smoker edifices that can be up to 45 m in height (inset photo). Venting supports dense and diverse biological communities. Seismic activity (red dots) (Wilcock *et al.*, 2009) and intrusion of melt beneath the seafloor results in profound transients in both heat flux and hydrothermal plumes which rise ~200 m above the seafloor (as shown in the three panels of potential temperature anomalies based on data collected over three years from 2004 to 2006) (Kellogg and McDuff, 2010).

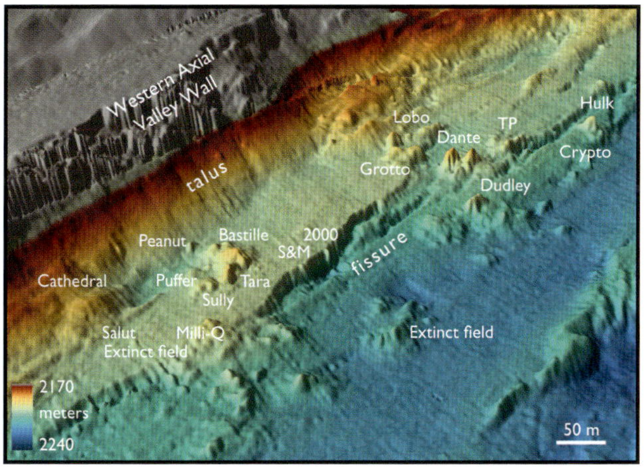

Figure 4.120 The Main Endeavour Field was the first field discovered along the Endeavour Segment – it was discovered in 1982 when a large dredge came up filled with massive sulfide. The field is also the best studied of the five fields and it, as well as the other fields, has been mapped to ~1–5 m bathymetric resolution. At this scale, the individual sulfide chimneys (e.g., Bastille, Grotto, Milli-Q) are easily delineated, as is the major northeast-trending fissure system that channels fluid flow.

Figure 4.121 The white smoker complex called Cathedral was discovered in 2000 on a talus rampart on the western axial valley wall in the Main Endeavour Field (Fig. 4.120). It hosted numerous slender and highly friable white smoker chimneys, reminiscent of pipe organs. The area was venting highly acidic fluids (pH 2) at temperatures up to 287 °C. It is thought that entrainment of seawater into the underlying talus and mixing with hydrothermal fluids resulted in precipitation of most of the metals within the rampart and the formation of white "smoke."

Figure 4.122 The 360 °C, multi-flanged structure called Milli-Q (Main Endeavour Field) rises 10 m above the surrounding seafloor. A microbial incubator, designed to investigate the upper temperature limit to life, is inserted into a drill hole. Similar to most chimneys in the field, the edifice is marked by numerous active and inactive flanges.

4.10.3 Sulfide chimneys

The highest temperature hydrothermal chimneys at Endeavour commonly host thick rinds of the copper-iron sulfide mineral chalcopyrite, lining the inner conduits. Though rarely directly measured due to inaccessibility, conduits in the largest chimneys may reach several tens of centimeters across (Fig. 4.127). The copper-rich minerals precipitate from the hottest hydrothermal fluids (generally hotter than 250 °C). Other high-temperature minerals include iron and zinc sulfides such as pyrite and sphalerite (Tivey and Delaney, 1986; Tivey *et al.*, 1999; Kristall *et al.*, 2006). The outer walls of the chimneys are highly complex, hosting multiple generations of smaller conduits, fractures, and mineralogy. Typically, they are highly porous, and almost sponge-like in appearance. Here, ingress of seawater into the chimney walls results in variable mixtures of minerals directly precipitated from hydrothermal fluids (pyrite, sphalerite,

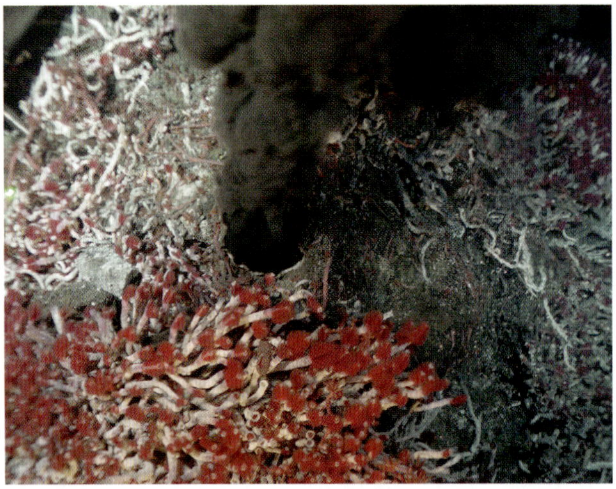

Figure 4.123 Several centimeter-wide orifices billowing black smoker fluids commonly cap larger sulfide edifices in the Main Endeavour Field. Numerous distinct outflow sites may occur on an individual chimney. This 380 °C vent on the structure Sully is surrounded by areas of diffuse flow that support thriving tubeworm (*Ridgeia piscesae*) colonies.

Figure 4.124 Diffuse flow around the monolith structure called Hulk in the Main Endeavour Field (see Fig. 4.120) supports lush tubeworm communities and yellow and white microbial mats. Yellow snails (*Buccinum thermophilum*) most commonly occur in areas of very weak or no flow. Field of view ~1 m.

chalcopyrite), and those largely derived from seawater, which typically contain sulfate (barite and anhydrite) (Figs. 4.126 and 4.128). These systems are highly dynamic and experience myriad fracturing events within the chimney walls, changes in temperatures and chemistry of the upwelling fluids, and periods of variable ingress of seawater (Kelley *et al.*, 2002). These processes result in the formation of anastomosing networks of variably mineralized channels, re-dissolution and precipitation of minerals, and multiple generations of mineral formation (Kristall *et al.*, 2006).

Flanges, which may reach sizes larger than the submersible *Alvin*, are a hallmark of the large black smoker chimneys at Endeavour (Delaney *et al.*, 1992; Robigou *et al.*, 1993; Kelley *et al.*, 2002) (Fig. 4.129). The flanges form when the walls of the black smoker edifices fracture and high-temperature hydrothermal fluids rush out of the cracks and rapidly mix with seawater.

Mixtures of hydrothermal fluids and seawater that flow through the top of the highly-porous flanges support dense tubeworm and limpet colonies. Rapid egress of the high-temperature, metal-rich hydrothermal fluids from chimney surfaces encases dense tubeworm colonies that grow on the outer chimney walls, rapidly killing the biological communities. Follow-on precipitation of sulfide- and sulfate-rich minerals during flange growth progressively incorporates the remains of tubeworms and results in their subsequent fossilization within the flange and chimney walls (Figs. 4.129–4.131) (Kristall *et al.*, 2006). Nutrient-rich fluids within the flange walls support dense microbial colonies (Hedrick *et al.*, 1992; Baross and Deming, 1995).

4.10.4 Extinct deposits

The Endeavour hydrothermal system is highly dynamic, as evidenced by individual extinct chimneys within vent fields and along the axial valley walls, as well as entire fields that are extinct, and well-preserved upflow zones located along the western axial valley wall. Only recently has some of the extinct material been dated, with the oldest sulfide deposits yielding ages of about 6000 years (Jamieson *et al.*, 2013). Within individual vent fields, extinct chimneys are common, as are highly oxidized sulfide blocks (Fig. 4.132) that mark chimney collapse and redistribution of localized upflow zones. During the history of Endeavour, entire fields have shut off, probably due to diking and/or seismic events. This is best evidenced in an area 90 m east of the Main Endeavour Field where a 500 m long zone of highly oxidized massive sulfide marks a past site of major hydrothermal upflow (Fig. 4.120).

Fossilized zones of upflow are rare in the Endeavour systems and have only rarely been found along MORs (Embley *et al.*, 1988). However, about 4 km south of Mothra, there is one of the best preserved upflow zones known. This area, known as the Stockwork Field, is located along the western axial valley wall and is 3.3 km above a highly reflective seismic feature interpreted to be

Figure 4.125 Small, isolated chimneys, which sprout from old blocks of sulfide intermixed with lava talus, typify vents in the Sasquatch Hydrothermal Field. Blue colonial ciliate mats (*Folliculinopsis* sp.) and small tubeworms are supported by localized hydrothermal flow. Spider crabs (middle right) commonly forage for food around the vents.

a magma chamber (Carbotte *et al.*, 2012; Kelley *et al.*, 2012). The area hosts numerous exposures of upflow zones. The most impressive fossilized system is a massive, continuous exposure, 200 m across and 100 m high. It is composed of variably altered and mineralized truncated pillow flows, interspersed with lenses of massive sulfide, which include blocks of monomineralic chalcopyrite (Figs. 4.133 and 4.134). Truncated pillows commonly exhibit pronounced mineralization by copper-rich minerals with the interstices between pillows coated or filled with red iron-oxide phases.

The base of the upflow zone is marked by steep talus slopes of variably altered pillow debris that has clearly come downslope from the upflow zones, trunks of chimneys, and variably oxidized massive sulfide deposits. Extinct sulfide chimneys adjacent to the axial wall reach up to 14 m in height, and are reminiscent in morphology to those at Mothra (Kelley *et al.*, 2001a; Glickson *et al.*, 2007). The slope base is delineated by an abrupt change from talus to tens of centimeters thick hydrothermal and pelagic sediment

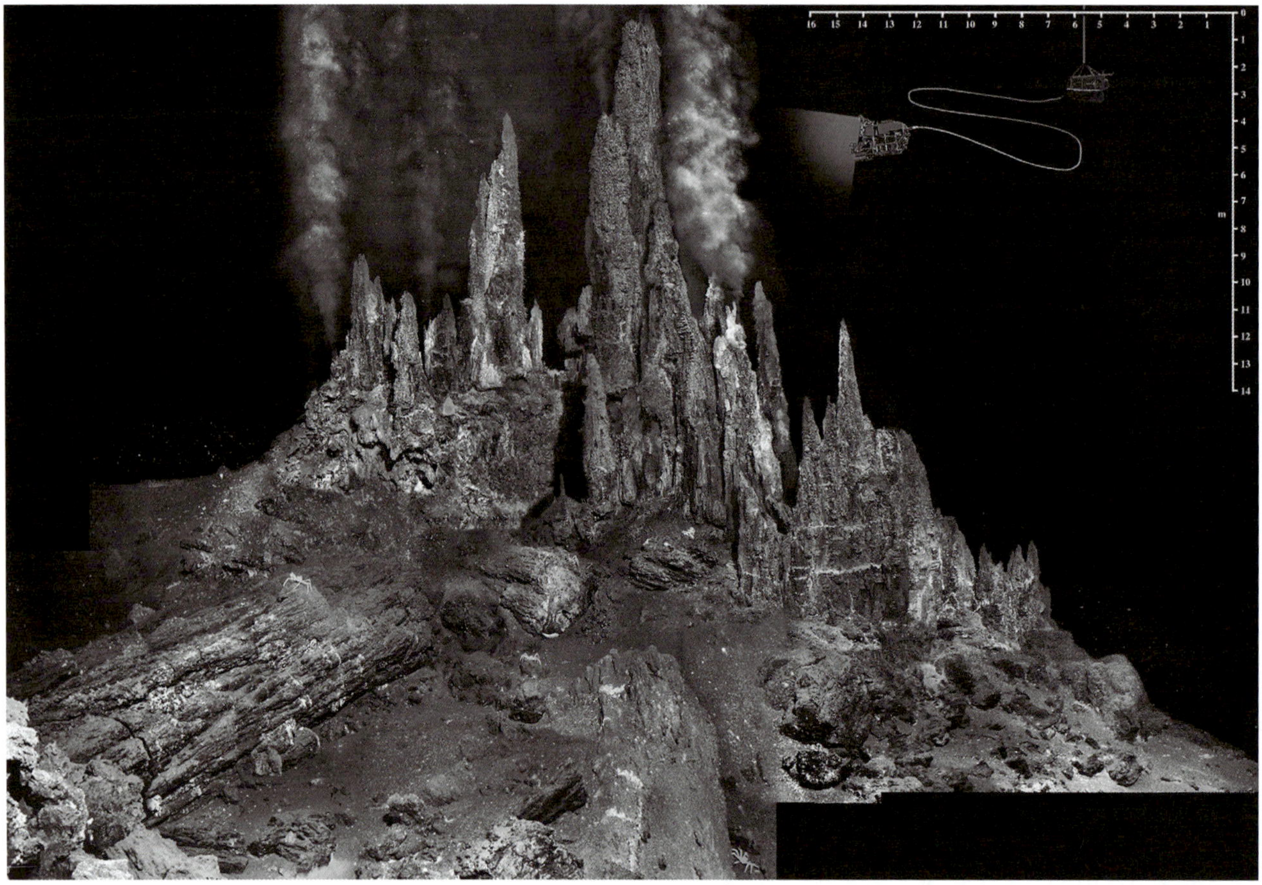

Figure 4.126 The Faulty Towers Complex is the largest array of chimneys in the Mothra Hydrothermal Field. It is composed of 14 active and inactive structures which reach 22 m in height. Nearly the entire surfaces of these structures are awash in diffusely venting fluids. The edifices are rooted at their base in a 6 m high steep rampart of massive sulfide and sulfide talus formed from collapse of the chimneys over time. Very large "logs" of sulfide occur on the west side of the field, marking past episodes of chimney collapse. They are underlain by a ~1 m thick layer of oxidized sulfide sediment deposited by both hydrothermal plume fall-out and chimney collapse. In 1998, the top ~2 m of the three southern chimneys – Finn, Roane and Phang, from left to right – were recovered intact during the University of Washington and American Museum of Natural History project called Edifice Rex. This mosaic is composed of 144 images taken with the ROV *Jason* (artificially added in for scale) in 1997 during a site characterization program led by the University of Washington in preparation for the sulfide recovery project.

deposits which completely mask the underlying geology. Water column studies and high-resolution bathymetry showing small chimney-like features suggest that there is a weakly venting system 200 m north and northeast of the Stockwork area.

4.10.5 Diking events, seismicity, and hydrothermal flow

A 20-year SOSUS (Sound Surveillance System) earthquake catalog (Dziak *et al.*, 2011) (see Chapter 1) shows that the Endeavour Segment lies in a region of elevated background seismicity that likely results from the reorganization of plate boundaries in the region (Wilcock *et al.*, 2002). The reorganization is due to the presence of an unstable triple junction at the northern end of the Juan de Fuca Ridge, which formed when the Explorer plate detached from the Juan de Fuca Plate along the Nootka transform fault at ~4 million years ago (Riddihough, 1984).

In addition to the elevated background level of seismicity, Endeavour has been punctuated by three major seismic swarm events during the past two decades, which have variably perturbed the hydrothermal systems (Fig. 4.135). Two elevated periods of seismic activity were recorded in June 1999 and January 2000 (Lilley *et al.*, 2003; Bohnenstiehl *et al.*, 2004), and another in February 2005 (Hooft *et al.*, 2010; Weekly *et al.*, 2013). The 1999 activity lasted 5 to 11 days and spanned the along-axis region of the shallow magma chamber reflectors. The January 2000 swarm was more limited in

Figure 4.127 In 1998, the top ~2 m of the 305 °C chimney called Finn was recovered by encasing the edifice in an open "cage" and by using an underwater chainsaw on the Canadian ROV *ROPOS* (see Fig. 4.128). Revisitation of Finn showed that it had regrown 2 m in the ensuing 72 hours! Organisms cultured from a sample of the new growth chimney in 1999 required temperatures of 121 °C and anaerobic conditions, marking the highest temperature organism then known (Kashefi and Lovely, 2003).

extent, of shorter duration, and did not show migration. In contrast to the 1999 and 2000 swarms, the February 2005 swarm began at the overlapping spreading center on the northern end of the Endeavour Segment (Hooft *et al.*, 2010; Weekly *et al.*, 2013).

4.10.6 The 1999–2000 diking event

Prior to 1999, there had been a long-standing chemical gradient between the northern and southern portions of the Main Endeavour Field. This had been interpreted to reflect feeding by two distinct limbs of upflow (Butterfield *et al.*, 1994; Delaney *et al.*, 1997; Kelley *et al.*, 2002). Fluid chemistry showed well-defined variation in salinity, carbon (CO_2 + CH_4), and temperature from north to south, with the highest salinity, lowest temperature fields occurring at Sasquatch and Mothra (~300 °C, 710 mmol/kg Cl) (Lilley *et al.*, 1993; Butterfield *et al.*, 1994; Delaney *et al.*, 1997; Lilley *et al.*, 2003).

The Main Endeavour Field hosted the long-standing record (two decades) for the highest vent fluid temperatures (up to 402 °C) on the segment (Delaney *et al.*, 1984), with many of the structures in the southern Bastille complex exhibiting both boiling and supercritical condensation (Fig. 4.18) (Butterfield *et al.*, 1994; Kelley *et al.*, 2002; Lilley *et al.*, 2003). However, beginning June 8, 1999, a diking event and resultant seismic swarm disrupted this gradient. This resulted in significant increases in gas concentrations and profound changes in venting temperature, intensity, and chemistry, which had been relatively stable since the discovery of vents here in 1982 (Delaney *et al.*, 1992).

The 1999 event caused dramatic increases in CO_2 concentrations and other magmatic gases, and produced elevated H_2 concentrations through increased high-temperature water–rock reactions (Lilley *et al.*, 2003; Proskurowski *et al.*, 2006). Many chimneys within the Main Endeavour Field increased in temperature by 15 °C, observations of boiling were common, and some vents emitted fluids that were nearly salt-free (Lilley *et al.*, 2003). Measurements of vent fluid temperatures at the Main Endeavour Field and Clam Bed site showed a tenfold increase in fluid output for at least 80 days following the event (Johnson *et al.*, 2000). This event was interpreted as a lateral dike injection (Johnson *et al.*, 2000; Lilley *et al.*, 2003; Bohnenstiehl *et al.*, 2004; Proskurowski *et al.*, 2004) (see Chapter 6). Modeling of pressure transients measured in Ocean Drilling Program boreholes east of the Endeavour Segment suggested a dike injection 40 km in length and depth extent of 3 km (Davis *et al.*, 2001). The southern Bastille complex (Fig. 4.120), which was venting 380 °C and nearly fresh fluids, is now dying, with numerous chimneys only weakly venting; temperatures at this site dropped to 230 °C by 2007 (Fig. 4.135).

Figure 4.128 The recovered section of the 305 °C black smoker Finn showing an inner channel several centimeters across, lined by the copper mineral chalcopyrite. Very rare to black smokers was the presence of significant clay deposits. The inner channel also hosted an anhydrite horn coated by chalcopyrite, indicating significant influx of seawater along a fracture, followed by fracture closure.

Figure 4.129 Flanges, such as the one shown here, are a hallmark of Endeavour chimneys. They form during outbreaks of hot hydrothermal fluids along fractures in the main trunks of the chimneys. Looking upward underneath a flange shows that they are convex and commonly host shallow reflecting pools of hot hydrothermal fluid that flows outward from the main edifice. The pools progressively overfill with hot, buoyant fluids that rise upward over the outer lip of the flange, causing outward growth. This flange on the structure Smoke and Mirrors within the Main Endeavour Field hosts 360 °C fluids and is 3 m across. Tubeworms, limpets, and palm worms dominate the area beneath the flange.

4.10.7 The 2005 diking event

In late January and late February 2005, the SOSUS arrays and an *in situ* seismic network recorded two complex swarm sequences that involved a north-to-south progression of seismicity over several days from 30 km northwest of the Endeavour Segment (Hooft *et al.*, 2010). The migration of earthquakes suggests that a dike propagated southward toward the center of the segment from a magma source to the north. Both swarms triggered increased rates of seismicity beneath the vent fields. The January swarm was concentrated between Salty Dawg and High Rise Fields (Weekly *et al.*, 2013), while the February swarm was concentrated further south between the High Rise and Main Endeavour Fields and coincided with a thermal perturbation to a diffuse vent site in the Mothra Field.

Immediately following the swarm sequences, seismicity almost ceased at both ends of the Endeavour Segment and decreased to a quarter of pre-swarm levels beneath the vent fields by the end of 2006 (Wilcock *et al.*, 2009). Beginning with the 1999 and 2000 earthquake swarms and ending with the 2005 swarm, it appears that these events marked the end of a 6-year episodic spreading event that cumulatively ruptured the entire segment and relieved the extensional stresses and magmatic pressures (Weekly *et al.*, 2013).

While the Main Endeavour Field is currently waning, water-column measurements show that High Rise, as of 2005, had the highest heat flux, and the temperatures of some chimneys had increased (Kellogg, 2011). Only two dives have been made at Salty Dawg since 2000, but a significant plume was detected in 2005 above the field. The plume could not have been sourced from either High Rise or Sasquatch (Fig. 4.119) (Kellogg and McDuff, 2010). Also, some vent fluid temperatures in 2006 increased, following a previous visit in 2000. Activity at Sasquatch has also increased since 2005; new chimneys have developed, while

Figure 4.130 A cross section of the hydrothermal edifice Roane from the Faulty Towers complex shows a classic example of early flange development. To the left is the gray-colored inner chimney wall, rich in iron and zinc sulfide minerals, and containing numerous small cavities. The white area to the right marks the development of a small flange, ~6 cm across, which hosted live tubeworms when it was brought to the surface in 1998 during the University of Washington and American Museum of Natural History Edifice Rex project. The white to gray minerals are rich in barium sulfate, with lesser iron sulfide. Tubeworms are progressively fossilized in these zones as warm to hot hydrothermal fluids leak out of the chimney walls and mix with seawater, resulting in mineral precipitation.

Figure 4.131 A cross section of the outermost portion of a chimney wall from Roane, showing progressively fossilized tubeworms. The outermost area still contains open tubes, while in the inner walls these tubes are completely filled and replaced by iron and zinc sulfides, sulfate minerals, and amorphous silica. The tubes are less than 0.5 cm across.

others have undergone dramatic growth (10 m in one year) (Kelley *et al.*, 2012, 2014). In contrast, no changes in either vent fluid chemistry or black smoker temperatures were measured at Mothra from 1996 to 2007; only weak seismic activity has been recorded beneath Mothra.

Figure 4.132 Blocks of highly oxidized massive sulfide commonly form skirts and talus ramps around the large sulfide edifices at Endeavour. The massive block in the bottom right portion of the image is ~2 m across and located in the Main Endeavour Field. Microbial communities are active in weathering–oxidation processes.

Figure 4.133 This preserved upflow zone along the western axial valley wall of the Endeavour Segment is part of the Stockwork Field (Fig. 4.116). Truncated pillow basalts are pervasively altered to copper and iron sulfide minerals, iron oxide, and clay. The outer portions of the upflow zone (right) are less altered basalts. Scale bar is 1 m across.

4.10.8 First real-time monitoring of geophysical, chemical, and biological processes

The Endeavour system has held the imagination of researchers and the public since its discovery in 1982. The massive sulfide edifices, billowing smokers, and rich animal communities truly make it one of the most amazing places on Earth. In part, because of its dynamic nature, and our inability to be there continually and directly observe rapid changes that have profound impacts on the thermal, chemical, and biological fluxes at this site, scientists decided to bring the Internet, with its significant observational power, to Endeavour.

This idea was first explored by John Delaney at the University of Washington in 1987, proposing an underwater

Figure 4.134 Bright green copper-halide minerals (atacamite) coat the outer surfaces of highly altered and mineralized pillow basalts. Some areas comprise zones of massive pyrite with several-centimeters-across angular blocks of copper sulfide (chalcopyrite). Pillow is ~1 m across.

cabled observatory at the scale of a tectonic plate – known then as NEPTUNE (Northeast Pacific Time-Series Underwater Experiment). In 2003, the Juan de Fuca Plate was chosen by the US science community as the site for the first regional cabled ocean observatory. This was in recognition that a representative suite of global ocean and seafloor processes occur in this area. That same year, NEPTUNE Canada, led by the University of Victoria, received funding for implementation of an 800 km cabled observatory with significant focus on Endeavour (Fig. 4.136). Both NEPTUNE Canada and the US system, known as the Regional Scale Nodes, are now in the water (see Section 4.11 and Chapter 9) (Kelley *et al.*, 2014).

In concert, these two high-power (8–10 kW) and high-bandwidth (10 Gb/s) expandable systems combine to be the first plate-scale experiment, with a total of 1700 km of fiber-optic cable, hundreds of seafloor sensors, and an array of instrumented water column moorings with robotic wire crawlers that extend to water depths of ~3 km. The ocean community is on the verge, for the first time, of being able to

Figure 4.135 The Main Endeavour Hydrothermal Field was profoundly disturbed by a diking event in 1999–2000. Almost all chimneys within the field increased in temperature by 15 °C, new chimneys started up, existing edifices grew several meters in one year, and microbial-generated flocculant material issued from new diffuse flow sites. Revisitation of the hydrothermal chimney Sully from 2000 to 2007 (shown here), and several other chimneys, shows that the southern portion of the field is waning with dramatic decreases in temperature, and some chimneys that had been venting fluids at 360–380 °C (e.g., Milli-Q; Fig. 4.122) are now extinct. The pressure housing in the 2005 image is about 0.75 m across.

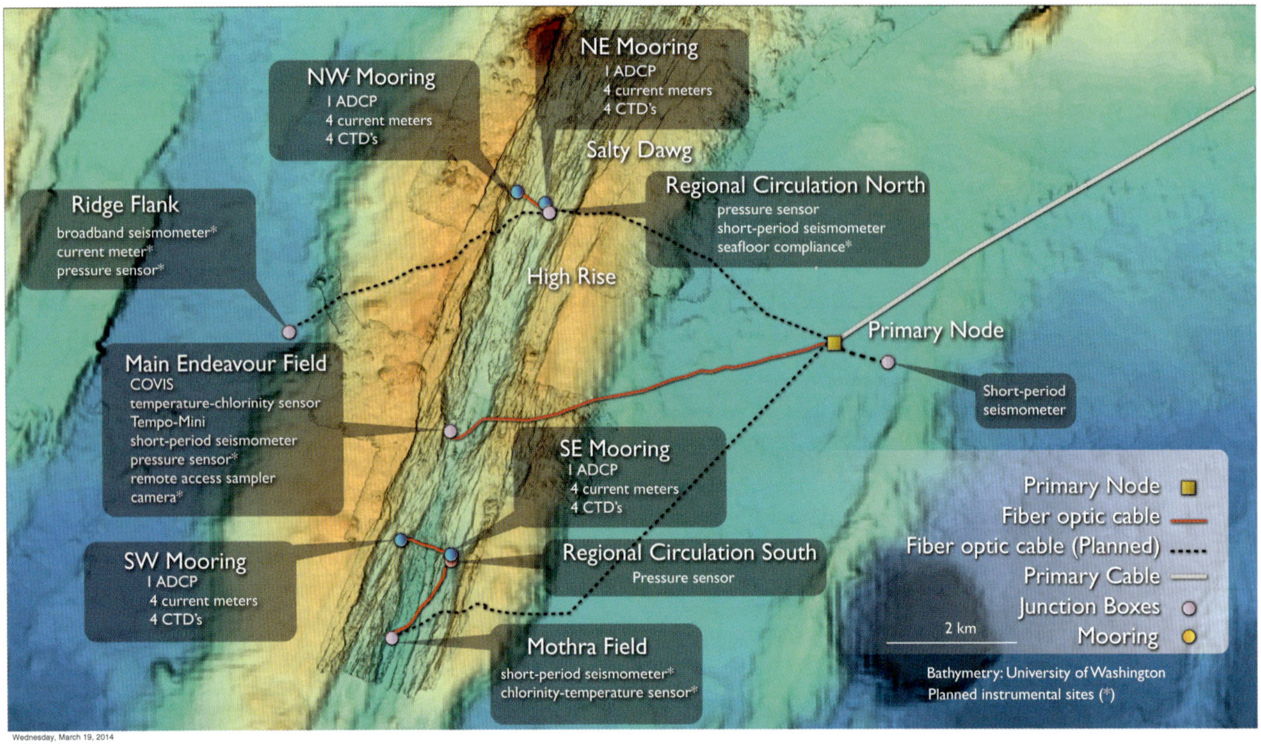

monitor and respond to events in real time at multiple sites on the Juan de Fuca Ridge. The cabled experiments on both the Endeavour Segment and Axial Seamount (see Section 4.11) will provide unprecedented access to these environments, facilitate the ability to conduct controlled perturbation experiments, and offer direct links to exciting seafloor environments to a global community of viewers and participants (Kelley *et al.*, 2014).

Figure 4.136 Installed and planned cable and sensor deployments at the Endeavour node of the NEPTUNE CANADA high-power, high-bandwidth cabled observatory (Kelley *et al.*, 2014). The system includes an array of regional cabled moorings to monitor flow of fluids along-axis, and an array of geophysical and chemical sensors, and cameras to examine linkages among seismic activity, fluid flow, and biological responses.

Axial Seamount

4.11 Robust volcanism on the Juan de Fuca Ridge

Axial Seamount is the most magmatically robust volcano on the intermediate-spreading Juan de Fuca Ridge. It is located 450 km off the Washington–Oregon coast. Its summit shoals to a water depth of 1.5 km, rising nearly 800 m above the surrounding seafloor (Fig. 4.137). Axial is the site of current volcanic activity associated with the Cobb–Eickelberg seamount chain. Its basalts are enriched in alkalis and highly incompatible trace elements, reflecting influence from a hotspot melt source (Chadwick *et al.*, 2005). It is underlain by a magma chamber at 1.4 km beneath the volcano's summit (West *et al.*, 2001; Kent *et al.*, 2003), which provides the heat for hydrothermal flow at numerous fields hosted within its caldera (Fig. 4.138).

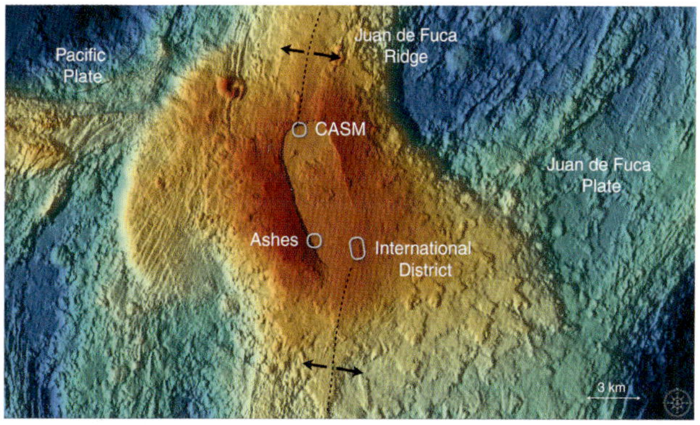

Figure 4.137 Axial Seamount is the most magmatically robust volcano on the Juan de Fuca Ridge, rising ~1000 m above the surrounding seafloor. Axial erupted in 1998 and in 2011. The volcano is dissected by the Juan de Fuca Ridge with the CoAxial Segment to the north and the Vance Segment to the south. A caldera, 3 km in diameter and 8 km long, hosts three distinct hydrothermal fields and numerous other sites of diffuse flow with isolated small chimneys.

Figure 4.138 Tubeworm- and limpet-covered sulfide chimney in the ASHES Vent Field near the southwestern caldera wall on Axial Seamount. The small, black chimney is very young (<1 year) and composed of fine-grained, highly friable sulfide minerals that include pyrite and zinc sulfide phases. Tubeworm casings are ~1 cm across.

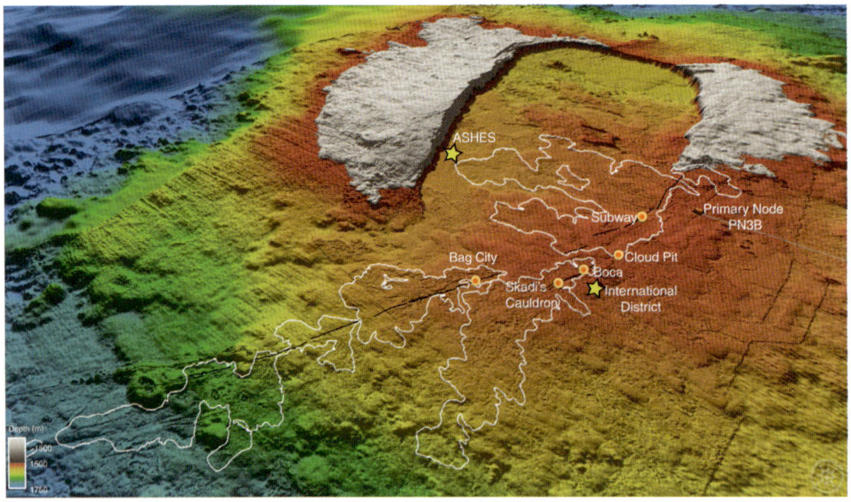

Figure 4.139 The summit of Axial Seamount was profoundly changed by the April 6, 2011 eruption. Numerous lava flows extended nearly 3 km across the caldera and flowed >8 km down the eastern fissure network (white outline). The flows were 3.5 to 17 m thick and were composed predominantly of lobate and pillow basalts (Caress *et al.*, 2012; Chadwick *et al.*, 2012). Many of the pre-existing channels were reutilized and are defined by an anastomosing network of variably filled collapse zones. The *en echelon* fissure system also hosted numerous snowblower sites (orange dots), which issued large plumes of microbial material and were enriched in CH_4.

Figure 4.140 The thicker parts of the 2011 lava flow were rapidly colonized by dense bacterial mats giving the flows an "old look." The microbial blooms result from expulsion of volatile-rich, warm fluids. Because the thicker flows stayed warm longer, fluid flow was likely more prolonged here. Green lasers from the ROV *ROPOS* are 10 cm apart at the surface of the flows.

Figure 4.141 Milky fluid vents from a skylight in a lobate flow four months after the 2011 eruption. This site is located along the eastern fissure system that hosted the eruption. The brown-orange color on the surface of the newly erupted basalts reflects a thick coating of microbial mats that colonized the surfaces of the flows shortly after eruption. The large microbially-covered cavity in the right-portion of the image is >1 m across.

Figure 4.142 Biological communities recolonized the Marker 33 diffuse flow site on the eastern fissure zone shortly after the eruption. Diffuse flow continued through the lobate flows. By August 2011, small tubeworms and limpets, as well as wispy filamentous bacteria, had colonized this area. The red polychaete *Vampiropolynoe embleyi* (up to 52 mm in length) also colonizes the new flow sites. This species was discovered at Axial after the 1998 eruption, grazing on bacterial mats (Marcus *et al.*, 2009).

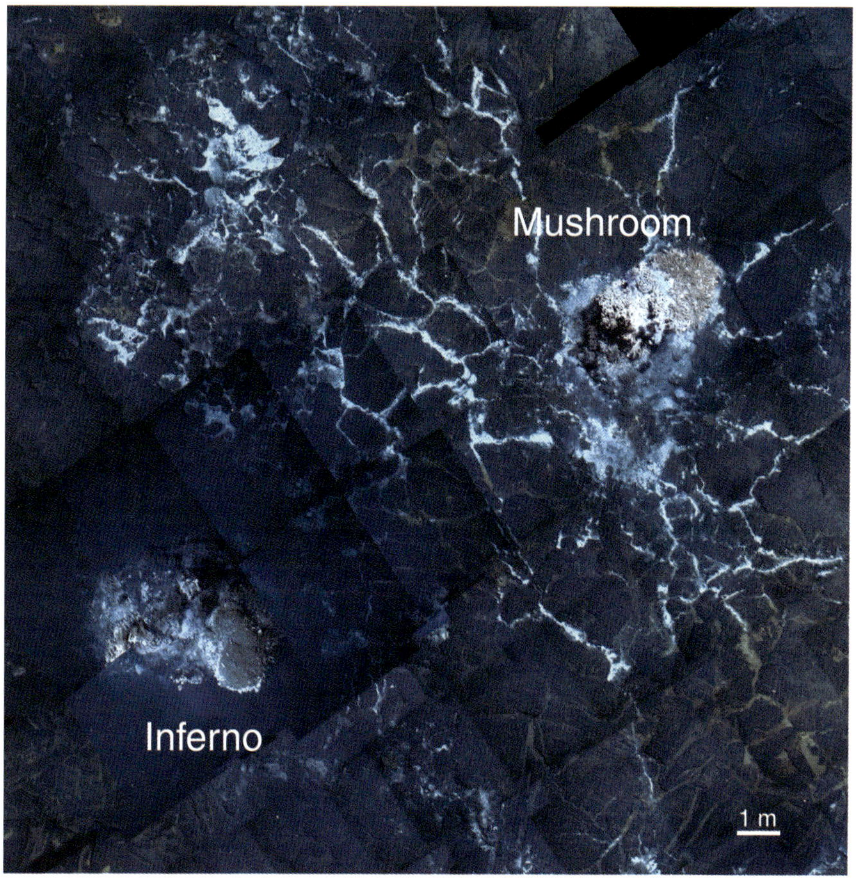

Figure 4.143 Down-looking mosaic of the ~4 m tall black smoker chimneys Inferno and Mushroom in the ASHES Hydrothermal Field. The chimneys are underlain by sheet and lobate flows. Low temperature, diffusely venting fluids issuing from radiating fractures in the basaltic seafloor support white, filamentous bacterial mats, limpets, and tubeworms. The two structures are 11 m apart. Images for this mosaic were taken with the ROV *Jason* flying ~5 m above the seafloor in 2010. Corresponding vertical mosaics of Inferno and Mushroom are shown in Figs. 4.145 and 4.146. Mushroom and the associated diffuse flow sites are now instrumented with chemical and temperature sensors and a high-definition camera as part of the Ocean Observatories Initiative cabled network installed in 2014.

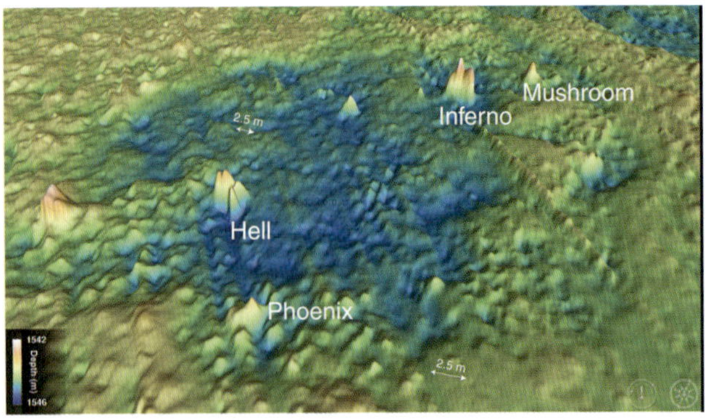

Figure 4.144 Meter-resolution bathymetry of the ASHES Hydrothermal Field on the southwest side of the caldera. The core of the field is covered in sheet flows (blue area), while the surrounding seafloor is composed predominantly of lobate flows.

Figure 4.145 Photomosaic of the 4 m tall, 250 °C black smoker Inferno, which hosts multiple orifices and is shrouded in dense colonies of tubeworms and limpets. The narrow chimneys on top are fine-grained sulfide and anhydrite and extremely friable.

4.11.1 Axial Seamount eruptions

Axial Seamount erupted in January/February 1998 and again on April 6, 2011. It is one of the few MOR sites where the full sequence of events associated with an underwater eruption has been documented (Fig. 4.139) (see Chapter 5). The beginning of the 1998 eruption was marked by an intense seismic swarm of 8247 earthquakes, which lasted 11 days and migrated 50 km from the summit caldera along the south rift zone (Dziak and Fox, 1990, 1999). A 3 km long sheet flow erupted along the southeastern part of the caldera, which was 500–800 m wide. Similar to other submarine diking and eruptive events, numerous sites of diffuse flow initiated, with some sites emitting abundant flocculant material and milky white fluid. The eruption partially or totally buried some pre-existing black smokers, and new diffuse flow sites supported nascent communities of tubeworms (Embley *et al.*, 1999).

Following the 1998 eruption, for several years the floor of the caldera at Axial Seamount rose at an average rate of ~16 cm/yr. Starting in 2010, however, this inflation rate increased to 34 cm/year and then to 50 cm from January to March 2011 (Chadwick *et al.*, 2012). On April 6 and 7, 3.0 and 3.7 magnitude earthquakes were recorded by land-based stations, and on April 6, >1000 earthquakes/day were recorded by ocean bottom seismometers (Dziak *et al.*, 2012). That same day the caldera floor rose 7 to 13 cm in 40–55 minutes. This was followed by a rapid deflation, marking the start of the eruption (Chadwick *et al.*, 2012). Similar to the 1998 event, the 2011 flow covered pre-existing diffuse flow sites, such as Marker 33, which remained active following the eruption. The toes of the flows, composed predominantly of pillow basalts, reached 17 m in thickness, with the thickest areas extensively covered by dense microbial mats (Fig. 4.140).

One of the most impressive consequences of the eruption was the development of numerous, spectacular snowblowers that formed along the eastern rift system, but which were also present at the southern termination of the caldera (Figs. 4.139 and 4.141). Most of the snowblowers were sourced in collapse pits several meters across and a few meters deep. Others issued from small skylights in lobate flows and from the interstices between lobate flows (Fig. 4.141). Thick, white microbial mats coated entire surfaces of the collapse zones and individual lava pillars. Significant egress of CH_4-rich diffuse fluids issued from the subsurface, forming dense clouds. The fluids were just

Figure 4.146 Photomosaic of the 4 m tall, black smoker Mushroom, which is covered in a very dense animal community. For a long-lived chimney, it is small in size and lacks a well-developed rampart of sulfide talus common to other chimneys in the Juan de Fuca system. Numerous fractures hosting diffuse flow radiate from its base (see Fig. 4.143).

above ambient temperature and contained massive amounts of flocculant and microbial material (Proskurowski *et al.*, 2012; Meyer *et al.*, 2013). Chemical gradients around the collapse pits resulted in zoning of microbial community structure: white mats dominated areas of most intense flow while brown-orange mats were abundant on the surfaces of adjacent basaltic flows (Fig. 4.141). Genomic and laboratory analyses show that the white flocculant was dominated by elemental sulfur that likely resulted from sulfide-oxidizing organisms. The issuing fluids and white floc also contained *Methanococcales* and anaerobic thermophilic archael groups, requiring that these organisms were sourced from the subsurface (Meyer *et al.*, 2013). In contrast, the orange mats contained iron-oxidizing bacteria, and lower concentrations of sulfur-oxidizing bacteria than the snowblowers, as well as organisms that colonized on settled plume particles, and multicellular debris (Meyer *et al.*, 2013). The snowblowers may serve as important nucleation sites for follow-on microbial and macrofaunal systems.

Three months after the eruption, the site was already colonized by biological communities including filamentous bacteria, tube and scale worms, limpets, and microbial material that continued to issue from the seafloor (Figs. 4.141 and 4.142). Significant temperature decreases (by up to 100 °C) were noted in many of the black smoker chimneys following the eruption, presumably from the drawdown of seawater to replace that displaced during formation of numerous snowblowers, which continued to vent several months after the eruption (Fig. 4.141). The snowblowers were predominantly located along the *en echelon* eastern rift system, extending down across the southern portion of the caldera.

4.11.2 Shallow hydrothermal vent sites

Axial Seamount hosts three major hydrothermal fields in addition to numerous sites of diffuse flow. The black smoker sites include (1) the ASHES (Axial Seamount Hydrothermal Expeditions) Field on the western side of the caldera (Butterfield *et al.*, 1990; Hammond, 1990) (Figs. 4.143–4.146); (2) CASM [Canadian American Seamount (Expedition)] at the northern edge of the caldera; and (3) the International District along the southeastern edge of the caldera (Figs. 4.137 and 4.139).

ASHES has been studied for over two decades, while many of the other sites have at least one decade of time-series measurements (Butterfield *et al.*, 1990, 2004). Chimneys at Axial are commonly small in size considering that they have been active for more than two decades, with many chimneys less than 5 m tall. The tallest chimney, El Guapo, is in the International District. It rises ~18 m above the seafloor (Figs. 4.147 and 4.148). In contrast, chimneys in the ASHES Vent Field are commonly 4–5 m tall (Figs. 4.143–4.146). ASHES also hosts sparse, fragile, small anhydrite chimneys that have been active for at least two decades (Fig. 4.149). The ASHES field is underlain by fractured sheet flows and lobate flows with diffuse venting issuing from many of the fractures (Fig. 4.143) (Kelley *et al.*, 2014). The outer portions of the chimneys are bathed in variable mixtures of hydrothermal fluid and seawater, which support dense animal and microbial communities.

The most active areas of venting are typically on the summits of the black smoker edifices, though some active orifices also occur near the base of the chimneys (Figs. 4.145–4.148). Actively forming black smokers form slender, highly

Figure 4.147 El Guapo – the handsome one – rises 18 m above the surrounding seafloor. The chimney is located in the International District Hydrothermal Field on the southeastern side of the summit caldera. At a depth of 1500 m, fluids issuing from this vent are boiling. The outer walls are covered in dense colonies of tubeworms, limpets, scale worms, and palm worms. White mats of filamentous bacteria also colonize areas of active flow.

friable, mottled black spires composed of mixtures of anhydrite and iron and zinc sulfides (pyrite, sphalerite, and marcasite) (Figs. 4.145 and 4.148). The highest temperature conduits also contain chalcopyrite, as well as other trace sulfide phases.

As the chimneys evolve, amorphous silica and barite form important mineral phases that give strength to the edifices. Chemical analyses of one of the chimneys showed that silica, zinc, and barium account for 65 wt% of the spire (Hannington and Scott, 1988). Many of the chimneys in ASHES have small ramparts of old sulfide debris (Fig. 4.146), while those in the International District are characterized by skirts of large blocks of sulfide debris. Numerous extinct chimneys also occur in the International District, some partially enclosed by lava flows. Sulfide structures in the International District are larger than those at ASHES (e.g., El Guapo) (Fig. 4.147).

4.11.3 Boiling fluids

High-temperature vent fluids issuing from chimneys within the Axial system exhibit a wide range of fluid compositions from gas-rich to gas-depleted and with salinities measured in the late 1980s ranging from 33% to 116% of normal seawater values (Butterfield *et al.*, 1990). The relatively shallow depth of the caldera below the ocean surface resulted in the common occurrence of boiling, at least until the April 2011 eruption. El Guapo, at a depth of about 1.5 km and a temperature of 340 °C, places the venting fluids precisely on the boiling curve as they exit its summit (Fig. 4.148).

Figure 4.148 The summit of El Guapo hosts numerous black smoker orifices. For many years the chimney produced fluids at ~349 °C. At a depth of 1500 m, this places the fluids on the boiling curve such that significant vapor was generated. On April 23, temperatures began to significantly drop in step functions and in August 2011 fluid temperatures had dropped to 210 °C, well off the boiling curve. The interpretation is that significant seawater was entrained into the subsurface to replenish venting from the snowblower sites and other diffuse sites on the newly erupted lava flows.

Figure 4.149 The anhydrite chimney called Virgin Mound is located in the ASHES Vent Field. It rises <1 m above the seafloor. Vent fluid temperatures reach 300 °C; historically it has had the highest dissolved gas concentrations and the lowest values for non-volatile elements. In 1990, this chimney had chloride concentrations of 176 mmol/kg versus 539 mmol/kg for seawater, consistent with subsurface boiling processes (Butterfield *et al.*, 1990).

Figure 4.150 Galatheid crab in diffuse flow site at the International District Vent Field on Axial Seamount. Dense, fine mats of filamentous bacteria coat the carapace. Also shown are snails and small limpets. The crab is only ~3 cm across.

Figure 4.151 Dense colonies of *Paralvinella palmiformis* worms and *vestimentiferan* tubeworms cover the outer surfaces of active chimneys at Axial. Limpets (white shells – *Lepetodrilus fucensis*), ~1 cm across, are also common. Sulfide worms typically colonize higher-temperature regions where exposed sulfide surfaces are visible.

Figure 4.152 Healthy tubeworms (*Ridgeia piscesae*) and dense limpet communities thrive in diffuse flow fluids on a chimney at Axial. Schillern patterns mark areas of robust diffuse flow.

Axial fluids contain extremely high concentrations of magmatic CO_2 (285 mmo/kg), consistent with the exceedingly high concentrations of CO_2 measured in basaltic glasses (Helo *et al.*, 2011). In addition, magmatic gases are also enriched during phase separation (boiling) near the seafloor, and during segregation of brines and vapor in the deeper subsurface as the fluids upwell from below (Butterfield *et al.*, 1990). Gas-rich fluids significantly impact the mineralogy of the chimneys. For example, Virgin Mound (Fig. 4.149) has been the most gas-rich chimney at Axial for

Figure 4.153 Octopi are relatively rare in the caldera floor of Axial Seamount. They likely feed on mollusks, such as clams, that thrive in sites of diffuse flow. This octopus (*Graneledone* sp.) is ~1 m long.

several years. It is an anhydrite-rich chimney that has remained small (<1 m high) for more than two decades (Butterfield *et al.*, 1990).

It is believed that the steep faults bounding the caldera walls and intersections with the north and south rift zones provide conduits for the buoyant, hot hydrothermal fluids (Butterfield *et al.*, 1990). In contrast, sites of diffuse flow are largely regulated by lava flow morphology. Venting occurs from long, wide fissures in sheet flows, pipe-like openings in zones of collapse or skylights, and along radiating fissures (Fig. 4.143). Significant diffuse flow also occurs out of the porous exterior chimney walls.

There are numerous diffuse flow sites on Axial, many of which occur along the southern portion of the caldera where the bounding fault has been covered by lava flows. The chemistry of the 3–78 °C fluids shows that they are a mixture of low-temperature fluids and fluids hotter than 350 °C, formed in the reaction zone at least 1 km beneath the seafloor (Butterfield *et al.*, 2004).

4.11.4 Eruptions and biological colonization

Vent species at Axial Seamount are largely similar to those found at other fields along the Juan de Fuca Ridge (Figs. 4.150–4.153). The outer portions of actively venting chimneys are covered in dense communities of tubeworms, palm and sulfide worms, and scale worms (Figs. 4.151 and 4.152). Thin tubeworms (*Ridgea piscesae*) and limpets (*Lepetodrilus fucensis*) are the most biomass abundant species, draping the outer walls of most chimneys. Unlike most non-sedimented hydrothermal fields along the Juan de Fuca Ridge, clams (*Calyptogena cf. pacifica*) are common in some of the basalt-hosted diffuse flow sites at Axial. White squat lobsters covered in fine microbial filaments are common in the International District, where they are associated with varied gastropods and abundant vent pycnogonid "spiders" (*Ammothea verenae* and *Sericosura venticola*) in the basalt-hosted diffuse flow sites (Fig. 4.150). Spider crabs are also present, though not in large abundances. The largest animals found at Axial are octopi, which likely feed on the mollusks around the vents (Fig. 4.153). Following the 1998 eruption, *R. piscesae* took up to three years to establish colonies at new diffuse vent sites. Within 30 months after the eruption, 60% of the known Axial species had colonized the new diffuse vent sites (Marcus *et al.*, 2009).

Axial is one of only a few sites in the world's oceans where decade-long studies have been completed on the evolution of microbial communities following a seafloor eruptive event. Hydrogen sulfide is the dominant energy source for microbial metabolism at Axial (Butterfield *et al.*, 2004). Surprisingly, genetic analyses of organisms in numerous diffuse vents show that the subseafloor microbial communities change from year to year and that they are distinct from site to site (Opatkiewicz *et al.*, 2009). Microdiversity in archaeal communities was less variable than in epsilon

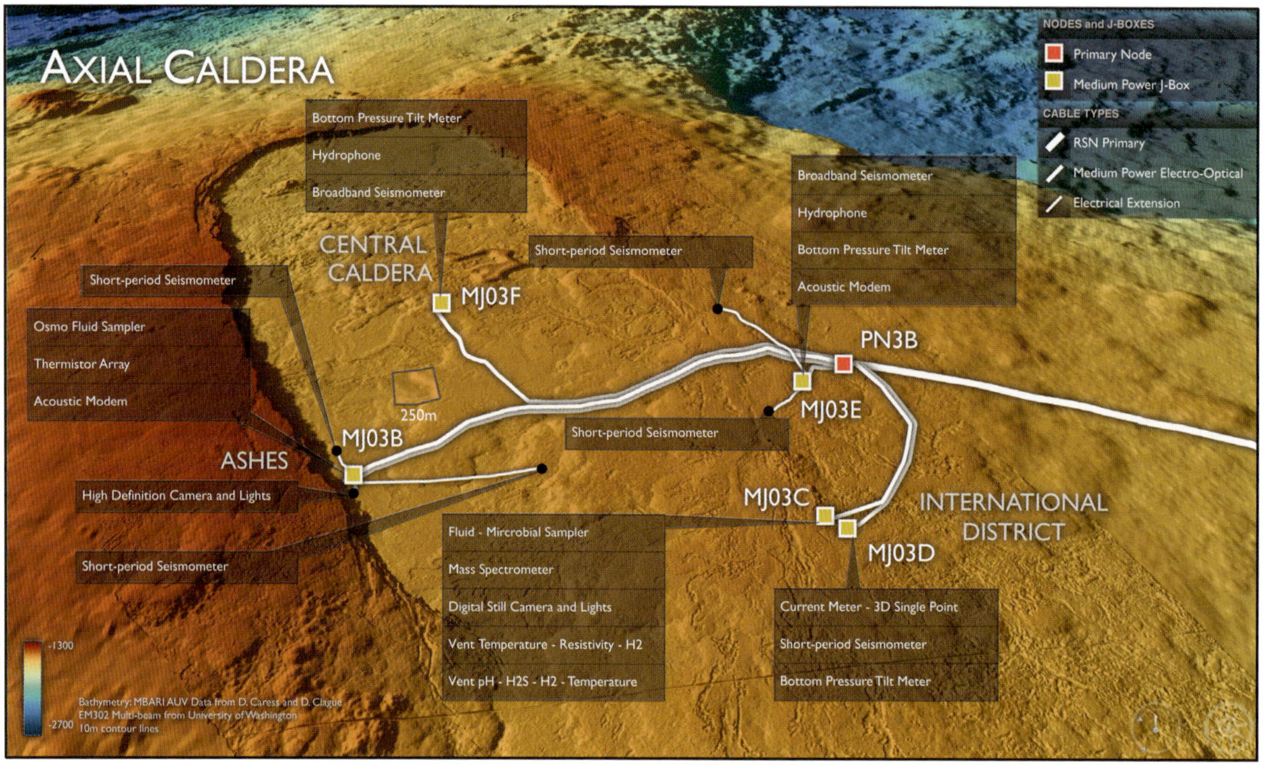

Figure 4.154 As part of the National Science Foundation's Ocean Observatories Initiative, Axial Seamount is now instrumented with high-power, high-bandwidth fiber-optic cables that connect to diverse sensor arrays. Extension cables, up to ~5 km in length, connect the primary node (PN3B) to a series of medium-powered junction boxes (e.g., MJ03C) that host the instruments (Kelley *et al.*, 2014). Two, ~2700 m tall instrumented moorings with robotic wire crawlers (not shown) are at the base of Axial. The cables provide real-time two-way communication and data flow from more than 300 miles offshore.

protobacterial communities between sites. Each site maintains its own distinct microbial community from year to year.

It is thought that persistence of diffuse flow at individual sites creates an environment, that allows isolation and stabilization of diverse microbial communities in the seafloor. A detailed, long-term study of one of the vents following the 1998 eruption showed that there was an increase in both the overall diversity of the microbes and also in microbes that are indigenous to the subseafloor (Huber *et al.*, 2003). Similar studies are underway in the black smokers at Axial to see whether each chimney is an island unto itself in a similar fashion to the diffuse flow sites.

4.11.5 Wiring an active submarine volcano

Axial Seamount is the site for the first US underwater cabled observatory, as part of the National Science Foundation's Ocean Observatories Initiative (Delaney and Kelley, 2014; Kelley *et al.*, 2014) (see Chapters 1 and 9). The high-bandwidth (10 Gb/s) and high-power (10 kV) fiber-optic cables installed at the summit and base of Axial Seamount are used to communicate with and power a diverse array of sensors installed in 2014 (Fig. 4.154). Real time high-definition video now provides unprecedented views of macrofaunal and microbial communities at ASHES. Chemical sensors and thermistor arrays will provide information on the environmental conditions in which the biological communities thrive. The instrumentation will also yield information on the impact of flow perturbations associated with eruptive and seismic events on biological communities.

Other installed sensors include mass spectrometers for fluid–volatile chemistry, broadband and short-period seismometers to monitor earthquake and magma migration activity, temperature and chemical probes in diffuse and black smoker sites, fluid and DNA samplers, and pressure-tilt meters for measurement of pre-eruptive inflation events and post-eruptive deflation. The *in situ* DNA samplers focus on filtering and preservation of time-series samples in diffuse flow sites. The instrument array at Axial, which is now fully deployed and operational, is the largest single *in situ* experiment in the global ocean focused on long-term measurements of underwater volcanoes with transmission of publically available real-time data and imagery back to shore.

Guaymas Basin

4.12 Venting at sedimented ridges

Sedimented spreading centers are found at divergent plate boundaries where high sedimentation rates coincide with active seafloor spreading. The chemistry of black smoker fluids at these sites typically contains hydrocarbons from the breakdown of organic matter (Figs 4.155 and 4.156). Commonly, the morphological character of these spreading centers is a rifted graben (Fig. 4.157), rather than a topographical high as in many MOR spreading centers.

Although these systems have not been studied as extensively as "bare-rock" MORs, two important impacts result from covering a spreading center with thick sediment sequences. At high sedimentation rates, sills commonly intrude into the sediments (see Chapter 3). The result is that these systems lack a well-developed basaltic cover at the seafloor (e.g., seismic Layer 2A). Instead, sediments overlay a complex of igneous sills and altered sediment, which grade at depth into wholly igneous crust. The second impact is that the upper crustal hydrologic system is influenced by thick sediments. The sediments tend to increase the fraction of conductive heat loss from the plate, even though crustal permeability at depth may be similar to that in the flanks of young bare-rock spreading centers.

There are two primary settings in which sedimented spreading systems are found: mature MORs that approach continental margins, such as at Middle Valley (Davis and Fisher, 1994; Wheat and Fisher, 2007) and the Escanaba Trough (Von Damm *et al.*, 2005) off the US Pacific Northwest; and young spreading centers that have recently (in the last few million years) completed rifting, of which the intermediate-spreading Guaymas Basin system (~6 cm/yr) in the Gulf of California is a prime example.

Figure 4.155 A typical assemblage of Guaymas Basin *Riftia* tubeworms near a small mound of hydrothermal sediment and minerals that can be 1–20 m high and hundreds of meters in area. The system is rich in hydrocarbons due to high organic concentrations in the sediments. Field of view is ~2.5 m.

Figure 4.156 *Riftia* tubeworms and scale worms (small, flat red worms) grazing on the outer surface of the tubes at Guaymas Basin. Field of view ~30 cm.

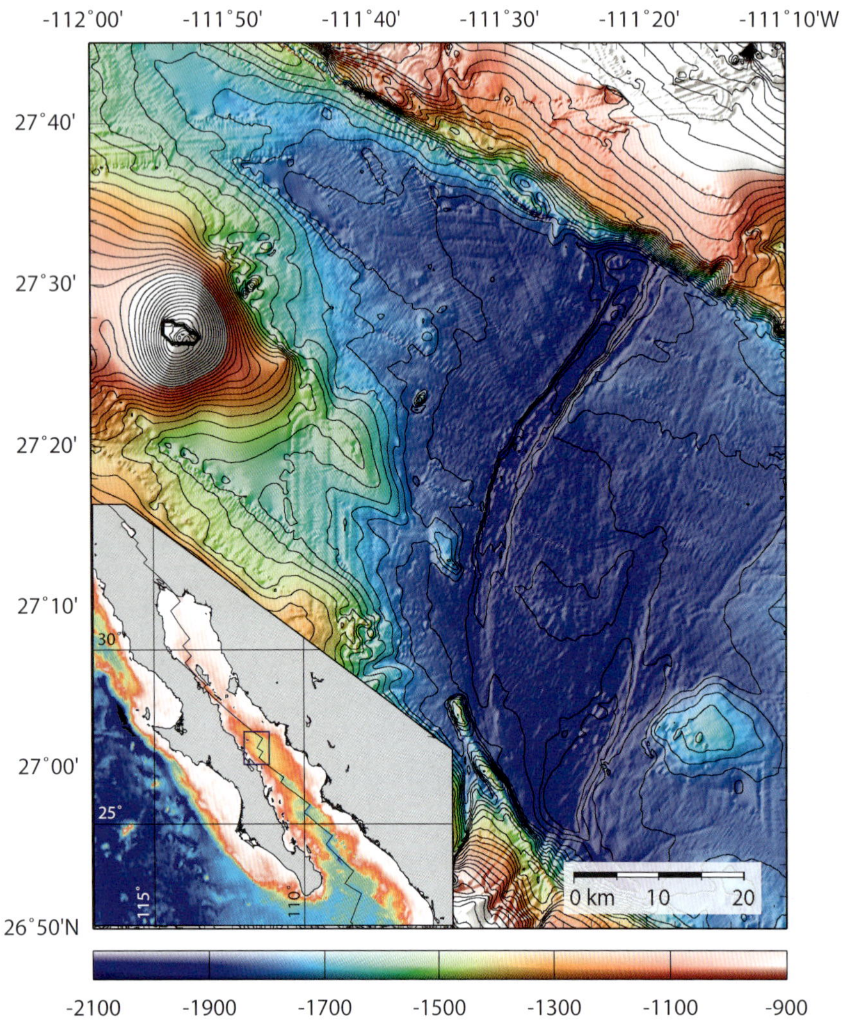

Figure 4.157 The Guaymas Basin is located in the Gulf of California (inset on left). The trough has formed during long-lived rifting of continental crust at the margin of western Mexico. Hydrothermal circulation through an organic-rich sedimentary sequence, over 500 m thick, results in the breakdown of organics at high temperature and formation of a variety of petroleum hydrocarbons. Diffuse flow of fluids at up to 200 °C releases hydrocarbons, short-chained organic acids, and ammonia from the sediments, which support particularly abundant and diverse communities of microbes (Bazylinski *et al.*, 1988; Teske *et al.*, 2007).

4.12.1 Guaymas Trough and venting

The first evidence for hydrothermal processes occurring in the Gulf of California came from heat-flow studies, helium anomalies in ocean bottom waters, and samples of hydrothermal deposits recovered by the US Navy submersible *Sea Cliff* (Lawver and Williams, 1979; Lupton, 1979; Williams *et al.*, 1979; Lonsdale *et al.*, 1980). The southern trough in the Gulf is a 30 km long curvilinear rift valley trending N-NE along its length. It overlaps with the northern trough for 20 km (Fig. 4.157).

Figure 4.158 Clear fluid issuing from one of the metal-poor vents in the southern trough of the Guaymas Basin. Field of view is ~1.5 m.

Vent discoveries at Guaymas were facilitated by detailed near-bottom investigations using the *Deep Tow* mapping and imaging system in 1980, when more than 20 hydrothermal plumes were discovered and over 120 sites of hydrothermal deposits imaged (Lonsdale *et al.*, 1980) (Figs. 4.158 and 4.159). Hydrothermal vents were discovered in the southern trough of the Guaymas Basin during *Alvin* diving in 1982 (see Section 4.1).

Active hydrothermal sites in Guaymas are concentrated along the base of rift valley walls, and along 2–25 m tall mounds that are typically steep-sided. The mounds are 10–50 m across and constructed in the center of the rift basin floor ~2 km below the ocean surface. Individual deposits generally strike NE–SW and form lineaments subparallel to the bounding rift walls. The alignment likely reflects permeability pathways for hydrothermal fluids sourced along faults that intersect recent sill intrusions and other heat sources beneath the graben floor. Chimneys are commonly built upon the mounds and range in height from less than a meter to above 20 m. They are either sulfide-rich or carbonate/sulfate-rich structures (Figs. 4.155 and 4.158–4.163). At least seven different actively venting sites have been discovered.

Guaymas Basin vents differ from most other vent sites because the hydrothermal fluids must pass through more than 500 m of organic-rich sediments before reaching the seafloor. Chemical reactions between vent fluids and the organic-rich sediments enhance metal sulfide precipitation. At elevated temperatures, the organics "cook," resulting in the formation of hydrocarbons (Whelan *et al.*, 1988). The metal content of the vent fluids is reduced by several orders of magnitude. This leads to enrichment of fluids in manganese relative to iron (Von Damm *et al.*, 1985); a characteristic

Figure 4.159 Delicate barite chimneys (~50 cm tall) at the top of a small Guaymas mound with a group of *Riftia* tubeworms occupying a diffuse flow area on the left side of the summit. The mound flank is covered in white to orange bacterial mat that grades into the surrounding organic-rich sediment at the base of the mound.

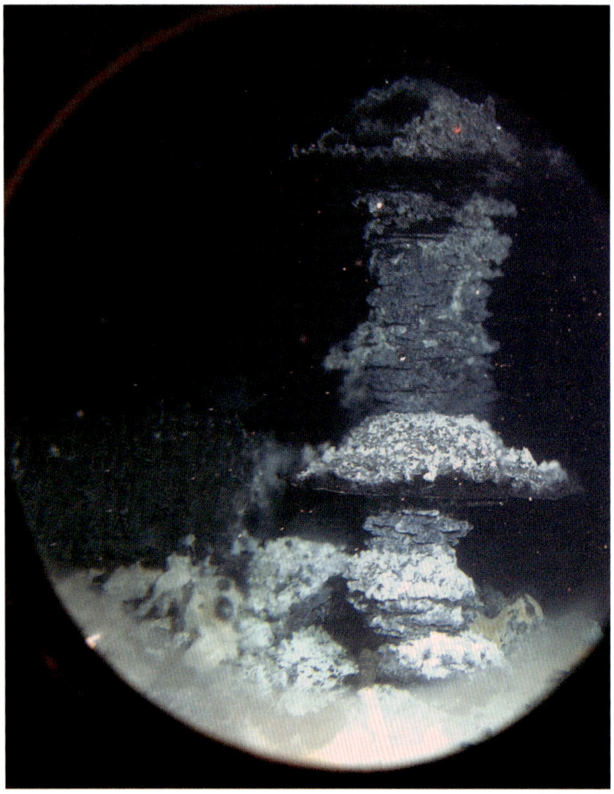

Figure 4.160 Typical "mushroom" vent structure at Guaymas as observed through *Alvin's* viewport. The "flange" morphology forms through diffuse and focused fluid flow issuing from cracks in the chimney so that mineralized ledges grow outward. Hydrothermal fluids commonly pool underneath the flanges. Base of structure is ~2 m across.

Figure 4.161 Classic black smoker chimney on a Guaymas mound. The chimney is ~3 m tall and has venting orifices along its side, as well as at the summit. Shimmering water diffusing through the outer wall can be seen around the base of the chimney, as well as >300 °C fluids issuing from the main vent.

Figure 4.162 Small, active black smoker at Guaymas forming a pyrrhotite-rich chimney (~1 m tall). *Alvin's* manipulator arm with a chemical sensor collects fluid chemistry data from the outer spongy surface of the chimney, as well as from the hottest fluid that issues from the chimney top. Mineral assemblages in Guaymas are commonly pyrrhotite-rich massive sulfide, with barite, calcite, anhydrite, and talc. Opaline silica forms amorphous, variably colored smoothed surfaces.

feature of Guaymas fluids compared to fluids in other vent fields. High carbonate and organic matter in the sediments buffer the pH of the vent fluids. Vent fluids from the Guaymas Basin and other sedimented systems (e.g., Middle Valley and Escanaba Trough) show a wide range in composition, but exhibit a higher pH (5.1 to 5.9 at 25 °C) and lower metal content than fluids formed in normal MOR settings (Fig. 4.158) (Von Damm, 1995).

Methane contents in Guaymas vent fluids range from 12 to 16 mM (Whelan, 1988), about two times greater than those of most bare-rock MOR vent sites. The high concentration and relatively low isotopic composition of CH_4 differ from those of most other MOR systems as well, where CH_4 originates from high-temperature reactions involving H_2 and CO_2, with minor contributions from mantle out-gassing (Claypool and Kvenvolden, 1983; Lilley *et al.*, 1993).

4.12.2 Microbial life in sedimented systems

Guaymas Basin is the only place known where cold seeps and sedimented hydrothermal vent systems are found in such close proximity. In 2009, these unique settings

Figure 4.163 *Alvin's* manipulator holding the sensing wand of an *in situ* chemical analyzer used to identify key components of the hydrothermal fluids, including H_2S and FeS, that are important sources of energy for organisms (Luther *et al.*, 2001). The exterior ledges of the hydrothermal mound deposits, seen here, commonly serve as pathways for fluids to reach the outer flanks of the mound where they deposit their mineral load.

Figure 4.164 A small Guaymas mound with shimmering water issuing from diffuse cracks on the left side of the summit. Bright white patches are likely anhydrite and broader, smooth areas of white to orange are bacterial mats.

Figure 4.165 Small chimneys (10–50 cm tall and ~5–10 cm in diameter) on the periphery of a Guaymas vent mound surrounded by carpets of white, yellow and orange bacterial mats.

contributed to Guaymas Basin being admitted into the International Marine Protected Area system organized by the United Nations. The distinctive chemistry of Guaymas vent fluids and sediments support a rich microbial diversity of unusual bacteria, Archaea, and eukaryotes (Figs. 4.164 and 4.165) (Teske *et al.*, 2007). Hydrothermal sediments at Guaymas sites contain highly diverse anaerobic thermophilic microorganisms, including methanogens and sulfate-reducing bacteria. Phylogenetically diverse thermophilic sulfate reducers have been isolated from hydrothermal vents in the Guaymas Basin (McKay *et al.*, 2012).

Abundant methanotrophic archaeal communities (that is, the organisms are able to metabolize CH_4) at the sediment surface reoxidize and incorporate CH_4 into their cellular biomass. The high eukaryotic diversity is associated with anoxic sediments and the overlying seawater. Bacterial populations in these sediments are primarily characteristic of anoxic, reduced, hydrocarbon-rich sedimentary habitats. Adaptation to anoxic environments is evidenced by specific affinity of environmental sequences to aerotolerant anaerobic species (that is, the organisms do not require oxygen during metabolic processes) superimposed against a background of widely distributed aerophilic and aerotolerant protists, some of which may migrate into and survive in the sediment. Because of the presence of both methanogenic and CH_4-oxidizing communities, the Guaymas Basin sediments are being used as a model system for studying anaerobic CH_4 cycling, with applicability to diverse hydrothermal and non-hydrothermal marine sediments and subsurface environments.

4.12.3 Fauna in hydrocarbon-rich environments

The first biological studies found previously unidentified infaunal assemblages in the Southern Trough of Guaymas Basin and compared them to the sedimented Escanaba Trough on the Gorda Ridge (Petrecca and Grassle, 1990;

Grassle and Petrecca, 1994). At Guaymas Basin, as in other sedimented chemosynthetic sites, hesionid polychaetes dominate with *Orseis grasslei* reaching densities of 2844 individuals per square meter. This polychaete is co-dominant in these ecosystems with the ampharetid (bristle worm) polychaete *Amphisamytha galapagensis* and the dorvilleid polychaete *Ophryotrocha akessoni*.

Dense thickets of the giant tubeworm *Riftia pachyptila*, which reach 1.5 m in length, are major sessile constituents at Guaymas vent sites (Figs. 4.155 and 4.163). Unlike those at all other Pacific vent sites, Guaymas *Riftia* are well known for their morphologically aberrant plume shape, which is highly elongate with frilly branchial lamellae. This feature initially suggested that this was either a different species or a population displaying a different morphology. Subsequent investigation of genetic variation revealed Guaymas, EPR, and Galápagos *Riftia* to be the same species. *Riftia* thickets at Guaymas provide habitat for a large array of organisms, among them a predominance of polynoid polychaetes and limpets. In addition, there are large mats of light orange and yellowish bacteria (*Beggiatoa*) (Fig. 4.165), several types of galatheid crabs (Fig. 4.166), and clams. The clams (*Calyptogena gigas*) are relatively small and are not biomass dominant as in other hydrothermal vent sites. Interestingly, at this site the clam *Vesicomya gigas* inhabits a sedimentary environment enriched with particulate mineral and organic matter, which may be ingested and partially responsible for the metal bioaccumulation observed in Guaymas vent fauna (Ruelas-Inzunza *et al.*, 2003).

Although many of the species in Guaymas Basin are similar to those found on the EPR (e.g., *Amphisamytha galapagensis* and the dorvilleid polychaete *Ophryotrocha akessoni*), there is only partial overlap of species listed among Guaymas and other EPR sites (Grassle, 1985). At least 14 of the more than 40 total species are considered endemic to Guaymas vents. They include polychaetes (6 species), nematodes (3), limpets (3), a copepod (1) and a species of actinostolid anemone (1). The majority of shared species are the diverse species of polychaetes and copepods that inhabit the EPR and Galápagos Rift as well. Thus, although there are faunal variations among vents, Guaymas Basin, EPR, and the Galápagos Rift are thought to form a single hydrothermal biogeographic province.

Figure 4.166 A lithodid crab captured by *Alvin's* manipulator.

East Pacific Rise

4.13 The fastest spreading center on Earth

Spreading rate has a profound impact on the extent and timing of magmatic activity beneath the MOR axis. Much of the northern and southern EPR spreads at fast rates, greater than 10 mm/yr near 10° N; near 30° S the EPR is the fastest spreading MOR, with rates of nearly 150 mm/yr. Depth of magma storage and frequency of eruptions have direct consequences for sustaining hydrothermal circulation along and across the ridge axis. These factors influence the morphology and distribution of vent chimneys along the EPR axis and their relationship to volcanic and tectonic features of the axial zone (Figs. 4.167–4.171), ultimately impacting the composition of the chemosynthetic communities (Macdonald and Fox, 1988; Haymon *et al.*, 1991).

Multichannel seismic imaging (Fig. 4.170) beneath several segments of the northern and southern EPR (Detrick *et al.*, 1987, 1993; Toomey *et al.*, 1994; Macdonald, 1998; MELT, 1998; Dunn *et al.*, 2000; Carbotte *et al.*, 2012; Canales *et al.*, 2012a) show that the crustal melt lens is relatively narrow (>500 m wide) and broadly continuous at scales of fifty to hundreds of kilometers. However, there are important small discontinuities at spatial scales of less than 10 km that match with transitions in crestal morphology and structure. Crustal magma bodies beneath the EPR crest occur at a depth of 1–2 km. The shallowest known occurrence is near

Figure 4.168 OBS black smoker vent at the EPR 21° N site in 2002, as imaged by *Alvin*. *Tevnia* and *Riftia* tubeworm assemblages occupy a flange on the edifice, which has three, small inactive sulfide chimneys. Field of view ~1.5 m.

Figure 4.167 Location of vent fields on the fast- to ultrafast-spreading EPR. Polymetallic sulfide chimneys (2–7 m tall) venting 350 °C fluids were first discovered at the intermediate-spreading EPR 21° N in 1979 (Spiess *et al.*, 1980).

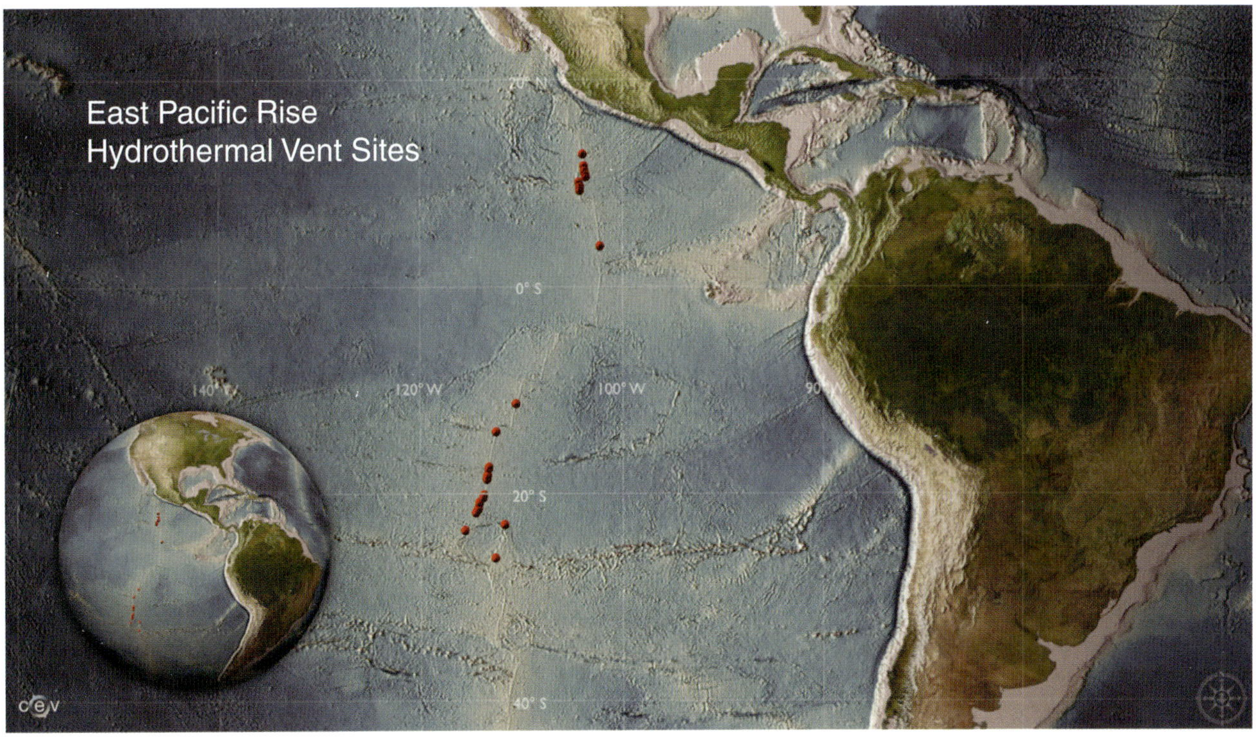

17.5° S, where a magma body was imaged 800 m below the seafloor (Detrick *et al.*, 1993). This is also a site where volcanic eruptions have occurred over the past few decades (See Chapter 5; Sinton *et al.*, 2002). The narrow zone above crustal magma chambers defines the area of active water–rock interaction and, as a result, most EPR hydrothermal vents occur in the narrow axial rift, predominantly within

Figure 4.169 Artist's depiction of the axial summit trough (AST), a primary morphological feature of many fast-spreading MORs. The trough is very narrow (~100–500 m wide), and is the origin for most of the volcanic eruptions. It also hosts nearly all of the high-temperature hydrothermal vents. This view includes many of the features observed at the EPR 9°50′ N area when first investigated with *Alvin* in the early to mid 1990s. The AST is also where primary eruptive fissures are located. The lava morphologies reflect rapid effusion of lava from the fissures, so they are mostly lobate and sheet flows. Hydrothermal chimneys are commonly found along the primary fissures or on benches on the collapsed wall of the AST.

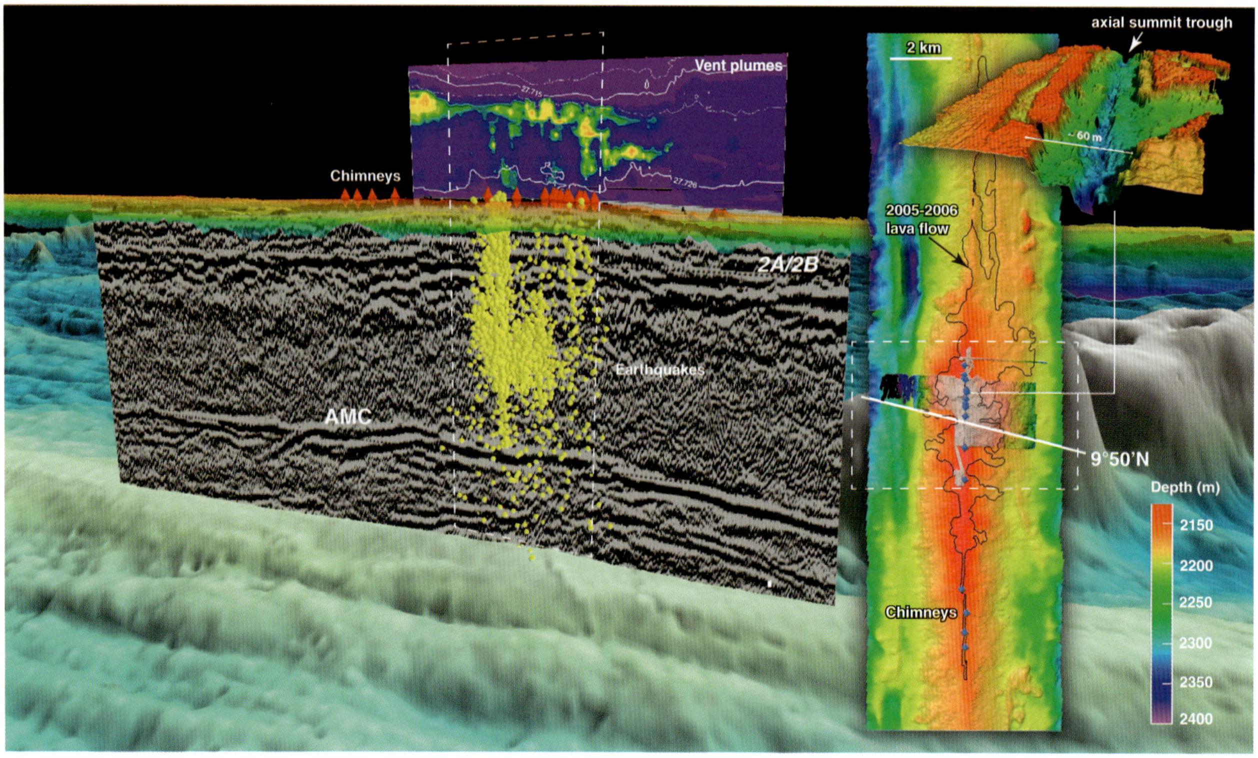

Figure 4.170 Compiled visualization of data sets from the East Pacific Rise Ridge 2000 Integrated Study Site (Fornari *et al.*, 2012). Right vertical inset displays ship-based EM300 bathymetry (25 m resolution) showing the axial high between 9°46′ N and 9°56′ N. A higher-resolution bathymetry data set (5 m resolution) collected in 2001 by the autonomous underwater vehicle *ABE* is overlain and shows greater details of the volcanic terrain. Also shown is the extent of flows produced during the 2005–2006 eruption (black line) (Soule *et al.*, 2007) and high-temperature vents (blue diamonds). To the left is a perspective view from the northeast. The EM300 bathymetry is elevated above a regional bathymetric map. Multichannel seismic reflection data collected in 2008 (Carbotte *et al.*, 2012) are shown relative to the seafloor. White labels show the crustal Layer 2A/B reflector and the top of the axial magma lens. Several hundred microearthquakes were recorded during 2003–2004 (Tolstoy *et al.*, 2008) (yellow dots) beneath the vent locations (red diamonds). A profile of turbidity recorded in late May 2006 is shown above the seafloor (Cowen *et al.*, 2007). The "eyeball" icon in the bathymetry inset shows the direction of the perspective view shown in the main figure. All data depicted are available at the Ridge 2000 data portal (http://www.marine-gio.org/index.php).

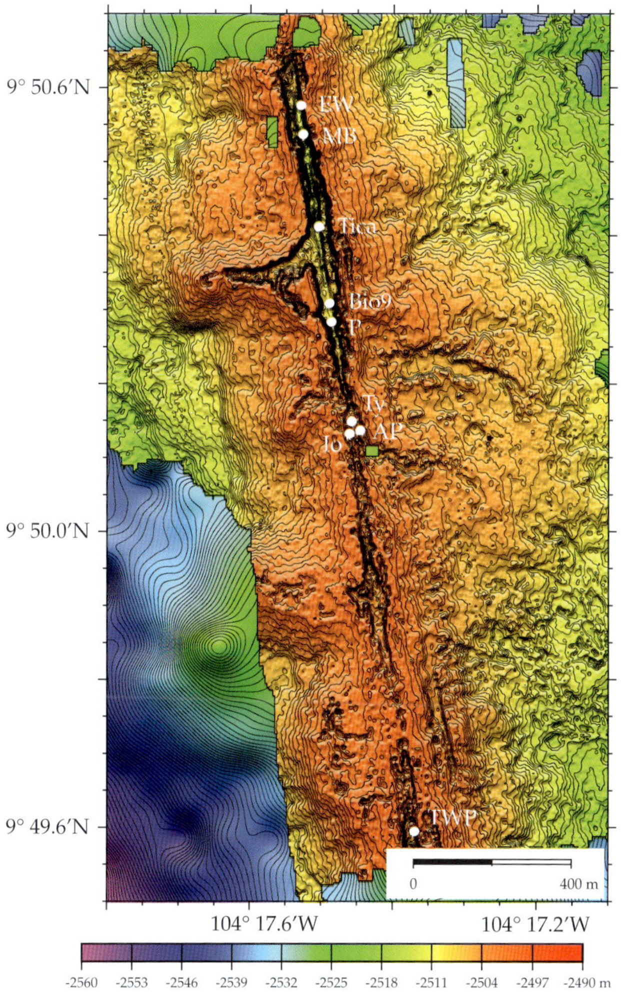

Figure 4.171 High-resolution bathymetry near 9°50′ N on the EPR showing the location of active vent sites. Volcanic and hydrothermal activity is narrowly focused within the ~40–100 m wide axial summit trough (Ferrini *et al.*, 2007).

Figure 4.172 Young black smoker chimneys, uncolonized by megafauna, near 9°50′ N on the EPR, 11 months after the 1991–1992 volcanic eruptions in this region. White bacterial mats coat the chimneys and their basal mound. Vent fluid temperatures were >390 °C. Field of view ~4 m.

or along the walls of the 100–500 m wide axial graben (Figs. 4.171 and 4.172) (Haymon *et al.*, 1991; Fornari *et al.*, 1998, 2004; Ferrini *et al.*, 2007).

Hydrothermal vent sites occur typically along the fast- to superfast-spreading EPR axis, fueled by small magma chambers (Fig. 4.170). Sites of volcanic eruptions, or above areas where magma has intruded, are the primary loci of venting. The seafloor manifestations of the eruption processes are primary fissures that can be more than 1 m wide, or as wide as 5–10 m (see Chapter 5). Large collapse areas occur where lava has drained back into the seafloor following eruptions (Figs. 4.169 and 4.171). Clusters of high- and low-temperature vents are typically found every 5–10 km along the EPR axis. They are commonly co-located with primary eruptive fissures that feed major volcanic eruptions, which occur roughly every 10–20 years (Fig. 4.171).

The hydrothermal plumbing system that develops at the fast- and superfast-spreading EPR is influenced primarily by the shallow depth to the magma lens, frequency of small melt bodies distributed along the crestal plateau, and frequent (decadal) repaving of the ridge axis and upper flanks by thin sequences of lava. The lavas in some places bury young hydrothermal vents and associated biological communities. These factors result in smaller hydrothermal chimneys than at intermediate- to slow-spreading systems. They also result generally in less voluminous deposits and rapid turnover of vent faunal assemblages that commonly follow a recognized pattern of community succession (Shank *et al.*, 1998; Mullineaux *et al.*, 2003, 2005; Van Dover, 2003; Sarrazin *et al.*, 2006).

The primary eruptive fissures and margins of the EPR axial graben provide well-established, vertical permeable pathways for hot fluids to ascend through the crust (Fig. 4.173), and supply hydrothermal vents (Figs. 4.174 and 4.175) (see Chapter 5). This pathway geometry leads to a distribution of 50–100 individual, active vents per 10–20 km of axis, and approximately the same number of inactive vents. Note that this is in significant contrast to intermediate-spreading systems such as the Endeavour, which hosts more than 800 chimneys over a distance of 15 km (Clague *et al.*, 2008; Kelley *et al.*, 2012, 2014; see Section 4.10).

Vent sites are most prevalent in areas where the EPR axis is shallow or the cross-sectional area is broad. These are both indicators of abundant and shallow magma supply (and thus heat to drive hydrothermal circulation). At 9°50′ N and near 17°30′ S, the vents are aligned in the axial graben

Figure 4.173 Hydrothermally altered basalt from the southern EPR showing extensive anhydrite along cracks in the rock.

Figure 4.174 Black smoker chimneys on the southern EPR (here at 17°37′ S) have diverse fluid compositions over small spatial areas. Detailed time-series studies at sites perturbed by magmatic emplacements show that vapor vents first, in the earliest stages after a magmatic/volcanic event, while the high chlorinity or "brine" liquid phase is expelled years later. On the southern EPR, both phases have been documented simultaneously at adjacent chimney structures venting >400 °C fluids at 2834 m depth (Brandon Vent). Field of view ~1.5 m.

and form fields or sites that each have one to five chimneys closely spaced within 10–50 m. Distances between sites are 100–500 m (Fig. 4.171).

The first high-temperature hydrothermal vents were discovered in 1979 (Francheteau *et al.*, 1979a; Spiess *et al.*, 1980) on the northern EPR near 21° N (Fig. 4.167). At 21° N, the MOR axis enters the Gulf of California and transitions from a fast-spreading center to a series of sediment-filled pull-apart basins spreading at intermediate rates, where magmatic activity is largely restricted to intrusions into hemi-pelagic sediment sequences because of density differences between the lava flows and overlying thick sediment section (see Section 4.12).

Figure 4.175 Vents along the southern ERP near 21° S were venting >350 °C fluids in 1998 when visited with *Alvin*. View outside the porthole is ~1 m.

Figure 4.176 Microbially produced flocculant material, forming snowblowers, issues from the subsurface following the 1991 9° 50′N eruption. This was the first observation of these features. Field of view ~2 m.

The 1979 discovery of vents at 21° N led to a series of primarily US and French cruises throughout the 1980s and early 1990s between ~8° N and 21° N along the northern EPR. This section of the global MOR in the eastern Pacific became a focal point for geological, geophysical, biological, and hydrothermal research where the first-order properties of fast-spreading crust were identified (Fig. 4.170) (Francheteau and Ballard, 1983; Hékinian *et al.*, 1983a,b; Fornari *et al.*, 2012). To better understand whether vent structures or fluid chemistry varied at different spreading rates, several international groups focused on discovering vents along the EPR using novel technologies that combined geological and hydrothermal plume mapping techniques to identify areas where venting was localized. French scientists produced some of the first detailed maps that revealed associations between hydrothermal and volcanic features near 12°50′ N. These included hydrothermal sites both on the ridge axis and on a large seamount located just east of the axis (Hékinian *et al.*, 1983a,b).

One of the first comprehensive near-bottom mapping programs directed at deducing vent distribution along the length of a ridge segment and the relationships between hydrothermal, volcanic, and tectonic features along the axis was carried out at the EPR between 9° N and 10° N in 1989 (Haymon *et al.*, 1991). That study led to a series of *Alvin* dives in 1991, which made the first discovery of a MOR submarine volcanic eruption and its impacts on hydrothermal venting (Haymon *et al.*, 1993) (Figs. 4.176 and 4.177).

Snowblowers, now known to be associated with seafloor eruptions, were also first documented here (Fig. 4.176). French and German vent exploration and research along the superfast-spreading southern EPR began in the mid-to-late 1980s. Japanese and US hydrothermal research along the southern EPR between about 5° S and 32° S began in the late 1990s, but relatively few cruises have occurred there in the past 10 years, largely due to the remoteness of that segment of the MOR.

4.13.1 Evolution of fast-spreading hydrothermal systems

Some of the first geochemical studies of high-temperature vent fluids were made at 21° N EPR. Results showed that end-member fluids contained low magnesium and sulfate, and abundant H_2S. The fluids have a pH of 3.5 and the sulfide-forming element concentrations show significant interfield variations, which are similar to metalliferous sediments on the ridge flanks (Von Damm, 1995) (see Section 4.14).

The 1991–1992 discoveries of very young lava, coupled with unusual and vigorous hydrothermal flow at the EPR 9°50′ N area less than one month after the eruptions (Fig. 4.174) (Haymon *et al.*, 1993; Rubin *et al.*, 1994), provided compelling evidence of fundamental linkages between ocean crust formation, phase separation of hydrothermal fluids, and alteration of the crust (Von Damm, 1995, 2000; Lilley *et al.*, 2003) and the impact that these processes have on biological communities (Figs. 4.178–4.185) (Shank *et al.*, 1998; Scheirer *et al.*, 2006; Mullineaux *et al.*, 2010). Complementary studies of volatile concentrations in erupted lavas showed that they have excess chlorine, suggesting contamination of erupted magmas through seawater interactions during eruption and by brines stored in shallow crustal fluid pathways and reservoirs (le Roux *et al.*, 2006; Soule *et al.*, 2006; Perfit *et al.*, 2003; Michael *et al.*, 2008).

Time-series changes in vent-fluid chloride indicate that the first fluid to be expelled in the immediate aftermath of an eruption is the "vapor" phase. Following volcanic

Figure 4.177 An intertwined community of spaghetti or acorn worms (*Saxipendium coronatum*) loosely attached to the broken lobate basalt at the outer periphery of a vent field near 9°50′ N. These worms are known as deposit feeders, scraping their proboscis over rock, as well as feeding by filtering food from the water column. The feeding end of this (highly stretchable) >3 m long worm is located at the snout or proboscis (lightest shade at the anterior end; center bottom). Gonads are visible as white ovoid bodies along the length of the worm. Field of view ~0.5 m.

eruptions and initial vapor-phase expulsion, some chimneys progressed to venting fluids with chlorinities greater than seawater much faster (taking 3 years or less) than others (e.g., 10 years) (Oosting and Von Damm, 1996). Other vents never made the transition, for example the Bio9 vent (Von Damm, 2000). Vent fluids sampled in 2004 and after the most recent 2005–2006 eruptions (Soule *et al.*, 2007), and in 2006–2008 (Foustoukos and Seyfried, 2007a,b) from the same vents first sampled in 1991, indicate a decrease in chloride content for fluids issuing from most vents, followed by non-monotonic increases in chloride contents leading up to the most recent eruptions.

Immediately after the eruptions, all 9°50′ N EPR vents sampled (starting in June 2006), including those that previously vented brines, were producing low-salinity, vapor phase-separated fluids. By late 2006, some of the vents began to return to their pre-eruption chemistries. The

Figure 4.178 Cirrate or "Dumbo" octopus (*Grimpoteuthis*) swims a few meters above the seafloor with fins extended upward, searching for food on the bottom. Their preferred prey includes mussels and clams (shown in background on the seafloor). With arm spans more than 3 m, these highly inquisitive animals are commonly attracted to submersibles.

Figure 4.179 Tubeworm communities on the 11 m tall Tubeworm Pillar show the spatial zonation of EPR vent habitats with *Tevnia* tubeworm (yellow stained tubes) occupying areas of more sulfide-rich fluids, temporally mixed with lower oxygen concentrations. The larger *Riftia* tubeworms (from middle to right side of image) displace *Tevnia* as vent flux and H_2S/temperature decrease over time. White brachyuran crabs and pink zoarcid fish are active in the communities. These sites can host more than several dozen species, including amphipods, gastropods, polychaetes, mussels, and shrimp. Brachyuran crabs graze on tubeworm tubes. An eelpout fish, known for eating tubeworm plumes and other polychaetes, is positioned amongst the tubes. Field of view ~1.5 m.

EPR hydrothermal response to magmatic activity and seafloor volcanism can manifest itself distinctly at each vent, even for vents located within only a few or at most tens of meters of each other, providing clear evidence of distinct and complex plumbing systems feeding the vent structures (Fornari *et al.*, 2004, 2012).

4.13.2 Relationships between EPR hydrothermal vents and biological communities

Exploration of the EPR between 12°50′ N and 21° N from 1979 to 1983 revealed endemic hydrothermal vent communities similar to those present along diffuse venting fissures at the Galápagos Spreading Center (GSC) (see Section 4.14) (Figs. 4.179–4.182). For example, at the EPR, vent communities are dominated by *Riftia* tubeworms, *Calyptogena* clams, and bathymodiolin mussels like those found at the GSC (Fig. 4.181), although a distinct suite of new species

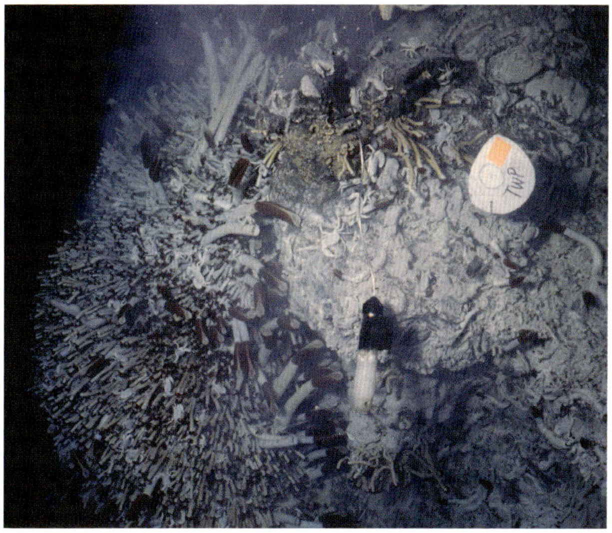

Figure 4.180 The upper surface of the 11 m tall Tubeworm Pillar (9°46.9′ N) discovered in 1989. A HOBO temperature logger [titanium wand visible with back and white cylindrical handle (center)] records the temperature of hydrothermal fluids emanating from this lava pillar. *Tevnia* tubeworms, many stained with iron oxides, surround the orifice. Once dominated almost exclusively by *Riftia* (1995), the later image of the pillar shown here is a mixture of *Riftia* and *Tevnia* tubeworms. Dynamic temporal changes in fluid chemistry and community structure have been associated with this structure, creating patchwork mosaics of various assemblages. Field of view ~2.5 m.

Figure 4.181 Panoramic view of the Choo Choo Train marker site hosting mussel assemblages, galatheid crabs, and white serpulid polychaete tubes (center left) along a collapsed sheet flow margin in the axial trough near 9°49.6′ N in 2000. Diffuse flow venting in this region diminished in the early 2000s and the large patch of *Riftia* (center), immediately adjacent to the site marker, became colonized by mussels. Field of view ~4.5 m.

Figure 4.182 Close-up of the Choo Choo Train *Riftia* community (shown in the center of Fig. 4.181) being outcompeted by bathymodiolin mussels for access to vent fluids. A cluster of amphipods (whitish haze) just above the *Riftia* plumes form dense swarms (~1000 individuals/liter). These amphipods, *Halice hesmonectes*, are pelagic crustaceans in the deep sea. Polypro rope holding the red and white marker is 0.6 cm wide. Field of view ~2 m.

Figure 4.183 Alvinellid polychaetes and their chimney-encrusting matrix of tubes inhabit EPR chimneys. These tubes, open at both ends, channel vent and non-vent fluids into and out of the tubes. *Alvinella pompejana* (red worms at center with extended buccal tentacles) have well-developed gills that are highly adapted for low oxygen conditions, They are deposit feeders, ingesting the free microbes present on the chimney surface. Field of view ~1 m.

was discovered inhabiting the surfaces of newly discovered high-temperature chimneys. The new communities were dominated by several species of vent-endemic tube-building polychaetes, notably the Pompeii worm, *Alvinella pompejana* (Fig. 4.183), and also included gastropod limpets, dirivultid copepods, and *Cyanograea* crabs.

Alvinella pompejana can reach lengths more than 13 cm and dwell in a matrix of chitonous tubes shared with other polychaetes (*Alvinella caudata*) on the hotter, less consolidated sections of smokers (Fig. 4.183). They feed on productive microbial communities and commonly position themselves with their posterior ends experiencing external temperatures as high as 80 °C (Johnson *et al.*, 1988a; Chevaldonne *et al.*, 1992; Cary *et al.*, 1998b; Desbruyères *et al.*, 1998). Simultaneously, their anterior end (with red tentacle-like gills surrounding their buccal or mouth region) is bathed in mostly ambient deep-seawater at 2.5 °C. As such, *Alvinella pompejana* is considered to be one of the most heat-tolerant animals on Earth (Cary *et al.*, 1998b). In general, similar types of black smoker communities are known at 21° N along the length of the EPR, south to about 23° S where alvinellid abundance is largely diminished and alvinocarid shrimp (*Chorocaris paulexa*) are the dominant fauna.

By 1990, observations during repeat visits to 21° N and 12°50′ N revealed dramatic changes in diffuse flow vent community composition and distribution. Partially dead beds of clams at 21° N between 1990 and 1992 reflected discontinuous vent emissions. In 1982, *Riftia* tubeworms were nearly eliminated at 21° N by intensive sampling, but then dramatically increased by 1990 (Fisher *et al.*, 1988). Between 1982 and 1990, the 12°50′ N vent sites experienced periods of venting inactivity coincident with sharp decreases in *Riftia* populations, increases in scavengers, and persistence of *Bathymodiolus* mussels. Reactivation of vent areas correlated with increases in a smaller siboglinid tubeworm, *Tevnia jerichonana*, and vent crab, *Bythograea thermydron*. Even though identifiable changes in faunal community structure were observed over short periods of time (months to years), distinguishing between naturally occurring faunal changes and the effects of human disturbances (extensive sampling and submersible maneuvering) was extremely difficult (Hessler *et al.*, 1985; Shank *et al.*, 1998).

The April 1991 discovery of newly formed hydrothermal vents in volcanically active areas of the EPR (Haymon *et al.*, 1993) presented a rare opportunity to document their biological and chemical evolution (Figs. 4.176 and 4.177). Biological, geochemical, and geological time-series studies over the past 20 years between 9°45′ N and 9°52′ N have provided a unique opportunity to examine the oceanographic processes that impact biological community establishment, and to define time scales over which communities develop, undergo structural/faunal changes, and eventually die.

By 2000, more than 90 vent-endemic species were documented at vent habitats along the EPR (Figs. 4.179–4.185). The approximately 16 vent areas between 9° 45′ N and 9° 51′ N contain both active and inactive black smoker chimneys. Diffuse flow venting occurs through cracks and fissures. In concert, these fluids support the full diversity of EPR community assemblages including vestimentiferan tube worms (*Riftia pachyptila* and *Tevnia jerichonana*), mussels (*Bathymodiolus thermophilus*), clams (*Calyptogena magnifica*),

Figure 4.184 Chimlets (1 m long) are iron oxyhydroxide stained. Bythograeid crabs transit over the structures' surface, scraping microbial material and predating on small polynoid polychaetes that inhabit this smoker.

Figure 4.185 Dense thickets of *Riftia pachyptila* along the EPR at 12°50′ N are commonly covered in iron oxide, the product of microbial activity and changing vent fluid chemistry. Note the spatial distribution of live versus dead tubeworms reflecting the spatial distribution of the venting activity. Live, densely packed tubeworms in the center and vacant tubes along the periphery suggest a recent change in the distribution of subsurface vent fluid supply and chemistry. Field of view ~2 m.

archaeogastropod limpets, alvinocarid shrimp, bythograeid crabs, serpulid worms, and anemones (Fig. 4.182).

The foundation of our current understanding of chemosynthetic organismal physiology and the unique chemoautotrophic and (endo)symbiotic relationships in these fauna stem from these well-characterized species. Comprehensive experiments in the 1980s revealed the physiology and metabolic functions of deep-sea vent fauna (Childress and Fisher, 1992). Physiological studies of the 1980s gave way to studies of biological community structure at the EPR vents (Shank *et al.*, 1998). Studies of larval dispersal (Mullineaux *et al.*, 2005), gene flow (Vrijenhoek, 2010), colonization dynamics associated with biological interactions and changes in chemical microhabitats (Luther *et al.*, 2001; Mullineaux *et al.*, 2005) and vent faunal distributions along various segments of the northern EPR (Van Dover, 2003) were also carried out.

Following the April 1991 eruption, biological community structure in this region underwent successional changes (see Figure 4.22). Communities progressed from syn-eruptive snowblower/microbial mats, to tubeworm dominance, to mussel dominance and increasing species richness (Shank *et al.*, 1998). These transitions followed a trend of decreasing temperatures, reduction in total sulfide concentrations, and decrease in hydrothermal flux over time. Post-eruption assemblages were characterized by massive microbial mats and a few, small bythograeid crabs. During the first three years post-eruption, tubeworms (*Tevnia jerichonana* and *Riftia pachyptila*) and large alvinellid polychaetes were the dominant community members (Fig. 4.185). At that time, large brachyuran and galatheid crabs, zoarcid fish, copepods, pardaliscid amphipods, and several limpet species were also part of the community composition.

By October 1994, small (under 10 mm) mussels and serpulid polychaetes were observed in areas adjacent to tubeworm colonies. The serpulid polychaetes and formerly sparsely distributed galatheid crabs had increased in abundance by November 1995. Mussels colonizing *Riftia* tubes were also initially observed in these areas in November 1995 (Shank *et al.*, 1998). In November 1997, vesicomyid clams (*Calyptogena magnifica*) were noted as a few sparse groups nestled in cracks adjacent to venting areas. Spaghetti worms were draped over inactive lava pillars peripheral to vent openings (Fig. 4.177). By December 2002, mussel populations had increased dramatically and the number of living tubeworms had declined precipitously. This was the setting when the 2005–2006 eruptions occurred, destroying many of the robust vent communities in the 9°50′ N region and resetting the stage for biological community development.

Microbial investigations on the EPR following the 1991–1992 eruptions expanded on early studies of chemoautotrophy (Wirsen *et al.*, 1986), leading to discoveries of new microbes with novel physiological and biochemical capabilities. Over the past decade, more than two dozen new microbial species have been detected or isolated, including ones that oxidize H_2 (Alain *et al.*, 2002), reduce nitrate to ammonia (Vetriani *et al.*, 2004a), reduce sulfur (Alain *et al.*, 2009), and are adapted to mercury exposure (Vetriani *et al.*, 2004b).

Microbes function in many ecological roles as producers, prey, re-mineralizers, and possibly as settlement cues for invertebrate larvae. East Pacific Rise studies have also revealed that microbial production is not necessarily constrained to vent sites, and may continue in the hydrothermal vent plume. Theory suggests that the latter may represent an important source of organic carbon to the deep ocean (McCollum, 2000) and field studies have supported that concept (Toner *et al.*, 2009).

After the 2005–2006 EPR eruptions in the 9°50′ N region, previous seafloor markers, biological communities, and ongoing faunal colonization experiments were completely covered by new lava. Relative to the 1991 eruption, the 2005–2006 event produced a greater volume of lava, including flows that extended nearly 3 km east of the axial rift (Soule *et al.*, 2007). As occurred in April 1991, white microbial mats surrounded the openings of diffuse-flow vents. Shimmering water and flocculent material were also abundant more than 6 months after the 2005–2006 eruptions. Among the microbial films were macrofaunal colonists, dominated by small (>1–40 mm length), sparse *Tevnia*. These occurred on basalt surfaces in vigorous diffuse flow and were attached to the sides and bottom surfaces of collected basalt. Among the recently settled *Tevnia* tubeworms at 9°49.8′ N, temperatures and H_2S concentrations were as high as 1.1 mmol/kg in 30 °C fluids, and two orders of magnitude higher H_2S (with maximum temperatures 10–20 °C higher) than measured one year earlier at this location (i.e., former Biomarker #82), when mussels were dominant at this site.

Galápagos Spreading Center

4.14 First vent discovery on a spreading center

The 1977 discovery of deep-sea hydrothermal vents and their associated animal communities at the Galápagos Spreading Center (GSC) transformed the oceanographic and biological sciences and ushered in a new era of scientific inquiry into life processes in extreme depth and chemical environments (Corlis *et al.*, 1979). Sites near 86° W, such as the Garden of Eden, Rose Garden, and Musselbed, were historic locations discovered using *Deep Tow* and *Alvin*, from which our early understanding of hydrothermal fluid chemistry, vent geological settings and chemosynthesis were derived (Figs. 4.186–4.193).

Hydrothermal vents at 86° W were discovered along young volcanic ridges constructed in the middle of the 2–4 km wide rift valley that defines the spreading axis at this intermediate-spreading (5–6 cm/yr) MOR (Figs. 4.188 and 4.189) (Ballard *et al.*, 1982). Galápagos vent sites host what we now know to be the iconic "textbook" vent communities – expansive vent fields inhabited by giant tubeworms, dinner-plate size clams, beds of mussels, bands of gastropod limpets, and alvinocarid shrimp (Figs. 4.190 and 4.191). The fields' peripheries were covered by extensive carpets of serpulid polychaetes, anemones, and dandelion siphonophores (Fig. 4.192).

Studies of Galápagos vents in the late 1970s and early 1980s provided the fundamental tenets of chemosynthesis in the deep sea, from physiology and intimate symbiosis to vent species interactions and community structure. For example, those early studies documented the processes of mussels outcompeting tubeworms via the removal of H_2S from vent fluids prior to the fluid reaching the worm plumes (Johnson *et al.*, 1988b) and complete physical displacement of *Riftia* tubes and plumes (Hessler *et al.*, 1988).

4.14.1 Vent activity

Hydrothermal activity at 86° W is characterized by diffusely flowing fluids emanating at temperatures between 3 °C and 13 °C (Edmond *et al.*, 1979b). The chemistry of the vent fluids sampled from 86° W in 1977 (Edmond *et al.*, 1979a,b; Lilley *et al.*, 1983) is typical of fluids associated with alteration of basalt at temperatures of ~300–500 °C (Gillis, 1995; Kelley *et al.*, 2002). Thus, relative to seawater, the vent fluid compositions exhibit enrichment in dissolved silica and selected cations (e.g., lithium, potassium, rubidium, calcium, strontium, barium, manganese), moderately reducing source conditions, and depletion in dissolved magnesium, oxygen, sulfate, and alkalinity.

Although the seafloor expressions of the source fluids were greatly diluted, high-precision analytical techniques allowed a faint hydrothermal signal to be recognized and quantified. These fluids were subsequently interpreted as a subsurface dilution of a single high-temperature hydrothermal end-member by cold seawater circulating in the upper crust (Edmond *et al.*, 1979a,b). Entirely consistent with the proposed seawater-mixing model, silica concentrations in the samples followed a linear relationship with respect to temperature. This simple mixing model was soon supported by the discovery of high-temperature (~350 °C) vent fluids at EPR, 21° N (Von Damm *et al.*, 1985), where end-member concentrations were remarkably similar to those extrapolated using the chemistry of the diffuse fluids at the GSC 86° W vents.

Figure 4.186 The fundamental discoveries of chemosynthesis, endosymbiosis, vent faunal physiology, and ecosystem development at deep-sea hydrothermal vents arose from studies at the iconic Rose Garden Vent Field shown here in a 1979-vintage 35 mm photograph taken from *Alvin*. Field of view ~1.5 m.

Figure 4.187 At the Rosebud vent site, *Riftia pachyptila* tubeworms (many tens of centimeters in length) colonize cracks in a sheet flow surface. Field of view ~0.5 m.

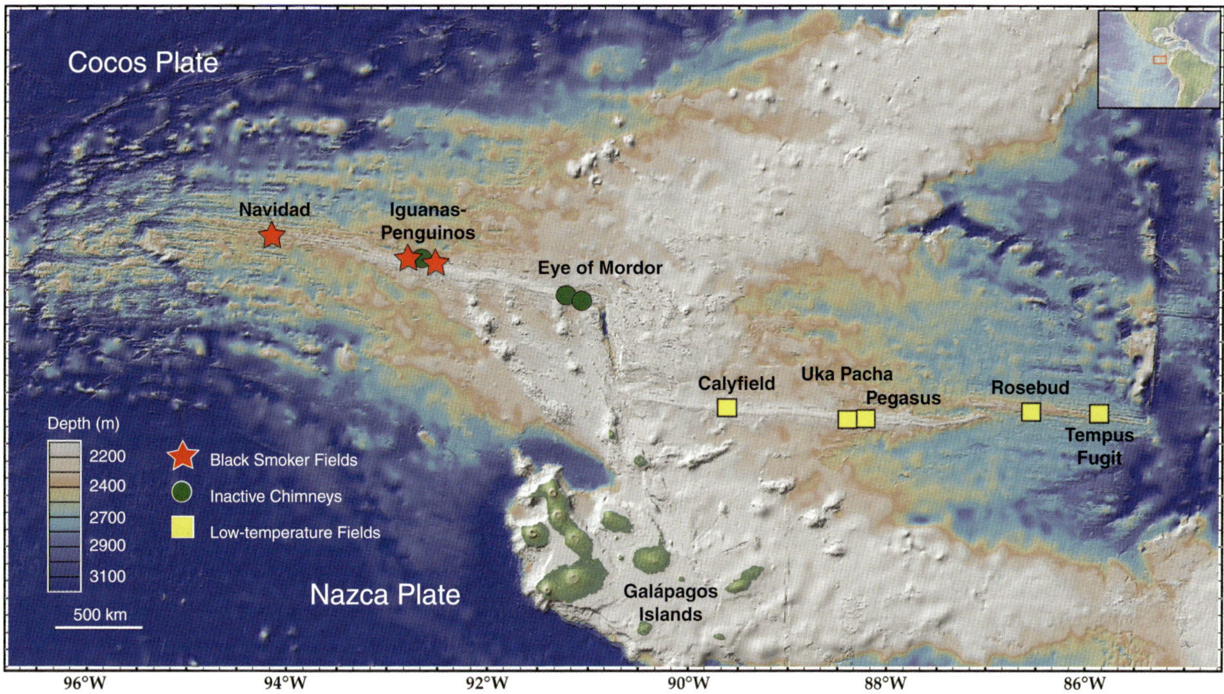

Figure 4.188 Regional distribution of high- and low-temperature vent fields along the intermediate-spreading GSC. Low-temperature vents were first discovered here in 1977. High-temperature fields were found near 91° W in 2005. At 91° W, the spreading center is not rifted and resembles a fast-spreading MOR with a shallow-depth axial high.

Figure 4.189 Locations of historic and iconic vent sites reported by Ballard *et al.* (1982) and Hessler *et al.* (1988) along the GSC near 86° W.

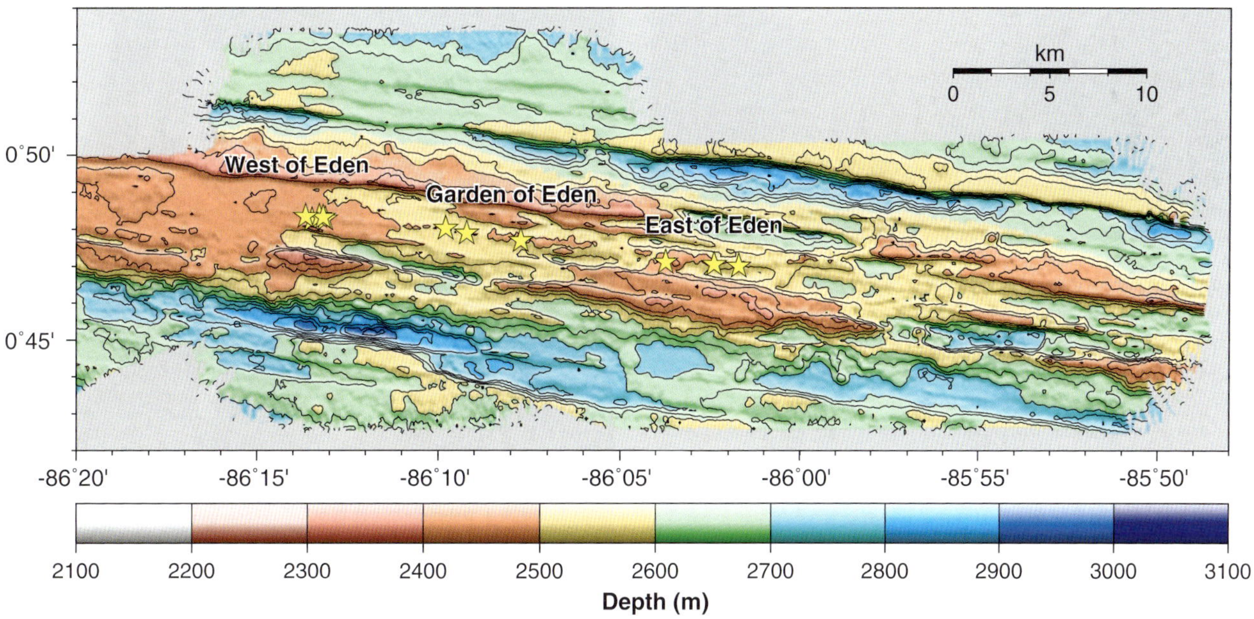

Although the Galápagos discoveries began in 1977, nearly 30 years later only a dozen vent sites, including inactive or extinct chimneys (Perfit *et al.*, 1999), had been discovered along this ridge, while active black smoker vents remained elusive. Until 2002, the majority of the eastern limb and the entire western limb of the GSC remained largely unexplored for hydrothermal activity.

4.14.2 Discovery of high-temperature vents

Since 2005, several expeditions have been devoted to exploring for the elusive high-temperature vents (Figs. 4.195 and 4.197–4.199) along the GSC to better understand the distribution of venting along the spreading center. Another goal was to understand the relationships of faunal

Figure 4.190 By 2002, the well-developed faunal communities documented years earlier at Rose Garden (Fig. 4.186) had apparently been buried by a new lava flow. Approximately 200 m northwest of Rose Garden, a new site, named Rosebud, at 2470 m depth, supported vent animals that were in the early stages of community development. Field of view ~1 m.

Figure 4.191 Small mussels grow in cracks on the surface of the Rosebud lava flow extending away from one of the main tubeworm assemblages. Brachyuran crabs are feeding among tubeworms and mussels. Small pink dots on the lava surface are anemones. Field of view ~30 cm.

Figure 4.192 Among the residents dominating the periphery of Galápagos Vent Fields are a colonial siphonophore, a relative of the Portuguese man-of-war. These animals were called "dandelions" by divers in *Alvin* in 1977, reflecting their delicate morphology. Their scientific name is *Thermopalia taraxaca*. These animals are particularly abundant along the margins of active fields (such as Tempus Fugit), along with alvinocarid shrimp, dead clams, and anemones. Field of view ~10 cm.

Figure 4.193 Pat Hickey piloting *Alvin* at the GSC during the dive program that discovered the Rosebud Vent Field in 2002.

communities both along the ridge and at the adjacent EPR (Figs. 4.190 and 4.191).

In December 2005, the GalAPAGoS Expedition (Galapágos Acoustical, Plumes, and Geobiological Surveys) explored an 800 km section of the GSC between 94.5° W and 89.5° W (Baker *et al.*, 2008; Haymon *et al.*, 2008). This area is where the topography of the spreading center axis is broad and inflated, reflecting abundant magma supply, similar to the typical topography for the fast-spreading EPR axis. Perhaps the primary reason for this dramatic change in axial topography along the GSC is that this segment of the ridge is closest to the mantle plume that created the Galápagos Islands, hence the increased

Figure 4.194 The majority of low-temperature sites on the Galápagos Rift are expansive and contain both live and dead *Calytogena* clams and Bathymodiolin mussels (e.g., Clambake and Oysterbed). Perhaps the largest active bivalve field is also the shallowest – the Calyfield Vent Field at 1670 m, near 89.5 W. The Calyfield hosts a large vesicomyid clam (*Calyptogena magnifica*), mussel (*Bathymodiolus thermophilus*), and endemic sponge (n. sp.) community that covers an approximate area of 60 m × 70 m. Extinct high-temperature sulfide chimneys were also discovered several hundred meters northeast of Calyfield. Field of view ~1 m.

magmatism, heat, and resulting shallow topography. During that cruise, several high-temperature vents were discovered using the towed imaging system *ARGO II*, but no samples of fluids or biota were recovered (Figs. 4.188 and 4.189).

In June and July 2011, the GALREX 2011 expedition mapped about 40,000 km^2 of seafloor and acquired a 400 km long hydrothermal plume survey of the unexplored eastern portion of the GSC from 101.3° W to 98.0° W (Shank *et al.*, 2012). Additional continuous hydrothermal plume surveys and multibeam transects were also completed between 89.33° W and 85.75° W. The plume surveys revealed at least 20 distinct water column anomalies along the eastern segment of the GSC, with venting concentrated in two distinct areas. The anomalies included a 50 km long (88.56° W–88.09° W) region of continuous, optically intense plumes rising as high as 250 m above the seafloor, and a suite of discontinuous weaker plumes near the site where Rose Garden had been located and the new Rosebud site (Shank *et al.*, 2003).

4.14.3 Rosebud and Calyfield Vent Fields

In May–June 2002, two new vent fields with markedly different community structures were discovered along the GSC using the AUV *ABE* and *Alvin* near 86°13′ W (Rosebud; Fig. 4.188) and 89°37′ W (Calyfield; Fig. 4.194) (Shank *et al.*, 2003). Based on the dive results at Rosebud, it is clear that the well-developed faunal communities documented 12 years earlier at Rose Garden had been buried by a new lava flow. Approximately 200 m northwest of Rose Garden a new vent site, Rosebud (Figs. 4.190–4.191) (2470 m depth), was supporting animal communities that were in the early stages of development. Detailed imaging and photomosaics of the Rosebud Field, which covered an area of approximately 60 m by 50 m, revealed that this site consists of four major venting areas. In 2004, these sites contained vestimentiferan tubeworms (the majority were less than 6 cm in length), linear rows of bathymodiolid mussels (average 1 cm in length) growing along cracks in the sheet lava surface, and adjacent carpets of amphianthid anemones

Figure 4.195 The shallowest part of the GSC is where the first discovery of active high-temperature black smoker venting was made. This digital image, collected using the *Medea* towed imaging platform, shows the Plumeria Vent Field. Plumeria was a cluster of more than six 12–14 m tall chimneys along an active fissure system. Other high-temperature black smoker vent fields were also discovered during this cruise in this shallowest part of the GSC. Field of view ~4 m.

Figure 4.196 High-definition ROV imaging in 2011 documented fresh lava flows and nascent vents hosting flocculent white microbial material on lava surfaces and mobile fauna at Uka Pacha and Pegasus Vent Fields at 88.2° W and 88.3° W, west of the overlapping spreading center. While these fields were less than 30 m × 40 m in size, they and the expansive areas of fresh basaltic lavas indicate the prevalence of recent volcanic activity throughout the segment. Field of view ~2.5 m.

(about 50 per m^2). A single assemblage of larger (longer than 10 cm) mussels was observed on the northeastern margin of the field. Vesicomyid clams (about 10 individuals), all less than 3 cm, were observed along cracks in the central sheet flow.

The sizes and distributions of the communities at Rosebud and the volcanic structures, freshness of lava surfaces and setting, all pointed to a volcanic eruption having occurred about two years prior to the 2002 diving expedition. The eruption paved over the Rose Garden site and provided the heat for rejuvenating the low-temperature hydrothermal flow that sustained Rosebud.

Figure 4.197 The summit of one of the more than six inactive chimneys over 30 m tall discovered along the rift valley of the GSC east of ~90° W in 2011. Field of view ~2 m.

Calyfield was also discovered in 2002 using the AUV *ABE* (see Chapter 2) and *Alvin* while exploring the shallow (~1900 m depth) non-rifted portion of the GSC (Fig. 4.188) west of the 86° W vent sites for high-temperature black smokers (Fig. 4.195). The axis of the GSC, west of 88° W, transitions from a well-developed rift valley east of this longitude that is several hundred meters deep and several kilometers wide, to a broad axial high that has nested axial fissures along the summit (Fig. 4.188).

Along the axis near 89°37′ W, at a water depth of 1690 m, an approximate 60 m × 60 m area of well-formed lava pillows and tubes was colonized by large vesicomyid clam (*Calyptogena magnifica*), mussel (*Bathymodiolus thermophilus*), and endemic sponge (n. sp.) communities (Fig. 4.194). In addition, extinct high-temperature sulfide chimneys were discovered several hundred meters northeast of Calyfield. Hydrothermal flow consisted of very low temperature fluids (less than 1 °C above ambient), but they contained sufficient H_2S to support long-lived bivalve communities estimated, based on their size, to be at least 10 years old.

4.14.4 Uka Pacha and Pegasus Vent Fields

Recently erupted lava flows were discovered near 88.3° W. The flows were spread over at least 14 km, and included several areas of vigorous diffuse venting. At two sites, white flocculent material – presumably microbial in origin – issued from the seafloor at snowblower vents (Fig. 4.196).

At two newly discovered vent fields, Uka Pacha and Pegasus, white microbial mats blanketed extensive areas along the base and sides of the axial graben (Shank *et al.*, 2012). Sessile fauna were not observed and mobile organisms were scarce, limited to copepods and shrimp, suggesting insufficient time since the creation of the vents for new colonists to arrive, similar to what has been observed at

Figure 4.198 Digital image taken by *Alvin* in 2010 of a black smoker hosting multiple small edifices indicating the short-term (months to <1 year) evolution of venting at the Iguanas site at 92° W. The structure shows active anhydrite deposition. Field of view ~2 m.

Figure 4.199 *Riftia* tubeworms, anemones, and small (<4 cm long) mussels (center lower right) recruiting along cracks adjacent to diffuse flow vent openings surrounded by large aggregations of live and dead clams at the Tempus Fugit Vent Field discovered in 2011 using *Little Hercules* and the *Okeanos Explorer*. Field of view ~1 m.

EPR vent sites near 9°50′ N after a volcanic eruption (Shank *et al.*, 1998). Extinct hydrothermal sulfide chimneys over 30 m tall (Fig. 4.197), remnants of an earlier time of sustained high-temperature venting, were discovered within 2 km of these active vents.

4.14.5 Tempus Fugit Vent Field

Three ROV dives at plume sites near 86° W discovered one of the largest low-temperature vent fields (120 m × 40 m) on the GSC (Figs. 4.188 and 4.198) (Shank *et al.*, 2012). Tempus Fugit was characterized by diffuse flow in a once massive clam bed, potentially more than 20 years old. The high abundance of juveniles among mature live vesicomyid clams, siboglinid tubeworms and mytilid mussels suggests multiple colonization events. Active colonization by recently colonized mussels occurred on relatively older pillow and lobate lavas that were ringed by large beds of dead clams. Clusters of scavenging dandelion siphonophores demarcated the site's periphery.

Observations to date suggest abundant and recent hydrothermal and volcanic activity on two adjacent segments of the GSC, spanning more than 200 km of the spreading axis. These findings reveal not only recent eruptive activity between 85° W and 89° W, but also that the rates of hydrothermal habitat turnover via eruption, dike intrusion, or cessation of venting may be considerably higher than previously expected along the GSC.

Central Indian Ridge

4.15 A new biogeographic province

A major driver for exploration of vents in the Indian Ocean was to test the hypothesis that the Indian Ocean Ridge system served as a pathway for exchange of vent organisms between the Pacific and Atlantic Oceans (Van Dover *et al.*, 2001). Mid-ocean ridge segments in the Indian Ocean range from ultraslow spreading (<1 to 2 cm/yr) along the Southwest and Southeast Indian Ridge to slow to intermediate (<6 cm/yr) spreading along the Central Indian Ridge (CIR), which extends north from the Rodriguez triple junction at 25°30′ S and 70° E where the Southwest and Southeast Ridges meet (Figs. 4.200 and 4.201). Hydrothermal plume exploration along these ridge segments began in the late 1980s with a German expedition (Hertzig *et al.*, 1986; Hertzig and Plüger, 1988; Plüger *et al.*, 1990), followed in the mid 1990s by French and Japanese cruises (Jean-Baptiste *et al.*, 1992; Gamo *et al.*, 1996) and British work along the CIR (Rudnicki *et al.*, 2002).

Active venting at the Kairei site, just north of the triple junction, was discovered by the Japanese in 2000 with their *Kaiko* ROV system (Gamo *et al.*, 2001; Hashimoto *et al.*, 2001). The Edmond vent site on the southern CIR was subsequently discovered by a US team in 2001 using CTD tow-yos to localize the area of the buoyant plume and the ROV *Jason* to follow seafloor features and locate the vents (Van Dover *et al.*, 2001).

4.15.1 Kairei Vent Field

The Kairei Vent Field (25°19.23′ S, 70°02.42′ E) is at a depth of 2415–2460 m, within a 20 km long rift valley segment that terminates in a 4.2 km deep basin at a non-transform discontinuity to the north and at the triple junction to the south (Fig. 4.201). The eastern rift-valley wall has a stair-step morphology with Kairei located on a 10°–30° slope on a step in the upper rift wall. The bench is nearly 1.8 km shallower than the rift valley floor and 700 m from the central axis of the rift valley. The main area of high-temperature venting (306–365 °C) occurs along a W-NW trend and extends 80 m along the rift wall, and is 30 m wide. Large accumulations of weathered sulfides and relict sulfide chimneys suggest that the site has been active for decades along the rift wall. Three mounds contain high-temperature black smokers. Low-temperature diffuse flow is mainly restricted to the sides of the black smoker complexes.

High-temperature fluids sampled from multiple orifices throughout the three active mounds (Figs. 4.202–4.204) have

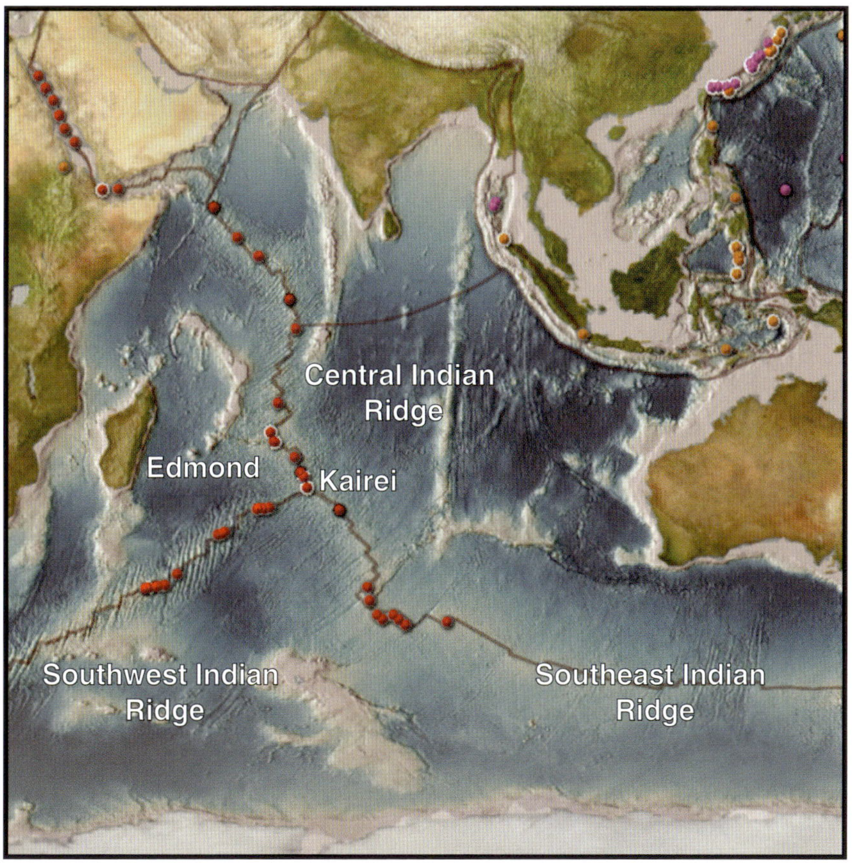

Figure 4.200 A major driver for exploration of vents in the Indian Ocean was to test the hypothesis that the Indian Ocean ridge system served as a pathway for exchange of vent organisms between the Pacific and Atlantic Oceans.

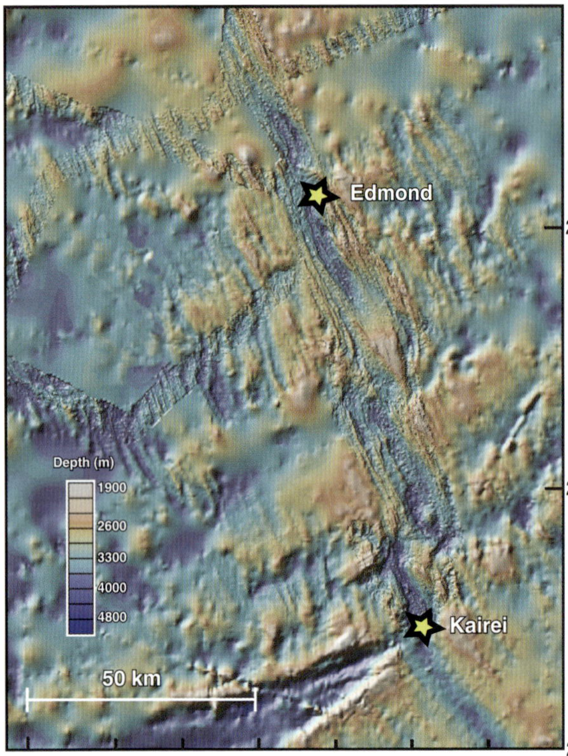

Figure 4.201 The Kairei and Edmond Vent Fields are located along the southern end of the 250 km long CIR. This bathymetric map focuses on the southernmost ridge segments before the CIR terminates at the Rodriguez triple junction where Kairei is located.

Figure 4.202 Black smoker vents at the Kairei Field, at 2415 to 2460 m water depth, are arranged in dense groups of 5–10 m tall structures along steep slopes of the eastern rift valley wall. Temperatures reach 365 °C and the vent fluids are acidic. Field of view ~3 m.

Figure 4.204 Faunal zonation is distinct at Kairei. Dense *Rimicaris* shrimp aggregations dominate the active sulfide chimney surfaces, while small clusters of *Bathymodiolus* mussels are commonly oriented along the basal margin of the edifices (seen here along the overhanging margin). Field of view ~3 m.

Figure 4.203 Most chimneys at Kairei in 2000 were venting from multiple orifices along the length of the main edifice. This orifice was formed when a sulfide sample was taken with the ROV *Jason* manipulator, creating an active vent at the base of one chimney. The diameter of each orifice was ~3 cm. Field of view ~12 cm.

identical chemistries with about 15% greater chlorinity than local ambient seawater (540 mmol/kg) and iron concentrations of about 5 mmol/kg (ambient seawater [Fe] is <1 nmol/kg) (Gallant and Von Damm, 2006). The vent fluids contain high concentrations of H_2 (8.5 mmol/kg), but do not have associated high values for CH_4 (0.2 mmol/kg) (Van Dover *et al.*, 2001).

Recent observations of the seafloor near Kairei from the Japanese submersible *Shinkai 6500* reveal olivine-rich plutonic rocks with olivine gabbro–troctolite–dunite assemblages (see Chapter 7) exposed within 15 km of the vent field. In addition, serpentinized ultramafic mantle rocks (see Chapter 8) occur on a core complex adjacent to the ridge segment where Kairei is located (Nakamura *et al.*, 2009). The core complex has been suggested to be a recharge zone for Kairei hydrothermal flow and a cause for H_2-rich vent fluids at this site, compared to the low H_2 fluids at Edmond (see below), which is hosted in normal MOR basalts (Nakamura *et al.*, 2009).

Animal communities at the Kairei field are highly zoned with dense aggregations of shrimp (*Rimicaris kairei*) congregating in 10–20 °C fluids, while anemones (*Marianactis*)

Figure 4.205 Dense aggregations of *Rimicaris kairei* reach up to 2500 individuals per square meter in ~30 °C fluids on active vents. Less numerous shrimp (shown here) commonly graze on inactive sulfide hosting white tufts of filamentous microbes. Morphologially and genetically distinct from *Rimicaris exoculata* on the MAR, *R. kairei* maintains a similar symbiosis with chemoautotrophic bacteria in their gills. Field of view ~6 cm.

Figure 4.206 A band of predatory *Marianactis* anemones is distributed along the apron of sulfide structures in the central Kairei Field. These anemones predate on shrimp. The accumulation of empty shrimp carapaces (molts) commonly forms dense puddles in low-lying areas, downslope from venting. Field of view ~0.5 m.

Figure 4.207 *Bathymodiolus marisindicus* mussels occur in small groups along rock margins. They are not extensive at Kairei, and even less common at the Edmond Field. The absence of byssus thread attachment points on shells indicates a notable lack of movement by these mussels. Field of view ~10 cm.

Figure 4.208 Large gastropods (e.g., *Alvinochoncha* sp. shown here at the center, and *Desbruyeresia* sp. at left) hesionid and orbiniid polychaetes in the upper right), and brachyuran crabs (*Austinograea rodriguezensis*) inhabit rocky surfaces in the mussel zone (upper and right) at the base of the black smoker complex at Kairei. Field of view ~15 cm.

populate the peripheral zones (Figs. 4.204–4.206). *Bathymodiolus marisindicus* mussels and gastropods occur in the transition zone and in small groups along broken rock margins (Figs. 4.207 and 4.208).

4.15.2 Edmond Vent Feld

The Edmond Vent Field (23°52.689′ S, 69°35.809′ E) is at a water depth of 3290–3320 m, approximately 160 km N-NW of Kairei (Figs. 4.200 and 4.209). Earlier hydrothermal plume work suggested the presence of active seafloor vents in this region. Discovery of the site involved interpreting the geological structure inferred from multibeam bathymetry and integrating the results of near-bottom water-column surveys to locate the seafloor origin of the hydrothermal plume signal.

The Edmond Field is at the northern end of a ridge segment on the eastern rift-valley wall, 6 km from the adjacent rift axis (Fig. 4.209). The vent field is on the upper section of the eastern rift-valley wall 3.3 km below the ocean surface, and measures 100 m × 90 m. It is on a small salient that extends south from the rift wall. The vents are 900 m deeper than those in the Kairei Vent Field. The seafloor structure at the non-transform discontinuity, which marks the northern end of the segment where Edmond is located, is characterized by a zone of cross-cutting, low-relief ridges and valleys that have orientations parallel, perpendicular,

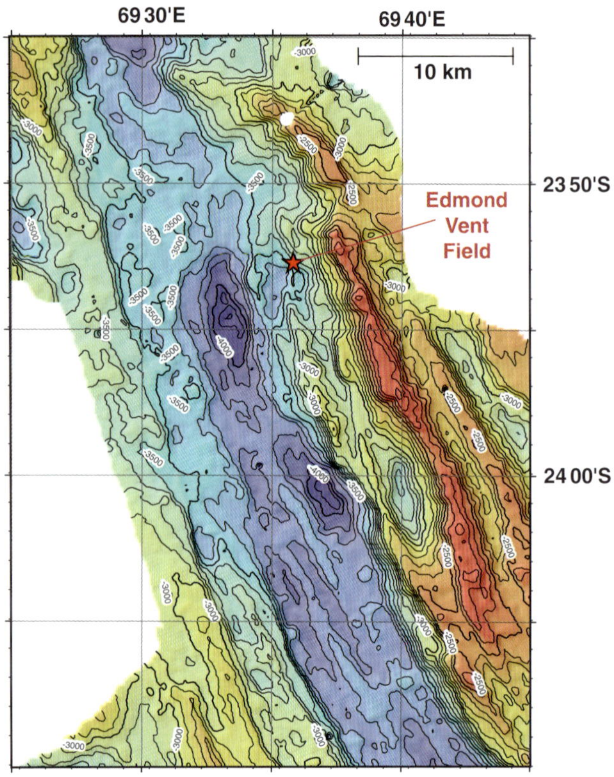

Figure 4.209 Multibeam bathymetric map showing the location of the Edmond Vent Field (red star) on a bench on the lower, eastern rift valley wall.

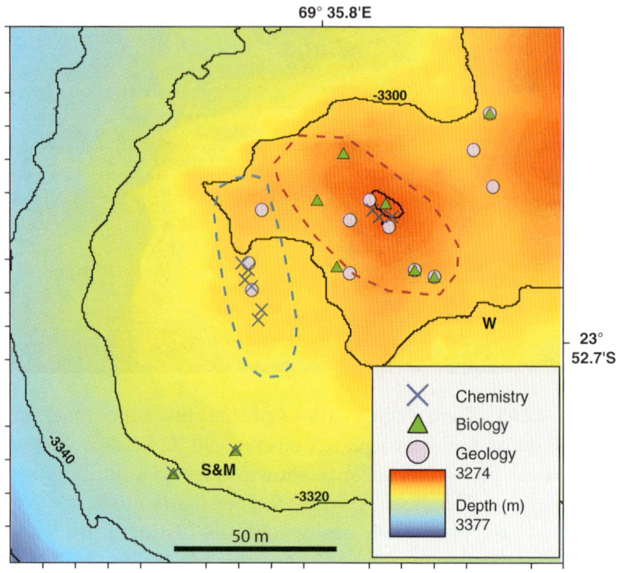

Figure 4.210 The distribution of high-temperature (circled in red dashed line) and lower-temperature focused venting (circled in blue dashed line) in the Edmond Field based on ROV *Jason* near-bottom multibeam surveys in 2001. Hydrothermal sediments and bacterial mats (labeled S&M) occur southwest of the high-temperature vents. Geological, fluid, and biological samples collected by the ROV *Jason* are also marked as shown by the key.

Figure 4.211 A black smoker in the Edmond Vent Field at 3278 m water depth hosts white anemones peripheral to the orifice. Venting at Marker A was from a ~50 cm tall chimney that merged into the western portion of the sulfide mound in the center of the field. Field of view ~1 m.

Figure 4.212 High-temperature beehive structures in the Edmond Vent Field support swarms of *Rimicaris* shrimp. No significant genetic differences were measured between *Rimicaris exoculata* in the Kairei and Edmond Fields. Field of view ~1.5 m.

and oblique to the N-NW trend of the rift valley. Similar cross-cutting ridge-crest structures are important in controlling sites of hydrothermal venting at the TAG Hydrothermal Field on the MAR (see Section 4.7).

Hydrothermal fluids collected from four smoker sites at Edmond (Figs. 4.210–4.212) have temperatures of 273–382 °C, and chlorinities at least 50% greater than ambient seawater, making them some of the hottest brines yet sampled. The high temperatures and chlorinities lead to more than 20 mmol/kg iron in the fluids – some of the highest concentrations yet measured (Gallant and Von Damm, 2006). The fluids are low in pH (about 3 measured at 25 °C and

Figure 4.213 Some of the vents at Edmond consist of small sulfide mounds, ~1–3 m tall and ~5–10 m across with delicate vent structures at their summits, including branching or candelabra structures as shown (water depth 3300 m). Diffuse fluids are venting from the small knob on the left side of the image. The knob is covered with white bacterial mat and a black beehive-looking structure. White anemones dominate the sessile fauna on these mounds and over much of the Edmond Field. Field of view ~15 cm.

Figure 4.214 Zoarcid eelpout fish (*Pachycara* sp.) thrive amongst anemones attached to a large inactive sulfide structure in the Edmond Vent Field. Field of view ~1 m.

Figure 4.215 The Kairei and Edmond Vent Fields appear highly similar in faunal dominance, given large aggregations of shrimp and anemones (shown here at Edmond). However, numerous faunal species are either scarce (e.g., limpets and mussels) or absent from Edmond (e.g., stalked barnacles). Field of view ~2 m.

1 atmosphere) and contain several mmol/kg of H_2S. Hence, in spite of the high iron concentrations, significant amounts of sulfide are transported and available for biological metabolic processes.

The Edmond Field consists dominantly of old disaggregated sulfide structures and massive sulfide talus, indicating that hydrothermal activity has been focused at this site over a long period of time, similar to the Kairei Vent Field. However, unlike Kairei, orange-brown, iron oxyhydroxide sediments are abundant at the Edmond Field, reaching several centimeters in thickness and coating many of the sulfide structures and much of the talus (Fig. 4.213).

Hydrothermal venting at Edmond exhibits a greater range of styles and chimney morphologies than at Kairei. High-temperature venting manifests as discrete clusters of large chimneys (up to 20 m tall and 2 m in diameter) with vigorous black smoker fluids emanating from multiple orifices. Within and between the clusters, smaller (up to 5 m high), branched candelabra-like structures with orifice diameters roughly 2–20 cm ornament the seafloor and discharge black smoker fluids at slower flow rates (Fig. 4.213). Beehive structures, similar to those described from other hydrothermal sites at slow- to fast-spreading MORs, are common at Edmond (Fig. 4.212). The beehives are gray, highly porous and bulbous structures, 2–50 cm in diameter, with clear fluid diffusing from the surfaces. The surfaces of the beehives are characteristically crenulated on the scale of centimeters, most likely due to variations in the rate of fluid discharge.

Diffuse flow is considerably more widespread at Edmond than at Kairei, where it is largely restricted to the flanks of the black smoker complexes. The presence of blocks of anhydrite at Edmond, which are absent at Kairei, indicates that there may be more subsurface seawater entrainment beneath vents at Edmond. Diffuse flow over large areas at Edmond lends support to extensive subseafloor mixing processes over a broad area, thereby providing abundant microhabitats for microbial activity, as evidenced by large bacterial mats observed at Edmond, but not at Kairei.

4.15.3 Kairei versus Edmond fauna

The macrofauna of the Edmond Field is, in general, a subset of that found at Kairei (Gamo *et al.*, 2001; Van Dover *et al.*, 2001; Van Dover, 2002; Gofreddi *et al.*, 2004). As at Kairei, the biomass of the Edmond Vent Field is dominated by peripheral beds of anemones (*Marianactis* sp.) on inactive sulfides and in areas of diffuse flow, and dense aggregations of shrimp (*Rimicaris kairei*) on active black smoker chimneys (Figs. 4.211 and 4.212). Other fauna include brachyuran crabs (*Austinograea rodriguezensis*), galatheid squat lobsters (*Munidopsis* n. sp.), a second shrimp species (*Mirocaris indica*), a large white polynoid scale worm (*Branchinotogluma* sp.), and large predatory gastropods (*Desbruyeresia marisindica* and *Provanna segonzaci*) (Fig. 4.208). A zoarcid fish species is shared by both fields (Fig. 4.214).

Despite the presence of visually similar habitats at Edmond, it is distinguished from Kairei by the complete absence or relative paucity of certain species that are components of the Kairei field. Although populations of hairy gastropods (*Alvinoconcha hessleri*) and highly dispersive mussels (*Bathymodiolus marisindicus*; Fig. 4.207) (Kyuno *et al.*, 2009) are not extensive at Kairei, they were even less abundant at Edmond. *Bathymodiolus marisindicus* at Edmond are distinctive both in their heavy coating of iron oxyhydroxides and in the poor condition of the tissues of the larger individuals. Species completely absent to date from Edmond, but observed at Kairei, are stalked neolepadin barnacles, scaly-foot gastropods, archinomid polychaetes, nemerteans, and turbellarians.

In 2004, limpets and small-coiled gastropods were scarce at Edmond, although they were commonly collected on inactive sulfide surfaces at Kairei. Small invertebrates including aplacophoran mollusks and small polychaetes (dorvilleids, orbiniids, hesionids), collected in association with hairy gastropods and mussels at Kairei, were not found in gastropod and mussel washings from Edmond. Unique to Edmond is a bed of tube-dwelling ampharetid polychaetes (provisionally *Glyphanostomum* sp.) in loose mineral deposits at the eastern extreme of the vent field.

The faunal differences observed between Edmond and Kairei are most likely related to contemporary habitat availability and microhabitat quality (fluid and rock chemistry) rather than to geographic or bathymetric isolation. The similarity of CIR vent fauna (Hashimoto *et al.*, 2001; Van Dover *et al.*, 2002) to extant vent-endemic genera (i.e., many new species from the CIR have been described within existing genera) supports the hypothesis that there has been significant historical communication of vent fauna between the Pacific and Atlantic Oceans via larval dispersal along MORs that span the breadth of the Indian Ocean.

Exploration of the Edmond and Kairei Fields highlights that there are still significant unexplored portions of the seafloor that remain open for discovery of new vent habitats. In the following chapters, the dynamic processes that form MOR volcanoes are discussed: they are tightly linked to the evolution of hydrothermal vents and their associated biological communities.

5 Submarine volcanism: fire beneath the sea

There are no manifestations of the forces of nature more calculated to inspire us with feelings of awe and admiration than volcanic eruptions. . .

E. Hull (1892)

5.1 Volcanoes on land and under the sea

Throughout recorded history, volcanic activity has had major impacts on societies and terrestrial ecosystems. But it has been only within the last 60 years that it has been recognized that the largest and most active volcanic province on Earth lies beneath the oceans. Over time, submarine eruptions produce the largest volume of volcanic products on Earth, but research on individual submarine eruptions is still in its infancy (Rubin *et al.*, 2012) (Fig. 5.1).

Figure 5.1 A pillow tube extruding down a near-vertical slope on West Mata Seamount, northern Lau Basin; photographed from the ROV *Jason* on Dive J2-420 during an eruption in May 2009. Lava has accumulated on the slope face in a folded, flow-deformed pillow pile. Field of view ~3 m (Resing *et al.*, 2011; Rubin *et al.*, 2012).

It is now understood that major volcanic eruptions such as those at Tambora, Krakatoa, Pinatubo, Vesuvius, Mt. Pelée, and Mt. St. Helens during the past few centuries – all violently explosive eruptions in island-arc and continental margin subduction environments – as well as effusive events such as the 1793 Lakagígar (Craters of Laki) eruption in Iceland, have had significant impacts upon ecosystems and humans and even affected short-term global climate conditions (Thordarson and Self, 2003). In contrast and hidden from view, the more continuous, markedly less explosive eruption of basaltic and in some cases more silica-rich lava along the global mid-ocean ridge (MOR) system, and intrusions of magma beneath it, produce the vast majority of Earth's oceanic crust (see Chapters 2 and 3). Unlike subaerial volcanism on Hawaii or Iceland, where active eruptions and lava flows can be monitored and studied with relative ease, information about deep-sea volcanic processes (except in rare cases) must be obtained using more complex instrumentation and remote-sensing methods. Flows and eruptions under the ocean surface are deduced using maps derived from hull-mounted multibeam and sidescan sonar, and more precise mapping at sub-meter resolution using near-bottom sonar imaging from autonomous underwater vehicles (AUVs) (see Chapter 1). The higher resolution data, combined with photographic images, detailed observations, and careful sampling of the solidified lava flows collected by human-occupied vehicles (HOVs) or remotely operated vehicles (ROVs) have all provided new and important perspectives about seafloor volcanic processes (Rubin *et al.*, 2012). These technologies now permit deep-ocean seafloor mapping to be conducted with detail similar to that possible on land. This chapter focuses on the volcanic carapace of the oceanic crust. These sequences of rocks emplaced on the crests and slopes of volcanoes along the MOR over time form the uppermost layer of the oceanic crust underlying all the ocean basins.

5.2 Submarine lavas

The primary differences between submarine lava flow morphologies (pillows, lobates, and sheets) are the size of individual flow units and their interconnectivity. Pillow flows extrude as ~1 m scale individual spherical or cylindrical tubes of lava that develop a glassy crust on all sides, generally inhibiting coalescence with neighboring pillows (Moore, 1975) (Fig. 5.2). Pillow flows are produced by the piling up of individual pillow lava lobes whereby the newest pillows are erupted from the top of the stack and flow outward a limited distance before freezing, resulting in the formation of steep-sided mounds, ridges, and haystacks (Chadwick *et al.*, 1998, 2001; Perfit and Chadwick, 1998; Embley *et al.*, 2000). Although, by necessity, some continuous pathways must exist within a pillow flow to deliver lava from the seafloor vent to the top or sides of an actively forming mound, the high relief of pillow flows and absence of collapse features that span multiple pillows suggest limited hydraulic connectivity.

Lobate lavas comprise individual flow units that are similar to pillows, but their form is more oblate and flow forms can span distances of ~2–10 m. The lobes develop a quenched crust on their top and sides, but are emplaced sufficiently rapidly that they can coalesce into one interconnected flow surface (Fig. 5.3). Lobate flows commonly inflate and are similar to terrestrial inflated pahoehoe sheet flows (Hon *et al.*, 1994). They produce relatively flat upper crusts that may later collapse if the molten interior of the flow subsides or trapped vapor from heated seawater condenses (Engels *et al.*, 2003; Perfit *et al.*, 2003). Flows that become ponded and later collapse are sometimes referred to

Figure 5.2 Pillow lava mound on the summit of Axial Seamount, Juan de Fuca Ridge, with well-developed bread-crust textures on pillows at the left side of the image. Lobate lava flows (darker lava) emplaced during the 2011 eruption on the right side of the image are flowing over and between the older pillows. Field of view ~3 m.

Figure 5.3 Coalesced lobate flows on the East Pacific Rise crest near 9°50′ N at 2550 m depth display low relief, lack of surface decorations, and are interfingered. This flow morphology suggests a relatively high effusion rate. Yellow staining on edges of the flows and near fractures is the result of low-temperature hydrothermal activity in regions where warm water vented from beneath the lava, which can accelerate alteration of the fresh glass crusts. Field of view ~3 m.

as "lava lakes," but lack a direct connection to a crustal magma supply and are more akin to lava ponds described in subaerial basaltic volcanic terranes (e.g., Wilson and Parfitt, 1993). Inflated lobate flows commonly display "lava pillars" in their collapsed interiors (Ballard *et al.*, 1979; Francheteau *et al.*, 1979; Gregg and Chadwick, 1996; Gregg *et al.*, 1996).

Submarine sheet flows erupt as a continuous sheet with an upper surface of uniform height, but they are typically modified into folded, lineated, jumbled, or hackly surface morphologies that reflect a variety of processes. These can include dynamic flow through topographic restrictions in the terrain the flow is advancing over; flow along channel margins, which can cause contortions in the surface of the flow; and repeated inflation/deflation episodes. Sheet flows are typically laterally extensive but can also be confined to axial valleys, graben, or the axial summit trough (AST). In some locations on fast-spreading MORs, sheet flows extend as far as ~3–5 km from the AST as a result of rapid, high-volume flow that forms lava channels (Sinton *et al.*, 2002; Soule *et al.*, 2005, 2007). Vertical sections of oceanic crust exposed along fault scarps (e.g., Karson *et al.*, 2002a,b; Stakes *et al.*, 2006) or revealed by drilling during the Integrated Ocean Drilling Program (IODP) (e.g., Tominaga and Umino, 2010) show that some sheet flows can attain thicknesses of a few meters and likely transition into what are known as "massive flows" (>3 m thick), which may have formed as lava ponds. A massive flow, known to be at least 75 m thick, drilled at IODP Site 1256 is proposed to represent a ponded flow that formed more than 3 km off-axis (Tominaga *et al.*, 2009).

Variations in lava flow type are observed on the ridge segment scale and over larger geographic areas. On most MOR, lavas extrude along linear eruptive fissures commonly found within the axial summit trough and axial valley, and can be from approximately one to tens of kilometers long. The eruptions can be sourced from melt lenses directly below the fissures or from magmatic dikes that travel along the ridge axis (Dzaik *et al.*, 1995; Delaney *et al.*, 1998). At fast-spreading ridges, it has been noted that lava at the ends of magmatic segments produces markedly more pillows than at the middle of the segments, where sheet flows and lobates predominate (White *et al.*, 2000). A similar increase in pillows has been observed at the ends of eruptive fissures, which commonly coincide with magmatic segment boundaries (Sinton *et al.*, 2002; Fundis *et al.*, 2010). At slow-spreading ridges, lavas erupted near the segment center, where ascending melts are focused within the shallow crust, have greater abundances of lobate and sheet flows relative to pillows than toward the segment ends, where melts are transported through lateral diking to eruption sites (Cann and Smith, 2005; Behn *et al.*, 2006). At the intermediate spreading rate Galápagos Spreading Center, melt supply systematically increases along-axis toward the intersection with the Galápagos hot spot, and the relative abundances of lobate and sheet flows to pillow flows is positively correlated with inferred melt supply (e.g., Behn *et al.*, 2004; White *et al.*, 2008).

5.3 Historical perspectives

Photographs and detailed studies of the MOR over the past few decades reveal the characteristic forms and the context of the lava flows, and the effect of key processes involved during eruptions. Much of what is presented in this chapter focuses on processes that influence the formation of volcanic features and different lava morphologies. The discussion includes how these deposits are used to interpret the timing and extent of eruptions along different portions of the MOR, determine the geochemical and petrogenetic processes operating in the sub-MOR crust and upper mantle, and how the time-integrated succession of lava flows builds the upper 5–10% of the oceanic crust.

Marine geologists have known of the existence of volcanic rocks on the seafloor since at least the 1870s, when submarine basalts were recovered from the deep sea by trans-Atlantic cable repair ships (Hall, 1876) and the *Challenger* Expedition (Murray and Renard, 1891). Early MOR photographic surveys and submersible expeditions in the 1960s and 1970s revealed a wide variety of volcanic landforms with great similarity to those in subaerial basaltic provinces such as Hawaii and Iceland (Ballard and Van Andel, 1977a). The diversity of processes associated with MOR volcanism has been studied in detail since the mid-1970s, when FAMOUS program investigators first made systematic observations and collected volcanic rocks from the Mid-Atlantic Ridge using *Alvin* (Fig. 5.4). Since then, significant advances have been made in the study of morphology, spatial extent, and eruptive processes leading to the emplacement of submarine lavas (Rubin *et al.*, 2012).

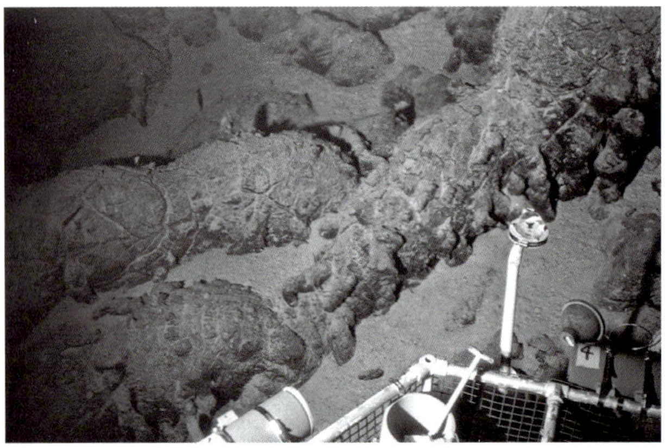

Figure 5.4 Single-frame 35 mm image taken from *Alvin* showing submarine pillow and tubular lava flows observed during the 1974 FAMOUS dives in the Mid-Atlantic Ridge rift valley. This was the first international, focused research expedition using deep-submergence vehicles to map and sample the MOR. Field of view ~3 m.

Despite decades of progress in understanding how the upper oceanic crust is generated, only recently have detailed investigations of the physical processes that lead to eruption, flow, and solidification of submarine lava yielded results. In fact, only in a few places have rudimentary data on the spatial dimensions of individual submarine eruptive lava flows been measured accurately (Fig. 5.5) (Embley *et al.*, 1995; Chadwick *et al.*, 2001, 2013; Sinton *et al.*, 2002; Soule *et al.*, 2007; Caress *et al.*, 2012; Colman *et al.*, 2012: Rubin *et al.*, 2012; Yeo *et al.*, 2012; Clague *et al.*, 2013).

Through detailed mapping, geochemical, and modeling studies of lava samples it is apparent that MOR lava can flow up to ~5 km outward from the axis and much greater distances along-axis at fast-spreading ridges (e.g., Soule *et al.*, 2007). At slow-spreading MORs, the seafloor lava unit commonly builds locally thick sections (tens to hundreds of meters in height) of pillow mounds, hummocks, and ridges of composite volcanic constructions. These commonly occur within the rift valley along the neovolcanic zone (e.g., Smith and Cann, 1990; Yeo *et al.*, 2012).

5.4 Direct observations of the lava units from tectonic windows and crustal drilling

Understanding the construction of the volcanic carapace of oceanic crust requires multiple lines of investigation. These include integration of geological and geochemical data on active MORs; study of exposures on tectonic escarpments that expose kilometer-thick sections of young ocean crust; analysis of lava sequences recovered in deep drilling cores; correlations to seismic structure and potential field studies from geophysical surveys; and studies of and comparisons with lavas sampled in ophiolites (Fig. 5.6). These types of data provide important information on the composition and construction of the uppermost oceanic crust across all spreading regimes (key areas where major studies have been conducted are summarized in Chapter 3). Perhaps the most compelling constraints on the internal structure of seafloor lava sequences that make up the oceanic crust come from crustal drilling and tectonic windows. These features comprise large fault scarps in transform faults or rift valley walls exposing vertical sections of upper crustal rocks that extend for hundreds to over a thousand meters.

Lava flow morphologies that preserve even delicate eruptive features in seafloor outcrops can be obscured by compaction and fracturing that increase downward in the unit. Nevertheless, many exposures in fault scarps preserve the distinctive round to oblate outlines of the pillow or tube forms and the characteristic radial fracturing that accompanies the cooling and contraction associated with those morphologies (Figs. 5.6 and 5.7).

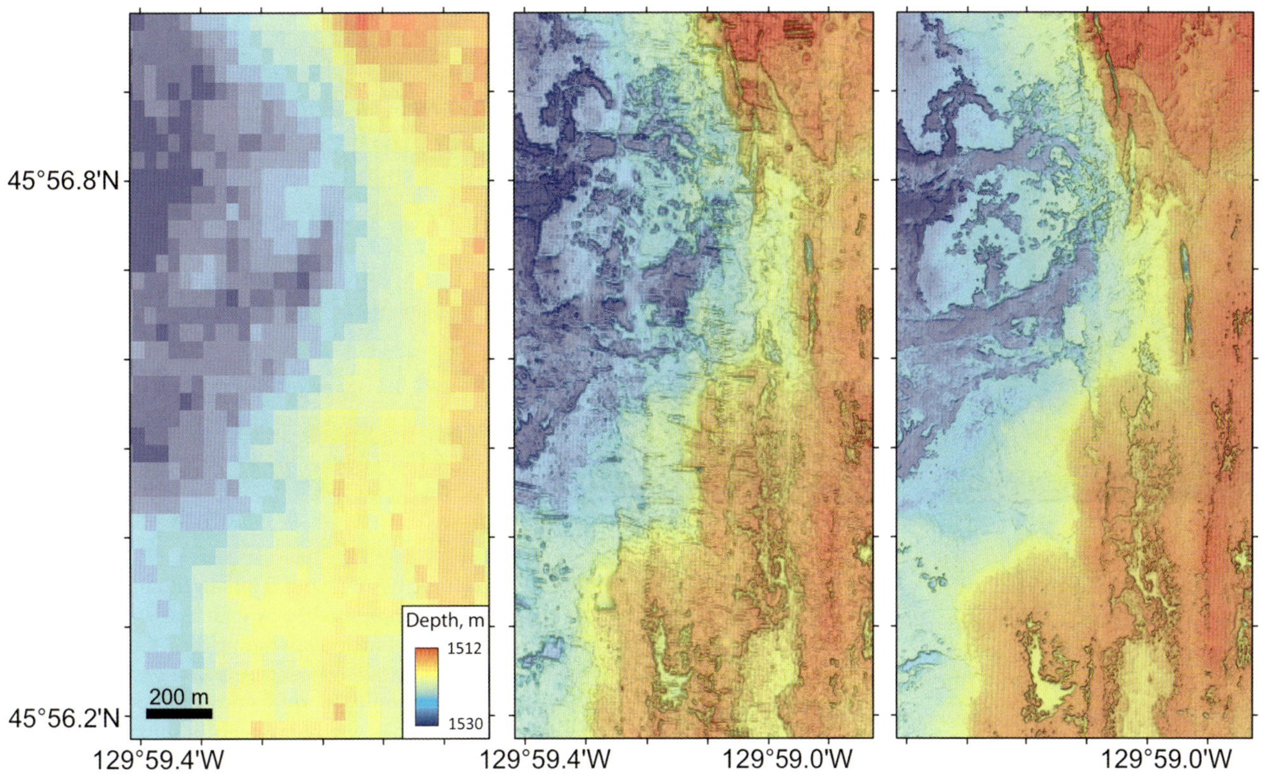

Figure 5.5 Comparison of the resolution of bathymetric mapping methods used to study the 1998 eruption of Axial Seamount (from Chadwick *et al.*, 2013). Each map shows precisely the same area in the southwest portion of the caldera where new flows erupted from fissures that extend south from the eastern margin of the caldera. (Left) Ship-based EM300 multibeam sonar shown at 25 m resolution. (Middle) ROV-mounted Imagenex 675 kHz scanning sonar image at 2 m resolution. (Right) MBARI AUV-based 200 kHz multibeam sonar at a resolution of 1 m (Caress *et al.*, 2012). The deeper areas in blue are lava channels within the caldera that transported lava west from fissures located along the eastern margin of the caldera floor. New, high-resolution images such as these provide data to better understand submarine volcanic emplacement processes.

Figure 5.6 Small mid-ocean ridge basalt (MORB) pillows (left) often form at the end of elongate lava cylinders. Pillow lava forms can wrap around each other as a result of the slow effusion rate and the cooling of the newly forming crust against the pre-existing surface. Manganese coatings that form on lava flows as they age obscure the glassy nature of the outer crusts. Outcropping pillow sequence in the Oman Ophiolite (middle). (Right) Fault scarp (bottom) exposing a cross section of pillow lava sequences at the Juan de Fuca Ridge near 45° N. Field of view across left and right images is ~3 m, middle view is ~5 m.

Figure 5.7 Tectonic fissures and faults in the upper oceanic crust and exposures in ophiolites provide cross sections of lava units that erupted near a spreading axis. Cross sections (left) of pillow and lobate lavas exposed along a fissure in the caldera of Axial Seamount, Juan de Fuca Ridge. Pillow lavas can also form in settings where lavas erupt into fresh water, for example, subglacial eruptions in Iceland (right). In both images note the distinctive forms of pillow and lobate lavas with rounded tops, cuspate bottoms, and radial cooling joints. Field of view in left image is ~6 m and in right photograph is ~5 m.

Drilling has provided samples through hundreds of meters of upper crustal lava units (Figs. 5.8 and 5.9), although commonly the recovery rate is low to moderate in the lava section because of the fragmental nature of the rocks and, in the youngest sequences, the lack of alteration that helps cement the unit. Boreholes in the ocean crust are important for *in situ* studies including logging using acoustical, geochemical, and microbial sensors. These techniques, however, provide only a narrow, one-dimensional perspective (depth), making precise identification of flow morphologies difficult. Recent advances in borehole seismic and acoustic imaging help provide three-dimensional context to interpret the drilling results.

Studies at tectonic windows and drilling in crust formed at intermediate to superfast rates show a range of thicknesses in lava flow units and lava morphologies (Karson, 2002). In general, pillow lavas dominate but lobate, sheet, and massive flows constitute a significant part of the section in some areas (Stakes *et al.*, 2006; Tominaga and Umino, 2010) (Fig. 5.9). The available borehole data show a generally consistent structure. From the top to the bottom of the drilled section the lava layers commonly follow a characteristic sequence. The uppermost lavas are largely intact, with only minor fracturing and jointing from radial cooling, and morphologies correspond well to features described from *in situ* and remote observations. The upper lavas grade downward into progressively more fractured and altered lavas, with compaction and fracturing that can be extreme locally, obscuring lava morphologies. Regardless of these effects, pillow fragments are commonly preserved at least locally. Identifiable lava flow surfaces, such as sheet flows and margins of oblate pillow and lobate flows (Fig. 5.10), progressively increase in dip downward. Dips are inconsistent in the upper lavas but become more consistently inward-dipping (toward the axis) in the lower half of the unit (Fig. 5.11). The proportion of dikes generally increases downward in the unit toward the underlying sheeted dike complex.

Figure 5.8 Drill cores of pillow basalts from the MOR collected by the *JOIDES Resolution* drill ship are laid out after recovery for preliminary inspection and analysis onboard the ship.

From observations of tectonic exposures on fault escarpments, dense high-level dike swarms tens to hundreds of meters across protrude upward into the lavas (see Chapter 6). In some places the dikes reach to within 100 m of the top of the lava unit. Dikes in the lower part of the unit dip steeply to moderately outward (away from the spreading

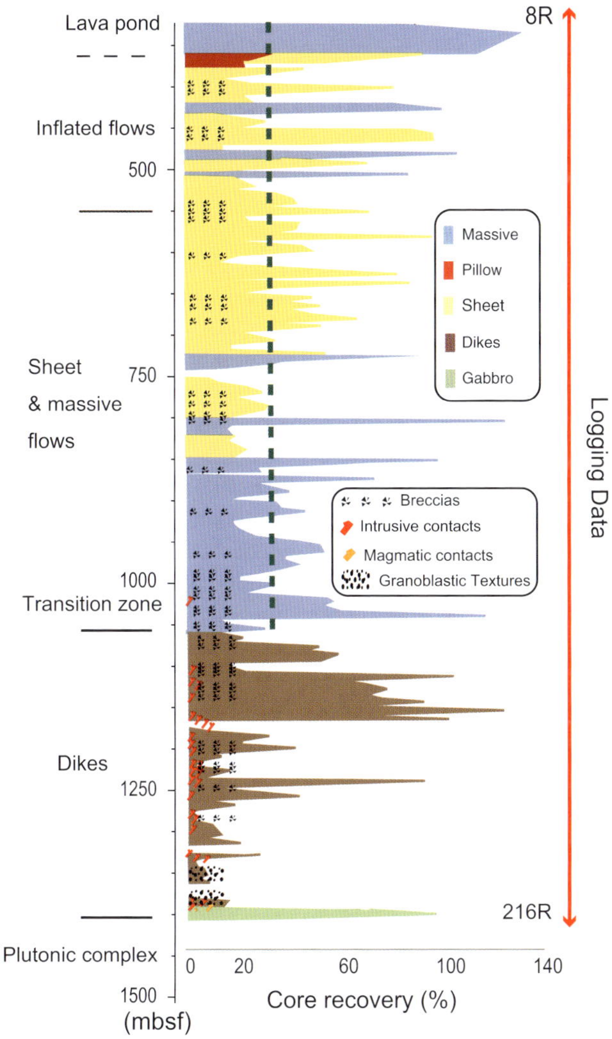

Figure 5.9 Diagrammatic representation of lithologies recovered by drilling in IODP Hole 1256D, based on logging results. This section corresponds to the logged interval from Cores 8R to 216R (red arrow). The dotted line indicates the average core recovery rate of 32% (from Tominaga and Umino, 2010). Vertical axis is meters below the seafloor.

Figure 5.10 Hydrothermally stained pillow and lobate lavas exposed on the inner rift of the Mid-Atlantic Ridge near the TAG hydrothermal site (see Chapter 4). Field of view ~2.5 m.

axis where they formed), and commonly cut lava flows at a high angle. Alteration observed both in boreholes and in samples from tectonic exposures generally increases downward in the unit and is highly heterogeneous, with low- to moderate-temperature hydrothermal assemblages in fractures and fault zones within the unit.

This relatively uniform internal structure in the lava unit of fast-spreading crust suggests a broadly consistent mode of construction in which magma supply is robust and lava erupts and spreads across and along the seafloor on the axis forming 1–3 m thick flows that are usually 10–20 km^2 in area.

The lava units at slow- to ultraslow-spreading ridges have been investigated in many places and appear to vary substantially in their thickness, relations with surrounding rock units, and internal structure (Karson *et al.*, 1987; Juteau *et al.*, 1990; Mevel *et al.*, 1991; Lawrence *et al.*, 1998; Blackman *et al.*, 2002). Tectonic windows and crustal drilling have documented thicknesses of tens to hundreds of meters, of mostly pillow lavas, in these settings. However, in only two localities – Vema Fracture Zone in the equatorial Atlantic (Auzende *et al.*, 1989) and King's Trough in the Northeast Atlantic (K. Klitgord, unpublished data) – have lavas been found to overlie a sheeted dike complex; similar to what has been observed in many ophiolites or faster-spreading oceanic crust.

Lava has also been found to overlie variably deformed and metamorphosed gabbroic rocks or peridotites (see Chapter 3 and Figs. 3.32 and 3.33), commonly in areas of extreme faulting (Karson, 1998) (see also Chapters 7 and 8). In the thickest lava units, hydrothermal metamorphism (Fig. 5.10) and fracturing increase downward.

Near-bottom studies of faulted lava units on the Mid-Atlantic Ridge (MAR) show that block tilting is common in the upper crust in slow-spreading environments. Lava flows and bedding in overlying pelagic sediment dip away from the ridge axis in several localities (Karson and Rona, 1990; Lawrence *et al.*, 1998; Blackman *et al.*, 2002). Paleomagnetic studies of dikes cutting the lavas also show that they have been tectonically rotated (Lawrence *et al.*, 1998). Lava units in the crust formed at relatively slow-spreading MORs appear to be missing in many places, notably in oceanic core complexes (MacLeod *et al.*, 2002; Smith *et al.*, 2006, 2008; Escartín *et al.*, 2008; Escartín and Canales, 2010).

The lava units in ophiolite complexes (see Chapter 2) provided the first indications of the thicknesses and types of eruptive products at MORs (Fig. 5.12). In most ophiolite

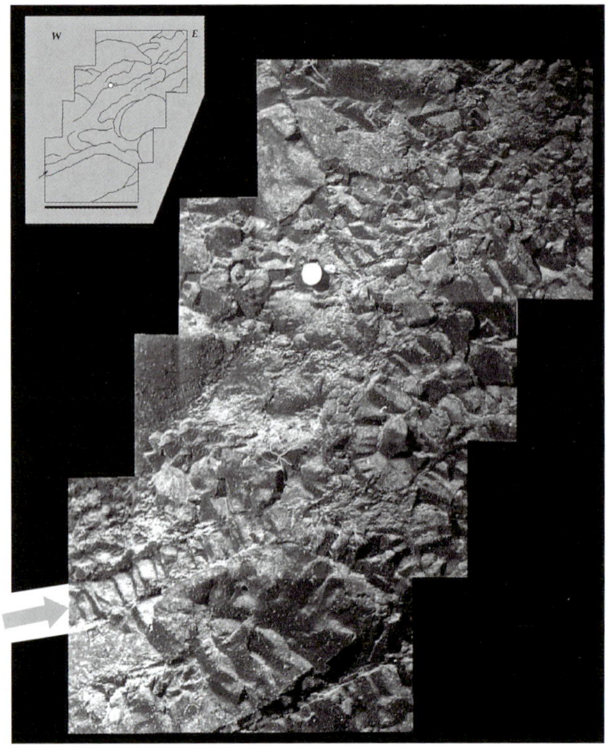

Figure 5.11 Photomosaic of a cliff face on the north wall of the Hess Deep Rift exposing a cross section of approximately 1 million years old (Ma) pillow lavas and thin sheet flows (gray arrow at lower left). Lava flow surfaces and the sheet flow dip toward the EPR axis where this crustal section was formed. Field of view across bottom of mosaic ~5 m.

complexes, these include a few hundred meters to more than a kilometer (reconstructed thicknesses) of mostly pillow lavas with fewer sheet flows (Dewey and Bird, 1971; Moores and Vine, 1971; Pallister and Hopson, 1981; Alabaster *et al.*, 1982). Pelagic sediments cap some of these lava units, providing a paleo-horizontal reference plane. Within many ophiolite sections, dikes intrude the lavas with increasing density toward the base of the unit and tend to cut the lava flow surfaces at high angles. In the absence of paleomagnetic or other independent constraints, reliable paleo-horizontal or vertical markers are not present, leading to uncertainties in reconstructions of this unit.

While the structure and composition of ophiolites have provided important context for understanding MOR lava units, it should be noted that spreading rates, spreading directions, and segment settings are not well constrained for most ophiolites. Consequently, caution is required in making comparisons between ophiolite lavas and MOR lava units. Many ophiolites have multiple lava units, some of which have geochemical characteristics normally associated with subduction-related lavas (see Chapters 2 and 3), thus calling into question the tectonic setting in which the entire complex might have formed (Miyashiro, 1973; Pearce and Cann, 1973; Robinson *et al.*, 1983; Pearce, 2003; MacLeod *et al.*, 2013).

Irrespective of the precise paleotectonic settings for ophiolites, it is clear that the lava units exposed in them represent submarine volcanic constructions. These constructs bear many similarities in both the flow morphology and stratigraphic position (with respect to dikes and plutonics) found at MORs and basaltic shield volcanoes,

Figure 5.12 Ancient (roughly 95 Ma) pillow lavas surrounded by red cherty sedimentary rocks crop out in an ophiolite in Oman. Exposures like this are remarkably similar to those seen along faults in modern oceanic crust.

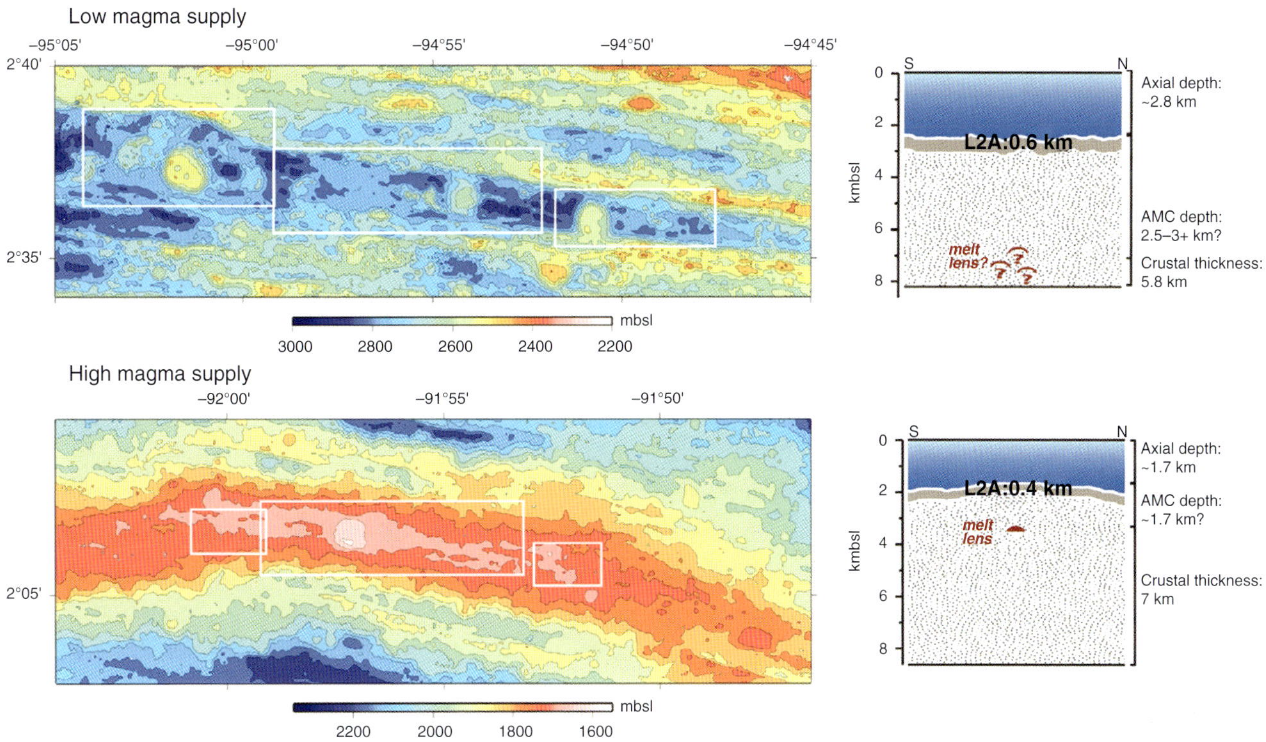

Figure 5.13 Bathymetric maps and schematic cross sections of low- and high-magma-supply study areas at the western Galápagos Spreading Center (GSC) (Colman *et al.*, 2012, Fig. 2). Low-magma-supply area (top), is characterized by a rifted axis and generally deeper axial depths. High-magma-supply area (bottom) has a more domed, inflated cross section, lacks an axial graben and the axial depth is hundreds of meters shallower. Cross sections (right top and bottom) summarize structural differences as defined by seismic data (Blacic *et al.*, 2004; Canales *et al.*, 2005). At low magma supply, the seafloor is deeper, the crust is thinner and melt lenses are deeper and not well imaged. At high magma supply the melt lens is clearly imaged and continuous. Depth keys refer to mbsl, meters below sea level; and kmbsl, kilometers below sea level.

hence they provide important opportunities that can help guide studies of oceanic crustal formation along the global MOR.

5.5 Seafloor volcanoes along the mid-ocean ridge

In subaerial volcanic terranes, studies of lava flow morphology, geochemistry, and structural relationships must be integrated to provide insights into the processes that generate magmas and control their eruption. As mentioned earlier, individual eruptive units have now been mapped in detail along several MOR segments, facilitated by improvements in AUVs, ROVs, sonar systems, and towed camera systems (see Chapter 1). For example, recent multidisciplinary studies completed along the Galápagos Spreading Center (GSC) provide excellent examples of the type of high-resolution, near-bottom mapping and sampling of individual eruptive units and lava fields required to fully understand submarine eruptions (Sinton *et al.*, 2002, 2010; White *et al.*, 2008; Colman *et al.*, 2012) (Fig. 5.13).

In addition, a few studies have begun to quantify the dynamics – including inferred effusion rate (i.e., effusion rate is volume flux per unit time, and flow rate is the local velocity of lava as it traverses the seafloor) – of submarine eruptions using novel geochemical and textural analyses (Michael *et al.*, 2008; Soule *et al.*, 2012; McClinton *et al.*, 2013), *in situ* sensors (Chadwick *et al.*, 2003, 2012), and repeat AUV surveys (Clague *et al.*, 2011, 2013; Caress *et al.*, 2012). Densely spaced lava sampling (samples taken less than 200 m apart), which can only be accomplished on the seafloor with HOVs and ROVs, is beginning to resemble the scale of sampling typically done at subaerial volcanoes. This fine-scale mapping, coupled with geochemical analysis of individual flows, has yielded important insights into understanding both mantle and crustal processes associated with oceanic volcanism (Perfit *et al.*, 1994, 2012; Rubin *et al.*, 2001a,b, 2012; Sinton *et al.*, 2002; Soule *et al.*, 2007, 2012; Goss *et al.*, 2010; Colman *et al.*, 2012).

While the products of most MOR eruptions are quite similar in composition and a variety of seafloor lava morphologies occur across the spectrum of MOR spreading rates, the modes of emplacement and the resulting volcanic constructs can be quite different at fast-spreading

Figure 5.14 Lobate lavas are common at the axis of the northern East Pacific Rise near 9°50′ N. The two lobes in the center-right foreground have flowed into and refilled a previously collapsed lobate flow. These are relatively young flows and have only a light dusting of pelagic carbonate sediment on their glassy surfaces. Field of view ~4 m.

centers compared to slower-spreading ridges. The key variable at different spreading rates is the magma supply and time-integrated mean effusion rate, which controls most of the gross variations observed in lava morphology. In general, at faster spreading rates and higher magma supply, a greater volume of magma over time is delivered to the spreading axis and results in the construction of a relatively uniform layer of extrusive rocks along and across axis (Fig. 5.14).

In contrast, along slower spreading ridges, lower magma supply and what is believed to be the more infrequent time-averaged effusion rate results in a highly variable three-dimensional thickness and structure of the lava unit. Evidence for this comes from observations on the distribution, spatial extent, and volume of volcanic constructs within the MAR rift valley, as mapped using multibeam and sidescan sonars, as well as detailed field observations from HOVs, ROVs, and AUVs (Smith and Cann, 1990, 1993; Yeo and Searle, 2013).

Submarine volcanic terranes have a broad range of morphologies, which in many cases mimic the types of features found in subaerial basaltic volcanic systems (Fig. 5.15). However, in submarine systems there are important differences in flow morphologies because eruptions occur at depths between 1.5 and 5 km, and are therefore influenced by high hydrostatic pressure created by the mass of overlying seawater.

The MOR axial zone, hot-spot-related ocean islands, arc volcanoes, and the vast number of widely distributed seamounts throughout the global ocean basins are typically constructed of thousands of small-volume eruptive units (order 0.1 km^3 and smaller). Some seamounts are built close to the MOR axis, suggesting areas where melt is channeled off-axis or where enhanced melt anomalies occur in the mantle beneath the global ridge crest leading to excess volcanism at these point sources. Some near-axis seamounts do form chains that extend outward from the MOR axis (e.g., Lamont Seamounts; Fornari *et al.*, 1984). However, traces of past magmatism are most obvious in mid-plate environments such as the Hawaiian Islands and other southern Pacific archipelagos supplied by hot spots (e.g., Line Islands, Marquesas; Staudigel *et al.*, 2010).

Although less common, very large eruptions (order 1 km^3 and larger) occur in each of these tectonic settings. Based on their volumes (equivalent of tens to hundreds of smaller eruptions), these very large submarine eruptions represent an important component of oceanic crustal construction. Some of the largest submarine eruptions formed as parts of rift-related flood basalts on volcanic rifted margins, oceanic plateaus, and ocean island chains created by hot spots and collectively are termed large igneous provinces (LIPs) (Coffin *et al.*, 2006; Sager *et al.*, 2013). Despite the large individual volumes on a global scale, lavas associated with LIPs amount to significantly less volume than lavas generated by persistent eruptions along MORs.

As was discussed in Chapter 3, the MOR neovolcanic zone is where much of the most recent eruptive activity is centered, approximately at the topographic axis of the ridge. At slow- to intermediate-spreading ridges, where accretion takes place beneath a broad rift valley, the neovolcanic zone

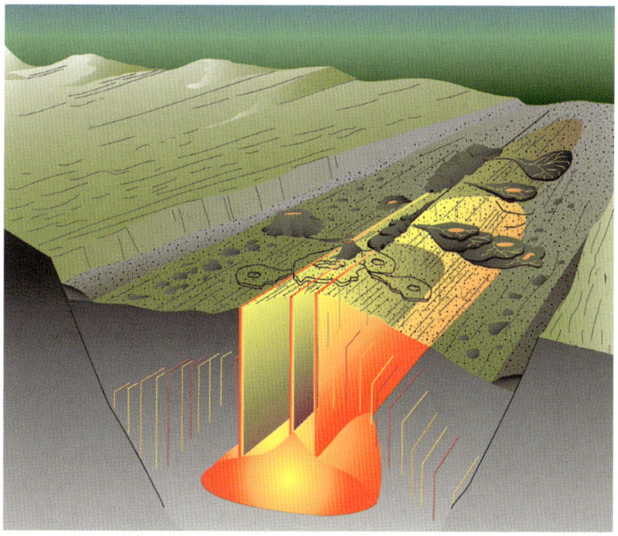

Figure 5.16 Illustration showing the arrangement of new lava flows and high (>100 m) rift valley walls at many intermediate- and slow-spreading MOR axial neovolcanic zones. At those sites, much of the volcanism is focused along fissures in the rift valley floor, often near the center of the valley within the bounding faults of the rift (from Smith and Cann, 1993).

Figure 5.15 An active Hawaiian lava flow from Kilauea's east rift zone, showing the formation of both smooth and ropy pahoehoe texture occurring during a breakout in the lobate surface due to a surge in the eruption. These complex field relationships are similar to what is observed in submarine volcanic terranes. Field of view ~2 m.

is located within the axial graben, which can be as wide as 5–10 km, hundreds of meters in relief and located at 2–6 km depth (Fig. 5.16). This contrasts with fast-spreading MORs that are characterized by a broad, elevated summit region of young lavas less than 5 km wide. At the axis, a narrow, shallow (typically <50 m deep) cleft (usually at depths 2.5–3 km below the ocean surface) marks the zone of repeated dike intrusion and coincides with the locus of eruptive fissures and high- and low-temperature hydrothermal vents (see Chapter 4) (Fig. 5.17).

Processes that build the architecture of the lava unit reflect the eruptive behavior of submarine lavas and their resulting morphology, and strongly influence its porosity and permeability. Because lava morphology is important for deducing eruption and emplacement processes, we present in detail the characteristics of different lava forms and their implications for building the upper crustal section.

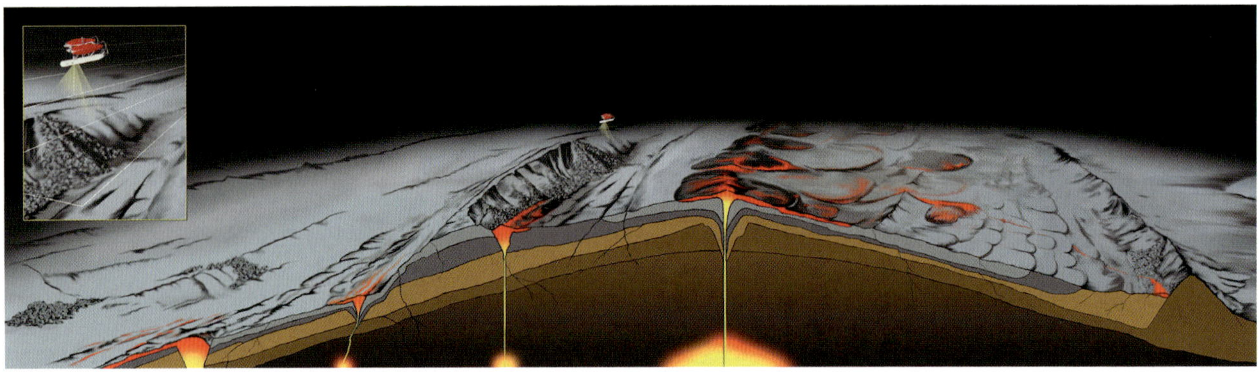

Figure 5.17 The low relief of a fast-spreading MOR, such as the East Pacific Rise near 9°50′ N, is dramatically different from rift valleys at slow-spreading ridges. At many fast- and intermediate-spreading MORS, the axis is marked by a narrow graben and fault scarps on the flanks are only 5–50 m high. At fast-spreading MORs, the axial terrane resembles a narrow cleft or trough [axial summit trough (AST)], which is 20–500 m wide, where the primary eruptive fissures and hydrothermal vents are located. At intermediate-spreading MORs the axial graben is usually 1–2 km wide, but commonly has a central cleft where volcanism and hydrothermal activity are focused. In these settings, some eruptions can also occur on the broad, elevated axial region (the "summit plateau"). Inset shows the AUV *ABE*, which collected high-resolution, near-bottom bathymetric data at the EPR axis near 9°28′ N and 9°50′ N in 2001 (Fornari *et al.*, 2004). *ABE* near-bottom bathymetric data were co-registered with high-resolution sidescan and with near-bottom magnetic measurements to understand MOR lava emplacement processes (Tivey, 2004).

5.6 Submarine lava morphologies

As in subaerial lava flows, the most important parameter that determines flow morphology is the effusion rate, the volume flux of lava issuing from a vent or fissure system during an eruption. The other key factors that control the development of lava morphology and transport across the seafloor include lava rheology (a function of composition, volatile content, temperature, vesicularity, and crystallinity), pre-existing regional slope and terrane roughness, and local slope variations (Wilson and Head, 1981; Bonatti and Harrison, 1988; Rowland and Walker, 1990; Griffiths and Fink 1992a,b; Gregg and Fink, 1995; Gregg *et al.*, 1996) (Fig. 5.18).

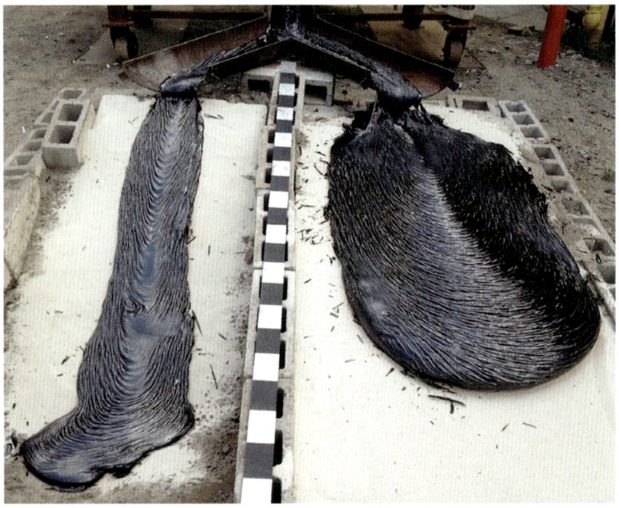

Figure 5.18 Experiments such as this one at Syracuse University (to model the flow structures developed in molten basaltic lava as it is delivered to flow surfaces with different slopes) have helped refine our ideas about the factors that influence the development of submarine lava morphologies. Slope on the left is 20°, resulting in a narrower, thinner, leveed flow relative to the flow on the right on a 5° slope, which is much wider and thicker. Composition, temperature, and pouring (effusion) rate are the same for both flows. Bar scale has 10 cm divisions (Karson and Wysocki, 2012; http://lavaproject.syr.edu).

Although explosive submarine eruptions have been documented recently on the MOR at depths greater than 1.5 km (Sohn *et al.*, 2008; Clague and Paduan, 2009; Helo *et al.*, 2011), the vast majority of mid-ocean ridge basalt (MORB) eruptions and lava flows sampled to date erupted under non-explosive and relatively quiescent conditions. Low-volatile contents (dominated by CO_2 and H_2O) in most MORB and the immense pressures (greater than 200–300 bars) at axial depths, limit explosive eruptions and production of submarine pyroclastic deposits (Dixon *et al.*, 1995). However, in some areas, the accumulation of gases in subseafloor magma chambers have apparently led to excess gas pressures and created eruptions that are sufficiently explosive to form glassy pyroclastic deposits and delicate bubble-wall and curved fragments of glass known as "limu o Pele" in the deep sea (Clague *et al.*, 2003, 2009; Sohn *et al.*, 2008; Clague and Paduan, 2009; Barreyre *et al.*, 2011; Helo *et al.*, 2011; Pontbriand *et al.*, 2012). Explosive eruptions appear to be common in submarine subduction zone settings including back-arc spreading centers, where gas contents in the lavas are commonly higher and volcanic edifices may be at shallower water depths (Head and

Figure 5.19 Contacts such as this one between young, tubular pillow lavas (on right) erupted at Axial Seamount on the Juan de Fuca Ridge in 2011 and older, sediment-covered lobate and pillow lavas beneath them (left side of photograph) are common when recent eruptions cover older lava flows along the MOR axis. Mixtures of younger and older lavas along contacts can be complex and difficult to discern as the lavas age, lose their glassy luster and pristine eruption surfaces, and are progressively covered with sediment. Field of view ~ 4 m.

Wilson, 2003; Wright *et al.*, 2006; Clague *et al.*, 2011; Resing *et al.*, 2011).

Recent studies of lava morphology including numerical modeling of the flows, where comprehensive near-bottom topography with resolution of less than a meter and continuous photographic imaging have been obtained before and after an eruption (Figs. 1.16, 1.33, 5.5 and 5.19), indicate that effusion rate is the primary controlling parameter in determining the dominant flow morphologies during MOR eruptions (Soule *et al.*, 2007; Fundis *et al.*, 2010). However, other factors such as slope and terrane roughness or pre-eruption topography also influence local flow transitions and specific volcanic features constructed during an eruption (Soule *et al.*, 2005; Fundis *et al.*, 2010; Caress *et al.*, 2012; Chadwick *et al.*, 2013; Clague *et al.*, 2013).

In a detailed study of the 2005–2006 eruption on the northern EPR, comprehensive analysis of imagery and topography before and after the eruption showed that pre-existing slope and other aspects of the terrain did not significantly affect the morphology of the most recently erupted lava flows (Fundis *et al.*, 2010). Additionally, petrologic and geochemical analyses of lava samples from that eruption suggested that the lava rheology remained very constant throughout the course of the eruption (Goss *et al.*, 2010). Thus, local flow rate (primarily controlled by lava effusion rate) was inferred to be the primary control determining which lava morphologies developed in the 2005–2006 EPR eruptions. In channels proximal to primary eruptive fissures where high flow velocities occurred, sheet flows were common. The morphologies transitioned downslope to lobate and pillow lavas at flow termini as velocities diminished (Fig. 5.20). The spatial distribution of morphologies was consistent with their distance from the primary eruptive fissures located within the narrow AST.

On fast-spreading ridges, lavas dominantly form lobate and sheet flows that vary from remarkably flat and thin (<10 cm thick) to ropy and jumbled or hackly (Hékinian *et al.*, 1985; Haymon *et al.*, 1991, 2008; Fornari *et al.*, 1998; Kurras *et al.*, 2000; Sinton *et al.*, 2002). Based on comparisons with subaerial basaltic lava flows (Hon *et al.*, 1994) and results of experiments with analog materials and the textures that are produced in the flow crusts, specific submarine flow morphologies have been correlated with laboratory-scaled effusion rates and slopes (Giffiths and Fink, 1992a,b; Gregg and Fink, 1995; Soule and Cashman, 2005).

At low effusion rates, pillow lavas extrude as individual spherical or cylindrical tubes of lava that cool and crust over on all sides, generally preventing coalescence with neighboring pillows (Moore, 1975; Ballard and Moore, 1977) (Figs. 5.21 and 5.22). The dominant lava morphology on slow-spreading ridges is bulbous pillow lava, which forms the building blocks for mound-like hummocks (typically less than 50 m high and under 500 m diameter), hummocky ridges (1–2 km long), or small circular seamounts (tens to hundreds of meters high and hundreds to thousands of meters in diameter). In some places these features coalesce to form axial volcanic ridges (AVRs) along the floor of the rift valley (Brown and Karson, 1988; Smith and Cann, 1990, 1993; Searle *et al.*, 2010; Yeo *et al.*, 2012; Yeo and Searle, 2013).

Pillow flows, mounds, and ridges are produced by the vertical accumulation of individual pillow lava lobes. As they grow, the newest pillows primarily erupt from the top of the stack and flow outward a limited distance, typically only a few meters, before solidifying. This process tends to produce steep-sided volcanic constructs such as flow fronts and "haystacks" or mounds (Fornari *et al.*, 1979; Chadwick *et al.*, 1991; Perfit and Chadwick, 1998) (Figs. 5.22 and 5.23).

The high-effusion-rate end of the eruptive spectrum includes morphologies formed when lava spreads out as a continuous sheet tens of meters across (Chadwick *et al.*, 1999; Soule *et al.*, 2005). In the middle of the spectrum are lobate flows (Figs. 5.3 and 5.14), which advance one lobe at a time. These flows form rapidly enough that the lobe edges may quickly coalesce back together into one interconnected flow, which can also inflate as the eruption progresses (much like "hummocky," "tube-fed," or "toey" pahoehoe on land (Wentworth and Macdonald, 1953; Swanson, 1973) (Fig. 5.15). In this way, lobate flows develop an upper crust that preserves the original lobate surface morphology, but which overlies a molten flow interior. The entire low-relief

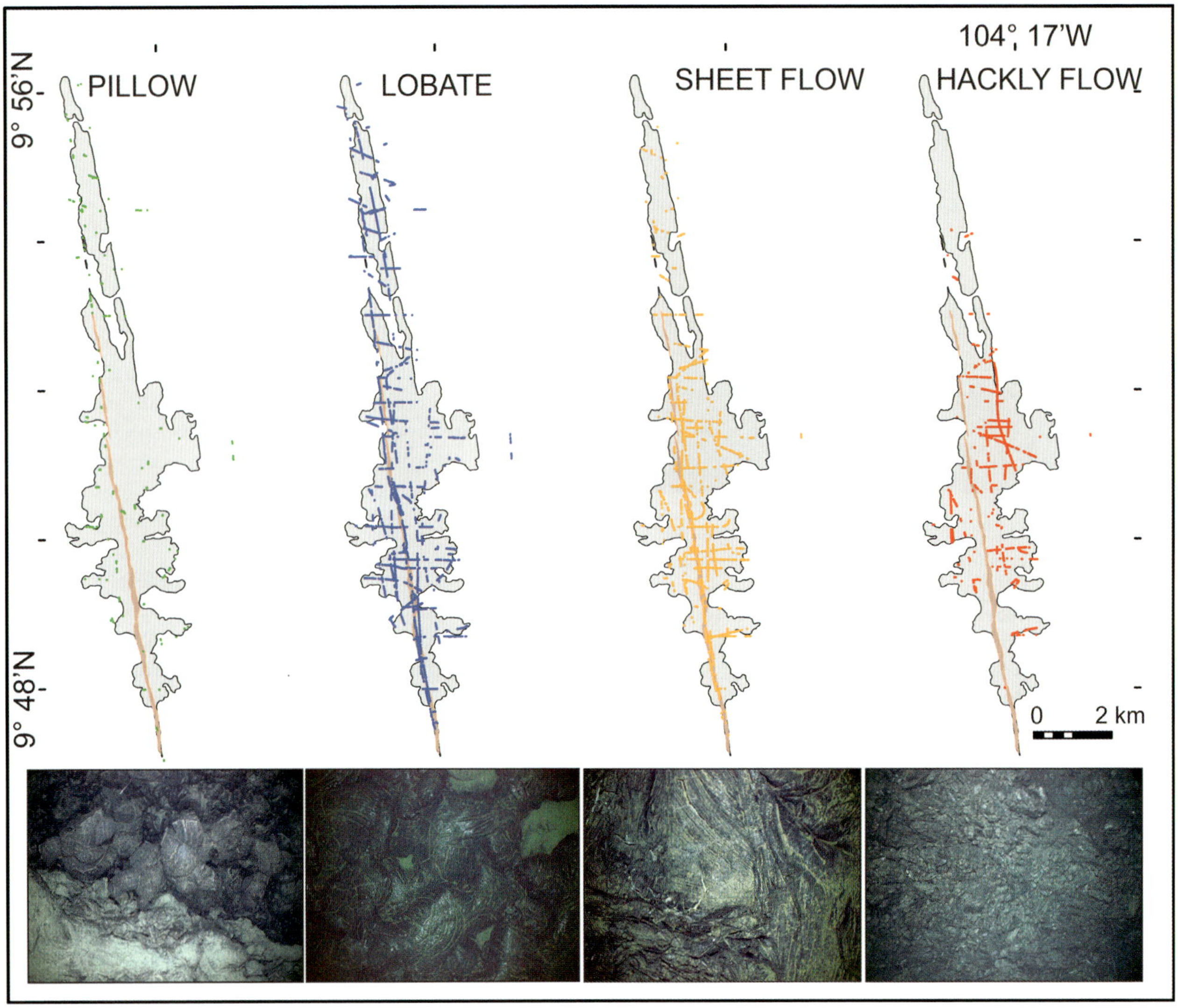

Figure 5.20 Maps showing the distribution of the four main types of lava morphology emplaced during the 2005–2006 EPR eruptions surveyed with the *TowCam* deep-sea digital camera system. Each point represents an analyzed image and the presence of that specific morphology type. The gray area shows the extent of 2005–2006 lava flows. *TowCam* images below each map represent "type" flow morphologies (from Fundis *et al.*, 2010).

lobate flow can be hundreds of meters to one or two kilometers long, with hydraulic connection via discontinuous networks of channels containing magma that flows through the interior (Chadwick *et al.*, 2013).

Lobate flows commonly spread out and then inflate upward, analogous to terrestrial inflated pahoehoe sheet flows (Hon *et al.*, 1994), producing relatively flat upper crusts (Fig. 5.14). These flows may later collapse (Fig. 5.24) by either lava drain-out toward the distal ends of flows, lava migration down primary eruptive fissures (drain back), or through vapor condensation and subsequent collapse of the outer lobate crust (Perfit *et al.*, 2003). Broad flows covering hundreds to thousands of square meters that become ponded and later collapse are referred to as "lava lakes," and are commonly associated with "lava pillars" (Figs. 5.25–5.27) (Ballard *et al.*, 1979; Francheteau *et al.*, 1979b; CYAMEX Scientific Team, 1981; Gregg and Chadwick, 1996; Gregg *et al.*, 1996, 2000; Chadwick *et al.*, 2003; Engels *et al.*, 2003; Clague *et al.*, 2013). The pillars were conduits where narrow plumes of vapor rose through the molten pond, quenching the magma surrounding the rising plumes and cooling their sides, which produced the pillars.

Sheet flows typically have low surface relief (under 0.5 m) and initially form as broad sheets or lava channels with a glassy upper crust over a molten interior. Sheet flows can be thin (less than 10 cm) glassy slabs, or massive flows a few meters thick with a thin glass veneer and more crystalline interiors (Fig. 5.28). Sheet flows normally are confined to axial valleys and the AST but may also extend hundreds of meters off-axis as low-relief channels. Delicate interior

Figure 5.21 Inflated, tubular pillows with well-developed bread-crust or "elephant hide" textures on Axial Seamount on the Juan de Fuca Ridge. The changing pattern of textures on the surface of the tube extending to the lower left of photo shows stages in pillow development as the extrusion from the original central form grew. Field of view ~2 m.

Figure 5.22 A pillow lava flow front, downslope from the Menez Gwen Hydrothermal Vent Field on the MAR near 37°51′ N at 900 m depth. Illumination of the terrain is enhanced by the high-intensity lighting on the MIR submersible because of the cinematography being conducted by James Cameron during the filming of the IMAX movie *Aliens of the Deep*. Field of view ~10 m.

surface textures are formed by magma–seawater interactions at high temperatures (~1000 °C) at the interface of voids and the interior surfaces of upper crusts of sheet and lobate flows. This process results in lava drips forming in vapor-supported cavities that commonly develop into collapse features (Engels *et al.*, 2003; Perfit *et al.*, 2003; Soule *et al.*, 2006, 2012) (Fig. 5.27). Larger collapse areas form in various MOR settings typically where eruptions have ponded in pre-existing depressions. Subsequent to ponding, the lava lakes drain away or back into an eruptive fissure, causing the upper crust to founder. This results in the floors of collapsed zones being covered by glassy talus slabs and high-standing fields of lava pillars (Francheteau *et al.*, 1979b; Gregg and Chadwick, 1996; Gregg *et al.*, 1996, 2000) (Figs. 5.26 and 5.27).

The upper crusts of sheet flows display a variety of textures ranging from smooth and very flat to lineated, ropy (with lava coils or whorls) (Lonsdale, 1977), to folded, jumbled, and hackly (Fig. 5.29). All of these textures may exist on single flow units. Variations in surface textures depend on several factors. These include underlying roughness or constraining boundaries such as faults, talus, edges

Figure 5.23 A video grab image of a "haystack" of pillow lavas and tubes ~5 m high, imaged by the MBARI *Tiburon* ROV on the Juan de Fuca Ridge (top); field of view ~4 m. These are common constructional features found on the ridges bounding the axial valley along the southern Juan de Fuca Ridge (Stakes *et al.*, 2006). They appear to be formed off-axis on top of extensive areas of pillow flows, probably during the final stages of eruption from point sources along eruptive vents. Bottom image shows a small "haystack" or "hornito" in Iceland; field of view ~10 m. These features are commonly interpreted to be rootless, meaning that they are not directly linked to eruptive fissures but more likely represent local accumulation of lava forms caused by blockage or other interference over a main flow conduit.

of adjacent lava flows, and increased viscosity due to cooling and crystallization. Textures are also impacted by changes in local flow rate due to variations in magmatic pressure at the vent because of constrictions or opening of flow channels, or a combination of these effects. Jumbled and hackly sheet flows are somewhat analogous to ʻaʻa lava texture on subaerial volcanoes (Fig. 5.30). ʻAʻa flows form by the physical disruption of the surface crust during transport, particularly when the viscosity and/or rate of shear strain increases as the flow progresses (Peterson and Tilling, 1980; Rowland and Walker, 1990; Soule and Cashman, 2005).

On fast-spreading MOR crests, sheet flows extend as far as 3–4 km from the AST as a result of rapid, high-volume flows that form lava channels (Soule *et al.*, 2005, 2007) (Figs. 5.31–5.33). Vertical sections of oceanic crust exposed along fault scarps (Karson *et al.*, 2002; Stakes *et al.*, 2006) or revealed by drilling during the Integrated Ocean Drilling Program (IODP) (Tominaga and Umino, 2010) show that some sheet flows are a few meters thick and likely transition into massive flows that are more than 3 m thick. These thicker flows may represent lava ponds. A massive, 75 m thick flow drilled at IODP Site 1256 may represent a ponded flow that formed more than 3 km off-axis due to rapid lava emplacement within a pre-existing depression (Tominaga *et al.*, 2009) (Fig. 5.9). Similar settings are common in subaerial volcanic terranes where escarpments 10–100 m high expose sections through massive flows emplaced during voluminous eruptions that accumulated in topographic depressions (Fig. 5.34).

In early investigations, it was commonly assumed that individual submarine flows comprise lavas with similar morphologies, for example, flattened, hollow, collapsed pillows (CYAMEX Scientific Team, 1981). However, more recent detailed studies of historic submarine eruptions have clearly shown that eruptive units or lava fields can consist of myriad morphologic types, similar to what is observed in

Figure 5.24 Edge of axial summit trough around 9°50′ N on the East Pacific Rise showing a lobate lava flow in upper right, and collapse pits and talus at the edges of the trough (in foreground and upper left). Curvilinear ledges in the axial summit trough are lava selvages created during the filling and drain out of lava in the axial trough. Field of view ~5 m.

Figure 5.25 Lava pillars (left), about 0.5 m diameter and 3–4 m tall, and archway formed by a 6 cm thick lobate crust suspended between the pillars. The top surface of the 1998 eruption (right) in the caldera of Axial Seamount, Juan de Fuca Ridge, showing bulbous and folded features created by the stretching and bending of the upper flow surface proximal to the tops of each lava pillar as the flow surface cooled (from Chadwick *et al.*, 2013).

Figure 5.26 A cross section (left) of a lava pillar collected from the EPR axis near 9°50′ N at 2505 m depth. The pillar formed during the 1991–1992 eruption. It shows a double channel configuration where two streams of seawater provided the nucleating cylinders of water around which the lava cooled as the axial lava pond inflated. The glass-lined interior of the two channels attests to the formation process. The distance across the long axis of the pillar is ~20 cm. Lava pillar (right) in the caldera of Axial Seamount that formed during the 2011 eruption showing the well-developed selvages on its side. These may represent still-stands of the ponded lava upper surface as it deflated during magma withdrawal that accompanied the waning phase of the eruption (pillar diameter ~20–30 cm).

subaerial flows. The 2005–2006 EPR eruption (Soule *et al.*, 2007; Fundis *et al.*, 2010) and the 2011 Axial Seamount eruption (Caress *et al.*, 2012; Chadwick *et al.*, 2013; Clague *et al.*, 2013) both provide excellent examples of lava flow fields that comprise diverse lava flow morphologies. Because of the excellent mapping done at both sites, the morphologies can be linked to causal dynamic processes related to distance from the eruptive fissures and pre-flow seafloor topography.

In submarine lava flows, lateral morphological transitions that are gradational typically imply changing flow conditions within the same eruptive episode, while more abrupt transitions displaying superposition of flow forms and/or relative ages commonly result from different

Figure 5.27 Lava pillar and archway at the EPR near 9°48′ N at 2520 m depth (left) showing lobate and sheet morphology on the top of the flow. The ~3 m distance between the base of the pillars and the flow top indicates the minimum thickness for the erupted lava flow that formed these features. Long (2–10 cm) drips formed within the interior of a lobate crust (right top) developed during emplacement of the flow indicate the presence of vapor-filled cavities that formed beneath lavas crusts during eruption. Interior surface of a deep-sea lobate lava crust (bottom right) collected by *Alvin* from the EPR with delicate, glassy drip structures (8 cm long).

eruptions (Fig. 5.35). Gradational changes are common near the outer margins or distal ends of a flow lobe where the volume of lava and flow rate decrease. Gradational transitions can occur from sheet to lobate flows, sheet to jumbled or hackly flows (Fig. 5.36), or from lobate to pillow flows, and from lobate to hackly flows.

Transitions from a sheet to a lobate morphology vary with respect to the length scales over which they occur, but typically they occur laterally over a few meters, implying pronounced variations in local flow rate. Contact relationships suggest sheet flows also form when lava issues from under a cooled lobate surface, perhaps reflecting a surge in magmatic pressure at the vent and an increase locally in the effusion rate. In some cases the surfaces of sheet lava formed in this setting are characterized by lineations, likely formed by the flow being raked against the partially cooled surface of the overlying lobate lava (Chadwick *et al.*, 1999).

Morphologic transitions from sheet into jumbled or hackly flows occur over a few meters as the flow's folded surface continues to deform during cooling, ultimately breaking the glassy surface crust because it exceeds its yield strength (Fig. 5.36). Transitions between lobate and pillow lavas can occur over distances of 1–10 m, and tend to occur toward the distal, flow front termini of lobate flows, where they are commonly characterized by one to a few pillows branching out of well-defined lobate flows. Lobate flows near these transitions can also be more "decorated" and have distinct lobes that are minimally coalesced. In some settings, the margins of lobate flows are entirely composed of pillowed fronts, reflecting the decrease in effusion rate and eruptive force as the eruption wanes; the pillowed flow fronts are the last features to form at distal sites during an eruption. These relationships are common at the edges of flows emplaced during the 2011 eruption at Axial Seamount on the Juan de Fuca Ridge (Chadwick *et al.*, 2013; Clague *et al.*, 2013).

In rare cases, lobate flows may transition into hackly flows, suggesting an abrupt increase in shear rate. These transitions tend to occur at the margins of lava channels (Soule *et al.*, 2005; Fundis *et al.*, 2010). Although individual flows in some places are characterized by lobate flows at the flow front rather than pillow lavas, the general trend is that lava morphologies associated with higher effusion rates (sheet, hackly, and flat-lying lobate flows) occur near the eruptive source. The lavas transition into morphologies that are typically associated with lower flow rates toward the distal ends of a flow.

High-silica lavas such as andesites, dacites, and rhyodacites are only rarely erupted along the MOR. They have been recovered near ridge–axis discontinuities such as ridge–transform intersections (RTIs) at the eastern GSC (Fornari *et al.*, 1983; Perfit *et al.*, 1983; Sinton *et al.*, 1983; Christie and Sinton, 1986; Thompson *et al.*, 1989), in back-arc basin settings such as the Lau Basin (Gill, 1976; Pearce *et al.*, 1994; Bezos *et al.*, 2009), and in overlapping spreading centers (Langmuir *et al.*, 1986; Wanless *et al.*, 2010, 2012). The occurrence of high-silica lavas at ridge–axis discontinuities is generally attributed to lower magma supplies at ridge segment ends or the "cold edge effect," where the propagation of a ridge through older and/or cooler lithosphere leads to increased cooling rates and greater extents of crystal fractionation in source magma bodies (Christie and Sinton, 1981; Perfit *et al.*, 1983; Sinton *et al.*, 1983; Wanless *et al.*, 2010). Such high-silica lavas can be difficult to identify on

the seafloor but typically form unusually large (greater than 2 m diameter) bulbous and elongate pillows, with well-defined surficial bread-crust textures (Fig. 5.37) or large angular blocky lavas (Fig. 5.38). In some cases these high-silica lavas are highly vesicular and microcrystalline, and in others they are composed entirely of glass; they rarely exhibit flow banding (Fig. 5.38). They commonly form pillow mounds and larger domes but rare andesite sheets have also been recognized on the southern Juan de Fuca Ridge (Stakes *et al.*, 2006).

Collapse pits are cavities in the surface of lobate flows that range in size from less than a meter to more than 5 m in relief and with areas of a few to tens of square meters (Fig. 5.39). The depressions form when the 2–10 cm thick surface crusts have collapsed, revealing water-filled voids beneath (Engels *et al.*, 2003; Fundis *et al.*, 2010). Collapse pits in seafloor lavas are common features that can form during flow inflation and post-eruption drain-back of lava into eruptive fissures or downflow through lava conduits as supply from the vent diminishes (Fornari *et al.*, 1998, 2004; Engels *et al.*, 2003).

5.7 Eruptions on the seafloor

Submarine lava flows that cover very large areas (tens to hundreds of square kilometers) present a paradox. Water has exceptionally high heat capacity, thermal conductivity, and enthalpy of vaporization. Thus it has been puzzling how submarine lava flows can travel great distances over low slopes before solidifying completely. But rapid crust

Figure 5.28 Top image shows the ODP Flow near 9°30′ N in the axial summit trough of the EPR, believed to have ponded within the trough during a fissure eruption in the 1980s. Pressure ridges within the flow suggest that it is at least 3–5 m thick. Drilling in ODP Hole 142 suggests that the flow may be as thick as 10–20 m. Galatheid crabs that populate this site may be feeding off bacteria in the milky hydrothermal fluids emanating from below the fissured flow (from Perfit *et al.*, 1994); field of view ~2 m. The second image from the top is a massive sheet flow exposed along a fissure on the Cleft Segment of the southern Juan de Fuca Ridge (Stakes *et al.*, 2006). Individual flow thicknesses are about 1 m in this image but in other places massive flows can be many meters thick. Second image from the bottom shows a collapsed sheet flow in the caldera of Axial Seamount, Juan de Fuca Ridge, showing the thin (10 cm thick) crust and the water-filled void beneath the flow created by magma withdrawal into the fissure that sourced the flow. Bottom image is a massive flow exposed in an eruptive fissure at the summit of Axial Seamount. Relict selvages of the flow surface stand out from the face of the exposure; field of view ~4 m.

Figure 5.29 Folded sheet flow (top) in the axial valley of the eastern Galápagos Spreading Center at the Rosebud Hydrothermal Vent Field near 86° W (Shank *et al.*, 2003). The folds in the lava surface are nearly entirely glass; field of view ~0.75 m. The eruption that created this flow is believed to have occurred sometime between 1992 and 2003, based on observations during *Alvin* diving programs to the area during that decade. Rat-tail fish is about 25 cm long. Folded sheet flow on the summit of Axial Seamount (middle); field of view ~1.5 m. Close-up of a hackly sheet flow surface on the summit of Axial Seamount, part of the 2011 eruption (bottom). All of the folds in this image are thin (under 0.5 cm) ribbons of glass that cooled on contact with seawater as the fast-moving flow traversed the seafloor. Note that the folds are well preserved, suggesting that the upper flow surface still behaved plastically during emplacement; field of view ~0.5 m.

formation during quenching of submarine flows provides a thermally efficient barrier to cooling and solidification of the molten interior. Land-based observations of basaltic volcanoes and numerical models of idealized submarine lava flows suggest that even a crust just a few centimeters thick provides sufficient thermal insulation to inhibit heat loss during flow. Thus, lava in sheet flows is able to travel 1–20 km across the seafloor (Gregg and Fornari, 1998). In addition, vapor created from degassing and superheated seawater trapped beneath flows may serve to enhance the mobility of some flows or portions of them by increasing local pore pressure and effectively decreasing the drag between the pre-existing terrain and the over-riding flow (Perfit *et al.*, 2003).

Figure 5.30 A historic-age folded sheet, ropy and lobate flow (left) emplaced within a pre-existing ʻaʻa lava terrane on the East Rift Zone of Kilauea Volcano, Hawaii. The young flow followed a shallow channel in the ʻaʻa flow as it was transported downslope from the eruptive fissure; field of view ~10 m. The Pacific Ocean and south coast of the Big Island of Hawaii are in the background. ʻAʻa lava flow in Iceland (right) that traversed about 1 km from its eruptive fissure (direction of flow was from lower left to center of image), and eventually transitioned into a lobate flow that poured into and overtopped older fissures along its path; field of view ~500 m.

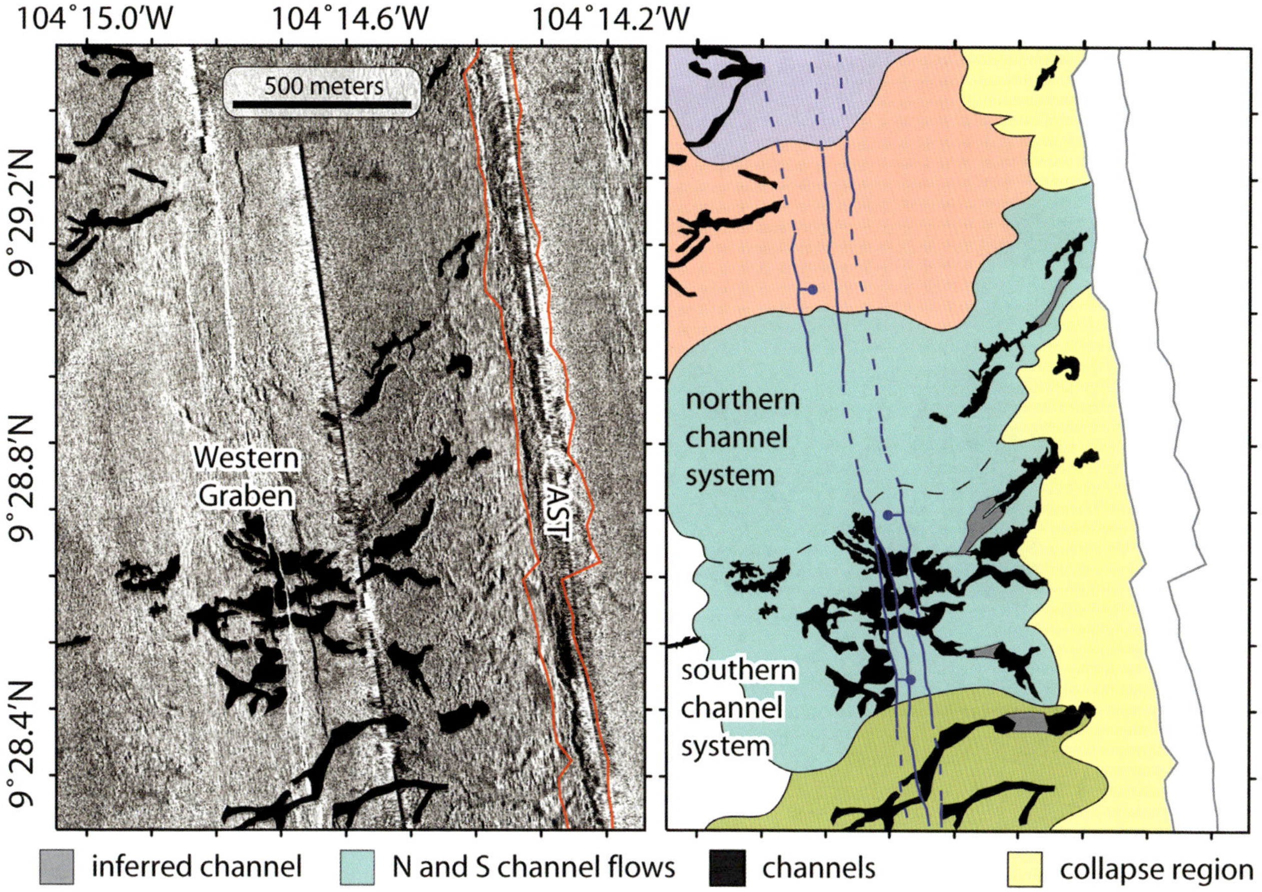

Figure 5.31 Sidescan sonar map (left) and geologic interpretation (right) of lava channels at the EPR 9°28′ N that flowed west of the axial summit trough (AST). Channels appear as dark (low reflectivity) sinuous patterns in the sonar because of their flat surfaces, compared to the rough/hummocky surfaces of the surrounding lava flows. These two channel systems are interpreted to represent eruptions separated by some time as the northern ones cover channels in the older, southern area. In the flows to the north of these two channel systems, other flow units (shown in red and purple) in the channels were destroyed by wholesale collapse of the lava flow that filled and overflowed the AST. The outline of the AST is shown in red on the sonar map and as a gray line on the interpreted geology.

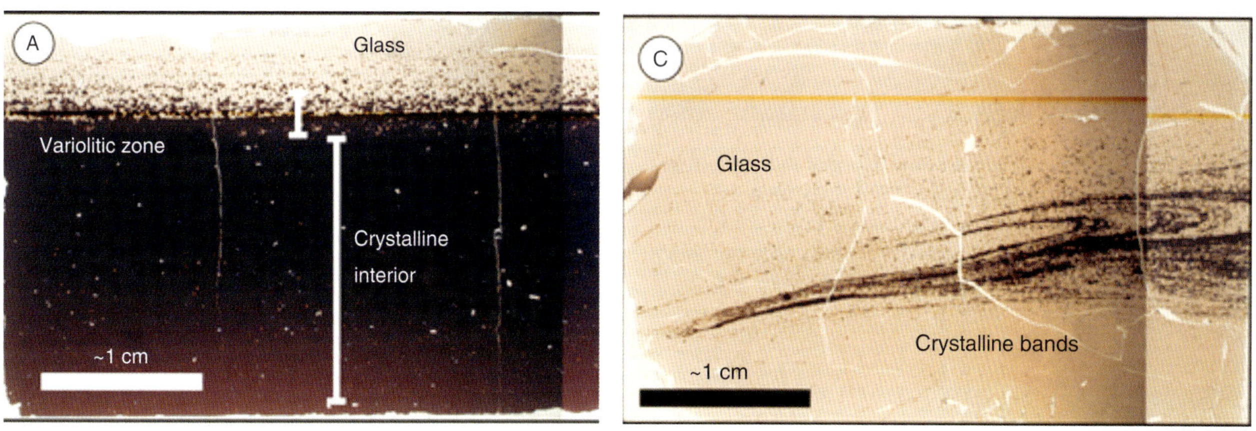

Figure 5.32 Photomicrographs of thin sections of EPR lobate lava sampled adjacent to a lava channel (left) and a folded upper portion of a sheet flow from within the lava channel (right). Mid-ocean ridge lavas commonly grade from outer crusts of aphyric glass into a variolitic zone of quenched crystals followed by a microcrystalline interior. The folded channel sample shows tortuous bands of quench crystals reflecting the plastic deformation of the lava as it flowed and cooled (from Soule *et al.*, 2005, Fig. 7).

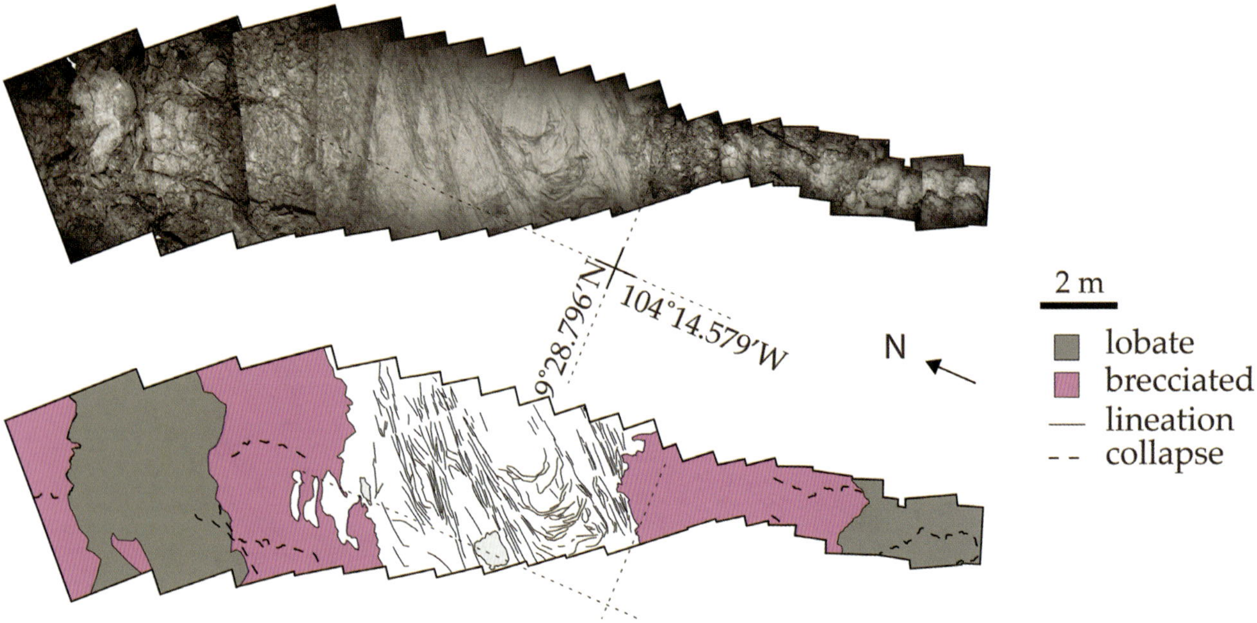

Figure 5.33 Photomosaic of digital-still, down-looking images collected using *Alvin* on the EPR near 9°29′ N on the west flank of the axis (top), and interpretative sketch of an area of extensive sheet flows where eruptions sourced by fissures within the AST erupted and flowed rapidly westward, forming extensive channelized flows (bottom) (from Soule *et al.*, 2005).

Figure 5.34 This massive sheet flow exposed in the Dettifoss Gorge in northern Iceland is about 25 m thick (note the person in blue shirt walking on the terrace in the middle of the image for scale), and has well-developed columnar joints between the chilled base and highly vesicular flow-top breccia.

Detailed studies of the 2005–2006 EPR eruptions allowed development of methods for quantifying critical parameters related to the thermal and dynamic history of submarine lava flows. Techniques including coupled geologic mapping (flow morphology), textural analysis (bubble growth and crystallization), and geochemical evolution of basaltic glasses (time-dependent volatile loss and fractionation) (Soule *et al.*, 2007, 2012) were used to test the thermal insulation hypothesis. Key to the success of these studies was an unequivocal knowledge of flow dimensions. In this case, the flow dimensions were documented by detailed mapping using a towed deep-sea camera (*TowCam*; Fornari, 2003), the *Alvin* submarine and the ROV *Jason*. Direct observations allowed identification of the contacts between old and new lava flows. These data provided the map-based geological context for collecting systematic, along-flow sample suites (Fig. 5.40).

In subaerial volcanic terranes, it is clear that rate of cooling limits lava flow length (Pinkerton and Wilson, 1994). Well-insulated, subaerial tube-fed flows in Hawaii have cooling rates of about 1 °C km^{-1} (Peterson and Swanson, 1974; Cashman *et al.*, 1994), whereas open-channel flows have cooling rates a factor of two to four times greater (Cashman *et al.*, 1999; Harris *et al.*, 2005). Recent estimates of submarine lava flow cooling rates derived from crystallization and volatile loss are close to those for subaerial tube flows (Paonita and Martelli, 2007, 2013; Soule *et al.*, 2012, 2013).

5.8 Mid-ocean ridge lava geochemistry and magmatic processes

Knowledge of the construction of the uppermost oceanic crust and the magmatism associated with accretionary phenomena is derived from geochemical and physical studies of exposed seafloor lavas. From a geochemical perspective,

Figure 5.35 A distinct transition between lava flows; a tubular pillow lava (0.75 m diameter) with well-developed bread crust surface texture flowed across an older lobate and pillow flow on the axis of the Galápagos Spreading Center. The site is near 91° W at the Calyfield Hydrothermal Vent Field site at a depth of 1950 m. Hydrothermal venting at this site is low temperature (only 1–2 °C above ambient) with low levels of H_2S supplying sufficient nutrients to sustain vesicomyid clam (white shells in the image) and bathymodiolid mussel communities. Open cavities within lobate flows and along the margins between lobes can provide pathways for hydrothermal fluids.

Figure 5.36 Abrupt transitions between flat sheet flow surface and ropy, folded, or crenulated sheet flow can occur frequently in submarine flows, and attest to the dynamic and fast movement of lava on the seafloor despite rapid chilling of the magma by the near-freezing water. Structures such as those observed in this image indicate that the lava remains plastic for a short period of time after extrusion in order to form the complex folded structures that typify these types of flows. This image is from the 2011 eruption of Axial Seamount. Field of view ~2 m.

Figure 5.37 (Top) Large andesite pillows (2–3 m diameter) on a near-vertical wall of a ridge along the northern rift zone of Axial Seamount at 1728 m depth. The higher viscosity of these silica-rich lavas, compared to most MOR basalts, results in elongate tubes and pillows that have distinctive lineated, bread-crust textured surfaces. Most MOR basalt pillows have ~1 m diameters. Center-right portion of the image shows a twisted "spill-out" of lava that froze when a tube broke open. The original bulbous tip of the tube is suspended by the twisted section of lava; field of view ~9 m. (Bottom) Large (2–3 m diameter) elongate tubular dacite lavas (right) from a constructional mound at the ridge–transform intersection between the Juan de Fuca Ridge and Blanco Transform Fault, 2219 m depth; field of view ~6 m.

Figure 5.38 Images of high-silica lavas from the East Pacific Rise and Iceland. A dacite lava (top) from the east limb of the 9° N overlapping spreading center imaged using the *TowCam* system showing prominent flow banding; field of view ~2 m. The banding and the blocky nature of the flow are uncommon in typical MOR basalts. A rhyolite lava (bottom) from Torfajökull, Iceland, shows banding that is similar to that observed at the EPR 9° N OSC in top image.

lavas erupted at MORs are a specific type of basalt known, for obvious reasons, as a mid-ocean ridge basalt or MORB for short. They are similar to basalts in other tectonic environments, but they have some unique chemical characteristics that distinguish them from lavas erupted on the continents, in island arcs, at intraplate ocean islands and from most oceanic seamounts (Engel *et al.*, 1965; Gast, 1968; Sun *et al.*, 1979). Petrological and geochemical studies of MORB (including more silica-rich differentiates) provide quantitative insights into fundamental igneous processes that lead to the formation of magmas deep within the mantle (see Chapters 3 and 8), and the chemical evolution of those liquids as they are emplaced to form the plutonic and volcanic rocks that comprise the oceanic crust. The chemical and volatile makeup and the constituent minerals that are formed from magmas originating beneath MORs provide clues about the composition of the mantle from which they were derived. This includes the temperature and pressures where the magmas formed and the variety of high-temperature processes that modify them as they are transported from depths greater than 100 km to shallow crustal magma chambers 1.5 to 3 km beneath the seafloor, and periodically to eruption sites on the MOR crest (Fig. 5.41).

A majority of MOR lavas have an upper crust of glass, a few centimeters thick, which forms when magmas erupt and freeze (quench) in the cold (2 °C) seawater (Fig. 5.32). This natural glass (Fig. 5.42), by definition, does not have a

Figure 5.39 View of a collapsed lobate lava field on the upper flanks of the East Pacific Rise axis near 9°50′ N showing one of the ocean-bottom seismometers deployed in 2004 to monitor small earthquakes on this section of the ridge. Data from OBSs have been key to identifying important aspects of submarine eruptions including the 2005–2006 eruptions at the EPR, as well as elucidating physical responses of the hydrothermal system to magmatic and tectonic activity (Tolstoy *et al.*, 2006; Soule *et al.*, 2007). Field of view ~10 m.

crystalline structure and commonly contains less than 10% (by volume) crystalline minerals, which may have been formed as the magma rose to the surface. Myriad chemical analyses are performed on lava samples from the MOR using instrumentation that determines everything from the major element contents (oxides over 1 wt%), to the concentrations of trace elements as low as a few parts per billion (ppb), to volatile concentrations (H_2O, CO_2, chlorine), and even the stable and radiogenic isotopic compositions [see White and Klein (2014) for a more thorough discussion]. Where possible, analysis of the glass is preferred and commonly used instead of whole-rock analyses because the glass is quite resistant to chemical/seawater alteration and considered to be equivalent to magma composition that has frozen.

5.9 Compositional variability of MORB

The characteristic chemical signatures of MOR lava are akin to a rock's DNA, in that they provide volcanologists with critical information about a sample's origin (i.e., its petrogenetic lineage) and history (Rubin and Sinton, 2007). Deciphering the origins and evolution of MORB suites is particularly important because they have relatively straightforward magmatic histories without the complications that arise when magmas must transit through thick, old continental crust or where mantle sources are contaminated by subducted sediments, such as found beneath island arcs. MORB range in composition from picrites with high magnesium (MgO) contents, to ferrobasalts and iron–titanium (FeTi) basalts containing lower MgO concentrations, and high concentrations of FeO and TiO_2. More rarely, there are silica-enriched lavas known as low-potassium basaltic andesites, andesites, dacites, and rhyodacites with low MgO (<4 wt%) and SiO_2 contents from 52 wt% to more than 70 wt% (Figs. 5.43 and 5.44).

A pronounced iron-enrichment trend with decreasing MgO contents (related to fractional crystallization as magmas cool) in suites of genetically related lavas is, in part, what classifies MORB as belonging to what is known as the "tholeiitic magmatic suite." Magmatic suites or lineages are groups of igneous rocks, commonly spatially and temporally

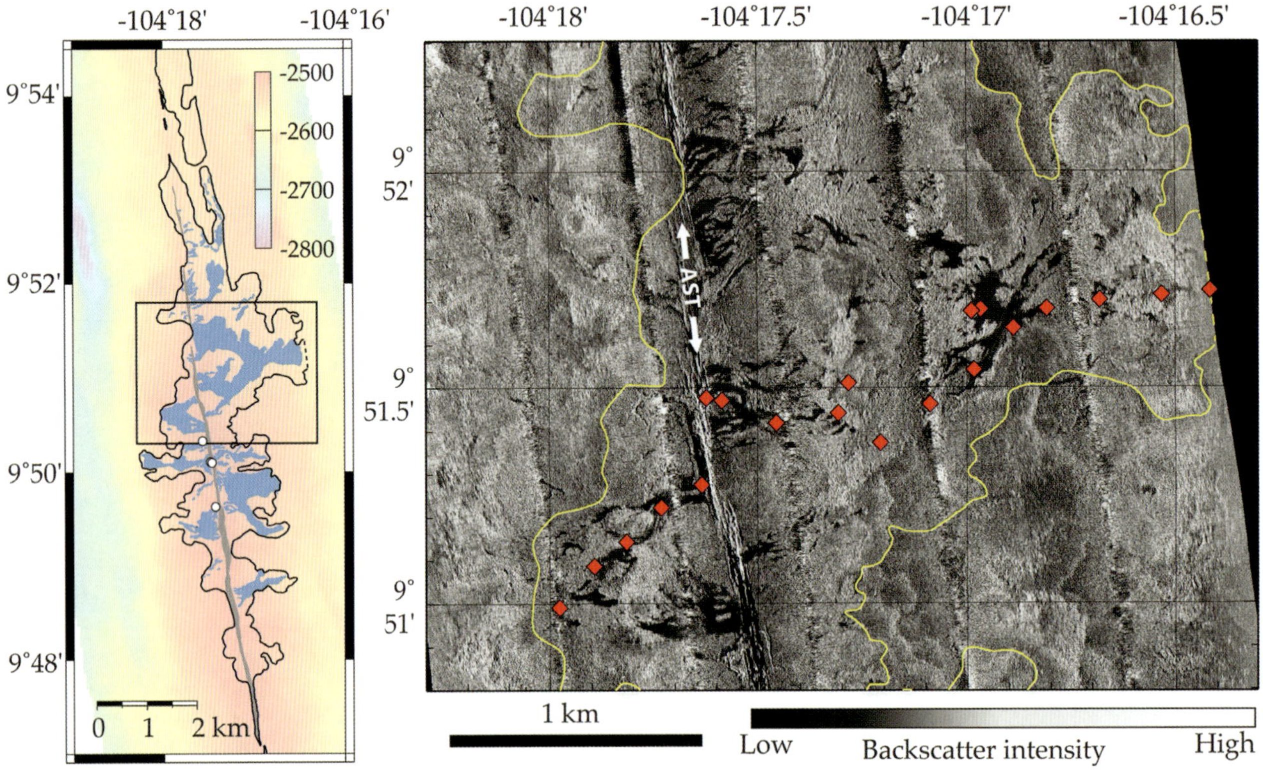

Figure 5.40 (Left) Map of the central portion of the 2005–2006 EPR eruption (outlined by black box) with channel systems (blue) comprising smooth sheet and hackly sheet flows (from Soule *et al.*, 2012, Fig. 2). Eruptive vents are located within the narrow AST (light gray). Box (out lined by black box) outlines the area shown in sidescan image to the right. Sidescan sonar backscatter image (right) of the eruption area shows the flow boundary (yellow line) and locations of lava samples (red diamonds) within the two channel systems. Samples collected along two flow transects, at 200 m intervals, were used in a detailed analysis of flow degassing (Soule *et al.*, 2012). Low backscatter intensities mark smooth sheet flows while higher backscatter indicates hackly sheet flows (Soule *et al.*, 2005).

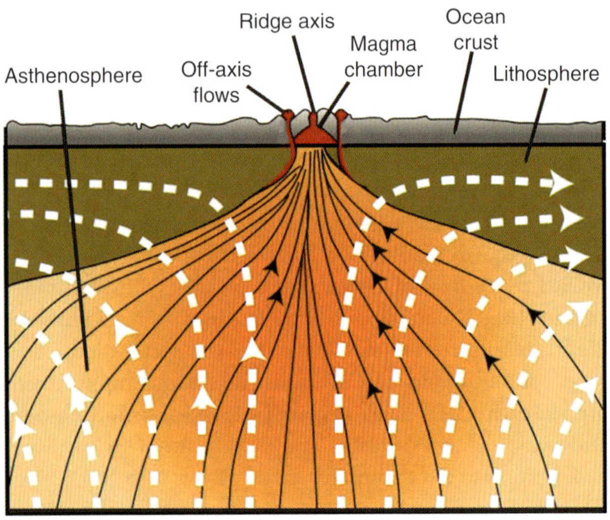

Figure 5.41 Illustration depicting a cross section through the lithosphere and upper mantle beneath a MOR showing an idealized view of the melting regime in the asthenosphere (from Perfit, 1999). Passive upwelling models of mantle flow suggest decompression partial melting of rising hot mantle (dashed lines) coincident with upwelling of magma (solid lines) in response to (rather than the cause of) plate divergence. The magma is concentrated at the ridge axis and forms magma chambers in both the upper mantle and the crust, but the melts converge from a wide area beneath the ridge. Some magma finds pathways away from the axis to form off-axis flows and near-axis seamounts.

related, that have chemical characteristics indicating they could have been "genetically related" or formed from the same or similar "parental melts." Hence they are part of a "family" of rocks that provide scientists with a history of their origins.

MORB are similar to tholeiitic basalts erupted on oceanic islands such as Iceland, the Galápagos, or Hawaii in that they are composed of about 50 wt% SiO_2 and lesser amounts of the oxides of magnesium, iron, calcium, and aluminum. However, MORB differ from ocean island basalts and oceanic arc basalts because they have different chemical characteristics (Sun *et al.*, 1979). MORB contain very low abundances of potassium (K_2O) and trace elements such as cesium, rubidium, barium, uranium, thorium, light rare earth elements (REE), and volatile elements – all of which are considered to be highly incompatible in typical mantle mineral assemblages during melting.

So-called "normal" mid-ocean ridge basalts, or N-MORB, are those with very low concentrations of highly incompatible elements. Because of their very low incompatible element abundances, N-MORB are considered to be "depleted" in trace elements that make up continental or ocean island volcanic rocks and consequently have low concentrations of trace elements such as potassium, rubidium, barium, lead, and uranium as well as the light rare earth elements (lanthanum, cerium, praseodymium, samarium) relative to the heavy rare earth elements. MORB containing slightly greater concentrations of highly incompatible elements, including potassium, are termed "enriched" or E-MORB (Fig. 5.44). In some localities, samples have been recovered that are even more depleted than N-MORB and are consequently classified as D-MORB.

The trace element characteristics of MORB lavas are commonly represented on normalized diagrams where the elemental abundances of each element are plotted and normalized to an accepted value of "primitive mantle" on a log scale (Fig. 5.45). An E-MORB pattern is characterized by a strong enrichment in the highly incompatible elements and a depletion in the middle to heavy REE, crossing over the REE patterns of the N- and D-MORB. Differences in the absolute abundances are commonly related to the amount of fractionation or mixing that a magma has experienced, whereas the slopes of the patterns are indicative of the source composition and/or extents of melting.

Figure 5.42 Hand specimens of MOR basalt lavas from the East Pacific Rise axis showing typical rock textures and surficial glass crust. Lobate flow crust (top) from ~9°50′ N (2545 m depth) with several large cavities that formed from vapor trapped within the flow during emplacement. Note the thin glassy crust on the upper surface. The very outermost glass is fragile and tends to spall off easily, whereas the glass beneath the exterior transitions into the microcrystalline rock below that typically adheres to the surface forming a turtle-shell-type texture. Folded and jumbled sheet lava samples (second from top) from ~9°29′ N (2566 m depth) also have finely textured outer glass surfaces, which commonly show the effects of stretching and folding before the lava completely quenched. In some cases these flows are so thin that the entire sample is glass. Lava finger (third from top) from the 9°7′ N OSC area where many pillowed flows were observed to have lava fingers as decorations on the flow forms, likely due to a late-stage surge in the eruptive pressure at the end of the eruption. The finger shown is about 12 cm long and is entirely composed of fresh glass through its diameter (~3–4 cm) as the end of the finger where it was broken off is entirely glassy. A piece of folded sheet lava from the Juan deFuca Ridge (bottom) collected by *Alvin* on *Dive* 2093 in 1988 has a thin glassy rind and microcrystalline interior suggesting the fold formed quickly.

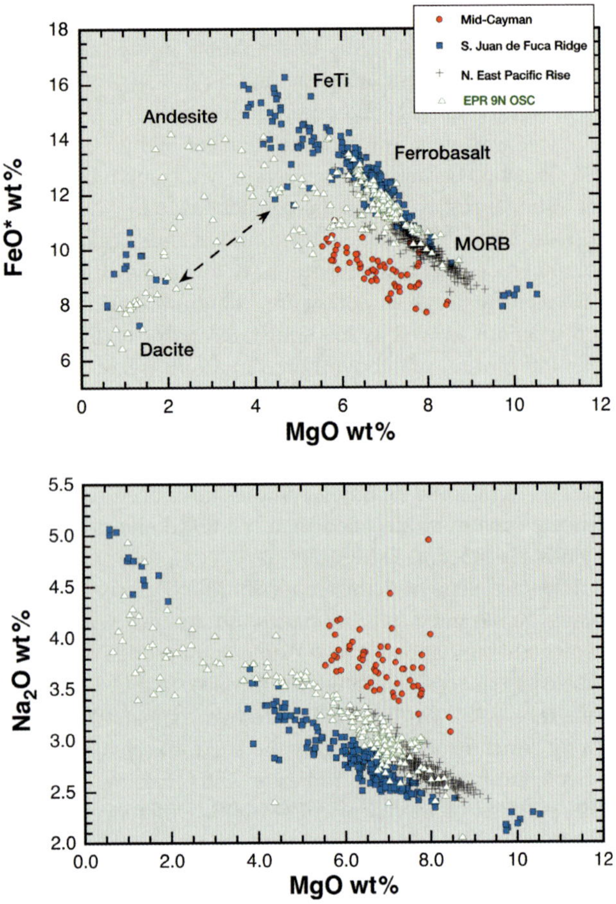

Figure 5.43 Major element diagram showing the variation of MgO wt% versus FeO wt% (top) and Na_2O (bottom) from three MORs that have different spreading rates. Volcanic rock names are shown adjacent to their typical range of compositions in top diagram. The general elemental trends with decreasing MgO are primarily due to fractional crystallization for typical MORB suites. Highly evolved lava compositions erupted at the 9° N OSC (Wanless *et al.*, 2010; green triangles) are a result of extreme fractional crystallization coupled with crustal assimilation along the propagating limb of the OSC, as well as a component of magma mixing between high-SiO_2 magmas and relatively evolved basaltic liquids. The variations in FeO and Na_2O at any given MgO reflect different extents and integrated depths of melting (see Klein and Langmuir, 1987) (data sources: northern East Pacific Rise 9°–10° N: Wanless *et al.*, 2010, 2012; Perfit *et al.*, various sources in Petdb; http://www.earthchem.org/petdbWeb/search/ref_info.jsp?singlenum=1548; http://www.earthchem.org/petdbWeb/search/set_ref_scrn2.jsp; Juan de Fuca Ridge: Stakes *et al.*, 2006; Mid-Cayman Spreading Center: Elthon *et al.*, 1995).

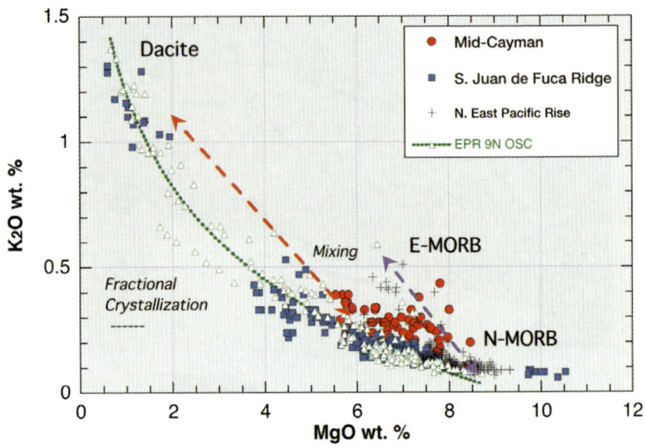

Figure 5.44 Variation diagram showing variations in K_2O wt% versus MgO wt%. Potassium is an incompatible element in MOR magmatic systems and as such exibits large increases in concentration with decreasing MgO as a consequence of fractional crystallization (typical fractional crystallization path shown as dotted green curve). Variations in K_2O at given MgO content reflect different parental MORB characteristics (e.g., N-MORB vs. E-MORB). These arise from different extents of mantle melting or mantle sources with different K_2O contents (and other incompatible elements). Some variability may also be due to mixing between N- and E-MORB melts (purple dashed line) or between highly evolved, low-MgO magmas and evolved basalts (red dashed line). Data sources are the same as those listed for Fig. 5.43.

Isotopic investigations have conclusively shown that values of the radiogenic isotopes of strontium, neodymium, hafnium, and lead in N-MORB are consistent with their incompatible element-depleted characteristics and indicate that this depletion occurred via one or more episodes of partial melting of upper mantle sources beginning more than 2 billion years ago (Arevalo and McDonough, 2010; Jenner and O'Neill, 2012; Gale *et al.*, 2013; White and Klein, 2014).

Overall, the geochemical characteristics of MORB are used to decipher the composition of the mantle source(s), how much mantle melting occurred, and at what temperatures and pressures (Langmuir *et al.*, 1992; Salters and Stracke, 2004; Putirka *et al.*, 2011). It has been clearly shown that different MOR segments have distinctive chemical characteristics that can be related to differences in regional mantle sources and melting (Schilling *et al.*, 1983; Michael, 1995; Arevalo and McDonough, 2010). This information allows researchers to better understand a fundamental process of plate tectonics and how Earth chemically evolved. On a much smaller scale, detailed geochemical investigations of lavas have documented the dominant effects of fractional crystallization and mixing on MORB geochemistry and the inherent short-term and spatial variability of MOR eruptions (e.g., Langmuir *et al.*, 1977, 1986; Frey *et al.*, 1993; Perfit *et al.*, 1996; Smith *et al.*, 2001; Bergmanis *et al.*, 2007; Rubin and Sinton, 2007; Waters *et al.*, 2011; Wanless *et al.*, 2012). Significant geochemical differences in closely spaced dikes from sheeted dike complexes (see Chapter 6), which presumably feed overlying lava flows, are consistent with the short-term temporal and spatial variability in MOR magmatic systems (Stewart *et al.*, 2005; Pollock *et al.*, 2009).

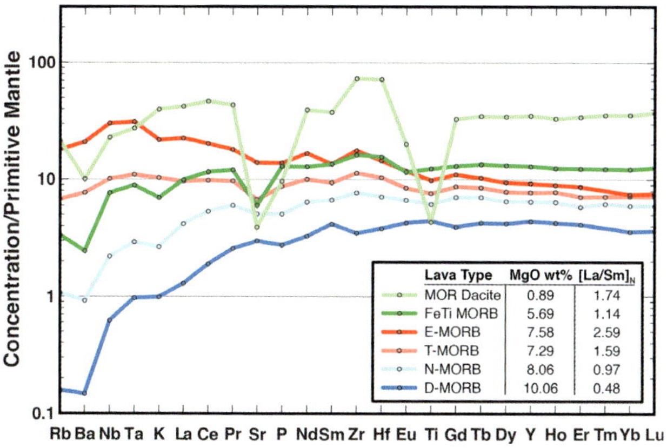

Figure 5.45 Incompatible trace element abundances are commonly used to chemically characterize different types of MORB and to provide information about magmatic processes and sources. Concentrations of incompatible elements (on x-axis) plotted on this primitive mantle-normalized diagram (values from Sun and McDonough, 1989) show the variations in trace element "patterns" for representative E-MORB (enriched), N-MORB (normal) and D-MORB (depleted). Transitional (slightly enriched) varieties are sometimes referred to as T-MORB. The degrees of enrichment in MORB show good correlations with K_2O contents (see Fig. 5.44) and K/Ti ratios. E-MORB have strong enrichments in elements that are the most incompatible during mantle melting (elements on far left) relative to the middle to heavy rare earth elements. For example, $(La/Sm)_N > 1$, may be the result of melting-enriched mantle source or smaller extents of melting. Highly differentiated N-MORB (low MgO) FeTi and dacite MORB exhibit overall enrichments of incompatible elements and some significant depletions (e.g., Sr, P, Eu, Ti). All samples shown are from the 8°–10° N segment of the EPR, including the D-MORB from the Siqueiros Transform (data from Perfit *et al.*, 1996; Sims *et al.*, 2005, Wanless *et al.*, 2010, 2012; Waters *et al.*, 2011).

It is these types of studies that allow scientists to decipher and model the physical and chemical processes that occur in the crust and upper mantle. Ultimately, with information obtained from the minerals in MORB (see Section 5.10), from samples from deeper parts of the oceanic lithosphere, and from ophiolites, more pieces of the oceanic crustal puzzle can be solved.

5.10 Crystals in MORB

The minerals present in MORB lavas can provide information about the temperatures and pressures of crystallization and compositional changes in magma composition. The most common minerals in MORB are olivine (Ol), plagioclase (Plag) and, less commonly, clinopyroxene (Cpx) (Table 5.1). Spinel (Sp) (chromium–aluminum–magnesium-rich varieties) is a common accessory phase in more magnesian lavas, where it is typically enclosed in olivine

Table 5.1 Compositions of common minerals in mid-ocean ridge basalts

Mineral	Composition/structural formula
Plagioclase feldspar (Pl)	$CaAl_2Si_2O_8$–$NaAlSi_3O_8$
Olivine (Ol)	Mg_2SiO_4–Fe_2SiO_4
Clinopyroxene (Cpx)	$Ca(Mg,Fe)Si_2O_6$
Pigeonite (Pig)	$(Mg,Fe)_2Si_2O_6$
Spinel (Sp)	$MgAl_2O_4$–$Fe_2Cr_2O_4$–$FeAl_2O_4$–$Mg(Fe,Cr,Al)_2O_4$
Titanomagnetite (TiMag)	Fe_2TiO_4–Fe_2O_4
Ilmenite (Ilm)	$FeTiO_3$

crystals. Magmatic sulfides also commonly exist in MORB as tiny round globules preserved in the glass or rock matrix. It is typical for either olivine or plagioclase to be the first mineral to crystallize in MORB – consistent with relatively low pressure conditions of crystallization (less than 0.5 GPa) at temperatures between 1170 and 1280 °C depending on the extent of differentiation (Bender *et al.*, 1978; Bryan, 1983; Grove *et al.*, 1992; Yang *et al.*, 1996). Calculated temperatures of likely primitive MORB melts (those in equilibrium with the mantle) have temperatures that range from around 1220 °C to just over 1400 °C (Putirka *et al.*, 2007; Kvassnes and Grove, 2008).

Olivine usually occurs as tiny, translucent, olive-green individual crystals. Plagioclase crystals are often seen as rectangular crystals with smooth, glassy to white surfaces and often in radiating crystal clusters with clinopyroxene and lesser olivine (Fig. 5.46). In the more chemically evolved MORB it is common to find plagioclase and augite together with a low-calcium pyroxene [pigeonite (Pig) or hypersthene (Opx)] and the oxides titanomagnetite (TiMag) and ilmenite (Ilm). High-silica MOR lavas may also contain sodic plagioclase, ferroaugite, and rare fayalitic olivine, hedenburgite, apatite, and zircon (Schmitt *et al.*, 2011). Unlike lavas in other tectonic environments, MORB typically contain very few phenocrysts (most below 40% by volume, but many below 10%) (Bryan, 1983; Davis and Clague, 1987). Textures (including grain size) vary depending on nucleation and crystallization rates – in part related to cooling rate. Hence, MOR lavas, which are quenched when erupted into seawater, commonly have relatively few, small phenocrysts in a glassy to very finely crystalline matrix (Figs. 5.32 and 5.46). Conversely, in the interiors of thick lava flows many of the rocks have time enough to become completely crystalline and hence can have totally crystalline (holocrystaline) textures similar to those observed in MOR dikes (see Chapter 6).

Some MORB contain xenocrysts or clots of crystals that appear to be foreign fragments of more coarsely crystalline rock (± glass), presumably entrained by the host magma as it transited through the developing crust prior to eruption. Several studies document gabbroic xenoliths in MORB from the active ridge axes of the Juan de Fuca Ridge and EPR (Dixon *et al.*, 1986; Davis and Clague, 1990; Smith *et al.*, 1994). Most, but not all, recovered xenoliths are medium-grained gabbros, composed of plagioclase + clinopyroxene and/or olivine and in some instances interstitial glass, minor orthopyroxene, and titanomagnetite.

Xenocrysts and minerals within the gabbroic xenoliths commonly are not in equilibrium with their host lava. In some cases they exhibit complex zoning patterns, which are akin to "growth rings," with different compositions reflecting changing conditions of crystallization. Disequilibrium textures between minerals are indicative of magma mixing (Bryan, 1983), assimilation or melt–rock interactions that most likely occur in the crystal-melt mush zone beneath ridge axes (Ridley *et al.*, 2006; Lissenberg and Dick, 2008; Kvassnes and Grove, 2008) (Figs. 5.41 and 5.47, and see Figs. 3.18 and 3.30 and Chapter 7).

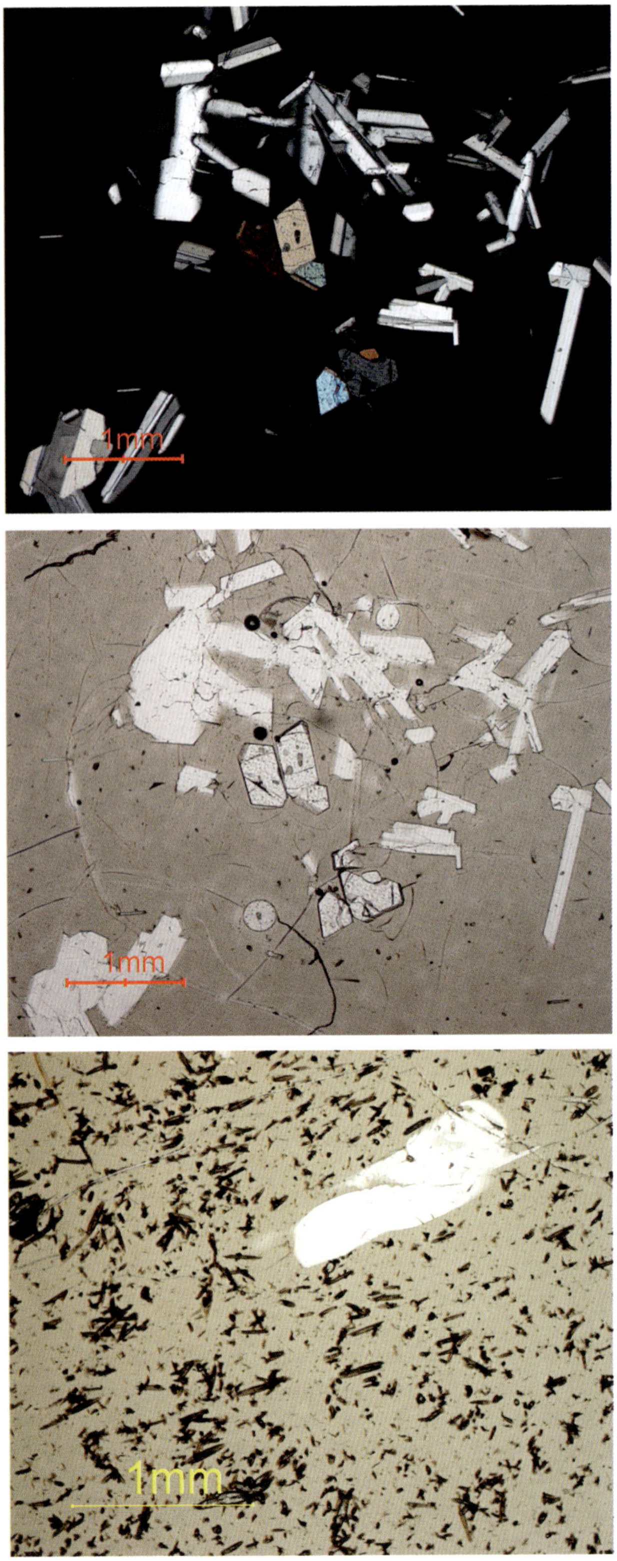

Figure 5.47 Gabbroic xenolith enclosed in MORB lava from the northern EPR showing glass and vesicles (open gas cavities) between plagioclase (white) and olivine (green) crystals. The textures and mineral compositions indicate that it formed deep in the mush zone deep in the crust below the axis before being incorporated in the melt that formed this basalt lava (width of sample ~8 cm).

Figure 5.46 Photomicrographs of thin sections of MORB taken using polarized light show the variety of mineral associations commonly present. Basalt T184-G5 (top) from the southern Juan de Fuca Ridge with crossed Nicol prisms, which causes non-crystalline glass surrounding mineral phases to appear black and minerals to exhibit different colors and optical properties as the sample is rotated. The middle image is the same sample and the same view in plane-polarized light. Elongate plagioclase (white to gray), and more textured equant olivine (colored) crystals are set in a tan basalt glass. Dacite lava T735-19 (bottom) displayed in plane-polarized light showing tiny microlites of clinopyroxene and magnetite set in tan glass with a larger, slightly resorbed, plagioclase crystal.

5.11 Generating MORB magma

Beneath both fast- and slow-spreading MORs, as the lithosphere spreads, basaltic melts are generated in a broad region (tens to hundreds of kilometers wide), where they accumulate and focus to feed the relatively narrow neovolcanic zone along the axis of the ridge (see Chapter 3). Primary MORB magmas are generated by decompression partial melting of peridotite (see Chapter 8) in the asthenospheric upper mantle (Fig. 5.41). The extent of melting, the depths of melting, melt extraction process, mineralogy, and bulk composition of the source or sources (for example, depleted harzburgite versus enriched pyroxenite), and magma mixing all influence the composition of resultant primitive magmas (Phipps Morgan, 1987; McKenzie and Bickle, 1988; McKenzie and O'Nions, 1991; Kinzler and Grove, 1992, 1993; Langmuir *et al.*, 1992; Spiegelman and Kelemen, 2003; Herzberg *et al.*, 2007; Stracke and Bourdon, 2008; Waters *et al.*, 2011).

Some of these parameters are determined by the thermal regimes that in turn can be related to crustal morphology and spreading rate (see Chapter 3). For example, at faster-spreading ridges, mantle upwelling is more rapid so that melting begins at greater depths and more melt can be generated during its ascent, resulting in a more continuous supply of magma to the spreading center (Langmuir *et al.*, 1992). Alternatively, slow-spreading ridges are expected to have a cooler melting regime, which provides a smaller volume of magma to a spreading center per unit time of extension. High melt supply at fast-spreading ridges leads to greater degrees of differentiation and homogenization of parental magmas in shallow magma chambers. At slower-spreading ridges where magma supply is low, episodic magma accumulation in deeper, poorly connected, and more isolated reservoirs produces less differentiated and more heterogeneous lavas (Rubin and Sinton, 2007).

During ascent from the mantle and cooling in the crust, primary mantle melts are subjected to a variety of physical and chemical processes such as fractional crystallization, magma mixing, crustal assimilation, and melt–rock interactions that modify and differentiate the original melt compositions (Langmuir *et al.*, 1992; Rubin *et al.*, 2009b). Consequently, primary melts are unlikely to erupt on the seafloor without undergoing some modification. Picritic lavas and magnesian glasses thought to represent possible primary basalts have only rarely been recovered from MOR environments. These compositionally more primitive lavas commonly occur in transform faults (Natland, 1989; Perfit *et al.*, 1996; Wendt *et al.*, 1999), or on near-axis seamounts (Fornari *et al.*, 1988; Allan *et al.*, 1989; Batiza *et al.*, 1990). Content of MgO in these basalts ranges from about 10% to over 15% by weight, and the lavas typically contain significant quantities of olivine crystals.

Chemical variations and elemental trends defined by spatially related suites of MOR lavas are primarily the result of progressive fractional crystallization of variable combinations and proportions of olivine, plagioclase, and clinopyroxene as the magma cools (Fig. 5.44). The compositional "path" that a magma takes, known as its liquid line of descent (LLD), can take different trajectories as a consequence of the order of crystallization and the different proportions of crystallizing phases, which are controlled by initial (and subsequent changes in) liquid composition, temperature, and pressure. In some MORB suites, linear elemental trends may result from mixing of primitive magmas with more evolved magmas that have evolved along an LLD. Suites of MORB glasses often define distinctive LLDs that match those determined by experimental crystallization of MORB at low to moderate pressures (Bender *et al.*, 1978; Grove *et al.*, 1992; Langmuir *et al.*, 1992; Yang *et al.*, 1996).

Much of the major element data from fast-spreading ridges such as the EPR are best explained by low-pressure [0.1 GPa or less, found at depths shallower than 3 km below the seafloor (bsf)] fractional crystallization. In contrast, at slow-spreading ridges such as the MAR, data seem to require unrealistically higher-pressure crystallization (0.3–1.0 GPa, which translates to depths of 10–30 km bsf) (Michael and Cornell, 1998; Herzberg, 2004; Eason and Sinton, 2009). This is consistent with other evidence suggesting that magmas at fast-spreading ridges evolve in a shallow crustal magma lens or chamber located only 1–3 km below the seafloor, while magmas at slow-spreading ridges evolve at significantly greater depths; possibly in the mantle lithosphere or at the crust–mantle boundary.

Volatile analyses of melt inclusions in olivine crystals separated from lavas recovered at the EPR and Juan de Fuca Ridge indicate that although most olivine crystallization occurred at pressures equivalent to the depth of the melt lens (1.5–3 km depth), over 25% crystallized at pressures consistent with formation within the lower crust and uppermost mantle (Wanless and Shaw, 2012).

MORB from normal, magmatically robust ridge segments typically exhibit minor-compositional variation due to relatively frequent magma recharge and mixing. However, in environments proximal to transforms, propagating rift tips, and overlapping spreading centers, highly differentiated FeTi basalts (basalts that are particularly enriched in iron and titanium) are commonly found, along with iron-rich andesites and dacites (Wanless *et al.*, 2010) (Figs. 5.43 and 5.44). This is largely due to extensive fractional crystallization coupled with some crustal assimilation (AFC, assimilation fractional crystallization) in crustal magma chambers as a consequence of the relatively cooler thermal regimes and magmatic processes associated with rift propagation.

5.12 Historic submarine eruptions

It has only been within the last 30 years that scientists have documented very young MOR submarine eruptions that occurred within historic time at fast- and intermediate-spreading ridges (Fig. 5.48) (Rubin *et al.*, 2012). The detection of hydrothermal event plumes along the Cleft Segment of the Juan de Fuca Ridge in the 1980s (Baker *et al.*, 1987) and repeat mapping of the MOR axis by ship-mounted multibeam sonar led to discovery of the first young deep-sea lava flows of known age, a series of pillow mounds up to 75 m high (Chadwick *et al.*, 1991, 2001; Embley *et al.*, 1991; Fox *et al.*, 1992). The first direct observation of the aftermath of a seafloor volcanic eruption that occurred in March 1991 was made by scientists diving in *Alvin* at the EPR at 9°50′ N (Haymon *et al.*, 1993), and comparing that site with observations made in 1989 using towed deep-sea cameras and sidescan sonar (Haymon *et al.*, 1991). Observations of newly dead and charred tubeworms strewn on top of and enveloped in fresh lava and diffuse vents spewing vast amounts of flocculated, sulfur-rich, microbial particles ("snow") into the water column suggested that an eruption had occurred within weeks prior to the diving (Fig. 5.49). Radiometric dating using Po–Pb systematics subsequently demonstrated that the observers had arrived just two to four weeks after the eruption (Rubin *et al.*, 1994). To date, no ongoing MOR eruptions have been witnessed directly, such as those happening at West Mata Volcano in the northern Lau Basin (Fig. 5.1) (Resing *et al.*, 2011).

The volcanic terrain created by the 1991 EPR eruption was not pristine, black lava as expected, but instead it was coated with white/gray gelatinous microbial matter (Haymon *et al.*, 1993) – something that has subsequently been observed at all sites of recent or ongoing deep-sea eruptions (Figs. 5.50 and 5.51). The EPR eruption continued with a second pulse in late 1991 to early 1992 determined from Po–Pb dating of young basaltic lavas recovered by *Alvin* (Rubin *et al.*, 1994). Within a year the microbial deposits coating the lava flows had been consumed by the animal community (Shank *et al.*, 1998) (see Chapter 4). A similar "snowblower" and young lava flow discovery at 17° S on the EPR was made in early 1993, although in this case the lava flow was not dated radiometrically and eruption timing constraints are less precise (see Sinton *et al.*, 2002).

In 1993, real-time hydroacoustic monitoring using the US Navy Sound Surveillance System (SOSUS) (see Chapter 1) in the Northeast Pacific became available (Fox *et al.*, 1994). Within the first month of acoustic monitoring, scientists at NOAA/PMEL detected an earthquake swarm on the CoAxial segment of the Juan de Fuca Ridge that appeared to have seismic characteristics typical of volcanic eruptions on land (harmonic "tremor" and migration of the epicenters related to dike intrusion and propagation along the spreading axis). This triggered

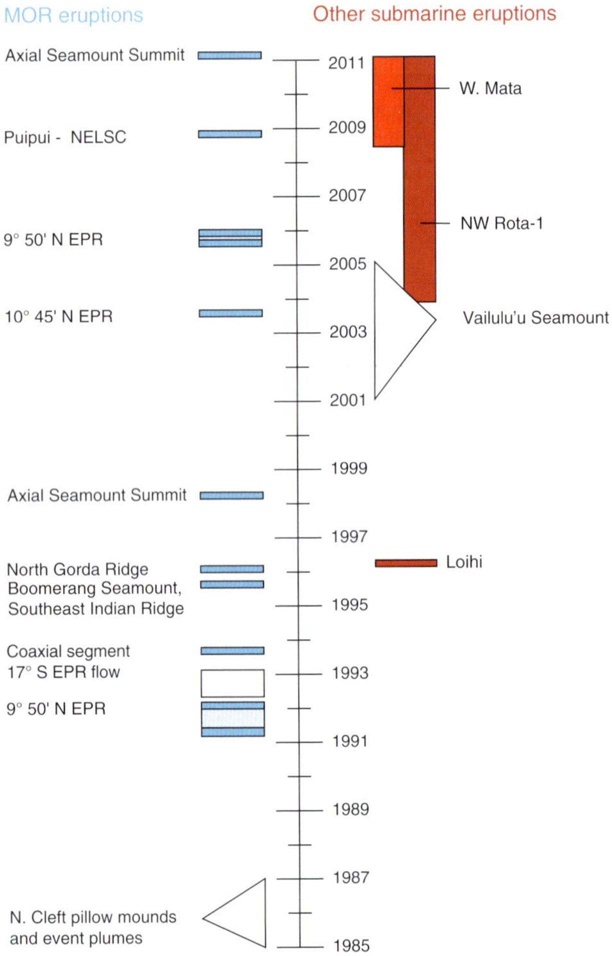

Figure 5.48 A timeline of historical, deep submarine eruption discoveries covering nearly 30 years since marine geologists began to have the sonar mapping technologies and deep-submergence vehicle access required to make precise pre- and post-eruption maps of seafloor terranes on the MOR, and the geochemical techniques required to quantitatively determine the age of young MORB lava (from Rubin *et al.*, 2012). Filled boxes correspond to eruptions with well-known dates and durations. Open boxes are for eruptions with less certain timing (see Rubin *et al.*, 2012, for details of methods used to determine eruption dates).

Figure 5.49 Freshly "cooked" tubeworms at the Tubeworm BBQ site in 1991, viewed from *Alvin* by observers witnessing the first discovery of the impacts of MOR submarine volcanic repaving of a hydrothermal vent community on the ridge axis at the EPR near 9°50′ N (Haymon *et al.*, 1993). Field of view ~1.5 m.

the first dedicated "response" expedition, which discovered event plumes, new hydrothermal vents, a still cooling lava flow (Embley *et al.*, 1995; Fox *et al.*, 1995; Delaney *et al.*, 1998) and two other previously unknown nearby young lava flows (Embley *et al.*, 2000) (Fig. 5.52).

Figure 5.50 *TowCam* image collected in May 2006 at the EPR near 9°49′ N just 6–8 months after the 2005–2006 eruptions at this site. White material is bacterial mat created by microbial blooms associated with the magmatic heat and hydrothermal circulation through the newly emplaced lava flows generated during the eruption. Field of view ~2 m.

Figure 5.51 *Alvin* photo taken in 1991 of small pillars at the margin of the EPR axial summit trough near 9°50′ N coated with microbial mat. During the eruption, hydrothermal fluids were seen emanating from the tops of pillars and many of the pillars were covered with microbial mat. This supports the supposition that pillars are formed in locations where hydrothermal fluids exit the seafloor below lava ponds and lakes. Field of view ~4 m.

Figure 5.52 (Top) CoAxial 1993 eruption showing 1993 flow (black lava) in contact with an older lava; field of view ~2 m. A close-up of the glassy tubular pillows that make up the large pillow ridge that formed during the CoAxial 1993 eruption (middle); field of view ~1 m. The new lava erupted in 1996 at Gorda Ridge (black lava in bottom image) and contact with older flows; field of view ~4 m.

The SOSUS (see Chapter 2) real-time hydroacoustic detection capability in the Northeast Pacific has led to multiple eruption discoveries and responses in that region over the past few decades (Perfit and Chadwick, 1998; Dziak *et al.*, 2007, 2011). In 1996, earthquake swarms were detected on the North Gorda Ridge (Fox and Dziak, 1998). A summit caldera eruption at Axial Seamount in 1998 (Dziak and Fox, 1999; Embley *et al.*, 1999) partially buried an instrument package containing pressure and temperature sensors at 1.5 km water depth (Fig. 5.53), leading to the first *in situ* record of the emplacement of a submarine lava flow. The instrument recorded lava flow inflation and drain-out phases (Fox *et al.*, 2001; Chadwick *et al.*, 2003), similar to eruptions on land (Hon *et al.*, 1994).

In 2006, the next discovery of a MOR eruption occurred when several ocean bottom seismometers (OBSs; see Chapter 1) still recording data were unable to release from

Figure 5.53 Illustration of the NOAA "rumbleometer" buried in 1998 Axial Seamount flow with the *ROPOS* ROV shown imaging the instrument. The inset photograph is an actual view of the rumbleometer buried in the flow (instrument is ~2 m on each side of the triangular base).

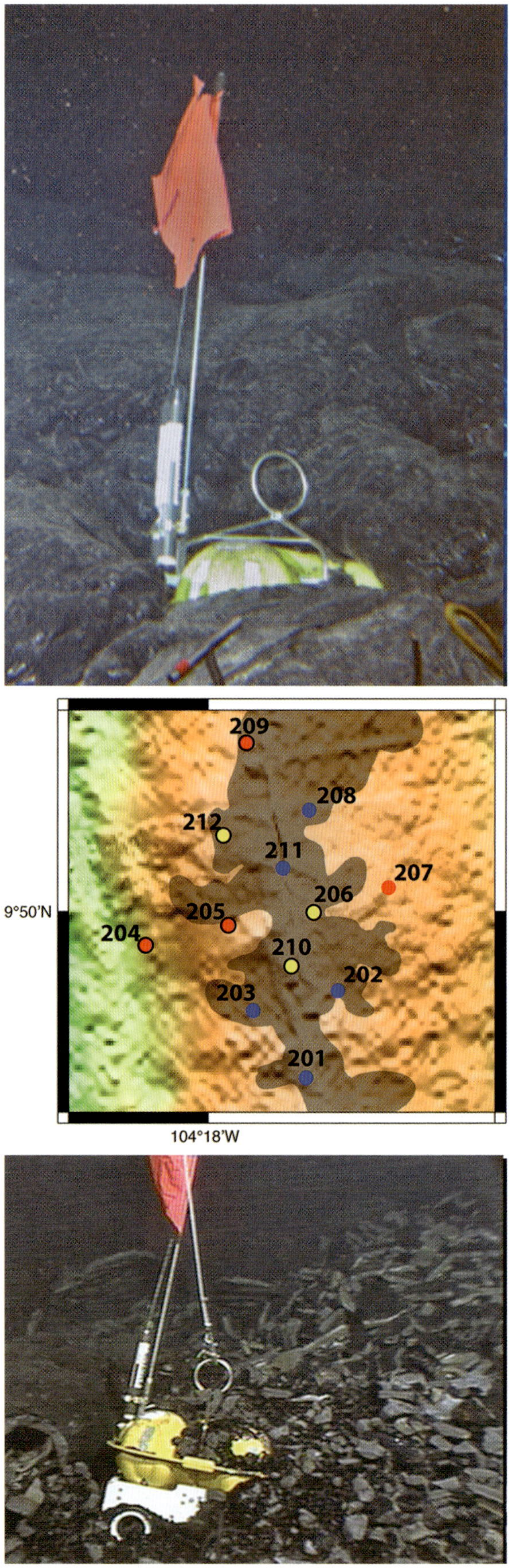

Figure 5.54 Deep-sea photographs of ocean bottom seismometers (top and bottom) deployed at the EPR in early 2005 that were trapped in newly erupted lava during the 2005–2006 eruption. Top image was taken by *Alvin* in June 2006 looking across the new flow surface. Bottom image was taken by ROV *Jason's* camera during recovery operations in March 2007. Map shows the location of the OBS array. Red symbols are OBSs recovered, blue symbols denote those lost to the erupted lava, and yellow symbols are the OBSs that were imaged by *TowCam* surveys. OBSs 212 and 206 were recovered, but data contained in the instruments were compromised by burning of cables by the lava.

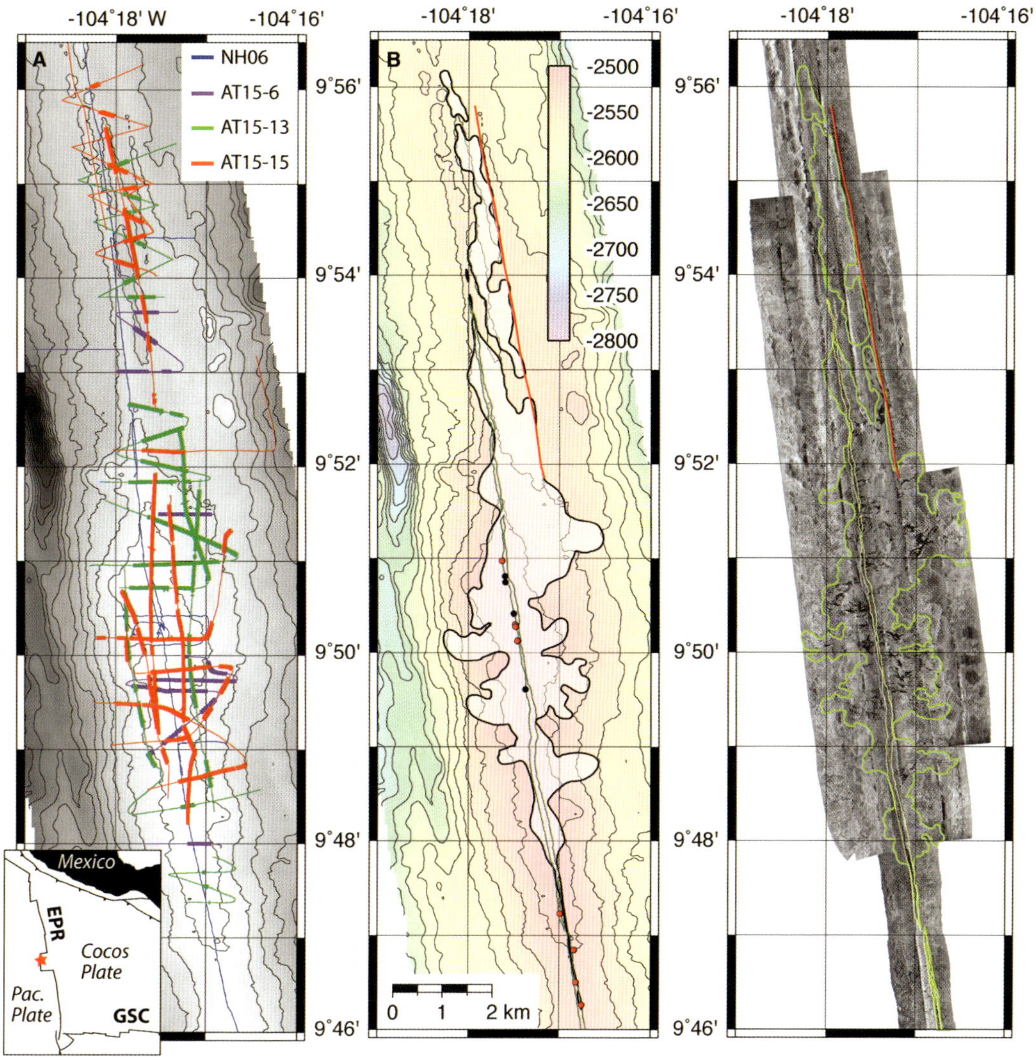

Figure 5.55 Seafloor imagery (e.g., Figs. 5.52 and 5.56), near-bottom bathymetry, and sidescan sonar data (right) were used to create a detailed map of the limits of the 2005–2006 EPR eruption (Soule *et al.*, 2007).

the seafloor at the 9°50′ N EPR site. This site was designated as an Integrated Study Site (ISS) for the NSF-sponsored Ridge 2000 Program (www.ridge2000.org).

Two subsequent "response" cruises discovered that recently emplaced lava flows had trapped several of the OBSs (Fig. 5.54), resulting in their inability to surface when acoustically released (Tolstoy *et al.*, 2006; Cowen *et al.*, 2007, Dziak *et al.*, 2007). This was the first documented example of a repeat historical eruption at the same MOR location and confirmed the theory that eruptions at fast-spreading MORs occur on average every 10–15 years (Perfit and Chadwick, 1998). It also marked the beginning of very detailed ROV and AUV surveys of newly erupted lava flows using high-resolution, near-bottom bathymetric, sidescan and photographic data from before and after an eruption (Figs. 5.55–5.57). The 2005–2006 EPR lava flows extend 19 km along the ridge axis and up to 3 km away from it into older terrain (greater than 20,000 to 60,000 years). The flow volume was roughly that of the 1991–1992 event in the same location, but it extended over a much larger area (Soule *et al.*, 2007; Fundis *et al.*, 2010). Observations from *Alvin* and a towed camera suggest that the dike feeding those eruptions may have extended within the upper crust for a distance of 10 to 20 km beyond that actual eruption site, which covered the AST area primarily between 9°48′ N and 9°56′ N (Soule *et al.*, 2007). However, seismic reflection data in this region of the EPR imaged three distinct melt lenses beneath the recent eruptions that coincided with transitions in lava composition, volume, and

Figure 5.56 Photomosaic of three *TowCam* images showing a tubular pillow and lobate flow of the 2005–2006 eruption on the EPR axis near 9°50′ N. Feld of view across each image ~4 m.

Figure 5.57 Three ages of lava on the EPR axis near 9°50′ N (left). Left lobe shows black, glassy lava emplaced during the 2005–2006 eruption, while the middle lobe of sediment-dusted lava is the oldest. Tubular pillow on right may be 1991–1992 lava; its age, based on the character of the outer crust and slight devitrification of the outer glass, is intermediate between the other two lava forms; field of view ~2 m. *TowCam* photo (right) of the contact between 2005–2006 lava ponded against the first inward facing fault east of the EPR axis at 9°53′ N. Fault height is about 3 m; field of view ~4 m.

Figure 5.58 Collapse pit in a lineated flow that transitions to both pillow and lobate forms along the margin of the flow. This flow is believed to be part of the early stages of the 2011 eruption at the summit of Axial Seamount. Field of view ~4 m.

morphology (Carbotte *et al.*, 2013). The data suggest the eruptions were associated with the vertical ascent of magma from lenses that are mostly physically isolated and coincide with fine-scale tectonic segmentation of the ridge axis.

The second example of a repeat eruption occurred in 2011 along the southeast rift zone of the summit caldera of Axial Seamount on the Juan de Fuca Ridge, where there was a well-documented eruption in 1998 (Chadwick *et al.*, 2013). New lava flows erupted from multiple, discontinuous fissures. Many of the fissures reoccupied those that were active during the 1998 eruption, but some new fissures also formed along the northern rim of the eastern caldera. Although the flows were discovered in much the same area of the summit caldera as the 1998 Axial eruption, high-resolution bathymetry and detailed mapping and observations using ROV and AUV surveys indicated the 2011 eruption was significantly more voluminous (Caress *et al.*, 2012; Chadwick *et al.*, 2012, 2013; Clague *et al.*, 2013) (Figs. 1.33 and 5.58). Geodetic surveys and *in situ* hydrophone data recorded a large deflation and a microearthquake swarm in the caldera in April 2011. It was not recognized in real-time SOSUS data due to recent reductions in system coverage. Nevertheless, three expeditions using a total of three ROVs, a mapping AUV, and four research vessels were able to respond and take advantage of the opportunity, employing excellent coordination to optimize mapping, sampling, and collection of complementary data sets throughout the Axial Seamount summit eruption area.

While not associated specifically with MOR eruptions, important discoveries of new seafloor eruptions at deep subduction zone and back-arc settings were made beginning in 2004 with ROV observations of pyroclastic eruptions at the 520 m deep summit of the Northwest Rota-1 Seamount in the Mariana Arc (Embley *et al.*, 2006; Chadwick *et al.*, 2008; Deardorff *et al.*, 2011).

In November 2008, two ongoing eruptions were serendipitously detected in the Northeast Lau Basin in the Southwest Pacific by water column anomalies, and were subsequently verified by ROV dives in May 2009, again as part of a coordinated multidisciplinary program associated with the NSF Ridge 2000 Program. The first one was an effusive eruption on the Northeast Lau spreading center (Rubin *et al.*, 2009a), which produced an event plume containing volcanic fragments (Baker *et al.*, 2011). The second one was at a nearby back-arc volcano named West Mata, characterized by an intense

Figure 5.59 A phase of the West Mata eruption imaged and sampled in 2009 using the ROV *Jason*, showing extensive water–magma interaction and resulting fragmental material caused by the explosive nature of the eruption. Yellow color in plume is sulfur produced during the eruption. Nozzle at right is the tip of a titanium gas-tight water sampler. Depth of the eruption site is about 1200 m. Field of view ~1 m.

particle plume and unusually high H_2 concentrations. ROV dives observed dramatic pyroclastic and active effusive pillow eruptions from two roughly 1200 m deep vents at the summit of West Mata volcano (Clague *et al.*, 2011; Resing *et al.*, 2011) (Fig. 5.59). Notably, the West Mata discovery allowed detailed, real-time observations of a deep submarine eruption to be made, including the first imagery of glowing hot lava flowing across the deep seafloor and the first sample of molten lava from the deep sea.

6 Dike intrusion and sheeted dike complexes

And the secret of it all – the secret of the earthquake, the secret of the "temple of fire," the secret of the ocean basin, the secret of the highland – is in the heart of the earth, forever invisible to human eyes.

R. A. Daly (1926)

6.1 Seafloor spreading: one meter at a time

On the seafloor, the most obvious manifestations of seafloor spreading along the mid-ocean ridge (MOR) system are young basaltic volcanic rocks and faults and fissures along the axis. At deeper levels, in the middle to lower crust, volumes of magma solidify to form coarse-grained gabbroic rocks. In between these regimes, magma from subsurface reservoirs is injected into the crust through narrow conduits. Basaltic magma passing through these channels is cooled by the surrounding upper crust and solidifies to form tabular structures called dikes. Steeply inclined dikes and sheet-like

Figure 6.1 Sheeted dike complexes are composed of parallel arrays of tabular basaltic intrusions, each about 1 m wide. In this outcrop from the Pito Deep Rift in the southeast Pacific, several well-defined dikes dip moderately to the southeast (right), away from the East Pacific Rise (EPR) spreading center where they were formed. Field of view ~4 m.

mafic intrusions in different orientations are well known from eroded volcanic terranes on land (Anderson, 1972; Walker, 1993). Some dikes feed lava flows at the surface, but many freeze in the subsurface before they can erupt.

At MORs the persistent supply of magma coupled with localized plate separation commonly creates a distinctive array of parallel, side-by-side intrusions termed "sheeted dike complexes" (Fig. 6.1). Dikes are intruded individually or in parallel families ("swarms") during magmatic events. Sheeted dike complexes are commonly composed of ~90% dikes, with minor amounts of lava or gabbroic rocks ("screens") between the dikes in some places. This distinctive geologic unit of the oceanic crust is sandwiched between the overlying lavas and underlying gabbroic rocks at crustal depths of roughly 500 to 2000 m beneath the seafloor (Fig. 6.2). Individual dikes are about 1 m wide, but they can be hundreds of meters tall and extend laterally for thousands of meters beneath the MOR axis.

Sheeted dike complexes are identified by their distinctive parallel structures, in some cases as regular as books on a shelf (Fig. 6.1). Individual dikes can be identified by their contacts (commonly chilled margins) against adjacent dikes, lava flows, or gabbroic rocks and by fracture patterns (columnar joints) that develop during cooling, contraction, and cracking of the injected magmatic material (Figs. 6.3 and 6.4).

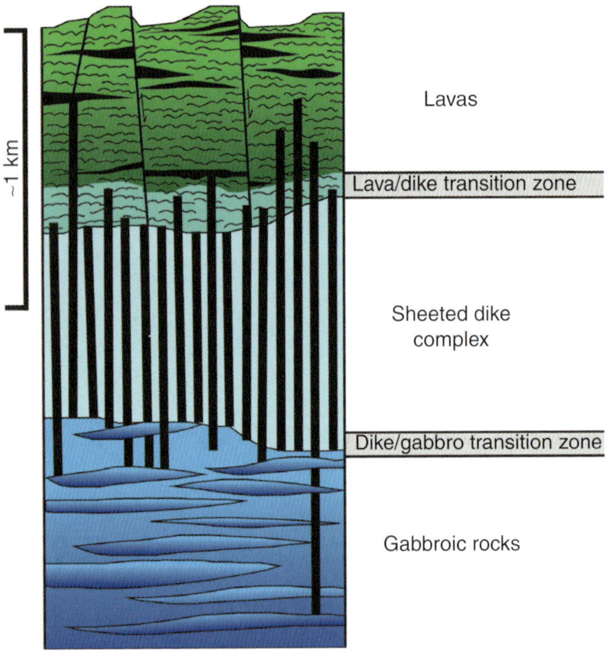

Figure 6.2 Sheeted dike complexes in the oceanic crust and ophiolites lie between the overlying lavas and the top of gabbroic material of the middle crust. The sheeted dikes are separated from the lavas by a transition zone composed of altered and fractured lower lavas cut by varying proportions of dikes. The thicknesses of the lavas, lava–dike transition zone and sheeted dikes are highly variable but typically a few hundred meters. The dike–gabbro transition zone has relatively low relief and marks the boundary between the rapidly chilled dikes and underlying gabbroic material crystallized from intrusive magma bodies.

Figure 6.3 Individual dikes have margins that can be traced continuously across outcrops where they have intruded surrounding crustal materials. Columnar joints, similar to those that form in thick lava flows and other igneous bodies are oriented at a high angle to the margins. Columnar joints terminate at edges of dikes, commonly giving each one a recognizable ladder-like pattern of fractures. A basaltic dike in Iceland ~2.5 m wide (above) illustrates this pattern. An entire individual column (~30 cm long) bounded by joints (bottom) was sampled with *Alvin's* manipulator from a dike (visible in background) on the north wall of Hess Deep Rift.

Figure 6.4 A chilled margin in the sheeted dike complex of the Oman Ophiolite (left) shows the typical dark, very fine-grained nature of the dike margin that grades into the lighter, coarser-grained, dike interior to the right. The dike chilled against another pre-existing dike on the left. Basaltic dikes have medium-grained, subophitic ("diabasic") textures with plagioclase laths partially enclosed by larger, more equant clinopyroxene crystals shown in the photomicrograph (upper right). The black groundmass is composed of very fine-grained crystals and glass. Rapid cooling produces very fine-grained textures (lower right) that grade into coarser material in the dike interior (right side of lower right image). Both photomicrographs shown in cross-polarized light. Field of view 4 mm for both images.

As discussed in Chapter 3, not all spreading systems have persistent magma supplies that could potentially produce crust with a sheeted dike complex, but even at MORs with very low magma budgets, dike intrusion probably plays a significant role in building the oceanic crust, at least near magmatic centers. In areas of low magma budget and major faulting, dikes commonly intrude deep crustal and even mantle rocks that have been uplifted to shallow levels. In these areas a continuous sheeted dike unit may never have formed as part of the spreading process. The absence of sheeted dikes may be an indication of spreading with a discontinuous magma supply (low magma budget) where plate separation is accommodated primarily by lithospheric stretching and faulting, rather than by magmatic construction.

Sheeted dike complexes are documented in oceanic crust formed at both fast and slow spreading rates. However, much of what is known about dikes and sheeted dike complexes in oceanic crust comes from a few studies of tectonic windows and deep drill holes, and from ophiolite complexes where extensive outcrops can be studied in detail (Karson, 2002) (Fig. 6.5). In extensional volcanic terranes such as Iceland, the Afar Rift in East Africa, and Hawaii, dike intrusion events have been documented and exposed dike swarms provide information that is relevant to understanding the nature of sheeted dike complexes in oceanic crust.

This chapter provides an overview of the characteristics found in individual dikes and sheeted dike complexes in oceanic crust, ophiolites, and related terranes, as well as the processes associated with dike intrusion in these environments. The images used to illustrate these processes show the diversity of dike-related features and their significance. Given the limited perspectives possible in seafloor outcrops, additional information from subaerial exposures is presented to provide a more complete picture of dike intrusion. The study of oceanic dikes provides important constraints on the composition and structure of the upper oceanic crust and how it forms at different types of spreading centers.

Figure 6.5 Sheeted dikes (dikes intruded side-by-side, making up more than 90% of the rock unit) are important components of ophiolites and oceanic crust. Vertical sheeted dikes in the Oman Ophiolite (left) are each about 1 m wide. Vertical sheeted dikes are also exposed on steep seafloor escarpments which reveal the structure of the upper oceanic crust (right), as in this example from the Pito Deep Rift. Field of view ~3 m in both images.

6.2 Dikes beneath the seafloor

Although sheeted dike complexes in the oceanic crust were anticipated from studies of ophiolites decades ago, this rock unit was not documented in the oceanic crust until the 1970s. Validation of the ophiolite model came from drilling through the lavas and into the underlying sheeted dikes at DSDP/ODP Hole 504B in the eastern Pacific (Alt *et al.*, 1986, 1996). However, only in the last decade was IODP Hole 1256D drilled through the entire thickness of the sheeted dike unit and into underlying gabbroic rocks (Wilson *et al.*, 2006). Sheeted dike complexes in the oceanic crust have also been described from exposures along tectonic windows in both the Atlantic and Pacific (Table 6.1).

Although extensive sheeted dike exposures have been found in relatively fast spreading (high magma budget) crust, there are very few documented examples of sheeted dikes in slower-spreading (lower magma budget) crust, despite many investigations of extensive seafloor escarpments and deep drill holes in slow-spreading crust (Table 6.1). Along escarpments of the Mid-Atlantic Ridge, for example, the exposures of dikes and sheeted dike complexes are very limited. Collectively, observations from these settings provide the context to understand the internal structure of sheeted dike complexes and how they are constructed at MORs with different spreading rates.

Theoretically, deep crustal drilling should provide only a limited perspective on sheeted dikes, as a vertical drill hole would not be expected to intersect many near-vertical dikes. But this is not the case. Numerous dike margins and igneous contacts are recovered in drill cores, suggesting that the majority of dikes are not truly vertical. Even though these rocks are prone to fracturing during the drilling process, so that recovery rates are commonly low and core materials are highly fractured, substantial vertical sections have been sampled. The drilled samples document important geologic relationships and also provide a basis for the interpretation of ophiolite structures and how they relate to seafloor dikes. Geophysical logging of drill holes provides a continuous record of down-hole physical

Table 6.1 Seafloor exposures of sheeted dike complexes (see Fig. 3.3 for key locations)

Location	Full rate (mm/yr)	Type of exposure	Notes	References
Slow-spreading crust				
Vema Transform Fault	2.4	Transform valley wall	Contacts above and below are faulted; dikes vertical	Auzende *et al.*, 1989
King's Trough	30	Intra-oceanic rift	Dikes vertical	Klitgord *et al.*, 1990
Intermediate-, fast-, and superfast-spreading crust				
Blanco Transform	54	Transform valley wall	Laterally extensive exposures, intensely faulted, no lower contact observed; dikes dip moderately outward	Juteau *et al.*, 1995; Naidoo, 1998; Tivey *et al.*, 1998; Karson *et al.*, 2002b
Hole 504B	60	DSDP/ODP drill hole	No lower contact reached; dikes fractured and dip moderately outward	Cann *et al.*, 1983; Adamson, 1985; Pariso and Johnson, 1989; Alt *et al.*, 1996
Hess Deep Rift	110	Upper scarp, propagating rift	Laterally extensive exposures; upper and lower contacts well preserved; dikes fractured and dip moderately outward	Francheteau *et al.*, 1990, 1992; Karson *et al.*, 1992, 2002a; Stewart *et al.*, 2002
Pito Deep Rift	142	Upper scarp, propagating rift	Laterally extensive exposures; upper and lower contacts preserved; dikes dip steeply outward	Francheteau *et al.*, 1988, 1994; Karson *et al.*, 2005; Chutas, 2007
Endeavour Deep Rift	200	Upper scarp, propagating rift	Local dense dike swarms; dikes weakly fractured; dip very steeply	Larson *et al.*, 2005
Hole 1256D	220	IODP drill hole	Upper and lower contacts penetrated; weakly fractured dikes, near vertical	Wilson *et al.*, 2006; Alt *et al.*, 2010; Umino *et al.*, 2013

properties, which helps to define dike margins. Borehole seismic experiments characterize the physical properties of the sheeted dike unit on the scale of tens of meters, providing an important link to broader-scale seismic refraction and reflection studies.

Tectonic windows, faulted escarpments that expose the upper oceanic lithosphere, can be tens or even hundreds of kilometers long and provide the most complete views and samples of sheeted dike complexes. Upper crustal cross sections at these sites are oriented parallel to spreading flow lines and represent millions of years of spreading history. The spreading rate, age relations, and contacts with other rock units in these sections are well constrained. Orientations of dikes in tectonic windows do not appear to have been greatly disturbed by post-spreading processes so that, for example, tilting of dikes during spreading can be evaluated. Structural, igneous, and metamorphic relationships remain basically intact. These data are used to infer the active processes beneath the spreading centers where they formed.

Along oceanic escarpments, dike margins can be difficult to discern because ferro-manganese hydroxide coatings obscure textural details such as chilled margins. This type of surface is typical of all but the very youngest rocks exposed on the seafloor. Thus, joint patterns, rather than dike contacts, are also used to help identify individual dikes (Figs. 6.1, 6.3, and 6.5). In addition, fracturing is commonly concentrated along the dike margins. Using these features as a guide, individual dikes can be traced upslope in outcrops for tens of meters (Karson *et al.*, 2002a).

Variations in dikes, sheeted dike units, and the structures associated with them highlight fundamental differences in the mode of crustal construction typical of crust formed at different spreading rates and different magma budgets. The simplest view of seafloor spreading might suggest that just below the basaltic lavas the oceanic crust would have a continuous horizontal array of vertical dikes extending all across the global seafloor. However, direct observations from the seafloor suggest a much more complicated architecture.

Traditionally, individual dikes in sheeted dike complexes are considered to have been intruded vertically beneath the MOR axis and to remain vertical as they are rafted off-axis during the spreading process. Many models of intermediate to fast seafloor spreading assume vertical dikes (e.g., Hooft *et al.*, 1996). Furthermore, in reconstructions of ophiolite complexes it is assumed that dikes provide a vertical, ridge-parallel, reference plane. So, documenting the internal structure of sheeted dike complexes *in situ* on the seafloor is critical to understanding the accretion of the upper oceanic crust and linking those characteristics to ophiolite analogs.

Surprisingly, dikes observed in sheeted dike complexes formed at intermediate to superfast spreading rates (Table 6.1) are not vertical. Instead, the dikes dip "outward" (that is, away from the spreading centers where they formed). Vertical dikes appear to occur only in very limited areas, mostly in high-level swarms where they have been injected upward into the overlying lavas (Karson, 2002; Karson *et al.*, 2002a; Larson *et al.*, 2005). The rare exposures of dikes and sheeted dikes in slow-spreading crust are also tilted, but they generally tip inwards (toward the axis), reflecting tectonic rotations of fault blocks in these settings.

The details of the key occurrences of sheeted dikes in the oceanic crust are outlined below. Table 6.1 summarizes the geologic context and internal structural features of sheeted dikes on the seafloor. The available data show that there is considerable variation in the internal structure of the sheeted dike complexes that may be related to spreading rate or magma supply (Karson, 2002).

6.2.1 Dikes and sheeted dikes in crust formed at fast to intermediate rates

Drilling at DSDP/ODP Hole 504B penetrated more than 1000 m of a sheeted dike complex in crust generated at ~65 mm/yr at the Costa Rica Rift (Figs. 6.6 and 6.7). Core recovery was very low (about 25% in lava units and 10% in sheeted dikes, Fig. 6.5), but downhole logging helps fill in the gaps in this section (Alt *et al.*, 1993). Even though this drill hole was vertical it intersected nearly 200 dikes that were identified on the basis of recovered dike margins, and magnetic and geochemical characteristics across igneous contacts, demonstrating that most of the dikes are not vertical (Anderson *et al.*, 1985; Becker *et al.*, 1988; Dilek *et al.*, 1996).

Assuming a typical dike width of about 1 m through the entire thickness of the sheeted dike unit, the average dip of the dike margins would be about 70°, in good agreement with the magnetically corrected dips of cored dike margins. Magnetic inclination measurements in core samples are scattered over a 30° to 40° range but they are skewed toward orientations indicating that they dip away (outward) from the Costa Rica Rift. This is consistent with the sense of tectonic rotation inferred from magnetic data in overlying lavas (Pariso and Johnson, 1989; Schouten and Denham, 2000).

In Hole 504B, two dike populations appear to be present: more gently dipping (average 68° outward dip) and steeper, cross-cutting dikes (average 82° outward dip). The more gently dipping dikes are inferred to have been tilted prior

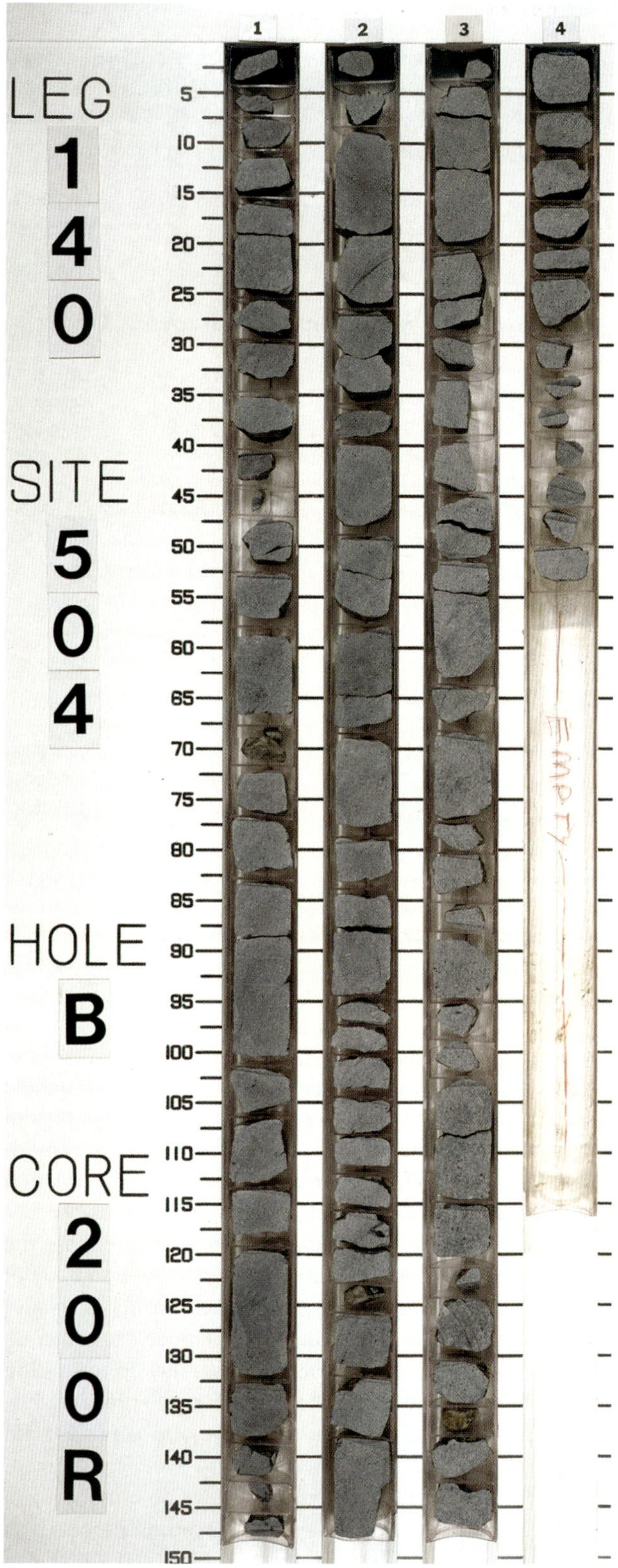

Figure 6.6 Material recovered in a 10 m long section of core in the sheeted dike complex of ODP Hole 504B. Note that fragments recovered do not necessarily fit together (Alt *et al.*, 1993). The recovery is about 50% for this section (the rest of the 10 m core was not recovered), much better than for most of the unit. This type of fragmented core material provided the first direct evidence of sheeted dikes in the oceanic crust. Scale on left is in centimeters.

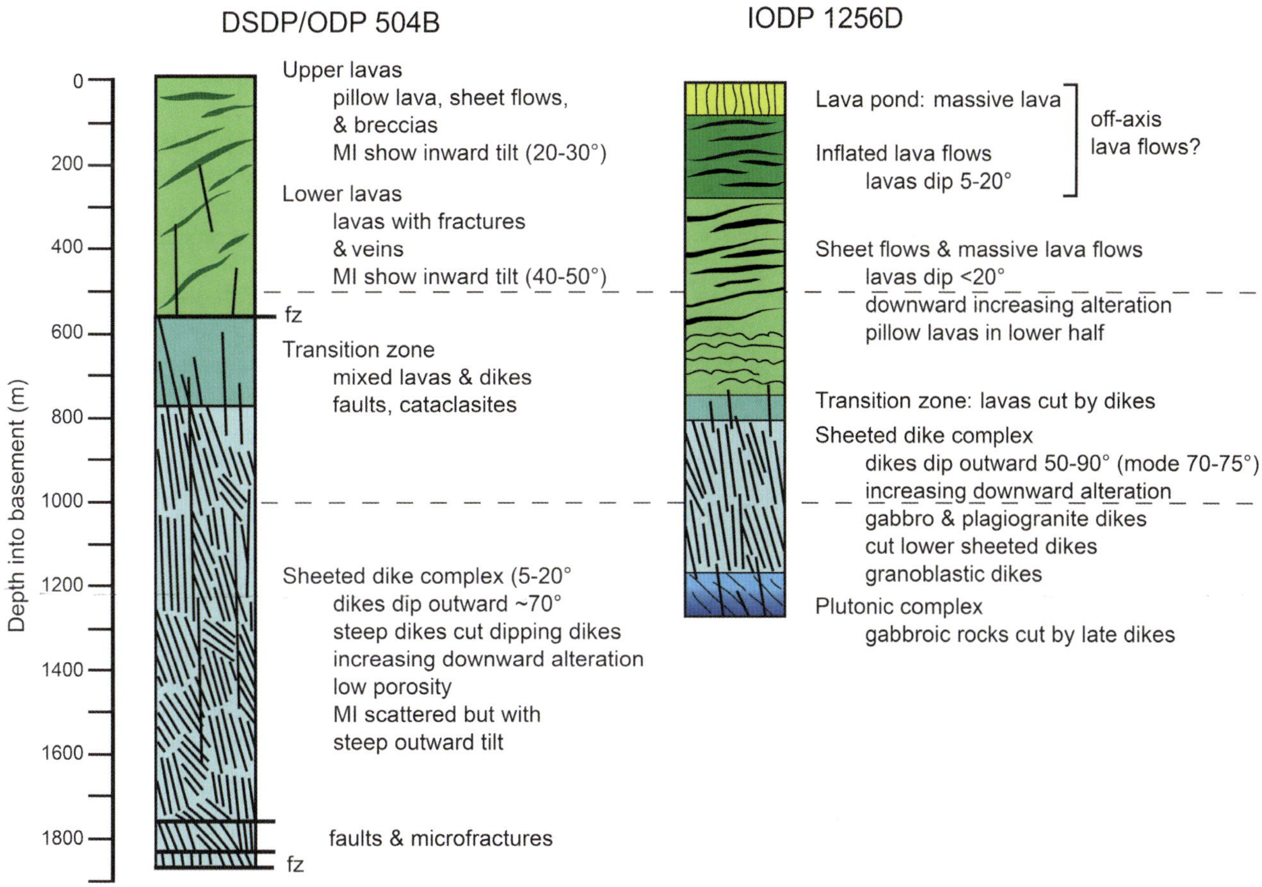

to the intrusion of the later, steeper dikes (Becker *et al.*, 1988). The later dikes may have been intruded near the axis or off-axis, but in either case, they show that the tilting of earlier dikes occurred during spreading.

IODP Hole 1256D drilled into superfast-spreading (220 mm/yr) oceanic crust in the eastern Pacific. It penetrated about 400 m of sheeted dikes and below that about 200 m of gabbroic rocks (Wilson *et al.*, 2006) (Fig. 6.7). Again, many discrete dikes were intersected, showing that they are commonly not vertical. Dips are typically 70°–75° away from the spreading center where they formed (Tartarotti *et al.*, 2006; Umino *et al.*, 2008). The sheeted dike complex at Hole 1256D is much thinner and less fractured than that drilled at Hole 504B. This difference is attributed to the very different spreading rates at which these crustal sections were generated (Wilson *et al.*, 2006).

Sheeted dike complexes have been documented in vertically and laterally continuous outcrops in tectonic windows at the Hess Deep Rift and the Pito Deep Rift. In these areas of fast-spreading crust, sheeted dike complexes are exposed nearly continuously for tens of kilometers along tectonic escarpments, corresponding to tens of thousands of years of spreading history at their respective spreading centers. Observations from widely spaced submersible dives suggest that these exposures may extend many kilometers further along the scarp as well. In these areas, HOVs and ROVs have been used to collect samples, make detailed

Figure 6.7 Vertical lithologic sequences in DSDP/ODP Hole 504B (left) and IODP Hole 1256D (right). These are the only two deep drill holes that reached the sheeted dike units of the upper oceanic crust (Alt *et al.*, 1996; Wilson *et al.*, 2006; Umino *et al.*, 2008). The thinner upper crustal sequence in Hole 1256D is believed to be related to accretion at a much higher spreading rate. Note that in both holes the sheeted dike complex is fractured and altered with dikes dipping outward (away from the spreading center where they formed). Magnetic inclination (MI) data from lavas at Hole 504B show that the lava pile has been tilted to dip inward by as much as 50° (Pariso and Johnson, 1989). Horizontal lavas at the top of the sections show that the tilting of dikes and lavas occurred during spreading.

observations, and to construct large-scale (tens of meters across) mosaics from digital still camera surveys to document critical structural relationships. In addition, structural measurements and oriented samples have been collected to provide quantitative data for evaluating the geometry and kinematics of deformation features (Fig. 6.8). Collectively, these approaches have provided important constraints on the nature of sheeted dike complexes and a context within which to consider more limited information derived from deep drill holes (Francheteau *et al.*, 1990; Karson *et al.*, 2002a, 2005).

Other exposures of dikes in crust generated at intermediate spreading rates are less extensive. At the Blanco

Figure 6.8 The Geocompass is a device for determining the orientations of geological structures on the deep seafloor from HOVs and ROVs. It provides quantitative structural data on outcrop structures and a means of collecting oriented samples for fabric and magnetic studies. Here one edge of the device is held flush against the margin of a tilted dike at the Pito Deep Rift with the manipulator on the Jason ROV. Field of view ~1 m.

Transform, separating the Juan de Fuca and Gorda Ridges, intensely fractured sheeted dikes are exposed beneath a thick (~1000 m) section of lavas (Juteau *et al.*, 1995; Tivey *et al.*, 1998; Karson *et al.*, 2002b). As in the locations described above, the dike margins dip outward. Dikes and dike swarms in lava units are also exposed at the Endeavour Deep Rift (Larson *et al.*, 2005) where the lava–dike transition zone resembles those found in other tectonic windows (see also Fig. 3.33).

The tectonic windows at Hess Deep and Pito Deep Rifts (Figs. 6.9–6.13) provide remarkable exposures of the internal structure of sheeted dike complexes as well as their upper and lower contacts with lavas and gabbroic rocks, respectively (Karson *et al.*, 2002a; Karson *et al.*, 2005). On these escarpments, individual dikes are not always clearly defined because of their complex forms and fracturing. Dikes commonly have lateral steps, bridges, horns, and related structures, suggesting that lateral dike injection as well as vertical intrusion is important in the construction of these units.

Similar to other seafloor locations, dikes in the sheeted dike complexes of these areas are not vertical. They dip outward ~70° at the Hess Deep Rift and 75–80° at the Pito Deep Rift (Fig. 6.10). In both cases, the lower the dip, the

Figure 6.9 The Pito Deep Rift is at the northward propagating tip of the fast-spreading southern EPR near the northwest corner of the Easter Microplate. Steep tectonic escarpments in two study areas (white boxes) shown in the bathymetric map (left) provide access to exposures of the upper oceanic crust. The sheeted dike unit as well as overlying lavas and underlying gabbroic rocks are well exposed and provide a window into the structure of the sheeted dike complex (right).

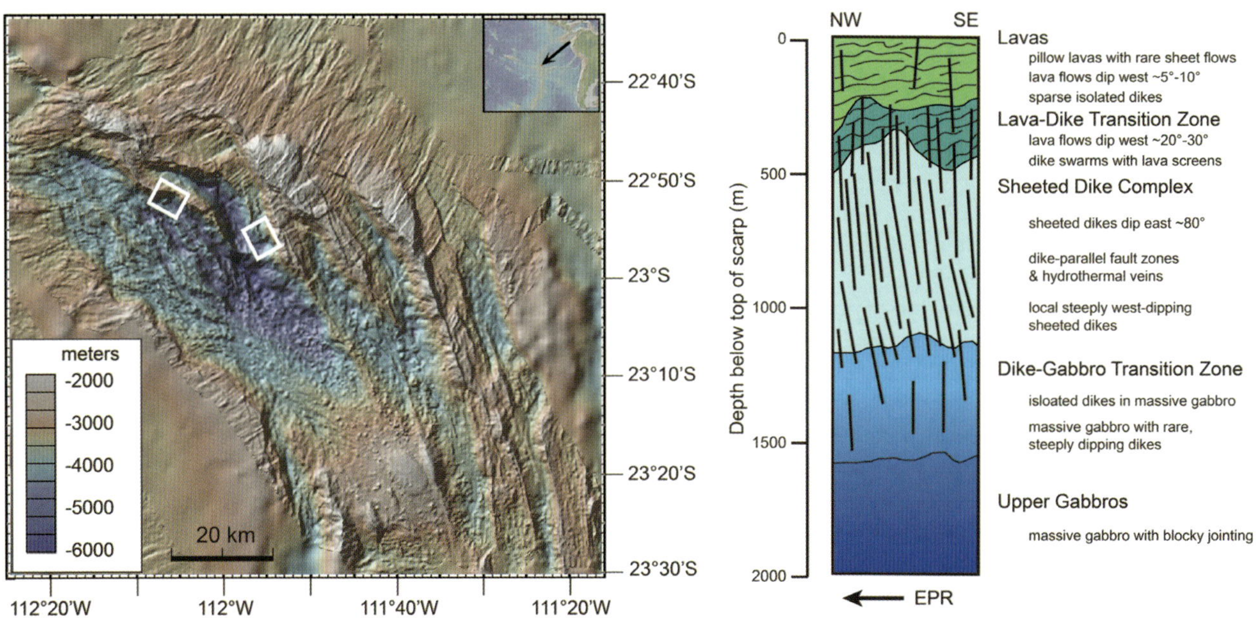

Figure 6.10 Spectacular exposures of sheeted dikes occur on near-vertical cliff faces in the Pito Deep Rift. These were formed about 3 million years ago at the southern EPR, at a spreading rate of ~142 mm/yr. The dikes, each about 1 m wide, dip moderately to steeply away from the spreading center where they formed. At this site, the EPR lies to the west (left in the image). Individual dikes are separated by coherent intrusive contacts (lines), faulted dike margins, and intensely deformed fault zones (f). The dikes are moderately fractured and jointed. This outcrop is very typical of the sheeted dike unit formed at relatively fast spreading rates. Field of view across the mosaic is ~40 m.

greater the degree of fracturing that occurs. Paleomagnetic studies show that the dikes were not intruded in these inclined orientations but instead were tectonically rotated from an original ridge-parallel, near-vertical orientation (Varga *et al.*, 2004; Horst *et al.*, 2011). In some outcrops, less fractured massive dikes cut across fractured material and across panels of tilted dikes (Fig. 6.12). Variable tilting and cross-cutting relations suggest that crustal construction involved a relatively continuous interval of spatially overlapping intrusion, fracturing, and rotation rather than two discrete episodes of dike intrusion before and after deformation.

Significant brittle deformation occurs within sheeted dike complexes. In some places, relatively intact sheeted dikes pass laterally or vertically into areas where brittle deformation is so intense that the shapes and boundaries of individual dikes are no longer visible (Fig. 6.13). Especially at the Hess Deep Rift, these types of intensely fractured swaths of fault breccia separate more coherent panels of parallel dikes tens to hundreds of meters across. Within groups of parallel dikes, fracturing and alteration is focused along dike margins (Fig. 6.14). Fracture networks and brecciated zones along dike margins serve as pathways for hydrothermal fluids, resulting in localized alteration and the

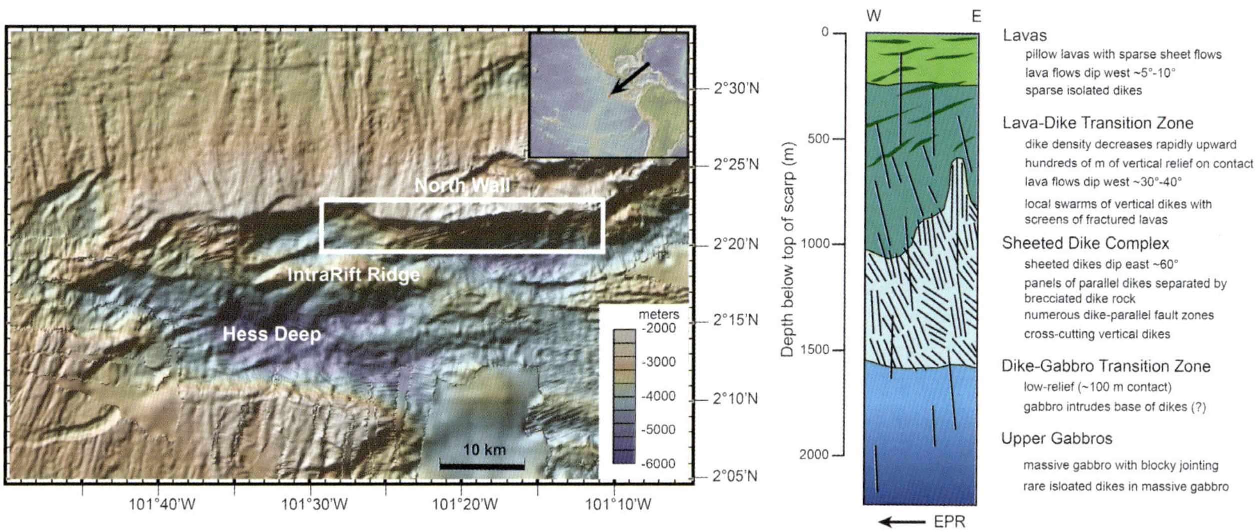

Figure 6.11 The Hess Deep Rift lies near the westward-propagating tip of the Cocos–Nazca plate boundary (left). The extensive east–west tectonic window exposes the upper crust created at the EPR that lies just to the west. Numerous HOV and ROV studies on the north wall of the rift (white box) provide details of the sheeted dike complex in this area, as summarized in the columnar section (right).

Figure 6.12 This large digital image mosaic of a vertical cliff face illustrates the complicated structure of the sheeted dike complex at the Hess Deep Rift. The upper third of the image shows a tilted panel of parallel dikes (arrows) dipping moderately away from the EPR spreading center where they formed. The dikes are moderately fractured and have dike-parallel faults and damage zones. A sinuous, steeply dipping dike in the middle lower half of the image (dashed outline) cuts the brecciated and tilted dikes, indicating that the faulting and tilting occurred during the formation of the sheeted dike complex. Field of view ~10 m (Karson *et al.*, 2002a).

deposition of mineralized veins. In some fault zones, layers of intensely deformed fault gouge, rich in clay-size material (Fig. 6.15), result in very low permeability and thus can effectively act as seals that restrict hydrothermal flow through the upper crust (Hayman and Karson, 2007).

Variably tilted and brecciated sheeted dikes are found in all drill holes and tectonic windows examined in crust formed at superfast to intermediate rates. In all cases the dikes dip away from the spreading centers where they formed (with overlying lavas dipping toward the axis). Different amounts of tilting and fracturing appear to correlate with spreading rate or local magma supply, with less fracturing and tilting at intermediate rates than at faster rates (Karson, 2002). The amount of tilting and fracturing also appear to correlate with the internal structure and thickness of the overlying lava unit (see Chapter 5).

Densely developed, ridge-parallel dikes, faults, and fissures in the upper oceanic crust create parallel zones of weakness that are exploited by spreading-related deformation. As crustal blocks are tilted, the blocks deform internally by slip on these types of surfaces. This style of deformation is referred to as "bookshelf faulting" because the resulting structure resembles the tilting of books on a shelf (Fig. 6.16). In this style of deformation many dikes are affected by slip along their margins and pre-existing faults are reactivated.

Sheeted dike complexes in ophiolites in some cases show significant deformation along dike-parallel fault zones, along with evidence of tilting during spreading (Harper, 1982; Varga, 1991). However, since the direction of tilting with respect to the spreading direction (or direction back to the spreading axis) is generally not known, it is not clear whether deformation in these ophiolites is related to inward block tilting and subaxial subsidence or back-tilting of blocks bounded by inward-dipping normal faults.

The internal structure of sheeted dike complexes has important implications for spreading processes. It is likely that dikes intruded along the axis of MORs are initially vertical and parallel to the spreading axis. This expectation is based on the general state of stress at spreading centers with the (horizontal) least compressive stress oriented normal to the spreading direction (Anderson, 1972). Thus, dikes intruded at the axis are used for reference markers as vertical, ridge-axis-parallel planes in the oceanic crust that can guide the interpretation of spreading structures in both oceanic crust and ophiolites.

However, not all dikes may be intruded in a vertical orientation. Mafic sheets resembling dikes (for example, cone sheets) are commonly intruded in non-vertical orientations near magmatic centers where the stress field is more complicated than in a simple rift setting. At both the Hess Deep and Pito Deep Rifts, weakly deformed dikes appear to have been intruded along the edges of previously tilted and fractured dikes, locally with low-angle cross-cutting relations. Thus, non-vertical dikes and mafic sheets need not have been tectonically rotated. Two lines of evidence demonstrate that the dikes have been tilted during the spreading process. First, overlying lavas, including sheet flows that must have formed in a subhorizontal orientation, dip inward. Second, paleomagnetic studies of oriented samples from both drill cores and oriented samples from tectonic windows show that the dipping dikes have been rotated to different extents from the orientations in which they were initially intruded. Paleomagnetic studies of dikes and lavas in DSDP/ODP Hole 504B also

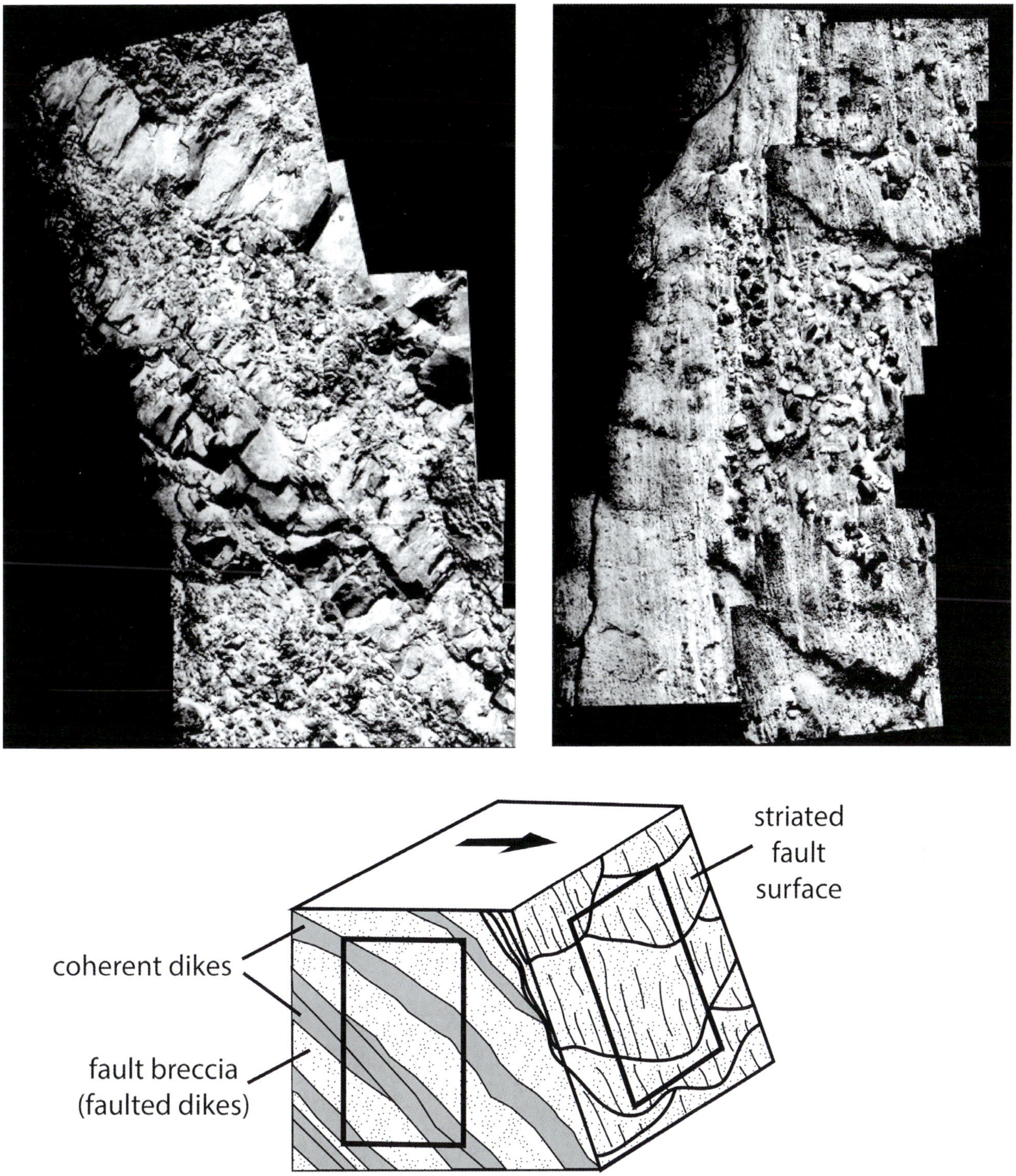

Figure 6.13 (Top left) Digital mosaic of a cliff face (view to north) showing an outcrop of massive subparallel dikes separated by intensely fractured dike rock. The dikes dip away from the EPR (located to west, left of image) where they formed. Field of view ~6 m. (Top right) View to the west of the same outcrop area showing a major fault surface, possibly an abyssal hill fault, cutting the sheeted dike unit. The fault zone is composed of shingled layers of striated fault breccia and cataclasite. The view is normal to the strike of the fault so dikes are not visible from this perspective. Field of view ~10 m. Schematic diagram (below) shows the relative orientations of left and right mosaics above. Black arrow shows spreading direction (Karson *et al.*, 2002a).

Figure 6.14 Dike-parallel fault zones in a vertical wall at the Hess Deep Rift (left) and in the Oman Ophiolite (right). Slip on zones of weakness along dike margins allows large crustal blocks to deform internally by "bookshelf faulting" in sheeted dike complexes. Field of view ~3 m (left) and ~8 m (right).

show rotations consistent with inward-dipping lavas and outward-dipping dikes. In all these cases, near-vertical cross-cutting dikes and flat-lying lavas at the top of the sections show that the tectonic rotations and fracturing occurred during crustal accretion. It is likely that this occurred immediately beneath spreading axes, but it is also possible that some of this material may have been added off-axis.

Anisotropy of magnetic susceptibility (AMS) studies on oriented samples from tectonic windows (Varga *et al.*, 2008; Horst *et al.*, 2014) and also from IODP Hole 1256D (Veloso *et al.*, 2014) show that magmatic flow in dikes from these areas involved both vertical and sub-horizontal transport. Considering the nearly continuous axial magma chamber (AMC) beneath fast-spreading ridges where these crustal sections formed, it is interesting that the magma plumbing system is not more uniform.

In all the settings discussed above, tectonic rotation of upper crustal units implies substantial (hundreds of meters) subsidence beneath the spreading axis. This is consistent with the occurrence of screens of pillow lava that erupted on the seafloor in the sheeted dike complex at crustal depths of hundreds of meters. Subaxial subsidence of the upper crustal units must be accommodated by magmatic intrusion and flow in the AMC and low-velocity zone (LVZ) regions (Fig. 6.17).

6.2.2 Dikes and sheeted dikes in slow-spreading crust

Despite extensive exposures of crust where sheeted dikes would be expected, outcrops of sheeted dike complexes in slow-spreading crust are rare. In fact, as noted in Chapter 5, a sheeted dike complex has been described in only two locations to date: the Vema Transform (Fig. 6.18) (Auzende *et al.*, 1989) and King's Trough, an abandoned rift in the eastern central Atlantic (Klitgord *et al.*, 1990). Although sheeted dikes appear to be rare in slow-spreading settings, individual basaltic dikes are commonly observed cutting lavas (Karson and Rona, 1990; Auzende *et al.*, 1994; Lawrence *et al.*, 1998). In these settings, the dikes dip steeply inward (toward the axis), consistent with back-tilting of

Figure 6.15 Fault breccia from the Pito Deep Rift has blocks of massive material surrounded by highly fractured and altered dike rock (left). Horizontal field of view ~3 m. Sample of fault gouge and fault breccia from a discrete fault zone from the Pito Deep Rift (middle). The block is bounded by smooth, curved, striated fault surfaces on either side. The block is ~20 cm long. Photomicrograph (shown in cross-polarized light) of fault breccia (right) with a clast of basaltic material (left side) in fine-grained fault gouge (dark material). Horizontal field of view ~4 mm.

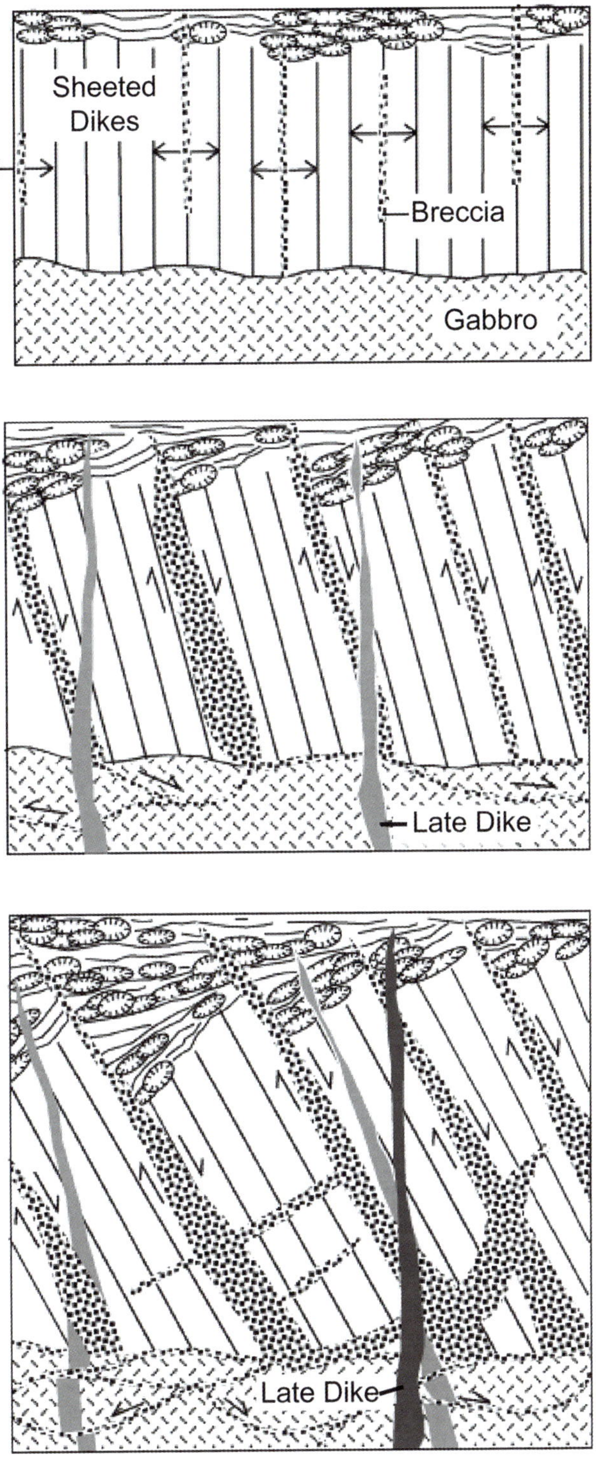

Figure 6.16 Sheeted dikes and adjacent rock units at the Hess Deep Rift show upper crustal structures with tilted dikes and lavas cut by later dikes and overlain by flat-lying lavas. Vertical dikes and horizontal lavas formed near the axis (top) are progressively tilted (middle and bottom) by slip on dike margins and spreading-related faults (bookshelf faulting) during the spreading process. Steeply dipping late dikes cut the tilted, altered and brecciated dikes and flat-lying lavas cap the section. The spreading axis is to the left in all of these sections (Varga *et al.*, 2004).

upper crustal blocks bounded by inward-dipping normal faults. Dikes intruding variably deformed and metamorphosed gabbroic and ultramafic rocks have been found in some seafloor outcrops in the rift valley walls at the MARK Area (Karson and Dick, 1983), the walls of the Kane Transform (Auzende *et al.*, 1994) and at the Atlantis Massif (Blackman *et al.*, 2002), and in drill holes (Dick *et al.*, 1991; Cannat *et al.*, 1995b; Kelemen *et al.*, 2007).

In slow-spreading environments, back-tilting of crustal blocks with dikes as well as overlying lava flows and pelagic sedimentary beds is well documented (Atwater and Mudie, 1973; Karson and Rona, 1990; Blackman *et al.*, 2001). Consequently, inward-dipping dikes appear to be associated with block rotations related to major normal faults near spreading centers. Paleomagnetic studies show that dikes cutting lavas on the median valley wall of the Mid-Atlantic Ridge at 23° N were intruded in various orientations and then tectonically rotated to different degrees (Lawrence *et al.*, 1998). Studies of some ophiolites interpret upper crustal structures as having inward-dipping (toward the axis) dikes and outward-dipping lavas, consistent with the structures described above for slow-spreading ridges (Harper, 1982; Varga, 1991).

6.2.3 Dikes and sills at sedimented MORs

In some places, sedimentary material derived from nearby continents buries MOR spreading centers. Oceanic crust generated in these settings has a distinctive upper crustal structure and thermal history, and the potential for exotic hydrothermal vents (see Chapter 4). Sedimented ridges include the Gulf of California (Einsele *et al.*, 1980; Einsele, 1982, 1985), parts of the Escanaba Trough of the Juan de Fuca system (Zierenberg *et al.*, 1993), and the Red Sea (Drake and Girdler, 1964; Lowell and Genik, 1972). Magma intruding the shallow crust in these areas commonly creates a network of steeply dipping dikes and more or less horizontal, bedding-parallel basalt sills in the sedimentary section (Fig. 6.19).

6.3 Petrology and geochemistry of sheeted dike complexes

Not unexpectedly, dikes in the oceanic crust have compositions and mineralogies similar to the overlying basaltic lavas they supplied to seafloor eruption sites from subcrustal magma chambers (Sparks, 1995; Pollock *et al.*, 2005, 2009; Sano *et al.*, 2011). Ubiquitous patchy, hydrothermal metamorphism of both the lower lavas and the sheeted dike complex complicates direct comparisons between the geochemistry of lavas and dikes in some instances. Where fresh, unaltered dikes have been sampled, however, there are some subtle differences between the composition of dikes and overlying lavas (e.g., Stewart *et al.*, 2003; Pollock *et al.*, 2009).

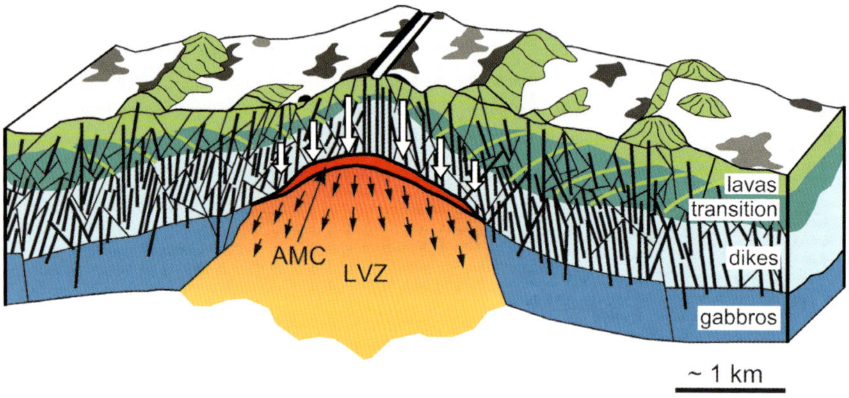

Figure 6.17 Tilting of crustal blocks in the sheeted dike complexes formed at superfast to intermediate spreading rates is related to subsidence and thickening of overlying lavas. Subaxial subsidence results in collapse and fragmentation of the sheeted dike complex and rotation of initially vertical dikes intruded beneath the axis. Deformation and subsidence in the sheeted dike complex requires mass redistribution in underlying (partially molten?) material of the AMC and underlying LVZ (after Karson, 2002).

Figure 6.18 Sheeted dikes have been described from only two very limited exposures in slow-spreading crust. One of these is the wall of the Vema Transform Fault where steeply dipping sheeted dikes crop out between lavas and gabbroic rocks (top). Field of view ~4 m. Outcrops on the eastern median valley wall in the SMARK Area (bottom), at 23° N on the MAR, include massive dikes cutting intensely brecciated lavas and dike material. The dikes dip inward toward the axis. Field of view ~2 m.

Dikes recovered from the oceanic crust are almost exclusively basaltic, medium-grained rocks comprising subequal amounts of plagioclase and pyroxene as the major mineral phases with a diabasic or subophitic texture (elongate plagioclase crystals partially enclosed by pyroxene crystals). Texturally, dikes typically are composed of holocrystalline, fine- to medium-grained material that ranges from being aphyric to containing sparse phenocrysts (Fig. 6.20).

Dikes tend to have coarser textures than lavas, as a consequence of their intrusive nature where cooling rates can be much slower than for lava extruded on the seafloor. Magma intruded as dikes into pre-existing sheeted dikes or other units can be rapidly chilled along the contacts to form microcrystalline to glassy margins (Fig. 6.3).

Thicker dikes that cool more slowly have coarser interiors, commonly with subophitic texture that characterizes many dikes (Fig. 6.19). However, only 30% of dikes from Pito Deep have subophitic texture (Pollock *et al.*, 2009). Olivine, if present, occurs as small phenocrysts in more magnesian samples but is typically altered to various low-temperature secondary minerals. The groundmass is composed of granular pyroxene (both pigeonite and augite) and iron–titanium oxides between tabular plagioclase. Rare spinel and sulfides are present in some basaltic dike rocks (Alt *et al.*, 1996; Bach *et al.*, 1996; Sano *et al.*, 2011).

Both basaltic dikes and associated lavas have been recovered at a few tectonic windows and from a few deep drilling sites. Where unaltered, dikes have major element compositions that, for the most part, overlap the composition of spatially associated lavas (Sparks, 1995; Kempton and Casey, 1997; Stewart *et al.*, 2005; Pollock *et al.*, 2009). In the Blanco Transform (Pollock *et al.*, 2005; Cordier *et al.*, 2007) the Pito Deep (Pollock *et al.*, 2009), and Hess Deep (Stewart *et al.*, 2005), dikes have a greater overall range in composition than nearby lavas (Fig. 6.21).

In some cases, for example dikes sampled by drilling in ODP holes 1256D and 504B, dikes can be – on average – relatively evolved compared to the overlying drilled lavas.

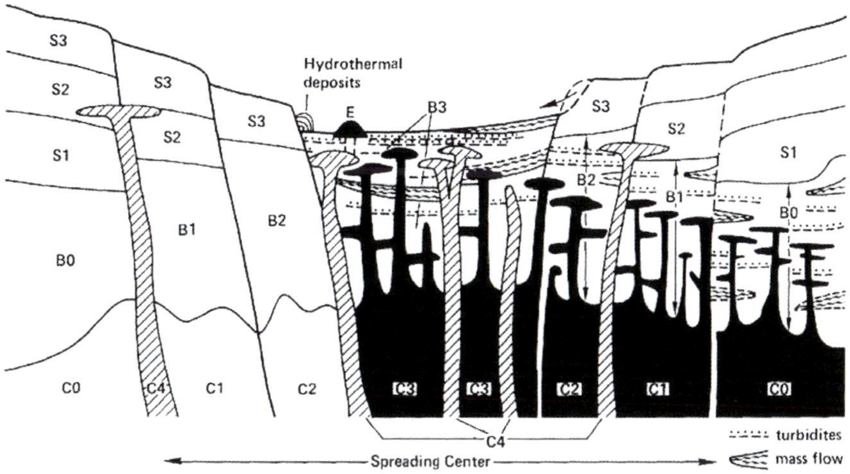

Figure 6.19 Dikes and sills intrude sedimentary rock units (S, B and C) that overlie the spreading center in the Guaymas Basin and drive hydrothermal systems. Shallow crustal intrusions in this type of sediment-covered MOR can result in distinctive hydrothermal vent systems (from Einsele, 1982).

Figure 6.20 Dike interiors can exhibit relatively coarse diabasic or subophitic textures where plagioclase (white and gray laths) is partially or wholly enclosed in clinopyroxene (green on left, gold and blue on right) as in this sample from the Mid-Cayman Spreading Center. Image on left shown in plane-polarized light and on the right in cross-polarized light. Field of view ~6 mm.

The effects of alteration/metamorphism in dikes at 504B are evident in the spread in compositions shown in Fig. 6.19. The systematic major element compositional variations in most other sheeted dike suites are consistent with low-pressure (less than 0.1 GPa) fractional crystallization similar to that proposed for many MOR lava suites (see Chapter 5).

At both Hess Deep and Pito Deep, a group of primarily low-FeO lavas has been distinguished from the main group of dikes. Together with their lower FeO, these lavas also have relatively low MgO and high Al_2O_3 and CaO (Fig. 6.20). These differences have been ascribed to plagioclase accumulation in the low-FeO lavas that may lead to the preferential eruption of these lavas due to their lower density (Stewart *et al.*, 2003; Stewart *et al.*, 2005; Pollock *et al.*, 2009; Sano *et al.*, 2011). The observation that dikes tend to have higher calculated liquid densities than overlying lavas has been used to propose that many dikes may not erupt and therefore have no volcanic equivalent. Hence the upper crust may act as a "density filter" (Sparks *et al.*, 1980; Hooft and Detrick, 1993).

Density differences might also promote horizontal magma transport along-axis, rather than simply vertical intrusion (Stewart *et al.*, 2005). In support of lateral intrusion, chemical analyses of closely spaced dikes show they can have significantly different compositions and in some cases were derived from different mantle sources (Stewart *et al.*, 2003; Pollock *et al.*, 2009). This has important implications for the association of recent lava flows and their underlying melt lenses, with lateral dike intrusion providing a mechanism for the distribution of magma along-axis (and lateral mixing of melts with different compositions). The heterogeneity in dikes from the Hess Deep and Pito Deep Rifts is also consistent with the relatively frequent (tens to thousands of years) changes in young axial lava compositions sampled along intermediate- to superfast-spreading ridges (Stewart *et al.*, 2005; Pollock *et al.*, 2009).

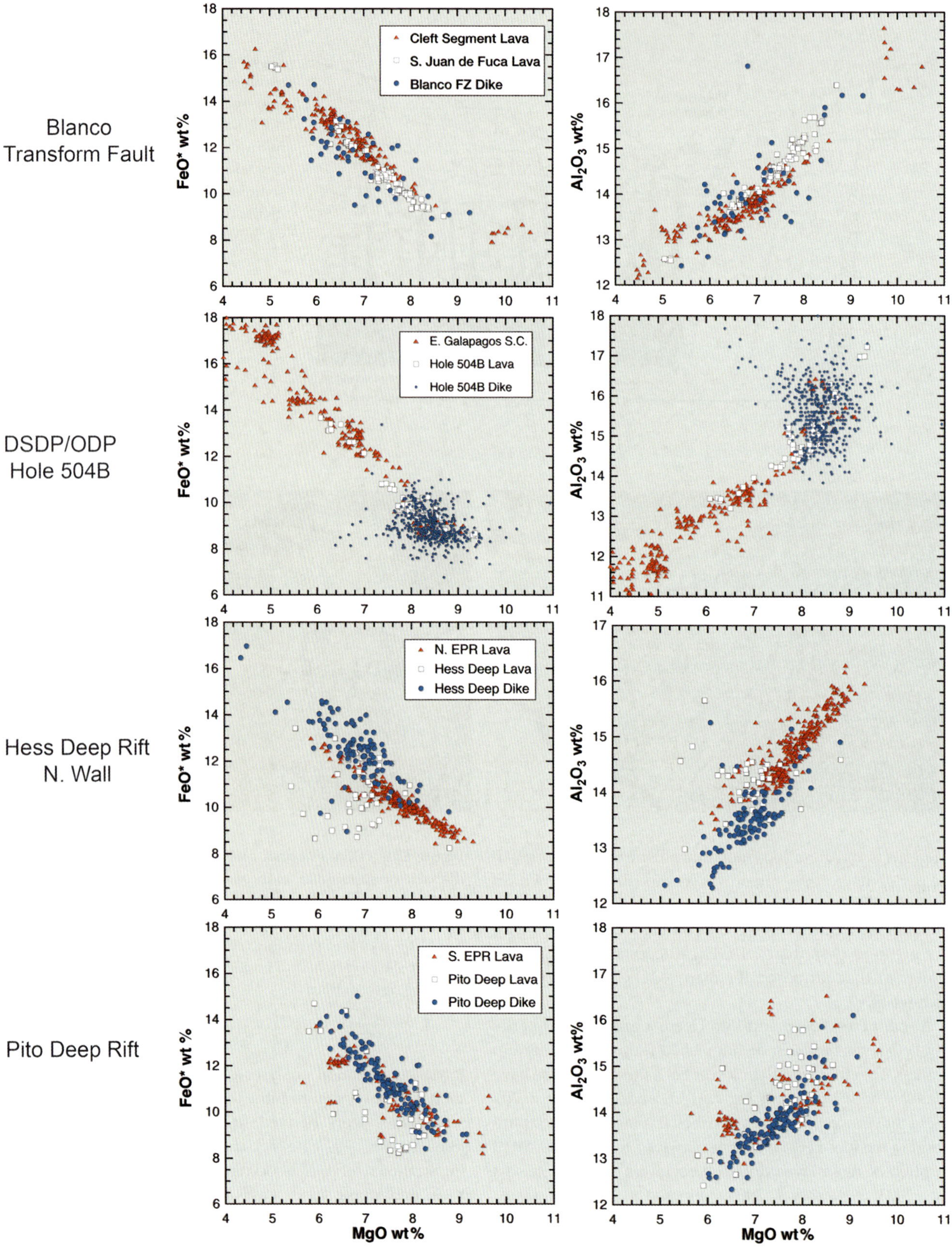

Figure 6.21 Dikes from sheeted dike complexes in the oceanic crust have compositions that are similar to those of spatially associated lavas. Subtle differences in geochemistry suggest that physical and magmatic processes in the crust may cause these distinctions. Comparisons between dikes (blue) and locally and regionally associated lavas (red and white symbols) are shown for four well-studied localities. Most of the systematic variation in FeO (left) and Al_2O_3 (right) with decreasing MgO is consistent with low-pressure fractional crystallization. The large spread in dike compositions at Hole 504B is due to variable amounts of alteration and low-grade metamorphism, but they also appear to have a more limited compositional range than lavas in the region. (Data from Alt *et al.*, 1996; Bach *et al.*, 1996; Stewart *et al.*, 2003; Pollock *et al.*, 2005; Stewart *et al.*, 2005; Wilson *et al.*, 2006; Cordier *et al.*, 2007; Pollock *et al.*, 2009; Sano *et al.*, 2011; and M. Pollock, personal communication, 2013).

6.4 Metamorphism in sheeted dikes

Deformation and metamorphism in dikes and the sheeted dike unit of the oceanic crust are closely related. Dikes are intruded into the brittle upper crust, where fracturing from thermal contraction and tectonic processes make the ingress of seawater or hydrothermal fluids possible. Fracture-controlled low-grade, hydrothermal metamorphism and veining are typical in these rock units. Given the shallow crustal level and relatively low temperatures, brittle deformation dominates; however, near the base of the sheeted dike unit, high-temperature ductile deformation may also be important.

As discussed in Chapter 4, hydrothermal circulation driven by magmatic heat from intruded dikes and deeper crystallizing magma bodies exploits faults and fractures in dike units as pathways for hot fluids to infiltrate and alter the upper crust (e.g., Lister, 1974; Mottl, 1983). Increasing temperature with depth in hydrothermal cells creates reaction zones in dikes and lavas with increasing metamorphic grade (and intensity) in which the primary mineralogies are replaced by the crystallization of secondary minerals. These mineralogic transformations and associated brecciation, veining, and sulfide mineralizations have been investigated in samples from scarps along tectonic windows (Gillis *et al.*, 2001; Heft *et al.*, 2008), cores drilled by the Ocean Drilling Program (Kelley *et al.*, 1995; Alt *et al.*, 1996, 2010), and outcrops in ophiolites (Elthon and Stern, 1978; Gillis and Robinson, 1988, 1990; Kelley *et al.*, 1992).

Alteration of the volcanic units in the upper oceanic crust typically occurs at temperatures below 100 °C, whereas metamorphism in the deepest parts of the dike complexes, near the interface with underlying plutonic rocks that crystallized in subaxial magma chambers, suggest temperatures as high as 750 °C (Manning *et al.*, 1996; Gillis and Roberts, 1999; Coogan *et al.*, 2006). Heated hydrothermal fluids exchange elements with dike rock along fine cracks and fissures, ultimately resulting in the complete recrystallization of some dikes. In these extreme cases the host rocks are enriched in sodium and magnesium (note higher MgO in dikes from Hole 504B in Fig. 6.21). Conversely, the hydrothermal fluids (350–550 °C) are typically enriched in sulfur, iron, calcium, copper, zinc, lead, and silicon (Seyfried *et al.*, 1988; Alt, 1995; Von Damm *et al.*, 1995) (see Fig. 4.6). Metamorphism of intact oceanic crust is most extensive toward the base of the sheeted dike complex. Because dynamic forces are typically not involved in this style of metamorphism the rocks can retain their original igneous textures or develop granular or even-grained textures known as "hornfels."

The mineral associations of altered material in basaltic lavas and sheeted dike complexes of oceanic and ophiolite rocks reflect the consequences of what has been termed "seafloor metamorphism," (Melson *et al.*, 1968; Miyashiro *et al.*, 1971; Spooner and Fyfe, 1973; Elthon and Stern, 1978). This type of metamorphism is distinct from those of the classic metamorphic terranes of the continents in a number of ways (see Chapter 7). First, because the oceanic crust is only 6–7 km thick, lithostatic pressures are less than ~200 MPa. These are much lower pressures than in other areas of active metamorphism (e.g., mountain belts) where crustal depths can be tens of kilometers. Unlike diverse continental rocks, the starting material (protolith) of oceanic metamorphic rocks is mafic igneous rock (basalt or gabbro). Finally, because it is a type of hydrothermal metamorphism requiring fluids penetrating through fractures to alter the rock, reactions do not completely transform the rock to a stable mineral assemblage that reflects a well-defined family of physical conditions (see Fig. 7.13). Chemical equilibrium is seldom achieved, even locally. So use of terms for metamorphic facies for oceanic rocks reflects mineralogy that may only be developed locally and be reflective of conditions that occurred over a short period of time.

At the lowest temperatures, commonly in the lower portions of the lavas and upper sections of dike complexes, greenschist assemblages prevail (Fig. 6.22). Typically, calcic plagioclase is replaced by sodic (albitic) plagioclase and olivine is recrystallized to chlorite or a chlorite–smectite mixture, and/or low-temperature amphibole (tremolite–actinolite). Pyroxene can be fairly resistant to alteration but is most commonly recrystallized to chlorite and tremolite–actinolite. Plagioclase may be replaced by prehnite, laumontite, epidote, and calcite. This assemblage has been referred to as the "actinolite facies" (Elthon and Stern, 1978). At deeper levels and higher temperatures, recrystallization produces amphibolite assemblages with calcic plagioclase, hornblende, and magnetite; titanomagnetite is replaced by titanite in most instances.

At the very highest temperatures (and pressures), most typical near the dike–gabbro transition, dikes can be completely recrystallized to fine-grained, granular metamorphic rocks known as granulites, consisting of equigranular calcic clinopyroxene (with or without orthopyroxene), intermediate composition plagioclase, magnetite, and ilmenite with or without quartz. These rocks typically have equigranular mosaic ("granoblastic") textures. Locally intense hydrothermal alteration involving large fluid fluxes can also produce veins and brecciated dike margins that contain abundant sulfide minerals such as pyrite, chalcopyrite, sphalerite, and anhydrite along with quartz (Alt *et al.*, 1996, 2010).

Faulting and fracturing in the sheeted dike complex near the spreading axis create permeable pathways for hydrothermal circulation related to hydrothermal vents on the seafloor (see Chapter 4). Stockwork systems (networks of cross-cutting and interconnected veins) provide evidence of the complex interplay between fracturing and hydrothermal flow at this level of the crust. These fractures ultimately become sealed by the deposition of hydrothermal mineral deposits and veins (Fig. 6.23).

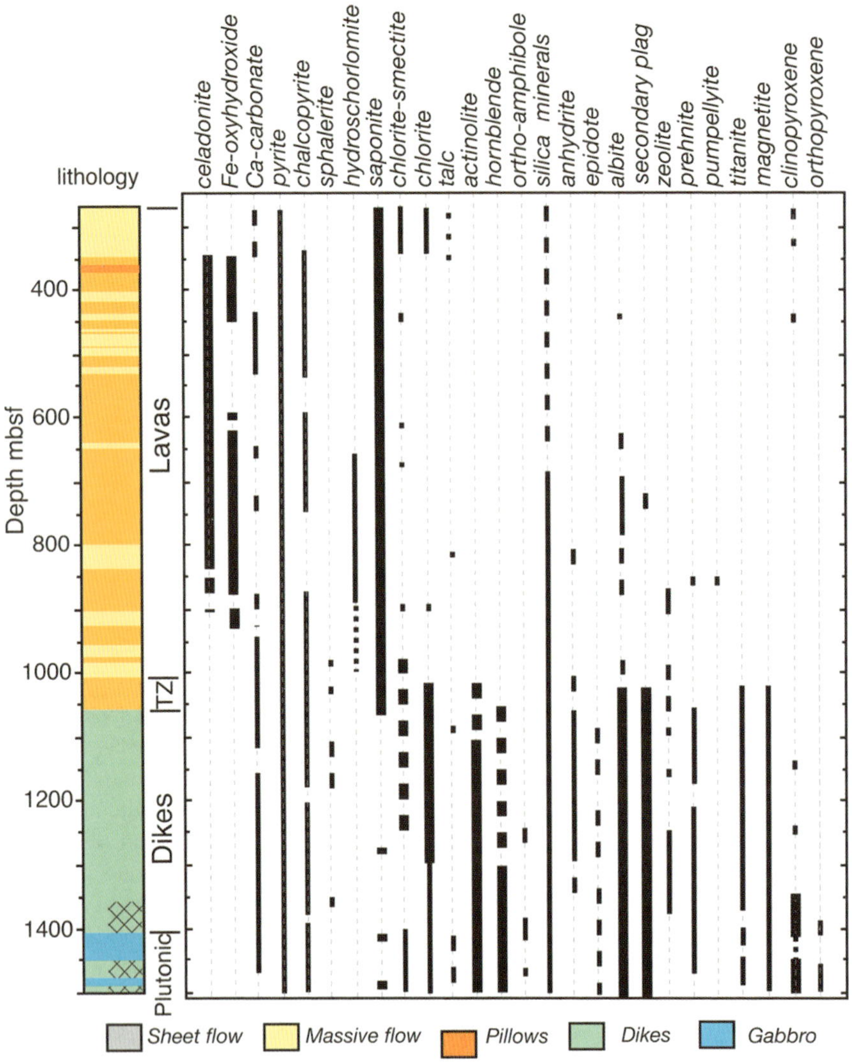

Figure 6.22 Sections of the upper oceanic crust have rarely been as completely sampled as in ODP Hole 1256D (also see Fig. 6.7). The various rock units sampled by drilling Hole 1256D are shown on the left side of the figure with the depths in meters below the seafloor (mbsf) where they were encountered. Vertical lines show the depth ranges at which secondary metamorphic minerals (listed across top of figure) are present. Major secondary mineral phases in recovered samples are shown as thick lines and minor and trace phases are shown as finer lines. Secondary minerals with higher temperature stabilities are more prevalent at depth. Crosshatching indicates where high-temperature granoblastic textures in dikes are present (from Alt *et al.*, 2010).

Figure 6.23 Faulting and alteration along dike margins at the Pito Deep Rift show moderately fractured dikes bounded by light-colored, highly altered fault breccia and fine-grained quartz veins (white). These relationships show how hydrothermal fluid flow is channeled through high-permeability damage zones such as faults and dike margins. Field of view ~3 m.

Intense alteration along fractures and fault zones is common, but locally, individual dikes have apparently channeled large-volume hydrothermal upflow. This results in the creation of individual dikes composed primarily of the metamorphic mineral epidote forming a characteristically pale green rock known as "epidosite" (Gillis and Robinson, 1990; Varga *et al.*, 1999; Gillis, 2002) (Fig. 6.24; see also Fig. 4.11).

6.5 Dikes and sheeted dike complexes on land

Overall, dikes have a tabular form and are commonly represented as simple planar features in textbooks and journal articles. However, most dikes have more complex forms, especially because of irregularities along their margins. In addition, the overall three-dimensional form of individual dikes is generally not well constrained because of their large dimensions (tens to thousands of meters) relative to even the most extensive outcrop exposures (few meters or tens of meters). Subaerial outcrops in exhumed magmatic terranes provide details of the shapes of dikes that reveal important constraints on their mode of origin. Most of these types of details have not been described in oceanic crustal dikes. This type of information could provide new constraints on subaxial magmatic plumbing systems and accretion processes at MORs. The recognition of sheeted dike complexes in ophiolites, in particular, has helped frame the current understanding of the roles of dikes and the dike intrusion processes in generating and modifying the oceanic crust.

Although the seafloor origin of ophiolites was recognized long ago (see Chapter 2), the sheeted dike complex was the last of the components to be recognized of what is now considered a complete, layered ophiolite complex (Anonymous, 1972). Early investigations of ophiolites often mapped dike complexes as parts of volcanic sequences. Perhaps the first recognition and understanding of the significance of a sheeted dike complex was by Barager (1954) working in the Betts Cove Ophiolite in Newfoundland, Canada, whose description below presaged later notions of seafloor spreading:

> *The volume which the dyke swarms occupy is extremely large but individually the intrusives are small (average about 3 feet). It does not seem reasonable that numerous small intrusives would be emplaced in a large opening if it were available at the time; rather, openings available at any one time were probably small but under the compulsion of persistent, slow-acting forces, additional openings with approximately the same orientation were, perhaps continuously being created.*

One of the first to correlate sheeted dikes in ophiolites with processes at MORs was Ian Gass (1968), based on his work in the Troodos Ophiolite. In the early 1970s, interest in ophiolites grew with the recognition of plate tectonics and seafloor spreading as a major global-scale process (see Chapter 2). Seminal papers by Moores and Vine (1971), Dewey and Bird (1971), and Church and Stevens (1971) interpreted ophiolite geology in terms of spreading processes (see Chapter 2), and sheeted dike complexes were an intuitively appealing part of this story (e.g., Robinson *et al.*, 2008).

Figure 6.24 Highly altered epidosite dikes in the sheeted dike complex of the Oman Ophiolite. The pale-green-colored dikes channeled intense hydrothermal flow in the upper crust. Surrounding dikes (gray to brown) are much less altered. Locally unaltered dikes intrude the epidosite dikes (dark gray). Field of view ~10 m.

6.5.1 General form and internal structure of dikes

Mafic dikes are easily identified in subaerial outcrops where they have intruded other rock types (Fig. 6.25). Their dark color contrasts strongly with the surrounding rocks that

Figure 6.25 Exhumed crustal sections near magmatic centers expose dikes and other shallow-level intrusions. Several basaltic dikes, each 2–3 m wide (dark gray to brown), cut light-colored granitic gneiss (left) showing dilational offsets where they intersect. Dense swarms of dikes create a tangle of intrusions on a cliff face more than 500 m high (right). Both images are from near the continent–ocean transition on the East Greenland Volcanic Rifted Margin.

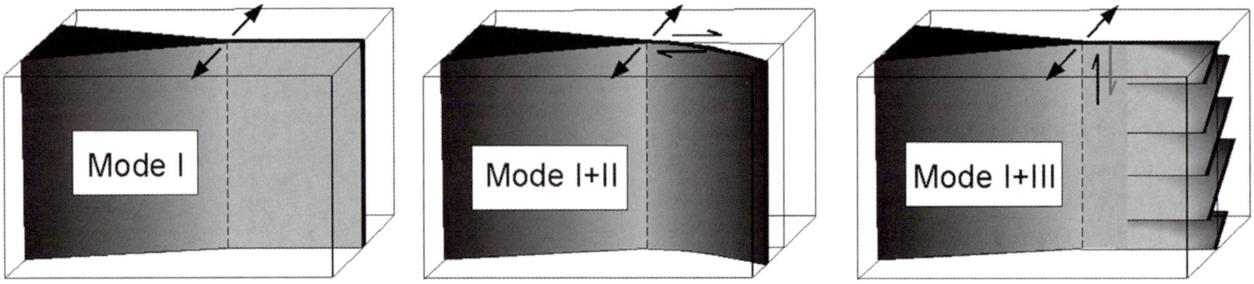

Figure 6.26 As dikes propagate (left to right in these diagrams), simple opening (Mode I, left) results in a simple planar intrusion. Dikes deviate from a planar form where shearing stresses (small arrows) occur during opening (large arrows). Shearing parallel to the propagation direction (Mode II, middle) can cause dikes to deflect from a straight propagation path. Shearing normal to the propagation direction (Mode III, right) results in a breakdown of the planar structure into *en echelon* segments. Dikes in nature commonly show all of these types of intrusion paths.

they intrude. In sheeted dike complexes, where dikes have been intruded along the margins or within the interiors of other dikes, it is more difficult to trace individual dikes and to recognize their rather subtle chilled margins in even the best outcrops.

Individual dikes commonly have subparallel margins with sharp contacts against adjacent country rocks and systematic, columnar joints at a high angle to the dike margins (e.g., Fig. 6.3). The spacing of these joints is proportional to the width of the dike, typically about one-fifth of the dike width. Margin-parallel, planar joints are common in some cases, especially near dike margins. Joints and fractures in country rocks are also parallel to dike margins and develop from tensile stresses associated with dike propagation (Pollard and Aydin, 1988).

The shapes of dikes reflect their mode of emplacement and interaction with the rock materials they intrude. These include both thermal and mechanical effects that are reflected in the rock textures, shapes, and other physical properties. Some features of dikes are the result of crack propagation processes caused by the injection of magma into the crust as a pressurized, fluid-filled crack. Other features reflect the flow of magma along the walls of the dike.

Dike intrusion in the crust is controlled by a number of physical parameters. These include the density of the magma, the magma pressure, the length of the dike, and the elastic properties of the surrounding crust. Basaltic magma is less dense than the surrounding solid basaltic crust that it intrudes, so it tends to rise buoyantly (Ryan, 1993; Rubin, 1995; Buck *et al.*, 1997). When the density of melt in a dike equals the density of the surrounding crust, the magma reaches its level of neutral buoyancy and may stall in its vertical ascent or propagate laterally (Sparks *et al.*, 1980; Hooft and Detrick, 1993; Ryan, 1993). However, magma pressure can drive an intrusion beyond this limit before it stalls because of cooling (Buck *et al.*, 1997).

Dikes are essentially large, pressurized cracks in the crust and, as such, the longer they are, the larger the stresses at their tips and the less stress is required for them to propagate horizontally or vertically. Crack propagation driven by dike intrusion can produce a range of different structures. In the simplest case, a dike propagates as a pure opening (Mode I) crack oriented at a high angle to the least compressive stress (Fig. 6.26) (Pollard and Aydin, 1988). Magma pressure allows the dike tip to propagate (vertically or laterally) ahead of the flowing magma. As they intrude laterally, dikes may encounter areas where the resistance to opening varies. In highly fractured, less resistant areas, dikes can continue to propagate. In stiffer areas, for example where there have been other recent dike intrusions, lateral propagation may be inhibited and dikes may instead intrude upward (or downward) and even erupt at the surface.

Where dikes propagate toward one another with small lateral offsets, shear stresses cause them to be deflected first away from and then toward one another (Figs. 6.27 and 6.28). The dikes may ultimately link up and can resemble lithosphere-scale cracks at overlapping spreading centers (see Chapter 3). Pre-existing fractures can also play a role in dike intrusion and the ultimate shape of a dike. The potential for fractures to open during dike intrusion is a function of the magma pressure in the dike relative to the maximum and minimum horizontal stresses (Jolly and Sanderson, 1995). Consequently, dikes show a wide range

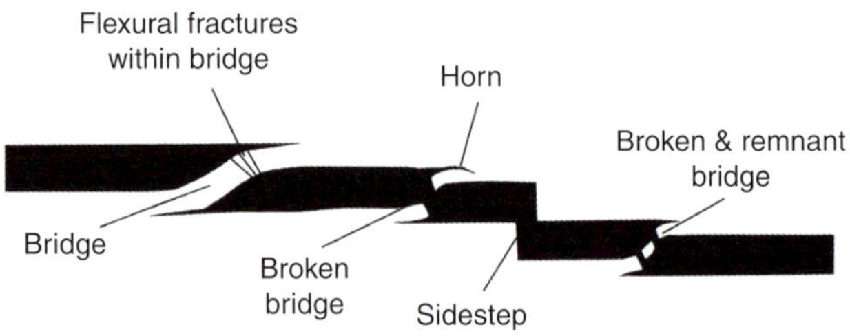

Figure 6.27 In detail, dikes are seldom simple, planar structures. Local stresses commonly result in the development of several types of irregularities and discontinuities illustrated above (after Nicholson and Pollard, 1985).

Figure 6.28 Dike segments (dark) intruding Precambrian metamorphic rocks (light) propagate toward one another resulting in slightly offset and overlapping segments with septa of country rocks (left). These mimic the plan-view arrangement of overlapping spreading centers (OSCs) (see Chapter 3) along MORs. Overlapping dike segments (right) taper near their tips and are linked by narrow bridges. Tiny horns protrude parallel to the main dike segments from terminations. Field of view of both images ~1 m.

of features that can be attributed the local stresses during dike intrusion (Figs. 6.27 and 6.28).

Many other types of contact relationships occur near dike margins as well (Fig. 6.29). Besides simple chilled margins these include irregular bulges, intruding figures or dikelets ("apophyses"), thermal erosion features, and partial melting effects. These types of features tend to make dike margins more complex and variable along-strike. For example, when magma flows through a dike, melting of the host rock can occur along dike margins, modifying the original shape of the margin and the composition of the magma in the dike.

Very fine-grained, and in some cases glassy, chilled margins form in dikes that are quenched against relatively cool host rocks in the shallow crust. However, at deeper crustal levels where dikes intrude hotter crustal material, chilled margins do not form and dikes may have very little grain-size change from interior to the margins. In oceanic crust and ophiolites, chilled margins are common where dikes intrude lavas or the sheeted dike complex. But where they intrude deeper-level gabbroic rocks they commonly lack chilled margins.

Various features found along the margins of dikes also shed light on magma transport mechanisms. For example, outcrop-scale studies of dikes commonly show distinctive striated surfaces on dike margins (Fig. 6.30) referred to as "hot slickensides," which are interpreted as the result of lateral magma transport (Varga *et al.*, 1998). These structures form as relatively cool, high-viscosity magma in the dike flows against the bounding host rocks. The resulting asymmetrical structures formed by shearing along the dike margin include striations, resembling those formed on fault surfaces. Other small-scale structures that indicate magmatic flow directions include aligned vesicles and crystals that were entrained in the magma (Fig. 6.30).

In addition, the alignment and clustering of magnetic particles in the cooling magma near dike margins can be detected and measured as the anisotropy of magnetic susceptibility (AMS) (Tauxe *et al.*, 1998; Canon-Tapia and Aubourg, 2004) (Fig. 6.31). Oriented samples from ophiolite or seafloor outcrops can be used to determine the magma transport direction in dikes using this technique (Staudigel *et al.*, 1992; Horst *et al.*, 2006; Varga *et al.*, 2008). The magnetization of dikes can also be used to determine whether dikes have been tectonically rotated after intrusion and to quantify that rotation. Both paleomagnetic rotation and AMS studies have been done on dikes in many different tectonic settings, including ophiolites (e.g., Staudigel *et al.*, 1992) and recently, on the seafloor (Varga *et al.*, 2008; Horst *et al.*, 2014; Veloso *et al.*, 2014).

In general, the interiors of dikes are massive and fine- to medium-grained. In some cases they include sparse or clustered crystals that may have been transported from a magma chamber. Banding produced by concentrations of crystals, vesicles, or entrained xenoliths develops in some dikes, parallel to dike margins.

6.5.2 Internal structure of dike swarms and sheeted dike complexes

Because dikes can transport magma both vertically and laterally, a sheeted dike complex in any cross section of oceanic crust can include interfingered dikes that originated in magma bodies at different crustal depths and possibly different locations along a spreading axis. In exposures of

Figure 6.29 Dike margins show a wide range of variations including simple planar chilled margins (top). In this case the dark dike (lower part of image) is chilled against the light-colored interior of another dike just above. Numerous tiny offshoots (apophyses) from a dike (middle) intrude granitic gneiss (light material above). Both images from the East Greenland Volcanic Rifted Margin. Irregular shapes indicating partial melting of granitic host rock (light) in South Africa (bottom).

the crust away from the MOR axis, off-axis magma bodies might also contribute dikes to this complex. Therefore the sheeted dike unit may be a composite of dikes intruded from different magma bodies at different depths and over a period of perhaps a few thousands years, making it extremely difficult to determine the precise provenance of magmas that contributed to constructing the unit.

Access to ophiolite complexes and other subaerial exposures in deeply eroded or glaciated volcanic terranes has provided important details of dike swarms. Dike swarms in Hawaii (Walker, 1986), Iceland (Gudmundsson, 1983; Helgason and Zentilli, 1985), the East Greenland Volcanic Rifted Margin (Klausen and Larsen, 2002), and the British Tertiary Igneous Province (Jolly and Sanderson, 1995) provide possible analogs for MORs. Studies in these areas show that dike swarms are constructed from multiple interfingered intrusions fed from a localized source such as a magma chamber or central volcano and that systematic variations in the dike assemblage occur along-strike (Fig. 6.32).

Near the magmatic centers, a variety of concentric and radial shallow-level intrusions form (Fig. 6.33). These intrusions are associated with the local stress field and inflation/deflation events (Anderson, 1972; Walker, 1993; Paquet *et al.*, 2007). In some areas, low-angle intrusions can be very densely developed similar to a sheeted dike complex. The geometry of these mafic sheets can be used to identify the locations of magma chambers (Walker, 1993; Siler and Karson, 2009; Burchardt *et al.*, 2011, 2013). These types of structures have not yet been identified in oceanic crust.

Along MOR spreading centers, magma chambers can be nearly continuous beneath the axis (e.g., EPR) or highly localized (e.g., MAR). Discrete central volcanoes are only found within the rift valleys of some relatively slow spreading MOR segments, commonly near where hot spots intersect ridge axes, for example the ultraslow-spreading Gakkel Ridge (Pontbriand *et al.*, 2012), Lucky Strike Seamount on the northern MAR (Humphris *et al.*, 2002), or Axial Seamount on the Juan de Fuca Ridge (Clague *et al.*, 2013).

Despite the ambiguities surrounding the provenance and reconstruction of structural features in ophiolites, the general processes that produce sheeted dike complexes are considered to be good analogs for spreading at MORs. Therefore, sheeted dikes in ophiolites and related terranes provide important glimpses of the details of these rock units (Figs. 6.34 and 6.35) that can guide future seafloor investigations.

In subaerial outcrops, weathering and grain-size variations help accentuate the margins of individual dikes. Detailed examinations of subaerial outcrops show that the internal structure and mode of construction of a sheeted dike complex can be very complicated (Fig. 6.35). For example, where young dikes intrude the interior of pre-existing dikes, the older dikes appear to have only one chilled margin. It is also common for younger dikes to intrude along pre-existing dike margins resulting in side-by-side chilled margins. Low-angle cross-cutting contacts occur in some places, creating the impression of a tangle of intrusive sheets (Figs. 6.24, 6.25, 6.35, and 6.36).

Figure 6.30 A basaltic dike (~1 m wide) cutting siltstones (upper left) in a steep cliff in Iceland has particularly well exposed margins. The lateral offset (arrow) suggests lateral intrusion. Fine, subhorizontal striations create "hot slickensides" on the margin of the same dike (upper right) consistent with horizontal magma transport. Well-defined "hot slickensides" on the margin of a dike in the Troodos Ophiolite (lower left) indicate inclined magma transport. Stretched vesicles along a dike margin in the Troodos Ophiolite (lower right) show low-angle magma transport. Part of black pen in lower right is ~5 cm long (lower two images from Varga *et al.*, 1998).

As in successions of lava flows or sills, columnar cooling joints may develop in a swarm of dikes resulting in joint patterns that transect multiple dike margins. Collectively, these types of complications make it difficult to unambiguously define the widths of individual dikes even in subaerial outcrops. The older dikes in a sheeted dike complex are especially problematic because of the difficulty in identifying both of their margins where separated by intervening dikes. Taking these uncertainties into consideration, it appears that most individual dikes

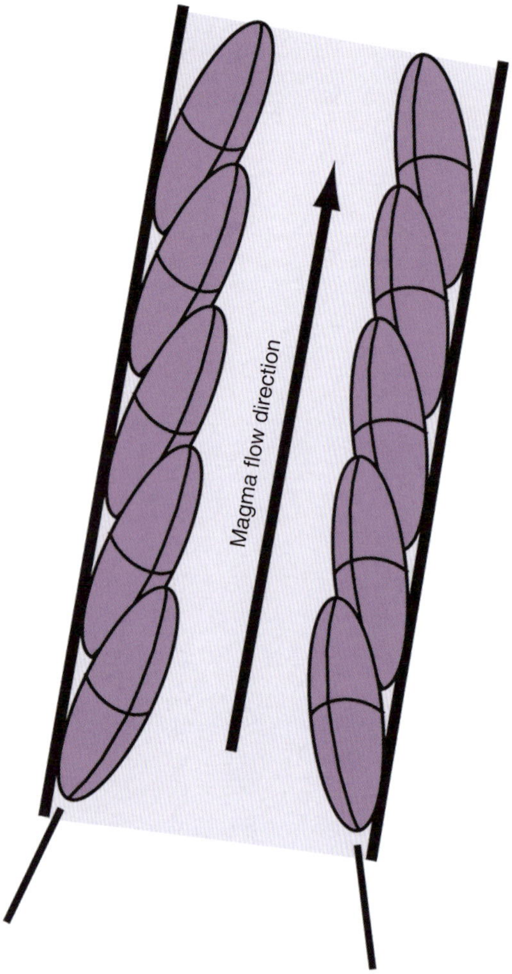

Figure 6.31 Anisotropy of magnetic susceptibility (AMS) reflects the alignment of elongated magnetic particles (represented here by ellipsoids) in dikes and other igneous bodies. In dikes, shearing of high-viscosity magma along dike margins commonly results in asymmetrical patterns that can be used to show the magma transport direction (after Tauxe *et al.*, 1998).

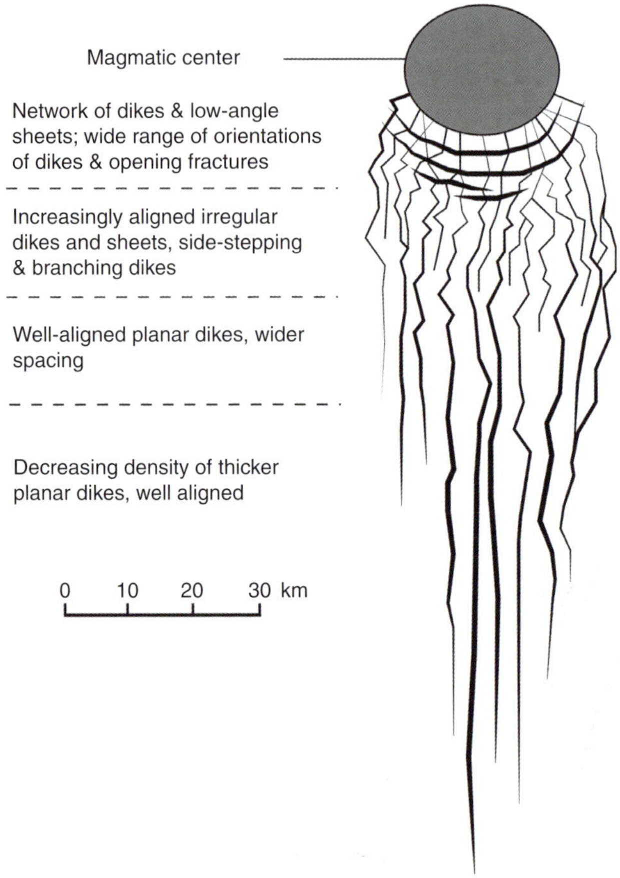

Figure 6.32 Exhumed dike swarms in continental magmatic systems show systematic along-strike variations in map view with respect to intrusive centers where they originate. Near intrusive centers, highly irregular dikes and intrusive sheets with various orientations (e.g., cone sheets) create a complex network of intrusions. Farther from the center, dikes are more planar and parallel and better aligned with local rift zones or fissure swarms. They are also more widely spaced and wider (after Jolly and Sanderson, 1995).

intruded into oceanic crust fall within the range of 0.5 to 1.5 m wide.

6.5.3 Contacts: linking the lava and gabbroic units

The upper and lower contacts of sheeted dike complexes are not clearly defined in either ophiolites or seafloor outcrops because dikes extend into both of the bounding units. Both upper and lower contacts appear to be "gradational" in the sense that the proportion of dikes decreases both upward into overlying lava units and, to a lesser extent, downward into underlying gabbroic rocks. Thus the vertical dimension of the sheeted dike unit increases during spreading, as dikes are injected both upward and downward during magmatic construction. The nature of these contacts helps to show how this process works. Although parts of these transition zones are visible in some seafloor escarpments, most of the important geological details are displayed in ophiolite complexes.

In a lava–dike transition zone, the upper limit of the sheeted dike complex consists of lavas cut by varying proportions of dikes (Fig. 6.37). Continuous exposures in both ophiolites and seafloor tectonic windows show that the transition zone is not a clearly defined, mappable rock unit. It is highly variable in terms of its internal structure, proportion of dikes and lavas, metamorphism, thickness, and relation to surrounding rock units.

In many places this transition zone, with 10–90% dikes, ranges from 100 to 700 m in thickness. The upper contact is defined in terms of a downward increase in the proportion of dikes relative to lavas. The density of fractures also commonly increases sharply in this interval. The proportion of dikes in this transition zone can vary rapidly laterally, with sharp contacts between sheeted dikes and overlying lower lava units with a very low proportion of dikes. High-level dike swarms (tens of meters across) protrude upward hundreds of meters into the lower lavas and in some places extend to within 100–200 m of the top of the lava unit.

Dikes cutting the lower lavas of oceanic crust show that lavas erupted on the seafloor must have subsided before being cut by later dikes feeding younger, overlying flows. The depth of the deepest lavas in the transition zone is a measure of the amount of subsidence that the lavas experienced during magmatic construction. In the few places where this has been documented, the subsidence ranges from about 300 m to as much as 1000 m. Such a dramatic subsidence of axial lavas is not reflected in the surface morphology, but might be masked by the rapid infilling of any depressions by younger lavas, especially at MORs with robust magma budgets.

Figure 6.33 Near magmatic centers, low-angle mafic sheets are common at shallow crustal levels. Stacks of sills (top) and inclined cone sheets are major components of central volcanoes in Iceland and other exhumed magmatic systems. Complex intersecting dikes and sills are common at shallow levels in the Troodos Ophiolite (bottom). Dikes and mafic sheets shown here are all one to a few meters thick.

Figure 6.34 Sheeted dikes (each 2–4 m wide) are exposed in a 30 m high cliff face on the East Greenland Volcanic Rifted Margin (top). Individual dikes can be distinguished by slightly different colors and by their patterns of columnar joints that terminate at dike margins. Sheeted dikes in ophiolite complexes are commonly rather poorly exposed and can be affected by post-spreading faulting and fracturing, as in the Bay of Islands Ophiolite Complex (middle). Three thin dikes (dark) are chilled throughout and cut the wider, lighter-colored dikes with coarser grain sizes in the Macquarie Island Ophiolite (bottom). Dikes that appear to have only one chilled margin (for example at the end of the hammer handle) show that later dikes intruded the interior of older ones.

Details of the lava–dike transition zone may provide important information on the width of the subaxial dike intrusion zone. For the theoretical end-member where dike intrusion is limited to a single dike width (each dike splits the previous one), the dike–lava transition zone would be a sharp contact with lavas directly overlying sheeted dikes. For a wider dike intrusion zone, dikes would intrude the lower lavas that had moved slightly off-axis to produce a lava–dike transition zone. The vertical distribution of dikes in the transition zone can be interpreted in terms of the shape and dimensions of the axial intrusion zone (Cann, 1974; Kidd, 1977). Available estimates of dike density variations from drill holes and tectonic windows in this transition zone (similar to that observed in ophiolites) suggest a very narrow dike intrusion zone, less than a kilometer wide, at least at fast-spreading MORs. This agrees well with the locus of recent eruptive and tectonic fissures at ridge axes, the width of the AMC reflector, and the locus of high-temperature hydrothermal venting along the spreading centers (Haymon *et al.*, 1991; Delaney *et al.*, 1992; Fornari *et al.*, 1998). However, the current paucity of systematic and thorough off-axis investigations limits our understanding of magmatism and hydrothermal processes away from MOR ridge axes.

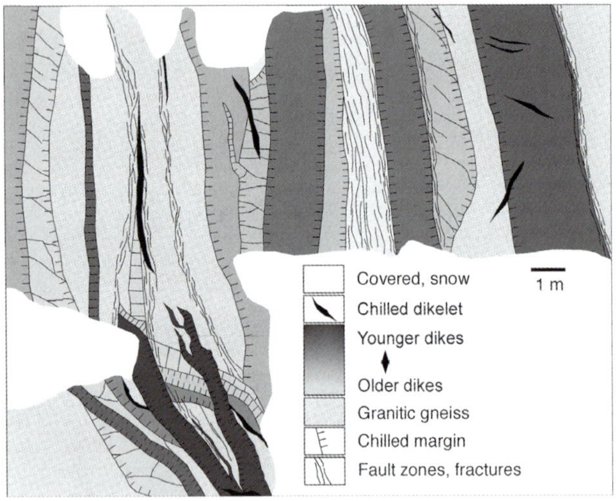

Figure 6.35 Sheeted dike complexes on land reveal complex intrusive relationships. In this map-view sketch of an outcrop on the East Greenland Volcanic Rifted Margin, dikes intrude along dike margins, split the interiors of dikes, and cross-cut one another at low angles. Slight differences in color and texture as well as traces of chilled margins make it possible to determine the sequence of intrusions.

Direct observations of gabbro–dike transition zones in seafloor tectonic windows, along with drilling in oceanic crust formed at relatively fast spreading rates, provide some constraints on the nature of the lower contact of sheeted dike complexes, but the details of this horizon remain unclear. Based on ophiolite analogs, at least three different types of relationships, or combinations of them, are likely to occur. Each of these must reconcile the following key observations found in both drill cores and tectonic windows: (1) The overlying upper crustal units have experienced intense fracturing and block rotations related to the subsidence and thickening of the overlying lava unit (see Chapter 5). (2) The uppermost gabbroic rocks are much less deformed than the overlying sheeted dikes. (3) The boundary appears to have relatively low vertical relief, less than 100 m, as it is traced across major escarpments.

Ophiolites and seafloor outcrops show diverse features at the gabbro–dike transition zone. These have been interpreted in a number of different ways, with implications for spreading processes (Fig. 6.38). In some ophiolites, notably some parts of the Troodos Ophiolite (Varga, 1991; Hurst *et al.*, 1994), the gabbro/sheeted dike contact is marked by

Figure 6.36 Dikes in sheeted dike complexes are not always parallel. Low-angle, cross-cutting relationships are common and may indicate changes in the local stress field or tilting during spreading as in an example from the Oman Ophiolite (left) and the Pito Deep Rift (right). Both of these are vertical outcrop faces showing steeply dipping dikes 1–1.5 m wide.

Figure 6.37 The upper contact of the sheeted dike complex is typically marked by an upward decreasing proportion of dikes cutting pillow lavas. In this example from the Troodos Ophiolite, individual near-vertical dikes and swarms of dikes (brownish) cut pillow lavas (gray). Dikes shown here are 1–2 m wide.

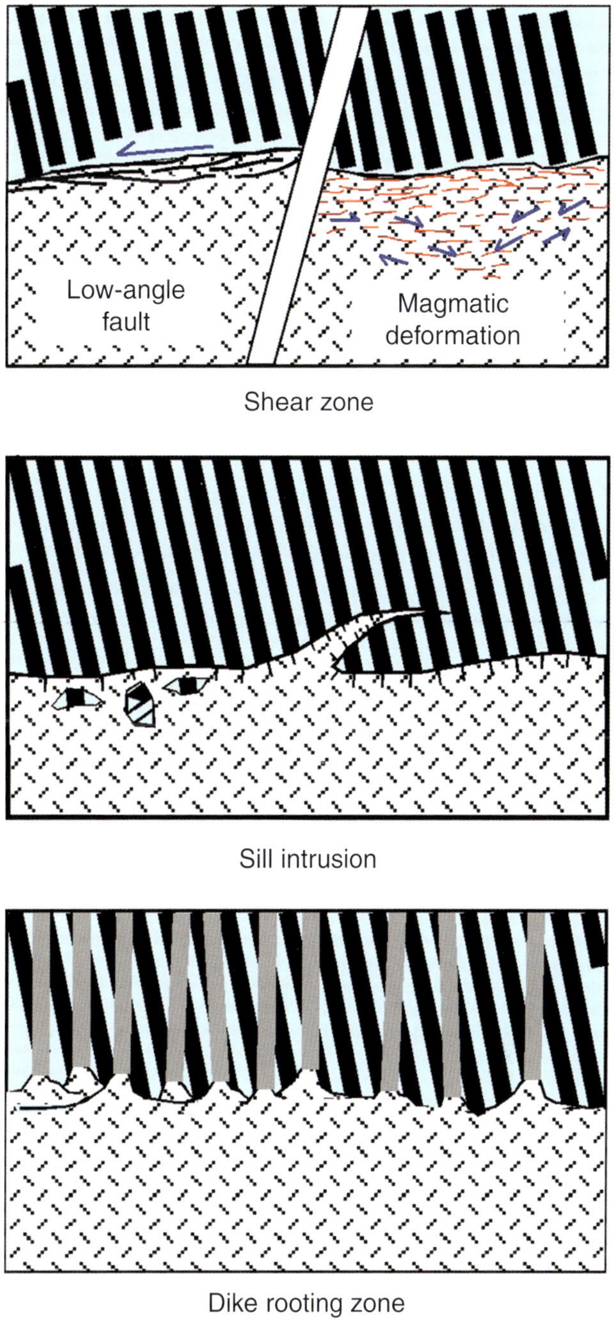

Figure 6.38 Observations from ophiolites and seafloor escarpments provide evidence of different geological relationships at the sheeted dike–gabbro contact. An abrupt change in the style of deformation from tilting and brittle deformation in the sheeted dikes to much less fractured gabbroic rocks below might be accommodated by a low-angle fault zone or by viscous deformation of underlying magmatic material (top). Sills or other volumes of gabbroic material may intrude along the base of the fractured sheeted dikes assimilating altered crustal material, creating xenoliths of dike rock, and producing contact metamorphic effects (middle). Dikes may "root" in the underlying gabbro unit where gabbroic material funnels upward into narrow dike conduits (bottom).

a major detachment fault and elsewhere by a more diffuse zone of intense fracturing (Agar and Klitgord, 1995). From this perspective, the contact appears to be related to brittle deformation. Alternatively, it is possible that brittle deformation and accretion of the upper crustal units is accommodated in the underlying gabbroic material by magmatic deformation (viscous flow). Near-bottom investigations along this contact in tectonic windows have not found evidence of faulting or localized magmatic deformation (Karson, 2002).

Gabbroic intrusions and intrusive contacts along the base of sheeted dike complexes are found in several ophiolite complexes (e.g., Gillis and Roberts, 1999). Commonly this contact zone is marked by gabbro to trondhjemite (or plagiogranite) net vein breccias and greenschist to granulite facies metamorphism (see Chapter 7). Evidence of stoping (magma that has worked its way up through existing rock), focused alteration, and cross-cutting dikes and veins are found at this level within ophiolites (Pallister and Hopson, 1981; Nicolas, 1989) as well as at some seafloor exposures (Coogan *et al.*, 2001, 2003).

Medium-grained, granoblastic metabasaltic rocks (hornfels) recovered by deep drilling suggest high-temperature contact metamorphism occurs at the contact (Gillis, 2008). The chemistry and textures of the uppermost gabbros in IODP Site 1256 suggest they represent an evolving fossilized melt lens that fed the overlying dikes and lavas (see Chapter 7) and that the granoblastic rocks represent relics of stoped and partially assimilated sheeted dikes (Koepke *et al.*, 2011).

In many parts of the Oman Ophiolite, the contact is relatively abrupt, occurring over a few tens of meters, and interpreted as a sharp boundary with underlying intrusive magma lenses (Boudier and Nicolas, 2011). However, in other parts of the Oman Ophiolite, this contact is described as "gradational," with coarse-grained gabbroic material funneling upward into localized, relatively fine-grained dikes (Fig. 6.38) (MacLeod and Yaouancq, 2000). The "varitextured gabbro" unit in the gabbro–dike transition zone has been likened to those rocks drilled at IODP Site 1256 (Koepke *et al.*, 2011). This sort of "rooting zone" might develop near the top of an AMC lens as a result of periodic hydrofracture of the overlying upper crustal lid.

6.6 Dike intrusion along spreading centers

The vertical partitioning of oceanic crust into discrete units of basaltic lavas with sparse dikes and a sheeted dike complex implies that most dikes do not reach the surface. Dike–lava relationships in exhumed cross sections of Icelandic crust show similar relationships (Gudmundsson, 1983, 1990). A consequence of subsurface intrusion of dikes is local horizontal extension and subsidence resulting from fissuring or faulting in overlying material.

As dikes intrude the brittle upper crust they wedge it apart. Dike intrusion creates a distinctive pattern of subsidence directly above the dike, and uplift to either side caused by horizontal compression as the dike is intruded (e.g., Rubin and Pollard, 1988). As crust in the subsurface is pushed apart by an intruding dike, the area above it must also be pushed apart. In this way, dikes that do not reach the surface cause extension in the crust above, resulting in normal faulting or fissuring immediately above the dike (Fig. 6.39). Dike intrusion can also cause slip on faults and fracturing in surrounding areas. The amount of horizontal extension in the crust above a dike can be related to its width and the depth of the dike tip beneath the surface (Rubin and Pollard, 1988; Rubin, 1992; Chadwick and Embley, 1998).

If the intrusion of a dike (or swarm of dikes) causes an amount of subsidence approximately equal to the dike width (Fig. 6.39), the formation of a sheeted dike complex across a 1 km wide intrusion zone would result in about 1000 m of subsidence. The subsided area could incrementally fill with lavas to build the upper oceanic crust (Karson, 2002). In other words, the subsidence required to accommodate the construction of the lava unit could simply be the result of subsurface dike intrusion.

With the identification of recent basaltic eruptions and related dike intrusion events on the seafloor (see Chapters 4 and 5), detailed investigations of the axial regions of spreading centers have been conducted (Fornari *et al.*, 1998; Embley *et al.*, 2000; Soule *et al.*, 2007; Fornari *et al.*, 2012). These document the changes in the structure of the axial summit trough and hydrothermal vents caused by dike intrusion (Embley *et al.*, 1995; Chadwick *et al.*, 1998, 2013; Fornari *et al.*, 1998, 2004; Soule *et al.*, 2007).

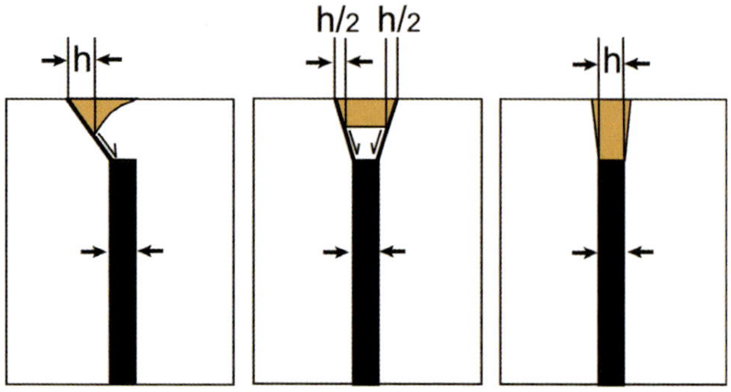

Figure 6.39 Where dikes do not reach the surface, extension equal to the dike width must also occur in the overlying crust and can result in localized subsidence (gold). Subsurface dikes may create half-graben, full-graben, or fissures (left to right) where the heave (h) of fault(s) approximately equals the dike width (black). The elastic properties of the crust and collapse of surface materials can decrease the surface effects.

Dike intrusion along MORs has a number of important consequences for the surrounding crust (see also Chapter 4). The intrusion of individual dikes or swarms of dikes during magmatic rifting events relieves the (least principal) compressive stress oriented at a high angle to a rift zone (Buck *et al.*, 2006). Tensional stresses ahead of propagating dike tips and compressive stresses adjacent to dikes can affect faulting and fissuring over distances of kilometers from the dike margin (Rubin, 1992; Curewitz and Karson, 1998).

As dikes propagate laterally, there are predictable changes in the mechanical and thermal effects on the surrounding crust (Fig. 6.40). Horizontal stresses related to dike intrusion scale with dike height, so that a 1000 m high dike could generate stresses sufficient to affect volcanism, faulting, and hydrothermal venting across the entire width of the axial summit region of the EPR (hundreds of meters). Dikes with greater vertical dimensions, originating in deeper magma sources, for example in the lower crust or upper mantle of slower spreading ridges, could affect the entire central portion of the rift valley (several kilometers wide) at slow-spreading ridges (Curewitz and Karson, 1998).

6.6.1 Dike intrusion events in spreading environments

If spreading were accommodated fully by dike intrusion, the frequency of dike intrusion events could be estimated directly from the spreading rate. At the highest spreading rates on Earth along the southern EPR (150 mm/yr) the average frequency of dike intrusion would be one meter-wide dike about every 5 years. At the slower rates of the MAR (20 mm/yr), a meter-wide dike would be expected about every 50 years. However, the frequency of historical dike intrusion events on MORs (see Chapter 5) shows that dike intrusion events are episodic, with bursts of extension

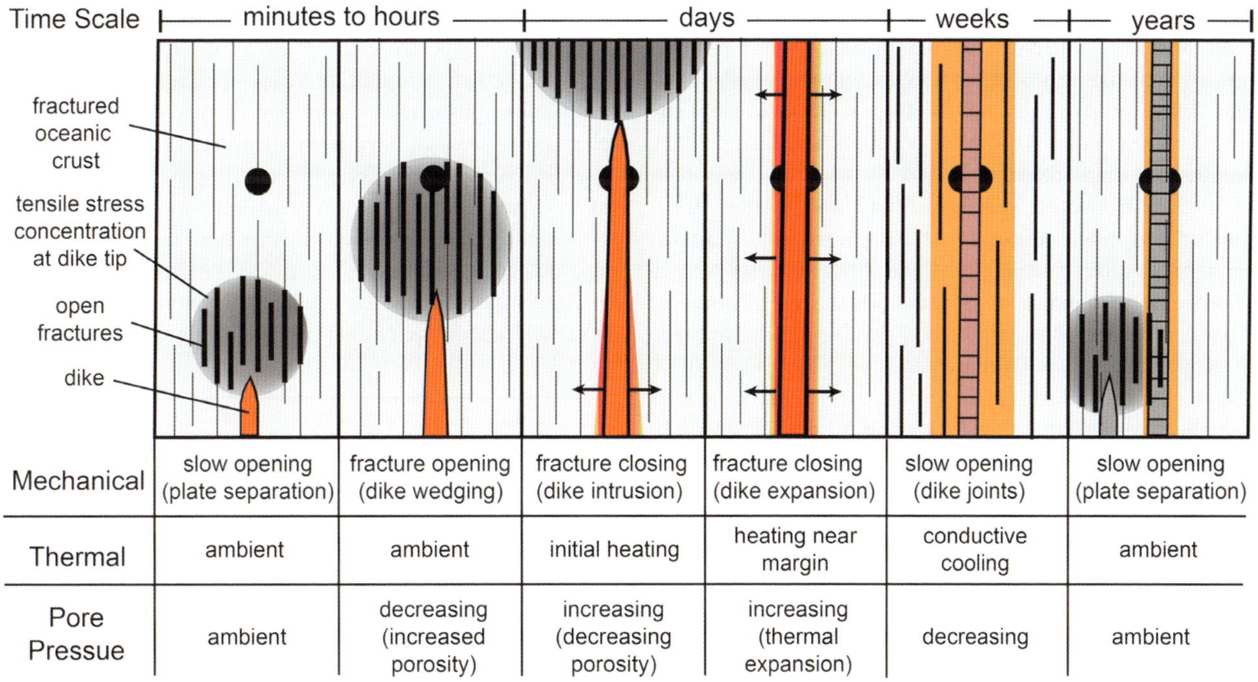

Figure 6.40 Dike intrusion (vertically or laterally) beneath a spreading axis is likely to have complicated mechanical, thermal, and pore pressure effects as the dike (red) approaches, reaches, and passes (panels left to right) a reference point in the crust (black dot). Active fracturing (bold lines) occurs in a region of elevated tensile stresses (dark gray) that moves ahead of the dike tip, resulting in locally increasing porosity and decreasing pore pressure. Fractures are closed by compressional stresses as the dike is intruded. Thermal effects (orange) are limited to narrow areas near the dike margin. Thermal expansion of crustal fluids and closing of pore space results in increasing pore pressure that may be relieved by opening of thermal contraction cracks. Later dike intrusion may repeat this cycle of processes (after Curewitz and Karson, 1998).

and magmatic construction that may match spreading rates only when averaged over hundreds or thousands of years.

Dike intrusion events on the seafloor have been documented in association with a number of spreading events on the MOR system (Dziak *et al.*, 1995, 2009, 2012c; Fox *et al.*, 1995, 2001; Hooft *et al.*, 2010; Rubin *et al.*, 2012) (see Fig. 5.48). Some of these events and their effects on volcanic and hydrothermal processes are described in Chapters 4 and 5. However, many details of dike intrusion events are best documented in subaerial magmatic rifts and these provide important analogs for processes along MORs (Einarsson and Brandsdóttir, 1980; Rubin and Pollard, 1988; Wright *et al.*, 2006).

Well-documented, historic dike intrusion events centered on the Krafla Central Volcano in Iceland show that multiple dikes were intruded for tens of kilometers along rift zones both to the north and south of this center (Fig. 6.41) (Einarsson and Brandsdóttir, 1980; Buck *et al.*, 2006; Hartley and Thordarsson, 2013). Field and petrological studies of eruptions related to the Askja Central Volcano suggest similar relationships with dominantly along-strike, lateral intrusion of dikes from a magmatic center (Fig. 6.42) (Hartley and Thordarsson, 2013). Magmatic activity in the Afar region of the East African Rift shows similar behavior with 13 dike intrusion events over six years (Fig. 6.43). Lateral dike intrusions in Afar extend for tens of kilometers along the rift zones causing widespread faulting and fissuring in the nearby crust (Wright *et al.*, 2006; Grandin *et al.*, 2010).

These studies emphasize the lateral intrusion of dikes, especially away from magmatic centers along rift zones. As such, they are probably good analogs for processes along MORs with discrete magmatic centers such as those present on relatively slow spreading ridges. At faster-spreading ridges, shallow AMCs are imaged almost continuously along spreading centers and the proportion of vertical to lateral magma transport is not known, although current concepts anticipate a greater proportion of vertical magma transport at these MORs.

6.7 Accretion of the upper oceanic crust from sheeted dikes

Limits on the dimensions of dikes intruded along MORs can be estimated using several different lines of evidence. First, the measured widths of dikes in ophiolites and seafloor outcrops are typically about 1 m. If MOR dikes are fed by the seismically imaged AMC, then they should be about 1–3 km in height, the typical depth of the AMC below the axial seafloor at fast-spreading MORs. Reflectors interpreted as magma bodies down to Moho depths might indicate deeper sources. Calculated pressures of equilibration of melt inclusions in crystals in some lavas also suggest dikes could come from as deep as ~7 km below the seafloor at the base of the crust (Wanless and Shaw, 2012). Finally, although dikes must have a vertical component of magma transport, seismic activity and related documented eruptions confirm that dike intrusion events in subaerial settings and along MORs can extend for tens of kilometers from magmatic sources. Thus, dikes in oceanic crust could have aspect ratios comparable to a knife blade or strip of ribbon with dimensions of 1:1000 to 1:10,000 (Fig. 6.44). Vertical intrusion of dikes from shallow magma chambers (especially at faster-spreading MORs) is probably an important process, but in some cases dikes may distribute magma along spreading axes over lateral distances that are much greater than the dimensions of seismically imaged AMC segments (Carbotte *et al.*, 2012; 2013) or individual lava flows (Soule *et al.*, 2007).

Although dike intrusion and the formation of sheeted dike complexes have long been considered to be important aspects of seafloor spreading, rapidly accumulating data from tectonic windows, crustal drilling, seafloor dike intrusion events, and subaerial analogs have begun to provide new perspectives on these processes. Instead of a simple, vertical array of planar intrusions, sheeted dike complexes appear to be composed of very complex assemblages of dikes that have been intruded both vertically and laterally during spreading events with periodicities of tens to thousands of years. Individual intrusion events have a profound impact on crustal construction, deformation, and hydrothermal processes.

At fast-spreading ridges, dike intrusion appears to be primarily a shallow crustal process in which magma from

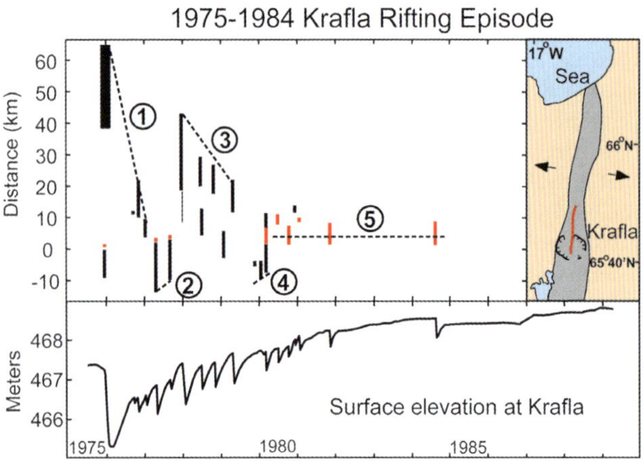

Figure 6.41 The distribution of seismic activity, surface observations, and inflation records at the Krafla Central Volcano in the northern rift zone of Iceland (upper right) show that dikes intruded for tens of kilometers along the rift zone both north and south of the volcanic center and its subaxial magma chamber. Intrusion events and eruptions correlate well with sudden deflations of Krafla (lower graph). Eruptions (red) occurred during only a few of the dike intrusion events, mostly close to Krafla and mostly after tectonic stresses were relieved by dike intrusion events (from Buck *et al.*, 2006).

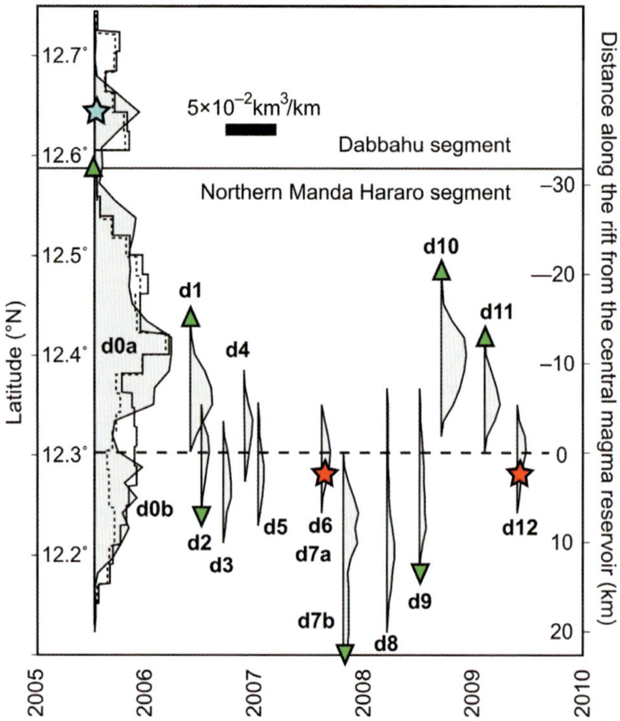

Figure 6.43 Summary of dike intrusion events in the Afar Rift Zone, East Africa, showing the lateral extent and volumes (gray areas show dike volumes per unit distance relative to black bar) of dike intrusion for 13 events (5–10 km depth) between 2006 and 2010 based on analysis of synthetic aperture radar data. Dikes intruded for tens of kilometers both to the north and south from the inferred magmatic center (dashed line) and into two different rift segments. Green triangles show the direction of migration of seismicity associated with intrusion events. Stars indicate eruptions (red, basalt; cyan, rhyolite) (Grandin *et al.*, 2010).

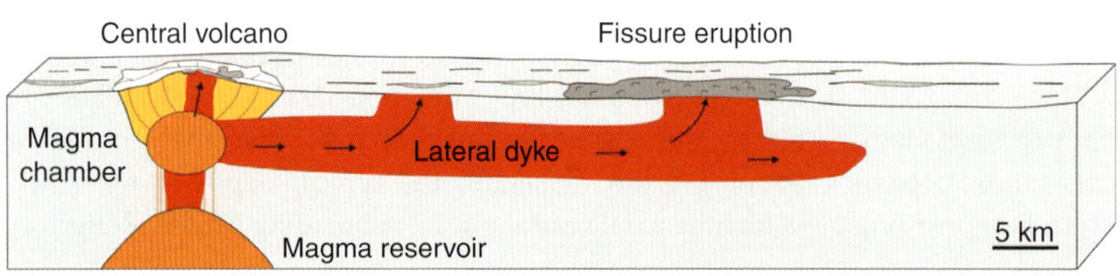

Figure 6.42 Schematic diagram summarizing the events of the 1874–1876 Askja volcano-tectonic episode. Magmas erupted at Askja Central Volcano in Iceland, and nearby fissures during this episode, evolved independently of each other in separate magma storage regions and arrived at the surface via separate plumbing systems (Hartley and Thordarsson, 2013).

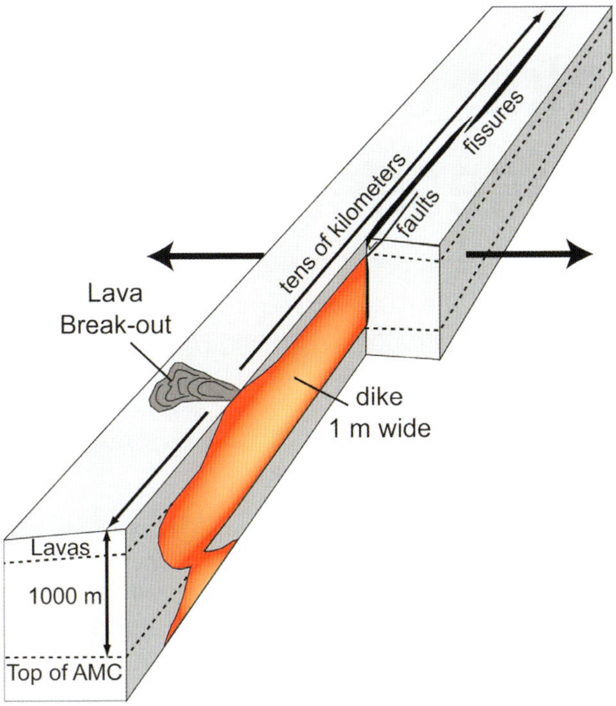

Figure 6.44 Schematic diagram showing the possible dimensions of dikes in oceanic crust. Dikes have aspect ratios similar to a ribbon or knife blade and can distribute magma and heat for tens of kilometers along spreading centers.

shallow AMCs is injected into the overlying crust, and periodically feeds surface lava flows. The sheeted dike complex is a crustal interval of rapid petrologic transitions linking the coarse-grained, slowly crystallized, gabbroic rocks at depth to fine-grained, rapidly cooled lavas at the seafloor. Metamorphic effects also increase rapidly downward as the top of the AMC and magmatic temperatures are approached. It is also a horizon where changes in physical properties result in rapid downward increases in seismic velocity structure. The internal structure of sheeted dike complexes consistently shows evidence of fracturing, crustal block rotation, and hydrothermal alteration during spreading that can be related to subaxial subsidence and thickening of the overlying lavas. The dikes dip moderately outward away from the spreading center where they formed (Fig. 6.45).

In contrast, at slower-spreading ridges, much less information is available and it is not clear to what extent sheeted dike complexes are developed. Sheeted dikes and dike swarms are likely to be present near magmatic centers or in areas of robust magmatism (high magma budget, like hot-spot influenced areas of the MAR such as Lucky Strike Seamount) but are clearly absent from areas of tectonic disruption or low magma budget, for example OCCs, transform faults and non-transform offsets. Dikes may have much deeper sources and therefore intrusions may affect much wider areas along MORs. Where observed, dikes dip inward, toward the spreading center, where they have been tilted by block rotations related to rifting (Fig. 6.45).

Sheeted dike complexes are a distinctive geological expression of seafloor spreading. Progress in unraveling these complex units will come from continued investigations of sheeted dikes *in situ* in the oceanic crust, especially taking advantage of tectonic windows in different spreading rate environments and using closely spaced drill holes coupled to studies in ophiolites and other analogous terranes. Monitoring dike intrusion events along MORs, through the use of cabled observatories (see Chapter 1) will also provide valuable information on dike intrusion processes that are fundamental to crustal construction and seafloor spreading.

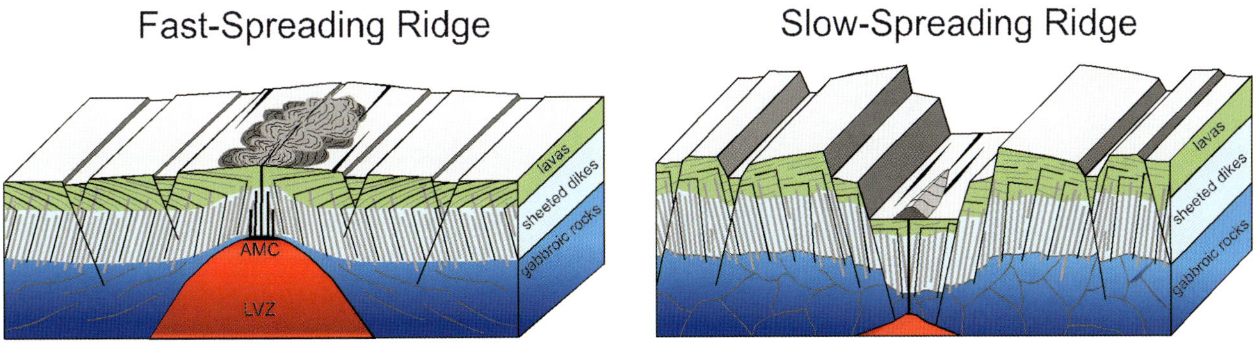

Figure 6.45 Where the magma budget is high, as in most parts of the EPR and other magmatic rifts, subaxial subsidence accommodates the thickening of the lava units (left). Subsidence is greatest near the axis and diminishes away from the axis, resulting in a pattern of dikes dipping "outward" or away from the axis and lavas dipping progressively more steeply "inward" with depth (after Pockalny and Larson, 2003). Where the magma budget is low, or variable, as on slow-spreading ridges (right), inward-dipping normal faults result in variable tilting of crustal blocks with inward-dipping dikes and outward-dipping lavas.

7 Gabbroic rocks: clues to magmatic processes

All the various kinds of granites which constitute the plutonic family are supposed to be of igneous or aqueo-igneous origin, and to have formed under great pressure, at considerable depth within the earth, or sometimes, perhaps, under a certain weight of incumbent ocean.

Sir Charles Lyell (1871)

7.1 The foundation of the oceanic crust

Mostly hidden beneath the carapace of basaltic lavas and sheeted dikes, the middle to lower portions of the oceanic crust are composed of a complex and varied assemblage of coarse-grained, gabbroic and ultramafic rocks (also known as "plutonic" rocks) (Fig. 7.1). These materials formed by slow cooling and crystallization of melts stored in magma chambers (mixtures of crystals and melt) located at depths of <7 km below the seafloor. At mid-ocean ridges (MORs) magma chambers are imaged using multichannel seismic techniques (see Chapter 1) as discrete lenses of melt and a

Figure 7.1 Coarse-grained gabbroic rocks make up most of the middle to lower oceanic crust. The structure and composition of this unit are critical to understanding magmatic processes and accretion at MOR spreading centers. Complex structures such as the cross-cutting volumes of layered gabbro from the Oman Ophiolite shown here reveal relationships at a scale and resolution that is not yet possible on the seafloor.

larger low-velocity zone (LVZ) in the crust beneath the ridge axis (see Chapter 3). Unlike crystal-poor lavas on the seafloor, these plutonic rocks are made up of crystalline minerals that reflect the physical and chemical conditions under which they formed. The compositions and textures of these crystalline rocks record histories that provide more detailed information about MOR magmatic and deformational processes than can be deduced from lavas and dikes. Mafic and ultramafic plutonic rocks constitute the bulk of the oceanic crust (>60% by volume). Despite their importance in the accretion of oceanic lithosphere, the magmatic and dynamic processes that create the plutonic lower crust largely remain a mystery because of the nature of exposures on the seafloor and the limited ability to perform detailed investigations of their structure and composition on the scale of meters to hundreds of meters.

The gabbroic and ultramafic rocks formed at the MOR are generally assumed to overlie upper mantle peridotites that have been depleted of their basaltic components by partial melting (see Chapter 8). It is unlikely that any of the primary mantle melts erupt on the seafloor, but rather undergo some extent of crystallization and crystal–melt interaction in axial magma chambers (AMCs) (Sinton and Detrick, 1992). Melt-rich parts of magma chambers form thin lenses concentrated near the base of the overlying sheeted dike complex (Chapter 6) and less commonly at deeper crustal levels near the spreading axis. Magma lenses in the crust supply melt to dikes, which in turn feed lavas that erupt on the seafloor.

The thickness of the plutonic assemblage, inferred from marine seismic studies and ophiolite complexes, is typically 4–5 km (see Chapter 2). Unfortunately, this part of the oceanic crust is largely hidden from view, with only limited exposures present at tectonic windows in oceanic crust and outcrops in ophiolites. On the seafloor, plutonic rocks are only exposed where major faulting has removed the overlying upper crustal units (Karson, 1998). Only the uppermost portions of the plutonic section, away from tectonic windows, have been penetrated by crustal drilling (Wilson *et al.*, 2006). The limited extent of seafloor exposures and difficulties in studying them *in situ* restrict the determination of internal structure of plutonic rock units and the processes by which they formed.

Despite this limited perspective from the oceans, it is clear that the gabbroic and related rocks from the seafloor are the most compositionally diverse and structurally complex part of oceanic crust. Ophiolite complexes provide an important perspective on the plutonic portion of the oceanic crust, but uncertainties in correlating ophiolites and oceanic crust formed at MOR hinder detailed comparisons. Given these constraints it should not be surprising that the middle to lower oceanic crust is also the most poorly understood section of the oceanic lithosphere.

Seismic velocities of gabbroic rocks are correlated with seismic Layer 3 of marine seismic refraction studies. Layer 3 is very continuous along the strike of the MOR in crust formed at fast spreading rates but highly variable in thickness and velocity structure for slower rate crust (see Chapter 3). These observations suggest that magmatic construction is a rather continuous process at faster spreading rates (or higher magma budgets) but, in contrast, spreading at lower rates or lower magma budgets creates a much more complicated and discontinuous middle to lower crustal assemblage. Thus, from a global perspective, the processes that construct the middle to lower oceanic crust are highly variable in time and space, and crustal structures generated under different tectonic and magmatic conditions may be quite distinct and not directly comparable.

The following sections provide an overview of gabbroic and related plutonic igneous rocks that formed during seafloor spreading under two end-member conditions: (1) relatively fast spreading MORs, with a more or less continuous magma supply; and (2) slow-spreading MORs with low, or discontinuous magma supplies. After reviewing the available information on gabbroic and related rocks from the seafloor, these concepts are placed within a framework of broader scale geological relationships and processes inferred from ophiolite complexes.

Some of most important progress in understanding the plutonic section of oceanic crust has come from a cross-pollination of ideas derived from studies of the seafloor

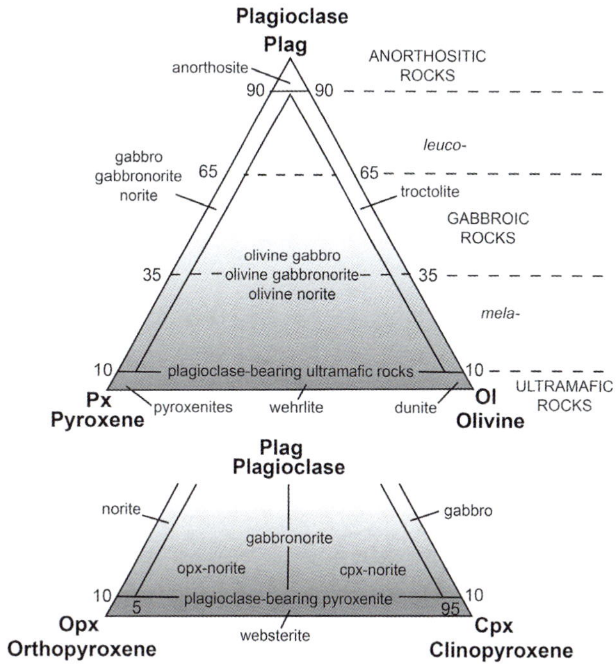

Figure 7.2 Classification of gabbroic and ultramafic rocks based on the modal proportions of the major phases plagioclase (Plag), olivine (Ol), and pyroxene (Px) including orthopyroxene (Opx) and clinopyroxene (Cpx). "Leuco-" and "melano-" are the prefixes given to light- and dark-colored gabbros determined by the proportions of plagioclase relative to mafic (dark-colored) minerals (after Streckeisen, 1976).

Table 7.1 Seafloor exposures of gabbroic and related rocks (see Fig. 3.3 for key locations)

Location	Full rate (mm/yr)	Type of exposure	Main rock types	Notes	References
Ultraslow to slow spreading rates					
Southwest Indian Ridge	15	OCC, drill core ODP 735B	Gabbro, ol-gabbro, troctolite, oxide-gabbro trondhjemite, metagabbro	Variable and locally intense synmagmatic deformation in >1 km drill hole	Dick *et al.*, 1990, 2000
Mid-Cayman Spreading Center	20	Rift valley walls, OCC	Gabbro, ol-gabbro, oxide-gabbro, metagabbro	Variable and locally intense hydrous deformation	Perfit, 1977; CAYTROUGH, 1979; Stroup and Fox, 1981; Elthon, 1987; Hayman *et al.*, 2011
FAMOUS Area	21	Drill core DSDP 334	Gabbro		Hodges and Papike, 1976
TAG Area	25	Rift valley wall	Gabbro		Tiezzi and Scott, 1980; Zonenshain *et al.*, 1989
Atlantis Massif Central Dome	24	OCC; drill core IODP UI309D	Gabbro, ol-gabbro, oxide-gabbro, wehrlite	Variable assemblage of gabbroic rocks in 1.4 km drill hole	Shipboard Scientific Party, 2005; Godard *et al.*, 2009
Atlantis Massif South Wall	24	OCC; sections on transform valley wall	Gabbro, oxide-gabbro	Small sills and plutons in peridotite	Miyashiro *et al.*, 1971; Blackman *et al.*, 2002; Schroeder and John, 2004; Boschi *et al.*, 2006
Atlantis Fracture Zone	24	Transform valley walls, OCC sections	Gabbro, ol-gabbro, metagabbro		Miyashiro *et al.*, 1971
MARK Area ridge–transform intersection	25	OCC on rift wall, drill cores ODP 921–924	Gabbro, ol-gabbro, gabbronorite, oxide gabbro	Gabbro, oxide gabbro, metagabbro	Karson and Dick, 1983; Mével *et al.*, 1991; Cannat *et al.*, 1995
MARK Area off-axis	25	OCC		Isolated gabbroic bodies in peridotite	Dick *et al.*, 2008
Kane Transform Fault	25	OCC; cross sections on transform wall	Gabbro, ol-gabbro, oxide-gabbro	Small plutons in peridotite	Auzende *et al.*, 1994
15° N MAR	22	OCC and rift walls	Gabbro	Small (100 m scale?) gabbro plutons in serpentinized peridotite	Cannat *et al.*, 1997; McLeod *et al.*, 2002; Kelemen *et al.*, 2007a
Vema Transform Fault	24	Transform and fracture zone valley walls	Gabbro	Between sheeted dikes and peridotites; fault contacts	Auzende *et al.*, 1989; Bonatti *et al.*, 2003
Romanche Fracture Zone		Fracture zone valley walls			Bonatti and Honnorez, 1976
Gorringe Bank	~25	Transpressional uplift of OCC	Gabbro	Between sheeted dikes and peridotites; later intrusions	Auzende *et al.*, 1978; Girardeau *et al.*, 1998
Intermediate to superfast spreading rates					
Mathematician Ridge	86	Abandoned spreading center	Gabbro, troctolite, ferro-gabbro	Dredged gabbros, peridotites and basalts	Vanko and Batiza, 1982; Batiza and Vanko, 1985

Table 7.1 (*cont.*)

Location	Full rate (mm/yr)	Type of exposure	Main rock types	Notes	References
Hess Deep Rift upper rift valley walls	110	Propagating rift wall	Gabbro, gabbronorite, ol-gabbro	Under regionally continuous sheeted dikes and lavas	Francheteau *et al.*, 1990; Hékinian *et al.*, 1993; Natland & Dick, 1996; Coogan *et al.*, 2002; Karson *et al.*, 2002a
Hess Deep Rift intra-rift ridge	110	Fault block on propagating rift wall	Gabbro, ol-gabbro	Depth in crust poorly constrained	Gillis *et al.*, 1993a; Hékinian *et al.*, 1993; Gillis *et al.*, 2013
Pito Deep Rift	142	Propagating rift wall	Ol-gabbro, gabbro	Under sheeted dikes and lavas	Francheteau *et al.*, 1994; Karson *et al.*, 2005; Perk *et al.*, 2007
Southwest Cocos Plate	220	Drill core IODP 1256D	Gabbro	Under lavas & sheeted dikes	Wilson *et al.*, 2006
Terevaka Transform Fault	113	Transform valley wall	Gabbro, oxide-gabbro		Francheteau *et al.*, 1988; Constantin *et al.*, 1996
Garrett Transform Fault	168	Transform valley wall	Gabbro, oxide-gabbro		Hébert *et al.*, 1983; Hékinian *et al.*, 1992

and on land. Relevant information on this topic comes from a wide range of tectonic perspectives and across a range of spatial scales including layered mafic intrusions as well as ophiolite complexes. Concepts pertaining to gabbroic to ultramafic oceanic rocks and the magmatic processes that create them have been continuously refined as new data have come to light from many different approaches.

Numerous models have been proposed for accretion of the middle to lower crust. The differences in these models come partly from various types of data used to construct them (for example, seismic, drilling results, near-bottom studies, and information from ophiolites) and from the significant variations in geology of the lower crust. Interpretations that led to the current models are a bit like the parable of a group of blind men trying to describe an elephant based on feeling different parts of the animal. The discussion below presents what is currently known about the structural, petrological, and geochemical aspects of the plutonic "elephant" and how these different lines of evidence are used to build models for construction of the oceanic lower crust.

7.2 Plutonic rock exposures on the seafloor

Typical gabbroic rock names are based on the proportions of the different major mineral phases that make up a sample (Fig. 7.2). Various magmatic processes that occur in the AMC or crystal mush zone, such as fractional crystallization coupled with crystal settling and accumulation, as well as secondary crystal–melt interactions, all serve to modify simple crystallization orders and textures. Crystal accumulation in a magma chamber forms cumulate plutonic rock types (see below) such as dunites rich in accumulated olivine, anorthosites rich in plagioclase, pyroxenites rich in pyroxenes, or combinations of these mineral phases as plagioclase-bearing ultramafic rocks, troctolite, gabbro, olivine gabbro, and gabbronorite. This range of compositions is commonly referred to collectively as gabbroic rocks.

Gabbroic and related plutonic rocks recovered from the seafloor have diverse mineralogic compositions and textures which hint at the range of compositions, structures, and processes that are well documented in ophiolites. Gabbroic rocks are widely exposed along major rift valley escarpments, oceanic core complexes (OCC), and fracture zones in ultraslow- to slow-spreading ridges (Chapter 3), but it is not clear whether these exposures are representative of the middle to lower crust away from these tectonic settings. A number of these types of exposures have been investigated with submersibles and drilling. In crust formed at superfast to intermediate spreading rates, faulting at the MOR axis or just off-axis usually does not penetrate deep into the crust. Gabbro exposures are very rare and limited to specific tectonic settings (Table 7.1), for example, along the faulted walls of propagating rifts. These too have been exploited by submersible studies and drilling. To date, only one drill site in superfast-spreading crust, IODP Hole 1256D, has penetrated the uppermost part (only 200 m) of the gabbroic assemblage beneath sheeted dikes in an area of seafloor far from major tectonic disruptions (Figs. 6.7 and 6.22). Given the very different spreading processes that characterize fast- and slow-spreading MORs, information acquired in one of these settings may not be directly applicable to another. In the metaphor employed above – the person at the plutonic "elephant's" foot may understand something quite different from the person touching the beast's head.

Not only do oceanic gabbros collectively have a wide range of textures and compositions, but it is also common for individual samples to be texturally heterogeneous on the scale of a few centimeters. In addition to the major rock types mentioned above (Fig. 7.2), assemblages of gabbroic rocks from the seafloor and in ophiolites typically include lesser amounts of more evolved plutonic rocks, which range from ferrogabbro and oxide gabbro to leucogabbro (see below). In rare instances silica-rich, highly evolved plutonic rocks (commonly classified as plagiogranites) have been recovered (Natland and Dick, 1996, 2009; Dick *et al.*, 2000). Plutonic rocks recovered by drilling during a number of ODP and IODP legs show that igneous compositional layering is common in gabbros from slow-spreading centers, but rare in fast-spread gabbroic crust (e.g., Gillis *et al.*, 2013).

7.2.1 Gabbros from fast to intermediate rate MORs

Extensive exposures of gabbroic rocks formed at relatively fast spreading rates are found only in a few places where major transform faults or propagating rifts provide "tectonic windows" into these mid-crustal materials (Figs. 7.3–7.7). None of these appear to provide vertically continuous exposures reaching more than about 1500 m – down to about the middle of the crust. At the Hess Deep (Fig. 6.11) and Pito Deep (Fig. 6.9) Rifts, over 1000 m thickness of gabbroic rocks crop out directly beneath upper crustal lavas and sheeted dikes (Karson *et al.*, 2002a, 2005). Thus, these exposures may provide more or less intact sections of the upper part of the oceanic crust in which the orientations of structures and crustal depth of samples can be constrained.

The extensive outcrops at the Hess Deep and Pito Deep Rifts are typically blocky and massive with widely spaced joints and rare faults or shear zones, making them easy to differentiate from overlying sheeted dikes (Fig. 7.3). Dikes within the gabbros commonly lack chilled margins and the distinctive cross-joints seen in higher-level dikes, making them difficult to trace in outcrops. The talus generated by the mass wasting of gabbroic outcrops is blocky and commonly consists mostly of large, angular boulders, very different from the talus of sheeted dikes that is much more tabular and smaller.

Figure 7.3 Large exposures (hundreds of meters vertically) of gabbroic rocks occur just below sheeted dike complexes in fast-spreading crust at the Hess Deep and Pito Deep Rifts. These images from the Hess Deep Rift show diverse outcrop expressions of this material. All images have a field of view of ~3 m and are from depths of 3000–4000 m.

Figure 7.4 Polished slab of gabbroic rock from below the sheeted dike complex at the Hess Deep Rift (left) shows the contacts between several different compositional and textural rock types. Photomicrographs (right) show textures in different parts of the rock in cross-polarized light (scale bar in top image is 1 cm; 1 mm in middle and bottom images).

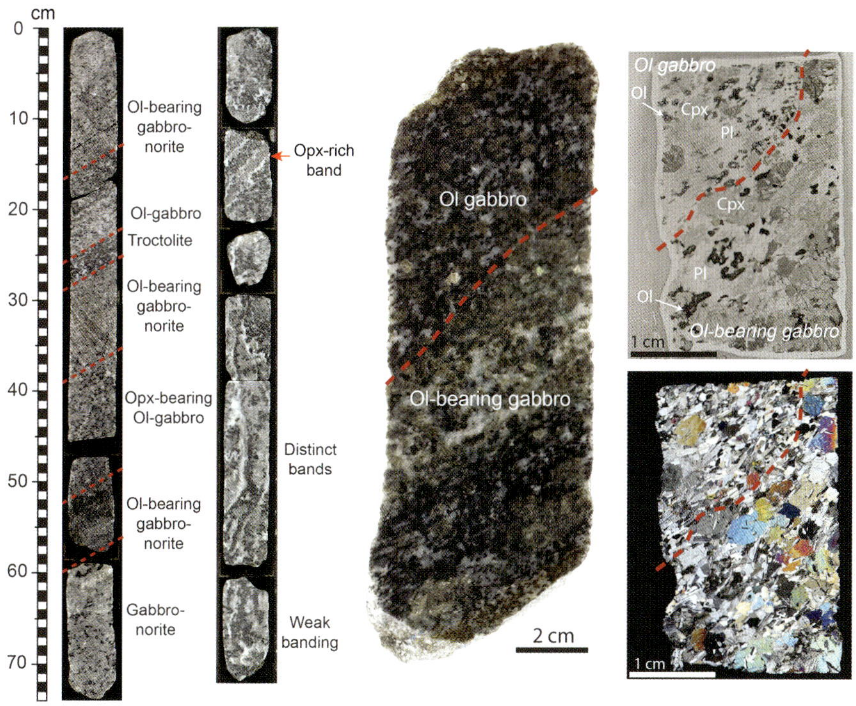

Figure 7.5 Sections of drill cores (left) from IODP Leg 345 on the Hess Deep IntraRift Ridge are interpreted as samples of the lower oceanic crust. Discrete banding and contacts between different rock types are similar to features found in gabbroic assemblages of ophiolite complexes. A piece of core (middle) shows the details of compositional layer contacts. Photomicrographs (right) show textural details in plane-polarized light (top) and cross-polarized light (bottom) (Gillis *et al.*, 2013).

Figure 7.6 Gabbroic rocks were recovered from below the sheeted dike complex at the Pito Deep Rift. Polished slab of layered gabbro (left) has pyroxene and olivine-rich layers (green) and plagioclase-rich layers (gray); scale bar is 1 cm long. Photomicrographs below show blocky igneous texture (upper right) and foliated gabbro (lower right) with bent albite crystal twinning, indicating weak crystal-plastic deformation. Both images in cross-polarized light and are at the same scale; scale bar is 5 mm long (photomicrographs from Perk *et al.*, 2007).

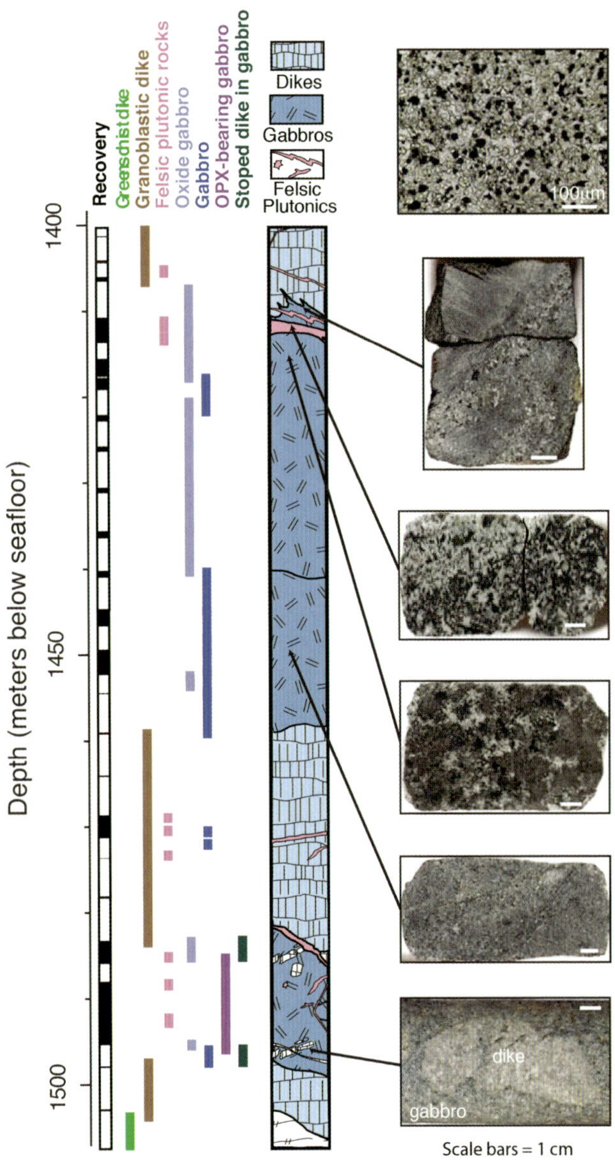

Figure 7.7 IODP Hole 1256D penetrated through the lavas and sheeted dikes of superfast-spreading crust in the eastern Pacific and recovered ~200 m of highly variable gabbroic material. The columnar section and ranges of rock types illustrate the diversity of this crustal section. Detailed photos of the core and photomicrographs to the right illustrate the contacts and textures in corresponding parts of the core. The black and white bar shows the actual amount of drill core recovered during drilling (Wilson *et al.*, 2006).

Samples from these tectonic windows (Figs. 7.4 to 7.7) show a wide range of compositions and textures, and in some cases cross-cutting relationships providing glimpses of a complex history of accretion. At the Hess Deep Rift, the upper gabbroic rocks sampled include gabbros, gabbronorites, ferrogabbros, and ferrodiorites, with minor amounts of plagiogranite (Natland and Dick, 1996, 2009). Samples recovered by drilling on the IntraRift Ridge, deeper in the rift valley, and thought to include deeper parts of the gabbro unit, include primitive layered gabbros. More evolved oxide gabbros and felsic veins that are present in the upper gabbroic sequence were not recovered in the deeper exposures (Gillis *et al.*, 2013). In contrast, even the uppermost gabbros from the Pito Deep Rift contain mafic, layered rocks including olivine gabbro to troctolite (Perk *et al.*, 2007).

Deeper-crustal gabbroic rocks and mantle peridotites crop out downslope on the IntraRift Ridge in the Hess Deep Rift valley (Francheteau *et al.*, 1990; Gillis *et al.*, 1993), but these units have not been spatially related to those shallower on the rift escarpments. Magnetic data on drill cores from these areas also suggest that the cores may come from a number of different rotated blocks that accumulated at the bottom of the rift valley (Gillis *et al.*, 2013). Gabbroic rocks recovered from this area include primitive (high-MgO) layered gabbros with a wide range of textures (Fig. 7.5) (Gillis *et al.*, 2013).

A single IODP drill hole (Hole 1256D) in superfast-spreading crust of the eastern Pacific penetrated about 200 m of gabbroic rock (Fig. 7.7) beneath about 1200 m of lavas and sheeted dikes formed by spreading at the East Pacific Rise (EPR) (Wilson *et al.*, 2006; Sano *et al.*, 2011). Samples from the recovered drill core show evidence of complex crystallization and contact metamorphism which probably occurred near the top of a magma chamber. Samples from this core show a wide range of compositions and textural variations (Fig. 7.7)

Apart from these samples from the Hess Deep and Pito Deep Rift tectonic windows and IODP Hole 1256D, the only *in situ* gabbroic materials from relatively fast-spreading crust come from dredging of two transform faults (Constantin *et al.*, 1996) and an abandoned rift (Vanko and Batiza, 1982) (Table 7.1). This limited sampling of the middle to lower crust of fast-spreading MORs makes it difficult to link that material to processes inferred from the extensive studies of lavas and geophysical data from these settings.

Rare fragments of gabbroic rocks ("xenoliths") captured and enclosed in some lava flows provide glimpses of the lower crust. Xenoliths in lavas from both fast- and intermediate-rate ridges provide a particularly useful perspective on the nature of the plutonic crust below active spreading centers and possible magma chamber processes. It is noteworthy that, despite the extensive studies of lavas and dikes from intermediate-rate ridges, xenoliths provide the only information on the gabbroic crust formed in these settings.

Xenoliths are found enclosed in moderately evolved mid-ocean ridge basalt (MORB) from the EPR (Ridley et al., 2006) and Juan de Fuca Ridge (Dixon et al., 1986; Davis and Clague, 1990; Smith et al., 1994) (Figs. 7.8 and 7.9). Many of these xenoliths are medium-grained gabbros containing

Figure 7.8 Gabbroic xenoliths enclosed in MORB from the Cleft Segment of the Juan de Fuca Ridge provide rare glimpses of the plutonic portions of the oceanic crust beneath an active spreading center. Aphyric host lava with medium-grained olivine gabbro (top left) and photomicrograph showing equigranular olivine and plagioclase with lesser clinopyroxene (top right). Photomicrographs of olivine-bearing gabbro xenolith (bottom left and right) showing small plagioclase crystals enclosed in a single large clinopyroxene oikocryst with minor quantities of interstitial glass. Images shown in cross-polarized light (right) and in plane-polarized light (left).

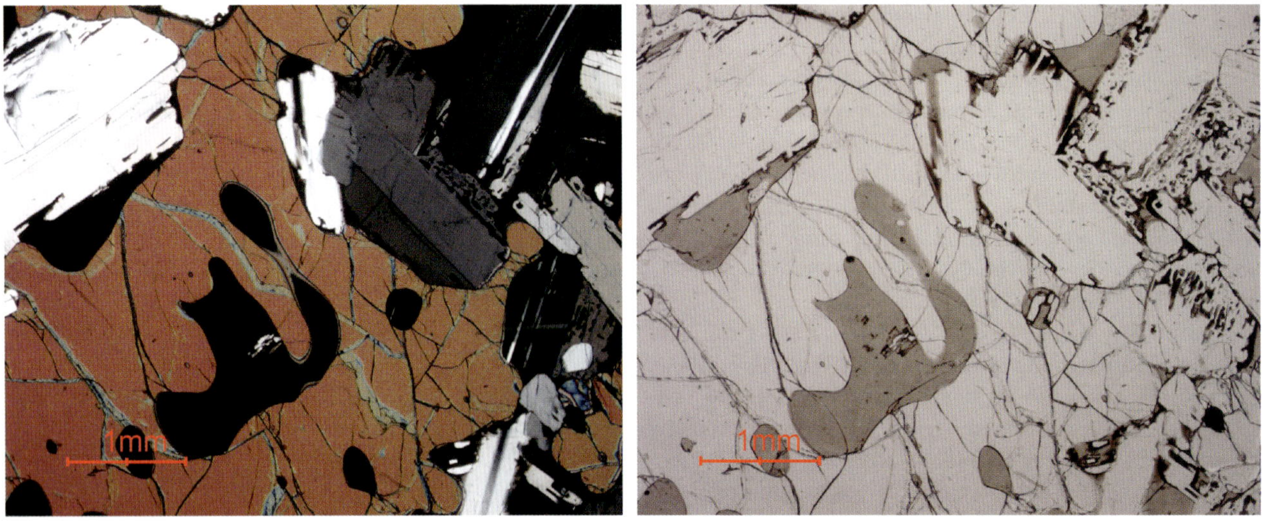

Figure 7.9 Photomicrographs of a troctolite xenolith from the northern EPR showing partially resorbed crystals of olivine and plagioclase enclosing and surrounded by glass (black on left, tan on right) that has a different composition than the host lava. Minor amounts of spinel and sulfide are also present. Plagioclase displays complex zoning and characteristic sieve texture on rims. Textures are indicative of changing melt compositions and crystal–melt interaction during crystallization in a subaxial mush zone (Ridley *et al.*, 2006). Image on left in cross-polarized light and on right in plane-polarized light.

interstitial glass (melt) between crystals (Fig. 7.9). Geochemical and textural data suggest they came from the seismic LVZ beneath the ridge axis, which is believed to represent a zone of crystal–liquid mush. Mineral textures, compositions, and zoning in the xenoliths suggest complex histories of crystal–melt reactions and magma mixing, in agreement with mineral trace element zoning patterns believed to be a consequence of interstitial melt migration within the crystal mush zone (Coogan et al., 2000; Kvassnes and Grove, 2008; Lissenberg and Dick, 2008; Lissenberg et al., 2013).

7.2.2 Gabbros from slow-spreading MORs

Exposures of gabbroic rocks are widespread in crust generated at slow to ultraslow spreading rates (<20 mm/yr) (Cannat *et al.*, 1997; Karson, 1998; Dick *et al.*, 2003). This may be a result of the occurrence of many large-offset fault zones that create kilometers of relief along rift and transform valley walls. Major, low-angle normal (detachment) faults on OCCs (see Chapter 3) provide some of the best-known exposures that have been investigated by dredging, deep crustal drilling, and near-bottom investigations with human-occupied vehicles (HOVs) and remotely operated vehicles (ROVs). These studies show that gabbro units range from continuous crustal exposures with dimensions of kilometers, both vertically and laterally, to much smaller bodies on the order of meters to tens of meters across in metamorphosed gabbros or peridotites. Drilling and closely spaced sampling in exposures of gabbroic rocks in a number of settings have provided samples that help constrain the scale, contacts, and composition of individual rock units (Fox and Stroup, 1981; Elthon, 1987; Cannat *et al.*, 1995a,b, 1997; Pedersen *et al.*, 1996; Dick *et al.*, 2000; Coogan *et al.*, 2002; Coogan, 2007; Perk *et al.*, 2007; Hayman *et al.*, 2011).

Mapping with deep-towed cameras, ROVs, and HOVs shows that gabbroic rocks commonly occur in outcrops with areas up to several kilometers across, ranging from massive, jointed rock to rugged angular cliffs (Figs. 7.10 and 7.11).

Figure 7.10 Gabbroic rocks such as these on the south wall of the Atlantis Massif are commonly exposed in very massive, jointed outcrops, making them very difficult to sample. Field of view ~3 m.

They can be crudely banded to strongly foliated, platy rocks commonly expressed as smooth dip-slopes (foliation essentially parallel to outcrop surface) in the exposures. Joints, fractures, and veins are spaced at intervals of meters except near fault zones. Foliated metagabbros tend to have distinct

Figure 7.11 Gabbroic rocks on OCCs show a range of outcrop-scale deformation structures. Intensely jointed gabbroic rocks on the wall of the Kane Transform Fault (cross section of an OCC) (upper left); complex pattern of foliation on horizontal surface of the Atlantis Bank OCC (upper right); coarse foliation and slabby form of metagabbros on the rift valley wall of the Kane RTI OCC (lower right); very platy, mylonitic metagabbros on the Kane RTI OCC (lower left). Note light-colored pelagic ooze highlights the deformation features. Field of view in all images is ~4 m. The inclined camera orientation makes the structures in the lower images appear much steeper than they are (~30° dip).

outcrop patterns with ledges and resistant ridges parallel to the metamorphic fabrics. Extensive faulted surfaces dipping 30°–50° have been followed upslope as much as one hundred meters. These are apparently the result of extensive sheet slides that exploit the fault surfaces and low-angle foliation planes.

Depressions that incise the walls and base of major escarpments are piled with talus blocks of gabbro and finer-grained debris from upslope. Talus at the base of outcrops made of foliated material has a distinctive tabular form reflecting the penetrative deformation fabric of the rock. Locally, huge boulders of massive gabbro up to a few tens of meters across have been observed; these can be difficult to distinguish from outcrops, especially if partially buried by rubble or pelagic sediment.

In general, these studies indicate that the gabbroic plutonic rock units of slow-spread crust are a highly heterogeneous collage of igneous units that have been assembled during deformation and metamorphism at spreading centers. Samples from massive outcrops and obtained by drilling include a range of compositions with blocky igneous textures (Fig. 7.12). Generally, less-deformed gabbroic rock masses cut somewhat older more deformed and metamorphosed units. Although the absolute ages of the rock bodies are not tightly constrained, the geological relations indicate a more or less continuous process of crustal deformation, hydration and metamorphism, and igneous activity during crustal accretion. Smaller gabbroic bodies are commonly hosted by variably deformed and metamorphosed mantle peridotites, resulting in very heterogeneous middle to lower crustal geological structures that contradict the traditional view of a simple layered structure of oceanic crust, which may be present in faster-spreading crust.

7.3 Deformation and metamorphism in the middle to lower oceanic crust

As in the sheeted dikes, deformation and metamorphism in the gabbroic section of the oceanic crust are closely related (see Chapter 6). Deformed and metamorphosed gabbroic rocks ranging from high-temperature gneissic and mylonitic

Figure 7.12 Gabbroic rocks from slow-spreading ridges can have highly variable textures and mineralogies. Photomicrograph mosaic of a typical gabbro (left) with blocky plagioclase and more irregular interstitial clinopyroxene. Varitextured gabbro (middle). Blocky gabbro (plane-polarized light) from the Vema Fracture Zone (top right). Blocky gabbro from the MAR at the Kane Transform (MARK) Area (middle right). Ferrogabbro with greenish hornblende and black oxide minerals (plane-polarized light, lower right) from the MARK Area. All images in cross-polarized light except top and bottom on right. Scale bars on all images are 0.5 cm.

metagabbros to low-temperature, brittle cataclasite and fault gouge have been recovered from the seafloor by dredging, crustal drilling, and near-bottom investigations (Figs. 7.13–7.16). Constraints on deformation and metamorphism of the middle to lower crust are fragmentary and come from widely separated areas. Detailed investigations of available materials have documented processes that have affected these rocks and attempted to link them to larger scale structures. However, because these materials all come from tectonic settings where the upper crust has been partially or entirely removed by faulting, the nature of the crust away from tectonically disrupted areas is not known with certainty. Ophiolite complexes once again provide a regional-scale view of these rocks and help to give a sense of how deformation and metamorphism affect the oceanic crust as it evolves.

Transform faults create major shear zones in the oceanic lithosphere. Nevertheless, theoretical models predict a complex history of vertical and horizontal displacements, variations in thermal effects, and igneous intrusion along these boundaries (Chapter 3) (DeLong *et al.*, 1979). Yet little is known of the details of the geological processes in these settings. Variably deformed and metamorphosed gabbroic rocks from transform valley walls, previously thought to represent transform fault processes, are now thought to be related to extensional shear zones related to detachment faults and OCCs. Samples from shear zones clearly related to transform faulting have not been recovered from the seafloor or by drilling. Major shear zones in ophiolites, however, provide analogs for the type of processes that may occur along oceanic transform faults.

At relatively fast spreading ridges, major deformation features that post-date accretion are not expected to affect the middle to lower crust (Chapter 3). The effects of faulting associated with abyssal hills, for example, are probably limited to relatively widely spaced faults in the upper crust (Macdonald *et al.*, 1996; Buck *et al.*, 2005). However, at middle to lower crustal levels, models for the accretion of this part of the crust demand substantial deformation related to the redistribution of magmatic material during spreading as it is added to the crust in magma chambers or smaller magma sills.

Historically, models for accretion on slow-spreading ridges have been adapted from fast-spreading models by scaling down magma chamber size and incorporating rift-related normal faults. However, over the past few decades, it has become clear that slow-spreading ridges include areas of extreme tectonic extension and modes of crustal construction that do not occur at faster spreading rates. Approximately half of the slow-spreading Mid-Atlantic Ridge plate boundary is affected by processes related to the formation of OCCs (see Chapter 3). This is clearly a fundamental aspect of seafloor spreading rather than a local phenomenon (e.g., Smith *et al.*, 2006). Studies at several OCCs provide important constraints on some

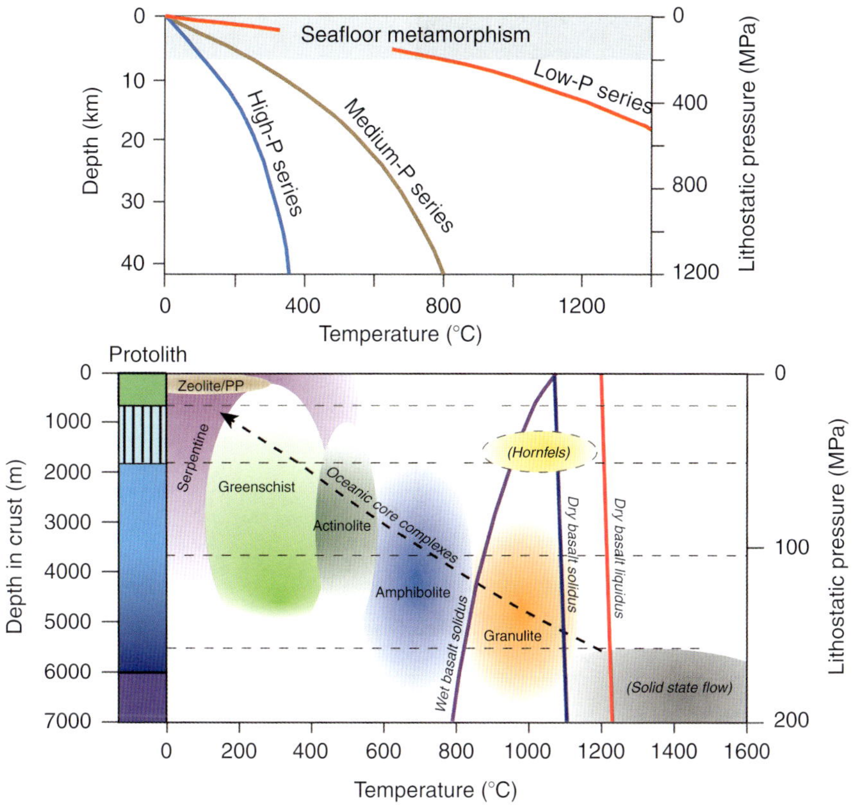

Figure 7.13 Metamorphic rocks from the oceanic crust and upper mantle reflect physical conditions that overlap with those of low-pressure metamorphic rocks of continental terranes (top) occurring in the shallow crust and commonly associated with contact metamorphism near igneous intrusions. The metamorphic facies indicated by the colored swaths (bottom) approximate the conditions associated with different rock units of the oceanic crust (column on left). Because temperatures and pressures generally increase downward in the crust, higher-grade assemblages occur in deeper level rocks and undergo greater degrees of plastic deformation producing strong fabrics. Rocks exposed at the seafloor and in drill cores typically show the effects of retrograde metamorphism (dashed line), reflecting age-related cooling of the crust or tectonic uplift to shallow crustal levels. Solidus curves (blue and purple) representing the temperatures above which dry and water-saturated (wet) basalt begin to melt are shown with the basalt liquidus where basalt becomes completely molten. Temperatures near the AMC (hornfels field) can be great enough to partially melt hydrated mafic basaltic rocks forming plagiogranite melts. Where mantle peridotites have been uplifted and hydrated, serpentine minerals form in a field overlapping with greenschist and lower-grade assemblages.

of the major processes involved in their evolution. In contrast, almost nothing is known of the processes affecting the middle to lower crust of slow-spreading systems away from OCCs, because of the lack of exposures of deep-level rocks and lack of penetration by deep drilling to date.

7.3.1 Oceanic crustal metamorphism

Metamorphic rocks from the oceanic crust and upper mantle have mineral assemblages and textures that reflect physical conditions that overlap with those of low-pressure metamorphic rocks of continental terranes occurring in the shallow crust, where they are commonly associated with contact metamorphic conditions around igneous intrusions. Metamorphism in oceanic crust occurs almost exclusively in mafic rocks at depths of <7 km and lithostatic pressures of less than ~200 MPa (Fig. 7.13). Because temperatures generally increase downward in the crust, higher-temperature (*T*) assemblages occur in deeper level rocks that also experience higher pressures (*P*). This leads to progressive changes in mineral content that reflect *P*–*T* conditions. Specific mineral assemblages (named for their characteristic minerals) form under specific *P*–*T* conditions (shown as colored fields on Fig. 7.13) defined by metamorphic reactions. Assemblages that reach chemical and textural equilibrium are referred to as metamorphic facies. The metamorphic facies represented by rocks recovered from the oceanic crust are similar to those of mafic continental rocks from low-pressure environments (Miyashiro, 1973).

The uppermost part of the lava unit ranges from fresh igneous assemblages to low-temperature zeolite to prehnite-pumpellyite (zeolite/PP) facies assemblages typical of continental burial metamorphism. Greenschist metamorphic assemblages are typical in the upper gabbros and include chlorite, actinolite, epidote, albite, and oxide minerals that replace the igneous mineral phases. Higher-temperature amphibolite (hornblende and calcic plagioclase) and even granulite (pyroxene and calcic plagioclase) assemblages are recovered locally. Contact metamorphism creates fine-grained, equigranular metamorphic rocks (hornfels) near the base of the sheeted dike units and is likely related to the intrusion of melt sills. Intense heating under static conditions in the thermal boundary layer near the top of the AMC may also result in partial melting of hydrated mafic basaltic rocks to create plagiogranite (Fig. 7.13) (Koepke *et al.*, 2008).

Figure 7.14 Deformed metagabbro (left) is cut by a complex vein with microbreccia healed by fine-grained, low-temperature metamorphic minerals (plane-polarized light). Low-temperature hydrothermal minerals concentrated along a fracture in otherwise fresh gabbroic rock (right) (cross-polarized light). Both images from the MARK Area. Field of view is 0.5 cm for both images.

Figure 7.15 Dark hornblende veins in lower gabbros just above the Moho in the Oman Ophiolite are produced by deep hydrothermal circulation.

Near spreading centers, fracture-controlled hydrothermal metamorphism is prevalent in the upper crust (lavas, dikes, and upper gabbros). In tectonic windows and ophiolites, many of the gabbro sections are only locally altered. Schistose metamorphic rocks occur in faults and shear zones cutting through any of the rock units. Deeper in the crust and upper mantle, anhydrous deformation produces strong fabrics. Rocks exposed at the seafloor and in drill cores typically show the effects of retrograde (progressively lower temperature, lower pressure, and more hydrous conditions) metamorphism (black lines with arrows in Fig. 7.13) reflecting age-related cooling of the crust or tectonic uplift to shallow crustal levels. Lower crustal and upper mantle rocks in OCC show high-temperature assemblages overprinted by conditions spanning a wide range of temperatures. In the mantle, anhydrous solid-state deformation occurs at depth beneath MORs. Where mantle peridotites have been uplifted and hydrated, serpentine minerals form in a field overlapping with greenschist and lower-grade assemblages.

7.3.2 Hydrothermal metamorphism

Hydrothermal vents are a direct result of seawater having been heated to near magmatic temperatures deep within the oceanic crust (see Chapter 4). When cold seawater penetrates the oceanic crust along fractures and faults it heats up to become hydrothermal fluid as it approaches magma bodies (e.g., dikes, melt lenses) and hot nascent crust that exists at depth beneath the MOR. The combination of increasing temperature and pressure with depth together with the chemically reactive nature of the hydrothermal fluids acts to change the mineralogy and chemistry of the oceanic crust. Depending on the magnitude of these

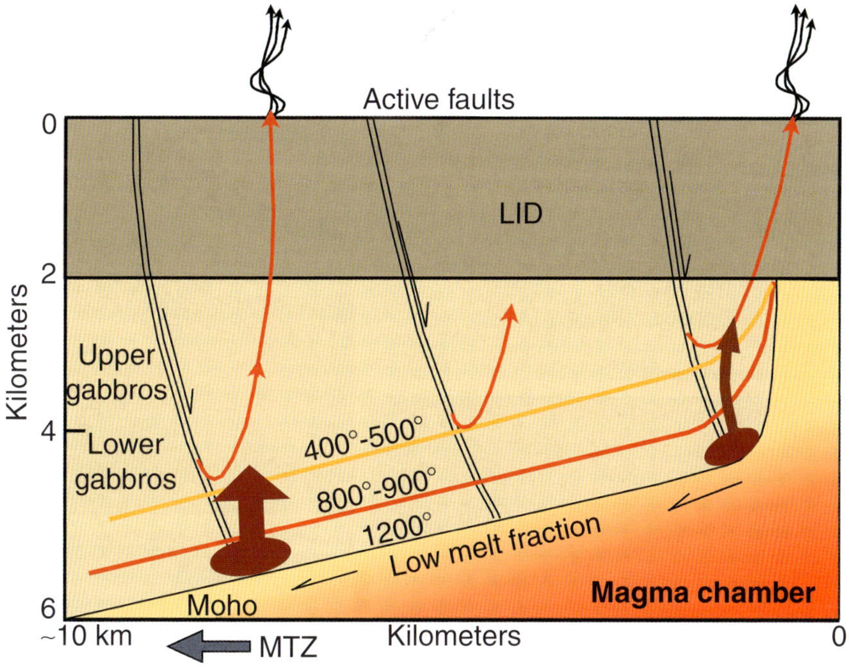

Figure 7.16 The distribution of metamorphic minerals and veins in the Oman Ophiolite is interpreted in terms of fault-controlled hydrothermal circulation driven by the heat of a subaxial magma. Black arrows: seawater downflow; red arrows: hydrothermal fluid upflow zones; brown: reaction zones; gray: upper crustal lid of lavas and dikes (from Bosch *et al.*, 2004).

physical and chemical factors and the original composition of the crustal materials, they are transformed into a variety of metabasalts and metagabbros. During this process, at temperatures between 350 °C and 550 °C, elements such as Cu, Zn, Fe, Pb, S, Ca and Si, are leached out of the rocks and are incorporated in the hydrothermal fluid – ultimately to be precipitated as metal-rich sulfides, anhydrite, and silica in hydrothermal vents and black smokers (see Chapter 4) (Mottl, 1983; Seyfried Jr. *et al.*, 1988; Von Damm *et al.*, 1995).

Even relatively fresh samples of seafloor gabbroic rocks commonly have haloes of metamorphic minerals, at least a few millimeters wide, reflecting localized effects of hydrothermal fluids. Pervasive fracturing and fluid interactions can result in the complete replacement of igneous minerals by lower-temperature, hydrous metamorphic minerals. Cross-cutting veins with hydrous phases are also common (Fig. 7.14). Where more extensive fracturing has occurred, metamorphism is also more widespread and pervasive. In many samples, mafic phases are replaced by amphiboles, chlorite, and other hydrous phases, with little obvious effect on the surrounding plagioclase.

Similar to other hydrothermal systems, the development of hydrous mineral assemblages in oceanic gabbroic rocks is highly heterogeneous both in its spatial development and in intensity (Manning *et al.*, 1996; Alt, 2004). Greenschist metamorphic assemblages are typical in the upper gabbros, with amphibolite and even granulite assemblages recovered locally. Contact metamorphism that creates hornfels occurs near the base of the sheeted dike units (Gillis and Coogan, 2002; Gillis, 2008; Koepke *et al.*, 2008) and is likely related to the intrusion of melt sills.

Studies of closely spaced samples along tectonic windows at the Hess Deep and Pito Deep Rifts show that fluid fluxes and temperatures of metamorphism in the sheeted dike units vary significantly on the scale of tens to hundreds of meters across the escarpments (Manning *et al.*, 1996; Gillis *et al.*, 2001, 2005; Alt, 2004; Heft *et al.*, 2008). Comparable investigations of gabbroic sequences away from tectonic windows have not yet been done. Ophiolite complexes show this same type of heterogeneous hydrothermal metamorphism, and high-temperature hydrothermal amphibole veins in the Oman Ophiolite (Fig. 7.15) suggest that fluid circulation at spreading centers penetrates to great depths beneath ridges (Fig. 7.16) (Bosch *et al.*, 2004; Koepke *et al.*, 2008).

7.3.3 Deformation and metamorphism in oceanic core complexes

Gabbroic rocks recovered from OCCs by HOVs, ROVs, and drilling reveal a complex evolution involving multiple episodes of high-temperature ductile deformation, hydration, metamorphism, and intrusion (Figs 7.17–7.20). Beneath the detachment faults, and in the underlying units, high-temperature crystal-plastic deformation has produced gabbroic gneisses and mylonites (Fig. 7.17) in major shear zones as much as a few hundred meters thick (Dick *et al.*, 2000). Cross-cutting relationships show that intrusion of gabbroic material occurred during and after the high-temperature deformation. Lower-temperature deformation, metamorphism, and fluid interactions in discrete shear zones are marked by talc–tremolite,

Figure 7.17 Photomicrographs of gabbroic material deformed at high temperatures (>800 °C) show evidence of crystal-plastic deformation and recrystallization. Mylonitic gabbro (left) with large, lens-shaped, porphyroclasts of clinopyroxene surrounded by finely recrystallized plagioclase and hornblende, white and green, respectively (upper left). Upper image shown in plane-polarized light; lower image in cross-polarized light. Photomicrographs on right show similar features. Finely recrystallized plagioclase enclosing stronger, and in some cases fractured and internally deformed, clinopyroxene (partially replaced by green amphibole) shown in plane-polarized light (top). Lower two images show details of crystallized plagioclase (gray) in deformed gabbroic rocks. Both shown in cross-polarized light. All images are from the MARK Area except upper right from the Vema Fracture Zone. Scale bars on all images are 1 cm.

serpentinite, and metagabbroic schists and cataclasites (Fig. 7.18). The lower-temperature deformation and metamorphic features overprint the higher-temperature fabrics, indicating a history of progressively lower-temperature deformation and metamorphism ranging from near magmatic conditions to cold, brittle faulting. This mode of spreading contrasts with the more continuous magmatic construction inferred for fast-spreading environments.

Serpentinization affects most of the ultramafic parts of OCCs and ophiolites (Chapter 8), but is also a local effect in olivine-bearing gabbroic rocks. This is typically manifest as static replacement of olivine and orthopyroxene by serpentine minerals and oxides. Serpentinization occurs below temperatures of 500 °C but mostly about 350 °C and even down to conditions found in the very shallow crust (Bach *et al.*, 2004; Früh-Green *et al.*, 2004; Boschi *et al.*, 2013). Gabbroic material adjacent to partially serpentinized areas is commonly intensely altered through interactions with calcium- and silica-rich fluids, resulting in chalky, white alteration of the gabbros (Fig. 7.21). These distinctive rocks, known as "rodingites," have been recovered from the seafloor (see Chapter 8) and are also very common in ophiolites (e.g., Boschi *et al.*, 2006).

7.4 Petrology, geochemistry, and magmatic processes in oceanic gabbros

The geochemistry of lavas and dikes (Chapters 5 and 6) are quite similar and provide information on the composition of the mantle from where they were derived and the processes that modified their chemical composition as they were transported to the seafloor. Much of the development of these rocks takes place in the magma chamber or chambers that underlie MOR crests and where they eventually crystallize to form gabbroic rocks. The paucity of minerals in lavas and evidence suggesting that magma mixing and crustal assimilation occur cloud our views of the origins of erupted magmas. The details of the magmatic evolution and diversification of primitive melts must be gleaned from the geochemistry and petrology of the plutonic rocks and minerals that make up the middle and lower crust.

Figure 7.18 Photomicrographs of gabbroic material affected by intense brittle deformation. Cataclasites have varying proportions of surviving lithic clasts and cross-cutting chloritic shear zones (left). Cataclasite with granular texture and mylonitic clasts (middle top). Lens-shaped clasts of mylonitic material in a schistose matrix (lower middle). Localized brittle deformation of coarse-grained gabbroic material (upper right). Intense hydrothermal alteration along fault zones produces fine-grained, quartz+chlorite rocks (lower right), All images from the MARK Area and shown in cross-polarized light. Scale bars on all images are 1 cm.

7.4.1 Petrology

Gabbroic rocks that make up the mid to lower oceanic crust comprise minerals that crystallized from basaltic magmas having compositions that commonly range from high-MgO picritic melts to FeO-rich ferrobasaltic melts (Fig. 7.22). Rarely there may be SiO_2-rich melts poor in FeO, TiO_2, CaO, and MgO and relatively enriched in alkalis that crystallize to form plagiogranites. In all cases the minerals present, and their compositions, are dependent not only on the bulk composition of the melt but also on the pressures, temperatures, and volatile contents as they crystallize. In the simplest case, gabbroic rocks are believed to crystallize from relatively anhydrous melts within the crust or uppermost mantle where pressures are less than 0.3–0.4 Gpa, although there is increasing evidence for even higher pressure (>0.7 GPa) crystallization of basaltic melts occurring in the mantle (Laukert *et al.*, 2013). Under relatively low pressure and dry conditions, olivine (rarely spinel), calcic plagioclase, clinopyroxene, orthopyroxene, and iron-titanium oxides are the most common mineral phases to crystallize – and typically in that order from evolving magmas. In highly evolved samples, amphibole can be a common phase along with ilmenite and magnetite and more rarely apatite and zircon (Coogan, 2014). Such samples are generally coarse-grained or even pegmatitic, presumably reflecting their high volatile contents.

Gabbroic xenoliths enclosed in young lavas (Figs. 7.8 and 7.9) are primarily composed of plagioclase + clinopyroxene or plagioclase + olivine (troctolite) with minor orthopyroxene and titanomagnetite (see Figs. 7.9 and 7.10). Most contain interstitial glass (melt) between crystals suggesting they came from the mush zone beneath the ridge axis. The chemistry of the minerals within these xenoliths commonly shows they are not in equilibrium with the lava in which they are enclosed or the interstitial glass indicative of complex histories of crystal–liquid reactions and magma mixing in the mush zone. Data from an EPR troctolite (Ridley *et al.*, 2006) in particular suggest the interaction of a number of different primitive melts with accumulating crystals before being incorporated by the host melt prior to eruption.

7.4.2 Gabbro geochemistry: deciphering the chemical fingerprints of magmatic processes

The diverse group of gabbroic rocks recovered from the oceans displays a much greater compositional variation than their spatially related dikes and lavas (Fig. 7.22).

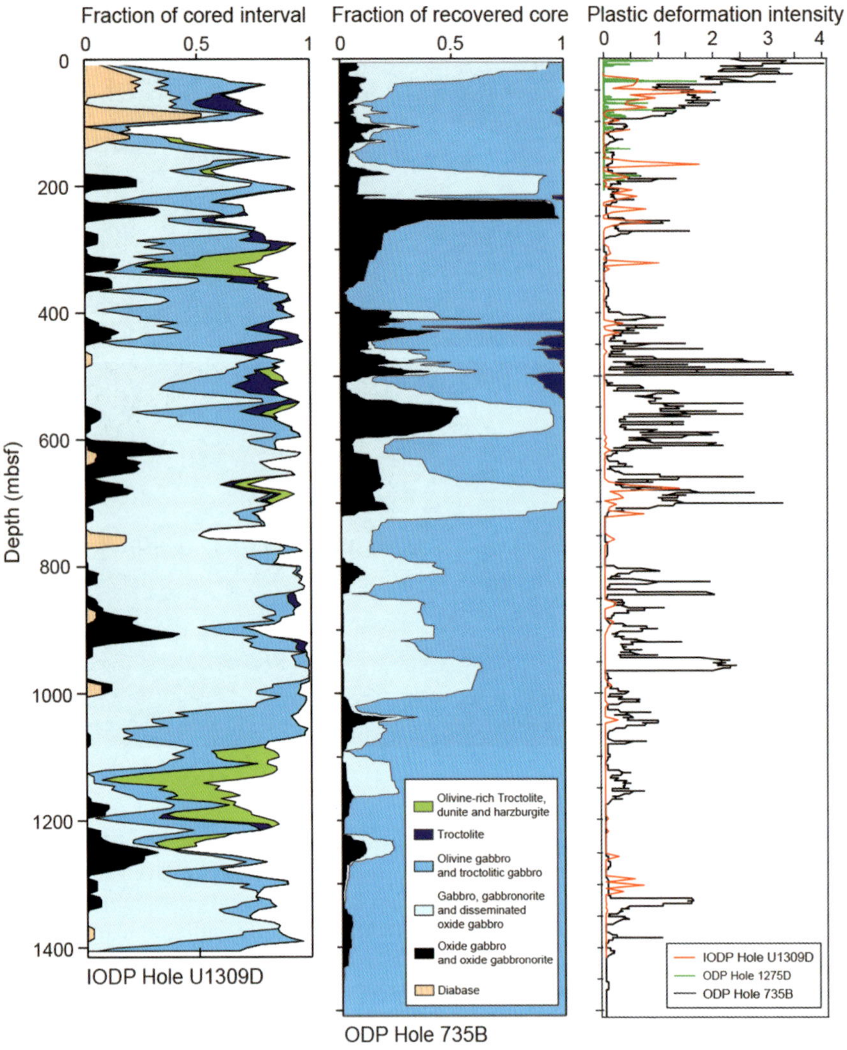

Figure 7.19 Gabbroic rocks drilled in OCCs from the Southwest Indian Ridge (ODP Hole 735B) and Mid-Atlantic Ridge (IDOP Hole U1309D) show a wide range of compositions and structures. Numerous intrusions ranging from centimeters to tens of meters thick cut variably deformed gabbroic rocks and serpentinized peridotite. From left to right, columns show lithologies in IODP Hole U1309D and ODP Hole 735B (colors indicate the relative abundance of major rock types recovered) and logs of crystal-plastic deformation intensity in these holes (IODP Hole 1275D for comparison). Note the very large range in deformation down-hole and large shear zones in Hole 735B (Ildefonse *et al.*, 2007). (See Fig. 2.20 for locations.)

Whereas the major element trends in the lava/glass compositions primarily reflect the evolution of primitive liquids via crystal fractionation (Plag + Ol ± Cpx), and in some cases magma mixing, the spread in gabbro data results from the accumulation of near-liquidus phases in variable proportions within a MOR magma chamber as well as the addition of intercumulus melts. For example, elemental trends toward high Al_2O_3 and low FeO and MgO contents in plutonic rocks reflect the accumulation of plagioclase in many of the troctolites, gabbros, and anorthositic gabbros, whereas more mafic samples with high MgO contents primarily show the effects of olivine accumulation (Fig. 7.22). The accumulation of plagioclase and olivine is consistent with the generation of relatively primitive MORB by the extraction of olivine (forming dunite bodies) and olivine gabbro cumulates from a range of potential primary magmas (Elthon *et al.*, 1995). Further, the liquid lines of descent for most MORB suites are consistent with the crystal fractionation of a gabbroic assemblage (Plag + Ol ± Cpx) and most gabbros are composed of near cotectic proportions of these phases (Coogan, 2014).

Petrological and geochemical studies of gabbros have, however, highlighted the complexities of basaltic magma evolution that are difficult to determine from lava geochemistry alone, in part due to efficient magma mixing before eruption. A common index of formation of mafic rocks and minerals is the Mg# ("Mg number") defined as [(100 × (Mg/(Mg + Fe)], where high Mg# corresponds to lower degrees of fractionation or higher pressures of crystallization. So it is surprising that clinopyroxene (and in rare cases orthopyroxene) with high Mg# (Gillis *et al.*, 2013) is common in gabbros even at shallow crustal levels (Perk *et al.*, 2007). Pyroxene with such a high Mg# is inconsistent with relatively low-pressure phase relations in MORB that predict clinopyroxene crystallization only occurs after plagioclase and olivine in magmas that have evolved to a significant extent. However, thin veins and bands of gabbroic rock recovered in abyssal peridotites from EPR transform faults also suggest deep crystallization may occur at fast-

Figure 7.20 Pieces of drill cores from OCCs sample a wide range of compositions and structures. Gneissic to mylonitic fabrics from ODP Hole 735B (left) and from ODP Hole 921 (middle) (Figs. 2.20 and 2.21). Altered troctolite (far right) from IDOP Hole U1309D show well-developed corona texture with secondary minerals rimming dark olivine grains.

Figure 7.21 Gabbroic dike in partially serpentinized peridotite in the Bay of Islands Ophiolite. The igneous minerals are replaced by a white aggregate of calcium silicate minerals including diopside, clinozoisite, chlorite, tremolite, and hydrogarnet.

spreading centers (Constantin *et al.*, 1996; Dick and Natland, 1996; Constantin, 1999).

Gabbroic rocks recovered from the seafloor exhibit a large range of trace element abundances, particularly the highly incompatible elements, including rare earth elements (REE), that correspond to their varied major element compositions. The trace element abundances partly reflect the concentrations in parental melts, but are also largely dependent on the proportion of cumulus phases (including accessory phases) and intercumulus melts. Chondrite-normalized REE patterns (Fig. 7.23) vary from very light REE depleted in troctolites to slightly light REE enriched ferrogabbros and plagiogranites (e.g., Coogan *et al.*, 2000; Dick *et al.*, 2000; Perk *et al.*, 2007; Godard *et al.*, 2009; Coogan, 2014). Some cumulate gabbros have low incompatible element and REE abundances with characteristic positive Sr and Eu anomalies – indicative of plagioclase accumulation (Fig. 7.23). The most evolved ferrogabbros exhibit elevated incompatible element contents and negative Sr and Eu anomalies indicative of plagioclase fractionation and consistent with their relatively low Al_2O_3 contents. Relatively few gabbros have trace element patterns that mimic those in dikes and lavas. Overall, the incompatible element compositions in gabbroic rocks vary widely between samples, partly because of the variable proportions of accumulated phases and amounts of interstitial melt, but also suggesting a variety of parental melts and or reactions with gabbroic rocks within the lower crust (Elthon *et al.*, 1995; Lissenberg *et al.*, 2013; Coogan, 2014).

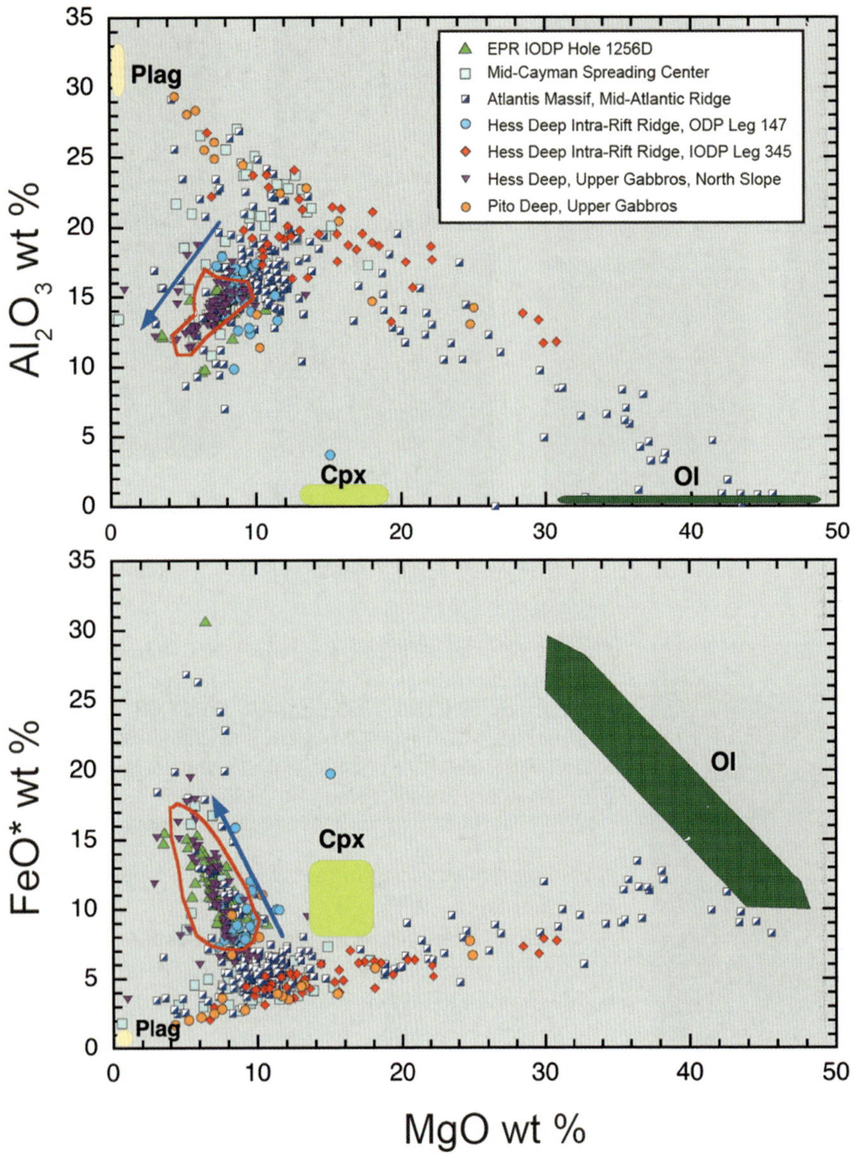

Figure 7.22 Major element variations in gabbros, from various ocean floor localities, compared to lavas and dikes from the Hess Deep Rift (Chapter 5). Overall, gabbros exhibit a much wider compositional range than lavas and dikes due to their cumulate nature. MgO-rich olivine gabbros and troctolites, such as the primitive gabbros from Hess Deep (Gillis *et al.*, 2013), have trends that project toward olivine (Ol) compositions (green solid field,) whereas high-Al_2O_3, low-MgO gabbros project toward plagioclase (Plag) compositions (orange field). Some more evolved upper crustal gabbros, IODP Hole 894 from Hess Deep and IODP Hole 1256D from the EPR, have compositions similar to lavas and dikes from Hess Deep (red outlined field). These more evolved gabbros, including ferrogabbros, follow the same differentiation trends observed in the lavas and dikes (blue arrow) due to low-pressure fractional crystallization. (Data from Natland and Dick, 1996; Pedersen *et al.*, 1996; Coogan *et al.*, 2002; Hanna, 2004; Perk *et al.*, 2007; Godard *et al.*, 2009; Neo *et al.*, 2009; Hayman *et al.*, 2011; Gillis *et al.*, 2013.)

Recent geochemical and petrologic studies of gabbros strongly suggest that rather than being simply related to the composition of mantle-derived melts and crystal accumulation, many of the unique compositional properties of gabbros result from crystal–melt interaction – a consequence of melt transport through the mush zone coupled with deformational processes (e.g., Natland and Dick, 2001; Gao *et al.*, 2007; Kvassnes and Grove, 2008; Dick *et al.*, 2010; Lissenberg *et al.*, 2013). A more detailed discussion of the use of major and trace elements in mineral phases to better understand the petrogenesis of oceanic gabbros is presented in Coogan (2014). Deciphering the tectonic and magmatic processes that create and affect the generation of oceanic gabbros appears to be the key to understanding the cryptic geochemical signals MORB provides and how, ultimately, the oceanic crust is formed.

7.5 Gabbroic rocks in ophiolites

As discussed above, the nature of seafloor outcrops and difficulty in accessing the deep oceanic crust have greatly limited the ability to investigate the lower half of the oceanic crust *in situ*. The general composition and processes responsible for the accretion of this part of the oceanic crust are inferred from a combination of results from marine geophysics and from ophiolites. These approaches are limited by the non-unique interpretation of geophysical data and by ambiguities related to the provenance of ophiolites. Counterparts of many features found in ophiolites have not yet been identified in the oceanic crust.

Despite their limitations, ophiolites nevertheless provide exceptional insights into oceanic crustal architecture and processes. The following sections highlight key features of

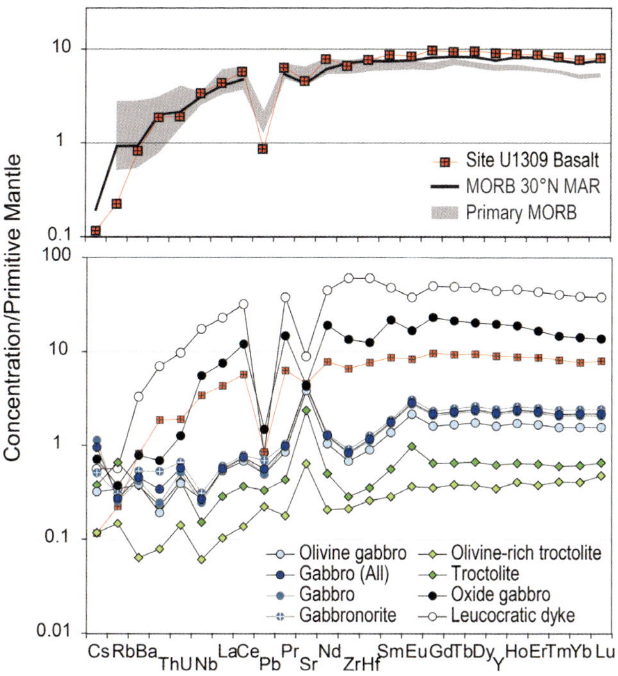

Figure 7.23 Primitive mantle-normalized trace element diagram showing the relative concentrations of incompatible trace elements in average gabbroic rock types (bottom) from the Atlantis Massif (IODP Site U1309) relative to spatially associated MORB (top) from the MAR (from Godard *et al.*, 2009). Incompatible trace element pattern for Site U1309 basalt (red) is also shown relative to the gabbros (lower plot), highlighting the differences between intrusive rocks and erupted basalts. The overall depletion in incompatible trace elements and anomalously high concentrations of Sr and Eu relative to the REE in most gabbroic rocks are indicative of their more primitive nature and accumulation of Ol and Plag in their origins, as emphasized in Fig. 7.21. Only more evolved gabbroic rocks (e.g., oxide gabbro) that have experienced Ol and Plag crystal fractionation have trace element contents and patterns similar to MORB. Estimated primary MORB compositions (gray field) (from Workman and Hart, 2005).

the gabbroic sections of ophiolites. These features provide windows into processes that may be important beneath MOR spreading centers, but are still beyond the resolution of current seafloor studies.

7.5.1 Sheeted dike–gabbro contact in ophiolites

The sheeted dike–gabbro contact on the seafloor and ophiolites is characterized by complex intrusive relationships, with gabbroic material intruding the base of the sheeted dike complex, but locally with dikes intruding the gabbros (Fig. 7.24). Faulting commonly disrupts this contact (Chapter 6). In IODP Hole 1256D, the sheeted dike–gabbro interface was drilled (Wilson *et al.*, 2006) and the lowermost dikes recovered in the cores show signs of high-temperature contact metamorphism and recrystallization that have been documented at the same level in ophiolites (Gillis and Coogan, 2002) and tectonic windows into fast-spreading crust (Gillis, 2008; Koepke *et al.*, 2008). Geological relationships near this contact show how the tops of subaxial magma chambers interact with the overlying rapidly cooled and hydrated upper crust. Magmatic heat alters the overlying sheeted dikes. Conversely, blocks of sheeted dike complex that are surrounded and melted by the magma ("assimilated") change the composition and crystallization processes in the magma. Given this setting, it is not surprising that there are many geochemical and structural complexities at this level in the oceanic crust.

Figure 7.24 Complex intrusive relationships occur at the top of the gabbro units of ophiolites. Left to right: coarse gabbro dikelets (light) intrude dark angular blocks of dike rock; plagiogranite (light) encloses angular blocks (xenoliths) of dike rock (dark); gabbro pegmatite (lighter colored band in middle of photo) cuts layered gabbro (all three from the Oman Ophiolite); and basaltic dike cuts lighter coarse-grained gabbro in the Macquarie Island Ophiolite.

Figure 7.25 The upper gabbros of ophiolites typically show patchy to layered textures with intercalated medium- and coarse-grained materials as in this outcrop from the Apennine Ophiolites. Field of view ~0.5 m.

7.5.2 Internal structure of gabbroic and related rocks in ophiolites

In many of the best-preserved layered ophiolite complexes, sections of gabbroic and lesser ultramafic plutonic rocks can be several kilometers thick and commonly occur as continuous rock units between overlying lavas and dikes and underlying mantle peridotites. This laterally continuous layered structure suggests a relatively constant magma supply during spreading, as postulated for fast-spreading ridges. In other ophiolite complexes, faulting has disrupted any simple layered structure that may have once existed. These are commonly referred to as "dismembered ophiolites." However, it is not clear whether the dismemberment occurred during mountain building after the oceanic crustal block was uplifted, or as an integral part of the spreading process as it was forming. Complex crustal structures and crustal sections lacking simple layering, documented in slow-spreading crust, suggest that some dismembered ophiolites may represent crust formed at slow-spreading ridges.

In the best-preserved ophiolite complexes, gabbroic and related rocks show physical and chemical characteristics that vary with depth. The upper and lower parts of the unit

Figure 7.26 Layering in the middle gabbros of ophiolites is commonly continuous across even large outcrops, for example in the Oman Ophiolite (top). The stone wall in the streambed is about 1.5 m high. Details of layering show that it is defined by changes in the proportions of minerals, changes in grain size, or both. Monomineralic layers are common (middle). Fine-scale layering with strong igneous foliations are also common (bottom).

are commonly very distinctive, with middle gabbros showing aspects of both of these. The gabbroic unit is commonly subdivided into petrologically and/or structurally distinct upper and lower parts (Casey *et al.*, 1981; Pallister and Hopson, 1981; Malpas *et al.*, 1987; Nicolas, 1989). The upper ~1000 m of the gabbroic unit is generally lithologically and compositionally heterogeneous, commonly containing more evolved lithologies such as gabbronorites and ferrogabbros. Networks of veins and small intrusive bodies of trondhjemite and tonalites (collectively known as plagiogranites) occur within the more abundant mafic gabbroic rocks. These rocks are also structurally heterogeneous on the outcrop scale, with local banding, flow foliation, pegmatitic veins, leucocratic net veins, and irregular grain-size variations (Fig. 7.25). Locally, the gabbroic material is cut by dikelets of plagiogranite to gabbro and fine-grained basaltic dikes (Fig. 7.24).

The diverse types of gabbros found in upper parts of the gabbroic unit of ophiolites have been described by various workers as isotropic, varitextured, and foliated. Isotropic gabbros typically are even-grained, two-pyroxene plutonic rocks with no mineral lineation or fabric. Magmatic flow fabrics in foliated gabbros are reconstructed as steep to vertical, and may reflect subsidence and vertical shearing of material intruded into shallow crustal magma lenses of the AMC (Boudier *et al.*, 1996; Boudier and Nicolas, 2011). Alternatively, the steep fabrics may be related to upward flow of gabbroic material feeding overlying dikes ("rooting") (Yaouancq and MacLeod, 2000). Important processes at this structural level include compaction, multiple intrusive events, magmatic flow, and *in situ* crystallization ("plating") against the solid roof or walls of the AMC. Texturally similar gabbroic rocks are exposed in tectonic windows and have been recovered in crustal drill holes (Fig. 7.4 and Table 7.1).

The composition and structure of the middle part of the gabbroic unit is not particularly distinct from material above and below. In general, the middle gabbros range from massive and homogeneous to well layered (Fig. 7.26). The rocks are described as "cumulates," which represent accumulated

Figure 7.27 The lower gabbros of ophiolites are typically strongly layered, as in this classic outcrop at Samrah Oasis in the Oman Ophiolite. The layering is the product of cumulate processes, magmatic deformation, and the intrusion of sills.

Figure 7.28 Details of the layering in the lower gabbros of ophiolites show grain-size and compositional variations between layers that may be sharp or gradational. Magmatic deformation and sill intrusion add to the complexity. Light-colored gabbroic sills with a very uniform, even-grained texture truncate variably deformed layering in the Oman Ophiolite (top). Variations in grain size define layering in gabbroic material in the Oman Ophiolite (middle). Sharply defined layering in dunite (brown) to feldspathic dunite and anorthosite (white) in the Bay of Islands Ophiolite (bottom).

Figure 7.29 Interlayered dunite, troctolite, gabbro, and anorthosite with cross-cutting anorthosite dikelets (white) are common in the lower part of the gabbro unit of the Bay of Islands Ophiolite.

crystals separated from magma as it cools and crystallizes. Cumulate layering defined by marked variations in the relative proportions of mineral phases, grain sizes, or both is typical (Fig. 7.26). Discrete volumes of layered material may correspond to magma lenses that differentiated more or less *in situ* with some amount of compaction and magmatic flow. Sills that may correspond to magma lenses are identified on the basis of low-angle cross-cutting relationships with host rocks as well as their mineral phase sequences and geochemical trends (Kelemen *et al.*, 1997b; Korenaga and Kelemen, 1997).

The lower part of the crustal assemblage in ophiolites, typically a few kilometers thick, mostly consists of strongly layered gabbro, olivine gabbro, troctolite and lesser anorthosite (Figs. 7.27–7.29). Ultramafic rocks interspersed with the gabbros include mainly dunites, wehrlites, and pyroxenites, which grade into gabbroic units as the proportions of plagioclase in them increase (Figs. 7.29 and 7.30). Igneous textures in these rocks are similar to those of the gabbroic material. Layered ultramafic rocks make up a substantial portion of this lower gabbroic unit in many ophiolites and may include

Figure 7.30 Layered wehrlite (brown) and clinopyroxenite (green-gray) cumulate rocks are common near the base of the plutonic section of many ophiolites, for example the Oman Ophiolite (top). Photomicrograph in cross-polarized light shows typical texture in clinopyroxenite layer from Bay of Islands Ophiolite. Scale bar is 1 mm.

Figure 7.31 Brown, smooth-weathering dunite in the Oman Ophiolite may form by a number of processes near the Moho transition zone.

a substantial amount of dunite (Fig. 7.31). Individual layers of ultramafic material a few centimeters to a few meters wide are common, but in some ophiolites thicknesses of hundreds of meters to more than a kilometer of ultramafic cumulate rocks exist, commonly with embedded lenses of gabbroic material. As discussed above, the presence and local abundance of these cumulate ultramafic rocks has important implications for the petrogenesis of MORB and the interpretation of the seismic Moho.

In some places meter-scale dikes and sills and larger plutonic bodies ranging from wehrlite to olivine gabbro intrude the gabbroic rocks (Karson *et al.*, 1983; Nicolas, 1989). Although these are most common in the lower part of the sequence, they can also be found in the upper gabbros. Chilled margins, apophyses, and xenoliths of gabbroic material occur at the contacts of these mafic intrusions with the surrounding gabbros (Fig. 7.32). These intrusions tend to occur in areas of magmatic deformation, resulting in complex cross-cutting relationships. The nature of seafloor outcrops generally does not permit these types of relationships to be identified.

In the lower gabbros, structures produced by deformation of partially molten material include foliations and lineations, pinch and swell structures, folds, and shear zones (Figs. 7.33 and 7.34). Outcrop features and igneous microstructures both show that these features are the result of magmatic deformation (for example crystal–mush). Most of the deformation is controlled by layer-parallel slip, probably reflecting significant differences in strength of layers with different amounts of interstitial melt. Local cross-cutting relations occur where deformation of partially molten assemblages results in mobilization of melt-rich material. Similar fabrics have been found in some oceanic gabbros. In extensive exposures of the Oman Ophiolite, some parts of the lower gabbro unit have been affected by anastomosing magmatic shear zones which separate lens-shaped bodies of layered gabbroic material (Fig. 7.1). These types of features

show that the base of the igneous crust can be a very dynamic environment as it is developing.

In other ophiolites, high-temperature, solid-state crystal-plastic deformation has also affected the lower gabbroic unit (Karson, 1977; Casey *et al.*, 1981, 1983; Girardeau and Nicolas, 1981). It is not clear whether the solid-state deformation overprints earlier magmatic deformation fabrics. Where solid-state deformation has occurred, the ultramafic material is significantly less deformed than surrounding or nearby, weaker, gabbroic rock. Thus, strain tends to be localized in gabbroic layers on an outcrop scale and in plagioclase on a hand-specimen to thin-section scale. Textures range from gneissic to mylonitic, reflecting locally large strains and layer-parallel displacements (Fig. 7.35). These relationships imply post-crystallization deformation that would be expected to occur at much higher stresses than for magmatic flow. Although crystal-plastic deformation of gabbroic rocks is common in slow-spread crust (see above) it is rare in gabbros from fast-spread crust (for example, see Fig. 7.2).

Fine-scale studies of ophiolites reveal geological details that provide important information on the process of melt extraction from the top of the mantle and channeling of magmas upward into the overlying gabbroic unit (Fig. 7.36). This area, termed the Moho transition zone (MTZ) is a region of complex interaction among residual upper mantle, melts passing through it, and deformational

Figure 7.32 Wehrlite to olivine gabbros with variable amounts of plagioclase (white) form discrete intrusive bodies hundreds of meters to centimeters across in many ophiolites. In the Bay of Islands Ophiolite these include fine-grained margins in spongy, poikilitic feldspathic wehrlite (upper left); xenoliths of gabbroic host rocks (white) enclosed in feldspathic wehrlite to lherzolite (upper right); small dikelets (dark) intruding surrounding gabbroic rocks (white) (lower right); and dikes and sills of dark wehrlite to feldspathic wehrlite cutting gabbroic rocks (lower left).

processes. Substantial variations occur at this level with respect to distance from centers of mantle upwelling where melt at the base of the crust is concentrated (Fig. 7.37).

7.5.3 Geological expressions of the Moho

The base of the plutonic igneous crust – the gabbro–peridotite contact – has not been penetrated by drilling nor is it exposed by faulting in oceanic crust. Inferences about this contact come from both marine seismic investigations (see Chapter 3) and studies of the lowermost sections of ophiolite complexes. Ophiolite complexes show a wide range of geological features at this structural level that may correspond to the seismic Moho, where seismic velocities increase to more than 8.0 km/s (see Chapter 3). This velocity increase and related seismic reflections can potentially be caused by a number of different geological relationships, so there is no unique interpretation of the Moho beneath the oceanic crust (Fig. 7.38) (Brocher *et al.*, 1985; Collins *et al.*, 1986; Mutter and Karson, 1992). Given the wide range of crustal construction processes at MORs, it is likely that the Moho varies from one place to another even within a limited range of spreading rates.

Figure 7.33 Magmatic deformation is common in the Oman Ophiolite. Deformation features such as folds, pinch and swell structures, and minor cross-cutting dikelets show that deformation took place before the unit was fully crystallized. Folds and lenses in layered dunite (brown) and clinopyroxenite (green) (top). Pinching and swelling (boudinage) of dark wehrlite layer in layered gabbro (bottom). Note that, despite the obvious deformation, the rocks do not show signs of significant crystal-plastic deformation.

In some parts of ophiolite complexes, igneous plutonic rocks with velocities of less than about 7.5 km/s directly overlie residual upper mantle peridotites, which would have seismic velocities of over 8.0 km/s (Table 2.1). The gabbroic rocks have crystallized from melts in subaxial magma chambers, but the upper mantle peridotites were never completely molten. Instead, the peridotites represent a complexly deformed volume of olivine and pyroxenes left behind as an unmelted residue from which basaltic melts have been extracted (see Chapter 8). Thus, this contact, referred to as the "petrologic Moho," represents a boundary between rock types not only with very different compositions, but also with very different geological histories. If a geologic contact of this type is developed over a lateral scale that can be detected in marine seismic experiments (kilometers), it could produce Moho refractions and reflections. However, important variations on this theme are found near the gabbro–peridotite contacts elsewhere in ophiolites, including, for example, layered sequences, lens-like intrusions, and ultramafic cumulate rocks, which suggest a range of geological relationships could produce very similar seismic responses (Fig. 7.38). Therefore, it is probable that the Moho (imaged seismically) has different geological/petrological significance in different areas (Karson *et al.*, 1984).

Significant variations occur from one ophiolite to another and even within the same ophiolite. In some parts of the Oman Ophiolite, this contact is relatively sharp (Fig. 7.37), with interlayered gabbroic and magmatic ultramafic rocks (wehrlites) directly overlying residual peridotites (see Chapter 8 for description and classification of ultramafic rocks). Elsewhere, magmatic deformation, melt impregnation, and sill intrusion within dunite units are common at this structural level. Significant variations in the composition and deformation structures of the MTZ appear to depend on the proximity to areas where previous focused mantle upwelling (diapirs) occurred (Jousselin *et al.*, 1998).

In contrast, in the Bay of Islands Complex in Newfoundland, volumes of dunite (100 m to >1000 m thick) host discontinuous "megalenses" of gabbroic and wehrlitic

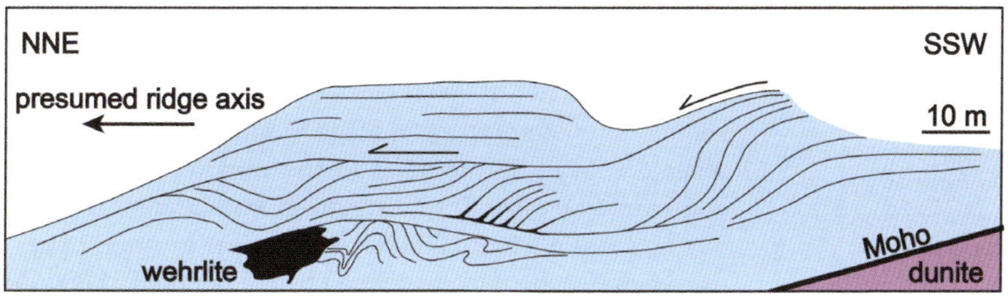

Figure 7.34 Sketch of structural relationships in gabbroic rocks just above the Moho transition at Wadi Marham in the Oman Ophiolite. This area is located near a center of mantle upwelling. Fine lines show the trace of layering, bold lines show magmatic shear zones with slip inferred to have the upper layers moving toward the paleo-ridge axis (after Jousselin *et al.*, 1998). The image shown in Fig. 7.1 is from just to the west (right) of the wehrlite body.

Figure 7.35 Platy dunite–wehrlite/pyroxenite mylonites at the Moho transition zone of the Bay of Islands Ophiolite (left). Outcrop is about 10 m high. Very strongly lineated wehrlite mylonite from same area (top right). Folded gabbro (white) and dunite to wehrlite (brown) also in the Bay of Islands Ophiolite (bottom right). Field of view ~1 m.

Figure 7.36 Partial melting of mantle rocks produces plagioclase-rich veins (white) that are progressively segregated and transported through the dunite and peridotite (dark) of the MTZ in the Oman Ophiolite. These images show diffuse melt veins (lower left) that merge to form larger veins and dikes (lower right) and dense networks (above) that transport melt from the MTZ to higher levels of the crust.

rocks, creating a transition zone that is in some cases kilometers thick (Casey *et al.*, 1981). Major variations in the layered structure occur over lateral distances of a few kilometers. Evidence of crystal-plastic deformation ranging from mild to mylonitic is widespread. Synthetic seismograms show that the scale and impedance contrasts of layering in mafic and ultramafic rocks can produce strong, continuous reflectors. Major reflectors comparable to the reflection Moho can occur more than 1000 m shallower than the top of the residual peridotites. Thus, the "petrologic Moho" and "seismic Moho" need not and generally do not coincide in this ophiolite (Collins *et al.*, 1986). This geologic/seismic mismatch will occur wherever there are large volumes of cumulate ultramafic rocks in the lower crust.

In still another variation, the Josephine Ophiolite in Oregon features a serpentinite–mylonite shear zone about 100 m thick at the gabbro–peridotite boundary (Harper, 1988). Only future deep drilling through oceanic crust is likely to demonstrate which of these ophiolite analogs or other relationships occur in the oceanic lithosphere.

In crust generated at slower spreading rates and lower magma budgets, the situation is even more complicated. The crust appears to be constructed from volumes of mafic gabbroic/ultramafic material intruding residual upper mantle rocks. Partial serpentinization lowers the density and seismic velocity of peridotites, diminishing the seismic distinction between gabbroic and ultramafic units. This geologically complex assemblage would have an average seismic velocity appropriate to the middle oceanic crust and would be difficult to distinguish from a more continuous and less altered gabbroic assemblage. In some places, the seismic Moho may be defined by the boundary between fresh and overlying partially serpentinized material. In this case, the seismic Moho would delineate the depth of hydrothermal alteration of the crust and have little or no significance for magmatic processes.

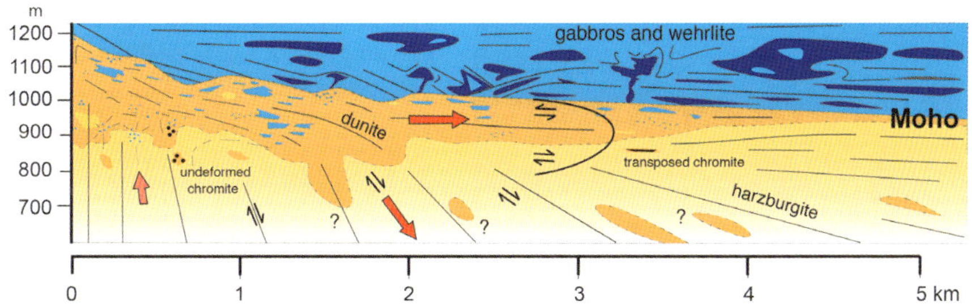

Figure 7.37 The Moho transition zone in the Oman Ophiolite (blue, gold, yellow) reveals the processes beneath spreading ridges and their variations with respect to subaxial mantle diapirs. Melt separated from the rising mantle is deformed along with mantle host rocks (Jousselin and Nicolas, 2000a).

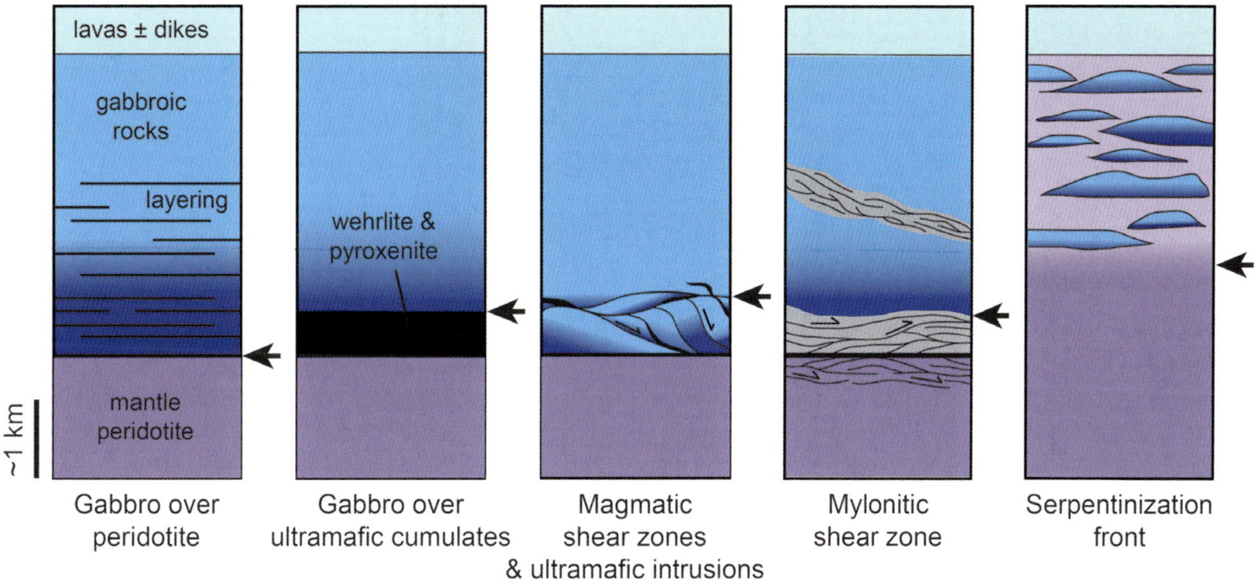

Figure 7.38 Geological relationships that could correspond to the seismic Moho at the base of the oceanic crust are highly variable in ophiolites. The arrow in each column marks the location of the seismic velocity change that would mark the Moho. The standard interpretation from marine reflection and refraction seismology (also found in some ophiolites) is represented in the section on the left. In some places major variation occurs over distances of only a few kilometers. The different structures include gabbroic rocks directly over peridotite or over ultramafic cumulate rocks, major magmatic or mylonitic shear zones, and gradations in serpentinization of mantle peridotites with or without gabbroic intrusions.

7.6 Building the middle to lower oceanic crust

During the mid twentieth century, investigations of layered mafic/ultramafic igneous intrusions in the continents were driven by the concentrations of valuable mineral deposits found in them, especially those including base metals, gold, and platinum group minerals. In these layered mafic intrusions, the sequence of crystallization of mineral phases and modes of accumulation are preserved; along with parallel studies of experimental petrology, these laid the foundation for modern igneous petrology (Wager and Brown, 1968; McBirney, 1975). In many early investigations, details of layering, mineral chemistry, and fine-scale textures suggested that they formed essentially as "magmatic sediments" within large magma chambers (Fig. 7.39).

These layered igneous rocks are composed of early formed crystals (one or more phases) that accumulated and were subsequently surrounded or at least partially enclosed by mineral phases that crystallized somewhat later. These components are typically referred to as "cumulus" and "post-cumulus" phases, respectively, and together make up "cumulate" plutonic rocks (Irvine, 1982). From the variations in mineralogy, textures, and layered sequences in plutonic rocks, changes in the composition and physical characteristics of the magma chamber were determined. In many places, the sequence of mineral assemblages and mineral compositions repeats in rhythmic or cyclic patterns, suggesting that the bulk composition of the magma chamber changed with fresh influxes of magma (e.g., Wager *et al.*, 1960; Wager and Brown, 1968; McBirney, 1975).

The interpretation of these cumulate rocks as "magmatic sediments" led to the assumption that the prominent layering and generally parallel alignment of elongated crystals must have originally formed in subhorizontal beds, just as in sedimentary rocks. The lateral continuity of individual

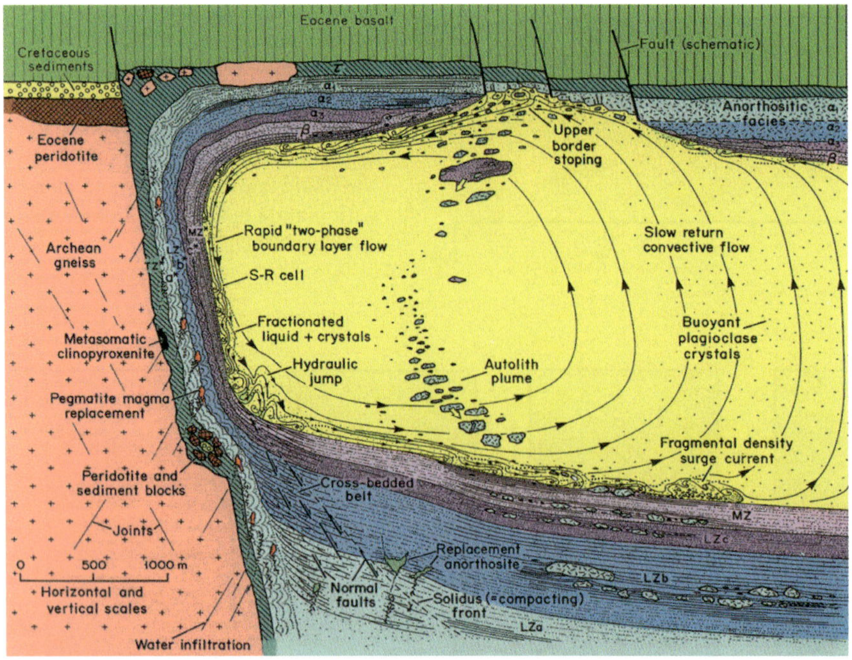

Figure 7.39 Schematic diagram showing major features and processes inferred for magma chambers that produced layered mafic intrusions such as the Skaergaard Intrusion in East Greenland, the Bushveld Intrusion in South Africa, and Stillwater Intrusion in Montana (from Irvine *et al.*, 1998).

compositional layers, even down to the centimeter scale, was found in some cases to be hundreds of meters to many kilometers. Accordingly, it was thought that the floor of magma chambers where this layering formed had similar lateral dimensions. Later investigations demonstrated that identical layering and shape fabrics could form even on vertical walls of intrusive bodies, demonstrating *in situ* crystallization as a potentially important magmatic process (Campbell, 1978). Further investigations have documented complex processes of segregation, compaction, and recrystallization in these types of rocks, casting further doubt on any simple interpretation of the formation of cumulate layering (e.g., Meurer and Boudreau, 1998a,b).

Before it was recognized that that ophiolites had faulted contacts with surrounding rocks and had been tectonically transported from their points of origin, many ophiolites were thought to be layered intrusive complexes that had been intruded into their present settings. Following the correlation of ophiolites with oceanic lithosphere, which was made in the early 1970s, magma chamber processes were linked to MOR magma chambers and processes of crustal accretion. Early models for accretion of the oceanic crust were based in part on cumulate processes observed in layered intrusions, and it was consequently anticipated that continuous magma chambers would be large at fast-spreading ridges and substantially smaller at slow-spreading ridges (Greenbaum, 1972; Cann, 1974; Fowler, 1976; Dewey and Kidd, 1977; Casey and Karson, 1981; Pallister and Hopson, 1981).

In the late 1980s, marine seismic reflection data documented the along- and across-axis extent of AMCs on fast-spreading MORs (Detrick *et al.*, 1987) (see Chapter 3), putting to rest the notion of very large subaxial magma chambers, even at fast-spreading ridges. Somehow, the accretion of several kilometers thickness of the middle to lower oceanic

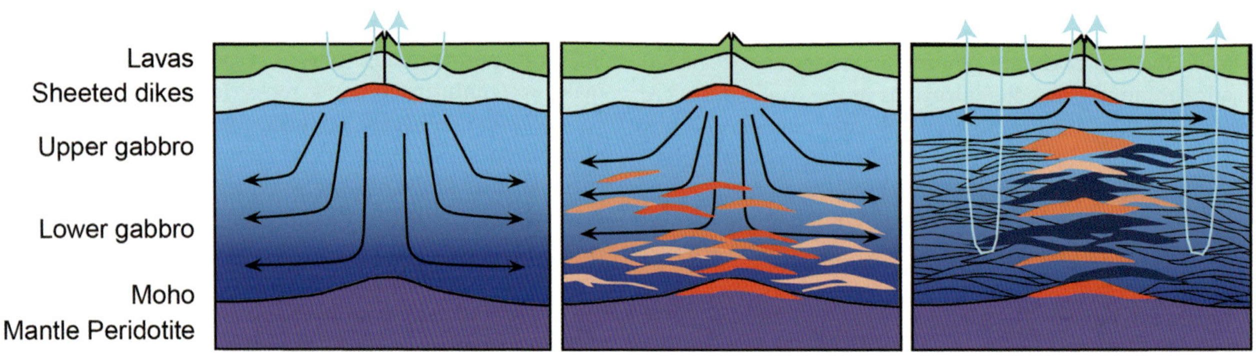

Figure 7.40 Recent models for accretion at relatively fast-spreading ridges reflect constraints from marine seismology, ophiolite studies, and ocean crust sampling. The "gabbro glacier" model (left) features downward and lateral magmatic flow from a shallow melt lens (after Quick and Denlinger, 1993). Accretion of the middle to lower crust mainly by sill intrusion (right; note gabbro glacier in upper gabbros) (Henstock *et al.*, 1993; Kelemen *et al.*, 1997b). Hybrid model (middle) with gabbro glacier in upper to middle crust and downward-increasing sill intrusions and superimposed deformation from mantle flow (Boudier *et al.*, 1996). Light blue lines indicate hydrothermal circulation.

crust had to be reconciled with magma lenses only tens of meters thick and not much more than 1000 m wide, perched above a low-velocity zone with a small percentage of interstitial melt (Sinton and Detrick, 1992).

This new view of MOR magma chambers spurred renewed efforts to find evidence for magma chambers preserved in ophiolites. The re-examination of ophiolite gabbros revealed that much of the outcrop-scale layering is the result of densely spaced sills on the order of meters to tens of meters thick with finer-scale cumulate layering (e.g., Kelemen *et al.*, 1997b). Hypersolidus (magmatic) and in some cases subsolidus (crystal-plastic) deformation tend to obscure the details of the sills and their internal layering. Although discrete sills are most obvious in the strongly layered rocks of the lower part of the gabbroic unit, they can also be found in the residual upper mantle beneath the main volume of plutonic rocks. The sills become much less distinct upward through the middle part of the gabbro unit. These observations, along with the marine seismic data (Chapter 3), suggest that the intrusion of magma lenses creating sills in the gabbroic unit is an important process from the upper mantle all the way up to the base of sheeted dike complexes in the upper crust.

At present, integrating available information from ophiolites, continental intrusions, rocks recovered from the seafloor, and marine seismic studies, does not lead to a uniquely constrained view of how the middle to lower crust forms. The major processes such as sill intrusion, solid-state flow, crustal assimilation, and chemical refining of ascending melts are well documented, at least locally, but how they integrate in time and space to build the bulk of the oceanic crust is not fully understood.

An interesting geometric/kinematic puzzle remains: how to build a more or less continuous, horizontally layered, middle and lower crust on the order of 5 km thick from densely spaced sills intruded in the very narrow (few kilometers wide) subaxial region (i.e., the AMC and LVZ) where magmatic processes are focused. The current generation of models seeks to reconcile the available data (Fig. 7.40). If most melt-rich lenses occur near the top of this volume, crustal accretion would require substantial downward and lateral flow to build the deeper parts of the gabbroic unit. If sill intrusion occurs through the full thickness of the middle to lower crust, then only limited flow is required.

Two end-member type models place different emphasis on sill intrusion as opposed to magmatic flow (Fig. 7.40). In one, the "gabbro glacier" model, melt is fed from the mantle to the upper crustal AMC or melt lens where it partially crystallizes. The resulting crystal mush subsides and

Figure 7.41 Model for development of vertical foliation in upper gabbros of the Oman Ophiolite. A strong fabric defined by the preferred orientation of elongate plagioclase and pyroxene crystals (photomicrographs below) develops as partially molten material initially formed on the floor of a magma lens subsides asymmetrically and deforms in a vertical shear zone as a crystal mush (after Nicolas *et al.*, 2009; Boudier and Nicolas, 2011).

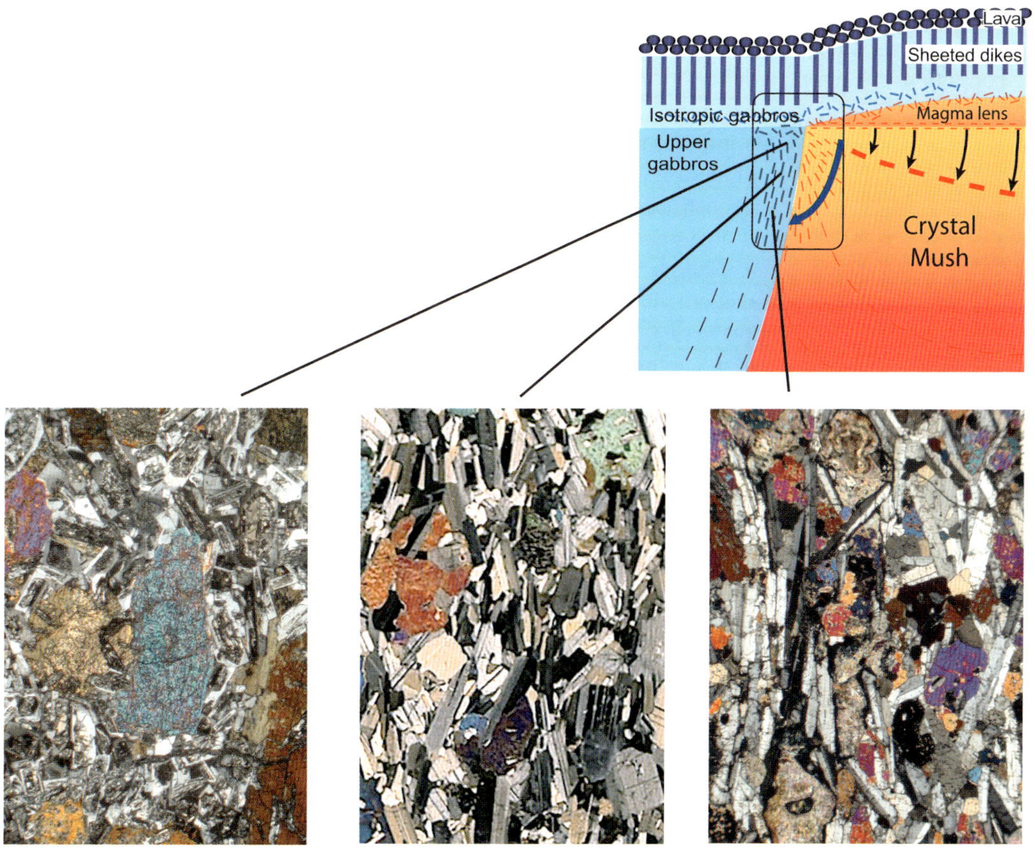

spreads laterally to form the middle to lower oceanic crust (Quick and Denlinger, 1993). Alternatively, in the "sheeted sill" model, melts crystallize in a series of sills as melt migrates from the mantle to the upper crust – the top of the AMC is simply the uppermost of these sills (Kelemen *et al.*, 1997b). One of the key differences between these models is how magmatic and hydrothermal processes are related. The models make important predictions for the internal structure of the middle to lower crust and heat transfer to the surface (Maclennan *et al.*, 2004, 2005).

Detailed studies in upper gabbros of the Oman Ophiolite provide important details of these types of processes (Nicolas *et al.*, 2009; Boudier and Nicolas, 2011). The evolution of fabrics in gabbroic rocks with depth beneath the sheeted dike complex suggests dramatic subsidence of the floor of magma lenses (Fig. 7.41). However, vertical fabrics elsewhere are interpreted as the result of upward flow of gabbroic material beneath the spreading axis (Yaouancq and MacLeod, 2000).

The diverse aspects of these two models, and hybrid models that incorporate aspects of both of them, illustrate the ambiguities in understanding how the middle to lower oceanic crust develops in a relatively continuous magmatic system present beneath fast-spreading ridges (Fig. 7.42). These models can be tested both with more direct geological investigations of oceanic rocks and with marine geophysical techniques. Crustal construction at slow spreading rates is likely to be much more variable in time and space and more complex because of the lower mantle temperatures and lower rates of mantle upflow compared to faster-spreading settings (Fig. 7.42). In areas of high magma budget (e.g., proximal to hot spots or segment centers), there may tend to be a more continuous magmatic system similar to those beneath faster-spreading ridges. However, seismic and geologic data from slow-spreading MORs indicate that many ridge segments must have very discontinuous magma supplies. Major faulting interrupts magmatic construction on time scales of 10^5 to 10^6 years. In these types of systems, magma bodies may be smaller and tend to have shorter life spans. They may also be subject to more interaction with different host rock types including hydrated upper crustal or mantle rocks and peridotites. Thus, the different modes of construction of the middle to lower crust in relatively fast- and slow-spreading environments are likely to have different petrologic/geochemical relationships (Rubin and Sinton, 2007).

The current view of the accretion of the middle to lower crust has evolved rapidly with the integration of marine seismic data and detailed investigations of gabbroic and related rocks in ophiolite complexes. As investigations in these areas continue to produce increasingly refined views of magmatic processes at spreading centers and "fossilized" magma chamber rocks on land in ophiolites and layered mafic intrusions, it will be important to bridge the gap between these approaches with detailed studies of oceanic gabbroic rocks *in situ*. Petrological and structural studies of major exposures at tectonic windows can be done with HOVs or ROVs, at essentially the same scale as investigations in ophiolite complexes. These exposures provide access to at least the upper third of the gabbroic units of the crust and could provide data with which to test the current generation of models. In addition, deep crustal drilling can provide important continuous vertical sampling in areas away from tectonic windows to help place magmatic processes in the context of spreading segment setting. Access to the deepest parts of the oceanic crust and upper mantle must await major drilling efforts that can reach the Moho.

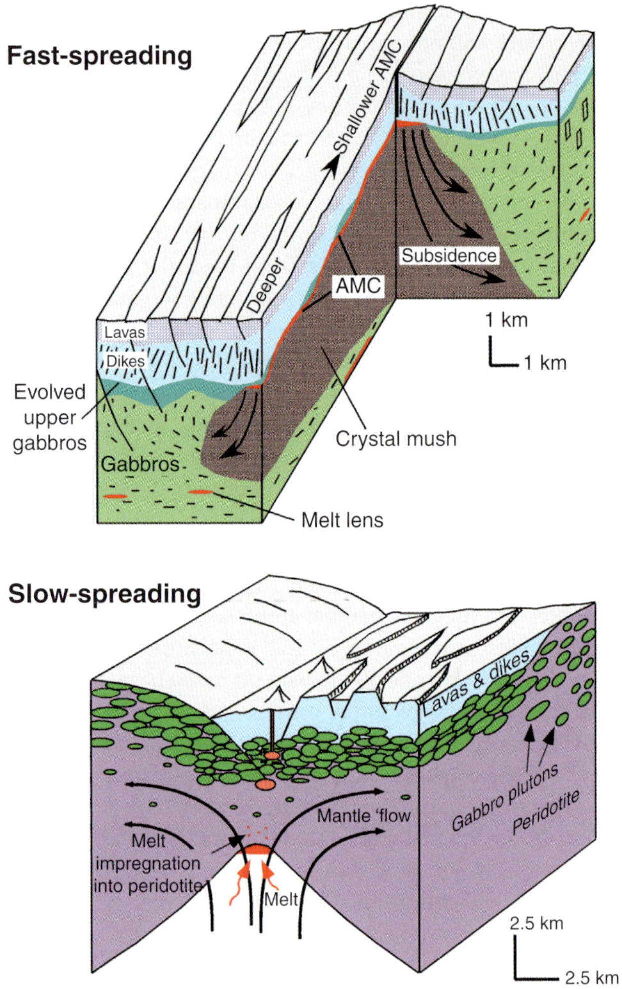

Figure 7.42 Schematic representations of magmatic processes inferred from gabbroic rocks from the seafloor and ophiolites emphasize the differences in the modes of accretion at different types of MORs. At fast-spreading ridges melt lenses and an extensive volume of crystal mush, represented by the LVZ, define a subaxial magma chamber from which gabbroic crust crystallizes. Melt lenses occur at various depths and as much as several kilometers away from the axis. Melts generated by partial melting in the underlying mantle pass through and react with material in the magma chamber. At slow-spreading ridges, magma chambers and melt lenses are more ephemeral and create a more discontinuous gabbroic complex made up of numerous discrete intrusions in the crust and upper mantle (Coogan, 2014).

8 Peridotites: windows into mantle processes

The lava floods and intriguing volcanoes tell us of the plasticity, mobility, of the deep interior of the globe.

R. A. Daly (1926)

8.1 Below the Moho: the oceanic mantle

Although peridotites are not common components of continental geology, large exposures of peridotites occur on the seafloor and as parts of ophiolite complexes. Unlike superficially similar coarse-grained gabbroic and ultramafic plutonic rocks that make up the middle to lower oceanic crust (see Chapter 7), peridotites are not igneous rocks in the

Figure 8.1 Earth's mantle is dominantly peridotite, a rock composed mostly of the mineral olivine. Ductile flow and partial melting of mantle peridotite produces the magmas from which the oceanic crust is constructed. The mosaic texture of the thin section of peridotite, as shown here in cross-polarized light, is the result of solid-state flow and recrystallization beneath a MOR spreading center. Field of view ~4 cm.

Figure 8.2 Samples of mantle peridotite recovered from the seafloor commonly have cross-cutting veins of pyroxene- or plagioclase-rich rock (left) interpreted as melt trapped in the mantle. Sample is about 20 cm across. Peridotite outcrops in ophiolites (right) show evidence of complex deformation of veins formed by melt interaction with the host rock during mantle flow. In this outcrop of mantle peridotite in the Bay of Islands Ophiolite, less deformed pyroxenite veins (dark) cut somewhat older, variably folded and sheared veins. Field of view ~2 m.

sense that they do not form from wholesale melts. Instead, they represent the residue left behind by partial melting of what was once more fertile upper mantle material from which basaltic melts (magmas) have been extracted. Peridotites commonly have metamorphic fabrics indicative of high-temperature, solid-state deformation (Fig. 8.1) related to the large-scale ductile flow in the mantle. The effects of this type of deformation are manifest on both macroscopic and microscopic scales. As such, they may be regarded as a special type of anhydrous metamorphic rock or "tectonite." Numerous veins and mineral segregations in peridotites mark the paths of melt that passed through and interacted with the flowing mantle beneath mid-ocean ridge (MOR) spreading centers (Fig. 8.2).

Exposures of peridotites on the seafloor (commonly referred to as "abyssal peridotites") and in ophiolite complexes provide important glimpses into the composition and structure of the upper mantle. They also yield important insights into magma generation processes that build the oceanic crust and provide the heat that drives hydrothermal systems. Extensive outcrops of peridotite, and its altered product, serpentinite (see below), are exposed on the seafloor by large-scale faulting and have been sampled by dredging, submersibles, and drilling. However, because of the processes that have created these exposures and intense serpentinization, it is difficult to constrain the spatial relations between different rock types, the orientations of structures, and other details on scales beyond that of individual samples. Ophiolite peridotites provide very large, accessible exposures of these mantle rocks, which can be examined, mapped, and sampled in the context of broader-scale (meters to kilometers) relationships, including their relationships to structurally overlying crustal rocks. Together, the perspectives from the seafloor and ophiolites provide windows into the rocks beneath the Moho. These data can be related to processes in the mantle inferred from marine geophysical studies (see Chapter 3). This chapter explores the nature of Earth's mantle beneath the oceanic crust through the use of geochemical and theoretical studies based on observations and samples of mantle peridotites.

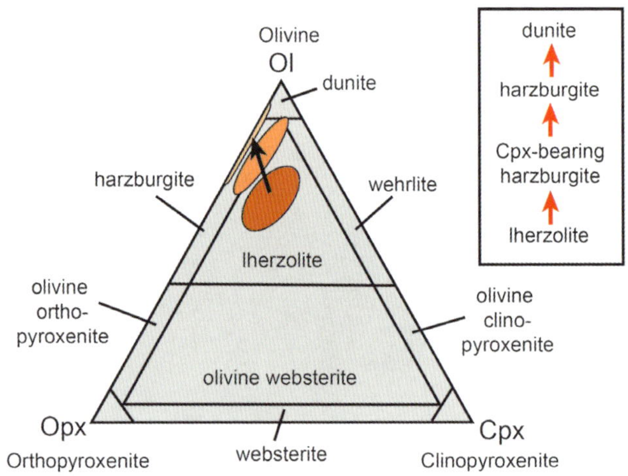

Figure 8.3 Ternary diagram showing the mineralogic (modal) composition of peridotites and depletion of mantle rocks by progressive melting (arrows). Mineral end-members are olivine (Ol), orthopyroxene (Opx), and clinopyroxene (Cpx). The colored areas show the main compositional fields of common peridotites that can be related by melting in the mantle. Primitive upper mantle material (lherzolite) is progressively depleted by melting to produce harzburgite and dunite. Pyroxenites and wehrlites commonly occur as dikes, veins, and lenses within these peridotite bodies.

8.2 Mantle rocks

Rock types in ophiolite and abyssal peridotite units include dominantly harzburgite, with lesser amounts of dunite, lherzolite, and pyroxenite – all primarily consisting of various proportions of the Mg-rich minerals olivine,

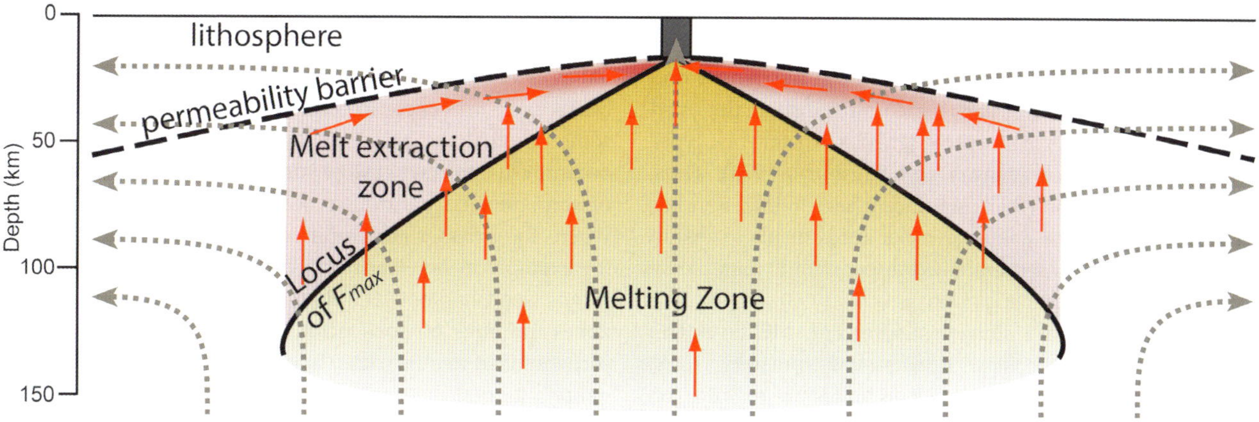

Figure 8.4 Diagrammatic representation showing a model of melt focusing, where melts rise buoyantly (red arrows) to the permeability barrier or thermal boundary layer (bold dashed line) at the base of the lithosphere where they may migrate upward to feed the axial magmatic system. These melts may also feed off-axis seamounts. Dotted flow lines show the path of solid asthenospheric mantle beneath the spreading center. As mantle material flows horizontally, it cools and increases in density and rigidity as it is transformed into oceanic lithosphere. Purple shaded area is the porous zone in the mantle where melting no longer occurs but from which melts can be extracted (Hébert and Montési, 2010).

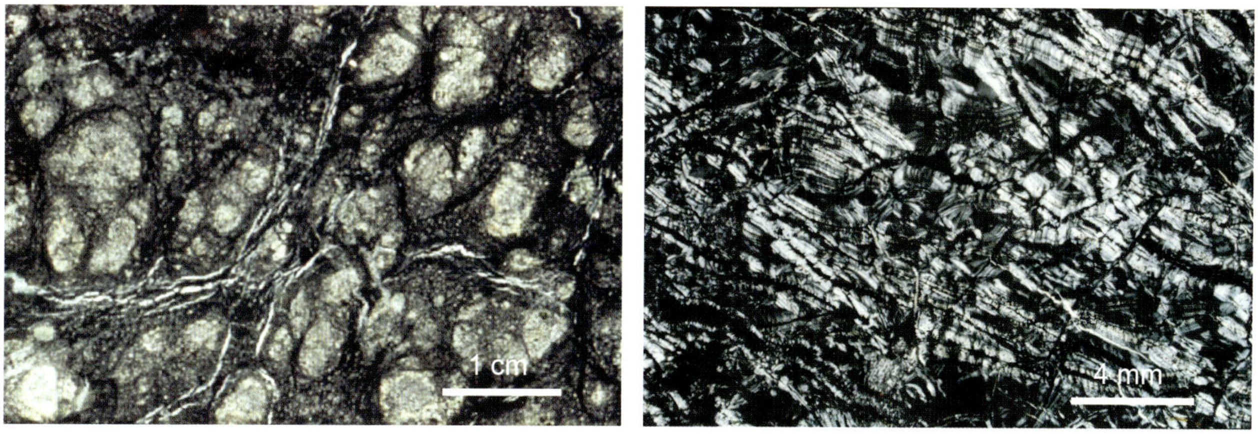

Figure 8.5 Hydration of mantle peridotites results in the conversion of olivine and orthopyroxene to serpentine minerals and related processes collectively referred to as serpentinization. Serpentinized peridotite from ODP Leg 153 (left) shows the evidence of mantle deformation in deformed pyroxene grains (pale green). Most of the olivine is replaced by serpentine minerals (black) and serpentine veins (white). In thin section (right) the colorful mosaic texture of anhydrous peridotite (Fig. 8.1) is replaced by gray "mesh-textured" serpentine.

orthopyroxene, and clinopyroxene (Fig. 8.3 and Table 5.1). Chrome spinel (chromite) typically occurs as disseminated grains in the peridotite, although in some places concentrations occur as variably deformed veins or layers or as lens-like "podiform" masses.

Peridotites are important because their geochemistry and their geological relationships in ophiolite complexes, and general considerations of seafloor spreading, make it possible to construct models for mantle processes beneath MORs (Fig. 8.4). In general, these involve the rise of buoyant upper mantle material and magma generation caused by decompression: a portion of this melt feeds crustal magma chambers. The residual, refractory mantle material rolls over to horizontal flow beneath diverging plates, where it becomes the solid foundation upon which the oceanic crust is constructed.

8.3 Peridotite exposures on the seafloor

Peridotites cannot be discussed without understanding how readily they can be altered. Where seawater interacts with peridotites at moderate to low temperatures (<500 °C) olivine and orthopyroxene react to form various serpentine minerals (O'Hanley, 1996) (Fig. 8.5). Nearly all peridotites from the seafloor and ophiolites show at least some signs of serpentinization as well as seafloor weathering. The "serpentinization" of peridotites results in profound changes in the chemistry and physical properties of peridotites (see below). Among the changes is a decrease in the density and seismic velocity of peridotites, which can substantially overlap with those of typical oceanic igneous crustal rock types (Fox *et al.*, 1973; Christensen and Salisbury, 1975). Thus, beneath the seafloor where significant serpentinization has occurred, subsurface gabbroic rocks are difficult to distinguish from partially serpentinized upper mantle material by seismic methods alone (e.g., Detrick *et al.*, 1993).

Peridotites are rare along fast-spreading MORs, but in lithosphere created at slow to ultraslow spreading rates, extensive areas of serpentinized peridotites are exposed on the seafloor. Overlying magmatic units have either been removed by faulting or did not form as part of the spreading process (Chapter 3). Here, the oceanic crust appears to be composed of serpentinized peridotite and the Moho might be defined by a metamorphic front separating fresh from altered peridotite. In these areas, substantial crustal-level deformation is required to accompany the uplift and stretching of the upper mantle associated with plate separation. This is basically the spreading process envisioned by Harry Hess (1962) before the birth of modern seafloor geology and plate tectonics (Chapter 2) – an old idea, rejuvenated by new data.

Exposures of partially serpentinized mantle peridotites on the seafloor occur in specific settings (Table 8.1) including major fault zones along slow- to ultraslow-spreading ridges, major transform faults and their non-transform fracture zone extensions, and major escarpments near the tips of propagating rifts. Exposures of mantle rocks formed at intermediate and faster rates are not only rare but also relatively small. In contrast, along ultraslow- to slow-spreading MORs, abyssal peridotites appear to be common at shallow structural levels in a number of different settings.

The most extensive exposures of serpentinized peridotites occur along oceanic fracture zones (e.g., Miyashiro *et al.*, 1969; Fox and Stroup, 1981; Bonatti, 1994). In areas of low magma budget, serpentinized peridotites crop-out along rift valley walls (Karson *et al.*, 1987; Cannat *et al.*, 1995b, 1997). Major exposures occur in oceanic core complexes and also in some anomalous areas of "smooth seafloor" found in oceanic crust formed at slow spreading rates. Some outcrops of serpentinized peridotites occur along spreading centers where the broader structural setting is poorly understood, for example in non-transform offsets along slow-spreading ridges (see Chapter 3). Drilling in areas away from fracture zones and other obvious tectonic features has also recovered serpentinized mantle rocks (Juteau *et al.*, 1990). These investigations suggest that serpentinites may be an important component of the upper oceanic crust at least in slow- to ultraslow-spreading environments. Serpentinite samples have been recovered by crustal drilling, providing information on this material at depth (Fig. 8.6). Overall, the rock types, fabrics, and structures found in abyssal peridotites recovered from the seafloor are essentially the same as those found in ophiolite complexes.

Historically, extensive outcrops of serpentinized ultramafic rock known as serpentinite massifs were commonly interpreted to have formed as serpentinite diapirs; that is, volumes of weak, low-density material that had moved

Figure 8.6 Summary of ODP drill holes into serpentinized peridotite (left). Legs and drill holes are labeled above, percentage of material recovered is shown schematically (purple) and in numbers below each column. Note that the only cores from fast-spreading areas are from ODP Leg 147. Sample of core from Leg 153 (Hole 920D) shows typical serpentine with pale green remnants of orthopyroxene, black serpentinized olivine, and white to black serpentine veins (right). Despite the serpentinization, the mantle flow fabric (dipping steeply to the right) is still visible. Width of core is 7 cm.

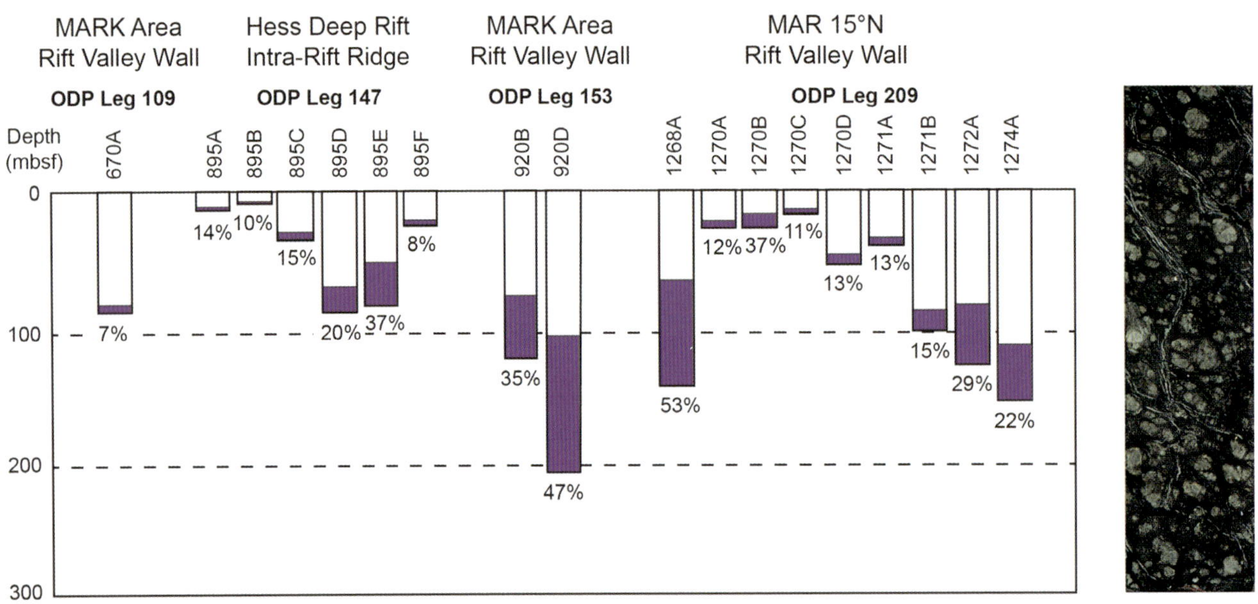

Table 8.1 Seafloor exposures of serpentinized peridotites (see Fig. 3.3 for key locations)

Location	Full rate (mm/yr)	Type of exposure	Main rock types	Notes	References
Ultraslow to slow spreading rates					
Gakkel Ridge		Rift valley and fracture zone walls	Serpentinized harzburgite	Extensive dredging	Dick *et al.*, 2003; Michael *et al.*, 2003
Southwest Indian Ridge	15	Rift valley and fracture zone walls	Serpentinized harzburgite, dunite, and lherzolite	Extensive dredging	Bonatti and Hamlyn, 1978; Dick, 1989; Dick *et al.*, 1991; Natland and Dick, 2001
Mid-Cayman Spreading Center	20	Rift valley walls, OCC	Serpentinized harzburgites	Related to very thin magmatic crust	CAYTROUGH, 1979; Stroup and Fox, 1981; Hayman *et al.*, 2011
MAR 45° N		Rift valley walls	Serpentinite	Dredges	Aumento and Loubat, 1971
North Central Atlantic	21	Drill core DSDP Hole 395	Serpentinites	Below basalts and breccias	Purdy *et al.*, 1978; Juteau *et al.*, 1990
Oceanographer Fracture Zone		Fracture zone valley walls and ridge–transform intersection	Serpentinized harzburgites	Directly below basalts at ridge–transform intersection	Fox *et al.*, 1976; OTTER, 1984, 1985
North Central Atlantic Non-Transform Offsets		Several small transform faults or non-transform offsets	Serpentinized harzburgites	Active hydrothermal venting in some of these areas	Lagabrielle *et al.*, 1998; Gracia *et al.*, 2000
Atlantis Massif South Wall		Transform valley wall, OCC section	Serpentinized harzburgite, dunite, meta-peridotites	Massive peridotite grades upward into a detachment shear zone	Miyashiro *et al.*, 1969; Blackman *et al.*, 2002; Schroeder and John, 2004; Boschi *et al.*, 2006
Atlantis Fracture Zone		Fracture zone walls	Serpentinized harzburgite and dunite	Dredges	Miyashiro *et al.*, 1969
MARK Area 23°15′ N		Detachment fault on rift wall; drill core ODP Legs 109 and 153	Serpentinized harzburgite, dunite, pyroxenite, breccias	Extensive exposures on the edge of a spreading segment	Karson *et al.*, 1987; Mével *et al.*, 1991; Cannat *et al.*, 1995a,b
West of MARK Area		OCC		Serpentinized harzburgite, dunite, feldspathic dunite	Dick *et al.*, 2006
Kane Transform Fault		Transform valley wall, OCC sections	Serpentinized peridotite, meta-peridotites	Complex faulting and minor gabbro intrusions	Purdy *et al.*, 1978; Auzende *et al.*, 1994
15°20′ N MAR	22	Rift valley wall, OCC	Serpentinized harzburgites, meta-peridotites	Detachment shear zone with gabbro bodies	Cannat *et al.*, 1999; McLeod *et al.*, 2002; Keleman *et al.*, 2007a
15°20′ N	22	Rift valley wall, OCC	Serpentinized harzburgite, dunite, lherzolite	Variably sheared serpentinites	Shipboard Scientific Party, 2002

Table 8.1 (*cont.*)

Location	Full rate (mm/yr)	Type of exposure	Main rock types	Notes	References
Vema Transform Fault	24	Transform valley wall	Serpentinized harzburgite	Below faulted contact with gabbro	Auzende *et al.*, 1989
Vema Fracture	24	Fracture zone	Serpentinized	Very long	Bonatti and Honnorez, 1976; Bonatti *et al.*, 2003
Intermediate to Fast and Superfast Spreading Rates					
Hess Deep Rift Intrarift Ridge	110	Propagating rift; intra-rift ridge	Serpentinite	Fault-bounded blocks	Francheteau *et al.*, 1990; Gillis *et al.*, 1994; Hekinian *et al.*, 1994
Pito Deep Rift	142	Propagating rift wall, ODP Leg 147, Nautile dives	Serpentinized harzburgite	Contacts not visible	Francheteau *et al.*, 1994
Terevaka Transform Fault	113	Transform valley wall	Serpentinized harzburgite		Francheteau *et al.*, 1988; Constantin *et al.*, 1996
Garrett Transform Fault	168	Transform valley wall	Serpentinized harzburgite, dunite	Gabbros form intusive dikes and pockets in peridotite. Metasomatic veins	Hébert *et al.*, 1983; Bideau *et al.*, 1991; Hékinian *et al.*, 1992; Constantin, 1999

buoyantly upward through major crustal fault zones and reached the seafloor in a process analogous to salt piercement structures in sedimentary terranes (Bonatti, 1976; Bonatti and Hamlyn, 1978; Francis, 1981). This process, however, has not been documented for serpentinites in any tectonic environment on the seafloor or continents.

In detail, the outcrop settings of serpentinized peridotites on the seafloor are extremely diverse. Many are on major fault scarps where weak, weathered serpentinites commonly create distinctive, rubbly exposures and extensive talus (e.g., OTTER, 1985). Some peridotites exposed on the seafloor near slow-spreading MORs are capped by basaltic lavas (Juteau *et al.*, 1990; Cannat, 1993; Cannat *et al.*, 1995b, 1997; Karson and Lawrence, 1997; Cipriani *et al.*, 2009), showing that, at least locally, oceanic crust does not have the idealized layered structure anticipated from marine seismology and ophiolite studies (see Chapter 2).

Abyssal peridotites are affected to varying degrees by the deformational processes that are responsible for their exposure on the seafloor. These relationships make it difficult to separate the features of the rocks developed in the mantle from those resulting from later processes, for example at crustal levels in transform faults and oceanic detachment faults. These factors also raise doubts about how representative peridotite samples from the seafloor may be in terms of the broader scale vertical and lateral composition of the oceanic upper mantle.

Peridotite outcrops on the seafloor have a wide range of expressions (Figs. 8.7 and 8.8). In some places they are massive with widely spaced planar joints, resembling outcrops of gabbroic rocks.

Elsewhere, serpentinites display very strong foliations and a platy form similar to shear zones in other rock types. Other areas superficially resemble lumpy, highly fragmented outcrops of pillow basalt (OTTER, 1984). Unlike exposures of gabbroic rocks, steep headwall scarps are less common in peridotites, especially where serpentinized. Instead, degraded, low-relief slopes with poorly preserved outcrops are typical. Pervasive development of fractures, veins, and other surfaces of low cohesion results in collapse of nearly all steep topography to produce an undulating rubbly surface. Like exposures of basalts, smooth, downslope lineated talus ramps form aprons on major exposures. Serpentinites tend to have numerous large, jagged blocks incorporated in the talus. Seafloor weathering commonly reduces exposures to friable masses of highly altered clay and oxide minerals.

In areas where detailed sampling has been done, the peridotites are moderately to extensively serpentinized and commonly deeply weathered in surface exposures (Fig. 8.9). Notable exceptions are remarkably fresh peridotites from Zagarbad Island in the Red Sea (Bonatti and Seyler, 1987) and peridotites dredged from the Arctic Gakkel Ridge (Hellebrand *et al.*, 2005; Liu *et al.*, 2008). Despite the alteration and weathering, surviving mantle minerals (relicts) show that abyssal peridotites are dominantly harzburgites and lesser lherzolites with rare pyroxenites, wehrlites, and dunites (Bonatti and Honnorez, 1976; Michael and Bonatti, 1985; Dick, 1989; Johnson and Dick, 1992).

In many seafloor serpentinite exposures, high-temperature deformation fabrics produced under mantle conditions are preserved. Typically these include strong

Figure 8.7 Seafloor outcrops of serpentinites showing a range of expressions. Upper left image shows very massive serpentinized harzburgite from the south wall of the Atlantis Massif along the Atlantis Transform Fault. Upper right image from nearby shows a shear zone cutting the serpentinite. Image on lower left from the MARK Area (23°15′ N on the Mid-Atlantic Ridge) has visible foliation, shear zones, and lumpy texture common in areas of deformed serpentinites. Image on the lower right is from the Vema Transform, showing a more blocky structure. See Fig. 3.3 for locations.

metamorphic foliations and lineations defined by the preferred dimensional and crystallographic orientations of olivine and lesser orthopyroxene with traces of clinopyroxene and spinel. Where oriented samples have been collected, these foliations are gently dipping (Boudier, 1978; Cannat *et al.*, 1990; Ceuleneer and Cannat, 1997), but may have been rotated significantly during faulting at crustal levels. Locally, millimeter- to centimeter-scale veins with concentrations of spinel, clinopyroxene, or plagioclase have dunite selvages and are interpreted as melt-reaction pathways (Casey, 1997; Kelemen *et al.*, 1997). Some samples have mylonitic fabrics indicative of extreme grain-size reduction and large strains developed in crustal shear zones (Melson *et al.*, 1972; Jaroslow *et al.*, 1996; Karson, 1998).

8.3.1 Peridotites from superfast to intermediate-rate spreading centers

Most geophysical constraints on mantle processes come from fast-spreading ridges such as the EPR; however, because of the high melt supply and relatively continuous igneous crust generated at these rates, exposures of mantle material are rare. The only known peridotite exposures are at a few large-offset transform faults and in the highly faulted axial regions of propagating rifts (Table 8.1). In both cases, it is not clear how crustal thickness variations and tectonic processes have combined to expose these upper mantle rocks. Samples of serpentinized abyssal peridotite have been recovered from widely spaced transects with submersibles, but the distance between samples does not provide substantial information on their geological relationships with surrounding material. One drill hole in the Hess Deep Rift, Hole 895S, penetrated 94 m of serpentinized peridotite and related rock types (Fig. 8.10) formed beneath the EPR about 1 Ma (Gillis *et al.*, 1993). The details of the nature of the mantle in relatively fast spreading environments is largely inferred from a combination of extrapolations from a few seafloor exposures and drill cores, geophysical data, and ophiolite analogs.

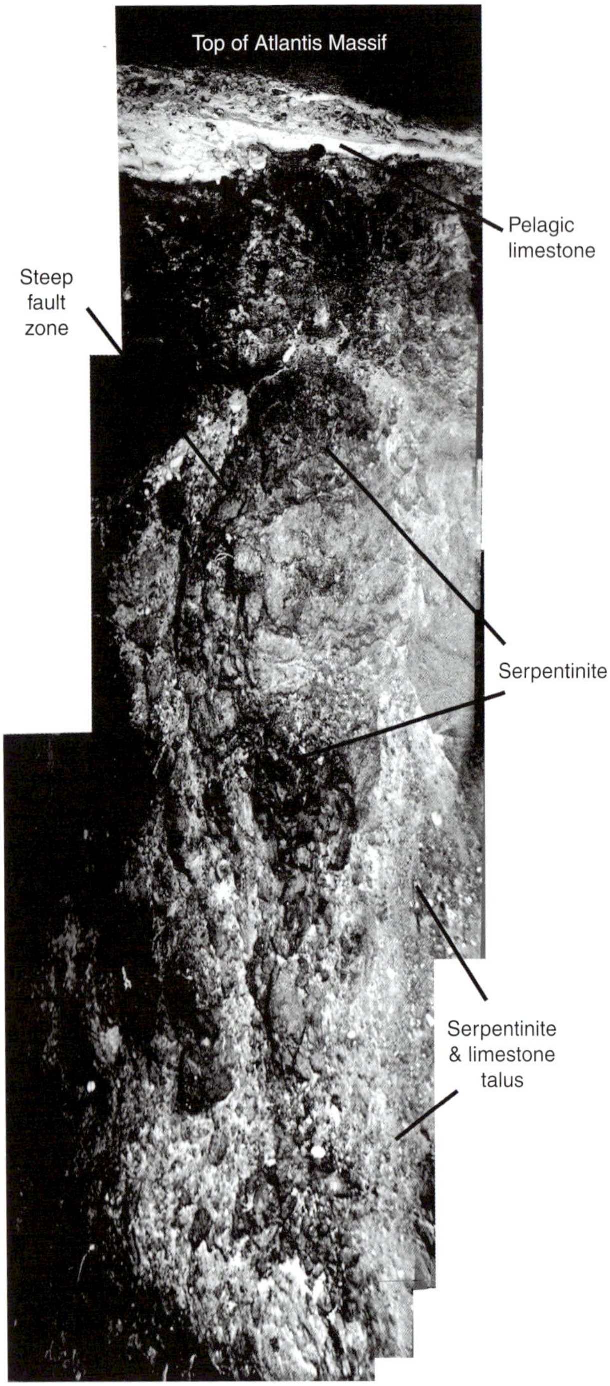

Figure 8.8 Digital image mosaic of a cliff face on the wall of the Atlantis Transform Fault which marks the south face of the Atlantis Massif. The massive peridotite is cut by numerous faults and fractures highlighted by a dusting of white, pelagic sediment. The mosaic field of view is about 25 m vertically.

8.3.2 Peridotites from slow- to ultraslow-spreading MORs

Extensive exposures of serpentinized peridotites occur along spreading centers with slower spreading rates and lower magma budgets. In some places relatively modest normal faulting has exposed abyssal peridotites beneath very thin magmatic units, or in places where there is no evidence that an extensive magmatic crust was ever present. These are believed to be regions where crust (defined by material with seismic velocity <8.0 km/s) consists largely of partially serpentinized peridotites (Brown and Karson, 1988; Cannat, 1993, 1996). In addition to oceanic core complexes (OCCs), serpentinized peridotites occur along the rift valley walls of some parts of the MAR (Karson *et al.*, 1987; Cannat *et al.*, 1997; Karson and Lawrence, 1997; Lagabrielle *et al.*, 1998). At present, it is not clear whether these areas are nascent OCCs or whether they represent some other mode of spreading in areas of thin, slowly spreading crust. Along slow-spreading ridges such as the MAR, the most extensive exposures of serpentinized peridotite appear to be mainly along or near major spreading discontinuities, such as fracture zones or "non-transform" offsets (Fig. 8.11). Some of the best-known exposures of mantle rocks in slow-spreading environments are highlighted in the discussion below.

Major fracture zones in slow-spreading oceanic crust have long been known to have major exposures of mantle peridotites and deep crustal rocks. These major exposures have been investigated by dredging as well as HOV and ROV studies (Miyashiro *et al.*, 1969; Engel and Fisher, 1975; Bonatti and Hamlyn, 1978; Fox and Stroup, 1981; Dick, 1989). Although many samples of serpentinized peridotite have come from these settings, the most systematic investigation of mantle rocks exposed along a fracture zone has been done at the Vema Fracture Zone (Fig. 8.12). A segment of the escarpment more than 300 km long exposes upper mantle rocks that formed during spreading over a time interval longer than 26 Ma, during which systematic variations in mantle melting and crustal generation have been documented (Bonatti *et al.*, 2003; Cipriani *et al.*, 2009). Results from this area show that variations in peridotite geochemistry and crustal thickness are laterally offset along the transform (to younger crustal age) with respect to corresponding signals in basalt geochemistry. This spatial separation of mantle and crustal geochemical markers is consistent with the interpretation that basaltic melts separated from parental mantle peridotite and erupted on the seafloor millions of years before the peridotite (moving at much slower, solid-state flow velocities) reached the surface. Further, the implied ascent rates for mantle material are significantly greater than the plate velocity, suggesting periodic rapid mantle upwelling (Bonatti *et al.*, 2003).

Several other fracture zone exposures have been investigated and sampled with submersibles including sites at the Oceanographer Transform Fault (OTTER, 1984), Kane Transform Fault (Auzende *et al.*, 1994), Vema Transform Fault (Auzende *et al.*, 1989) and the Atlantis Massif (Blackman *et al.*, 2002) (see Fig. 3.3 for locations). In nearly all cases, contact relations with nearby rock units are not visible or have been tectonically disrupted.

While it may not be surprising that mantle peridotites are exposed in areas of large-scale faulting, such as oceanic

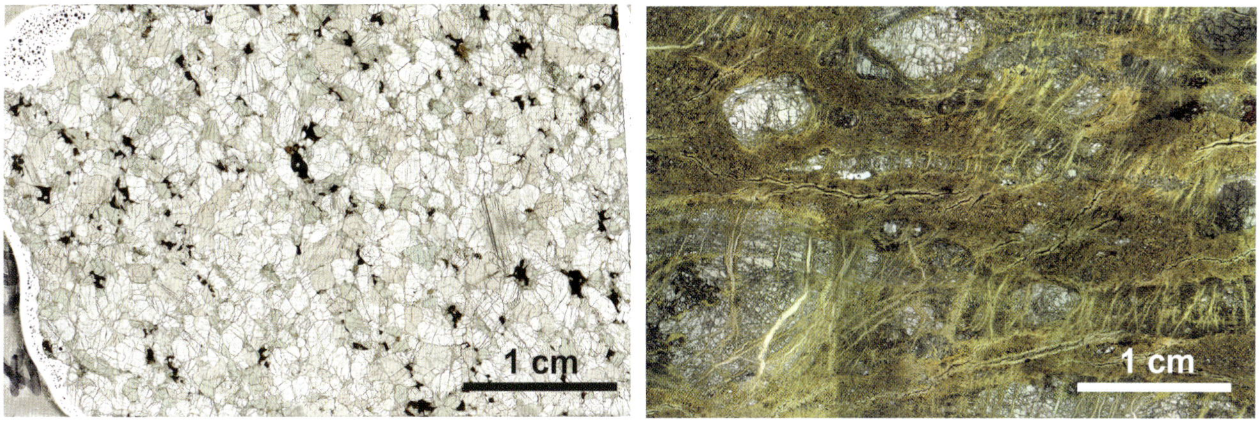

Figure 8.9 Samples of oceanic peridotites are highly variable in texture and degree of serpentinization. The sample on the left, from Zagarbad Island, has a mosaic texture of fresh olivine (greenish) and chromite (black) and is almost completely free of alteration. The sample on the right, from the Vema Fracture Zone, is more typical of samples recovered from the seafloor. It is highly serpentinized (brown, green, and yellow material), but partially altered light-colored grains of orthopyroxene preserve the foliation developed during mantle flow. Both photomicrographs taken in plane-polarized light.

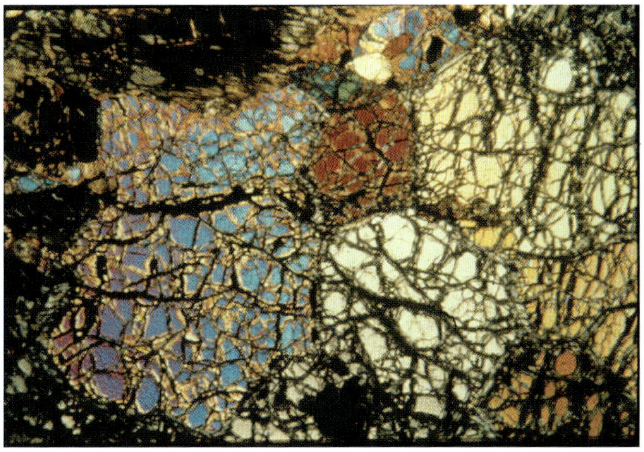

Figure 8.10 Partially serpentinized peridotite from the ODP Hole 895D at the Hess Deep Rift, one of the rare locations from which upper mantle material from a fast-spreading ridge has been recovered. Field of view 7 mm.

fracture zones or OCC, some other exposures are more enigmatic. For example, serpentinized abyssal peridotite outcrops have been found along small spreading discontinuities referred to as non-transform offsets (NTOs) along the MAR between 38° N and 34° N. These have lateral offsets of the spreading axis of only ~15–30 km and age offsets of 1–3 Ma (Gràcia *et al.*, 2000). Seismic investigations show that the crust is anomalously thin in these areas relative to along-axis segment centers (Canales *et al.*, 2000). One example is the Saldana Massif on the MAR, a roughly equant massif in a non-transform offset between the FAMOUS and AMAR study areas with evidence of past hydrothermal venting (Barriga *et al.*, 1998). Peridotite exposures in these areas of limited faulting show that magmatic components of the crust must be very thin or discontinuous in these areas, despite the robust magma supply expected proximal to the Azores hot spot.

As noted above, some OCCs apparently develop in regions of thin or discontinuous magmatic crust with limited evidence of spreading-related igneous activity (see Chapter 3). The footwalls (lower plates) of these OCCs are dominated by serpentinized abyssal peridotites with varying proportions of gabbroic intrusions (MacLeod *et al.*, 2002; Cannat *et al.*, 2006; Dick *et al.*, 2008; Cheadle and Grimes, 2010).

Atlantis Massif, at 30° N on the Mid-Atlantic Ridge is one of the most intensely investigated OCCs. The south wall of

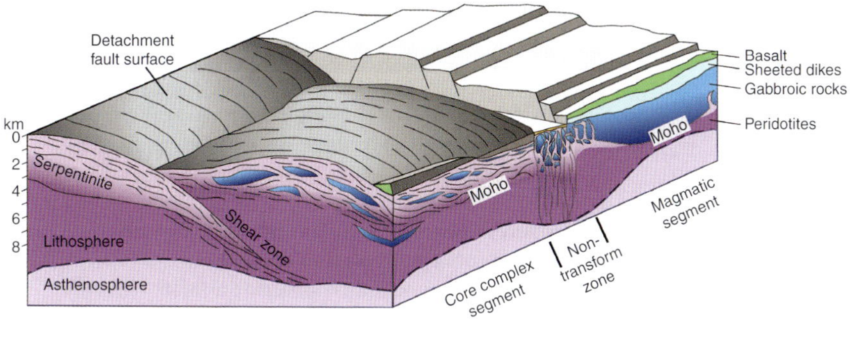

Figure 8.11 The crust along slow- to ultraslow-spreading ridges is highly heterogeneous and variable in thickness along spreading centers. Deep crustal and upper mantle rocks are exposed by extreme tectonic extension in oceanic core complexes and also in relatively cool areas such as transform faults and non-transform offsets.

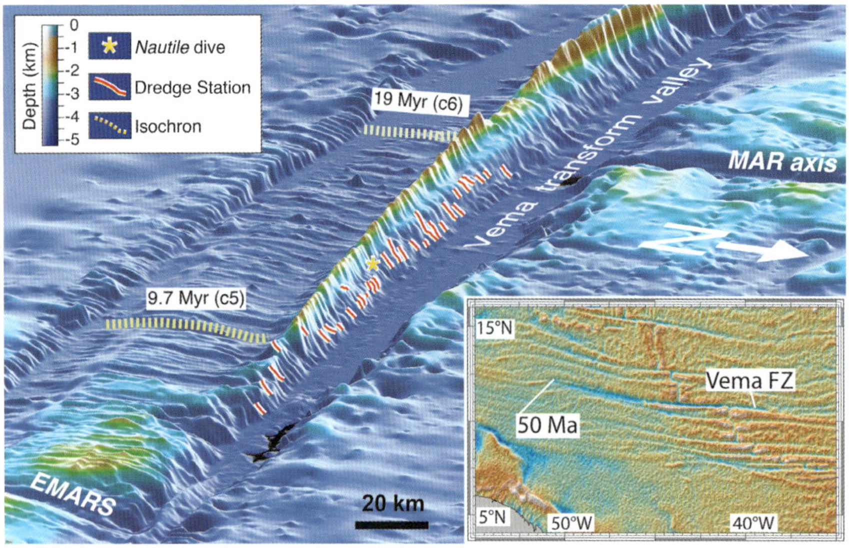

Figure 8.12 Peridotite samples collected along the wall of the Vema Fracture Zone using HOVs and dredges (red) represent a 10 Ma record of mantle processes related to spreading on the Mid-Atlantic Ridge (Bonatti *et al.*, 2003). Later studies have extended sampling to 26 Ma over 300 km along the fracture zone. EMAR-Eastern MAR axis. 9.7 and 19 Myr ages determined by magnetic anomalies c5 and c8, respectively.

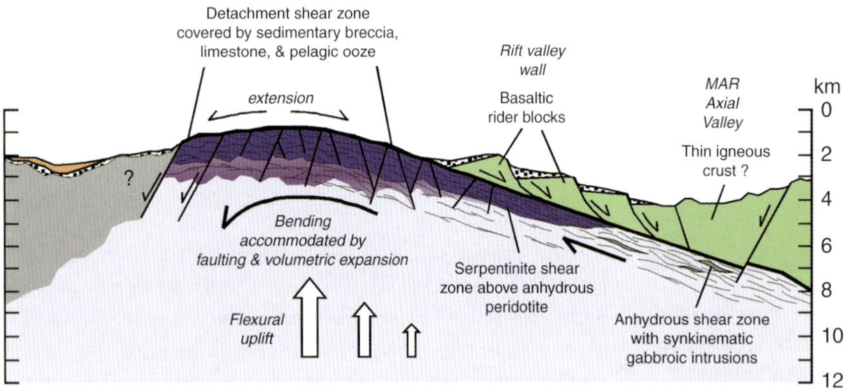

Figure 8.13 This schematic cross section of the Atlantis Massif OCC shows the serpentinite detachment shear zone bending over to an arched form that is subhorizontal at the summit of the massif. Fault-bounded rider blocks of basaltic pillow lavas lie above the fault zone (Karson *et al.*, 2006a).

the massif along the Atlantis Transform Fault is a major escarpment exposing a vertical section dominated by serpentinized peridotites at least 4 km thick (Blackman *et al.*, 2002; Karson *et al.*, 2006) (Fig. 8.13). Only a few kilometers away from this escarpment, ODP Hole U1309D penetrated ~1400 m of gabbroic rocks and minor amounts of ultramafic material (Shipboard Scientific Party, 2005), revealing how heterogeneous the crust can be in these settings.

In the steep south wall of Atlantis Massif, massive serpentinized peridotites with lesser gabbroic intrusions grade upward into a 100 m thick, detachment shear zone dominated by sheared serpentinites (Figs. 8.13 and 8.14) which arch across the top of this OCC (Karson *et al.*, 2006). Kinematic indicators show that this is a normal-sense shear zone with displacement of the hanging wall (upper plate) toward the MAR axis.

Crystal-plastic deformation and metamorphism of ultramafic and mafic material in the shear zone have produced amphibolite-facies mylonites overprinted by lower-temperature fault rocks, including talc–tremolite schists and exotic metasomatized mafic rocks (Schroeder and John,

Figure 8.14 The detachment fault surface at the top of the Atlantis Massif is a 100 m thick serpentinite shear zone superimposed on rocks with earlier, high-temperature deformation fabrics. The top of the serpentinite shear zone is shown here, where it is overlain by a sedimentary breccia (whitish cap rock) composed of serpentinite and basalt clasts in a pelagic carbonate matrix about 1 m thick. Field of view ~3 m.

2004; Boschi *et al.*, 2006). Very coarse grained sedimentary breccias with clasts of serpentinites and lesser basalts blanket this assemblage and are in turn covered with beds of pelagic limestone (Fig. 8.15). The Lost City Hydrothermal Field, located at the crest of this major escarpment, is an active, serpentinite-hosted vent system (Chapter 4) (Kelley *et al.*, 2001b, 2007). Hydrothermal fluids have deposited carbonate veins in fractures near the vent field providing an analog for similar rock types found in some ophiolites referred to as "ophicalcites."

Detachment shear zones have also been investigated in peridotite-dominated OCCs at the 15°40′ N area on the MAR (MacLeod *et al.*, 2002; Kelemen *et al.*, 2007) and 20 km west of the MAR rift valley near the Kane Transform Fault (Dick *et al.*, 2008). Mechanically weak serpentinites and related rock types, such as talc schists, may play an important role in localizing deformation along detachment shear zones in OCCs dominated by peridotites.

The distinctive assemblages of rock units found in association with serpentinites on OCCs can also be found in some ophiolite complexes (Abbate and Bortolotti, 1980; Lagabrielle and Cannat, 1990; Dilek *et al.*, 1991; Nicolas *et al.*, 1999). Some of these areas have been interpreted as oceanic crust and upper mantle that had been highly faulted and dismembered following spreading. It now appears that this class of ophiolites may preserve structures related to OCC formation along slow- to ultraslow-spreading ridges.

Spectacular examples of arched, corrugated detachment fault surfaces in serpentinites occur along the median valley wall of the MAR (Figs. 3.20 and 3.21) in some places, demonstrating that mantle material can be exhumed by faulting essentially at the spreading axis (MacLeod *et al.*, 2002; Smith *et al.*, 2006). Elsewhere, low-angle fault zones in serpentinite bordering the MAR median valley (Fig. 8.16) suggest that low-angle slip may occur in this weak material without large-scale rotations of a fault zone that was once more steeply dipping (Karson and Lawrence, 1997). The presence of lava flows directly overlying the serpentinites once again indicates that the oceanic crust in areas of low magma budget does not necessarily have the thick sequence of igneous crustal material expected along MORs with higher magma budgets.

Drilling has penetrated peridotites along the MAR and in some shallow crustal sites in crust produced along the MAR previously (Fig. 8.8). These include holes drilled during

Figure 8.15 Sedimentary rocks atop the Atlantis Massif OCC. Lower images show serpentinite clast sedimentary breccias directly on top of the detachment shear zone on the south wall (lower left) and in large, jumbled outcrops on top of the central dome of the massif (lower right). The breccias are overlain by white pelagic limestone (upper left) through a gradational unit with dark serpentinite clasts in pink carbonate matrix (upper right). Photomicrograph (upper right) shown in plane-polarized light. Field of view ~2 cm. Outcrop images have fields of view of ~3 m.

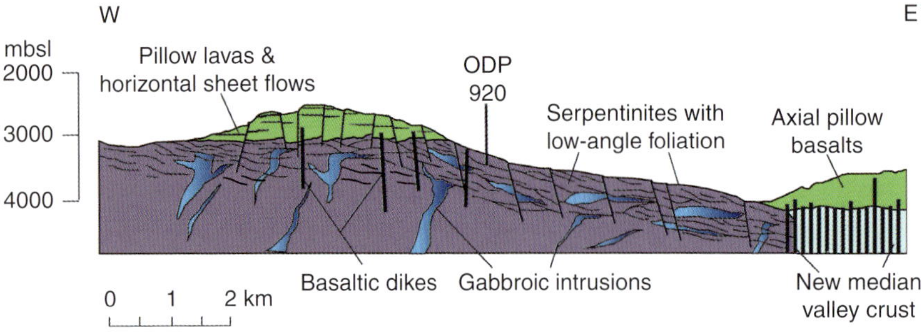

Figure 8.16 Serpentinites occur along a low-angle shear zone on the edge of the median valley of the Mid-Atlantic Ridge at 23°15′ N. The serpentinite is locally cut by steeply dipping dikes and overlain by flat-lying basaltic lavas (Karson and Lawrence, 1997). ODP Hole 920 was drilled in the serpentinite outcrop represented here during Leg 153 (Fig. 8.6).

ODP Leg 153 at 23°15′ N (Cannat *et al.*, 1995a) and Leg 209 near 15° N (Kelemen *et al.*, 2007), both OCC settings. Some of these holes penetrated 100–200 m of variably serpentinized peridotite with sparse dikes and gabbro bodies (Fig. 8.6).

Ultraslow-spreading ridges, such as the Southwest Indian Ridge, Mid-Cayman Spreading Center, and Knipovich and Mohns Ridges in the Arctic, are expected to represent an extreme endmember in terms of low magma budget predicted from theoretical models, crustal thermal structure, and basalt geochemistry (e.g., Dick *et al.*, 2003). Ultraslow ridges commonly are divided into magmatic and amagmatic segments that may or may not be related to tectonic segmentation and have extensive exposures of emplaced abyssal peridotite interspersed with volcanic centers (Dick *et al.*, 2003; Michael *et al.*, 2003; Pedersen *et al.*, 2010a; Hayman *et al.*, 2011; Laukert *et al.*, 2013). Although most areas have only been sampled by dredging, it is clear that extensive areas of the seafloor along this type of spreading axis are dominated by serpentinized peridotite (Fig. 8.17). Details of the fault structure and geologic setting of peridotites in these ultraslow-spreading ridges have yet to be defined, but the morphology of some of these areas indicates that they are probably exposed on OCCs.

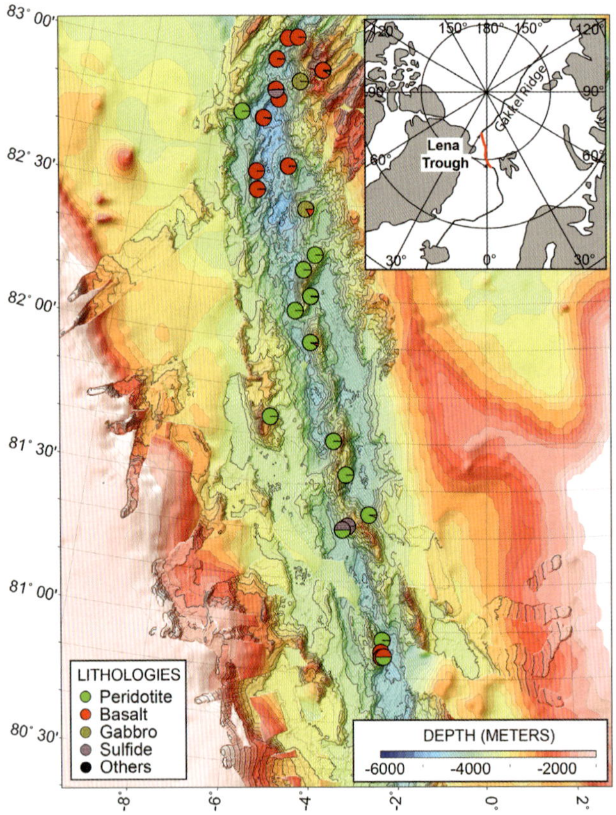

Figure 8.17 Widespread exposures of abyssal peridotites along spreading centers in the Arctic (Lena Trough) and other ultraslow-spreading ridges suggest basalt and other magmatic units may be absent from the oceanic crust over long sections of this ultraslow-spreading MOR system (Laukert *et al.*, 2013).

8.4 Serpentinization on the seafloor

Hydrothermal reactions in peridotites are distinctly different from those found in mafic host rocks typical of most seafloor hydrothermal systems (Chapter 4). Low-temperature hydration of anhydrous peridotites, serpentinization, is widespread and has dramatic effects on the chemical and physical properties of mantle peridotites (O'Hanley, 1996; Früh-Green *et al.*, 2004). It is likely that serpentinization of peridotites is ubiquitous in the heterogeneous and commonly very thin crust of slow- to ultraslow-spreading systems. An important consequence of serpentinization includes a decrease in density relative to anhydrous peridotite. One of the most profound changes that can occur during serpentinization is that the volume of the rock can increase by up to 40%. This may be crucial to hydrothermal flow because it can lead to both closing and opening of cracks (O'Hanley, 1996). Serpentinization also results in a dramatic weakening of peridotite, which probably results in fault localization (Escartín *et al.*, 1997, 2001).

The mineralogical and geochemical details of serpentinization are complex and involve the transformation of olivine to serpentine and related minerals as well as changes in fluid reactants (see, O'Hanley, 1996; Beard and Hopkinson, 2000; Früh-Green *et al.*, 2004; Morishita *et al.*, 2009; Boschi *et al.*, 2013). Static (versus dynamic) alteration of abyssal peridotites at temperatures from ~300 °C to

Figure 8.18 Photomicrograph of a weakly serpentinized peridotite (cross-polarized light) dredged from the Gakkel Ridge, an ultraslow-spreading ridge in the Arctic Ocean. Field of view 6 mm.

500 °C results in the replacement of olivine by mesh- to ribbon-textured serpentine consisting of lizardite ± chrysotile (varieties of serpentine) together with lesser amounts of Fe-brucite, magnetite, chlorite (± rare antigorite and FeNi alloy) (Figs. 8.5 and 8.9).

Serpentinization of orthopyroxene commonly results in the formation of "bastite," orthopyroxene crystals replaced by serpentine. Bastites retain the grain-scale textural characteristics of the original orthopyroxene (Fig. 8.19). When clinopyroxene is present in lherzolites and Cpx-bearing harzburgites, it is largely unaffected by serpentinization. Mg–Al-rich spinel may be recrystallized to Al-poor chromites and spinel and ilmenite may be partially replaced by titanite (a Ca–Ti silicate). It is common for relict grains of olivine to remain in the cores of serpentinized grains or for secondary minerals to form pseudomorphs of the minerals they replace, providing evidence of higher-temperature mantle-flow fabrics (Figs. 8.18–8.21).

In higher-temperature, hydrous, mafic/ultramafic shear zones, complex fluid exchange results in metasomatic reactions and exotic rock types (Boschi *et al.*, 2006). These include mylonitic to schistose rocks rich in talc, tremolite/actinolite, and chlorite (± serpentine) (Fig. 8.21).

Hydrothermal processes within mantle rocks can be important environmentally. For example, one of the key serpentine minerals is chrysotile – the dangerous form of asbestos that was commonly used as a fire retardant. In numerous ophiolites, variably altered peridotite outcrops host significant aquifer systems used as sources for drinking water, and alteration of some ultramafic rocks may concentrate economically important metals. Along major fault zones in seafloor serpentinites and in ophiolite peridotites, active serpentinization in the subsurface creates high-pH fluids that precipitate carbonate both in the subsurface as

Figure 8.19 Serpentinization of mantle peridotites results in the partial replacement of olivine and orthopyroxene by different forms of serpentine minerals. A cut slab (~15 cm across) of serpentinized harzburgite (top) shows black areas of olivine replaced by mesh-textured serpentine and bands of coarse, light-colored serpentinized orthopyroxene crystals replaced by serpentine, which mimics the orthopyroxene crystal structure (bastite grains). Photomicrographs (below) show bastites in mesh serpentine cut by dark serpentine veins in plane-polarized light (left) and a bastite grain in cross-polarized light, which retains kinked and distorted crystal structure from upper mantle deformation (right). Field of view in both photomicrographs is 4 mm.

Figure 8.20 Deformed serpentinites and related rocks appear to be common along detachment faults in oceanic core complexes. These examples show mylonitic fault rocks from the Atlantis Massif OCC (top pair) with dark serpentinized peridotite enclosed in finely recrystallized serpentine and talc (left) and talc schist (right); fault rocks from the MARK Area (middle and lower pairs) with cataclastic serpentinite fault rock with colorful remnants of pyroxene (middle left), serpentine extension reins (middle right), shear zone (lower left), and detail of multiple cross-cutting serpentine veins (black and gray) (lower right). All images shown in cross-polarized light. Field of view is 4 cm for all images except the lower right, which is 4 mm.

veins and at the surface as blocky deposits and chimneys (Neal and Stranger, 1983; Kelley *et al.*, 2001c; Früh-Green *et al.*, 2004; Kelley *et al.*, 2005). In the Oman Ophiolite, ephemeral high-pH springs precipitate thin (millimeters) carbonate crusts. Surficial sedimentary deposits are commonly cemented by carbonate derived from these fluids and small carbonate growths develop in fractures in the peridotite (Fig. 8.22).

As noted above, hydrothermal venting in peridotite massifs (see Chapter 4) has created a distinctive family of structures. Although recognized for many years in ophiolites, similar features have only recently been described from the seafloor. In ophiolites of the Alps and Apennines in particular, assemblages known as "ophicalcites" are common (Treves and Harper, 1994). These include variably sheared red to green serpentinites cut by numerous cross-cutting white to pink carbonate veins and carbonate-cemented breccias. The remarkable array of patterns and textures related to faulting, vein growth, fluid

Figure 8.21 Highly deformed and metamorphosed fault rocks from the Atlantis Massif detachment shear zone also show evidence of complex fluid interactions. Partially serpentinized peridotite and gabbro mylonites are cut by talc shear zones and veins (white) (top). Photomicrographs shown in cross-polarized light (bottom) show complexly deformed mylonitic to schistose fault rocks rich in talc, tremolite, and serpentine minerals. Pinch and swell structure (boudinage) in mylonite with carbonate veins (light) cutting dark serpentine lenses (left) and tight folds and lenses of mylonitic material fine-grained schist (right). Field of view is 4 cm for both images.

over-pressurization, and sedimentary processes make this a prized ornamental stone (Fig. 8.23). The surrounding geology shows that these structures formed in major, tectonically active, seafloor shear zones in peridotites. Similar assemblages of serpentinized peridotite with calcite veins and sedimentary breccias occur near the Lost City Hydrothermal Vent Field (Chapter 4) (Figs. 8.15 and 8.24).

Localized deformation and metamorphism occur in mantle peridotites in association with major tectonic features. The observed geological relationships provide important insights to the chemical and mechanical roles of peridotites and serpentinites in these settings. The extreme contrast in strength of variably serpentinized peridotites and spatially associated crustal rocks commonly results in serpentinite mélanges (mixtures) in fault zones. In these areas, angular to lensoid blocks of partially serpentinized peridotite, gabbro, and dike rocks are enclosed in schistose serpentinite with a phaccoidal fabric (Fig. 8.25). Serpentinite mélanges are common in ophiolite complexes and many seafloor samples may come from these types of assemblages.

8.5 Peridotites in ophiolites

Peridotites are a fundamental component of complete ophiolite complexes (Anonymous, 1972) (see Chapter 2) and the minerals and features these rocks contain provide windows into mantle processes beneath the oceanic crust. The details of features in these rocks and the processes they

Figure 8.22 Active serpentinization of peridotites is reflected in the formation of carbonate crusts in springs within the Oman Ophiolite (top) and in carbonate crust growing from transmissive fractures in peridotite outcrops (bottom).

imply provide important constraints on mantle processes at scales that are beyond detection by current geophysical methods and are not observable in seafloor outcrops.

Especially in the large, layered ophiolite complexes (Fig. 8.26), the composition and structures in peridotite units vary smoothly on the scale of kilometers (Davies, 1971; Boudier and Coleman, 1981; Girardeau and Nicolas, 1981; Casey *et al.*, 1983; Nicolas *et al.*, 1988). These units are dominated by harzburgite and lherzolite and cut by a complex array of variably deformed, centimeter-scale veins and larger volumes of dunite and pyroxenites marking areas where melt interacted with the deforming mantle asthenosphere (Figs. 8.2 and 8.27). These heterogeneities are variably deformed by the same processes as affect the host peridotite, so they appear to have been generated as the mantle material was actively rising and deforming beneath a spreading center. Locally, variably deformed lenses of mafic to ultramafic material meters to hundreds of meters thick, probably originating as sills, are isolated within this material. These lenses are composed of strongly layered and variably deformed gabbro, wehrlite, and pyroxenite, and represent volumes of melt that were trapped in the residual mantle. Networks of veins and large dunite bodies up to hundreds of meters across occur in some places (Kelemen *et al.*, 1992) (Figs. 8.27–8.30). These features are concentrated near the top of the mantle section, at the base of the crustal units in a region of deformation and melt segregation termed the Moho transition zone (MTZ). The composition and distribution of dunite and gabbro veins in this region of the mantle provide constraints on models of melt generation, segregation, and transport beneath spreading centers (Fig. 8.31).

In general, strong foliations and weaker lineations lie in a plane that is nearly parallel to the contact with magmatic crustal rocks (fossil Moho) and other boundaries that separate the crustal units, assumed to have originally formed in a subhorizontal orientation. Lineations are oriented at a high angle to the strike of dikes and other spreading-related features, indicating flow away from a spreading axis at this structural level.

In the Oman Ophiolite, large-scale (tens of kilometers across) diapiric structures are described by patterns of foliations and lineations in the peridotite (Fig. 8.32). Crustal structures near the crest of these features show evidence of localized melt enrichment and "impregnation" of residual mantle material by basaltic melts (Ceuleneer *et al.*, 1988; Nicolas, 1989; Jousselin *et al.*, 1998). These structures are potential analogs for the seismically imaged areas of focused mantle upwelling and melt generation on the EPR (Jousselin *et al.*, 2003; Toomey *et al.*, 2007).

8.6 Peridotite petrology and geochemistry

The mineralogic and bulk-rock composition of peridotites and the composition of the individual mineral phases provide information critical to understanding the sources and processes involved in the generation of the volcanic and plutonic rocks that make up the oceanic crust. Multiple igneous processes occur in the magmatic system beneath MORs (see Chapters 5–7). These processes modify primary melt compositions and make it difficult to decipher both the composition and physical conditions (e.g., depths and extents of partial melting) in the sub-ridge mantle.

Figure 8.23 Hydrothermally altered serpentinites with networks of white calcite veins (ophicalcites) are quarried for decorative building materials in the northern Apennines (left and middle). These types of assemblages are commonly overlain by serpentinite-clast sedimentary breccias with a carbonate matrix. The breccias (lower right), in some cases meters thick (upper right), overlie serpentinites (bottom of outcrop). Field of view is ~10 m in upper right image and 0.5 m in lower right image.

Figure 8.24 Light-colored carbonate veins and fracture fillings in dark-colored serpentinites near the Lost City Hydrothermal Field on the Atlantis Massif. Field of view ~3 m.

Figure 8.25 Serpentinite mélanges are made up of phaccoidal (lens shaped) slices of serpentinite bounded by shear zones (top). In some places blocks of massive serpentinite and oceanic crustal rocks are enclosed in a highly sheared matrix of serpentinite (bottom).

Unfortunately, nearly all mantle peridotites recovered from the ocean floor are moderately to highly altered (~40–100% by volume). Although serpentinization is the primary alteration process, complex textures and mineral assemblages are present in most serpentinized peridotites, reflecting multiple phases of alteration, deformation, veining, and weathering from their birth in the mantle to their current exposure on the seafloor (Früh-Green *et al.*, 2004). The effects of these processes make it difficult to ascertain the exact compositions and primary mineralogies of abyssal peridotites. Attempts have been made at determining protolith compositions, mineralogies, and modal abundances by focusing on relatively immobile elements (e.g., Niu, 2004; Morishita *et al.*, 2009) and/or the identification of pseudomorphs of primary mineral phases (e.g., Dick, 1989). Analysis of minor and trace elements in primary and relict igneous phases in peridotites provides important information regarding melts that were in equilibrium with the rocks and extents of melting that samples have experienced (Michael and Bonatti, 1985; Cannat, 1993; Niu and Hékinian, 1997; Constantin, 1999; Hellebrand *et al.*, 2002).

Figure 8.26 Large exposures of oceanic upper mantle rocks can be found in the Oman Ophiolite. Detailed studies of these types of exposures help put samples from seafloor outcrops and drill cores in a broader context of mantle flow and melt generation and also help guide the interpretation of geophysical data. Field of view in the foreground is ~300 m.

Figure 8.27 Peridotite outcrops in ophiolite complexes show a wide range of structures and heterogeneities, but on the scale of tens of meters they tend to be more uniform overall. Strong banding is produced by the foliation and lineation in the rock fabric and also by highly deformed dunite (pale, smooth layers) and pyroxenite (dark, resistant layers) bands. Resistant, coarse-grained pyroxene crystals stand out on the surfaces. Variably deformed cross-cutting veins indicate that they were formed during deformation in the mantle. Clockwise from top left the outcrops of peridotite are from the Macquarie Island, Bay of Islands, Troodos, and Oman Ophiolites.

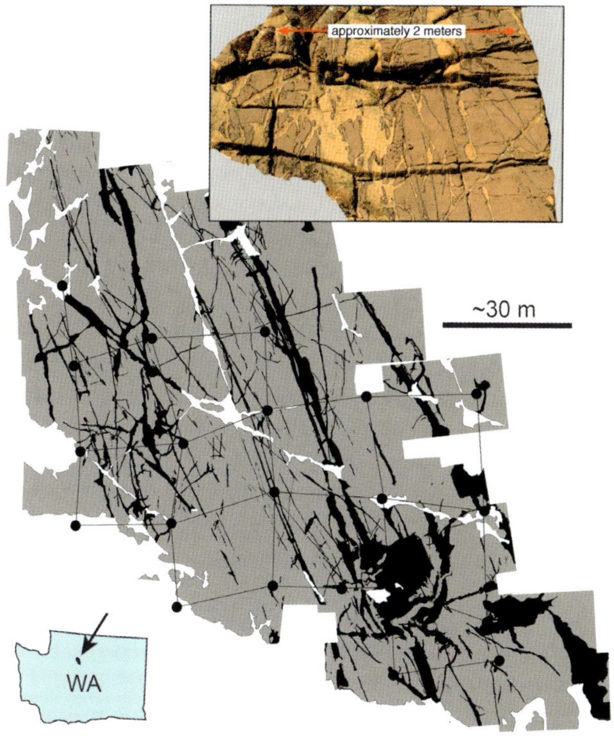

Figure 8.28 Map of dunite pods and veins (black) in the Ingalls Ophiolite marking the paths of melt that migrated through the peridotite (gray). Upper panel shows a detail of the large map area with dunite veins (light) in harzburgite (dark). Inset shows location in Washington State (Kelemen *et al.*, 2000).

Figure 8.29 Large-scale (hundreds of meters across) dunite bodies (light) in peridotite (dark) in the Oman Ophiolite are concentrated near the top of the unit, near the geological expression of the Moho transition zone (MTZ).

Figure 8.30 Gabbroic veins in peridotites represent concentrations of melt trapped in the mantle. Examples here are from the Oman Ophiolite (top) and the Bay of Islands Ophiolite (bottom). In both cases gabbroic dikes (a few centimeters to tens of centimeters wide) with white plagioclase and gray clinopyroxene cut dunite to harzburgite (brown) mantle rocks. Hammer (top) and coin (bottom) provide scales.

8.6.1 Peridotite mineralogy and phase chemistry

Estimates of primary modal mineralogies in abyssal peridotites indicate most are harzburgites (containing 70–80 vol% olivine) and less common lherzolites, dunites, wehrlites, and pyroxenites (Fig. 8.3). All comprise olivine (Ol), orthopyroxene (Opx), clinopyroxene (Cpx), and spinel (Sp) in variable proportions (Fig. 8.33) (Bodinier and Godard, 2014). Plagioclase is present in some samples but commonly proposed to be secondarily formed by melts that infiltrate peridotites after they have been partially melted. Similarly, cross-cutting lenses of plagioclase- and/or clinopyroxene-rich peridotites likely represent late-stage interstitial melts (Constantin, 1999; Laukert *et al.*, 2013). Greater proportions of olivine relative to the other phases are indicative of samples that have experienced greater extents of partial melting and are therefore considered more refractory (Fig. 8.34) (Niu, 2004). Typical abyssal peridotites are more depleted in elements found in basaltic magmas than those comprising typical fertile lherzolites found in continental mountain ranges (orogenic peridotites) that mostly contain >10% Cpx (Fig. 8.33).

The chemical compositions of individual minerals, particularly trace elements in clinopyroxene (rather than whole-rock compositions that are liable to alteration), are

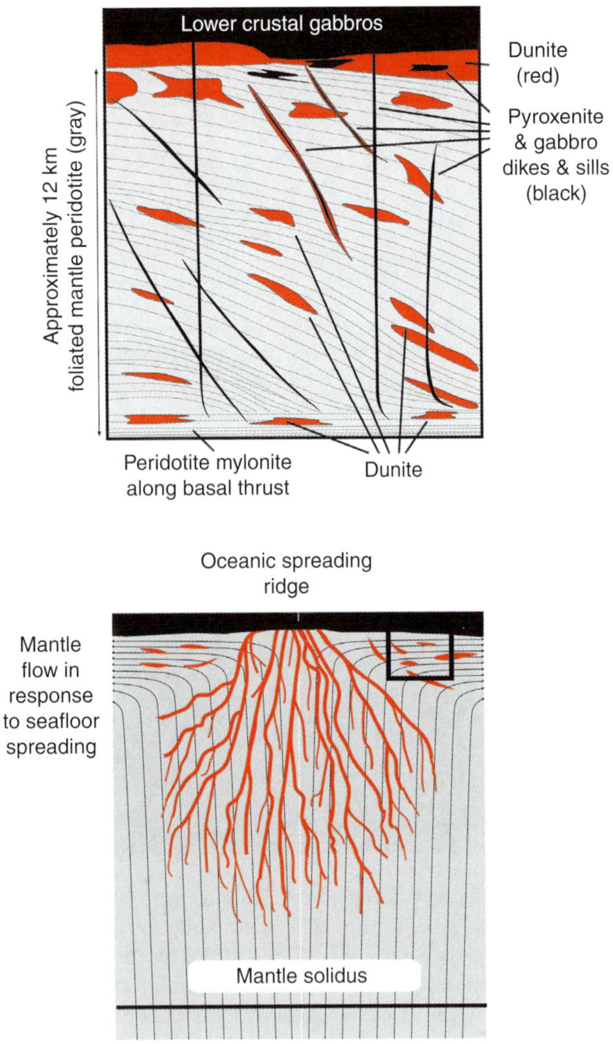

Figure 8.31 Melt flow and segregation beneath MOR spreading centers can be inferred from the internal structure of ophiolite peridotites. Upper panel shows a schematic representation of peridotites in the Oman Ophiolite. Lower panel shows one possible model for the melt paths through the melting regime beneath a spreading center. Box in lower illustration shows relative location of structures found in the Oman section (Kelemen *et al.*, 2000).

commonly used to determine the relative fertility (enriched versus depleted source relative to presumed primitive mantle composition) of abyssal peridotites (i.e., Johnson *et al.*, 1990; Constantin, 1999; Hellebrand *et al.*, 2002; Laukert *et al.*, 2013). Depleted, refractory harzburgites commonly contain olivine and pyroxene(s) with high Mg# [100 × Mg/(Mg + Fe)], low Ti, and high concentrations of Ni and Cr respectively. Spinels with high Mg# and Cr# [100 × Cr/(Cr + Al)] are also an indication that these ultramafic rocks represent residues after high degrees of partial melting involved in producing basaltic magmas. Peridotites that have been refertilized by secondary magmatic processes (e.g., melt impregnation, metasomatism) commonly have olivine and pyroxene with lower Mg# and Ni, and lower Cr and Ti in pyroxene.

8.6.2 Peridotite geochemistry

Overall, abyssal peridotites are characterized by having high MgO contents (~36–48 wt%) and relatively low SiO_2 contents (~49–42 wt%) that vary inversely with MgO content. Contents of Al_2O_3, CaO, and to lesser extents Na_2O and TiO_2 exhibit negative correlations with MgO (Niu, 2004; Bodinier and Godard, 2014). In addition, abyssal peridotites contain extremely low, but quite variable, concentrations of lithophile trace elements that are incompatible in mantle phases. Depletions in incompatible trace elements correspond to the more refractory nature of abyssal peridotites relative to many ultramafic rocks found in other tectonic environments (see review in Bodinier and Godard, 2014). There is general consensus that the depletions in major and trace elements are a consequence of either decompression partial melting beneath ridge crests as basaltic melts are removed or more complicated scenarios involving previous melting events coupled with more recent reenrichment or "refertilization." There is ample evidence to support the latter for even the most refractory samples. For example, slight light rare earth element (REE) enrichments relative to the middle and heavy REE suggest post-melting magmatic or metasomatic refertilization (Niu, 2004; Bodinier and Godard, 2014).

Comparisons between the mineral modes in abyssal peridotites and bulk-rock chemistry show that increased MgO concentrations and decreased contents of oxides such as Al_2O_3, CaO, and Na_2O correlate with increasing proportions of Ol and decreased proportions of Opx and Cpx. The changes can be modeled by variable extents of melting (up to 25%) of a fertile lherzolite (Fig. 8.34) (Niu, 2004). Although the compositions of many abyssal and ophiolite peridotites can be modeled by single batch melting events in the spinel peridotite field, analysis of REE in clinopyroxene crystals suggest some have undergone near-fractional polybaric decompression melting beginning in the garnet stability field of the upper mantle (Johnson, 1998). The REE in the cores of clinopyroxene-bearing peridotites also support the hypothesis that the light REE enrichments, and in some cases high field strength elements (HFSE), observed in some bulk samples reflect secondary enrichment/refertilization rather than the original compositional characteristics.

Dunite samples, generally considered to be the most refractory peridotites, may actually represent reaction channels where rising basaltic melts react with pyroxenes leaving residues of olivine (Kelemen *et al.*, 1992) (Fig. 8.19). Likewise, petrographic evidence suggests some seemingly olivine-rich harzburgites have had olivine added as melt percolates along grain boundaries (Niu, 2004). Some plagioclase- and/or Cpx-bearing abyssal peridotites appear to have been impregnated with more fertile melts below ridge crests (Dick and Bullen, 1984; Elthon *et al.*, 1992; Dick and Natland, 1996; Niu and Hékinian, 1997; Constantin, 1999; Niu, 2004). Similarly, some lherzolites and Cpx-harzburgites

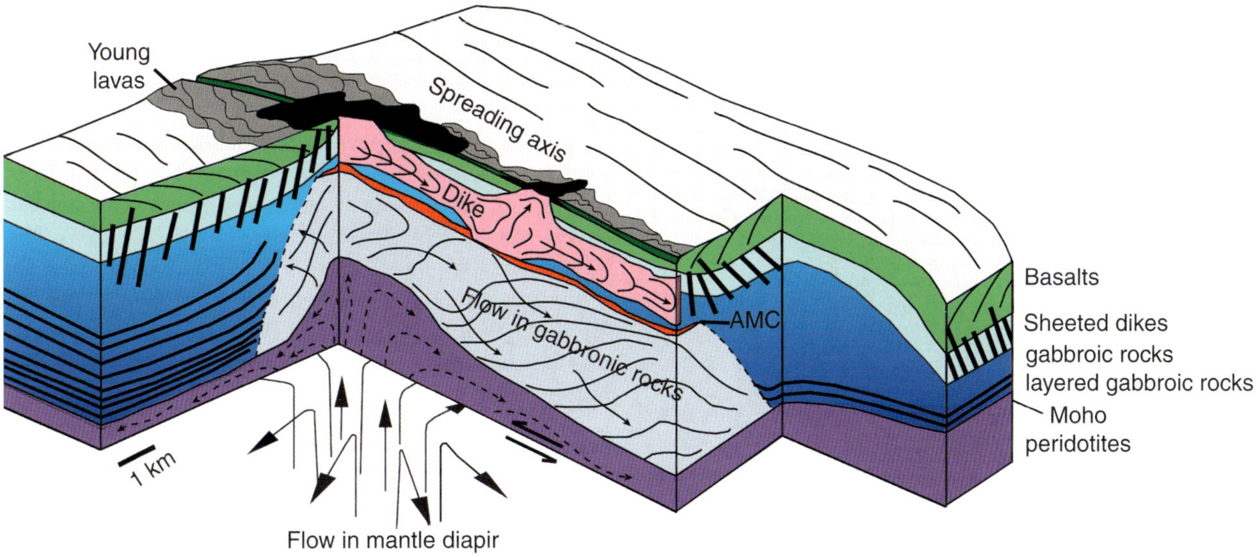

Figure 8.32 Schematic diagram of subaxial mantle upwelling at a mantle diapir and along-axis variations based on studies of the Oman Ophiolite (after Nicolas, 1989). Thin lines and arrows in the intruding dike (pink) and gabbroic unit (gray) illustrate the geometry of magmatic flow in those units.

in ophiolites may have been formed by refertilization from melts that ponded along the thermal boundary layer at the base of the lithosphere away from spreading axes (Godard *et al.*, 2000).

Relatively rare pyroxenites that typically occur as cross-cutting veins and dikes or as deformed bands and lenses in peridotite foliation exhibit wide compositional variations (Fig. 8.2). In some cases, they appear to be basaltic melts that crystallized at high pressures in the mantle and likely represent enriched components derived from recycled oceanic crust (Bodinier and Godard, 2014). It is these pyroxenites that are believed to provide the low-temperature "enriched component" in the largely refractory oceanic mantle. However, pyroxenites from the ultraslow-spreading Lena Trough (see Fig. 8.17) appear to have

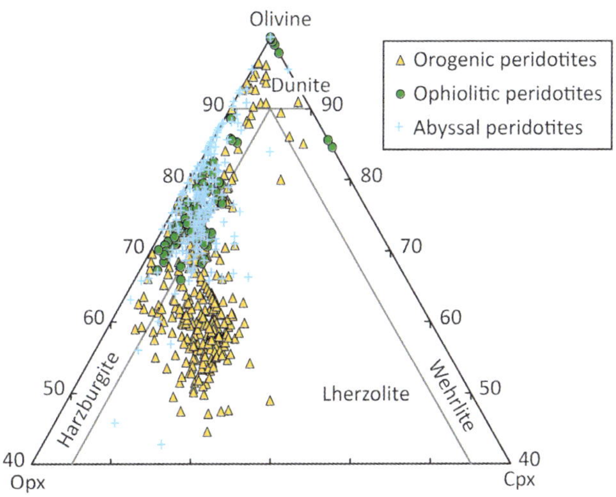

Figure 8.33 Modal mineral proportions of orogenic, ophiolitic, and abyssal peridotites classified based on the olivine–clinopyroxene (Cpx)– orthopyroxene (Opx) diagram of Streckeisen (1976). Abyssal peridotites (185 samples from both MOR and back-arc spreading centers) contain much less Cpx than orogenic peridotites as a consequence of their residual nature after variable extents of prior partial melting. The most depleted samples have the greatest proportion of olivine (from Bodinier and Godard, 2014).

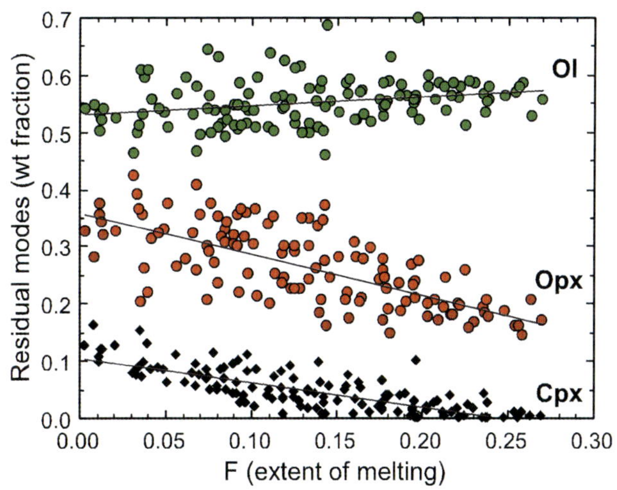

Figure 8.34 Observed and calculated modal abundances of olivine (Ol), orthopyroxene (Opx), and clinopyroxene (Cpx) in abyssal peridototes relative to estimated extents of melting (F) (Niu, 2004). Note that the proportion of Ol remaining in peridotite samples increases and Cpx is eliminated from the assemblage (lherzolite becomes more refractory harzburgite) with greater extents of melting. See Niu (2004) for details of the calculations and sample database.

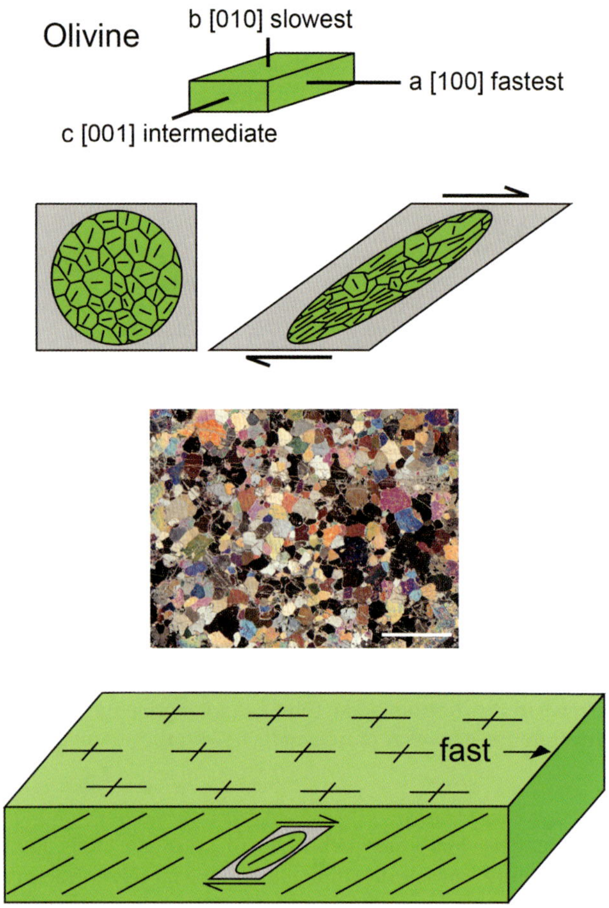

Figure 8.35 Deformation of olivine leads to the lattice preferred orientations (LPO) in peridotites. Because olivine crystals are anisotropic, the LPO results in seismic anisotropy in deformed peridotites. For horizontal shearing, as might be expected at the base of lithosphere plates, the slowest direction is vertical and the fastest direction is oriented in the shearing direction. Measurements of mantle seismic anisotropy show the direction of plate movement relative to the underlying mantle.

formed by assimilation–fractional crystallization of primitive, depleted MORB melts at high pressures as these melts reacted with refractory harzburgites (Laukert *et al.*, 2013).

8.7 Evolution of the oceanic mantle lithosphere

The current understanding of mantle processes beneath MORs comes from an integration of information from marine geophysics, geochemistry of seafloor peridotite samples, and ophiolite geology. As the mantle rises beneath spreading centers and spreads laterally, it cools and dramatically increases in strength and rigidity (Fig. 8.4). Here, weak asthenosphere is transformed into strong lithosphere as it cools without changing composition. This strong upper mantle material gives lithosphere plates their strength and rigidity and makes plate tectonics possible.

The impacts of different mantle process can be detected acoustically. For example, the movement of mantle material is recorded by crystal-plastic deformation of olivine, which results in seismic anisotropy (directional variation in velocity) of compressional and shear waves (Christensen, 1984; Kendall, 1994; Blackman *et al.*, 1996) (Fig. 8.35). This is caused by high-temperature crystal-plastic deformation at a grain scale that occurs when crystals deform by slip on specific crystallographic planes. With even moderate shearing, the constituent olivine crystals in peridotites develop strong lattice preferred orientations of deformed crystals.

Based on deformational relationships, the direction of mantle flow near MORs should be indicated by seismic anisotropy. In general, experiments show that the highest velocities are oriented at high angles to MOR plate boundaries, indicating mantle flow away from the plate boundary as plates diverge (Wolfe and Solomon, 1998). Because plates are decoupled from the underlying, weaker, mantle asthenosphere, plate movements need not be directly related to asthenosphere flow. Oblique angles between plate boundaries and inferred mantle flow directions are observed at some MORs (Fig. 8.36), suggesting more complex flow kinematics related to changing poles of rotation for spreading centers and the influence of along-axis flow near hot spots (Toomey *et al.*, 2007). Vertical, axis-parallel melt segregations near spreading axes are expected to result in slow mantle velocities in the melt delivery region.

8.8 Melt generation and transport

There are two end-member models of melt flow and focusing toward a spreading axis (Fig. 8.37): porous flow in response to pressure gradients and matrix shear (e.g., McKenzie, 1984; Phipps Morgan, 1987; Spiegelman and McKenzie, 1987), and active or buoyancy-driven flow forming a high-porosity decompaction channel (Sparks and Parmentier, 1991; Spiegelman, 1993). In general, the rise of asthenospheric mantle is regarded as a passive process wherein weak, melt-rich mantle rises in response to plate separation, driven by plate diving forces. In some areas, the buoyantly rising asthenosphere may play a more active mechanical role, by helping to push the plates apart. Mantle flow paths and the role of asthenosphere can interact in different ways to influence the geochemistry of melts that are generated (Spiegelman and Reynolds, 1999; Spiegelman and Kelemen, 2003) (Fig. 8.37).

Mantle melting occurs as a result of adiabatic decompression of asthenospheric mantle upwelling in response to plate separation at MORs beginning at depths between ~120 km and 75 km (Fig. 8.38). Subaxial temperatures are believed to be the major control on the composition of primary melts because they determine the extents of partial melting; however, variations in mantle source composition

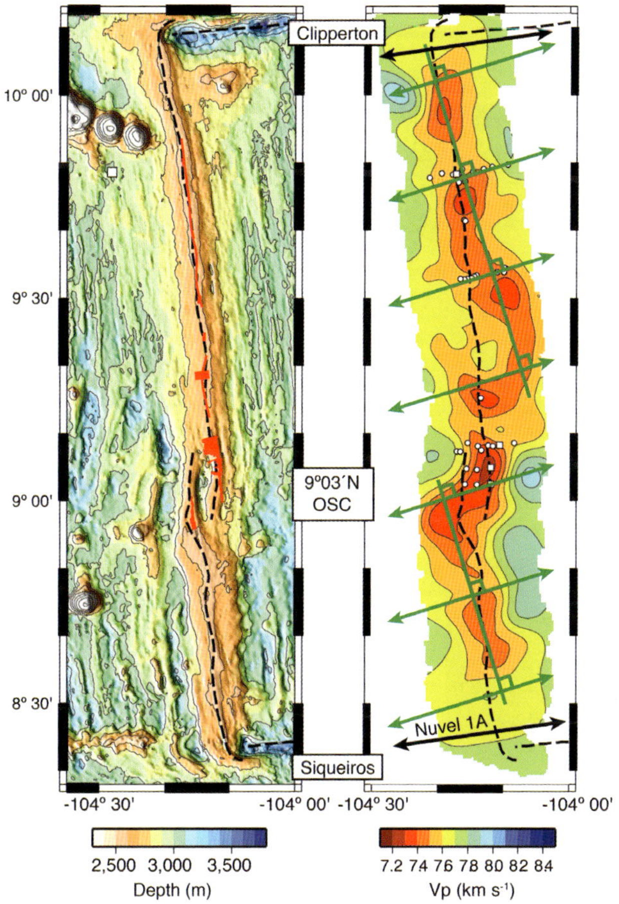

Figure 8.36 The fast direction of compressional seismic waves (green arrows on right side) in the mantle near the EPR is oblique to the spreading direction (nearly E–W), showing that the underlying mantle asthenosphere is not moving at 90° to the plate boundary. Nuvel 1A (DeMets *et al.*, 1990) (bottom black arrow on right map) is the direction of plate movement predicted from a model of global plate movements (Toomey *et al.*, 2007).

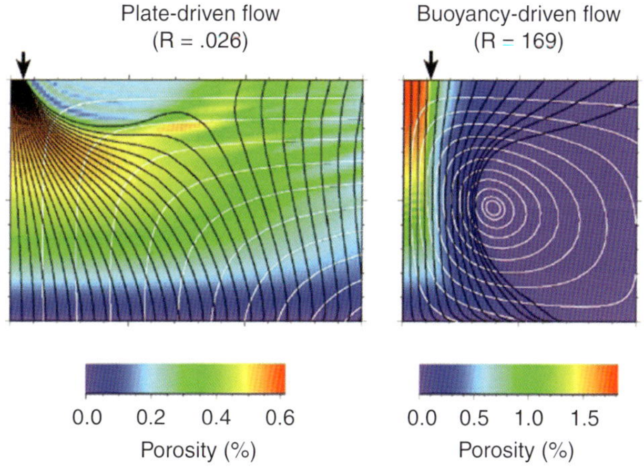

Figure 8.37 Numerical models of mantle upwelling and melting for MOR spreading centers. The left panel shows the porosity distribution beneath a ridge crest as a result of passive upwelling caused by plate separation which results in a broad region of upwelling and melting. The right panel is for "active" buoyancy-driven flow which promotes more rapid upwelling in a narrow region beneath the axis. The diagrams are vertical cross sections of the right half of the melting region beneath a MOR crest, scaled to depths of ~80 km. Arrows mark the width of the "crustal accretion region," defined as the zone over which approximately 85% of the melt produced within the box is added to the crust. Black lines show trajectories of melt flow. White lines show the solid-state flow of the mantle (Spiegelman, 1996). In both models, melt tends to migrate along grain boundaries at low porosity but accumulate into larger more porous channels beneath the ridge crest. The maximum extent of melting used in the model is 20%.

and lithology can also play a major role (Langmuir *et al.*, 1992; Katz and Weatherley, 2012). Higher initial mantle temperatures will result in a greater amount of melt generation for a given mantle composition and ascent rate (e.g., White and McKenzie, 1989). Other processes such as melt extraction, melt transport and mixing, and melt–rock reaction can also significantly influence the composition of primary mantle melts as they traverse the uppermost mantle (e.g., Kelemen *et al.*, 1997). Consequently, unraveling the processes and degree and depth of melting that form primary melts for the plethora of MORB types is not straightforward.

Melting occurs in a broad region (>400 km wide) and over a depth range of maybe as much as 150 km beneath the fast-spreading EPR (Melt Seismic Team, 1998; Key *et al.*, 2013) (Fig. 3.42) to shallower levels (~80 km) in slow-spreading environments (White *et al.*, 2001). More restricted regions of melting are expected beneath slower-spreading ridges because of lower mantle temperatures and slower rates of ascent (White and McKenzie, 1989). An exception would be at axial thermal anomalies such as Iceland or the Galápagos hot spots.

The geochemistry of peridotite samples from the seafloor and ophiolites indicates that 5–20% melting (by volume) is typical (Dick, 1989; Johnson *et al.*, 1990; Johnson and Dick, 1992). As partial melting occurs at depth, aggregated melts, by virtue of their lower viscosity and greater buoyancy, move upward more rapidly than the ascending mantle residue. Theoretical considerations supported by outcrop structures show that the melts segregate and become channelized (Kelemen *et al.*, 1997) (Fig. 8.24). As they migrate upward, they interact with surrounding, previously depleted mantle peridotite (Kelemen *et al.*, 1992) resulting in modification of both the melts and the surrounding wall rocks (Fig. 8.38).

Evidence of interactions between ophiolite peridotites and melts includes lenses, dikes, and veins of pyroxenite, plagioclase-bearing peridotites, and gabbroic material ranging from olivine gabbro to ferrogabbro, representing areas of concentrated melt. These are commonly separated from host peridotites by dunite margins. Diffuse areas of melt concentrations pass upward into progressively larger and more numerous concentrations of gabbroic material. This material is

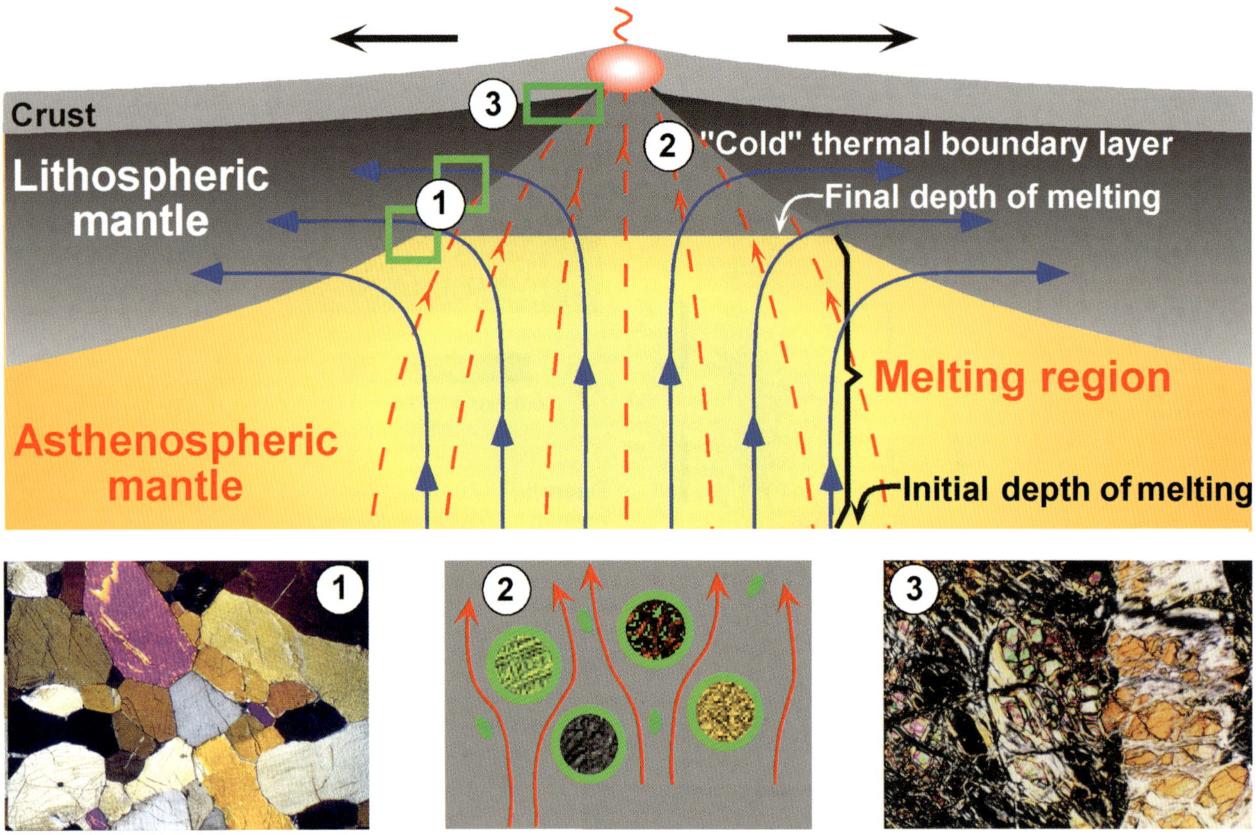

Figure 8.38 Processes and locations in the melting regime beneath a MOR. Images 1–3 (below) correspond to petrologic relationships that characterize regions highlighted in the schematic illustration above. Generalized motion of the solid mantle is represented by blue lines/arrows and melt trajectories are shown as red dashed lines focusing toward the ridge axis. Melting begins where the temperature of upwelling mantle crosses the solidus of peridotite and ceases when the temperature falls below the solidus at lower pressures. The cooler lithosphere is bounded by the thermal boundary layer (TBL) between the base of the crust and the final depth of melting. Above the melting region, melts (red arrows) rise within the solid residue of mantle (1) where they can interact with the depleted peridotites and potentially refertilized peridotites (2). At shallow levels beneath the crust, hydrothermal circulation results in serpentinization of many abyssal peridotites (3). Unaltered peridotites (1) from deeper levels are unlikely to be sampled in the oceans (after Niu, 2004).

interpreted to be melts that were trapped in the mantle. Most of the melt generated must have escaped the melting regime and fed magma bodies at higher structural levels. Deformation of melt concentrations in peridotites results in complexly layered rock in the MTZ (Jousselin *et al.*, 2012) (see Chapter 7). Although the details of outcrop relationships are not clear, samples of peridotites with trapped melt veins have also been recovered from outcrops on the seafloor (Table 8.1 and Fig. 8.2) (Constantin, 1999).

Even in the fastest spreading regimes, melt delivery to the surface is not focused in a continuous, ridge-parallel plane. Mantle flow and melting in the mantle, and hence magmatic crustal construction, becomes focused in discrete, 10–20 km wide, diapir-like centers beneath the EPR (Fig. 8.36) (Toomey *et al.*, 2007). Seismic imaging of these centers shows that they are not always centered directly beneath the surface expression of the EPR axis and can be located as much as 10 km off-axis. It is not yet clear whether these mantle upwelling centers dictate the location of the plate boundary and/or discontinuities or vice versa. These features may correspond to the diapiric structures mapped in the peridotites of the Oman Ophiolite and inferred to have formed both beneath the axis and in some cases off-axis during spreading (Jousselin *et al.*, 1998; Jousselin and Nicolas, 2000a,b; Nicolas and Boudier, 2000). Variations in the thickness and deformation of the overlying crustal section of the ophiolite are correlated with proximity to these zones of upwelling and melt production (Fig. 8.31).

An important but commonly overlooked point is that melt can move much more rapidly than the surrounding olivine-rich mantle material. The mantle asthenosphere moves at about the same rate as plate separation, whereas melt, especially if in channels or hydraulic fractures, can move orders of magnitude more rapidly and therefore reach crustal levels before its mantle source material is added to the base of the lithosphere. The result is that, in a vertical section of oceanic lithosphere, the crustal magmatic rocks are unlikely to be directly related to the mantle peridotites that they overlie. Further complications will arise if components of along-axis mantle flow or magma transport occur.

Mantle peridotites are the source of melts that generate the igneous oceanic crust. The major rock units discussed in earlier chapters (lavas, dikes, and gabbros) are petrogenetically related to this material. Having arrived at the mantle, the journey downward through the oceanic crust has reached its natural conclusion. Although the rock units of the oceanic crust and upper mantle have been presented from top to bottom, magma from which the oceanic crust is constructed mainly moves in the opposite direction. Magma migrates from the upper mantle through the peridotites and into crustal-level magma chambers, where it may crystallize and react with other rocks before being injected as dikes into the shallow crust and, in some cases, erupting as lavas on the seafloor. The composition of the mantle rocks and melting and crystallization processes determine the details of the composition of the overlying crust, which in turn will influence the chemistry of hydrothermal fluids and the water–rock reactions that occur beneath the seafloor at the MOR crest. Mantle melting determines the magma budget along MORs and hence the nature of the oceanic crust in different tectonic environments. It is likely that the special chemical and mechanical properties of serpentinites derived from mantle peridotites play a major role in the tectonics of slow- to ultraslow-spreading ridges. Unraveling the geochemical and structural complexities of these rocks and the processes by which they are generated is crucial to a comprehensive understanding of seafloor spreading and how oceanic crust is generated.

9 Future research

The pioneer spirit is still vigorous within this nation. Science offers a largely unexplored hinterland for the pioneer who has the tools for his task. The rewards of such exploration both for the Nation and the individual are great. Scientific progress is one essential key to our security as a nation, to our better health, to more jobs, to a higher standard of living, and to our cultural progress.

V. Bush (July 1945)

As head of the US Office of Scientific Research and Development, Vannevar Bush, a leading scientist at the Massachusetts Institute of Technology, developed strategic technologies that helped the Allies win World War II. His quote is taken from the 1945 report to President Franklin Roosevelt: *Science, The Endless Frontier*, a seminal document that promoted the establishment of the National Research Foundation, which became the National Science Foundation in 1950.

His words in that report remain as true today as when they were delivered. They helped direct the past decades of investigations and set the stage for the future.

The responsibility for the creation of new scientific knowledge – and for most of its application – rests on that small body of men and women who understand the fundamental laws of nature and are skilled in the techniques of scientific research. We shall have rapid or slow advance on any scientific frontier depending on the number of highly qualified and trained scientists exploring it.

Significant scientific advances since 1950 have established the founding theories of plate tectonics and seafloor spreading and other fundamental Earth and oceanographic phenomena. The list includes mid-ocean ridge earthquakes and magmatic events; complex geological processes that create oceanic lithosphere; the biogeochemistry of hydrothermal vent ecosystems; chemosynthetic organisms; and chemical and microbial processes within the oceanic lithosphere. Our expanded knowledge of the mid-ocean ridge (MOR) system, the global ocean, and the history and evolution of oceanic crust might suggest that not much remains to be discovered.

But that is far from true.

We have touched on many of the crucial processes that occur in MOR environments in this book, but there are still many unanswered questions about fundamental phenomena associated with magma generation in the mantle, melt transport and storage in the upper oceanic lithosphere, volcanism, hydrothermal circulation, and biological processes in all these dynamic settings. We continue to seek a deeper understanding of how oceanic crust develops through the wide range of spreading processes along the global MOR, including the precise nature of lower crustal and mantle rocks, faulting, seafloor eruptions, and evolution of oceanic lithosphere as it ages. Like all complex areas of science, each new increment of information gained has raised new questions, with exciting implications.

How does knowing about MORs and ocean processes help us solve key problems facing global society today? The answer to that important question, and many related issues, is not simple. As Dr. Bush suggested, however, the need to advance knowledge and pursue scientific questions for the benefit of society is crucial to the advancement of the human species. The deep seafloor represents a vast portion of the Earth's surface, yet the exploration and investigation of this enormous region and its potential are still in their infancy. This pursuit can engender a renewed sense of responsibility for the global community in the twenty-first century for stewarding planet Earth and its unique land and ocean ecosystems.

The global ocean is crucial to life on this planet. It is the engine that drives climate variation over time spans of centuries to thousands of years. Human input to that engine, in the form of carbon released into the atmosphere and oceans, represents a key challenge facing the global society in this century and beyond. That carbon input profoundly influences the processes that regulate climate and global ocean circulation and chemistry. Indeed, some scientists suggest that we have entered a new epoch of geologic time called the Anthropocene, in which humankind has the dominant impact on the planet. How human influence on our global oceans plays out over the next few decades will determine the quality of life on our planet for generations to come.

By 2050, it is estimated that there will be another 2 billion people on this planet, which will further stress marine resources and demand informed decisions on how to manage them. One example is underwater mining of massive sulfides generated at hydrothermal vents (Hoagland *et al.*, 2010; Collins *et al.*, 2013a,b; Chapter 4). These systems contain high-grade ores and trace metals that may be critical to some emerging technologies and may become economically viable to mine in the future. It will take an informed community of scientists, policymakers, and the public to determine the best way forward on the possible use and preservation of these resources and the ecosystems, both macro-biological and microbial, that they host.

Much still needs to be done to comprehensively survey the global ocean seafloor at a resolution that is meaningful to truly assess its economic potential. Less than 15% of the global seafloor has been surveyed with multibeam bathymetry, and less than 1% has been surveyed with near-bottom methods that provide crucial links between physical seafloor features at sub-meter resolution and related geological, chemical, and biological processes.

The concepts of plate tectonics and seafloor spreading have revolutionized Earth and ocean sciences. Those discoveries paved the way for an entirely new approach to how planet Earth works. This has resulted in more accurate models to guide exploration and exploitation of resources such as minerals and hydrocarbons in the twentieth–twenty-first centuries. The study of MORs also revolutionized our ideas about life and the conditions under which it can thrive, survive, and expire, on Earth, as well as other planets in our solar system and elsewhere in the universe. Most people are not likely to ascribe this geological and oceanographic revolution in knowledge to improvements that accrued to human society, but we hope that in some small way the information presented here will help spread the word that it has.

Building a holistic understanding of the entire Earth system requires numerous scientific investigations. These include focused studies on human health and the oceans, fisheries and aquaculture, and exploration of all types of energy and mineral resources. These efforts will require that public and government officials recognize the need to re-prioritize and elevate funding of basic scientific research in the Earth and ocean sciences and nurture scientific education and collaboration.

Equally important is continued support for technical innovation and for seagoing research and drilling vessels, environmental sampling, and myriad ocean observatories at both local and basin scales. To advance monitoring and develop response capabilities to events in the ocean, we will need more sophisticated sensors, cabled and uncabled instrument networks, and an array of progressively advanced undersea vehicles, both human-piloted and robotic. Such systems are critical to observe and sample the details of processes on relevant temporal and spatial scales that span microseconds to decades and micrometers to hundreds of kilometers.

Technological advances will enable discoveries and facilitate our understanding of linkages between disparate phenomena in the oceans and within Earth's lithosphere. But it will also take human intelligence to recognize those linkages and explore them. We will need parallel progress in developing increasingly sophisticated iterative models that incorporate this new wealth of data so that we can have better predictive capabilities for volcanic eruptions, earthquakes, methane release from gas hydrate deposits, and biological responses to changing climate, for example.

What important future research topics will guide and expand our understanding of mid-ocean ridge processes beyond today's concepts?

Fundamental gaps remain in our knowledge of the volume and flux of heat and chemical energy from Earth's mantle through the lithosphere and into the deep ocean. These processes rely on complex transfer phenomena occurring both in solid or partially molten rock and in fluids beneath MORs. Similar phenomena also occur at oceanic subduction zones where oceanic lithosphere is recycled into the mantle. The rheological behavior and chemical composition of oceanic lithosphere that is subducted in part governs both the size and location of large seismic events along subducting continental margins. Why has the frequency of very large earthquakes increased over the past five decades? Clearly we need to develop better models of subduction zone earthquake behavior and to assess the historical record of those events in both spatial and temporal domains, as well as new strategies to monitor continental margins where megathrusts are expected to occur.

The hydrology of the oceanic lithosphere at different plate settings, such as transform faults and subduction zones, also remains equally vague. Yet, knowledge of fluid flow pathways and storage mechanisms at all levels within the oceanic lithosphere has important ramifications for how to relate seismicity and rock alteration at different boundaries. There are hints that vast "aquifer" systems exist within the oceanic crust, yet the exact nature and impact on chemical transfer and biological systems in the subsurface remain elusive (Davis and Elderfield, 2004). Indeed, research shows that seamounts transported off-axis may be hydrothermal "breathing holes" for millions of years, acting as both downwelling and upwelling zones (Fisher *et al.*, 2003a,b). We know that these processes integrate biogeochemically to provide energy sources to sustain the unique macrobiological and microbial communities occurring at MORs and within the oceanic lithosphere as it evolves. However, we are just beginning to understand the global variability and distribution of these communities, their unique metabolic energetics, and how they contribute to primary production throughout the global ocean and within the oceanic crust.

Of all the discoveries at MORs, perhaps the most profound was the finding in 1979 of single-celled life living off magmatic gases. Since that pivotal discovery, there has been exponential growth in interdisciplinary research focused on the subseafloor biosphere and the role that submarine volcanoes play in sustaining these novel life forms, their metabolic and genetic diversity, and evolution. Emerging technologies, largely driven by the human genome project, have revolutionized our ability to characterize the genetic makeup of microbes and their metabolic activities. Studies of MOR environments are starting to provide hints that microbial diversity is significantly higher than that documented by traditional 16S rRNA studies (e.g., Huber *et al.*, 2003; Brazelton *et al.*, 2010). We are only starting to get a glimpse into this rare biosphere hosted in the lithosphere and within vents (Sogin *et al.*, 2006), yet it may have profound importance in understanding how we, as a species, got here. These novel organisms may also be important in the development of new catalysts for industrial applications and in biomedical and pharmaceutical applications (e.g., National Research Council, 1999; Wilcock *et al.*, 2004; Girguis and Holden, 2012).

We have also just begun to discover how oceanic crustal generation is manifested at slow- to ultraslow-spreading ridges, which currently make up nearly half of the global MOR system. The structural responses of large sections of the upper lithosphere to very slow accretion rates can lead to the formation of oceanic core complexes. The unique biogeochemical reactions occurring between exposed lower crustal and mantle rocks and circulating fluids in these settings promise to reveal much about the origins of early life on Earth and perhaps on other planets (Martin *et al.*, 2008). There exists only cursory knowledge of serpentinization processes in MOR and rifted margin settings. Are there other vent sites like Lost City, providing novel geochemical environments for unique hydrothermal ecosystems in the virtually unexplored sections of the global MOR in the South Atlantic, Indian, Arctic, and Southern Oceans? The implications for "Lost City-type" venting are seminal for understanding first-order structural and geochemical processes within large volumes of the lithosphere.

There is no shortage of excellent conceptual models and field and laboratory approaches to explore these environments, but we need new, specialized technologies, and a decades-long commitment to investigate these remote and often difficult-to-access regions of the MOR. We need new multidisciplinary programs, such as IODP, OOI, US RIDGE, and Ridge 2000, and the other international MOR programs of the past few decades, to investigate myriad processes occurring within the oceanic lithosphere, because often one phenomenon influences another. New insights are likely to come from coordinated shallow and deep crustal drilling programs at targeted areas, where detailed geophysical investigations can be linked with a series of drill holes providing access to the lithospheric interior. This will permit direct crustal sampling with simultaneous measurements and monitoring of physical, chemical, and biological fluxes and phenomena within a well-constrained volume of MOR.

The current investment over the past decade in ocean observing systems, both on the seafloor and in the water column, will stimulate different, quantitative approaches to monitoring and sensing plate boundary processes. Real-time monitoring systems similar to OOI will provide high power to sensors on the seafloor and in the water column and direct connection to the global Internet. In conjunction with more traditional observational and sampling expeditions to specific study areas, new observatory sensors will yield data that will undoubtedly affirm or revise current models of processes associated with crustal accretion and subduction, as the lithosphere is cycled from its origin at the MOR to its transformation back into the mantle, as well as the life that these systems host.

Techniques used in future research will be guided by advances in robotic technologies coupled with miniaturization and automation of sensing systems. Specific sensors will target key chemical compounds essential to sustaining metabolic processes at the MOR crest and within the upper lithosphere. New vehicle and observational systems will allow direct observations of ephemeral processes that have been missed by traditional oceanographic methods. No one system is likely to satisfy all the needs for the spectrum of environments or monitoring/sampling tasks to be carried out. Hence, the approach developed over the past 25 years of new multi-vehicle operations within a nested survey strategy will continue to deliver significant results. But the resolution and rate of data acquired are likely to surpass anything we have experienced to date.

Over the next decade the ocean and the dynamic geology, chemistry, and biology occurring within and beneath it will be increasingly important to our global society. Discoveries on many fronts within the ocean and below the seafloor are within our grasp. This is an exciting time to be a marine scientist.

Abbreviations

Please note that abbreviations for some universities, institutions and organizations have been used throughout to save space. They are defined as follows:

BEA	(Bureau d'Enquêtes et d'Analyses pour la sécurité de l'aviation civile)
CSSF	(*Canadian Scientific Submersible Facility*)
IFREMER	(Institut français de recherche pour l'exploitation de la mer)
JAMSTEC	(Japan Agency for Marine-Earth Science and Technology)
IODP	(International Ocean Drilling Program)
LDEO	(Lamont-Doherty Earth Observatory)
MARUM	(Center for Marine Environmental Sciences)
MBARI	(Monterey Bay Aquarium Research Institute)
NSF	(National Science Foundation)
NOAA	(National Oceanic and Atmospheric Administration)
OOI	(Ocean Observing Initiative)
UNC	(University of North Carolina)
UNH	(University of New Hampshire)
URI	(University of Rhode Island)
SIO	(Scripps Institution of Oceanography)
UW	(University of Washington)
WHOI	(Woods Hole Oceanographic Institution)

Figure credits

P.1 Delaney, J. R. and Kelley, D. S.; UW, CSSF-ROV *ROPOS*; NSF/OOI/UW.
P.2 ©WHOI-NDSF; Lutz, R.; Rutgers University; Lange, W.; WHOI.
P.3 Delaney, J. R., Kelley, D. S. and Elend, M.; UW, NSF/OOI/UW.
P.4 NOAA-NGDC base map (http://www.ngdc.noaa.gov).
P.5 Karson, J.; Syracuse University.
P.6 Delaney, J. R. and Kelley, D. S.; UW, CSSF-ROV *ROPOS*; NSF/OOI/UW.
P.7 Kelley, D. S., Delaney, J. R. and Elend, M.; UW; CSSF-ROV *ROPOS*; W. M. Keck Foundation/UW.
P.8 Delaney, J. R. and Kelley, D. S.; UW, CSSF-ROV *ROPOS*; NSF/OOI/UW.
1.1 ©UW; Photo by Mitch Elend.
1.2 ©WHOI; Photo by C. German.
1.3 ©The Challenger Society.
1.4 from Thomson, C. W. (1877); Fig. 17.
1.5 ©National Geographic Society; Illustration by E. Bostelmann.
1.6 ©Lamont-Doherty Earth Observatory.
1.7 Fornari, D.; WHOI.
1.8 from Sandwell and Smith (1997); Fig. 1.
1.9 ©WHOI; Illustration by E. Paul Oberlander.
1.10 Fornari, D.; WHOI; Marine Geoscience Data System www.marine-geo.org.
1.11 ©SIO; Marine Physical Laboratory.
1.12 Harpp, K.; Colgate University; University of Hawaii, Hawaii Mapping Research Group (HMRG).
1.13 Edwards, M.; University of Hawaii, Hawaii Mapping Research Group (HMRG).
1.14 Applegate, B.; SIO; Swan, D.; Honolulu Star Bulletin.
1.15 Soule, S.A.; WHOI.
1.16 Soule, S.A.; WHOI.
1.17 ©WHOI; Woods Hole Oceanographic Archives.
1.18 ©WHOI; Woods Hole Oceanographic Archives; US Navy.
1.19 Fornari, D.; WHOI.
1.20 Fisher, C.; Pennsylvania State University; Yoerger, D.; WHOI.
1.21 ©WHOI; Triptych by Dave Grey.
1.22 Escartín, J.; IPGP; IFREMER.
1.23 ©JAMSTEC.
1.24 P. P. Shirshov Institute of Oceanography; http://www.ocean.ru.
1.25 ©WHOI; Photo by Lisa Smith.
1.26 ©WHOI; Photo by D. Fornari.
1.27 ©MBARI; Photos by Todd Walsh; MBARI.
1.28 Kelley, D. S.; UW; CSSF-ROV *ROPOS*.
1.29 ©IFREMER.
1.30 ©WHOI; Photos by D. Fornari (left) and C. German (right); WHOI.
1.31 ©WHOI; Illustration by E. Paul Oberlander.
1.32 ©MBARI; Photos by Todd Walsh; MBARI.
1.33 ©MBARI; Caress, D.; MBARI.
1.34 Tivey, M.; WHOI (Tivey and Johnson, 2002).
1.35 ©WHOI; REMUS Group; Purcel, M.; WHOI.
1.36 ©BEA, BEA (2012).
1.37 ©WHOI; Photo by Tim Shank; WHOI.
1.38 ©Charile Arneson.
1.39 Carbotte, S.: LDEO; illustration LDEO.
1.40 Carbotte, S.: LDEO; (from Carbotte *et al.*, 2012).
1.41 ©WHOI; WHOI OBS Group and IRIS; top photo by A. Gardner; WHOI.
1.42 ©WHOI; Illustration by E. Paul Oberlander.
1.43 ©WHOI; Smith, D.; WHOI; Smith (2004) Oceanus.
1.44 ©IODP; ©JAMSTEC.
1.45 ©IODP.
1.46 Delaney, J. R. and Kelley, D. S.; UW; Center for Environmental Visualization; NSF/OOI/UW.
1.47 Delaney, J. R. and Kelley, D. S.; UW; Center for Environmental Visualization; NSF/OOI/UW.
1.48 ©WHOI; Illustration by E. Paul Oberlander.
1.49 Delaney, J. R. and Kelley, D. S.; UW; Center for Environmental Visualization; NSF/OOI/UW.
1.50 Delaney, J. R. and Kelley, D. S.; UW; Center for Environmental Visualization; NSF/OOI/UW.
2.1 D. Fornari, WHOI; GeoMapApp http://www.geomapapp.org and Ryan *et al.* (2009).
2.2 Delaney, J. R., Barga, R. S. and Stoermer, M.; UW; Center for Environmental Visualization.
2.3 J. Karson, Syracuse University; after Oceanus graphic.
2.4 J. Karson, Syracuse University.
2.5 Heezen, B. G. and Tharp, M. (1977).
2.6 J. Karson, Syracuse University, after Heezen, B. C. (1960).
2.7 J. Karson, Syracuse University.
2.8 Müller, R. D., *et al* (2008).
2.9 Isaaks, B., Oliver, J. and Sykes, L. R. (1968).
2.10 J. Karson, Syracuse University; GeoMapApp http://www.geomapapp.org and Ryan *et al.*, 2009.
2.11 J. Karson, Syracuse University and M. Perfit, University of Florida.
2.12 J. Karson, Syracuse University, after Ewing, M. A. and Engel, L. (1962).

2.13 J. Karson, Syracuse University, after White, R. S., McKenzie, D. P. and O'Nions, R. K. (1992).
2.14 J. Karson, Syracuse University.
2.15 D. Fornari, WHOI.
2.16 J. Karson, Syracuse University.
2.17 J. Karson, Syracuse University.
2.18 after C. MacLeod, V. Ferrini and D. Schillington (2010) (right); D. Fornari, WHOI (left).
2.19 J. Karson, Syracuse University; GeoMapApp http://www.geomapapp.org and Ryan *et al.* (2009).
2.20 Expedition 335 Scientists (2011), Integrated Ocean Drilling Program.
2.21 Expedition 335 Scientists (2011), Integrated Ocean Drilling Program.
2.22 Karson, J. A., Cannat, M. Miller, D. J. and Elthon, D. (1997).
2.23 J. Karson, Syracuse University.
2.24 J. Karson, Syracuse University.
2.25 J. Karson, Syracuse University, after Aubouin, J. (1965). and Hess, H. H. (1962).
2.26 J. Karson, Syracuse University.
2.27 J. Karson, Syracuse University, after Twiss, R. J. and Moores, E. M. (1992).
3.1 T. Orr, US Geological Survey.
3.2 Global bathymetry data from E-Topo2 satellite altimetry, MGDS database and University of New Hampshire's Center for Coastal and Ocean Mapping Joint Hydrographic Center. Globe perspectives produced using Fledermaus imaging software from QPS (http://www.qps.nl). Images produced by D. Fornari, WHOI.
3.3 Kelley, D. S.; Center for Environmental Visualization, University of Washington.
3.4 data source: GeoMapApp http://www.geomapapp.org and Ryan *et al.* (2009).
3.5 after Perfit, M. R. (2003).
3.6 data source: GeoMapApp http://www.geomapapp.org and Ryan *et al.* (2009).
3.7 data source: GeoMapApp http://www.geomapapp.org and Ryan *et al.* (2009).
3.8 after Macdonald, K. C. (1998a).
3.9 courtesy of S. Carbotte, Lamont–Doherty Earth Observatory.
3.10 after Mutter, J. C. and Karson, J. A. (1992).
3.11 from Wang *et al.* (2014).
3.12 Kelley, D. S.; Center for Environmental Visualization, University of Washington.
3.13 from Soule *et al.* (2005).
3.14 from Stakes *et al.* (2006).
3.15 after Cannat *et al.* (1999), Barreyre *et al.* (2012), Escartín *et al.* (2001a), Combier (2007), Ondreas *et al.* (2009).
3.16 data source: GeoMapApp http://www.geomapapp.org and Ryan *et al.* (2009).
3.17 after Lawrence *et al.* (1998).
3.18 J. Karson, Syracuse University.
3.19 after Kelley *et al.* (2012).
3.20 data source: GeoMapApp http://www.geomapapp.org and Ryan *et al.* (2009).
3.21 from MacLeod *et al.* (2009).
3.22 from deMartin *et al.* (2007).
3.23 from Escartín and Canales (2010).
3.24 J. Karson, Syracuse University.
3.25 J. Karson, Syracuse University.
3.26 data source: GeoMapApp http://www.geomapapp.org and Ryan *et al.* (2009).
3.27 J. Karson, Syracuse University.
3.28 from Hey, R. N. (2004) and data source: GeoMapApp http://www.geomapapp.org and Ryan *et al.* (2009).
3.29 from Hey, R. N. (2004) and data source: GeoMapApp http://www.geomapapp.org and Ryan *et al.* (2009).
3.30 from Christeson *et al.* (2012).
3.31 after IODP Leg 335 (2012).
3.32 J. Karson; WHOI, NDSF-DSV Jason ©WHOI; NSF, Pito Deep Cruise, 2003.
3.33 J. Karson, Syracuse University.
3.34 from Dunn *et al.* (2000).
3.35 from Han *et al.* (2014).
3.36 from Canales *et al.* (2009).
3.37 from Singh *et al.* (2006).
3.38 from Canales *et al.* (2005).
3.39 from Canales *et al.* (2003).
3.40 from Banda *et al.* (1992) (top); Reston *et al*, (1999) (bottom).
3.41 after Karson (1998).
3.42 after Key *et al.* (2013).
3.43 from Dunn, R. A. and Toomey, D. R. (1997).
3.44 J. Karson, Syracuse University.
3.45 J. Karson, Syracuse University.
4.1 Delaney, J. R.; Kelley, D. S.; UW; and Girguis, P.; MBARI; ©MBARI-Tiburon; W. M. Keck Foundation/UW/MBARI.
4.2 Delaney, J. R., Kelley, D. S., Elend, M.; UW; ©CSSF-ROV *ROPOS*; W. M. Keck Foundation/UW/University of Victoria.
4.3 from Spiess *et al.* (1980).
4.4 Robigou, V.; Ocean et Terra Studio.
4.5 Kelley, D. S.; UW.
4.6 Courtesy of Tivey, M. K. WHOI; ©WHOI Graphics.
4.7 NASA NSSDC Photo Gallery.
4.8 Kelley, D. S.; UW.
4.9 from Wilcock *et al.* (2009).
4.10 Kelley, D. S.; UW.
4.11 Kelley, D. S.; UW.
4.12 Kelley, D. S.; UW.
4.13 Delaney, J. R. and Kelley, D. S.; UW, ©CSSF-ROV *ROPOS*; NSF/OOI/UW.
4.14 Kelley, D. S. and Delaney, J. R.; UW.
4.15 Kelley, D. S.; UW.
4.16 Von Damm, K. UNH, and Lilley, L. UW, NDSF-DSV *Alvin*, ®WHOI.
4.17 Kelley, D. S.; UW.
4.18 Delaney, J. R. and Kelley, D. S.; UW.

4.19 from Lupton *et al.* (2006).
4.20 from Lupton *et al.* (2006).
4.21 from Summit, M. (2000).
4.22 from Shank *et al.* (1998).
4.23 after Kelley *et al.* (1998).
4.24 Delaney, J. R. and Kelley, D. S.; UW; ©CSSF-ROV *ROPOS*; NSF/OOI/UW.
4.25 after Escartin *et al.* (2008) and Kelley and Shank (2010). Logachev Type 2 image was acquired by the ROV *Quest 4000*, from ©MARUM, University of Bremen.
4.26 Kelley, D. S.; UW; NDSF-DSV *Alvin*; ©WHOI.
4.27 Delaney, J. R. and Kelley, D. S.; UW; NDSF-ROV *Jason*; WHOI; W. M. Keck Foundation/UW.
4.28 ©Cameron, J., Earthship Productions; MIR.
4.29 Lutz, R.; Rutgers, U. and Lange, W; WHOI; NDSF-DSV *Alvin*; ©WHOI; NSF.
4.30 Sohn, R. and Humphris, S.; WHOI; NDSF-ROV Jason, *Alvin*, ©WHOI; NSF.
4.31 Shank, T.; WHOI and Cook, J.; ©WHOI Graphics.
4.32 Pedersen, R.; ©University of Bergen; ROV *Bathysaurus*, Research Council of Norway.
4.33 Pedersen, R.; ©University of Bergen; ROV *Bathysaurus*, Research Council of Norway.
4.34 from Pedersen *et al.* (2010b).
4.35 from Pedersen *et al.* (2010b).
4.36 Pedersen, R. and Denny, A.; University of Bergen.
4.37 Pedersen, R.; ©University of Bergen; ROV *Bathysaurus*, Research Council of Norway.
4.38 from Pedersen *et al.* (2010a).
4.39 Pedersen, R.; ©University of Bergen; ROV *Bathysaurus*, Research Council of Norway.
4.40 Pedersen, R.; ©University of Bergen; ROV *Bathysaurus*, Research Council of Norway.
4.41 ©Cameron, J.; Earthship Productions; MIR.
4.42 ©Cameron, J.; Earthship Productions; MIR.
4.43 Fornari, D; WHOI; Bathymetry data from Marine Geoscience Data System, Lamont-Doherty Earth Observatory.
4.44 ©Cameron, J.; Earthship Productions; MIR.
4.45 ©Cameron, J.; Earthship Productions; MIR.
4.46 ©Cameron, J.; Earthship Productions; MIR.
4.47 ©Cameron, J.; Earthship Productions; MIR.
4.48 ©Cameron, J.; Earthship Productions; MIR.
4.49 from Humphris *et al.* (2002).
4.50 from Barreyre *et al.* (2012).
4.51 ©Cameron, J.; Earthship Productions; MIR.
4.52 ©Cameron, J.; Earthship Productions; MIR.
4.53 ©Cameron, J.; Earthship Productions; MIR.
4.54 ©Cameron, J.; Earthship Productions; MIR.
4.55 ©Cameron, J.; Earthship Productions; MIR.
4.56 ©Cameron, J.; Earthship Productions; MIR.
4.57 ©Cameron, J.; Earthship Productions; MIR.
4.58 Desbruyères, D.; IFREMER, HOV *Nautile*; ©IFREMER-Nautile.
4.59 Desbruyères, D.; IFREMER, HOV *Nautile*; ©IFREMER-Nautile.
4.60 Courtesy of Escartín, J.; IPG-Paris and IFREMER.
4.61 Desbruyères, D.; IFREMER, HOV *Nautile*; ©IFREMER-Nautile.
4.62 Desbruyères, D.; IFREMER, HOV *Nautile*; ©IFREMER-Nautile.
4.63 Desbruyères, D.; IFREMER, HOV *Nautile*; ©IFREMER-Nautile.
4.64 Desbruyères, D.; IFREMER, HOV *Nautile*; ©IFREMER-Nautile.
4.65 Desbruyères, D.; IFREMER, HOV *Nautile*; ©IFREMER-Nautile.
4.66 Desbruyères, D.; IFREMER, HOV *Nautile*; ©IFREMER-Nautile.
4.67 Kelley, D. S. and Center for Environmental Visualization; UW.
4.68 Kelley, D. S. and Elend, M., UW, URI, ROV *Hercules*; NOAA Ocean Exploration.
4.69 After Boschi *et al.* (2008).
4.70 Kelley, D. S.; UW, and Yoerger, D. A.; WHOI; WHOI-AUV *ABE*; NSF.
4.71 Kelley, D. S. UW; URI-ROV *Hercules*. IFE, URI-IAO, Lost City Science Party, and NOAA; NOAA Ocean Exploration.
4.72 after Karson *et al.* (2006).
4.73 Kelley, D. S. UW; URI-ROV *Hercules*, IFE, URI-IAO, Lost City Science Party, and NOAA Ocean Exploration.
4.74 Kelley, D. S. UW; URI-ROV *Hercules*, IFE, URI-IAO, Lost City Science Party, and NOAA Ocean Exploration.
4.75 Kelley, D. S. UW; URI-ROV *Hercules*, IFE, URI-IAO, Lost City Science Party, and NOAA Ocean Exploration.
4.76 Kelley, D. S. UW; URI-ROV *Hercules*, IFE, URI-IAO, Lost City Science Party, and NOAA Ocean Exploration.
4.77 Kelley, D. S., and Elend, M., UW; URI-ROV *Hercules*, and NOAA Ocean Exploration.
4.78 Kelley, D. S. UW; URI-ROV *Hercules*, IFE, URI-IAO, Lost City Science Party, and NOAA Ocean Exploration.
4.79 Kelley, D. S. UW; URI-ROV *Hercules*, IFE, URI-IAO, Lost City Science Party, and NOAA Ocean Exploration.
4.80 Kelley, D. S. and Elend, M., UW; URI-ROV *Hercules*; UW; URI-ROV Hercules, and NOAA Ocean Exploration.
4.81 Kelley, D. S. UW; URI-ROV *Hercules*, IFE, URI-IAO, Lost City Science Party, and NOAA Ocean Exploration.
4.82 From Kelley *et al.* (2007).
4.83 Kelley, D.S. UW; URI-ROV *Hercules*, IFE, URI-IAO, Lost City Science Party, and NOAA Ocean Exploration.

4.84 Kelley, D.S. UW; URI-ROV *Hercules*, IFE, URI-IAO, Lost City Science Party, and NOAA Ocean Exploration.
4.85 Kelley, D.S. UW; URI-ROV *Hercules*, IFE, URI-IAO, Lost City Science Party, and NOAA Ocean Exploration.
4.86 Kelley, D.S. UW; URI-ROV *Hercules*, IFE, URI-IAO, Lost City Science Party, and NOAA Ocean Exploration.
4.87 Kelley, D.S. UW; URI-ROV *Hercules*, IFE, URI-IAO, Lost City Science Party, and NOAA Ocean Exploration.
4.88 Kelley, D.S. UW; URI-ROV *Hercules*, IFE, URI-IAO, Lost City Science Party, and NOAA Ocean Exploration.
4.89 Kelley, D. S., and Center for Environmental Visualization; UW.
4.90 Sohn, R., Humphris, S., Canales, J. P., Roman, C.; ©WHOI.
4.91 Sohn, R. and Humphris, S.; WHOI; NDSF-ROV *Jason*, DSRV *Alvin*, ©WHOI; NSF.
4.92 Karson, J.; Syracuse University.
4.93 from Tivey, M. K. (2007).
4.94 Sohn, R. and Humphris, S.; WHOI; NDSF-ROV *Jason*, *Alvin*, ©WHOI; NSF.
4.95 Sohn, R. and Humphris, S.; WHOI; NDSF-ROV *Jason*, *Alvin*, ©WHOI; NSF.
4.96 Sohn, R. and Humphris, S.; WHOI; NDSF-ROV *Jason*, *Alvin*, ©WHOI; NSF.
4.97 Sohn, R. and Humphris, S.; WHOI; NDSF-ROV *Jason*, *Alvin*, ©WHOI; NSF.
4.98 Sohn, R. and Humphris, S.; WHOI; NDSF-ROV *Jason*, *Alvin*, ©WHOI; NSF.
4.99 from Petersen *et al.* (2009); ROV *MARUM-QUEST*; MARUM - Center for Environmental Sciences, University of Bremen, Germany. ©MARUM.
4.100 Borowsk, C.; MPI Bremen; ROV *Jason 2*; ©SPP1144, German Science Foundation.
4.101 Bathymetry collected during HYDROMAR-IV cruises with the R/V *Meteor* and R/V *S. Merian* between 2004 and 2007 (Petersen *et al.*, 2009).
4.102 ROV *MARUM-QUEST*; MARUM - Center for Environmental Sciences, University of Bremen, Germany. ©MARUM.
4.103 reprinted from Petersen, S., Kuhn, K., Kuhn, T., Augustin, N., Hékinian, R., Franz, L. and Borowski, C., The geological setting of the ultramafic-hosted Logatchev hydrothermal field (14°45′ N, Mid-Atlantic Ridge) and its influence on massive sulfide formation. *Lithos*, 112, p. 47, Copyright (2009), with permission from Elsevier.
4.104 reprinted from Petersen, S., Kuhn, K., Kuhn, T., Augustin, N., Hékinian, R., Franz, L. and Borowski, C., The geological setting of the ultramafic-hosted Logatchev hydrothermal field (14°45′ N, Mid-Atlantic Ridge) and its influence on massive sulfide formation. *Lithos*, 112, p. 52, Copyright (2009), with permission from Elsevier.
4.105 ROV *MARUM-QUEST*; MARUM - Center for Environmental Sciences, University of Bremen, Germany. ©MARUM.
4.106 Borowski, C.; MPI Bremen; ROV *Jason 2*; ©SPP1144, German Science Foundation.
4.107 MARUM - Center for Marine Environmental Research; University of Bremen, Germany; ROV *Quest*; ©MARUM.
4.108 MARUM - Center for Marine Environmental Research; University of Bremen, Germany; ROV *Quest*; ©MARUM.
4.109 Fornari, D. Woods Hole Oceanographic Institution, Bathymetry data from Marine Geoscience Data System, Lamont-Doherty Earth Observatory.
4.110 after German *et al.* (2008).
4.111 MARUM - Center for Marine Environmental Research; University of Bremen, Germany; ROV *Quest*; ©MARUM.
4.112 MARUM - Center for Marine Environmental Research; University of Bremen, Germany; ROV *Quest*; ®MARUM.
4.113 MARUM - Center for Marine Environmental Research; University of Bremen, Germany; ROV Quest; ®MARUM.
4.114 MARUM - Center for Marine Environmental Research; University of Bremen, Germany; ROV Quest; ®MARUM.
4.115 Kelley, D. S. and Delaney, J. R.; UW.
4.116 Kelley, D. S. and Delaney, J. R.; UW.
4.117 Delaney, J. R., Kelley, D. S. and Elend, M.; UW; ©MBARI-ROV *Tiburon*; W. M. Keck Foundation/UW/MBARI.
4.118 Delaney, J. R., Kelley, D. S. and Elend, M.; UW; ©CSSF-ROV *ROPOS*; W. M. Keck Foundation/UW.
4.119 From Kelley *et al.* (2012).
4.120 From Kelley *et al.* (2012).
4.121 Kelley, D. S., Delaney, J. R. and Elend, M.; UW; WHOI-DSRV *Alvin*; ©WHOI; NSF.
4.122 Delaney, J. R., Kelley, D. S. and Elend, M.; UW; ©CSSF-ROV *ROPOS*; W. M. Keck Foundation/UW.
4.123 Delaney, J. R., Kelley and D. S.; UW; ©CSSF-ROV *ROPOS*; W. M. Keck Foundation/UW.
4.124 Delaney, J. R. and Kelley, D. S., UW; WHOI-DSRV *Alvin*; NSF.
4.125 Delaney, J. R., Kelley, D. S. and Elend, M.; UW; and Girguis, P.; ©MBARI-ROV *Tiburon*; W. M. Keck Foundation/UW/MBARI.
4.126 Delaney, J. R., Kelley, D. S. and Elend, M.; UW; WHOI-ROV *Jason*; UW, American Museum of Natural History, New York.
4.127 Delaney, J. R., Kelley, D. S., Elend, M., Mathez, E., UW; CSSF-ROV *ROPOS*; UW/American Museum of Natural History, New York.

4.128 Delaney, J. R. and Kelley, D. S.; UW.
4.129 Delaney, J. R. and Kelley, D. S., UW, and Juniper, K.; University of Quebec-Montreal; ©CSSF-ROV *ROPOS*. W. M. Keck Foundation/UW/University of Victoria.
4.130 Delaney, J. R. and Kelley, D. S.; UW.
4.131 Delaney, J. R. and Kelley, D. S.; UW.
4.132 Kelley, D. S., UW; WHOI-DSRV *Alvin*; ©WHOI; NSF.
4.133 Kelley, D. S., Delaney, J. R. and Elend, M.; UW; ©CSSF-ROV *ROPOS*; W. M. Keck Foundation/UW.
4.134 Kelley, D. S. and Delaney, J. R.; UW; ©CSS -ROV *ROPOS*; W. M. Keck Foundation/UW.
4.135 after Kelley *et al.* (2012).
4.136 Kelley, D. S. and Delaney, J. R.; Center for Environmental Visualization, UW.
4.137 Kelley, D. S. and Delaney, J. R.; UW.
4.138 Delaney, J. R. and Kelley, D. S.; UW; ©CSSF-ROV *ROPOS*; NSF/OOI/UW.
4.139 Kelley, D. S. and Delaney, J. R.; UW.
4.140 Delaney, J. R. and Kelley, D. S.; UW; ©CSSF-ROV *ROPOS*; NSF/OOI/UW.
4.141 Delaney, J. R. and Kelley, D. S.; UW; ©CSSF-ROV *ROPOS*; NSF/OOI/UW/CSSF.
4.142 Delaney, J. R. and Kelley, D. S.; UW; ©CSSF-ROV *ROPOS*; NSF/OOI/UW.
4.143 Delaney, J. R., Kelley, D. S. and Elend, M.; UW; WHOI-ROV *Jason*; NSF/OOI/UW.
4.144 Delaney, J. R., Kelley, D. S. and Elend, M.; UW; WHOI-ROV *Jason*; NSF/OOI/UW.
4.145 Delaney, J. R., Kelley, D. S. and Center for Environmental Visualization; UW; WHOI-ROV *Jason* NSF/OOI/UW.
4.146 Delaney, J. R., Kelley, D. S. and Elend, M.; UW; WHOI-ROV *Jason*; NSF/OOI/UW.
4.147 Delaney, J. R., Kelley, D. S. and Elend, M.; UW; WHOI-ROV *Jason*; NSF/OOI/UW.
4.148 Delaney, J. R. and Kelley, D. S.; UW; ©CSSF-ROV *ROPOS*; NSF/OOI/UW.
4.149 Delaney, J. R. and Kelley, D. S.; UW; ©WHOI-ROV *Jason*; NSF/OOI/UW.
4.150 Delaney, J. R. and Kelley, D. S.; UW; ©CSSF-ROV *ROPOS*; NSF/OOI/UW.
4.151 Delaney, J. R. and Kelley, D. S.; UW; ©CSSF-ROV *ROPOS*; NSF/OOI/UW.
4.152 Delaney, J. R. and Kelley, D. S.; UW; ©CSSF-ROV *ROPOS*; NSF/OOI/UW.
4.153 Delaney, J. R. and Kelley, D. S.; UW; WHOI-ROV *Jason*; NSF/OOI/UW.
4.154 Delaney, J. R. and Kelley, D.S.; Center for Environmental Visualization; UW.
4.155 Teske, A.; UNC; NDSF-DSV Alvin; ©WHOI; NSF.
4.156 Cary, C. University of Delaware; NDSF-DSV *Alvin*; ©WHOI; NSF.
4.157 Soule, A., WHOI.
4.158 Cary, C. University of Delaware; NDSF-DSV *Alvin*; ©WHOI; NSF.
4.159 Teske, A.; UNC; NDSF-DSV *Alvin*; ©WHOI; NSF.
4.160 Teske, A.; UNC; NDSF-DSV *Alvin*; ©WHOI; NSF.
4.161 Teske, A.; UNC; NDSF-DSV *Alvin*; ©WHOI; NSF.
4.162 Cary, C. University of Delaware; NDSF-DSV *Alvin*; ©WHOI; NSF.
4.163 Cary, C. University of Delaware; NDSF-DSV *Alvin*; ©WHOI; NSF.
4.164 Teske, A.; UNC; NDSF-DSV *Alvin*; ©WHOI; NSF.
4.165 Teske, A.; UNC; NDSF-DSV *Alvin*; ©WHOI; NSF.
4.166 Cary, C. University of Delaware; NDSF-DSV *Alvin*; ©WHOI; NSF.
4.167 Kelley, D. S., UW.
4.168 Von Damm, K.; UNH; NDSF-DSV *Alvin*; ©WHOI, NSF.
4.169 ©WHOI Graphics Paul Oberlander artist.
4.170 after Fornari *et al.* (2012).
4.171 Fornari, D., Soule, A., WHOI.
4.172 Lutz, R. Rutgers University; NDSF-DSV *Alvin*; ©WHOI; NSF.
4.173 Tivey, M. K.; WHOI; NDSF-DSV *Alvin*; ©WHOI; NSF.
4.174 Von Damm, K.; UNH, and Lilley, L.; UW; NDSF-DSV *Alvin*; ©WHOI; NSF.
4.175 Von Damm, K.; UNH, and Lilley, L.; UW; NDSF-DSV *Alvin*; ©WHOI; NSF.
4.176 Haymon, R., University of California at Santa Barbara, and Fornari, D., NDSF-DSV *Alvin*; ©WHOI; NSF.
4.177 Lutz, R. Rutgers University; NDSF-DSV *Alvin*; ©WHOI; NSF.
4.178 Lutz, R. Rutgers University; NDSF-DSV *Alvin*; ©WHOI; NSF.
4.179 Lutz, R. Rutgers University; NDSF-DSV *Alvin*; ©WHOI; NSF.
4.180 Lutz, R. Rutgers University; NDSF-DSV *Alvin*; ©WHOI; NSF.
4.181 Lutz, R. Rutgers University; NDSF-DSV *Alvin*; ©WHOI; NSF.
4.182 Lutz, R. Rutgers University; NDSF-DSV *Alvin*; ©WHOI; NSF.
4.183 Cary, C.; University of Delaware; NDSF-DSV *Alvin*; ©WHOI; NSF.
4.184 Cary, C. University of Delaware; NDSF-DSV *Alvin*; ©WHOI; NSF.
4.185 Cary, C.; University of Delaware; NDSF-DSV *Alvin*; ©WHOI; NSF.
4.186 Edmond, J., MIT; and Hessler R., SIO; NDSF-DSV *Alvin*; ©WHOI; NSF.
4.187 Shank, T. M.; WHOI, NDSF-DSV *Alvin*; ©WHOI; NSF, NOAA Ocean Exploration.
4.188 Shank, T. M., WHOI.
4.189 after Shank, T., Fornari, D., Yoerger, D., Humphris, S., Bradley, A.S. *et al.* (2003).
4.190 Shank, T. M.; WHOI, NDSF-DSV *Alvin*; ©WHOI; NSF, NOAA Ocean Exploration.
4.191 Shank, T. M.; WHOI, NDSF-DSV *Alvin*; ©WHOI; NSF, NOAA Ocean Exploration.

4.192 Shank, T. M.; WHOI, NDSF-DSV *Alvin*; ©WHOI; NSF, NOAA Ocean Exploration.
4.193 Shank, T. M.; WHOI; URI-ROV *Little Hercules*; NOAA Ocean Exploration.
4.194 Shank, T. M.; WHOI, NDSF-DSV *Alvin*; ©WHOI; NSF, NOAA Ocean Exploration.
4.195 Haymon, R. University of California at Santa Barbara, NSF, NOAA Ocean Exploration WHOI, NDSF, DSV *Medea*.
4.196 Shank, T. M.; WHOI; URI-ROV *Little Hercules*; NOAA Ocean Exploration.
4.197 Shank, T. M.; WHOI; URI-ROV *Little Hercules*; NOAA Ocean Exploration.
4.198 Sinton, J.; University of Hawaii; NDSF-DSV *Alvin*; ©WHOI; NSF.
4.199 Shank, T. M.; WHOI; URI-ROV *Little Hercules*; NOAA Ocean Exploration.
4.200 Kelley, D. S., and Center for Environmental Visualization, UW.
4.201 Fornari, D., WHOI; Bathymetry data from Marine Geoscience Data System, Lamont-Doherty Earth Observatory of Columbia University.
4.202 Van Dover, C.; Duke; NDSF-ROV *Jason*; ©WHOI; NSF.
4.203 Van Dover, C.; Duke; NDSF-ROV *Jason*; ©WHOI; NSF.
4.204 Van Dover, C.; Duke; NDSF-ROV *Jason*; ©WHOI; NSF.
4.205 Van Dover, C.; Duke; NDSF-ROV *Jason*; ©WHOI; NSF.
4.206 Van Dover, C.; Duke; NDSF-ROV *Jason*; ©WHOI; NSF.
4.207 Van Dover, C.; Duke; NDSF-ROV *Jason*; ©WHOI; NSF.
4.208 Van Dover, C.; Duke; NDSF-ROV *Jason*; ©WHOI; NSF.
4.209 Fornari, D., WHOI.
4.210 after Gallant, R. M. and Von Damm, K. L. (2006).
4.211 Van Dover, C.; Duke; NDSF-ROV *Jason*; ©WHOI; NSF.
4.212 Van Dover, C.; Duke; NDSF-ROV *Jason*; ©WHOI; NSF.
4.213 Van Dover, C.; Duke; NDSF-ROV *Jason*; ©WHOI; NSF.
4.214 Van Dover, C.; Duke; NDSF-ROV *Jason*; ©WHOI; NSF.
4.215 Van Dover, C.; Duke; NDSF-ROV *Jason*; ©WHOI; NSF.
5.1 Resing, J., Rubin, K., NOAA; U. Hawaii; NSF; WHOI-NDSF-ROV *Jason*.
5.2 Kelley, D. S., UW; CSSF -ROV *ROPOS*.
5.3 ©WHOI; WHOI-NDSF-*Alvin*; Lange, W., WHOI.
5.4 ©WHOI; WHOI-NDSF-*Alvin*.
5.5 from Chadwick *et al.* (2013), G3ed supplementary materials.
5.6 ©WHOI; WHOI-NDSF-*Alvin*; Karson, J.; Syracuse University (middle); ROV *Tiburon* ©MBARI (right).
5.7 Kelley, D. S., UW; CSSF-ROV *ROPOS*; Karson, J., Syracuse University (right).
5.8 ©IODP.
5.9 from Tominaga and Umino (2010).
5.10 Kelley, D. S., UW; CSSF -ROV *ROPOS*.
5.11 Karson, J., Syracuse University; WHOI; WHOI-NDSF-ROV *Jason*.
5.12 Karson, J., Syracuse University.
5.13 from Colman *et al.* (2012), G3ed Fig. 2.
5.14 ©WHOI; WHOI-NDSF-*Alvin*; Lange, W., WHOI.
5.15 Fornari, D., WHOI.
5.16 ©WHOI; from Humphris and Tivey (1993).
5.17 ©WHOI; from Tivey (2004); Oceanus
5.18 Karson, J., Syracuse University.
5.19 Kelley, D. S., UW; CSSF-ROV *ROPOS*.
5.20 from Fundis *et al.* (2010).
5.21 Kelley, D. S., UW; CSSF-ROV *ROPOS*.
5.22 ©Cameron, J., Earthship Productions; MIR.
5.23 Perfit, M., UF, ROV *Tiburon* ©MBARI (top); Karson, J., Syracuse University (bottom).
5.24 ©WHOI; WHOI-NDSF-*Alvin*; Lange, W., WHOI.
5.25 from Chadwick *et al.* (2009) (left); Kelley, D. S., UW; CSSF-ROV *ROPOS* (right).
5.26 Fornari, D., WHOI; Kelley, D. S., UW; CSSF-ROV *ROPOS*.
5.27 ©Cameron, J., Earthship Productions; MIR (left); Fornari, D., WHOI (top right); Perfit, M., University of Florida (bottom right).
5.28 Perfit, M., UF, ROV *Tiburon* ©MBARI (second from top); Fornari, D.; WHOI-NDSF-*Alvin* (top); Kelley, D., UW; CSSF-ROV *ROPOS* (bottom two).
5.29 Shank, T.; WHOI-NDSF-*Alvin* (top); Kelley, D., UW; CSSF-ROV *ROPOS* (bottom two).
5.30 Karson, J., Syracuse University (right); Fornari, D., WHOI (left).
5.31 from Soule *et al.* (2005).
5.32 from Soule *et al.* (2005).
5.33 from Soule *et al.* (2005).
5.34 Karson, J., Syracuse University.
5.35 ©WHOI; WHOI-NDSF-*Alvin*; Shank, T., WHOI.
5.36 Kelley, D. S., UW; CSSF-ROV *ROPOS*.
5.37 ROV *Doc Rickets*, ©MBARI (top); Perfit, M. R., UF, ROV Tiburon ©MBARI (bottom).
5.38 Perfit, M., UF (bottom); Fornari, D., WHOI (top).
5.39 ©Cameron, J., Earthship Productions; MIR.
5.40 from Soule *et al.* (2012).
5.41 from Perfit (1999).
5.42 Perfit, M., University of Florida.
5.43 Perfit, M., University of Florida.
5.44 Perfit, M., University of Florida.
5.45 Perfit, M., University of Florida.
5.46 Perfit, M., University of Florida.
5.47 Perfit, M., University of Florida.
5.48 from Rubin *et al.* (2012).
5.49 from Haymon *et al.* (1993); WHOI-NDSF-*Alvin*.
5.50 Fornari, D., WHOI.
5.51 ©WHOI; WHOI-NDSF-*Alvin*.
5.52 Chadwick, W., NOAA; OSU; Embley, R., NOAA; Fornari, D., WHOI.

5.53 Chadwick, W., NOAA; OSU.
5.54 Fornari, D., WHOI.
5.55 from Soule *et al.* (2007).
5.56 Fornari, D. and Shank T., WHOI.
5.57 Fornari, D., WHOI.
5.58 Kelley, D. S., UW; CSSF-ROV *ROPOS*.
5.59 Resing, J., Embley, R., NOAA; Chadwick, W., NOAA OSU; Rubin, K., University of Hawaii; NOAA; NSF; WHOI-NDSF-ROV *Jason*.
6.1 J. Karson, Syracuse University.
6.2 J. Karson , Syracuse University.
6.3 J. Karson, Syracuse University.
6.4 J. Karson, Syracuse University.
6.5 J. Karson, Syracuse University.
6.6 from Alt, J. C., Kinoshita, H., Stokking, L. B. and Michael, P. J. (eds.) (1993).
6.7 after Alt, J., Laverne, C., Vanko, D. A. *et al.* (1996); Umino, S., Crispini, L., Tartarotti, P. *et al.* (2008); and Wilson, D. S., Teagle, D. A. H., Alt, J. A. *et al.* (2006).
6.8 J. Karson, Syracuse University.
6.9 data source: GeoMapApp http://www.geomapapp.org and Ryan *et al.* (2009); and Karson, J. A., Francheteau, J., Gee, J. S. *et al.* (2005).
6.10 J. Karson, Syracuse University.
6.11 data source: GeoMapApp http://www.geomapapp.org and Ryan *et al.* (2009); and Karson, J. A., Klein, E. M., Hurst, S. D. *et al.* (2002).
6.12 Karson, J. A., Klein, E. M., Hurst, S. D. *et al.* (2002).
6.13 J. Karson, Syracuse University.
6.14 J. Karson, Syracuse University.
6.15 J. Karson, Syracuse University.
6.16 Varga, R. J., Karson, J. A. and Gee, J. S. (2004).
6.17 Karson, J. A., Klein, E. M., Hurst, S. D. *et al.* (2002).
6.18 Karson, J. A. (1998).
6.19 Einsele, G. (1982).
6.20 M. Perfit, University of Florida.
6.21 M. Perfit, University of Florida.
6.22 from Alt, J. C., Laverne, C., Coggon, R. M. *et al.* (2010).
6.23 J. Karson, Syracuse University.
6.24 J. Karson, Syracuse University.
6.25 J. Karson, Syracuse University.
6.26 J. Karson, Syracuse University.
6.27 after Nicholson, R. and Pollard, D. D. (1985).
6.28 J. Karson, Syracuse University.
6.29 J. Karson, Syracuse University.
6.30 Varga, R. J., Gee, J. S., Staudigel, H. and Tauxe, L. (1998).
6.31 after Tauxe, L., Gee, J. S. and Staudigel, H. (1998).
6.32 after Jolly, R. J. H. and Sanderson, D. J. (1995).
6.33 J. Karson, Syracuse University, and M. Perfit, University of Florida.
6.34 J. Karson, Syracuse University.
6.35 J. Karson, Syracuse University.
6.36 J. Karson, Syracuse University.
6.37 M. Perfit, University of Florida.
6.38 J. Karson, Syracuse University.
6.39 J. Karson, Syracuse University.
6.40 after Curewitz, D. and Karson, J. A. (1998).
6.41 Buck, W. R., Einarsson, P. and Brandsdóttir, B. (2006).
6.42 Hartley, M. E. and Thordarson, T. (2013).
6.43 Grandin, R., Socquet, A., Jacques, E. *et al.* (2010).
6.44 J. Karson, Syracuse University.
6.45 after Pockalny, R. A. and Larson, R. L. (2003).
7.1 J. Karson, Syracuse University.
7.2 Streckeisen, A. (1976).
7.3 J. Karson, Syracuse University.
7.4 J. Karson, Syracuse University.
7.5 J. Karson, Syracuse University, after Gillis, K. M., Snow, J. E., Klaus, A. *et al.* (2013).
7.6 Perk, N. W., Coogan, L. A., Karson, J. A. and Klein, E. M. (2007).
7.7 J. Karson, Syracuse University, after Wilson, D. S., Teagle, D. A. H., Alt, J. A. *et al.* (2006).
7.8 Dixon, J. E., Clague, D. A. and Eissen, J.-P. (1986) (top pair); M. Perfit, University of Florida (bottom pair).
7.9 Ridley, W. I., Perfit, M. R., Fornari, D. J. and Smith, M. C. (2006).
7.10 Karson, J. A., Früh-Green, G. L., Kelley, D. S. *et al.* (2006).
7.11 J. Karson, Syracuse University; Dick, H. J. B. (1998). and Karson, J. A. (1998).
7.12 J. Karson, Syracuse University.
7.13 J. Karson, Syracuse University.
7.14 J. Karson, Syracuse University.
7.15 J. Karson, Syracuse University.
7.16 after Bosch, D., Jamais, M., Boudier, F. *et al.* (2004).
7.17 J. Karson, Syracuse University.
7.18 J. Karson, Syracuse University.
7.19 Ildefonse *et al.* (2007) and Dick, H. J. B., Natland, J. H., Alt, J. C. *et al.* (2000).
7.20 J. Karson, Syracuse University.
7.21 J. Karson, Syracuse University.
7.22 M. Perfit, University of Florida.
7.23 Godard, M., Awaji, S., Hansen H.-E. *et al.* (2009). and Workman and Hart (2005)
7.24 J. Karson, Syracuse University.
7.25 J. Karson, Syracuse University.
7.26 J. Karson, Syracuse University.
7.27 J. Karson, Syracuse University.
7.28 J. Karson, Syracuse University.
7.29 J. Karson, Syracuse University.
7.30 J. Karson, Syracuse University.
7.31 J. Karson, Syracuse University.
7.32 Karson, J. A., Elthon, D. L. and Delong, S. E. (1983).
7.33 J. Karson, Syracuse University.
7.34 after Jousselin, D., Nicolas, A. and Boudier, F. (1998).
7.35 J. Karson, Syracuse University.
7.36 J. Karson, Syracuse University.
7.37 Jousselin, D. and Nicolas, A. (2000).
7.38 J. Karson, Syracuse University.
7.39 Irvine, T. N., Anderson, J. C. Ø. and Brooks, C. K. (1998).

7.40 Kelemen, P. B., Koga, K. and Shimizu, N. (1997), Boudier, F., Nicolas, A. and Ildefonse, B. (1996) and Quick, J. E. and Denlinger, R. P. (1993).
7.41 Boudier, F. and Nicolas, A. (2011). Nicolas, A., Boudier, F. and France, L. (2009).
7.42 Coogan, L. A. (2014).
8.1 J. Karson, Syracuse University.
8.2 J. Karson, Syracuse University.
8.3 J. Karson, Syracuse University, after Streckeisen, A. (1976).
8.4 Hébert, L. B. and Montési, L. G. J. (2010).
8.5 Karson, J. A. and Lawrence, R. M. (1997).
8.6 J. Karson, Syracuse University.
8.7 J. Karson, Syracuse University and Auzende, J. M., Bideau, D., Bonatti, E. *et al.* (1990) (bottom right). Karson, J. A. and Lawrence, R. M. (1997). Karson, J. A., Früh-Green, G. L., Kelley, D. S. *et al.* (2006).
8.8 J. Karson, Syracuse University, digital mosaic from NDSF *Jason* images, Lost City Cruise (2003).
8.9 Images courtesy of A. Cipriani.
8.10 Dick, H. J. B. and Natland, J. H. (1996).
8.11 J. Karson, Syracuse University.
8.12 Bonatti, E., Ligi, M., Brunelli, D. *et al.* (2003).
8.13 J. Karson, Syracuse University.
8.14 Karson, J. A., Früh-Green, G. L., Kelley, D. S. *et al.* (2006).
8.15 left images: NOAA OE Lost City Cruise (2005), D. Kelley chief scientist; upper right: J. Karson, Syracuse University; lower right: Blackman, D. K., Karson, J. A., Kelley, D. S. *et al.* (2002).
8.16 after Karson, J. A. and Lawrence, R. M. (1997).
8.17 after Laukert, G., Von Der Handt, A., Hellebrand, E. *et al.* (2013).
8.18 courtesy of P. Michael, University of Tulsa.
8.19 J. Karson, Syracuse University.
8.20 J. Karson, Syracuse University.
8.21 J. Karson, Syracuse University, NDSF *Jason* sample, Lost City Cruise (2003), D. Kelley and J. Karson co-chief scientists; photomicrographs J. Karson, Syracuse University.
8.22 J. Karson, Syracuse University.
8.23 J. Karson, Syracuse University.
8.24 Hercules digital image, NOAA OE Lost City Cruise (2005), D. Kelley chief scientist; UW; URI-ROV *Hercules*, IFE, URI-IAO, Lost City Science Part, and NOAA Ocean Exploration.
8.25 J. Karson, Syracuse University.
8.26 J. Karson, Syracuse University.
8.27 J. Karson, Syracuse University.
8.28 Kelemen, P. B., Braun, M. and Hirth, G. (2000).
8.29 J. Karson, Syracuse University.
8.30 J. Karson, Syracuse University.
8.31 Kelemen, P. B., Braun, M., and Hirth, G. (2000).
8.32 J. Karson, Syracuse University; after Nicolas, A. (1989).
8.33 after Bodinier, J. L. and Godard, M. (2014).
8.34 after Niu, Y. (2004).
8.35 J. Karson, Syracuse University.
8.36 after Toomey, D. R., Jousselin, D., Dunn, R. A., Wilcock, W. S. D. and Detrick, R. S. (2007).
8.37 Spiegelman, M. and Reynolds, J. R. (1999).
8.38 Modified from Niu, Y. (2004).

References

Abbate, E. and Bortolotti, V. (1980). Apennine ophiolites: A peculiar oceanic crust. *Ofioliti*, 1, 59–96.

Abbot, D. H. and Hoffman, S. E. (1984). Archean plate tectonics revisited: 1. Heat flow, spreading rate, and the age of subducting oceanic lithosphere and their effects on the origin and evolution of continents. *Tectonics*, 3, 429–448.

Acton, G. D., Stein, S. and Engeln, J. F. (1991). Block rotation and continental extension in Afar: A comparison to oceanic microplate systems. *Tectonophysics*, 10, 501–526.

Adamson, A. C. (1985). Basement lithostratigraphy, Deep Sea Drilling Project Hole 504B. In Anderson, R. N., Honnorez, J., Becker, K. *et al.* (eds.), *Initial Reports of the Deep Sea Drilling Project*. Washington, DC: US Government Printing Office, pp. 121–127.

Agar, S. M. and Klitgord, K. D. (1995). A mechanism for decoupling within the oceanic lithosphere revealed in the Troodos ophiolite. *Nature*, 374, 232–238.

Aghaei, O., Nedimovic', M. R., Carton, H. *et al.* (2014). Crustal thickness and Moho character of the fast-spreading East Pacific Rise from 9°42′ N to 9°57′ N from poststack-migrated 3-D MCS data. *Geochemistry, Geophysics, Geosystems*, 15, 634–657, doi:10.1002/2013GC005069.

Alabaster, T., Pearce, J. A. and Malpas, J. G. (1982). The volcanic stratigraphy and petrogenesis of the Oman ophiolite complex. *Contributions to Mineralogy and Petrology*, 81, 168–213.

Alain, K., Callac, N., Guégan, M. *et al.* (2009). Nautilia abyssai sp. Nov., a thermophilic, chemolithoautotrophic, sulfur-reducing bacterium isolated from an East Pacific Rise hydrothermal vent. *International Journal of Systematic Evolutionary Microbiology*, 60, 1182–1186.

Alain, K., Querellou, J., Lesongeur, F. *et al.* (2002). *Caminibacter hydrogeniphilus* gen. nov., sp nov., a novel thermophilic, hydrogen-oxidizing bacterium isolated from an East Pacific Rise hydrothermal vent. *International Journal of Systematic and Evolutionary Microbiology*, 52, 1317–1323.

Allan, J. F., Batiza, R., Perfit, M. R., Fornari, D. J. and Sack, R. O. (1989). Petrology of lavas from the Lamont Seamount Chain and adjacent East Pacific Rise, 10° N. *Journal of Petrology*, 30, 1245–1298.

Allen, D. E. and Seyfried, W. E. Jr. (2004). Serpentinization and heat generation: Constraints from Lost City and Rainbow hydrothermal systems. *Geochimica et Cosmochimica Acta*, 68, 1347–1354.

Alt, J. C. (1995). Subseafloor processes in mid-ocean ridge hydrothermal systems. In Humphris, S., Zierenberg, R., Mullineau, L. and Thomson, R. (eds.), *Physical, Chemical, Biological, and Geological Interactions within Seafloor Hydrothermal Systems. Geophysical Monograph 91*, Washington, DC: American Geophysical Union, pp. 178–193.

Alt, J. C. (2004). Alteration of the upper oceanic crust: Mineralogy, chemistry and processes. In Davis, E. E. and Elderfield, H. (eds.), *Hydrogeology of the Oceanic Lithosphere*. Cambridge: Cambridge University Press, pp. 495–533.

Alt, J. C. and Teagle, D. A. H. (2000). Hydrothermal alteration and fluid fluxes in ophiolites and oceanic crust. In Dilek, Y., Moores, E., Elthon, D. and Nicolas, A. (eds.), *Ophiolites and Oceanic Crust: New Insights from Field Studies and the Ocean Drilling Program. Geological Society of America Special Papers, 349*. Boulder, CO: Geological Society of America, pp. 273–282.

Alt, J. C., Honnorez, J., Laverne, C. and Emmermann, R. (1986). Hydrothermal alteration of a 1 km section through the upper oceanic crust, Deep Sea Drilling Project Hole 504B: Mineralogy, chemistry, and evolution of seawater-basalt interactions. *Journal of Geophysical Research*, 91, 10309–10335.

Alt, J. C., Kinoshita, H., Stokking, L. B. and Michael, P. J. (eds.) 1993. *Proceedings of the Ocean Drilling Program, Initial Reports*, Leg 148, College Station, TX: Ocean Drilling Program.

Alt, J. C., Laverne, C., Coggon, R. M. *et al.* (2010). Subsurface structure of a submarine hydrothermal system in oceanic crust formed at the East Pacific Rise, ODP/IODP Site 1256. *Geochemistry, Geophysics, Geosystems*, 11, Q10010, doi:10.1029/2010GC003144.

Alt, J., Laverne, C., Vanko, D. A. *et al.* (1996). Hydrothermal alteration of a section of upper oceanic crust in the eastern equatorial Pacific: A synthesis of results from Site 504 (DSDP Legs 69, 70, and 83, and ODP Legs 111, 137, 140, and 148). In Alt, J. C., Kinoshita, H., Stokking, L. B. and Michael, P. J. (eds.), *Proceedings of the Ocean Drilling Program, Scientific Results, Leg 148*. College Station, TX: Ocean Drilling Program, pp. 417–434.

Anderson, E. M. (1972). *The Dynamics of Faulting and Dyke Formation with Applications to Britain*. New York: Hafner Publishing Co.

Anderson, R. N., Honnorez, J., Becker, K. *et al.* (1982). DSDP Hole 504B, the first reference section over 1 km through Layer 2 of the oceanic crust. *Nature*, 300, 589–594.

Anderson, R. N., Honnorez, J., Becker, K. *et al.* (1985). *Initial Reports of the Deep Sea Drilling Project, Leg 92*. Washington, DC: US Government Printing Office.

Anonymous (1972). Penrose Field Conference on Ophiolites. *Geotimes*, 17, 24–25.

Arevalo, R., Jr. and McDonough, W. (2010). Chemical variations and regional diversity observed in MORB. *Chemical Geology*, 271, 70–85, doi:10.1016/j.chemgeo.2009.12.013.

Arnulf, A. F., Harding, A. J., Kent, G. M., Singh, S. C. and Crawford, W. (2014). Constraints on the shallow velocity structure of the Lucky-Strike Volcano, Mid-Atlantic Ridge, from downward continued multi-channel streamer data. *Journal of Geophysical Research*, doi:10.1002/2013JB010500.

Atwater, T. M. and Mudie, J. (1973). Detailed near-bottom geophysical study of the Gorda Rise. *Journal of Geophysical Research*, 78, 8665–8686.

Aubouin, J. (1965). *Geosynclines*. Amsterdam: Elsevier.

Augustin, N., Lackschewitz, K. S., Kuhn, T. and Devey, C. W. (2008). Mineralogical and chemical mass changes in mafic and ultramafic rocks from the Logatchev hydrothermal field (MAR 15° N). *Marine Geology*, 256, 18–29.

Aumento, F. and Loubat, H. (1971). The Mid-Atlantic Ridge near 45° N: Serpentinized ultramafic intrusions. *Canadian Journal of Earth Sciences*, 8, 631–663.

Auzende, J. M., Bideau, D., Bonatti, E. *et al.* (1989). Direct observation of a section through slow-spreading oceanic crust. *Nature*, 337, 726–729.

Auzende, J. M., Cannat, M., Gente, P. *et al.* (1994). Observation of sections of oceanic crust and mantle cropping out on the southern wall of the Kane Fracture Zone (N. Atlantic). *Terra Nova*, 6, 143–148.

Auzende, J. M., Olivet, J. L., Charvet, J. *et al.* (1978). Sampling and observation of oceanic mantle and crust on Gorringe Bank. *Nature*, 273, 45–59.

Baba, K., Chave, A. D., Evans, R. L., Hirth, G. and Mackie, R. L. (2006). Mantle dynamics beneath the East Pacific Rise at 17 S: Insights from the Mantle Electromagnetic and Tomography (MELT) experiment. *Journal of Geophysical Research*, 111(B2), doi:10.1029/2004JB003598.

Bach, W., Erzinger, J., Alt, J. C. and Teagle, D. A. H. (1996). Chemistry of the lower sheeted dike complex, Hole 504B (Leg 148): Influence of magmatic differentiation and hydrothermal alteration. In Alt, J. C., Kinoshita, H., Stokking, H. and Michael, P. J. (eds.), *Proceedings of the Ocean Drilling Program, Scientific Results, Leg 148*. College Station, TX: Ocean Drilling Program, pp. 39–56.

Bach, W., Garrido, C. J., Paulick, H. and Rosner, M. (2004). Seawater–peridotite interactions: first insights from ODP Leg 209, MAR 15° N. *Geochemistry, Geophysics, Geosystems*, 5, Q09F26, doi:10.1029/2004GC000744.

Bachraty, C., Legendre, P. and Desbruyères, D. (2009). Biogeographic relationships among deep-sea hydrothermal vent faunas at global scale. *Deep-Sea Research, Part I*, 156, 1371–1378.

Baker, E. T. (2009) Relationships between hydrothermal activity and axial magma chamber distribution, depth, and melt content. *Geochemistry, Geophysics, Geosystems*, 10(6), doi:10.1029/2009GC002424.

Baker, E. T. and German C. R. (2004). On the global distribution of hydrothermal vent fields. In German, C. R., Lin, J. and Parsons, L. M. (eds.), *Mid-Ocean Ridges: Hydrothermal Interactions Between the Lithosphere and Oceans*. Geophysical Monograph 148, Washington, DC: American Geophysical Union, pp. 245–266.

Baker, E. T. and Lupton, J. E. (1990). Changes in submarine hydrothermal ^{3}He/heat ratios as an indicator of magmatic/tectonic activity. *Nature*, 346, 556–558.

Baker, E. T., Cormier, M. -H., Langmuir, C. H. and Zavala, K. (2001). Hydrothermal plumes along segments of contrasting magmatic influence, 15°20′ –18°30′ N, East Pacific Rise: Influence of axial faulting. *Geochemistry, Geophysics, Geosystems*, 2(9), doi:10.1029/2000GC000165.

Baker, E. T., Feely, R. A., Mottl, M. J. and Sansone, F. J. (1994). Hydrothermal plumes along the East Pacific Rise, 8° 40′ to 11° 50′ N: Plume distribution and relationship to the apparent magmatic budget. *Earth and Planetary Science Letters*, 128, 1–17.

Baker, E. T., Fox, C. G. and Cowen, J. P. (1999). *In situ* observations of the onset of hydrothermal discharge during the 1998 submarine eruption of Axial Volcano, Juan de Fuca Ridge. *Geophysical Research Letters*, 26, 3445–3448.

Baker, E. T., Haymon, R. M., Resing, J. A. *et al.* (2008). High-resolution surveys along the hotspot-affected Galápagos Spreading Center: 1. Distribution of hydrothermal activity. *Geochemistry, Geophysics, Geosystems*, 9, Q09003, doi:10.1029/2008GC002028.

Baker, E. T., Hey, R. N., Lupton, J. E. *et al.* (2002). Hydrothermal venting along Earth's fastest spreading center: East Pacific Rise, 27.5°–32.3°S. *Journal of Geophysical Research*, 107(B7), 1–14, doi: 10.1029/2001JB000651.

Baker, E. T., Lupton, J. E., Resing, J. A. *et al.* (2011). Unique event plumes from a 2008 eruption on the Northeast Lau Spreading Center. *Geochemistry, Geophysics, Geosystems*, 12, Q0AF02, doi:10.1029/2011GC00372.

Baker, E. T., Massoth, G. J. and Feely, R. A. (1987). Cataclysmic hydrothermal venting on the Juan de Fuca Ridge. *Nature*, 329, 149–151.

Baker, E. T., Walker, S. L., Embley, R. W. and de Ronde, C. E. J. (2012). High-resolution hydrothermal mapping of Brothers Caldera, Kermadec Arc. *Economic Geology and the Bulletin of the Society of Economic Geologists*, 107(8), 1583–1593.

Ballard, R. D. (1985). The Titanic: Lost and Found – Introduction. *Oceanus*, 28, 4.

Ballard, R. D. and Moore, J. G. (1977). *Photographic Atlas of the Mid-Atlantic Ridge Rift Valley*. New York: Springer-Verlag.

Ballard, R. D. and van Andel, Tj. H. (1977a). Project FAMOUS: Operational techniques and American submersible operations. *Geological Society of America Bulletin*, 88, 495–506.

Ballard, R. D. and van Andel, Tj. H. (1977b). Morphology and tectonics of the inner rift valley at lat. 36°50′ N on the Mid-Atlantic Ridge. *Geological Society of America Bulletin*, 88, 507–530.

Ballard, R. D., Francheteau, J., Juteau, T., Rangan, C. and Normark, W. (1981). EPR at 21°N: Volcanic, tectonic and hydrothermal processes of the central axis. *Earth and Planetary Science Letters*, 55, 1–10.

Ballard, R. D., Holcomb, R. T. and van Andel, Tj. H. (1979). The Galapagos Rift at 86°W: 3. Sheet flows, collapse pits and lava lakes of the rift. *Journal of Geophysical Research*, 84, 4507–5422.

Ballard, R. D., van Andel, Tj. H. and Holcomb, R. T. (1982). The Galapagos Rift at 86° W. 5. Variations in volcanism, structure, and hydrothermal activity along a 30 km segment of the rift valley. *Journal of Geophysical Research*, 87, 1149–1161.

Banda, E., Ranero, C. R., Danobeitia, J. J. and Rivero, A. (1992). Seismic boundaries of the eastern central Atlantic Mesozoic crust from multichannel seismic data. *Geological Society of America Bulletin*, 104, 1,340–1,349.

Barager, W. R. A. (1954). *Betts Pond Area, Burlington Peninsula, Newfoundland. Newfoundland Department of Mines, Open File Report*. St. John's Newfoundland, Canada.

Barany, I. and Karson, J. A. (1989). Basaltic breccias of the Clipperton Fracture Zone: sedimentation and tectonics in a fast-slipping oceanic transform. *Geological Society of America Bulletin*, 101, 204–220.

Baranzangi, M. and Dorman, J. (1969). World seismicity maps compiled from ESSA, Coast and Geodetic Survey, epicenter data, 1961–1968. *Bulletin of the Seismological Society of America*, 59, 369–380.

Barker, A. K., Coogan, L. A. and Gillis, K. M. (2010). Insights into the behaviour of sulphur in mid-ocean ridge axial hydrothermal systems from the composition of the sheeted dyke comlex at Pito Deep. *Chemical Geology*, 275, 105–115.

Barnes, I. and O'Neil, J. R. (1969). The relationship between fluids in some fresh alpine-type ultramafics and possible modern serpentinization, Western United States. *Bulletin of the Geological Society of America*, 80, 1947–1960.

Baross, J. A. and Deming, J. W. (1995). Growth at high temperatures: Isolation and taxonomy, physiology, and ecology. In Karl, D. M. (ed.), *The Microbiology of Deep-Sea Hydrothermal Vents*. Boca Raton, FL: CRC Press, pp. 169–217.

Baross, J. A. and Hoffman, S. E. (1985). Submarine hydrothermal vents and associated gradient environments as sites for the origin and evolution of life. *Origins of Life*, 15, 327–345.

Barreyre, T., Escartín, J., Garcia, R. *et al.* (2012). Structure, temporal evolution, and heat flux estimates from the Lucky Strike deep-sea hydrothermal field derived from seafloor image mosaics. *Geochemistry, Geophysics, Geosystems*, 13, 4Q04007, doi:10.1029/2011GC003990.

Barreyre, T., Soule, S. A. and Sohn, R. A. (2011). Dispersal of volcaniclasts during deep-sea eruptions: Settling velocities and entrainment in buoyant seawater plumes. *Journal of Volcanology and Geothermal Research*, 205(3), 84–93, doi 10.1016/j.jvolgeores.2011.05.006.

Barriga, F., Fouquet, Y., Almeida, A. *et al.* (1998). Discovery of the Saldanha hydrothermal field on the FAMOUS segment of the MAR (36°30′ N). *Eos, Transactions of the American Geophysical Union*, 79, 67.

Bascom, W. (1959). The Mohole. *Scientific American*, 200(4), 41–49.

Batiza, R. and Vanko, D. A. (1986). Petrologic evolution of large failed rifts in the eastern Pacific: Petrology of volcanic and plutonic rocks from the Mathematician Ridge area and Guadalupe Trough. *Journal of Petrology*, 26, 564–602.

Batiza, R., Niu, Y. L. and Zayac, W. C. (1990). Chemistry of seamounts near the East Pacific Rise: Implications for the geometry of subaxial mantle flow. *Geology*, 18, 1122–1125.

Batuev, B. N., Krotov, A. G., Makrov, V. F. et al. (1994). Massive sulfide deposits discovered and sampled at 14°45′ N Mid-Atlantic Ridge. *Bridge Newsletter*, 6, 6–10.

Bazylinski, D. A., Farrington, J. W. and Jannasch, H. W. (1988). Hydrocarbons in surface sediments from a Guaymas Basin hydrothermal vent site. *Organic Geochemistry*, 12(6), 547–558.

Beard, J. S. and Hopkinson, L. (2000). A fossil, serpentinization-related hydrothermal vent, Ocean Drilling Program Leg 173, Site 1068 (Iberia Abyssal Plain): Some aspects of mineral and fluid chemistry. *Journal of Geophysical Research*, 105(B7), 16,527–16,539.

Becker, K., Sakai, H., Merrill, R. B. *et al.* (1988). *Proceedings of the Ocean Drilling Program, Initial Reports, Leg 111*. College Station, TX: Ocean Drilling Program.

Behn, M. D., Buck, W. R. and Sacks, I. (2006). Topographic controls on dike injection in volcanic rift zones. *Earth and Planetary Science Letters*, 246(3–4), 188–196, doi:10.1016/j.epsl.2006.04.005.

Behn, M. D., Sinton, J. M. and Detrick, R. S. (2004). Effect of the Galápagos hotspot on seafloor volcanism along the Galápagos Spreading Center (90.9°–97.6° W). *Earth and Planetary Science Letters*, 217(3), 331–347.

Bender, J. F., Hodges, R. N. and Bence, A. E. (1978). Petrogenesis of basalts from the Project FAMOUS area: Experimental study from 0 to 15 kbars. *Earth and Planetary Science Letters*, 41, 277–302.

Benthos, I. and Smith, P. F. (Compiler) (1984). *Underwater Photography, Scientific and Engineering Applications*. New York: Van Nostrand Reinhold.

Bergman, E. A. and Solomon, S. C. (1984). Source mechanisms of earthquakes near mid-ocean ridges from body waveform inversion: Implications for the early evolution of oceanic lithosphere. *Journal of Geophysical Research*, 89, 11,415–11,441.

Bergmanis, E. C., Sinton, J. M. and Rubin, K. H. (2007). Eruptive history and magma reservoir dynamics on the East Pacific Rise at 17°30′ N. *Geochemistry, Geophysics, Geosystems*, 8(12), doi:10.1029/2007GC001 742.

Bezos, A., Escrig, S., Langmuir, C. H., Michael, P. J. and Asimow, P. D. (2009). Origins of chemical diversity of back-arc basin basalts: A segment-scale study of the Eastern Lau Spreading Center. *Journal of Geophysical Research*, 114(B6), doi:10.1029/2008JB005924.

Bickle, M. J. *et al.* (2011). Illuminating Earth's Past, Present, and Future: The International Ocean Discovery Program, Science Plan for 2013–2023. www.iodp.org/Science-Plan-for-2013-2023/.

Bischoff, J. L. and Rosenbauer, R. J. (1989). Liquid-vapor relations for the system NaCl–H_2O: Summary of the P-T-x surface from 300 °C to 500 °C. *American Journal of Science*, 289, 217–248.

Blacic, T. M., Ito, G., Canales, J. P., Detrick, R. S. and Sinton, J. (2004). Constructing the crust along the Galapagos Spreading Center 91.3W-95.5W: Correlation of seismic layer 2A with axial magma lens and topographic characteristics. *Journal of Geophysical Research*, 109, B10310, 19, doi:10.1029/2004JB003066.

Blackman, D. K., Cann, J. R., Janssen, B. and Smith, D. K. (1998). Origin of extensional core complexes: Evidence from the Mid-Atlantic Ridge at Atlantis fracture zone. *Journal of Geophysical Research*, 103, 21,315–21,334.

Blackman, D. K., Karson, J. A., Kelley, D. S. *et al.* (2002). Geology of the Atlantis Massif (Mid-Atlantic Ridge, 30°N): Implications for the evolution of an ultramafic oceanic core complex. *Marine Geophysical Researches*, 23, 443–469.

Blackman, D. K., Kelley, D. S., Karson, J. A. and Shipboard Scientific Party (2001). Seafloor mapping and sampling of the MAR 30°N Oceanic Core Complex-MARVEL (Mid-Atlantic Ridge Vents in Extending Lithosphere) 2000. *InterRidge Newsletter*, 10, 33–36.

Blackman, D. K., Kendall, J., Dawson, P. R., Wenk, H., Boyce, D. and Morgan, J. P. (1996). Teleseismic imaging of subaxial flow at mid-ocean ridges: Traveltime effects of anisotropic mineral texture in the mantle. *Geophysical Journal International*, 127(2), 415–426.

Blondel, P. (2009). *Handbook of Sidescan Sonar*. Springer-Praxis.

Bodinier, J. L. and Godard, M. (2014). *Orogenic, Ophiolitic*, and Abyssal Peridotites, Treatise on Geochemistry, 2nd edn. Elsevier, pp. 103–151.

Bohnenstiehl, D. R., Dziak, R. R., Tolstoy, M., Fox, C. G. and Fowler, M. (2004). Temporal and spatial history of the 1999–2000 Endeavour Segment seismic series, Juan de Fuca Ridge. *Geochemistry, Geophysics, Geosystems*, 5, Q09003, doi:10.1029/2004GC000735.

Bonatti, E. (1976). Serpentinite protrusions in the oceanic crust. *Earth and Planetary Science Letters*, 32, 107–113.

Bonatti, E. (1994). The Earth's mantle below the oceans. *Scientific American*, 270(3), 44–51.

Bonatti, E. and Hamlyn, P. R. (1978). Mantle uplifted blocks in the western Indian Ocean. *Science*, 201, 249–252.

Bonatti, E. and Harrison, C. G. A. (1988). Eruption styles of basalt in oceanic spreading ridges and seamounts: Effect of magma temperature and viscosity. *Journal of Geophysical Research*, 93, 2967–2980.

Bonatti, E. and Honnorez, J. (1976). Sections of the Earth's crust in the Equatorial Atlantic. *Journal of Geophysical Research*, 81, 4104–4116.

Bonatti, E. and Seyler, M. (1987). Crustal underplating and evolution in the Red Sea rift: Uplifted gabbro/gneiss crustal complexes on Zabargad and Brothers Islands. *Journal of Geophysical Research*, 92, 12,803–12,821.

Bonatti, E., Ligi, M., Brunelli, D. *et al.* (2003). Mantle thermal pulses below the Mid-Atlantic Ridge and temporal variation in the formation of oceanic lithosphere. *Nature*, 423, 499–505.

Bosch, D., Jamais, M., Boudier, F. *et al.* (2004). Deep and high-temperature hydrothermal circulation in the Oman Ophiolite: Petrological and isotopic evidence. *Journal of Petrology*, 45(6), 1181–1208, doi:10.1093/petrology/egh010.

Boschi, C., Bonatti, E., Ligi, M. *et al.* (2013). Serpentinization of mantle peridotites along an uplifted lithospheric section, Mid-Atlantic Ridge at 11° N. *Lithos*, 178, 3–23.

Boschi, C., Dini, A., Früh-Green, G. L. *et al.* (2008). Isotopic and element exchange during serpentinization and metasomatism at the Atlantis Massif: Insights from B and Sr isotope data. *Geochimica et Cosmochimica Acta*, 72, 1801–1823.

Boschi, C., Früh-Green, G. L., Delacour, A., Karson, J. A. and Kelley, D. S. (2006). Mass transfer and fluid flow during detachment faulting and development of an oceanic core complex. *Geochemistry, Geophysics, Geosystems*, 7(1), doi:10.1029/2005GC001074.

Boudier, F. (1978). Microstructural study of three peridotite samples drilled at the western margin of the Mid-Atlantic Ridge. In Melson, W. G., Rabinowitz, P. *et al.* (eds.), *Initial Reports of the Deep Sea Drilling Project, Leg 45*. Washington, DC: US Government Printing Office, pp. 603–608.

Boudier, F. and Coleman, R. G. (1981). Cross section through the peridotite in the Samail Ophiolite southeastern Oman Mountains. *Journal of Geophysical Research*, 86, 2573–2592.

Boudier, F. and Nicolas, A. (2011). Axial melt lenses at oceanic ridges: A case study in the Oman Ophiolite. *Earth and Planetary Science Letters*, 304, 313–325.

Boudier, F., Nicolas, A. and Ildefonse, B. (1996). Magma chambers in the Oman Ophiolite: Fed from the top or bottom? *Earth and Planetary Science Letters*, 144, 239–250.

Bougault, H., Charlou, J. -L., Fouquet, Y. *et al.* (1993). Fast and slow spreading ridges: Structure and hydrothermal activity, ultramafic topographic highs, and CH_4 output. *Journal of Geophysical Research*, 98, 9643–9651.

Bowen, A. D., Yoerger, D. R., Taylor, C. *et al.* (2009). The Nereus hybrid underwater robotic vehicle. *Underwater Technology*, 28, 79–89.

Brazelton, W. J. and Baross, J. A. (2009). Abundant transposases encoded by the metagenome of a hydrothermal chimney biofilm. *The ISME Journal*, 3, 1420–1424.

Brazelton, W. J., Ludwig, K. A., Sogin, M. L. *et al.* (2010). Archaea and bacteria with surprising microdiversity show shifts in dominance over 1,000-year time scales in hydrothermal chimneys. *Proceedings of the National Academy of Sciences, USA*, 107, 1612–1617, doi:1073/pnas.0905369107.

Brazelton, W. J., Mehta, M. P., Kelley, D. S. and Baross, J. A. (2011). Physiological differentiation within a single-species biofilm fueled by serpentinization. *mBio*, 2(4):e00127-11, doi:10.1128/mBio.00127-11.

Brazelton, W. J., Schrenk, M. O., Kelley, D. S. and Baross, J. A. (2006). Methane and sulfur metabolizing microbial communities dominate in the Lost City hydrothermal vent ecosystem. *Applied Environmental Microbiology*, 72, 6257–6270.

Brocher, T. M., Karson, J. A. and Collins, J. A. (1985). Seismic stratigraphy of the oceanic Moho based on ophiolite models. *Geology*, 13, 62–65.

Brongniart, A. (1813). Essai de classification minéralogique des roches mélanges. *Journal des Mines*, XXXIV, 190–199.

Brown, J. R. and Karson, J. A. (1988). Variations in axial processes on the Mid-Atlantic Ridge: The median valley of the MARK area. *Marine Geophysical Researches*, 10, 109–138.

Bryan, W. B. (1983). Systematics of modal phenocryst assemblages in submarine basalts: Petrologic implications. *Contributions to Mineralogy and Petrology*, 83, 62–74.

Buck, W. R. (1991). Modes of continental lithospheric extension. *Journal of Geophysical Research*, 96, 20,161–20,178.

Buck, W. R., Carbotte, S. M. and Mutter, C. Z. (1997). Controls on extrusion at mid-ocean ridges. *Geology*, 25, 935–938.

Buck, W. R., Einarsson, P. and Brandsdóttir, B. (2006). Tectonic stress and magma chamber size as controls on dike propagation: Constraints from the 1975–1984 Krafla rifting episode. *Journal of Geophysical Research*, 111, B12404, doi:10.1029/2005JB003879.

Buck, W. R., Lavier, L. L. and Poliakov, A. (2005). Modes of faulting at mid-ocean ridges. *Nature*, 434, 719–723.

Burchardt, S., Tanner, D. C., Troll, V. R., Krumbholz, M. and Gustafsson, I. E. (2011). Three-dimensional geometry of concentric intrusive sheet swarms in the Geitafell and the Dyrfjöll Volcanoes, eastern Iceland. *Geochemistry, Geophysics, Geosystems*, 12, Q0AB09, doi:10.1029/2011GC003527.

Burchardt, S., Troll, V. R., Mathieu, L., Emeleus, H. C. and Donaldson, C. H. (2013). Ardnamurchan 3D cone-sheet architecture explained by a single elongate magma chamber. *Science Reports*, 3, doi:10.1038/srep02891.

Burke, K. and Kidd, W. S. F. (1978). Were Archean continental gradients much steeper than those of today? *Nature*, 272, 240–241.

Bush, V. (1945). Science: The Endless Frontier. Washington, DC: US Government Printing Office.

Butterfield, D. A., Jonasson, I. R., Massoth, G. J. *et al.* (1997). Seafloor eruptions and evolution of hydrothermal fluid chemistry. *Philosophical Transactions of the Royal Society of London*, 355, 369–386.

Butterfield, D. A., Massoth, G. J., McDuff, R. E., Lupton, J. E. and Lilley, M. D. (1990). The geochemistry of hydrothermal fluids from ASHES vent field, Axial Seamount, Juan de Fuca Ridge: Subseafloor boiling and subsequent fluid-rock interaction. *Journal of Geophysical Research*, 95, 12,895–12,922.

Butterfield, D. A., McDuff, R. E., Mottl, M. J. *et al.* (1994). Gradients in the composition of hydrothermal fluids from the Endeavour segment vent field: Phase separation and brine loss. *Journal of Geophysical Research*, 99, 9561–9583.

Butterfield, D. A., Roe, K. K., Lilley, M. D. *et al.* (2004). Mixing, reaction and microbial activity in the subseafloor revealed by temporal and spatial variation in diffuse flow vents at Axial Volcano. In Wilcock, W. S. D., DeLong, E. F., Kelley, D. S., Baross, J. A. and Cary, S. C. (eds.), *Subseafloor Biosphere at Mid-Ocean Ridges*. Geophysical Monograph 144, Washington, DC: American Geophysical Union, pp. 269–290.

Campbell, I. H. (1978). Some problems with the cumulus theory. *Lithos*, 11, 311–323.

Canales, J. P., Carton, H., Carbotte, S. M. *et al.* (2012a). Network of off-axis melt bodies at the East Pacific Rise. *Nature Geoscience Letters*, doi: 10.1038/NGEO1377.

Canales, J. P., Carton, H., Mutter, J. C. *et al.* (2012b). Recent advances in multichannel seismic imaging for academic research in deep oceanic environments. *Oceanography*, 25(1), 113–115, doi.org/10.5670/oceanog.2012.09.

Canales, J. P., Detrick, R. S., Carbotte, S. M. *et al.* (2005). Upper crustal structure and axial topography at intermediate spreading ridges: Seismic constraints from the southern Juan de Fuca Ridge. *Journal of Geophysical Research*, 110, B12104, doi:10.1029/2005JB003630.

Canales, J. P., Detrick, R. S., Lin, J., Collins, J. A. and Toomey, D. R. (2000). Crustal and upper mantle seismic structure beneath the rift mountains and across a nontransform offset at the Mid-Atlantic Ridge (35°N). *Journal of Geophysical Research*, 105, 2699–2719.

Canales, J. P., Detrick, R. S., Toomey, D. R. and Wilcock, W. S. D. (2003). Segment-scale variations in the crustal structure of 150–300 kyr old fast spreading oceanic crust (East Pacific Rise, 8°15′ N–10°5′ N) from wide-angle seismic refraction profiles. *Geophysical Journal International*, 152, 766–794, doi:10:1046/j.1365-246X.2003.10885.x.

Canales, J. P., Nedimović, M. R., Kent, G. M., Carbotte, S. M. and Detrick, R. S. (2009). Seismic reflection images of near-axis melt sill within the lower crust at the Juan de Fuca Ridge. *Nature*, 460, 89–93, doi:10.1038/nature08095.

Canales, J. P., Singh, S. C., Detrick, R. S. *et al.* (2006). Seismic evidence for variations in axial magma chamber properties along the southern Juan de Fuca Ridge. *Earth and Planetary Science Letters*, 246, 353–366.

Canales, J. P., Sohn, R. A. and deMartin, B. J. (2007). Crustal structure of the Trans-Atlantic Geotraverse (TAG) segment (Mid-Atlantic Ridge, 26°10′ N): Implications for the nature of hydrothermal circulation and detachment faulting at slow spreading ridges. *Geochemistry, Geophysics, Geosystems*, 8(8), doi:10.1029/2007GC001629.

Canales, J. P., Tucholke, B. E., Xu, M., Collins, J. A. and DuBois, D. L. (2008). Seismic evidence for large-scale compositional heterogeneity of oceanic core complexes. *Geochemistry, Geophysics, Geosystems*, 9, Q08002, doi:10.1029/2008GC002009.

Cann, J. R. (1974). A model for oceanic crustal structure developed. *Geophysical Journal of the Royal Astronomical Society*, 39, 169–187.

Cann, J. R. and Smith, D. K. (2005). Evolution of volcanism and faulting in a segment of the Mid-Atlantic Ridge at 25° N. *Geochemistry, Geophysics, Geosystems*, 6(9), doi:10.1029/2005GC000954.

Cann, J. R., Blackman, D. K., Smith, D. K. *et al.* (1997). Corrugated slip surfaces formed at ridge-transform intersections on the Mid-Atlantic Ridge. *Nature*, 385, 329–332.

Cann, J. R., Langseth, M. G., Honnorez, J. *et al.* (1983). *Initial Reports of the Deep Sea Drilling Project, Leg 69*. Washington, DC: US Government Printing Office.

Cannat, M. (1993). Emplacement of mantle rocks in the sea floor at mid-ocean ridges. *Journal of Geophysical Research*, 98, 4163–4172.

Cannat, M. (1996). How thick is the magmatic crust at slow spreading oceanic ridges? *Journal of Geophysical Research*, 101, 2847–2857.

Cannat, M., Briais, A., Deplus, C. *et al.* (1999). Mid-Atlantic Ridge–Azores hotspot interactions: Along-axis migration of a hotspot-derived magmatic pulse 14 to 4 myrs ago. *Earth and Planetary Science Letters*, 173, 257–269, doi:10.1016/S0012-821X(99)00234-4.

Cannat, M., Juteau, T. and Berger, E. (1990). Petrostructural analysis of the Leg 109 serpentinized peridotites. In Detrick, R. S., Bryan, W. B., Honnorez, J. and Juteau, T. (eds.), *Proceedings of the Ocean Drilling Program, Scientific Results, Leg 109*. College Station, TX: Ocean Drilling Program, pp. 47–57.

Cannat, M., Karson, J. A., Miller, D. J. *et al.* (1995a). *Proceedings of the Ocean Drilling Program, Initial Reports, Leg 153*. College Station, TX: Ocean Drilling Program.

Cannat, M., Karson, J. A., Miller, D. J. *et al.* (1995b). Probing the plutonic foundation of the Mid-Atlantic Ridge. *Eos, Transactions of the American Geophysical Union*, 76, 129–133.

Cannat, M., Lagabrielle, Y., Coutures, N. D. *et al.* (1997). Ultramafic and gabbroic exposures at the Mid-Atlantic Ridge: Geological mapping in the 15°N region. *Tectonophysics*, 279, 193–214.

Cannat, M., Mével, C., Maia, M. *et al.* (1995c). Thin crust ultramafic exposures and rugged faulting patterns at the Mid-Atlantic Ridge (22°–24°N). *Geology*, 23, 49–52.

Cannat, M., Sauter, D., Mendel, V. *et al.* (2006). Modes of seafloor generation at a melt-poor ultraslow-spreading ridge. *Geology*, 34, 605–608, doi:10.1130/G22486A.1.

Canon-Tapia, E. and Aubourg, C. (2004). Anisotropy of magnetic susceptibility of lava flows and dykes: A historical account. In Martin-Hernandez, F. (ed.), *Magnetic Fabric: Methods and Applications*. Boulder, CO: Geological Society of America, pp. 205–225.

Carbotte, S. M., Canales, J. P., Nedimović, M. R., Carton, H. and Mutter, J. C. (2012). Recent seismic studies at the East Pacific Rise 8°20′ –10°10′ N and Endeavour Segment: Insights into mid-ocean ridge hydrothermal and magmatic processes. *Oceanography*, 25(1), 100–112, doi:org/10.5670/oceanog.2012.08.

Carbotte, S. M., Marjanovic, M., Carton, H. *et al.* (2013). Fine-scale segmentation of the crustal magma reservoir beneath the East Pacific Rise. *Nature Geoscience*, 6, 866–870, doi:10.1038/NGEO1933.

Carbotte, S. M., Mutter, J. C. and Xu, L. (1997). Contribution of volcanism and tectonism to axial and flank morphology of the southern East Pacific Rise 17°10′ – 17°40′ S from a study of layer 2A geometry. *Journal of Geophysical Research*, 102, 10,165–10,184.

Carbotte, S. M., Solomon, A. and Ponce-Correa, G. (2000). Evaluation of morphological indicators of magma supply and segmentation from a seismic reflection study of the East Pacific Rise 15°30′ –17°N. *Journal of Geophysical Research*, 105, 2737–2759.

Caress, D. W., Clague, D. A., Paduan, J. B. *et al.* (2012). Repeat bathymetric surveys at 1-metre resolution of lava flows erupted at Axial Seamount in April 2011. *Nature Geoscience*, 5(7), 483–488, doi:10.1038/NGEO1496.

Cary, S. C., Fisher, C. R. and Felbeck, H. (1998a). Mussel growth supported by methane as sole carbon and energy source. *Science*, 240, 78–80.

Cary, S. C., Shank, T. and Stein, J. (1998b). Worms bask in extreme temperatures. *Nature*, 391, 545–546.

Casey, J. F. (1997). Comparison of major- and trace-element geochemistry of abyssal peridotites and mafic plutonic rocks with basalts from the MARK region of the Mid-Atlantic Ridge. In Karson, J. A., Cannat, M., Miller, D. J. and Elthon, D. (eds.), *Proceedings of the Ocean Drilling Program, Scientific Results, Leg 153*. College Station, TX: Ocean Drilling Program, pp. 181–242.

Casey, J. F. and Karson, J. A. (1981). Magma chamber profiles from the Bay of Islands ophiolite complex. *Nature*, 292, 295–301.

Casey, J. F., Dewey, J. F., Fox, P. J., Karson, J. A. and Rosencrantz, E. (1981). Heterogeneous nature of the oceanic crust and upper mantle: A perspective from the Bay of Islands Ophiolite. In Emiliani, C. (ed.), *The Oceanic Lithosphere*. New York: Wiley, pp. 305–338.

Casey, J. F., Karson, J. A., Elthon, D. L., Rosencrantz, E. and Titus, M. (1983). Reconstruction of the geometry of accretion during formation of the Bay of Islands Ophiolite Complex. *Tectonics*, 2, 509–528.

Cashman, K. V., Mangan, M. T. and Newman, S. (1994). Surface degassing and modifications to vesicle size distributions in active basalt flows. *Journal of Volcanology and Geothermal Research*, 61, 45–68.

Cashman, K. V., Thornber, C. and Kauahikaua, J. P. (1999). Cooling and crystallization of lava in open channels, and the transition of pahoehoe lava to 'A'a. *Bulletin Volcanologique*, 61, 306–323.

CAYTROUGH (1979). Geological and geophysical investigation of the Mid-Cayman Rise spreading center: Initial results and observations. In Talwani, M., Harrison, C. G. and Hayes, D. E. (eds.), *Deep Drilling Results in the Atlantic Ocean: Ocean Crust*. Maurice Ewing Series, 2, Washington, DC: American Geophysical Union, pp. 66–93.

Cavanaugh, C. M., Levering, P. R., Maki, J. S., Mitchell, R. and Lidstrom, M. E. (1987). Symbiosis of methylotrophic bacteria and deep-sea mussels. *Nature*, 325, 346–347.

Ceuleneer, G. and Cannat, M. (1997). High-temperature ductile deformation of Site 920 peridotites. In Karson, J. A., Cannat, M., Miller, D. J. and Elthon, D. (eds.), *Proceedings of the Ocean Drilling Program, Scientific Results, Leg 153*. College Station, TX: Ocean Drilling Program, pp. 23–34.

Ceuleneer, G., Nicolas, A. and Boudier, F. (1988). Mantle flow patterns at an oceanic spreading centre: The Oman peridotites record. *Tectonophysics*, 151, 1–26.

Chadwick, J., Perfit, M., Ridley, I. *et al.* (2005). Magmatic effects of the Cobb hot spot on the Juan de Fuca Ridge. *Journal of Geophysical Research Letters*, 110, B03101, doi:10.1029/2003JB002767.

Chadwick, W. W., Jr. (1997). Is the "axial summit caldera" on the East Pacific Rise really a "graben"? *Ridge Events*, 8(1), 8–11.

Chadwick, W. W., Jr. and Embley, R. W. (1998). Graben formation associated with recent dike intrusions and volcanic eruptions on the mid-ocean ridge. *Journal of Geophysical Research*, 103, 9807–9826.

Chadwick, W. W., Jr., Cashman, K. V., Embley, R. W. *et al.* (2008). Direct video and hydrophone observations of submarine explosive eruptions at NW Rota-1 Volcano, Mariana Arc. *Journal of Geophysical Research*, 113, B08S10, doi:10.1029/2007JB005215.

Chadwick, W. W., Jr., Clague, D. A., Embley, R. W. *et al.* (2013). The 1998 eruption of Axial Seamount: New insights on submarine lava flow emplacement from high-resolution mapping. *Geochemistry, Geophysics, Geosystems*, 14(10), 3939–3968, doi:10.1002/ggge.20202.

Chadwick, W. W., Jr., Embley, R. W. and Fox, C. G. (1991). Evidence for volcanic eruption on the southern Juan de Fuca Ridge between 1981 and 1987. *Nature*, 350, 416–418.

Chadwick, W. W., Jr., Embley, R. W. and Fox, C. G. (2003). Quantitative constraints on the growth of submarine lava pillars from a monitoring instrument that was caught in a lava flow. *Journal of Geophysical Research*, 108(B11), 2534, doi:2510.1029/2003JB002422.

Chadwick, W. W., Jr., Embley, R. W. and Shank, T. M. (1998). The 1996 Gorda Ridge eruption: Geologic mapping, sidescan sonar, and SeaBeam comparison results. *Deep Sea Research II*, 45, 2547–2569.

Chadwick, W. W., Jr., Gregg, T. K. P. and Embley, R. W. (1999). Submarine lineated sheet flows: A unique lava morphology formed on subsiding lava ponds. *Bulletin of Volcanology*, 61, 194–206.

Chadwick, W. W., Jr., Nooner, S. L., Butterfield, D. A. and Lilley, M. D. (2012). Seafloor deformation and forecasts of the April 2011 eruption at Axial Seamount. *Nature Geoscience*, 5, 474–477, doi: 10.1038/NGEO1464.

Chadwick, W. W., Jr., Scheirer, D. S., Embley, R. W. and Johnson, H. P. (2001). High-resolution bathymetric surveys using scanning sonars: Lava flow morphology, hydrothermal vent and geologic structure at recent eruption sites on the Juan de Fuca Ridge. *Journal of Geophysical Research*, 106, 16,075–16,100.

Charlou, J. L., Dmitried, L., Bougault, H. and Needham H. D. (1988). Hydrothermal CH_4 between 12°N and 15°N over the Mid-Atlantic Ridge. *Deep-Sea Research Part A*, 35, 121–131.

Charlou, J. L., Donval, J. P., Douville, E. *et al.* (2000). Compared geochemical signature and evolution of Menez Gwen (37°50′ N) and Lucky Strike (37°17′ N) hydrothermal fields, south of the Azores Triple Junction on the Mid-Atlantic Ridge. *Chemical Geology*, 171, 49–75.

Charlou, J. L., Donval, J. P., Fouquet, Y., Jean-Baptiste, P. and Holm, N. (2002). Geochemistry of high H_2 and CH_4 vent fluids issuing from ultramafic rocks at the Rainbow hydrothermal field (36°14′ N, MAR). *Chemical Geology*, 191, 345–359.

Charlou, J. L., Donval, J. P., Jean-Baptiste, P. and Dapoigny, A. (1996) Gases and helium isotopes in high temperature solutions sampled before and after ODP Leg 158 drilling a TAG hydrothermal field (26°N, MAR). *Geophysical Research Letters*, 23, 3491–3494.

Charlou, J. L., Donval, J. P., Konn, C., Ondreas, H. and Fouquet, Y. (2010). High production and fluxes of H2 and CH4 and evidence of abiotic hydrocarbon synthesis by serpentinization in ultramafic-hosted hydrothermal systems on the Mid-Atlantic Ridge.

In Rona, P., Devey, C. and Murton, B. (eds.), *Diversity of Hydrothermal Systems on Slow Spreading Ocean Ridges*. Geophysical Monograph 188, Washington, DC: American Geophysical Union, pp. 265–296.

Cheadle, M. J. and Grimes, C. (2010). Structural geology: To fault or not to fault. *Nature Geoscience*, 3, 454–456, doi:10.1038/ngeo910.

Chevaldonné, P., Desbruyères, D. and Childress, J. J. (1992). Some like it hot... and some even hotter. *Nature*, 359, 593–594.

Childress, J. J. and Fisher, C. R. (1992). The biology of hydrothermal vent animals: Physiology, biochemistry, and autotrophic symbiosis. *Oceanography and Marine Biology*, 30, 337–441.

Christensen, N. I. (1978). Ophiolites, seismic velocities and oceanic crustal structure. *Tectonophysics*, 47, 131–157.

Christensen, N. I. (1984). The magnitude, symmetry and origin of upper mantle anisotropy based on fabric analyses of ultramafic tectonites. *Geophysical Journal of the Royal Astronomical Society*, 76, 89.

Christensen, N. I. and Mooney, W. D. (1995). Seismic velocity structure and composition of the continental crust: A global view. *Journal of Geophysical Research*, 100, 9761–9788.

Christensen, N. I. and Salisbury, M. H. (1975). Structure and composition of the lower oceanic crust. *Reviews of Geophysics and Space Physics*, 13, 57–86.

Christensen, N. I. and Salisbury, M. H. (1982). Lateral heterogeneity in the seismic structure of the oceanic crust inferred from velocity studies in the Bay of Islands ophiolites, Newfoundland. *Geophysical Journal of the Royal Astronomical Society*, 68, 675–688.

Christeson, G. L., Karson, J. A. and McIntosh, K. D. (2010). Mapping of seismic layer 2A/2B boundary above the sheeted dike unit at intermediate-spreading crust exposed near the Blanco Transform. *Geochemistry, Geophysics, Geosystems*, 11, Q03015,, doi: 10.1029/2009 GC002864.

Christeson, G. L., Kent, G. M., Purdy, G. M. and Detrick, R. S. (1996). Extrusive thickness variability at the East Pacific Rise 9°–10°N: Constraints from seismic techniques. *Journal of Geophysical Research*, 101, 2859–2783.

Christeson, G. L., McIntosh, K. D. and Karson, J. A. (2007). Inconsistent correlation of seismic layer 2a and lava layer thickness in oceanic crust. *Nature*, 445, 418–421, doi:10.1038/nature05517.

Christeson, G. L., Morgan, J. V. and Warner, M. R. (2012). Shallow oceanic crust: Full waveform tomographic images of the seismic layer 2A/2B boundary. *Journal of Geophysical Research*, 117, B05101, doi:10.1029/2011 JB008972.

Christeson, G. L., Purdy, G. M. and Fryer, G. J. (1992). Structure of young upper crust at the East Pacific Rise near 9°30′ N. *Geophysical Research Letters*, 19, 1045–1048.

Christie, D. M. and Sinton, J. M. (1981). Evolution of abyssal lavas along propagating segments of the Galápagos spreading center. *Earth and Planetary Science Letters*, 56, 321–335.

Christie, D. M. and Sinton, J. M. (1986). Mantle processes in the generation of MORB from the 95°W Galápagos propagating rift area. *Contributions to Mineralogy and Petrology*, 94, 274–288.

Church, W. R. and Stevens, R. K. (1971). Early Paleozoic complexes of the Newfoundland Appalachians as mantle–oceanic crust sequences. *Journal of Geophysical Research*, 76, 1460–1466.

Chutas, L. A. (2007). *Structures in Upper Oceanic Crust: Perspectives from Pito Deep and Iceland*. Masters thesis, Duke University.

Ciccarelli, F. D., Doerks, T., von Mering, C. *et al.* (2006). Toward automatic reconstruction of a highly resolved tree of life. *Science*, 311, 1283–1286.

Cipriani, A., Bonatti, E., Brunelli, D. and Ligi, M. (2009). 26 million years of mantle upwelling below a segment of the Mid Atlantic Ridge: The Vema Lithospheric Section revisited. *Earth and Planetary Science Letters*, 285, 87–95.

Clague, D. A. and Paduan, J. B. (2009). Submarine basaltic volcanism. In Cousens, B. and Piercey, S. J. (eds.), *Submarine Volcanism and Mineralization: Modern Through Ancient*. Geological Association of Canada, pp. 41–60, doi:10.1029/2011GC003791.

Clague, D. A., Caress, D. W., Thomas, H. *et al.* (2008). Abundance and distribution of hydrothermal chimneys and mounds on the Endeavour Ridge determined by 1-m resolution AUV multibeam mapping surveys. *Eos, Transactions of the American Geophysical Union*, 89(53), V41B-2079.

Clague, D. A., Davis, A. S. and Dixon, J. E. (2003). Submarine strombolian eruptions along the Gorda mid-ocean ridge. In White, J. D. L., Smellie, J. L. and Clague, D. A. (eds.), *Explosive Subaequeous Volcanism.* Geophysical Monograph 140, Washington, DC: American Geophysical Union, pp. 1–128.

Clague, D. A., Dreyer, B. M., Paduan, J. B. *et al.* (2013). Geologic history of the summit of Axial Seamount, Juan de Fuca Ridge. *Geochemistry, Geophysics, Geosystems*, 14, 4403–4443, doi:10.1002/ggge.20240.

Clague, D. A., Paduan, J. B. and Davis, A. S. (2009). Widespread strombolian eruptions of mid-ocean ridge basalt. *Journal of Volcanology and Geothermal Research*, 180, 171–188.

Clague, D. A., Paduan, J. B., Caress, D. W., Thomas, H. and Chadwick, W. W., Jr. (2011). Volcanic morphology of West Mata Volcano, NE Lau Basin, based on high-resolution bathymetry and depth changes. *Geochemistry, Geophysics, Geosystems*, 12(11), doi:10.1029/2011GC 003791.

Clark, A. C. (1991) In Mackay, A. L. (ed.), *A Dictionary of Scientific Quotations*. Boca Raton, FL: CRC Press.

Clark, M. R., Rowden, A. A., Schlacher, T. *et al.* (2010). The ecology of seamounts: Structure, function and human impacts. *Annual Review of Marine Science*, 2, 253–278.

Claypool, G. E. and Kvenvolden, K. A. (1983). Methane and other hydrocarbon gases in marine sediments. *Annual Review of Earth and Planetary Sciences*, 11, 299–327.

Cochran, J. R., Goff, J. A., Malinverno, A. *et al.* (1993). Morphology of a 'superfast' mid-ocean ridge crest and flanks: The East Pacific Rise, 7–9° S. *Marine Geophysical Researches*, 15(1), 65–75.

Coffin, M. F., Duncan, R. A., Eldholm, O. *et al.* (2006). Large igneous provinces and scientific ocean drilling: Status quo and a look ahead. *Oceanography*, 19(4), 150–160, doi.org/10.5670/oceanog.2006.13.

Colaco, A., Blandin, J., Cannat, M. *et al.* (2011). MoMAR-D: A technological challenge to monitor the dynamics of the Lucky Strike vent ecosystem. *ICES Journal of Marine Science*, 68(2), 416–424, doi:10.1093/icesjms/fsq075.

Colaco, A., Dehairs, F. and Desbruyères, D. (2002). Nutritional relations of deep-sea hydrothermal fields at the Mid-Atlantic Ridge: A stable isotope approach. *Deep-Sea Research I*, 49, 395–412.

Coleman, R. G. (1981). Tectonic setting for ophiolite obduction in Oman. *Journal of Geophysical Research*, 86, 2497–2508.

Collins, J. A., Brocher, T. M. and Karson, J. A. (1986). Two-dimensional seismic reflection modeling of the oceanic crust/mantle transition in the Bay of Islands ophiolite complex. *Journal of Geophysical Research*, 91, 12,520–12,538.

Collins, J. A., Tucholke, B. E. and Canales, J. P. (2001). Structure of Mid-Atlantic Ridge megamullions from seismic refraction experiments and multichannel seismic reflection profiling. *Eos, Transactions of the American Geophysical Union*, 82, F1100.

Collins, P. C., Croot, P., Carlsson, C. *et al.* (2013a). A primer for the environmental impact assessment of mining at seafloor massive sulfide deposits. *Marine Policy*, 42, 198–209.

Collins, P. C., Kennedy, R., Copley, J. *et al.* (2013b). VentBase: Developing a consensus among stakeholders in the deep-sea regarding environmental impact assessment for deep-sea mining. *Marine Policy*, 42, 334–336.

Colman, A., Sinton, J. M., White, S. M. *et al.* (2012). Effects of variable magma supply on mid-ocean ridge eruptions: Constraints from mapped lava flows along the Galápagos Spreading Center. *Geochemistry, Geophysics, Geosystems*, 13, Q08014, doi:10.1029/2012GC004163.

Connelly, D. P., Germa, C. R., Asada, M. *et al.* (2007). Hydrothermal activity on the ultra-slow spreading southern Knipovich Ridge. *Geochemistry, Geophysics, Geosystems*, 8, Q08013, doi:10.1029/2007GC001652.

Constantin, M. (1999). Gabbroic intrusions and magmatic metasomatism in harzburgites from the Garrett transform fault: Implications for the nature of the mantle–crust transition at fast-spreading ridges. *Contributions to Mineralogy and Petrology*, 136, 111–130.

Constantin, M., Hékinian, R., Bideau, D. and Hébert, R. (1996). Construction of the oceanic lithosphere by magmatic intrusions: Petrologic evidence from plutonic rocks formed along the fast-spreading East Pacific Rise. *Geology*, 24(8), 731–734.

Coogan, L. A. (2007). The lower oceanic crust. In Turekian, K. and Holland, H. D. (eds.), *Treatise on Geochemistry*. Oxford: Pergamon, pp. 1–45, doi:10.1016/B978-0080 43751-4/00230-3.

Coogan, L. A. (2014). The lower oceanic crust. In Holland, H. and Turekian, K. (eds.), *Reference Module in Earth Systems and Environmental Sciences, from Treatise on Geochemistry (Second Edition), Vol. 4: The Crust*. Oxford: Elsevier, pp. 497–541, doi.org/10.1016/B978-0-08-095975-7.00316-8.

Coogan, L. A., Gillis, K. M., MacLeod, C. J., Thompson, G. M. and Hékinian, R. (2002). Petrology and geochemistry of the lower oceanic crust formed at the East Pacific Rise and exposed at Hess Deep: A synthesis and new results. *Geochemistry, Geophysics, Geosystems*, 3(11), 8604, doi:10.1029/2001/GC000230.

Coogan, L. A., Howard, K. A., Gillis, K. M. *et al.* (2006). Chemical and thermal constraints on focused fluid flow in the lower oceanic crust. *American Journal of Science*, 306, 389–427, doi:10.2475/06.2006.01.

Coogan, L. A., Kempton, P. D., Saunders, A. D. and Norry, M. J. (2000). Melt aggregation within the crust beneath the Mid-Atlantic Ridge: Evidence from plagioclase and clinopyroxene major and trace element compositions. *Earth and Planetary Science Letters*, 176, 245–257.

Coogan, L. A., Mitchell, N. C. and O'Hara, M. J. (2003). Roof assimilation at fast-spreading ridges: An investigation combining geophysical, geochemical and field evidence. *Journal of Geophysical Research*, 108(B1), 2002, doi:10.1029/2001JB001171.

Coogan, L. A., Wilson, R. N., Gillis, K. M. and MacLeod, C. J. (2001). Near-solidous evolution of oceanic gabbros: Insights from amphibole geochemistry. *Geochimica et Cosmochimica Acta*, 65, 4339–4357.

Cordier, C., Caroff, M., Juteau, T. *et al.* (2007). Bulk-rock geochemistry and plagioclase zoning in lavas exposed along the northern flank of the Western Blanco Depression (Northeast Pacific): Insight into open-system magma chamber processes. *Lithos*, 99, 289–311.

Corliss, J. B., Dymond, J., Gordon, L. I. *et al.* (1979). Submarine hydrothermal springs on the Galápagos Rift. *Science*, 203, 1073–1083.

Cousteau, J.-Y. (1953). *The Silent World*. New York: Harper & Brothers.

Cowen, J. P., Betram, M. A., Baker, E. T. *et al.* (1998). Geomicrobial transformations of manganese in Gorda Ridge event plumes. *Deep-Sea Research II*, 45, 2713–2738.

Cowen, J. P., Bertram, M. A., Wakeham, S. *et al.* (2001). Biogeochemical cycling in hydrothermal plumes. In Halbach, P. E., Tunnicliffe, V. and Hein, J. R. (eds.),

Energy and Mass Transfer in Marine Hydrothermal Systems. Berlin: Dahlem University Press, pp. 303–316.

Cowen, J. P., Fornari, D. J., Shank, T. M. *et al.* (2007). Volcanic eruptions at East Pacific Rise crest near 9°50′ N. *Eos, Transactions of the American Geophysical Union*, 88(7), 81–82.

Cox, A. (1969). Geomagnetic reversals. *Science*, 163, 000–000.

Cravo, A., Foster, P., Almeida, C., Bebianno, M. J. and Company, R. (2008). Metal concentrations in the shell of *Bathymodiolus azoricus* from contrasting hydrothermal vent fields on the Mid-Atlantic Ridge. *Marine Environmental Research*, 65, 338–348.

Crawford, W. C., Webb, S. C. and Hildebrand, J. A. (1999). Constraints on melt in the lower crust and Moho at the East Pacific Rise, 9° 48′ N, using seafloor compliance measurements. *Journal of Geophysical Research*, 104(B2), 2923–2939.

Crone, T. J., Wilcock, W. S., Barclay, A. H. and Parsons, J. D. (2006). The sound generated by mid-ocean ridge black smoker hydrothermal vents. *PLoS ONE*, 1(1), doi:10.1371/journal.one.0000133.

Croon, M. B., Cande, S. C. and Stock, J. M. (2010). Abyssal hill deflections at Pacific–Antarctic ridge–transform intersections. *Geochemistry, Geophysics, Geosystems*, 11 (11), Q11004, doi:10.1029/2010GC003236.

Curewitz, D. and Karson, J. A. (1998). Geological consequences of dike intrusion at mid-ocean ridge spreading centers. In Buck, W. R., Delaney, P. T., Karson, J. A. and Lagabrielle, Y. (eds.), *Faulting and Magmatism at Mid-Ocean Ridges*. Washington, DC: American Geophysical Union, pp. 117–136.

Curtis, A. C., Wheat, C. G., Fryer, P. and Moyer, C. L. (2013). Mariana forearc serpentinite mud volcanoes harbor novel communities of extremophilic Archaea. *Geomicrobiology Journal*, 30, 430–441.

Cuvelier, D., Sarrazin, J., Colaco, A. *et al.* (2009). Distribution and spatial variation of hydrothermal faunal assemblages at Lucky Strike (Mid-Atlantic Ridge) revealed by high resolution video image analysis. *Deep-Sea Research*, 1(156), 2026–2040.

CYAMEX Scientific Team (1981). First manned submersible dives on the East Pacific Rise at 21°N (Project RITA): General results. *Marine Geophysical Researches*, 4, 345–379.

Daly, R. A. (1926). *Our Mobile Earth*. New York: Charles Scribner's Sons.

Davies, A., Roberts J. M. and Hall-Spencer, J. M. (2007). Preserving deep-sea natural heritage: emerging issues in offshore conservation and management. *Biological Conservation*, 138, 299–312.

Davies, H. L. (1971). *Peridotite-Gabbro-Basalt Complex in Eastern Papua: An Overthrust Plate of Oceanic Mantle and Crust*. Bureau of Mineral Resources, 128, Geology and Geophysics, Australia.

Davis, A. S. and Clague, D. A. (1987). Geochemistry, mineralogy, and petrogenesis of basalt from the Gorda Ridge. *Journal of Geophysical Research*, 92, 10,467–10,483.

Davis, A. S. and Clague, D. A. (1990). Gabbroic xenoliths from the northern Gorda Ridge: Implications for magma chamber processes under slow-spreading centers. *Journal of Geophysical Research*, 95, 10,885–10,906.

Davis, E. E. and Elderfield, H. (2004). *Hydrogeology of the Oceanic Crust*. Cambridge: Cambridge University Press.

Davis, E. E. and Fisher A. T. (1994). On the nature and consequences of hydrothermal circulation in the Middle Valley sedimented rift: Inferences from geophysical and geochemical observations, Leg 139. In Mottl, M. J., Davis, E. E., Fisher, A. T. and Slack, J. F. (eds.), *Proceedings of the Ocean Drilling Program, Scientific Results, Leg 139*. College Station, TX: Ocean Drilling Program, pp. 695–717.

Davis, E. E., Wang, K., Thompson, R. E., Becker, K. and Cassidy, J. F. (2001). An episode of seafloor spreading and associated plate deformation inferred from crustal fluid pressure transients. *Journal of Geophysical Research*, 106, 21,953–21,963.

Deardorff, N. D., Cashman, K. V. and Chadwick, W. W., Jr. (2011). Observations of eruptive plumes and pyroclastic deposits from submarine explosive eruptions at NW Rota-1, Mariana Arc. *Journal of Volcanology and Geothermal Research*, 202(1–2), 47–59, doi:10.1016/j.jvolgeores.2011.01.003.

Dechaine, E. G., Bates, A. E., Shank, T. M. and Cavanaugh, C. M. (2006). Off-axis symbiosis found: Characterization and biogeography of bacterial symbionts of *Bathymodiolus* mussels from Lost City hydrothermal vents. *Environmental Microbiology*, 8(11), 1902–1912.

Delacour, A., Früh-Green, G. L., Bernasconi, S. M., Schaeffer, P. H. and Kelley, D. S. (2008a). Carbon geochemistry of serpentinites at the Lost City Hydrothermal System (30°N, MAR). *Geochimica et Cosmochimica Acta*, 72, 3681–3702.

Delacour, A., Früh-Green, G. L., Frank, M., Gutjahr, M., and Kelley, D. S. (2008b). Sr- and Nd-isotope geochemistry of the Atlantis Massif (30°N, MAR): Implications for fluid fluxes and lithospheric heterogeneity. *Chemical Geology*, 254, 19–35.

Delaney, J. R. and Barga, R. S. (2009). A 2020 vision for ocean science. In Hey, T., Tansley, S. and Tolle, K. (eds.), *The Fourth Paradigm Data-Intensive Scientific Discovery*. Microsoft Research, pp. 27–38.

Delaney, J. R. and Kelley, D. S. (2014). Understanding the planetary life support system: Next generation science in the ocean basins. In Favali, P., de Santis, A. and Beranzoli, L. (eds.), *Seafloor Observatories: A New Vision of the Earth from the Abyss*. Springer Praxis.

Delaney, J. R., Kelley, D. S., Lilley, M. D. (1998). The quantum event of oceanic crustal accretion: Impacts of diking at mid-ocean ridges. *Science*, 281, 222–230.

Delaney, J. R., Kelley, D. S., Lilley, M. D. (1997). The Endeavour hydrothermal system I: Cellular circulation above an active cracking front yields large sulfide structures, "fresh" vent water, and hyperthermophile archaea. *RIDGE Events*, 8, 11–19.

Delaney, J. R., Kelley, D. S. and Mathez, E. A. (2001). "Edifice Rex" sulfide recovery project: Analyses of submarine hydrothermal, microbial habitat. *Eos, Transactions of the American Geophysical Union*, 82(6), 67, 72–73.

Delaney, J. R., McDuff, R. E. and Lupton, J. E. (1984). Hydrothermal fluid temperatures of 400 °C on the Endeavour Segment, northern Juan de Fuca Ridge. *Eos, Transactions of the American Geophysical Union*, 65(45), 973.

Delaney, J. R., Mogk, D. W. and Mottl, M. J. (1987a). Quartz-cemented breccias from the Mid-Atlantic Ridge: Samples of a high-salinity hydrothermal upflow zone. *Journal of Geophysical Research*, 92, 9175–9192.

Delaney, J. R., Robigou, V., McDuff, R. E. and Tivey, M. K. (1992). Geology of a vigorous hydrothermal system on the Endeavour Segment, Juan de Fuca Ridge. *Geophysical Research Letters*, 97, 19,663–19,682.

Delaney, J. R., Spiess, F. N., Solomon, S. C. *et al.* (1987b). Scientific rationale for establishing long-term ocean bottom observatory/laboratory systems. In Teleki, P. *et al.* (eds.), *Marine Minerals: Resource Assessment Strategies.* Springer, pp. 389–411.

DeLong, S. E., Dewey, J. F. and Fox, P. J. (1977). Displacement history of oceanic fracture zones. *Geology*, 5, 199–202.

DeLong, S. E., Dewey, J. F. and Fox, P. J. (1979). Topographic and geologic evolution of fracture zones. *Journal of the Geological Society of London*, 136, 303–310.

deMartin, B. J., Sohn, R. A., Canales, J. P. and Humphris, S. E. (2007). Kinematics and geometry of active detachment faulting beneath the Trans-Atlantic Geotraverse (TAG) hydrothermal field on the Mid-Atlantic Ridge. *Geology*, 35, 711–714, doi:10.1130/G23718A.1.

DeMets, C., Gordon, R. G., Argus, D. F. and Stein, S. (1990). Current plate motions. *Geophysical Journal International*, 101, 425–478.

Desbruyères, D., Alayse-Danet, A.-M., Ohta, S. and the Scientific Parties of Biolau and Starmer Cruises (1994). Deep-sea hydrothermal communities in Southwestern Pacific back-arc basins (the North Fiji and Lau Basins): Composition, micro-distribution and food web. *Marine Geology*, 116, 227–242.

Desbruyères, D., Almeida, A., Biscoito, M. *et al.* (2000). A review of the distribution of hydrothermal vent communities along the northern Mid-Atlantic Ridge: Dispersal vs. environmental controls. *Hydrobiologia*, 440, 201–216.

Desbruyères, D., Biscoito, M. Caprais, J. C. *et al.* (2001). Variations in deep-sea hydrothermal vent communities on the Mid-Atlantic Ridge near the Azores plateau. *Deep-Sea Research Part I*, 48, 1325–1346.

Desbruyères, D., Chevaldonné, P, Alayse, A. M. *et al.* (1998). Biology and ecology of the 'Pompeii worm' (*Alvinella pompejana* Desbruyères and Laubier), a normal dweller of an extreme deep-sea environment: A synthesis of current knowledge and recent developments. *Deep-Sea Research*, 45, 383–422.

Desbruyères, D., Segonzac, M. and Bright, M. (2006). *Handbook of Deep-Sea Hydrothermal Vent Fauna: Denisia*, Vol. 18, Linz-Dornach, Austria: Oberösterreichischen Landesmuseen.

Detrick, R. S., Buhl, P., Vera, E. *et al.* (1987). Multichannel seismic imaging of a crustal magma chamber along the East Pacific Rise. *Nature*, 326, 35–41.

Detrick, R. S., Collins, J., Stephen, R. and Swift, S. (1994). In situ evidence for the nature of the seismic layer 2/3 boundary in oceanic crust. *Nature*, 370, 288–290.

Detrick, R. S., Harding, A. J. Kent, G. M. *et al.* (1993a). Seismic structure of the southern East Pacific Rise. *Science*, 259, 499–503.

Detrick, R. S., White, R. S. and Purdy, G. M. (1993b). Crustal structure of North Atlantic fracture zones. *Reviews of Geophysics*, 31, 439–458.

Devey, C. W., Lackschewitz, K. S. and Baker, E. (2005). Hydrothermal and volcanic activity found on the Southern Mid-Atlantic Ridge. *Eos, Transactions of the American Geophysical Union*, 86, 209–216.

Dewey, J. F. and Bird, J. M. (1971). Origin and emplacement of the ophiolite suite: Appalachian ophiolites in Newfoundland. *Journal of Geophysical Research*, 76, 3179–3206.

Dewey, J. F. and Kidd, W. S. F. (1977). Geometry of plate accretion. *Geological Society of America Bulletin*, 88, 960–968.

Dick, H. J. B. (1989). Abyssal peridotites, very slow spreading ridges and ocean ridge magmatism. In Saunders, A. D. and Norry, M. J. (eds.), *Magmatism in the Ocean Basins*. Special Publication 42, London: Geological Society of London, pp. 77–105.

Dick, H. J. B. and Bullen, T. (1984). Chromian spinel as a petrogenetic indicator in abyssal and alpine-type peridotites and spacially associated lavas. *Contributions to Mineralogy and Petrology*, 86, 54–76.

Dick, H. J. B. and Natland, J. H. (1996). Late-stage melt evolution and transport in the shallow mantle beneath the East Pacific Rise. In Mével, C., Gillis, K. M., Allan, J. F. and Meyer, P. S. (eds.), *Proceedings of the Ocean Drilling Program, Scientific Results, Leg 147*. College Station, TX: Ocean Drilling Program, pp. 439–538.

Dick, H. J. B., Lin, J. and Schouten, H. (2003). An ultraslow-spreading class of ocean ridge. *Nature*, 426, 405–412.

Dick, H. J. B., Lissenberg, C. J. and Warren, J. M. (2010). Mantle melting, melt transport, and delivery beneath a slow-spreading ridge: The paleo-MAR from 23°15′ N to 23°45′ N. *Journal of Petrology*, 51(1–2), 425–467, doi:10.1093/petrology/egp088.

Dick, H. J. B., Meyer, P., Bloomer, S. *et al.* (1991). Lithostratigraphic evolution of an in situ section of oceanic layer 3. In von Herzen, R. P. and Robinson, P. T. (eds.), *Proceedings of the Ocean Drilling Program, Scientific Results, Leg 118*. College Station, TX: Ocean Drilling Program, pp. 439–515.

Dick, H. J. B., Natland, J. H., Alt, J. C. et al. (2000). A long in situ section of the lower oceanic crust: Results of ODP Leg 176 drilling at the Southwest Indian Ridge. *Earth and Planetary Science Letters*, 179, 31–51.

Dick, H. J. B., Schouten, H., Meyer, P. S. et al. (1991). Tectonic evolution of the Atlantis II fracture zone. *In Proceedings of the Ocean Drilling Program, Scientific Results, Leg 118*. pp. 359–398.

Dick, H. J. B., Tivey, M. A. and Tucholke, B. E. (2008). Plutonic foundation of a slow-spreading ridge segment: The oceanic core complex at Kane megamullion, 23°30′ N, 42°20′ W. *Geochemistry, Geophysics, Geosystems*, 9, Q05014, doi:10.1029/2007GC001645.

Dietz, R. S. (1961). Continental and ocean basin evolution by spreading of the sea floor. *Nature*, 190, 854–857.

Dilek, Y., Harper, G. D., Pezard, P. A. and Tartarotti, P. (1996). Structure of the sheeted dike complex in Hole 504B (Leg 148). In Alt, J. C., Kinoshita, H., Stokking, L. B. and Michael, P. J. (eds.), *Proceedings of the Ocean Drilling Program, Scientific Results, Leg 148*. College Station, TX: Ocean Drilling Program, pp. 229–243.

Dilek, Y., Moores, E. M., Delaloye, M. and Karson, J. A. (1991). Amagmatic extension and tectonic denudation in the Kizildag ophiolite, southern Turkey: Implications for the evolution of Neotethyan oceanic crust. In Peters, T., Nicolas, A. and Coleman, R. G. (eds.), *Ophiolite Genesis and Evolution of Oceanic Lithosphere*. Dordrecht: Kluwer, pp. 485–500.

Dixon, J. E., Clague, D. A. and Eissen, J. P. (1986). Gabbroic xenoliths and host ferrobasalt from the southern Juan de Fuca Ridge. *Journal of Geophysical Research*, 91(B3), 3795–3820.

Dixon, J. E., Stolper, E. M. and Holloway, J. R. (1995). An experimental study of water and carbon dioxide solubilities in mid-ocean ridge basaltic liquids. Part I: Calibration and solubility models. *Journal of Petrology*, 36(6), 1607–1631.

Douville, J. L., Charlou, E. H., Oelkers, P. et al. (2002). The rainbow vent fluids (36°14′ N, MAR): The influence of ultramafic rocks and phase separation on trace metal content in Mid-Atlantic Ridge hydrothermal fluids. *Chemical Geology*, 184, 37–48.

Drake, C. L. and Girdler, R. W. (1964). A geophysical study of the Red Sea. *Geophysical Journal of the Royal Astronomical Society*, 8(5), 473–495.

Dunkelman, T. J., Karson, J. A. and Rosendahl, B. R. (1988). Structural style of the Turkana Rift, Kenya. *Geology*, 16, 258–261.

Dunn, R. A. and Toomey, D. R. (1997). Seismological evidence for three-dimensional melt migration beneath the East Pacific Rise. *Nature*, 388, 259–262.

Dunn, R. A., Toomey, D. R. and Solomon, S. C. (2000). Three-dimensional seismic structure and physical properties of the crust and shallow mantle beneath the East Pacific Rise at 9°30′ N. *Journal of Geophysical Research*, 105(B10), 23,537–22,555.

Dziak, R. P. and Fox, C. G. (1990). Long-term seismicity and ground deformation at Axial Volcano, Juan de Fuca Ridge. *Geophysical Research Letters*, 26, 3641–3644.

Dziak, R. P. and Fox, C. G. (1999). The January 1998 earthquake swarm at Axial Volcano, Juan de Fuca Ridge: Hydroacoustic evidence of seafloor volcanic activity. *Geophysical Research Letters*, 26, doi:10.1029/1999GL002332.

Dziak, R. P., Baker, E. T., Shaw, A. M. et al. (2012a). Flux measurements of explosive degassing using a yearlong hydroacoustic record at an erupting submarine volcano. *Geochemistry, Geophysics, Geosystems*, 13, Q0AF07, doi:10.1029/2012GC004211.

Dziak, R. P., Bohnenstiehl, D. R. and Smith, D. K. (2012b). Hydroacoustic monitoring of oceanic spreading centers: Past, present, and future. *Oceanography*, 25(1), 116–127.

Dziak, R. P., Bohnenstiehl, D. R., Cowen, J. P. et al. (2007). Rapid dike emplacement leads to eruptions and hydrothermal plume release during seafloor spreading events. *Geology*, 35, 579–582.

Dziak, R. P., Bohnenstiehl, D. R., Matsumoto, H. et al. (2009). January 2006 seafloor-spreading event at 9°50′ N, East Pacific Rise: Ridge dike intrusion and transform fault interactions from regional hydroacoustic data. *Geochemistry, Geophysics, Geosystems*, 10, Q06T06, doi:10.1029/2009GC002388.

Dziak, R. P., Fox, C. G. and Schreiner, A. E. (1995). The June-July 1993 seismo-acoustic event at CoAxial segment, Juan de Fuca Ridge: Evidence for a lateral dike injection. *Geophysical Research Letters*, 22, 135–138.

Dziak, R. P., Hammond, S. R. and Fox, C. G. (2011). A 20-year hydroacoustic time series of seismic and volcanic events in the Northeast Pacific Ocean. *Oceanography*, 24, 280–293, doi:10.5670/oceanog.2011.79.

Dziak, R. P., Haxel, J. H., Bohnenstiehl, D. R. et al. (2012c). Seismic precursors and magma ascent before the April 2011 eruption at Axial Seamount. *Nature Geosciences*, 5, 478–482.

Eason, D. E. and Sinton, J. M. (2009). Lava shields and fissure eruptions of the Western Volcanic Zone, Iceland: Evidence for magma chambers and crustal interaction. *Journal of Volcanology and Geothermal Research*, 186, 331–348.

Edmond, J. M., Measures, B., Mangum, B. et al. (1979a). On the formation of metal-rich deposits at ridge crests. *Earth and Planetary Science Letters*, 46, 19–30.

Edmond, J. M., Measures, C., McDuff, R. E. et al. (1979b). Ridge crest hydrothermal activity and the balances of the major and minor elements in the ocean: The Galápagos data. *Earth and Planetary Science Letters*, 46, 1–18.

Edmonds, H. N., Michael, P. J., Baker, E. T. et al. (2003). Discovery of abundant hydrothermal venting on the ultraslow-spreading Gakkel ridge in the Arctic Ocean. *Nature*, 421, 252–256.

Edwards, M. H., Fornari, D. J., Malinverno, A., Ryan, W. B. F. and Madsen, J. (1991). The regional tectonic fabric of the East Pacific Rise from 12°50′ N to 15°10′ N. *Journal of Geophysical Research*, 96, 7995–8017.

Ehlmann, B. L., Mustard, J. F. and Murchie, S. L. (2010). Geologic setting of serpentine deposits on Mars. *Geophysical Research Letters*, 37, L06201, doi:10.1029/2010GL042596.

Einarsson, P. and Ericksson, J. (1982). Earthquake fractures in the districts land and Rangarvellir in the South Iceland Seismic Zone. *Jokull*, 32, 113–119.

Einarsson, P. and Brandsdóttir, B. (1980). Seismological evidence for lateral magma intrusion during the July 1978 deflation of the Krafla volcano in NE-Iceland. *Journal of Geophysics*, 47, 160–165.

Einsele, G. (1982). Mechanism of sill intrusion into soft sediment and expulsion of pore water. In Curray, J. R. and Moore, D. G. (eds.), *Initial Reports of the Deep Sea Drilling Project, Leg 64, Part 2*. Washington, DC: US Government Printing Office, pp. 1169–1176.

Einsele, G. (1985). Basaltic sill-sediment complexes in young spreading centers: genesis and significance. *Geology*, 13, 249–252.

Einsele, G., Gieskes, J., Curray, J. *et al.* (1980). Intrusion of basaltic sills into highly porous sediments, and resulting hydrothermal activity. *Nature*, 283, 441–445.

Elthon, D. (1987). Petrology of gabbroic rocks from the Mid-Cayman Rise spreading center. *Journal of Geophysical Research*, 92, 658–682.

Elthon, D. and Stern, C. (1978). Metamorphic petrology of the Sarmiento Ophiolite Complex, Chile. *Geology*, 6, 464–468.

Elthon, D., Ross, D. K. and Meen, J. K. (1995). Compositional variations of basaltic glasses from the Mid-Cayman Rise spreading center. *Journal of Geophysical Research*, 100(B7), 12,497–12,512, doi:10.1029/94JB02777.

Elthon, D., Stewart, M. and Ross, D. K. (1992). Compositional trends of minerals in oceanic cumulates. *Journal of Geophysical Research*, 97, 15,189–15,200.

Embley, R. W. and Lupton, J. E. (2004). Diking, event plumes, and the subsurface biosphere at mid-ocean ridges. In Wilcock, W. D., Kelley, D. S., Baross, J. A., DeLong, E. and Cary, C. (eds.), *The Subseafloor Biosphere at Mid-Ocean Ridges*. Geophysical Monograph 144, Washington, DC: American Geophysical Union, pp. 75–97.

Embley, R. W., Chadwick, W. W., Jr., Baker, E. T. *et al.* (2006). Long-term eruptive activity at a submarine arc volcano. *Nature*, 441, 494–497.

Embley, R. W., Chadwick, W. W., Jr., Clague, D. A. and Stakes, D. (1999). 1998 eruption of Axial Volcano: Multibeam anomalies and sea-floor observations. *Geophysical Research Letters*, 26, 3425–3428.

Embley, R. W., Chadwick, W. W., Jr., Jonasson, I. R., Butterfield, D. A. and Baker, E. T. (1995). Initial results of the response to the 1993 CoAxial event: Relationships between hydrothermal and volcanic processes. *Geophysical Research Letters*, 22, 143–146, doi:.org/10.1029/94GL02281.

Embley, R. W., Chadwick, W. W., Jr., Perfit, M. R. and Baker, E. T. (1991). Geology of the northern Cleft segment, Juan de Fuca Ridge: Recent eruptions, seafloor spreading, and the formation of megaplumes. *Geology*, 19, 771–775.

Embley, R. W., Chadwick, W. W., Jr., Perfit, M. R., Smith, M. C. and Delaney, J. F. (2000). Recent eruptions on the CoAxial segment of the Juan de Fuca Ridge: Implications for mid-ocean ridge accretion processes. *Journal of Geophysical Research*, 105, 16,501–16,525.

Embley, R. W., Jonasson, I. R., Perfit, M. R. *et al.* (1988). Submersible investigation of an extinct hydrothermal system on the Galápagos Ridge: Sulfide mounds, stockwork zone, and differentiated lavas. *Canadian Mineralogist*, 26, 517–539.

Embley, R., Chadwick, W. W., Jr., Baker, E. T. *et al.* (2006). Long-term eruptive activity at a submarine arc volcano. *Nature*, 441, 494–497.

Engel, A. E. J., Engel, C. G. and Havens, R. G. (1965). Chemical characteristics of oceanic basalts and the upper mantle. *Geological Society of America Bulletin*, 76(7), 719–734.

Engel, C. G. and Fisher, R. L. (1975). Granitic to ultramafic rock complexes of the Indian Ocean ridge system western Indian Ocean. *Geological Society of America Bulletin*, 86, 1553–1578.

Engels, J. L., Edwards, M. H., Fornari, D. J., Perfit, M. R and Cann, J. R. (2003). A new model for submarine volcanic collapse formation. *Geochemistry, Geophysics, Geosystems*, 4(9), doi:10.1029/2002GC000483.

Escartín, J. and Canales, J. P. (2010). Chapman Conference on Detachments in Oceanic Lithosphere: Deformation, Magmatism, Fluid Flow and Ecosystems (Conference report). *Eos, Transactions of the American Geophysical Union*, 92, 31, doi:10.1029/2011EO040003.

Escartín, J., Cannat, M., Pouliquen, G., Rabain, A. and Lin, J. (2001a). Crustal thickness of V-shaped ridges south of the Azores: Interaction of the Mid-Atlantic Ridge (36°–39° N) and the Azores hot spot. *Journal of Geophysical Research*, 106, 21,719–21,735, doi:10.1029/2001JB000224.

Escartín, J., García, R., Delaunoy, J. *et al.* (2008a). Globally aligned photomosaic of the Lucky Strike hydrothermal vent field (Mid-Atlantic Ridge, 37°18.5′ N): Release of georeferenced data, mosaic construction, and viewing software. *Geochemistry, Geophysics, Geosystems*, 9(12), doi:10.1029/2008GC002204.

Escartín, J., Hirth, G. and Evans, B. (1997). Effects of serpentinization on the lithospheric strength and the style of normal faulting at slow-spreading ridges. *Earth and Planetary Science Letters*, 151, 181–190.

Escartín, J., Hirth, G. and Evans, B. (2001b). Strength of slightly serpentinized peridotites: Implications for the tectonics of oceanic lithosphere. *Geology*, 29, 1023–1026.

Escartín, J., Mevel, C., MacLeod, C. J. and McCaig, A. M. (2003). Constraints on deformation conditions and the origin of oceanic detachments: The Mid-Atlantic Ridge core complex at 15°45′ N. *Geochemistry, Geophysics, Geosystems*, 4(8), doi:1029/2002GC000472.

Escartín, J., Smith, D. K., Cann, J. R. *et al.* (2008b). Central role of detachment faults in accretion of slow-spreading oceanic lithosphere. *Nature*, 455, 790–794, doi:10.1038/nature07333.

Escartín, J., Soule, S. A., Fornari, D. J. *et al.* (2007). Interplay between faults and lava flows in construction of the upper oceanic crust: The East Pacific Rise crest 9° 25′ –9° 58′ N. *Geochemistry, Geophysics, Geosystems*, 8(6), doi:10.1029/2006GC001399.

Ewing, M. A. and Engel, L. (1962). Seismic shooting at sea. *Scientific American*, 5, 116–126.

Ewing, M. A., Vine, A. C. and Worzel, J. L. (1946). Photography of the ocean bottom. *Journal of the Optical Society of America*, 36, 307.

Ferrini, V. L., Fornari, D. J., Shank, T. M. *et al.* (2007). Sub-meter bathymetric mapping of volcanic and hydrothermal features on the East Pacific Rise Crest at 9° 50′ N. *Geochemistry, Geophysics, Geosystems*, 8, Q01006, doi:10.1029/2006GC001333.

Ferrini, V. L., Tivey, M. K., Carbotte, S. M., Martinez, F. and Roman, C. N. (2008). Variable morphologic expression of volcanic, tectonic, and hydrothermal processes at six hydrothermal vent fields in the Lau Back-arc Basin. *Geochemistry, Geophysics, Geosystems*, 9, Q07022, doi:10.1029/2008GC002047.

Fisher, A. T. (1998). Permeability within basaltic oceanic crust. *Reviews in Geophysics*, 36, 143–182.

Fisher, A. T. and Wheat, G. (2010). Seamounts as conduits for massive fluid, heat, and solute fluxes on ridge flanks. *Oceanography Magazine*, 23(1), 74–87.

Fisher, A. T., Davis, E. E. and Becker, K. (2008). Borehole-to-borehole hydrologic response across 2.4 km in the upper oceanic crust: Implications for crustal-scale properties. *Journal of Geophysical Research*, 113, doi:10.1029/2007JB005447.

Fisher, A. T., Davis, E. E., Hutnak, M. *et al.* (2003a). Hydrothermal recharge and discharge across 50 km guided by seamounts on a young ridge flank. *Nature*, 421, 618–621.

Fisher, A. T., Stein, C. A., Harris, R. N. *et al.* (2003b). Abrupt thermal transition reveals hydrothermal boundary and role of seamounts within the Cocos Plate. *Geophysical Research Letters*, 30(11), 1550, doi:1510.1029/2002GL016766.

Fisher, C. R., Childress, J. J., Arp, A. J. *et al.* (1988). Microhabitat variation in the hydrothermal vent mussel, *Bathymodiolus thermophiles*, at the Rose Garden vent on the Galápagos Rift. *Deep-Sea Research Part A. Oceanographic Research Papers*, 35, 1769–1791, doi.org/10.1016/0198-0149(88) 90049-0.

Fontaine, F. and Wilcock, W. S. D. (2006). Dynamics and storage of brine in mid-ocean ridge hydrothermal systems. *Journal of Geophysical Research*, 111, 1–16, B06102, doi:10.1029/2005JB003866.

Fornari, D. J. (2003). A new deep-sea towed digital camera and multi-rock coring system. *Eos, Transactions of the American Geophysical Union*, 84, 69 and 73.

Fornari, D. J. (2004). Realizing the dreams of da Vinci and Verne. *Oceanus*, 42, 1–5.

Fornari, D. J., Beaulieu, S. E., Holden, J. F., Mullineaux, L. S. and Tolstoy, M. (2012b). Introduction to the Special Issue: From RIDGE to Ridge 2000. *Oceanography*, 25(1), 12–17.

Fornari, D. J., Gallo, D. G., Edwards, M. H. *et al.* (1989). Structure and topography of the Siqueiros transform fault system: Evidence for the development of intra-transform spreading centers. *Marine Geophysical Researches*, 11, 263–299.

Fornari, D. J., Haymon, R. M., Perfit, M. R., Gregg, T. K. P. and Edwards, M. H. (1998a). Axial summit trough of the East Pacific Rise 9°–10°N: Geological characteristics and evolution of the axial zone on fast spreading mid-ocean ridges. *Journal of Geophysical Research*, 103, 9827–9855.

Fornari, D. J., Humphris, S. E. and Perfit, M. R. (1997). Deep submergence science takes a new approach. *Eos, Transactions of the American Geophysical Union*, 78, 402 and 408.

Fornari, D. J., Humphris, S. E. *et al.* (1996). LUSTRE '96 (LUcky STRike Exploration) R/V Knorr Cruise 145–19, Multidisciplinary investigations of hydrothermal vents on Lucky Strike Seamount and the tectonic and volcanic structure of the Mid-Atlantic Ridge rift valley between 37°10′ -25′ N: Near-bottom studies using the DSL-120kHz Sonar, ARGO-II and ROV Jason: Final Cruise Report, WHOI Technical Report. Woods Hole Oceanographic Institution.

Fornari, D. J., Malahoff, A. and Heezen, B. C. (1979). Visual observations of the volcanic micromorphology of Tortuga, Lorraine and Tutu seamounts, and petrology and chemistry of ridge and seamount features in and around the Panama Basin. *Marine Geology*, 31(1), 1–30.

Fornari, D. J., Perfit, M. R., Allan, J. F. *et al.* (1988). Geochemical and structural studies of the Lamont seamounts: Seamounts as indicators of mantle processes. *Earth and Planetary Science Letters*, 89, 63–83.

Fornari, D. J., Perfit, M. R., Malahoff, A. and Embley, R. (1983). Geochemical studies of abyssal lavas recovered by DSRV ALVIN from eastern Galápagos Rift,

Inca Transform and Ecuador Rift: 1. Major element variations in natural glasses and special distribution of lavas. *Journal of Geophysical Research*, 88, 10,519–10,529.

Fornari, D. J., Ryan, W. B. F. and Fox, P. J. (1984). The evolution of craters and calderas on young seamounts: Insights from Sea MARC I and Sea Beam sonar surveys of a small seamount group near the axis of the East Pacific Rise at ~ 10°N. *Journal of Geophysical Research*, 89, 11,069–11,083.

Fornari, D. J., Shank, T., Von Damm, K. L. *et al.* (1998b). Time-series temperature measurements at high-temperature hydrothermal vents, East Pacific Rise 9° 49′ –51′ N: Monitoring a crustal cracking event. *Earth and Planetary Science Letters*, 160, 419–431.

Fornari, D. J., Tivey, M. A., Schouten, H. *et al.* (2004). Submarine lava flow emplacement at the East Pacific Rise 9°50′ N: Implications for uppermost oceanic crust stratigraphy and hydrothermal fluid circulation. In German, C. R., Lin, J. and Parson, L. M. (eds.), *Mid-Ocean Ridges: Hydrothermal Interactions Between the Lithosphere and Oceans*. Geophysical Monograph 148, Washington, DC: American Geophysical Union, pp. 187–218.

Fornari, D. J., Von Damm, K. L., Bryce, J. G. *et al.* (2012a). The East Pacific Rise between 9°N and 10°N: Twenty-five years of integrated, multidisciplinary oceanic spreading center studies. *Oceanography*, 25(1), 18–43, doi.org/10.5670/oceanog.2012.02.

Forsyth, D. (1992). Geophysical constraints on mantle flow and melt generation beneath mid-ocean ridges. In Phipps Morgan, J., Blackman, D. K. and Sinton, J. M. (eds.), *Mantle Flow and Melt Generation at Mid-Ocean Ridges*. Washington, D C: American Geophysical Union, pp. 1–66.

Fouquet, Y., Charlou, J.-L., Costa, I. *et al.* (1995). Atlantic lava lakes and hot vents. *Nature*, 377, 201.

Fournier, R. O. (1987). Conceptual models of brine evolution in magmatic-hydrothermal systems. In Decker, R. W., Wright, T. L. and Stauffer, P. H., *Volcanism in Hawaii*. US Geological Survey Professional Paper, 1350, pp. 1487–1506.

Foustoukos, D. I. and Seyfried, W. E., Jr. (2004). Hydrocarbons in hydrothermal vent fluids: The role of chromium-bearing catalysts. *Science*, 304, 1002–1005.

Foustoukos, D. I. and Seyfried, W. E., Jr. (2007a). Fluid phase separation processes in submarine hydrothermal systems. In Liebscher, A. and Heinrich, C. A. (eds.), *Fluid-Fluid Interactions*. Mineralogical Society of America, pp. 213–233.

Foustoukos, D. I. and Seyfried, W. E., Jr. (2007b). Quartz solubility in the two-phase and critical region of the NaCl-KCl-H_2O system: Implications for submarine hydrothermal vent systems at 9°50′ N East Pacific Rise. *Geochimica et Cosmochimica Acta*, 71, 186–201

Fowler, C. M. R. (1976). Crustal structure of the Mid-Atlantic Ridge crest at 37°N. *Geophysical Journal of the Royal Astronomical Society*, 47, 459–491.

Fox, C. G. (1990). Consequences of phase separation on the distribution of hydrothermal fluids at ASHES vent field, Axial Volcano, Juan de Fuca Ridge. *Journal of Geophysical Research*, 95, 12,923–12,926.

Fox, C. G. and Dziak, R. P. (1998). Hydroacoustic detection of volcanic activity on the Gorda Ridge, February–March 1996. *Deep-Sea Research Part II: Topical Studies in Oceanography*, 45(12), 2513–2530.

Fox, C. G., Chadwick, W. W., Jr. and Embley, R. W. (1992). Detection of changes in ridge-crest morphology using repeated multibeam surveys. *Journal of Geophysical Research*, 97, 11,149–11,162.

Fox, C. G., Chadwick, W. W., Jr. and Embley, R. W. (2001). Direct observation of a submarine volcanic eruption from a sea-floor instrument caught in a lava flow. *Nature*, 412, 727–729.

Fox, C. G., Dziak, R. P., Matsumoto, H. and Schreiner, A. E. (1994). Potential for monitoring low-level seismicity on the Juan de Fuca Ridge using military hydrophone arrays. *Marine Technology Society Journal*, 27(4), 22–30.

Fox, C. G., Radford, W. E., Dziak, R. P. *et al.* (1995). Acoustic detection of a seafloor spreading episode on the Juan de Fuca Ridge using military hydrophone arrays. *Geophysical Research Letters*, 22(2), 131–134.

Fox, P. J. and Gallo, D. G. (1986). The geology of the North Atlantic transform plate boundaries and their aseismic extensions. In Vogt, P. R. and Tucholke, B. E. (eds.), *The Geology of North America: The Western North Atlantic Region*. DNAG Series, Boulder, CO: Geological Society of America, pp. 157–172.

Fox, P. J. and Gallo, D. G. (1989). Transforms of the eastern central Pacific. In Winterer, E. L., Hussong, D. M. and Decker, R. W. (eds.), *The Eastern Pacific and Hawaii*. Boulder, CO: Geological Society of America, pp. 111–124.

Fox, P. J. and Stroup, J. B. (1981). The plutonic foundation of the oceanic crust. In Emiliani, C. (ed.), *The Oceanic Lithosphere*. New York: Wiley, pp. 119–218.

Fox, P. J., Detrick, R. S. and Purdy, G. M. (1980). Evidence for crustal thinning near fracture zones: Implications for ophiolites. In Panayiotou, A. (ed.), *Proceedings of the Ophiolite Symposium*. Nicosia, Cyprus: Geological Survey Department, pp. 161–168.

Fox, P. J., Schreiber, E. and Peterson, J. (1972). Compressional wave velocities in basalt and altered basalt recovered during Leg 14. In Hayes, D. E., Pimm, A. C. *et al.* (eds.), *Initial Reports of the Deep Sea Drilling Project, Leg 14*. Washington, DC: US Government Printing Office, pp. 773–775.

Fox, P. J., Schreiber, E. and Peterson, J. J. (1973). The geology of the oceanic crust: Compressional wave velocities of oceanic rocks. *Journal of Geophysical Research*, 78, 5155–5172.

Fox, P. J., Schreiber, E., Rowlett, H. and Mccamy, K. (1976). The geology of the Oceanographer Fracture Zone: A model for fracture zones. *Journal of Geophysical Research*, 81, 4117–4128.

Francheteau J. and Ballard, R. D. (1983). The East Pacific Rise near 21°N, 13°N, and 20°S: Inferences for along-strike variability of axial processes of the Mid-Ocean Ridge. *Earth and Planetary Science Letters*, 64, 93–116.

Francheteau, J., Armijo, R., Cheminee, J. L. *et al.* (1990). 1 Ma East Pacific Rise oceanic crust and uppermost mantle exposed by rifting in Hess Deep (equatorial Pacific Ocean). *Earth and Planetary Science Letters*, 101, 281–295.

Francheteau, J., Armijo, R., Cheminee, J. L. *et al.* (1992). Dyke complex of the East Pacific Rise exposed in the walls of Hess Deep and the structure of the upper oceanic crust. *Earth and Planetary Science Letters*, 111, 109–121.

Francheteau, J., Armijo, R., Cogné, J. P. *et al.* (1994). Submersible observations of the Easter microplate and its boundary. *Eos, Transactions of the American Geophysical Union*, 75, 582.

Francheteau, J., Juteau, T. and Rangan, C. (1979b). Basaltic pillars in collapsed lava-pools on the deep ocean-floor. *Nature*, 281, 209–211.

Francheteau, J., Needham, H. D., Choukroune, P. *et al.* (1979a). Massive deep sea sulfide ore deposits discovered on the East Pacific Rise. *Nature*, 277, 523–528.

Francheteau, J., Patriat, P., Segoufin, J. *et al.* (1988). Pito and Orongo fracture zones: The northern and southern boundaries of the Easter microplate (southeast Pacific). *Earth and Planetary Science Letters*, 89, 363–374.

Francheteau, J., Yelles-Chaouche, A. and Craig, H. (1987). The Juan Fernandez microplate north of the Pacific-Nazca-Antarctic plate junction at 35°S. *Earth and Planetary Science Letters*, 86, 253–286.

Francis, T. J. G. (1981). Serpentinization faults and their role in the tectonics of slow spreading ridges. *Journal of Geophysical Research*, 86, 11,616–11,622.

Frey, F. A., Walker, N., Stakes, D., Hart, S. R. and Nielsen, R. (1993). Geochemical characteristics of basaltic glasses from the AMAR and FAMOUS axial valleys, Mid-Atlantic Ridge (36°–37° N): Petrogenetic implications. *Earth and Planetary Science Letters*, 115(1), 117–136.

Fricke, H., Giere, O., Stetter, K. *et al.* (1989). Hydrothermal vent communities at the shallow subpolar Mid-Atlantic Ridge. *Marine Biology*, 102, 425–429.

Früh-Green, G. L., Connolly, J. A. D., Kelley, D. S., Plas, A. and Grobety, B. (2004). Serpentinization of oceanic peridotites: Implications for geochemical cycles and biological activity. In Wilcock, W. D., Kelley, D. S., Baross, J. A., DeLong, E. and Cary, C. (eds.), *The Subseafloor Biosphere at Mid-Ocean Ridges*. Geophysical Monograph 144, Washington, DC: American Geophysical Union, pp. 119–135.

Früh-Green, G. L., Kelley, D. S., Bernasconi, S. M. *et al.* (2003). 30,000 years of hydrothermal activity at the Lost City Vent Field. *Science*, 302, 495–498.

Fryer, P., Wheat, C. G. and Mottl, M. J. (1999). Mariana blueschist mud volcanism: Implications for conditions within the subduction zone. *Geology*, 27, 103–106.

Fundis, A. T., Soule, S. A., Fornari, D. J. and Perfit, M. R. (2010). Paving the seafloor: Volcanic emplacement processes during the 2005–2006 eruptions at the fast-spreading East Pacific Rise, 9°50″ N. *Geochemistry, Geophysics, Geosystems*, 11(8), Q08024, doi:10.1029/2010GC003058.

Fustec, A., Desbruyères, D. and Laubier, L. (1988). Estimation de la biomase des peuplements associes aux sources hydrothermales profoundes de la dorsale du Pacifique oriental a 13°N. Biologie et ecologie des sources hydrothermales. *Oceanologica Acta*, 8, 15–22.

Gablina, I. F., Semkova, T. A., Stepanova, T. V. and Gor'kova, N. V. (2006). Diagenetic alteration of copper sulfides in modern ore-bearing sediments of the Logatchev-1 hydrothermal field (Mid-Atlantic Ridge 14°45′ N). *Lithology and Mineral Resources*, 41, 27–44.

Gale, A., Dalton, C. A., Langmuir, C. H., Su, Y. and Schilling, J.-G. (2013). The mean composition of ocean ridge basalts. *Geochemistry, Geophysics, Geosystems*, 14, 489–518, doi:10.1029/2012GC004334.

Gallant, R. M. and Von Damm, K. L. (2006). Geochemical controls on hydrothermal fluids from the Kairei and Edmond vent fields, 23°–25°S, Central Indian Ridge. *Geochemistry, Geophysics, Geosystems*, 7(6). Q06018; doi:10.1029/2005GC001067.

Gallo, D. G., Fox, P. J. and Macdonald, K. C. (1986). A Sea Beam investigation of the Clipperton transform fault: The morphotectonic expression of a fast slipping transform boundary. *Journal of Geophysical Research*, 91, 3455–3467.

Gamo, T., Chiba, H., Yamanaka, T. *et al.* (2001). Chemical characteristics of newly discovered black smoker fluids and associated hydrothermal plumes at the Rodriguez Triple Junction, Central Indian Ridge. *Earth and Planetary Science Letters*, 193, 371–379.

Gamo, T., Nakayama, E., Shitashima, K. *et al.* (1996). Hydrothermal plumes at the Rodriguez Triple Junction, Indian Ridge. *Earth and Planetary Science Letters*, 142, 261–270.

Gao, Y., Hoefs, J., Hellebrand, E., Von Der Handt, A. and Snow, J. E. (2007). Trace element zoning in pyroxenes from ODP Hole 735B gabbros: Diffusive exchange or synkinematic crystal fractionation? *Contributions to Mineralogy and Petrology*, 152, 429–442, doi:10.1007/s00410-006-0158-4.

Garces, M. and Gee, J. S. (2007). Paleomagnetic evidence of large footwall rotations associated with low-angle faults at the Mid-Atlantic Ridge. *Geology*, 35, 279–282.

Gardner, J. V., Armstrong, A. A., Calder, B. R. and Beaudoin, J. (2014). So, how deep is the Mariana Trench? *Marine Geodesy*, 42, doi:10.1080/01490419.2013.837849.

Garry, W. B., Gregg, T. K. P., Soule, S. A. and Fornari, D. J. (2006). Formation of submarine lava channel textures: Insights from laboratory simulations. *Journal of Geophysical Research*, 111, B03104, doi:10.1029/2005JB003796.

Gass, I. G. (1968). Is the Troodos massif of Cyprus a fragment of Mesozoic ocean floor? *Nature*, 220, 39–42.

Gast, P. W. (1968). Trace element fractionation and the origin of tholeiitic and alkaline magma types. *Geochimica et Cosmochimica Acta*, 32, (10), 1057–1086.

Gebruk, A. V., Chevaldonne, P., Shank, T., Lutz, R. A. and Vrijenhoek, R. C. (2000). Deep-sea hydrothermal vent communities of the Logatchev area (14°45′ N, Mid-Atlantic Ridge): Diverse biotopes and high biomass. *Journal of the Marine Biological Association*, 80, 383–393.

Gebruk, A. V., Pimenov, N. V. and Savvichev, A. S. (1993). Feeding specialization of bresiliid shrimps in the TAG site hydrothermal community. *Marine Ecology Progress Series*, 98, 247–253.

Gente, P., Dyment, J., Maia, M. and Goslin, M. (2003). Interaction between the Mid-Atlantic Ridge and the Azores hot spot during the last 85 Myr: Emplacement and rifting of the hot spot-derived plateaus. *Geochemistry, Geophysics, Geosystems*, 4(10), 8514 doi:10.1029/2003GC000527.

German, C. R., Bennett, S. A., Connelly, D. P. *et al.* (2008a). Hydrothermal activity on the southern Mid-Atlantic Ridge: Tectonically- and volcanically-controlled venting at 4–5°S. *Earth and Planetary Science Letters*, 273, 332–344.

German, C. R., Thurnherr, A. M., Knoery, J. *et al.* (2010). Heat, volume and chemical fluxes from submarine venting: A synthesis of results from the Rainbow hydrothermal field, 36°N MAR. *Deep-Sea Research Part I*, 57, 518–527.

German, C. R., Yoerger, D. R., Jakuba, M. *et al.* (2008b). Hydrothermal exploration with the Autonomous Benthic Explorer. *Deep-Sea Research Part 1*, 55, 203–219.

Gill, J. B. (1976). Composition and age of Lau Basin and Ridge volcanic rocks: Implications for evolution of an interarc basin and remnant arc. *Geological Society of America Bulletin*, 87(10), 1384–1395.

Gillis, K. M. (1995). Controls on hydrothermal alteration in fast-spreading oceanic crust. *Earth and Planetary Science Letters*, 134, 473–489.

Gillis, K. M. (2002). The root zone of an ancient hydrothermal system exposed in the Troodos ophiolite, Cyprus. *Journal of Geology*, 110, 57–74.

Gillis, K. M. (2008). The roof of an axial magma chamber: A hornfelsic heat exchanger. *Geology*, 36(4), 299–302, doi:10.1130/G24590A.1.

Gillis, K. M. and Coogan, L. A. (2002). Anatectic migmatites from the roof of an ocean ridge magma spreading centre. *Journal of Petrology*, 43, 2075–2095.

Gillis, K. M., and Roberts, M. D. (1999). Cracking at the magma-hydrothermal transition: Evidence from the Troodos Ophiolite, Cyprus. *Earth and Planetary Science Letters*, 169, 227–244.

Gillis, K. M. and Robinson, P. T. (1988). Distribution of alteration zones in the upper oceanic crust. *Geology*, 16, 262–266.

Gillis, K. M. and Robinson, P. T. (1990). Patterns and processes of alteration in the lavas and dikes of the Troodos Ophiolite, Cyprus. *Journal of Geophysical Research*, 95, 21,523–21,548.

Gillis, K. M., Coogan, L. A. and Pedersen, R. B. (2005). Strontium isotope constraints on fluid flow in the upper oceanic crust at the East Pacific Rise. *Earth and Planetary Science Letters*, 232, 82–94.

Gillis, K. M., Mével, C., Allan, J. *et al.* (1993a). *Proceedings of the Ocean Drilling Program, Initial Reports, Leg 147*. College Station, TX: Ocean Drilling Program.

Gillis, K. M., Muehlenbachs, K., Stewart, M., Gleeson, T. and Karson, J. A. (2001). Fluid flow patterns in fast-spreading East Pacific Rise crust exposed at Hess Deep. *Journal of Geophysical Research*, 106, 26,311–26,329.

Gillis, K. M., Snow, J. E., Klaus, A. *et al.* (2013). Primitive layered gabbros from fast-spreading lower oceanic crust. *Nature*, doi:10.1038/nature12778.

Gillis, K. M., Thompson, G. and Kelley, D. S. (1993b). A view of the lower crustal component of hydrothermal systems at the Mid-Atlantic Ridge. *Journal of Geophysical Research*, 98, 19,597–19,619.

Girardeau, J. and Nicolas, A. (1981). Structures in two of the Bay of Islands (Newfoundland) ophiolite massifs: A model for oceanic crust and upper mantle. *Tectonophysics*, 77, 1–34.

Girardeau, J., Cornen, G., Beslier, M.-O. *et al.* (1998). Extensional tectonics in the Gorringe Bank rocks, Eastern Atlantic Ocean: Evidence of an oceanic ultra-slow mantellic accreting centre. *Terra Nova*, 10, 330–336.

Girguis, P. R. and Holden, J. F. (2012). On the potential for bioenergy and biofuels from hydrothermal vent microbes. *Oceanography*, 25(1), 213–217, doi:org/10.5670/oceanog.2012.20.

Glennie, K. W., Boeuf, M. G. A., Hughes Clark, M. W. *et al.* (1973). Late Cretaceous nappes in Oman mountains and their geologic evolution. *American Association of Petroleum Geologists Bulletin*, 57, 5–27.

Glickson, D. A., Kelley, D. S. and Delaney, J. R. (2007). Geology and hydrothermal evolution of the Mothra Hydrothermal Field, Endeavour Segment, Juan de Fuca Ridge. *Geochemistry, Geophysics, Geosystems*, 8(6), Q06010, doi:10.1029/2007GC001588.

Godard, M., Awaji, S., Hansen H.-E. *et al.* (2009). Geochemistry of a long in-situ section of intrusive slow-spread lithosphere: Results from IODP Site U1309 (Atlantis Massif, 30°N Mid-Atlantic Ridge). *Earth and Planetary Science Letters*, 279, 110–122.

Godard, M., Dautria, J.-M. and Perrin, M. (2003). Geochemical variability of the Oman ophiolite lavas: Relationship with spatial distribution and paleomagnetic directions. *Geochemistry, Geophysics, Geosystems*, 4(6), doi:2002GC000452.

Godard, M., Jousselin, D. and Bodinier, J.-L. (2000). Relationships between geochemistry and structure beneath a paleo-spreading centre: A study of the mantle section in the Oman Ophiolite. *Earth and Planetary Science Letters*, 180, 133–148.

Goff, J. A., Fornari, D. J., Cochran, J. R., Keeley, C. and Malinverno, A. (1993). Wilkes transform system and "nannoplate". *Geology*, 21(7), 623–626.

Goffredi, S. K., Warén, A., Orphan, V. J., Van Dover, C. L. and Vrijenhoek, R. C. (2004). Novel forms of structural integration between microbes and a hydrothermal vent gastropod from the Indian Ocean. *Applied Environmental Microbiology*, 70(5), 3082–3090, doi:10.1128/AEM.70.5.3082-3090.2004.

Goss, A. R., Perfit, M. R., Ridley, W. I. *et al.* (2010). Geochemistry of lavas from the 2005–2006 eruption at the East Pacific Rise, 9°46′ N–9°56′ N: Implications for ridge crest plumbing and decadal changes in magma chamber compositions. *Geochemistry, Geophysics, Geosystems*, 11, Q05T09, doi:10.1029/2009GC00297.

Goud, M. R. and Karson, J. A. (1985). Tectonics of short-offset, slow-slipping transform zones in the FAMOUS area, Mid-Atlantic ridge. *Marine Geophysical Researches*, 7, 489–514.

Gràcia, E., Charlou, J. C., Radford-Knoery, J. and Parson, L. M. (2000). Non-transform offsets along the Mid-Atlantic Ridge south of the Azores (38°N–34°N): Ultramafic exposures and hosting of hydrothermal vents. *Earth and Planetary Science Letters*, 177, 89–103.

Grandin, R., Socquet, A., Jacques, E. *et al.* (2010). Sequence of rifting in Afar, Manda-Hararo Rift, Ethiopia, 2005–2009: Time-space evolution and interactions between dikes from interferometric synthetic aperture radar and static stress change modeling. *Journal of Geophysical Research*, 115(B10), doi:10.1029/2009JB000815.

Grassle, J. F. (1985). Hydrothermal vent animals: Distribution and biology. *Science*, 229, 713–717.

Grassle, J. F. and Petrecca, R. (1994). Macrofaunal response to artificial enrichments and depressions in a deep-sea habitat. *Journal of Marine Research*, 52, 345–369.

Grassle, J. F., Berg, C. J., Childress, J. J. *et al.* (1979). Galápagos '79: Initial findings of a deep-sea biological quest. *Oceanus*, 22(2), 2–10.

Greenbaum, D. (1972). Magmatic processes at ocean ridges and evidence from the Troodos massif, Cyprus. *Nature Physical Science*, 238, 18–21.

Greenberg, D. S. (1964). Mohole: The project that went awry. *Science*, 143, 115–119.

Gregg, T. K. P. and Chadwick, W. W. Jr. (1996). Submarine lava-flow inflation: A model for the formation of lava pillars. *Geology*, 24, 981–984.

Gregg, T. K. P. and Fink, J. H. (1995). Quantification of submarine lava-flow morphology through analog experiments. *Geology*, 23, 73–76.

Gregg, T. K. P. and Fornari, D. J. (1998). Long submarine lava flows: Observations and results from numerical modeling. *Journal of Geophysical Research*, 103, 27,517–27,531.

Gregg, T. K. P., Fornari, D. J., Perfit, M. R., Haymon, R. M. and Fink, J. H. (1996). Rapid emplacement of a mid-ocean ridge lava flow: The East Pacific Rise at 9°46′ – 51′ N. *Earth and Planetary Science Letters*, 144, E1–E7.

Gregg, T. K. P., Fornari, D. J., Perfit, M. R., Ridley, W. I. and Kurz, M. D. (2000). Using submarine lava pillars to record mid-ocean ridge eruption dynamics. *Earth and Planetary Science Letters*, 178, 195–214.

Griffiths, R. W. and Fink, J. H. (1992a). Solidification and morphology of submarine lavas: A dependence on extrusion rate. *Journal of Geophysical Research*, 97, 19,729–19,737.

Griffiths, R. W. and Fink, J. H. (1992b). The morphology of lava flows in planetary environments: Predictions from analog experiments. *Journal of Geophysical Research*, 97, 19,739–19,748.

Grimes, C. B., John, B. E., Cheadle, M. J. and Wooden, J. L. (2008). Protracted construction of gabbroic crust at a slow spreading ridge: Constraints from $^{206}Pb/^{238}U$ zircon ages from Atlantis Massif and IODP Hole U1309D (30°N, MAR). *Geochemistry, Geophysics, Geosystems*, 9(8), Q08012, doi:10.1029/2008GC002063.

Grove, T. L., Kinzler, R. J. and Bryan, W. B. (1992). Fractionation of mid-ocean ridge basalts. In Phipps Morgan, J., Blackman, D. K. and Sinton, J. (eds.), *Mantle Flow and Melt Generation at Mid-Ocean Ridges*. Washington, DC: American Geophysical Union, pp. 281–310, doi:org/10.1029/GM071p0281.

Gudmundsson, A. (1983). Form and dimensions of dykes in eastern Iceland. *Tectonophysics*, 95, 295–307.

Gudmundsson, A. (1990). Emplacement of dikes, sills and crustal magma chambers at divergent plate boundaries. *Tectonophysics*, 176, 257–275.

Haase, K. M., Koschinsky, A. and Petersen, S. (2009). Diking, young volcanism and diffuse hydrothermal activity on the southern Mid-Atlantic Ridge: the Lilliput field at 9°33′ S. *Marine Geology*, 266, 52–64.

Haase, K. M., Petersen, S., Koschinsky, A. *et al.* (2007). Young volcanism and related hydrothermal activity at 5°S on the slow-spreading southern Mid-Atlantic Ridge. *Geochemistry, Geophysics, Geosystems*, 8(11), Q11002, doi:10.1029/2006GC001509.

Hall, M. (1876). Note upon a portion of basalt from Mid-Atlantic. *Mineralogical Magazine*, 1(1), 1–3, doi:10.1180/minmag.1876.001.1.03.

Hammond, S. R. (1990). Relationships between lava types, seafloor morphology, and the occurrence of hydrothermal venting in the ASHES vent field of

Axial Volcano. *Journal of Geophysical Research*, 95, 12,875–12,893.

Han, S., Carbotte, S. M., Carton, H. *et al.* (2014). Architecture of on- and off-axis magma bodies at EPR 9°37′ – 40′ N and implications for oceanic crustal accretion. *Earth and Planetary Science Letters*, 390, 31–44.

Hanna, H. D. (2004). *Geochemical Variations in Basaltic Glasses from an Incipient Rift and Upper Level Gabbros from Hess Deep, Eastern Equatorial Pacific*. M.S. thesis, Duke University.

Hannington, M. D. and Scott, S. D. (1988). Mineralogy and geochemistry of a hydrothermal silica-sulfide-sulfate spire in the caldera of Axial Seamount, Juan de Fuca Ridge. *Canadian Mineralogist*, 26, 603–625.

Hannington, M. D., de Rhonde, C. E. J. and Petersen, S. (2005). Sea-floor tectonics and submarine hydrothermal systems. In Hedquist, J. W., Thompson, J. F. H., Goldfarb, R. J. and Richards, J. R. (eds.), *Economic Geology*, One Hundredth Anniversary Volume, 1905–2005. Littleton, CO: Society of Economic Geologists, pp. 111–142.

Hannington, M. D., Galley, A. G. Herzig, P. M. and Petersen, S. (1998). Comparison of the TAG mound and stockwork complex with Cyprus-type massive sulfide deposits. In Herzig, P. M., Humphris, S. E., Miller, D. J., and Zierenberg, R. A. (eds.), *Proceedings of the Ocean Drilling Program, Scientific Results Leg 158*. College Station, TX: Ocean Drilling Program, pp. 389–415.

Hannington, M. D., Jonasson, I. R., Herzig, P. M. and Petersen, S. (1995). Physical and chemical processes of seafloor mineralization at mid-ocean ridges. In Humphris, S., Zierenberg, R., Mullineau, L. and Thomson, R. (eds.), *Seafloor Hydrothermal Systems: Physical, Chemical, Biological, and Geological Interactions*. Geophysical Monograph 91, Washington, DC: American Geophysical Union, pp. 115–157.

Hannington, M., Herzig, P. Stoffers, P. *et al.* (2001). First observations of high-temperature submarine hydrothermal vents and massive anhydrite deposits off the north coast of Iceland. *Marine Geology*, 177, 199–220.

Harding, A. J., Kent, G. M. and Orcutt, J. A. (1993). A multichannel seismic investigation of upper crustal structure at 9°N on the East Pacific Rise: Implications for crustal accretion. *Journal of Geophysical Research*, 98, 13,925–13,944.

Harding, A. J., Orcutt, J., Kappus, M. *et al.* (1989). The structure of young oceanic crust at 13°N on the East Pacific Rise from expanding spread profiles. *Journal of Geophysical Research*, 94, 12,163–12,196.

Harper, G. D. (1982). Evidence for large-scale rotations at spreading centers from the Josephine Ophiolite. *Tectonophysics*, 82, 25–44.

Harper, G. D. (1988). Episodic magma chambers and amagmatic extension in the Josephine ophiolite. *Geology*, 16, 831–834.

Harris, A., Bailey, J., Calvari, S. and Dehn, J. (2005). Heat loss measured at a lava channel and its implications for down-channel cooling and rheology. In Manga, M. and Ventura, G. (eds.), *Kinematics and Dynamics of Lava Flows*. Boulder, CO: Geological Society of America, Special Paper 396, pp. 125–146.

Hart, S. R., Staudigel, H., Koppers, A. A. P. *et al.* (2000). Vailulu'u undersea volcano: The new Samoa. *Geochemistry, Geophysics, Geosystems*, 1(12), 1056, doi:10.1029/2000GC000108.

Hartley, M. E. and Thordarsson, T. (2013). The 1874–1876 volcano-tectonic episode at Askja, North Iceland: Lateral flow revisited. *Geochemistry, Geophysics, Geosystems*, 14, 2286–2309, doi:10.1002/ggge.20151.

Hashimoto, J., Ohta, S., Gamo, T. *et al.* (2001). Hydrothermal vents and associated biological communities in the Indian Ocean. *InterRidge News*, 10, 21–22.

Hawkins, J. W. (2003). Geology of supra-subduction zones: Implications for the origin of ophiolites. In Dilek, Y. and Newcomb, S. (eds.), *Ophiolite Concept and the Evolution of Geological Thought*. Boulder, CO: Geological Society of America, pp. 227–268.

Hayman, N. W. and Karson, J. A. (2007). Faults and damage zones in fast-spread crust exposed on the north wall of the Hess Deep Rift: Conduits and seals in seafloor hydrothermal systems. *Geochemistry, Geophysics, Geosystems*, 8(10), Q10002, doi:10.1029/2007GC001623.

Hayman, N. W., Grindlay, N. R., Perfit, M. R. *et al.* (2011). Oceanic core complex development at the ultraslow spreading Mid-Cayman Spreading Center. *Geochemistry, Geophysics, Geosystems*, 12(3), Q0AG02, doi:10.1029/2010GC003240.

Haymon, R. M. (1983). Growth history of black smoker hydrothermal chimneys. *Nature*, 301, 695–698.

Haymon, R. M., Fornari, D. J., Edwards, M. H. *et al.* (1991). Hydrothermal vent distribution along the East Pacific Rise crest (9°09′ – 54′ N) and its relationship to magmatic and tectonic processes on fast-spreading mid-ocean ridges. *Earth and Planetary Science Letters*, 104, 513–534

Haymon, R. M., Fornari, D. J., Von Damm, K. L. *et al.* (1993). Volcanic eruption of the mid-ocean ridge along the East Pacific Rise crest at 9°42–52′ N: Direct submersible observations of seafloor phenomena associated with an eruption event in April, 1991. *Earth and Planetary Science Letters*, 119, 85–101.

Haymon, R. M., Koski, R. A. and Abrams, M. J. (1989). Hydrothermal discharge zones beneath massive sulfide deposits mapped in the Oman ophiolite. *Geology*, 17, 531–535.

Haymon, R. M., Koski, R. A. and Sinclair, C. (1984). Fossils of hydrothermal vent worms from cretaceous sulfide ores of the Samail Ophiolite, Oman. *Science*, 223(4643), 1407–1409, doi: 10.1126/science.223.4643.1407.

Haymon, R. M., White, S. M., Baker, E. T. *et al.* (2008). High-resolution surveys along the hot spot-affected Galápagos Spreading Center: 3. Black smoker discoveries and the implications for geological controls on hydrothermal activity. *Geochemistry, Geophysics, Geosystems*, 9 (12), 1–30.

Head, J. W. III and Wilson, L. (2003). Deep submarine pyroclastic eruptions: Theory and predicted landforms and deposits. *Journal of Volcanology and Geothermal Research*, 121(3), 155–193.

Head, J. W., Hiesinger, H., Ivanov, M. A. *et al.* (1999). Possible ancient oceans on Mars: Evidence from Mars orbiter laser altimeter data. *Science*, 286, 2134–2137.

Hébert, L. B. and Montési, L. G. J. (2010). Generation of permeability barriers during melt extraction at mid-ocean ridges. *Geochemistry, Geophysics, Geosystems*, 11, Q12008, doi:10.1029/2010GC003270.

Hébert, R., Bideau, D. and Hékinian, R. (1983). Ultramafic and mafic rocks from the Garrett Transform Fault near 13°30′S on the East Pacific Rise: Igneous petrology. *Earth and Planetary Science Letters*, 65, 107–125.

Hedrick, D. B., Pledger, R. J., White, D. C. and Baross, J. A. (1992). In situ microbial ecology of hydrothermal vent sediments. *FEMS Microbial Ecology*, 101, 1–10.

Heezen, B. C. (1960). The rift in the ocean floor. *Scientific American*, 203, 98–110.

Heezen, B. C. and Hollister, C. D. (1971). *The Face of the Deep*. New York: Oxford University Press.

Heezen, B. C. and Tharp, M. (1977). *World Ocean Floor*. Palisades, NY: Office of Naval Research, US Navy.

Heezen, B. C., Tharp, M. and Ewing, M. (1959). *The Floors of the Oceans. 1. The North Atlantic*. Special Paper 65, Boulder, CO: Geological Society of America, pp. 1–126.

Heft, K., Gillis, K. M., Pollock, M. A., Karson, J. A. and Klein, E. M. (2008). Constraints on the nature of axial hydrothermal systems from the sheeted dike complex exposed at Pito Deep. *Geochemistry, Geophysics, Geosystems*, 9(5), doi:10.1029/2007GC001926.

Heirtzler, J. R. and LePichon, X. (1974). FAMOUS: A plate tectonics study of the genesis of the lithosphere. *Geology*, June, 273–274.

Hékinian, R., Auzende, J. M., Francheteau, J. *et al.* (1985). Offset spreading centers near 12°53′ N on the East Pacific Rise: Submersible observations and composition of the volcanics. *Marine Geophysical Researches*, 7, 359–377.

Hékinian, R., Bideau, D., Cannat, M., Francheteau, J. and Hébert, R. (1992). Volcanic activity and crust-mantle exposure in the ultrafast Garrett Transform Fault near 13°18′ S in the Pacific. *Earth and Planetary Science Letters*, 108, 259–275.

Hékinian, R., Bideau, D., Francheteau, J. *et al.* (1993). Petrology of the East Pacific Rise crust and upper mantle exposed in Hess Deep (Eastern Equatorial Pacific). *Journal of Geophysical Research*, 98, 8069–8094.

Hekinian, R. M., Fevrier, F., Avedik, P. *et al.* (1983a). East Pacific Rise near 13°N: Geology of new hydrothermal fields. *Science*, 219, 1321–1324.

Hékinian, R., Francheteau, J., Renard, V. *et al.* (1983b). Intense hydrothermal activity at the axis of the East Pacific Rise near 13°N: Submersible witnesses the growth of sulfide chimney. *Marine Geophysical Research*, 6(1), 1–14, doi:10.1007/BF00300395.

Helgason, J. and Zentilli, M. (1985). Field characteristics of laterally emplaced dikes: Anatomy of exhumed Miocene dike swarm in Reydarfjordur, eastern Iceland. *Tectonophysics*, 115, 247–274.

Hellebrand, E., Snow, J. E., Mostefaoui, S. and Hoppe, P. (2005). Trace element distribution between orthopyroxene and clinopyroxene in peridotites from the Gakkel Ridge: A SIMS and NanoSIMS study. *Contributions to Mineralogy and Petrology*, 150(5), 486–504.

Hellebrand, E., Snow, J. E. and Mühe, R. (2002). Mantle melting beneath Gakkel Ridge (Arctic Ocean): Abyssal peridotite spinel compositions. *Chemical Geology*, 182, (2), 227–235.

Helo, C., Longpre, M.-A., Shimizu, N., Clague, D. A. and Stix, J. (2011). Explosive eruptions at mid-ocean ridges driven by CO_2-rich magmas. *Nature Geoscience*, 4, 260–263, doi:10.1038/NGEO1104.

Henstock, T. J., Woods, A. W. and White, R. S. (1993). The accretion of oceanic crust by episodic sill intrusion. *Journal of Geophysical Research*, 98, 4143–4161.

Hersey, J. B. (ed.) (1967). *Deep-Sea Photography*. Baltimore: The Johns Hopkins University Press.

Herzberg, C. (2004). Partial crystallization of mid-ocean ridge basalts in the crust and mantle. *Journal of Petrology*, 45(12), 2389–2405.

Herzberg, C., Asimow, P. D., Arndt, N. *et al.* (2007). Temperatures in ambient mantle and plumes: Constraints from basalts, picrites, and komatiites. *Geochemistry, Geophysics, Geosystems*, 8, Q02006, doi:10.1029/2006GC001390.

Herzig, P. M. and Plüger, W. L. (1988). Exploration for hydrothermal activity near the Rodriguez Triple Junction, Indian Ocean. *Canadian Mineralogist*, 26, 721–736.

Hess, H. H. (1954). Geological hypotheses and the Earth's crust under the Oceans. *Proceedings of the Royal Society, A*, 222, 341–348.

Hess, H. H. (1955). Serpentinites, orogeny and epeirogeny. In *Crust of the Earth: A Symposium*. Boulder, CO: Geological Society of America, Special Paper 62, pp. 391–408.

Hess, H. H. (1962). History of ocean basins. In Engel, A. E. J., Hall, H. L. and Leonard, B. F. (eds.), *Petrologic Studies: A Volume in Honor of A. F. Buddington*. Boulder, CO: Geological Society of America, pp. 599–620.

Hess, H. H. (1965). Mid-oceanic ridges and tectonics of the sea floor. In Whittard, W. F. and Bradshaw, R. (eds.), *Submarine Geology and Geophysics, Proceedings of the*

Seventeenth Symposium of the Colston Research Society. London: Butterworths, pp. 317–334.

Hessler, R. R., Smithey, W. M., Boudrias, M. A. *et al.* (1988). Temporal changes in megafauna at the Rose Garden hydrothermal vent, Galápagos Rift, eastern tropical Pacific. *Deep-Sea Research*, 35, 1681–1709.

Hessler, R. R., Smithey, W. M. and Keller, M. A. (1985). Spatial and temporal variation of giant clams, tube worms, and mussels at deep-sea hydrothermal vents. *Bulletin of the Biological Society of Washington*, 6, 411–428.

Hey, R. N. (1977). A new class of pseudofaults and their bearing on plate tectonics: A propagating rift model. *Earth and Planetary Science Letters*, 37, 321–325.

Hey, R. N. (2004). Propagating rifts and microplates at mid-ocean ridges. In Selley, R. C. (ed.), *Encylopedia of Geology*. London: Academic Press, pp. 396–405.

Hey, R. N., Baker, E., Bohnenstiehl, D. *et al.* (2004). Tectonic/volcanic segmentation and controls on hydrothermal venting along Earth's fastest seafloor spreading system, EPR 27°–32°S. *Geochemistry, Geophysics, Geosystems*, 5,(12, Q12007, doi:10.1029/2004GC000764.

Hey, R. N., Kleinrock, M. C., Miller, S. P., Atwater, T. M. and Searle, R. C. (1986). Sea Beam/Deep-Tow Investigation of an active oceanic propagating rift system, Galápagos 95.5°W. *Journal of Geophysical Research*, 91, 3369–3393.

Hey, R. N., Naar, D. F., Kleinrock, M. C. *et al.* (1985). Microplate tectonics along a superfast seafloor spreading system near Easter Island. *Nature*, 317, 320–325.

Hey, R. N., Sinton, J. M. and Duennebier, F. K. (1989). Propagating rifts and spreading centers. In Winterer, E. L., Hussong, D. M. and Decker, R. W. (eds.), *The Geology of North America: The Eastern Pacific Ocean and Hawaii*. Boulder, CO: Geological Society of America, pp. 161–176.

Hoagland, P., Beaulieu, S., Tivey, M. A. *et al.* (2010). Deep-sea mining of seafloor massive sulfides. *Marine Policy*, 34, 728–732.

Hodges, F. N. and Papike, J. J. (1976). DSDP site 334: Magmatic cumulates from oceanic layer 3. *Journal of Geophysical Research*, 81, 4135–4151.

Holden, J. F., Summit, M., Baross, J. A. (1998). Thermophilic and hyperthermophilic microorganisms in 3–30 °C fluids following a deep-sea volcanic eruption. *FEMS Microbiology Ecology*, 25, 33–41.

Holdsworth, R. E., Van Diggelen, E. W. E., Spiers, C. J. *et al.* (2011). Fault rocks from the SAFOD core samples: Implications for weakening at shallow depths along the San Andreas Fault, California. *Journal of Structural Geology*, 33(2), 132–144, doi:10.1016/j.jsg.2010.11.010 .

Holm, N. G. and Baltscheffsky, H. (2011). Links between hydrothermal environments, pyrophosphate, Na^+ and early evolution. *Origins of Life and Evolution of the Biosphere*, 41, 483–493.

Holm, N. G. and Charlou, J. L. (2001). Initial indications of abiotic formation of hydrocarbons in the Rainbow ultramafic hydrothermal system, Mid-Atlantic Ridge. *Earth and Planetary Science Letters*, 191, 1–8.

Hon, K., Kauahikaua, J. P., Denlinger, R. and Mackay, K. (1994). Emplacement and inflation of pahoehoe sheet flows: Observations and measurements of active lava flows on Kilauea Volcano, Hawaii. *Geological Society of America Bulletin*, 106(3), 351–370.

Hooft, E. E. and Detrick, R. S. (1993). The role of density in the accumulation of basaltic melts at mid-ocean ridges. *Geophysical Research Letters*, 20, 423–426.

Hooft, E. E., Detrick, R. S. and Kent, G. M. (1997). Seismic structure and indicators of magma budget along the southern East Pacific Rise. *Journal of Geophysical Research*, 102, 27,319–27,340.

Hooft, E. E., Detrick, R. S., Toomey, D. R., Collins, J. A. and Lin, J. (2000). Crustal thickness and structure along three contrasting spreading segments of the Mid-Atlantic Ridge, 33.5°–35°N. *Journal of Geophysical Research*, 105, 8205–8226.

Hooft, E. E., Patel, H., Wilcock, W., Becker, K. *et al.* (2010). A seismic swarm and regional hydrothermal and hydrologic perturbations: The northern Endeavour segment, February 2005. *Geochemistry, Geophysics, Geosystems*, 11, Q12015, doi:10.1029/2010GC003264.

Hooft, E. E., Schouten, H. and Detrick, R. S. (1996). Constraining crustal emplacement processes from the variation in seismic layer 2A thickness at the East Pacific Rise. *Earth and Planetary Science Letters*, 142, 289–310.

Horita, J. and Berndt, M. E. (1999). Abiogenic ethane formation and isotopic fractionation under hydrothermal conditions. *Science*, 285, 1055–1057.

Horst, A. J., Varga, R. G., Gee, J. S. and Karson, J. A. (2006). Magnetic remanence and anisotropy of magnetic susceptibility of dikes from super-fast spread crust exposed at Pito Deep Rift. *Transactions of the American Geophysical Union*, 87, T51C–1546.

Horst, A. J., Varga, R. J., Gee, J. S. and Karson, J. A. (2011). Paleomagnetic constraints on constructional deformation of superfast-spread oceanic crust exposed at Pito Deep Rift. *Journal of Geophysical Research*, 116(12), doi:10.1029/2011JB008268.

Horst, A. J., Varga, R. J., Gee, J. S. and Karson, J. A. (2014). Diverse magma flow directions during construction of sheeted dike complexes at fast- to superfast-spreading centers. *Earth and Planetary Science Letters*, 408, 119–131.

Hosford, A., Tivey, M. A., Matsumoto, T. *et al.* (2003). Crustal magnetization and accretion at the Southwest Indian Ridge near the Atlantis II fracture zone, 0–25 Ma. *Journal of Geophysical Research*, 108, doi:1029/2001JB 000604.

Houtz, R. E. (1976). Seismic properties of layer 2A in the Pacific. *Journal of Geophysical Research*, 81, 6321–6340.

Huber, J. A., Butterfield, D. A. and Baross, J. A. (2003). Bacterial diversity in a subseafloor habitat following a deep-sea volcanic eruption. *FEMS Microbiology Ecology*, 42, 393–409.

Hull, E. (1892). *Volcanoes: Past and Present*. New York: Charles Scribner's Sons and London: Walter Scott.

Humphris, S. E. and Cann, J. R. (2000). Constraints on the energy and chemical balances of the modern TAG and ancient Cyprus seafloor sulfide deposits. *Journal of Geophysical Research*, 105, 28,477–28,488.

Humphris, S. E. and Kleinrock, M. C. (1996). Detailed morphology of the TAG active hydrothermal mound: Insights into its formation and growth. *Geophysical Research Letters*, 23, 3443–3446.

Humphris, S. E. and Tivey, M. K. (2000). A synthesis of geological and geochemical investigations of the TAG hydrothermal field: Insights into fluid-flow and mixing processes in a hydrothermal system. In Dilek, Y., Moores, E., Elthon, D. and Nicolas, A. (eds.), *Ophiolites and Oceanic Crust: New Insights from Field Studies and the Ocean Drilling Program*. Special Paper, 349, Boulder, CO: Geological Society of America, pp. 213–235.

Humphris, S. E., DeMenocal, P. B., Edwards, K. J., Fisher, A. T. and Saffer, D. (2011). The need for scientific ocean drilling. *Eos, Transactions of the American Geophysical Union*, 92(10), 83–84.

Humphris, S. E., Fornari, D. J., Scheirer, D. S., German, C. R. and Parson, L. (2002). Geotectonic setting of hydrothermal activity on the summit of Lucky Strike Seamount (37°17′ N, Mid-Atlantic Ridge). *Geochemistry, Geophysics, Geosystems*, 3(8), doi: 10.1029/2001GC000284.

Humphris, S. E., Herzig, P. M. *et al.* (1996). *Proceedings of the Ocean Drilling Program, Initial Reports Leg, 158*. College Station, TX: Ocean Drilling Program.

Hurst, S. D., Moores, E. M. and Varga, R. J. (1994). Structure and geophysical expression of the Solea Graben, Troodos Ophiolite Cyprus. *Tectonics*, 13, 139–156.

Hussenoeder, S. A., Detrick, R. S., Kent, G. M., Schouten, H. and Harding, A. J. (2002a). Fine-scale seismic structure of young upper crust at 17°20′ S on the fast spreading East Pacific Rise. *Journal of Geophysical Research*, 107(B8), doi:10.1029/2001JB001688.

Hussenoeder, S. A., Kent, G. M. and Detrick, R. S. (2002b). Upper crustal seismic structure of the slow spreading Mid-Atlantic Ridge, 35 degrees N: Constraints on volcanic emplacement processes. *Journal of Geophysical Research*, 107(B8), 2156, doi:10.1029/2001JB001691.

Hynes, A. (1975). Comment on "The Troodos ophiolite was probably formed in an island arc" by A. Miyashiro. *Earth and Planetary Science Letters*, 25, 213–216.

Ildefonse, B., Blackman, D. K., John, B. E. *et al.* (2007). Oceanic core complexes and crustal accretion at slow-spreading ridges. *Geology*, 35(7), 623–626, doi:10.1130/G23531A.1.

Irvine, T. N. (1982). Terminology for layered intrusions. *Journal of Petrology*, 23, 127–162.

Irvine, T. N., Anderson, J. C. Ø. and Brooks, C. K. (1998). Included blocks (and blocks with blocks) in the Skaergaard intrusion: Geologic relations and the origins of rhythmic modally graded layers. *Geological Society of America Bulletin*, 110(11), 1398–1447.

Isaaks, B., Oliver, J. and Sykes, L. R. (1968). Seismology and the new global tectonics. *Journal of Geophysical Research*, 73, 5855–5899.

Jambon, A. (1994). Earth degassing and large-scale geochemical cycling of volatile elements. In Caroll, M. R. and Holloway, J. R. (eds.), *Volatiles and Magmas.* Reviews in Mineralogy, 30, Washington DC: Mineralogy Society of America, pp. 479–517.

Jamieson, J. W., Hannington, M. D., Clague, D. *et al.* (2013). Sulfide geochronology along the Endeavour Segment of the Juan de Fuca Ridge. *Geochemistry, Geophysics, Geosystems*, 14, 2084–2099, doi:10.1002/ggge.20133.

Jannasch, H. W. (1995). Microbial interaction with hydrothermal fluids. In Humphris, S. E., Zierenberg, R. A., Mullineaux, L. S. and Thomson, R. E. (eds.), *Seafloor Hydrothermal Systems: Physical, Chemical, Biological and Geological Interactions*. Geophysical Monograph 91, Washington, DC: American Geophysical Union Press, pp. 273–296.

Jaroslow, G. E., Hirth, G. and Dick, H. J. B. (1996). Abyssal peridotite mylonites: Implications for grain-size sensitive flow and strain localization in the oceanic lithosphere. *Tectonophysics*, 256, 17–37.

Jean-Baptiste, P., Mantisi, F., Pauwells, H., Grimaud, D. and Patriat, P. (1992). Hydrothermal ^{3}He and manganese plumes at 19°29′ S on the Central Indian Ridge. *Geophysical Research Letters*, 19, 1787–1790.

Jenner, F. E. and O'Neill, H. S. C. (2012). Analysis of 60 elements in 616 ocean floor basaltic glasses. *Geochemistry, Geophysics, Geosystems*, 13, Q02005, doi:10.1029/2011GC004009.

Johnson, H. P., Hutnak, M., Dziak, R. P. *et al.* (2000). Earthquake-induced in a hydrothermal system on the Juan de Fuca Ridge. *Nature*, 407, 174–177.

Johnson, H. P., Tivey, M. A., Bjorklund, T. A. and Salmi, M. S. (2010). Hydrothermal circulation within the Endeavour Segment, Juan de Fuca Ridge. *Geochemistry, Geophysics, Geosystems*, 11(5), Q05002, doi.org/10.1029/2009GC002957.

Johnson, K. S., Childress, J. J. and Beehler, C. L. (1988a). Short-term temperature variability in the Rose Garden hydrothermal vent field: An unstable deep-sea environment. *Deep-Sea Research Part 1*, 35, 1711–1721.

Johnson, K. S., Childress, J. J., Hessler, R. R. *et al.* (1988b). Chemical and biological interactions in the Rose Garden [eastern Pacific Ocean] hydrothermal vent field, Galápagos spreading center. *Deep-Sea Research Part 1.* 35, 1723–1744.

Johnson, K. T. M. (1998). Experimental determination of partition coefficients for rare earth and high-field-strength elements between clinopyroxene, garnet, and basaltic melt at high pressures. *Contributions to Mineralogy and Petrology*, 133, 60–68.

Johnson, K. T. M. and Dick, H. J. B. (1992). Open system melting and temporal and spatial variation of peridotite and basalt at the Atlantis II fracture zone. *Journal of Geophysical Research*, 97, 9219–9241.

Johnson, K. T. M., Dick, H. J. B. and Shimizu, N. (1990). Melting in the oceanic upper mantle: An ion microprobe study of diopsides in abyssal peridotites. *Journal of Geophysical Research*, 95, 2661–2678.

Jolly, R. J. H. and Sanderson, D. J. (1995). Variations in the form and distribution of dykes in the Mull swarm, Scotland. *Journal of Structural Geology*, 17, 1543–1557.

Jousselin, D. and Nicolas, A. (2000a). Structural mapping of an off-axis mantle diapir in the Mansah region (Sumail massif in the Oman ophiolite). *Marine Geophysical Researches*, 21, 243–257.

Jousselin, D. and Nicolas, A. (2000b). The Moho transition zone in the Oman ophiolite: Relation with wehrlites in the crust and dunites in the mantle. *Marine Geophysical Researches*, 21, 229–241.

Jousselin, D., Dunn, R. and Toomey, D. R. (2003). Modeling the seismic signature of structural data from the Oman Ophiolite: Can a mantle diapir be detected beneath the East Pacific Rise? *Geochemistry, Geophysics, Geosystems*, 4(7), 8610, doi:10.1029/2002GC000418.

Jousselin, D., Morales, L. F. G., Nicolle, M. and Stephant, A. (2012). Gabbro layering induced by simple shear in the Oman Ophiolite Moho transition zone. *Earth and Planetary Science Letters*, 331–332, 55–66.

Jousselin, D., Nicolas, A. and Boudier, F. (1998). Detailed mapping of a mantle diapir below a paleo-spreading center in the Oman ophiolite. *Journal of Geophysical Research*, 103, 18,153–18,170.

Juniper, S. K. and Sarrazin J. (1995). Interaction of vent biota and hydrothermal deposits: Present evidence and future experimentation. In Humphris, S., Zierenberg, R., Mullineau, L. and Thomson R., *Seafloor Hydrothermal Systems: Physical, Chemical, Biological, and Geological Interactions*. Geophysical Monograph 91, Washington, DC: American Geophysical Union, pp. 178–193.

Juniper, S. K. and Tunnicliffe, V. (1997). Crustal accretion and the hot vent ecosystem. *Philosophical Transactions of the Royal Society A: Mathematical, Physical and Engineering Sciences*, 355, 459–474.

Juniper, S. K., Martineu, P., Sarrazin, J. and Gelinas, Y. (1995). Microbial-mineral floc associated with nascent hydrothermal activity on CoAxial Segment, Juan de Fuca Ridge. *Geophysical Research Letters*, 22, 179–182.

Juteau, T., Bideau, O., Dauteuil, G. *et al.* (1995). A submersible study in the western Blanco Fracture Zone, N.E. Pacific: Lithostratigraphy, magnetic structure and magmatic and tectonic evolution during the last 1.6 Ma. *Marine Geophysical Researches*, 17, 399–430.

Juteau, T., Cannat, M. and Lagabrielle, Y. (1990). Serpentinized peridotites in the upper oceanic crust away from transform zones: A comparison of the results of previous DSDP and ODP legs. In Detrick, R. S., Honnorez, J., Bryan, W. B. and Juteau, T. (eds.), *Proceedings of the Ocean Drilling Program, Scientific Results, Leg 81*. College Station, TX: Ocean Drilling Program, pp. 303–308.

Kadar, E., Costa, V. and Santos, R. S. (2006). Distribution of micro-essential (Fe, Cu, Zn) and toxic (Hg) metals in tissues of two nutritionally distinct hydrothermal shrimps. *Science of the Total Environment*, 358, 143–150.

Kappel, E. and Ryan, W. B. F. (1986). Volcanic episodicity and a non-steady state rift valley along the Northeast Pacific spreading centers: Evidence from Sea MARC I. *Journal of Geophysical Research*, 91, 13,925–13,940.

Kappel, E. S. and Normark, W. R. (1987). Morphometric variability within the axial zone of the southern Juan de Fuca Ridge: Interpretation from Sea MARC II and Sea MARC I, and deep-sea photography. *Journal of Geophysical Research*, 92, 11,291–11,302.

Karl, D. M. (1995). *The Microbiology of Deep-Sea Hydrothermal Vents*. Boca Raton, FL; New York: CRC Press, pp. 1–299.

Karson, J. A. (1977). *The Geology of the Northern Lewis Hills, Western Newfoundland*. Ph.D. dissertation, State University of New York at Albany.

Karson, J. A. (1987). Factors controlling the orientations of dykes in ophiolites and oceanic crust. In Halls, H. C. and Fahrig, W. F. (eds.), *Mafic Dyke Swarms*. Special Paper 34 Ottawa: Geological Association of Canada, pp. 229–241.

Karson, J. A. (1990). Seafloor spreading on the Mid-Atlantic Ridge: Implications for the structure of ophiolites and oceanic lithosphere produced in slow-spreading environments. In Malpas, J., Moores, E. M., Panayioutou, A. and Xenophontos, C. (eds.), *Ophiolites and Oceanic Crustal Analogues: Proceedings of the Symposium "Troodos 1987"*. Nicosia, Cyprus: Geological Survey Department, pp. 125–130.

Karson, J. A. (1998). Internal structure of oceanic lithosphere: A perspective from tectonic windows. In Buck, W. R., Delaney, P. T., Karson, J. A. and Lagabrielle, Y. (eds.), *Faulting and Magmatism at Mid-Ocean Ridges*. Geophysical Monograph 106, Washington, DC: American Geophysical Union, pp. 177–218.

Karson, J. A. (1999). Geological investigation of a lineated massif at the Kane transform: Implications for oceanic core complexes. *Philosophical Transactions of the Royal Society, A*, 357, 713–740.

Karson, J. A. (2002). Geologic structure of uppermost oceanic crust created at fast- to intermediate-rate spreading centers. *Annual Review of Earth and Planetary Sciences*, 30, 347–384.

Karson, J. A. and Dewey, J. F. (1978). Coastal Complex, western Newfoundland: An Early Ordovician oceanic fracture zone. *Geological Society of America Bulletin*, 89, 1037–1049.

Karson, J. A. and Dick, H. J. B. (1983). Tectonics of ridge-transform intersections at the Kane Fracture Zone, 24°N on the Mid-Atlantic Ridge. *Marine Geophysical Researches*, 6, 51–98.

Karson, J. A. and Lawrence, R. M. (1997). Tectonic setting of serpentinite exposures on the western median valley wall of the MARK Area in the vicinity of Site 920. In Karson, J. A., Cannat, M., Miller, D. J. and Elthon, D. (eds.), *Proceedings of the Ocean Drilling Program, Scientific Results, Leg 153*. College Station, TX: Ocean Drilling Program, pp. 5–22.

Karson, J. A. and Rona, P. A. (1990). Block-tilting, transfer faults and structural control of magmatic and hydrothermal processes in the TAG Area, Mid-Atlantic Ridge 26°N. *Geological Society of America Bulletin*, 102, 1635–1645.

Karson, J. A., Collins, J. A. and Casey, J. F. (1984). Geologic and seismic velocity structure of the crust/mantle transition in the Bay of Islands Ophiolite Complex. *Journal of Geophysical Research*, 89, 6129–6138.

Karson, J. A., Elthon, D. L. and DeLong, S. E. (1983). Ultramafic intrusions in the Lewis Hills Massif, Bay of Islands ophiolite complex, Newfoundland: Implications for igneous processes at oceanic fracture zones. *Geological Society of America Bulletin*, 94, 15–29.

Karson, J. A., Francheteau, J., Gee, J. S. *et al.* (2005). Nested-scale investigation of tectonic windows into super-fast spread crust exposed at the Pito Deep Rift, Easter Microplate, SE Pacific. *InterRidge News*, 14, 5–8.

Karson, J. A., Früh-Green, G. L., Kelley, D. S. *et al.* (2006a). Detachment shear zone of the Atlantis Massif core complex, Mid-Atlantic Ridge, 30° N. *Geochemistry, Geophysics, Geosystems*, 7(6), Q06016, doi:10.1029/2005GC001109.

Karson, J. A., Hurst, S. D. and Lonsdale, P. (1992). Tectonic rotations of dikes in fast-spread oceanic crust exposed near Hess Deep. *Geology*, 20, 685–688.

Karson, J. A., Klein, E. M., Hurst, S. D. *et al.* (2002a). Structure of uppermost fast-spread oceanic crust exposed at the Hess Deep Rift: Implications for subaxial processes at the East Pacific Rise. *Geochemistry, Geophysics, Geosystems*, 3, 1002, doi: 10.1029/2001GC000155.

Karson, J. A., Thompson, G., Humphris, S. E. *et al.* (1987). Along-axis variations in seafloor spreading in the MARK area. *Nature*, 328, 681–685.

Karson, J. A., Tivey, M. A. and Delaney, J. R. (2002b). Internal structure of uppermost oceanic crust along the western Blanco Transform Scarp: Implications for subaxial accretion and deformation at the Juan de Fuca Ridge. *Journal of Geophysical Research*, 107(B9), 2181, doi:10.1029/2000JB000051.

Karson, J. A., Williams, E. A., Früh-Green, G. L. *et al.* (2006b). Detachment shear zone on the Atlantis Massif Core Complex, Mid-Atlantic Ridge 30°N. *Geochemistry, Geophysics, Geosystems*, 7, (6), doi:10.1029/2005GC001109.

Kashefi, K. and Lovley, D. R. (2003). Extending the upper temperature limit for life. *Science*, 301, 934.

Katz, R. F. and Weatherley, S. M. (2012). Consequences of mantle heterogeneity for melt extraction at mid-ocean ridges. *Earth and Planetary Science Letters*, 335–336, 226–237.

Kay, M. (1951). *North American Geosynclines*. Memoir 48, Boulder, CO: Geological Society of America.

Kelemen, P. B., Braun, M. and Hirth, G. (2000). Spatial distribution of melt conduits in the mantle beneath oceanic spreading ridges: Observations from the Ingalls and Oman ophiolites. *Geochemistry, Geophysics, Geosystems*, 1(1), 1005, doi:10.1029/1999GC000012.

Kelemen, P. B., Hirth, G., Shimizu, N., Spiegelman, M. and Dick, H. J. B. (1997a). A review of melt migration processes in the adiabatically upwelling mantle beneath oceanic spreading ridges. *Philosophical Transactions of the Royal Society of London A*, 355, 283–318.

Kelemen, P. B., Kikawa, E., Miller, D. J. and Shipboard Scientific Party (2007). Leg 209 summary: Processes in a 20-km-thick conductive boundary layer beneath the Mid-Atlantic Ridge, 14°–16°N. In Kelemen, P. B., Kikawa, E. and Miller, D. J. (eds.), *Proceedings of the Ocean Drilling Program, Leg 209*. College Station, TX: Ocean Drilling Program, pp. 1–33, doi:10.2973/odp.proc.sr.209.001.2007.

Kelemen, P. B., Koga, K. and Shimizu, N. (1997b). Geochemistry of gabbro sills in the crust/mantle transition zone of the Oman ophiolite: Implications for the origin of the oceanic lower crust. *Earth and Planetary Science Letters*, 146, 475–488.

Kelemen, P. B., Quick, J. E. and Dick, H. J. B. (1992). Formation of harzburgite by pervasive melt/rock reaction in the upper mantle. *Nature*, 358, 635–641.

Kelley, D. S. (1996). Methane-rich fluids in the oceanic crust. *Journal of Geophysical Research*, 101, 2943–2962.

Kelley D. S. and Delaney, J. R. (1987). Two-phase separation and fracturing in mid-ocean ridge gabbros at temperatures greater than 700 °C. *Earth and Planetary Science Letters*, 83, 53–66.

Kelley, D. S. and Früh-Green, G. (1999). Abiogenic methane in deep-seated mid-ocean ridge environments: Insights from stable isotope analyses. *Journal of Geophysical Research*, 104, 10,439–10,460.

Kelley, D. S. and Früh-Green, G. (2001). Volatile lines of descent in submarine plutonic environments: Insights from stable isotope and fluid inclusion analyses. *Geochimica et Cosmochimica Acta*, 65, 3325–3346.

Kelley, D. S. and Shank, T. M. (2010). Hydrothermal systems: A decade of discovery in slow-spreading centers. In Rona, P. A., Devey, C. W., Dyment, J. and Murton B. J. (eds.), *Diversity of Hydrothermal Systems on Slow-Spreading Ocean Ridges*. Geophysical Monograph 188, Washington, DC: American Geophysical Union, pp. 369–407, doi: 10.1029/2010GM000945.

Kelley, D. S., Baross, J. A. and Delaney, J. R. (2002). Volcanoes, fluids, and life in submarine environments. *Annual Review of Earth and Planetary Sciences*, 30, 385–491.

Kelley, D. S., Carbotte, S. M, Clague, D. A. *et al.* (2012). Endeavour Segment of the Juan de Fuca Ridge: One of the most remarkable places on Earth. *Oceanography*, 25 (1), 44–61, doi.org/10.5670/oceanog.2012.03.

Kelley, D. S., Delaney, J. R. and Juniper, S. K. (2014). Establishing a new era of submarine volcanic observatories: Cabling Axial Seamount and the Endeavour Segment of the Juan de Fuca Ridge. *Marine Geology*, 352, 426–450.

Kelley, D. S., Delaney, J. R. and Yoerger, D. A. (2001a). Geology and venting characteristics of the Mothra Hydrothermal Field, Endeavour Segment, Juan de Fuca Ridge. *Geology*, 29, 959–962.

Kelley, D. S., Früh-Green, G. L. and Lilley, M. D. (2004). Volatiles in mid-ocean ridge environments: Food for life. In Wilcock, W. D., Kelley, D. S., Baross, J. A., DeLong, E. and Cary, C. (eds.), *The Subseafloor Biosphere at Mid-Ocean Ridges*. Geophysical Monograph 144, Washington, DC: American Geophysical Union, pp. 167–189.

Kelley, D. S., Früh-Green, G. L., Karson, J. A. and Ludwig, K. A. (2007). Lost City hydrothermal field revisited. *Oceanography*, 20(4), 90–99.

Kelley, D. S., Gillis, K. M. and Thompson G. (1993). Fluid evolution in submarine magma-hydrothermal systems at the Mid-Atlantic Ridge. *Journal of Geophysical Research*, 98, 19,579–19,596.

Kelley, D. S., Karson, J. A., Blackman, D. K. *et al.* (2001b). An off-axis hydrothermal field discovered near the Mid-Atlantic Ridge at 30°N. *Nature*, 412, 145–149.

Kelley, D. S., Karson, J. A., Blackman, D. K. *et al.* (2001c). Discovery of Lost City: An off-axis, peridotite-hosted, hydrothermal field near 30°N on the Mid-Atlantic Ridge. *RIDGE Events Newsletter*, 11, 3–9.

Kelley, D. S., Karson, J. A., Früh-Green, G. L. *et al.* (2005). A serpentinite-hosted ecosystem: The Lost City Hydrothermal Field. *Science*, 307, 1428–1434.

Kelley, D. S., Lilley, M. D., Lupton, J. E. and Olson, E. J. (1998). Enriched H_2, CH_4 and ^3He concentrations in hydrothermal plumes associated with the 1996 Gorda Ridge Eruptive Event. *Deep Sea Research II, Event Detection and Response: Gorda Ridge*, 45, 2665–2682.

Kelley, D. S., Robinson, P. T. and Malpas, J. (1992). Processes of brine generation and circulation in the oceanic crust: Fluid inclusion evidence from Troodos ophiolite, Cyprus. *Journal of Geophysical Research*, 97, 9307–9322.

Kelley, D. S., Vanko, D. A. and Gu, C. (1995). Fluid evolution in oceanic crustal layer 2: Fluid inclusion evidence from the sheeted dike complex, ODP Hole 504B, Costa Rica Rift. In Stokking, L., Dick, H. J. B. and Erzinger, J. (eds.), *Proceedings of the Ocean Drilling Program, Scientific Results, Leg 148*. College Station, TX: Ocean Drilling Program, pp. 191–197.

Kellogg, J. P. (2011). *Temporal and Spatial Variability of Hydrothermal Fluxes Within a Mid-Ocean Ridge Segment*. Ph.D. thesis, University of Washington.

Kellogg, J. P. and McDuff, R. E. (2010). A hydrographic transient above the Salty Dawg hydrothermal field, Endeavour segment, Juan de Fuca Ridge. *Geochemistry, Geophysics, Geosystems*, 11, doi:10.1029/2010GC003299.

Kempton, P. D. and Casey, J. F. (1997). Petrology and geochemistry of crosscutting diabase dikes, Sites 920 and 921. In Karson, J. A., Cannat, M., Miller, D. J. and Elthon, D. (eds.), *Proceedings of the Ocean Drilling Program, Scientific Reports, Leg 153*. College Station, TX: Ocean Drilling Program, pp. 363–377.

Kendall, J.-M. (1994). Teleseismic arrivals at a mid-ocean ridge: Effects of mantle melt and anisotropy. *Geophysical Research Letters*, 21(4), 301, doi: 10.1029/93GL02791.

Kent, G. M. (1998). Melt to mush variations in crustal magma properties along the ridge crest at the southern East Pacific Rise. *Nature*, 394, 874–878.

Kent, G. M., Harding, A. J. and Orcutt, J. A. (1993). Distribution of magma beneath the East Pacific Rise between the Clipperton transform and the 9°17′ N Deval from forward modeling of common depth point data. *Journal of Geophysical Research*, 98, 13,945–13,969.

Kent, G. M., Harding, A. J. and Orcutt, J. A. (2000). Evidence from three-dimensional seismic reflectivity images for enhanced melt supply beneath mid-ocean-ridge discontinuities. *Nature*, 406, 614–618.

Kent, G. M., Harding, A. J., Orcutt, J. A. *et al.* (1994). Uniform accretion of oceanic crust south of the Garrett transform at 14°15′ S on the East Pacific Rise. *Journal of Geophysical Research*, 99(B5), 9097–9116.

Kent, G., Harding, A., Babcock, J. *et al.* (2003). A new view of 3-D magma chamber structure beneath Axial Seamount and Coaxial Segment: Preliminary results from the 2002 multichannel seismic survey of the Juan de Fuca Ridge. *Eos, Transactions of the American Geophysical Union*, 84(46), Fall Meeting Supplement, Abstract B12A0544.

Key, K., Constable, S., Liu, L. and Pommier, A. (2013). Electrical image of passive mantle upwelling beneath the northern East Pacific Rise. *Nature*, 495, 499–502, doi:10.1038/nature11932.

Kidd, R. G. W. (1977). A model for the process of formation of the upper oceanic crust. *Geophysical Journal of the Royal Astronomical Society*, 50, 149–183.

Kingsley, C. (1863). *The Water-Babies, A Fairy Tale for a Land Baby*. New York: Houghton Mifflin.

Kinzler, R. J. and Grove, T. L. (1992). Primary magmas of mid-ocean ridge basalts, 1. Experiments and methods. *Journal of Geophysical Research*, 97, 6885–6906.

Kinzler, R. and Grove, T. (1993). Corrections and further discussion of the primary magmas of mid-ocean ridge basalts, 1 and 2. *Journal of Geophysical Research*, 98, 22,339–22,347.

Kivelson, M. G., Khurana, K. K., Russell, C. T. *et al.* (2000). Galileo magnetometer measurements: A stronger case for a subsurface ocean at Europa. *Science*, 289, 1340–1343.

Klausen, M. B. and Larsen, H. C. (2002). East Greenland coast-parallel dike swarm and its role in continental breakup. In Menzies, M. A., Klemperer, S. L., Ebinger, C. J. and Baker, J. (eds.), *Volcanic Rifted Margins*. Special Paper 362, Boulder, CO: Geological Society of America, pp. 133–158.

Klein, E. M., White, S., Nunnery, A. *et al.* (2013). Seafloor photo-geology and sonar terrain modeling at the 9°N overlapping spreading center, East Pacific Rise. *Geochemistry, Geophysics, Geosystems*, 14(12), 5146–5170, doi:10.1002/2013GC004858.

Kleinrock, M. C. and Hey, R. N. (1989). Migrating transform zone and lithospheric transfer at the Galápagos 95.5°W propagator. *Journal of Geophysical Research*, 94, 13,859–13,878.

Kleinrock, M. C. and Humphris, S. E. (1996a). Structural asymmetry of the TAG rift valley: Evidence from a near-bottom survey for episodic spreading. *Geophysical Research Letters*, 23, 3439–3442.

Kleinrock, M. C. and Humphris, S. E. (1996b). Structural control on sea-floor hydrothermal activity at the TAG active mound. *Nature*, 382, 149–153.

Klinkhammer, G., Rona, P., Gleaver, M. and Elderfield, H. (1985). Hydrothermal manganese plumes in the Mid-Atlantic Ridge valley. *Nature*, 314, 727–731.

Klitgord, K., Casey, J. F., Agar, S. and Cruise Participants (1990). *Cruise Report of Russian Mir Dives at Kings Trough* (unpublished report).

Koepke, J., Christie, D. M., Dziony, W. *et al.* (2008). Petrography of the dike-gabbro transition at IODP Site 1256 (equatorial Pacific): The evolution of the granoblastic dikes. *Geochemistry, Geophysics, Geosystems*, 9, Q07O09, doi:10.1029/2008GC001939.

Koepke, J., France, L., Müller, T. *et al.* (2011). Gabbros from IODP Site 1256, equatorial Pacific: Insight into axial magma chamber processes at fast spreading ocean ridges. *Geochemistry, Geophysics, Geosystems*, 12, doi: 10.1029/2011GC003655.

Komai, T., Giere, O. and Segonzac, M. (2007). New record of alvinocaridid shrimps (Crustacea: Decapoda: Caridea) from hydrothermal vent fields on the southern Mid-Atlantic Ridge, including a new species of the genus *Opaepele*. *Species Diversity*, 12, 237–253.

Kong, L. S., Solomon, S. C. and Purdy, G. M. (1992). Microearthquake characteristics of a mid-ocean ridge along-axis high. *Journal of Geophysical Research*, 97, 1659–1685.

Kongsrud, J. A. and Rapp, H. T. (2012). *Nicomache (Loxochona) lokii* sp. nov. (Annelida: Polychaeta: Maldanidae) from the Loki's Castle vent field: An important structure builder in an Arctic vent system. *Polar Biology*, 35, 1612–170.

Konn, C., Charlou, J. L., Donval, J. P. *et al.* (2009). Hydrocarbons and oxidized organic compounds in hydrothermal fluids from Rainbow and Lost City ultramafic-hosted vents. *Chemical Geology*, 258, 299–314.

Korenaga, J. and Kelemen, P. B. (1997). Origin of gabbro sills in the Moho transition zone of the Oman ophiolite: Implications for magma transport in the lower oceanic crust. *Journal of Geophysical Research*, 102, 27,729–27,749.

Koschinsky, A., Billings, A., Devey, C. *et al.* (2006). Discovery of new hydrothermal vents on the Southern Mid-Atlantic Ridge (4°–10°S) during cruise M68/1. *InterRidge News*, 15, 9–15.

Koschinsky, A., Garbe-Schonberg, D., Sander, S. *et al.* (2008). Hydrothermal venting at pressure-temperature conditions above the critical point of seawater, 5°S on the Mid-Atlantic Ridge. *Geology*, 36, 615–618.

Krasnov, S. G., Cherkashev, G. A., Stepanova, T. V. *et al.* (1995). Detailed geological studies of hydrothermal fields in the North Atlantic. In Parson, J., Walker, C. L. and Dixon, D. R. (eds.), *Hydrothermal Vents and Processes*. Special Publication 87 London: Geological Society of London, pp. 43–64.

Kristall, B., Kelley, D. S., Hannington, M. D. and Delaney, J. R. (2006). Growth history of a diffusely venting sulfide structure from the Juan de Fuca Ridge: A petrological and geochemical study. *Geochemistry, Geophysics, Geosystems*, 7(7), Q07001, doi:10.1029/2005GC001166.

Kristall, B., Nielsen, D., Hannington, M. D., Kelley, D. S. and Delaney, J. R. (2011). Chemical microenvironments within sulfide structures from the Mothra Hydrothermal Field: Evidence from high-resolution zoning of trace elements. *Chemical Geology*, 290, 12–30.

Kuhn, T., Alexander, B., Augustin, N. *et al.* (2004). The Logatchev hydrothermal field revisited: Preliminary results of the R/V *Meteor* cruise Hydromar I (M60/3). *InterRidge News*, 13, 1–4.

Kuhn, T., Herzig, P. M., Hannington, M. D., Garbe-Schönberg, D. and Stoffer, P. (2003). Origin of fluids and anhydrite precipitation in the sediment-hosted Grimsey hydrothermal field north of Iceland. *Chemical Geology*, 202, 5–12.

Kurras, G. J., Fornari, D. J., Edwards, M. H., Perfit, M. R. and Smith, M. C. (2000). Volcanic morphology of the East Pacific Rise crest 9°49′ –52′ : Implications for volcanic emplacement processes at fast-spreading mid-ocean ridges. *Marine Geophysical Researches*, 21, 23–41, doi:10.1023/A:1004792202764.

Kuzenetsov, V., Cherkashev, G., Lein, A. *et al.* (2006). ^{230}Th/U dating of massive sulfides from the Logatchev and Rainbow hydrothermal fields (Mid-Atlantic Ridge). *Geochronometria*, 25, 51–56.

Kvassnes, A. J. and Grove, T. L. (2008). How partial melts of mafic lower crust affect ascending magmas at oceanic ridges. *Contributions to Mineralogy and Petrology*, 156(1), 49–71.

Kyuno, A., Shintaku, M., Fujita, Y. *et al.* (2009). Dispersal and differentiation of deep-sea mussels of the genus *Bathymodiolus* (Mytilidae, Bathymodiolinae). *Journal of Marine Biology*, doi:10.1155/2009/625672.

Lagabrielle, Y. and Cannat, M. (1990). Alpine Jurassic ophiolites resemble the modern central Atlantic basement. *Geology*, 18, 319–322.

Lagabrielle, Y., Bideau, D., Cannat, M., Karson, J. A. and Mével, C. (1998). Ultramafic-mafic plutonic rock suites exposed along the Mid-Atlantic Ridge (10°N–30°N): Symmetrical-asymmetrical distribution and implications for seafloor spreading processes. In Buck, W. R., Delaney, P. T., Karson, J. A. and Lagabrielle, Y. (eds.), *Faulting and Magmatism at Mid-Ocean Ridges*. Geophysical Monograph 106, Washington, DC: American Geophysical Union, pp. 153–176.

Lagabrielle, Y., Le Moigne, J., Maury, R. C., Cotton, J. and Bourgois, J. (1994). Volcanic record of the subduction of an active spreading ridge, Taito Peninsula (southern Chile). *Geology*, 22, 515–518.

Lalou, C., Brichet, J. S., Krasnov, T. *et al.* (1996). Initial chronology of a recently discovered hydrothermal field at 14°45′ N, Mid-Atlantic Ridge. *Earth and Planetary Science Letters*, 144, 483–490.

Lalou, C., Reyss, J. L. and Brichet, E. (1998). Age of sub-bottom sulfide samples at the TAG active mound. In Herzig, P. M. et al. (eds.), *Proceedings of the Ocean Drilling Program, Scientific Results, Leg 158*. College Station, TX: Ocean Drilling Program, pp. 111–117.

Lalou, C., Reyss, J. L., Brichet. E., Rona, P. A. and Thompson, G. (1995). Hydrothermal activity on a 10^5-year scale at a slow-spreading ridge, TAG hydrothermal field, Mid-Atlantic Ridge 26°N. *Journal of Geophysical Research*, 100, 17,855–17,862.

Lang, S. Q., Butterfield, D. A., Schulte, M., Kelley, D. S. and Lilley, M. D. (2010). Elevated concentrations of formate, acetate and dissolved organic carbon found at the Lost City Hydrothermal Field. *Geochimica et Cosmochimica Acta*, 74, 941–952.

Lang, S. Q., Früh-Green, G. L., Bernasconi, S. M. *et al.* (2012). Microbial utilization of abiogenic carbon and hydrogen in a serpentinite-hosted system. *Geochimica et Cosmochimica Acta*, 92, 82–99.

Langmuir, C. H. and Bender, J. F. (1984). The geochemistry of oceanic basalts in the vicinity of transform faults: Observations and implications. *Earth and Planetary Science Letters*, 69(1), 107–127.

Langmuir, C. H. and Broecker, W. (2012). *How to Build a Habitable Planet: The Story of Earth from the Big Bang to Humankind*. Princeton, NJ: Princeton University Press.

Langmuir, C. H., Bender, J. F. and Batiza, R. (1986). Petrologic and tectonic segmentation of the East Pacific Rise, 5°30′ –14°30′ N. *Nature*, 322, 422–426.

Langmuir, C. H., Bender, J. F., Bence, A. E., Hanson, G. N. and Taylor, S. R. (1977). Petrogenesis of basalts from the FAMOUS area: Mid-Atlantic Ridge. *Earth and Planetary Science Letters*, 36, 133–156.

Langmuir, C. H., Humphris, S. E., Fornari, D. J. *et al.* (1997). Hydrothermal vents near a mantle hot spot: The Lucky Strike vent field at 37°N on the Mid-Atlantic Ridge. *Earth and Planetary Science Letters*, 148, 69–91.

Langmuir, C. H., Klein, E. M. and Plank, T. (1992). Petrological systematics of mid-ocean ridge basalts: Constraints on melt generation beneath ocean ridges. In Phipps Morgan, J., Blackman, D. K. and Sinton, J. (eds.), *Mantle Flow and Melt Generation at Mid-Ocean Ridges*. Geophysical Monograph 71, Washington, DC: American Geophysical Union, pp. 183–280.

Larson, B. I., Olson, E. J. and Lilley, M. D. (2009). Parameters of subsurface brines and hydrothermal processes 12–15 months after the 1999 magmatic event at the Main Endeavor Field as inferred from in situ time series measurements of chloride and temperature. *Journal of Geophysical Research*, 114, doi:10.1029/2008JB005627.

Larson, R. L., Popham, C. T. and Pockalny, R. A. (2005). Lithologic and structural observations at Endeavor Deep and their implications for the accretion process at fast to ultra-fast spreading rates. *Eos, Transactions of the America Geophysical Union*, 85(52), T33D–5094.

Laukert, G., Von Der Handt, A., Hellebrand, E. *et al.* (2013). High-pressure reactive melt stagnation recorded in abyssal pyroxenites from the ultraslow-spreading Lena Trough, Arctic Ocean. *Journal of Petrology*, 55, 427–458.

Lavier, L., Buck, W. R. and Poliakov, A. N. B. (1999). Self-consistent rolling-hinge model for the evolution of large-offset low-angle faults. *Geology*, 27, 1127–1130.

Lawrence, R. M., Karson, J. A. and Hurst, S. D. (1998). Dike orientations, fault-block rotations, and the construction of slow spreading oceanic crust at 22°40′ N on the Mid-Atlantic Ridge. *Journal of Geophysical Research*, 103, 663–676.

Lawver, L. A. and Williams, D. L. (1979). Heat flow in the central Gulf of California. *Journal of Geophysical Research*, 84, doi:10.1029/0JGREA0000840000B7003465000001.

le Roux, P. J., Shirey, S. B., Hauri, E. H., Perfit, M. R. and Bender, J. F. (2006). The effects of variable sources, processes and contaminants on the composition of northern EPR MORB (8–10°N and 12–14°N): Evidence from volatiles (H_2O, CO_2, S) and halogens (F, Cl). *Earth and Planetary Science Letters*, 251, 209–231.

Lein, A. Y., Peresypking, V. I. and Simoneit, B. R. T. (2003). Origin of hydrocarbons in hydrothermal sulfide ores in the Mid-Atlantic Ridge. *Lithology and Mineral Resources*, 38, 383–393.

Leitch, E. C. (1984). Island arc elements and arc-related ophiolites. *Tectonophysics*, 106, 177–203.

Lilley, M. D., Baross, J. A. and Gordon, L. I. (1983). Reduced gases and bacteria in hydrothermal fluids:

The Galápagos spreading center and 21°N East Pacific Rise. In Rona, P. A., Widenfalk, L. and Bostrom, K. (eds.), *Hydrothermal Processes at Seafloor Spreading Centers*. New York: Plenum, pp. 411–447.

Lilley, M. D., Butterfield, D. A., Lupton, J. E. and Olson E. J. (2003). Magmatic events can produce rapid changes in hydrothermal vent chemistry. *Nature*, 422, 878–881.

Lilley, M. D., Butterfield, D. A., Olsen, E. J. *et al.* (1993). Anomalous CH_4 and NH_4^+ concentrations at an unsedimented mid-ocean ridge hydrothermal system. *Nature*, 364, 45–47.

Lilley, M. D., Landsteiner, M. C., McLaughlin, E. A. *et al.* (1995). Real-time mapping of hydrothermal plumes on the Endeavour Segment of the Juan de Fuca. *Eos, Transactions of the American Geophysical Union*, 76, 420.

Lin, J. and Phipps Morgan, J. (1992). The spreading rate dependence of three-dimensional mid-ocean ridge gravity structure. *Geophysical Research Letters*, 19, 13–16.

Lissenberg, C. J. and Dick, H. J. B. (2008). Melt–rock reaction in the lower oceanic crust and its implications for the genesis of mid-ocean ridge basalt. *Earth and Planetary Science Letters*, 271(1), 311–325.

Lissenberg, C. J., MacLeod, C. J., Howard, K. A. and Goddard, M. (2013). Pervasive reactive melt migration through fast-spreading lower oceanic crust (Hess Deep, equatorial Pacific Ocean). *Earth and Planetary Science Letters*, 361, 436–447.

Lister, C. R. B. (1974). On the penetration of water into hot rock. *Geophysical Journal of the Royal Astronomical Society*, 39, 465–509.

Lister, C. R. B. (1983). The basic physics of water penetration into hot rock. In Rona, P. A., Widenfalk, L. and Bostrom, K. (eds.), *Hydrothermal Processes at Seafloor Spreading Centers*. New York: Plenum, pp. 141–176.

Liu, C. Z., Snow, J. E., Hellebrand, E. *et al.* (2008). Ancient, highly heterogeneous mantle beneath Gakkel Ridge, Arctic Ocean. *Nature*, 452(7185), 311–316.

Lizarralde, D., Soule, S. A., Seewald, J. S. and Proskurowski, G. (2011). Carbon release by off-axis magmatism in a young sedimented spreading centre. *Nature Geoscience*, 4, doi:10.1038/NGEO1006.

Lonsdale, P. (1977a). Abyssal pahoehoe with lava coils at the Galápagos rift. *Geology*, 5, 147–152.

Lonsdale, P. (1977b). Clustering of suspension-feeding macrobenthos near abyssal hydrothermal vents at oceanic spreading centers. *Deep-Sea Research*, 24, 857–863.

Lonsdale, P. (1977c). Structural geomorphology of a fast-spreading rise crest: The East Pacific Rise crest at 3°25′ S. *Marine Geophysical Researches*, 3, 251–293.

Lonsdale, P. (1985). Linear volcanoes along the Pacific-Cocos plate boundary, 9°N to the Galápagos triple junction. *Tectonophysics*, 116, 255–279.

Lonsdale, P. and Spiess, F. N. (1980). Deep-tow observations at the East Pacific Rise, 8°45′ N, and some interpretations. *Initial Reports of the Deep Sea Drilling Project, Leg 54*. Washington, DC: US Government Printing Office, pp. 43–62.

Lonsdale, P., Bischoff, J. L., Burns, V. M., Kastner, M. and Sweeney, R. E. (1980). A high temperature hydrothermal deposit on the sea bed at a Gulf of Mexico spreading center. *Earth and Planetary Science Letters*, 44, 463–472.

Lourenço, N., Miranda, J. M., Luis, J. F. *et al.* (1998). Morpho-tectonic analysis of the Azores volcanic plateau from a new bathymetric compilation of the area. *Marine Geophysical Researches*, 20, 141–156.

Love, B. A., Resing, J. A., Cowen, J. P. *et al.* (2008). Methane, manganese, and helium in hydrothermal plumes following volcanic eruptions on the East Pacific Rise near 9°50′ N. *Geochemistry, Geophysics, Geosystems*, 9(6), Q06T01, doi:10.1029/2008GC002104.

Lowell, J. D. and Genik, G. J. (1972). Sea-floor spreading and structural evolution of southern Red Sea. *American Association of Petroleum Geologists Bulletin*, 56, 247–259.

Lowell, R. P., Rona, P. A. and Von Herzen, R. P. (1995). Seafloor hydrothermal systems. *Journal of Geophysical Research*, 100, 327–352.

Ludwig, K. A., Kelley, D. S., Butterfield, D. A., Nelson, B. and Früh-Green, G. (2006). Formation and evolution of carbonate chimneys at the Lost City Hydrothermal Field. *Geochimica et Cosmochimica Acta*, 70, 3625–3645.

Ludwig, K. A., Shen, C.-C., Kelley, D. S., Cheng, H. and Lawrence, R. L. (2011). U-Th systematics and ^{230}Th ages of carbonate chimneys at the Lost City Hydrothermal Field. *Geochimica et Cosmochimica Acta*, 75, 1869–1888.

Lupton, J. E. (1979). Helium^3 in the Guaymas basin: Evidence for injection of mantle volatiles in the Gulf of California. *Journal of Geophysical Research*, 84, 7446–7452, doi:10.1029/JB084iB13p07446.

Lupton, J. E. (1995). Hydrothermal plumes: Near and far field. In Humphris, S., Zierenberg, R., Mullineau, L. and Thomson, R. (eds.), *Seafloor Hydrothermal Systems: Physical, Chemical, Biological, and Geological Interactions*. Geophysical Monograph 91, Washington, DC: American Geophysical Union, pp. 317–346.

Lupton, J. E., Baker, E. T. and Massoth, G. J. (1999). Helium, heat, and the generation of hydrothermal event plumes at mid-ocean ridges. *Earth and Planetary Science Letters*, 171, 343–350.

Lupton, J. E., Baker, E. T. and Massoth, G. J. (2000). Reply to comment by M. R. Palmer and G. G. J. Ernst on "Helium, heat, and the generation of hydrothermal event plumes at mid-ocean ridges" by J. E. Lupton, E. T. Baker, and G. J. Massoth. *Earth and Planetary Science Letters*, 180, 219–222.

Lupton, J. E., Butterfield, D. A., Lilley, M. D. *et al.* (2006). Young, submarine venting of liquid carbon dioxide on a Mariana arc volcano. *Geochemistry, Geophysics, Geosystems*, 7(8), Q08007, doi:10.1029/2005GC001152.

Luther, G. W. III, Rozan, T. F., Taillefert, M. *et al.* (2001). Chemical speciation drives hydrothermal vent ecology. *Nature*, 410, 813–816.

Lyell, C. Sir. (1871). Student's Elements of Geology, London: John Murray.

Macdonald, K. C. (1982). Mid-ocean ridges: Fine scale tectonic, volcanic and hydrothermal processes within the plate boundary zone. *Annual Review of Earth and Planetary Sciences*, 10, 155–190.

Macdonald, K. C. (1998a). Exploring the global Mid-Ocean Ridge: A quarter century of discovery. *Oceanus*, 41, (1).

Macdonald, K. C. (1998b). Linkages between faulting, volcanism, hydrothermal activity and segmentation on fast spreading centers. In Buck, W. R., Delaney, P. T., Karson, J. A. and Lagabrielle, Y. (eds.), *Faulting and Magmatism at Mid-Ocean Ridges*, Geophysical Monograph 106. Washington, DC: American Geophysical Union, pp. 27–58.

Macdonald, K. C. and Fox, P. J. (1983). Overlapping spreading centers: New accretion geometry on the East Pacific Rise. *Nature*, 302, 55–57.

Macdonald, K. C. and Fox, P. J. (1988). The axial summit graben and cross-sectional shape of the East Pacific Rise as indicators of axial magma chambers and recent volcanic eruptions. *Earth and Planetary Science Letters*, 88, 119–131.

Macdonald, K. C. and Luyendyk, B. P. (1977). Deep-tow studies of the structure of the Mid-Atlantic Ridge crest near 37° N (FAMOUS). *Geological Society of America Bulletin*, 88, 621–636.

Macdonald, K. C. and Luyendyk, B. P. (1985). Investigation of faulting and abyssal hill formation on the flanks of the East Pacific Rise (21°N) using *Alvin*. *Marine Geophysical Researches*, 7, 515–535.

Macdonald, K. C., Fox, P. J., Alexander, R. T., Pockalny, R. and Gente, P. (1996). Seafloor topography: A faulted or volcanic origin? *Nature*, 380, 125–129.

Macdonald, K. C., Fox, P. J., Miller, S. *et al.* (1992). The East Pacific Rise and its flanks 8–18° N: History of segmentation, propagation and spreading direction based on Seamarc-II and Sea Beam studies. *Marine Geophysical Researches*, 14, 299–344.

Macdonald, K. C., Scheirer, D. S. and Carbotte, S. M. (1991). Mid-ocean ridges: Discontinuities, segments and giant cracks. *Science*, 253, 986–994.

Macdonald, K., Luyendyk, B. P., Mudie, J. D. and Spiess, F. N. (1975). Near-bottom geophysical study of the Mid-Atlantic Ridge median valley near lat. 37°N: Preliminary observations. *Geology*, 3, 211–215.

Maclennan, J., Hulme, T. and Singh, S. C. (2004). Thermal models of oceanic crustal accretion: Linking geophysical, geological and petrological observations. *Geochemistry, Geophysics, Geosystems*, 5, doi:10.1029/2003GC000605.

Maclennan, J., Hulme, T. and Singh, S. C. (2005). Cooling the lower oceanic crust. *Geology*, 5, 357–360, doi:10.1130/G21207.

MacLeod, C. J. and Yaouancq, G. (2000). A fossil melt lens in the Oman ophiolite: Implications for magma chamber processes at fast spreading ridges. *Earth and Planetary Science Letters*, 176, 357–373.

MacLeod, C. J., Escartín, J., Banerjee, D. *et al.* (2002). Direct geological evidence for oceanic detachment faulting: The Mid-Atlantic Ridge, 15°45′ N. *Geology*, 30(10), 879–882, doi: 10.1130/0091-7613(2002)030087.

MacLeod, C. J., Lissenberg, C. J. and Bibby, L. E. (2013). "Moist MORB" axial magmatism in the Oman ophiolite: The evidence against a mid-ocean ridge origin. *Geology*, 41(4), 459–462, doi:10:1130/G33904.1.

MacLeod, C. J., Searle, R. C., Murton, B. J. *et al.* (2009). Life cycle of oceanic core complexes. *Earth and Planetary Science Letters*, 287(3–4), 333–344.

Magde, L. S., Barclay, A. H., Toomey, D. R., Detrick, R. S. and Collins, J. A. (2000). Crustal magma plumbing within a segment of the Mid-Atlantic Ridge, 35°N. *Earth and Planetary Science Letters*, 175, 55–67.

Malinverno, A. and Cowie, P. (1993). Normal faulting and the topographic roughness of mid-ocean ridge flanks. *Journal of Geophysical Research*, 98, 17,921–17,940.

Malinverno, A., Hildebrand, J., Tominaga, M. and Channell, J. E. T. (2012). M-sequence geomagnetic polarity time scale (MHTC12) that steadies global spreading rates and incorporates astrochronology constraints. *Journal of Geophysical Research*, 117, (B06104), doi: 10.1029/2012JB009260.

Malpas, J., Calon, T. and Xenophontos, C. (1987). Plutonic rocks of the Troodos Ophiolite. In *Field Excursion Guidebook, Troodos'87 Symposium, Ophiolites and Oceanic Lithosphere*, Nicosia, Cyprus: Geological Survey Department, pp. 158–181.

Manning, C. E., MacLeod, C. J. and Weston, P. E. (2000). Lower crustal cracking front at fast spreading ridges: Evidence from the East Pacific Rise and Oman ophiolite. In Dilek, Y., Moores, E., Elthon, D. and Nicolas, A. (eds.), *Ophiolites and Oceanic Crust: New Insights from Field Studies and the Ocean Drilling Program*. Special Paper 349, Boulder, CO: Geological Society of America, pp. 261–272.

Manning, C. E., Weston, P. E. and Mahon, K. I. (1996). Rapid high temperature metamorphism of the East Pacific Rise gabbros from Hess Deep. *Earth and Planetary Science Letters*, 144, 123–132.

Marcon, Y., Sahling, H., Borowski, C. *et al.* (2013). Megafaunal distribution and assessment of total methane and sulfide consumption by mussel beds at Menez Gwen hydrothermal vent, based on geo-referenced photomosaics. *Deep-Sea Research Part 1*, 175, 93–109.

Marcus, J., Tunnicliffe, V. and Butterfield, D. A. (2009). Post-eruptive succession of macrofunal communities at diffuse flow hydrothermal vents on Axial Volcano, Juan de Fuca Ridge, Northeast Pacific. *Deep-Sea Research Part II*, 56, 1586–1598.

Marques, A. F. A., Barriga, F., Chavagnac, V. and Fouquet, Y. (2006). Mineralogy, geochemistry, and Nd isotope composition of the Rainbow hydrothermal field, Mid-Atlantic Ridge. *Mineralium Deposita*, 41, 52–67.

Marques, A. F. A., Barriga, F. J. A. S. and Scott, S. D. (2007). Sulfide mineralization in an ultramafic-rock hosted seafloor hydrothermal system: From serpentinization to the formation of Cu-Zn-(Co)-rich massive sulfides. *Marine Geology*, 245, 20–39.

Marques, A. F. A., Scott, S. D., Gorton, M. P., Barriga, F. J. A. S. and Fouquet, Y. (2009). Pre-eruption history of enriched MORB from the Menez Gwen (37°50′ N) and Lucky Strike (37°17′ N) hydrothermal systems, Mid-Atlantic Ridge. *Lithos*, 112, 18–39.

Martin, W., Baross, J. A., Kelley, D. S. and Russel, M. J. (2008). Hydrothermal vents and the origin of life. *Nature Reviews Microbiology*, 6, 805–814, doi:10.1038/nrmicro 1991, 1–10.

Martins, I., Colaco, A., Dando, P. R. *et al*. (2008). Size-dependent variations on the nutritional pathway of *Bathymodiolus azoricus* demonstrated by a C-flux model. *Ecological Modeling*, 217, 59–71.

McBirney, A. R. (1975). Differentiation of the Skaergaard intrusion. *Nature*, 253, 691–694.

McCaig, A. M., Cliff, R. A. Escartín, J. Fallick, A. E. and MacLeod, C. J. (2007). Oceanic detachment faults focus very large volumes of black smoker fluids. *Geology*, 35, 935–938.

McCaig, A. M., Delacour, A. Fallick, A. E. Castelain, T. and Früh-Green, G. L. (2010). Detachment fault control on hydrothermal circulation systems: Interpreting the subsurface beneath the TAG Hydrothermal Field using the isotopic and geological evolution of oceanic core complexes in the Atlantic. In Rona, P. A., Devey, C. W., Dyment, J., Murton, B. J. (eds.), *Diversity of Hydrothermal Systems on Slow Spreading Ocean Ridges*. Geophysical Monograph 108, Washington, DC: American Geophysical Union, pp. 207–240.

McCarthy, J., Mutter, J. C., Morton, J. L., Sleep, N. H. and Thompson, G. A. (1988). Relict magma chamber structures preserved within the Mesozoic North Atlantic crust? *Geological Society of America Bulletin*, 100, 1423–1436.

McClinton, T., White, S. M., Colman, A. and Sinton, J. M. (2013). Reconstructing lava flow emplacement processes at the hotspot-affected Galápagos Spreading Center, 95°W and 92°W. *Geochemistry, Geophysics, Geosystems*, 14, 2731–2756, doi 10.1002/ggge.20157.

McCollom, T. M. (2000). Geochemical constraints on primary productivity in submarine hydrothermal plumes. *Deep-Sea Research Part 1: Oceanographic Research Papers*, 47, 85–101.

McCollom, T. M. (2007a). Geochemical constraints on sources of metabolic energy for chemolithoautotrophy in ultramafic-hosted deep-sea hydrothermal systems. *Astrobiology*, 7, 933–950.

McCollom, T. M. (2007b). Observational, experimental, and theoretical constraints on carbon cycling in mid-ocean ridge hydrothermal systems. In Lowell, R. P., Seewald, J. S., Metaxas, A. and Perfit, M. R. (eds.), *Magma to Microbes: Modeling Hydrothermal Processes at Ocean Spreading Centers*. Geophysical Monograph 178, Washington, DC: American Geophysical Union, pp. 193–213.

McCollom, T. M. and Bach, W. (2009). Thermodynamic constraints on hydrogen generation during serpentinization of ultramafic rocks. *Geochimica et Cosmochimica Acta*, 73, 856–875.

McCollom, T. M. and Seewald, J. S. (2001). A reassessment of the potential for reduction of dissolved CO_2 to hydrocarbons during serpentinization. *Geochimica et Cosmochimica Acta*, 65, 3769–3778.

McCollom, T. M. and Seewald, J. S. (2003). Experimental constraints on the hydrothermal reactivity of organic acids and acid anions: I. Formic acid and formate. *Geochimica et Cosmochimica Acta*, 67, 3625–3644.

McCollom, T. M. and Seewald, J. S. (2006). Carbon isotope composition of organic compounds produced by abiotic synthesis under hydrothermal conditions. *Earth and Planetary Science Letters*, 243, 74–84.

McKay, L. J., MacGregor, B. J., Biddle, J. F. *et al*. (2012). Spatial heterogeneity and underlying geochemistry of phylogenetically diverse orange and white *Beggiatoa* mats in Guaymas Basin hydrothermal sediments. *Deep-Sea Research Part 1*, 67, 21–31.

McKenzie, D. (1984). The generation and compaction of partially molten rock. *Journal of Petrology*, 25, 713–765.

McKenzie, D. and Bickle, M. J. (1988). The volume and composition of melt generated by extension of the lithosphere. *Journal of Petrology*, 29, 625–679.

McKenzie, D. and O'Nions, R. K. (1991). Partial melt distributions from inversion of rare earth element concentrations. *Journal of Petrology*, 32, 1021–1091.

Melson, W. G., Hart, S. R. and Thompson, G. (1972). St. Paul's Rocks, Equatorial Atlantic: Petrogenesis, radiometric ages, and implications for sea-floor spreading. In Shagan, R. (ed.), *Studies in Earth and Space Sciences: The Hess Volume*. Boulder, CO: Geological Society of America, pp. 241–272.

Melson, W. G., Thompson, G. and Van Andel, Tj. H. (1968). Volcanism and metamorphism in the mid-Atlantic Ridge, 22°N latitude. *Journal of Geophysical Research*, 73(18), 5925–5941, doi:10.1029/JB073i018p 05925.

Melt Seismic Team (1998). Imaging deep seismic structure beneath the mid-ocean ridge: The MELT experiment. *Science*, 280, 1215–1218.

Menard, H. W. (1964). *Marine Geology of the Pacific*. New York: McGraw Hill.

Menard, H. W. (1967). Extension of Northeastern-Pacific fracture zones. *Science*, 155, 72–74.

Menard, H. W. and Atwater, T. (1969). Origin of fracture zone topography. *Nature*, 222, 1037–1040.

Meurer, W. P. and Boudreau, A. E. (1998a). Compaction of igneous cumulates part I: Geochemical consequences for cumulates and liquid fractionation trends. *Journal of Geology*, 106(3), 281–292.

Meurer, W. P. and Boudreau, A. E. (1998b). Compaction of igneous cumulates part II: Compaction and the development of igneous foliations. *Journal of Geology*, 106(3), 293–304.

Mével, C., Cannat, M., Gente, P. *et al.* (1991). Emplacement of deep crustal and mantle rocks on the West Median Valley Wall of the MARK Area (M.A.R. 23°N). *Tectonophysics*, 190, 31–53.

Mével, G., Faidy, C. and Prieur, D. (1996). Distribution, activity, and diversity of heterotrophic nitrifiers originating from East Pacific deep-sea hydrothermal vents. *Canadian Journal of Microbiology*, 42, 162–171.

Meyer, J. L., Akerman, N. H., Proskurowski, G. and Huber, J. A. (2013). Microbial characterization of post-eruption "snowblower" vents at Axial Seamount, Juan de Fuca Ridge. *Frontiers in Microbiology*, 4, doi:10.3389/frmicb.2013.00153.

Michael, P. J. (1995). Regionally distinctive sources of depleted MORB: Evidence from trace elements and H_2O. *Earth and Planetary Science Letters*, 131(3), 301–320.

Michael, P. J. and Bonatti, E. (1985). Peridotite compositions from the North Atlantic: Regional and tectonic variations and implications for partial melting. *Earth and Planetary Science Letters*, 73, 91–104.

Michael, P. J. and Cornell, W. C. (1998). Influence of spreading rate and magma supply on crystallization and assimilation beneath mid-ocean ridges: Evidence from chlorine and major element chemistry of mid-ocean ridge basalts. *Journal of Geophysical Research*, 103 (B8), 18,325–18,356.

Michael, P. J., Langmuir, C. H., Dick, H. J. B. *et al.* (2003). Magmatic and amagmatic seafloor generation at the ultraslow-spreading Gakkel ridge, Arctic Ocean. *Nature*, 423, 956–961, doi:10:1038/nature01704.

Michael, P. J., Perfit, M. R., Palke, A., Fornari, D. J. and Soule, S. A. (2008). Observations of CO_2 exsolution in submarine lava flows: The 2005–06 eruption on the East Pacific Rise, 9°50′ N. *Eos, Transactions of the American Geophysical Union*, 89, V21B–2106.

Mittelstaedt, E., Escartín, J., Gracias, N. *et al.* (2012). Quantifying diffuse and discrete venting at the Tour Eiffel vent site, Lucky Strike hydrothermal field. *Geochemistry, Geophysics, Geosystems*, 13(4), Q04008, doi:10.1029/2011GC003991.

Miyashiro, A. (1972). Metamorphism and related magmatism in plate tectonics. *American Journal of Science*, 272, 629–656.

Miyashiro, A. (1973a). *Metamorphism and Metamorphic Belts*. New York: John Wiley and Sons.

Miyashiro, A. (1973b). The Troodos ophiolite complex was probably formed in a volcanic arc. *Earth and Planetary Science Letters*, 19, 218–224.

Miyashiro, A. (1975a). Classification, characteristics, and origin of ophiolites. *Journal of Geology*, 83, 249–281.

Miyashiro, A. (1975b). Discussion of "Origin of the Troodos and other ophiolites: A reply to Hynes". *Earth and Planetary Science Letters*, 25, 217–222.

Miyashiro, A. (1975c). Origin of the Troodos and other ophiolites: A reply to Moores. *Earth and Planetary Science Letters*, 25, 227–235.

Miyashiro, A., Shido, F. and Ewing, M. (1969). Composition and origin of serpentinites from the Mid-Atlantic Ridge near 24° and 30°N lat. *Contributions to Mineralogy and Petrology*, 23, 117–127.

Miyashiro, A., Shido, F. and Ewing, M. A. (1971). Metamorphism in the Mid-Atlantic Ridge near 24°N and 30°N. *Philosophical Transactions of the Royal Society of London, Series A*, 268, 589–603.

Moore, J. G. (1975). Mechanism of formation of pillow lava. *American Scientist*, 63, 269–277.

Moore, J. G. and Fabbi, B. P. (1971). An estimate of the juvenile sulfur content of basalt. *Contributions to Mineralogy and Petrology*, 33, 118–127.

Moores, E. M. (1969). *Petrology and Structure of the Vourinos Ophiolite Complex, Northern Greece*. Special Paper 118, Boulder, CO: Geological Society of America, pp. 3–66.

Moores, E. M. (1975). Discussion of "Origin of the Troodos and other ophiolites: A reply to Hynes" by Akiho Miyashiro. *Earth and Planetary Science Letters*, 25, 223–226.

Moores, E. M. (1982). Origin and emplacement of ophiolites. *Reviews of Geophysics and Space Physics*, 20, 735–760.

Moores, E. M. and Vine, F. J. (1971). The Troodos massif, Cyprus, and other ophiolites as oceanic crust: Evaluation and implications. *Philosophical Transactions of the Royal Society of London A*, 268, 443–466.

Moores, E. M., Kellogg, L. H. and Dilek, Y. (2000). Tethyan ophiolites, mantle convection, and tectonic "historical contingency:" A resolution of the "Ophiolite Conundrum". In Dilek, Y., Moores, E. M., Elthon, D. and Nicolas, A. (eds.), *Ophiolites and Oceanic Crust: New Insights from Field Studies and the Ocean Drilling Program*. Boulder, CO: Geological Society of America, Special Paper 349, pp. 3–12.

Morishita, T., Hara, K., Nakamura, K. *et al.* (2009). Igneous, alteration and exhumation processes recorded in abyssal peridotites and related fault rocks from an oceanic core complex along the Central Indian Ridge. *Journal of Petrology*, 50(7), 1299–1325.

Morris, A., Gee, J. S., Pressling, N. *et al.* (2009). Footwall rotation in an oceanic core complex quantified using reoriented Integrated Ocean Drilling Program core samples. *Earth and Planetary Science Letters*, 287, 217–228.

Morris, E., Detrick, R. S., Minshull, T. A. *et al.* (1993). Seismic structure of oceanic crust in the Western North Atlantic. *Journal of Geophysical Research*, 98, 13879–13903.

Mottl, M. J. (1983). Metabasalts, axial hot springs, and the structure of hydrothermal systems at mid-ocean ridges. *Geological Society of America Bulletin*, 94, 161–180.

Mottl, M. J., Komor, S. C., Fryer, P. and Moyer C. L. (2003). Deep-slab fluids fuel extremophilic Archaea on a Mariana forearc serpentinite mud volcano: Ocean Drilling Program Leg 195. *Geochemistry, Geophysics, Geosystems*, 4(11), doi:10/10.1029/2003GC000588.

Mount, V. S. and Suppe, J. (1987). State of stress near the San Andreas fault: Implications for wrench tectonics. *Geology*, 15, 1143–1146.

Müller, R. D., Sdrolias, M., Gaina, C. and Roest, W. R. (2008). Age, spreading rates, and spreading asymmetry of the world's ocean crust. *Geochemistry, Geophysics, Geosystems*, 9, Q04006, doi.org/10.1029/2007GC001743.

Mullineaux, L. S., Adams, D. K., Mills, S. W. and Beaulieu, S. E. (2010). Larvae from afar colonize deep-sea hydrothermal vents after a catastrophic eruption. *Proceedings of the National Academy of Sciences*, 107, 7829–7834.

Mullineaux, L. S., Mills, S. W., Sweetman, A. K. *et al.* (2005). Spatial structure and temporal variation in larval abundance at hydrothermal vents on the East Pacific Rise. *Marine Ecology Progress Series*, 293, 1–16.

Mullineaux, L. S., Peterson, C. H., Micheli, F. and Mills, S. W. (2003). Successional mechanism varies along a gradient in hydrothermal fluid flux at deep-sea vents. *Ecological Monographs*, 73, 523–542.

Müntener, O. (2010). Serpentine and serpentinization: A link between planet formation and life. *Geology*, 38, 959–960.

Murray, J. A. and Renard, F. (1891). *Report on Deep-Sea Deposits, Based on Specimens Collected During the Voyage of H.M.S. Challenger in the Years 1872–76*. London: HM Stationery Office.

Murton, B. J., Baker, E.T., Sands, C.M., and German, C.R., (2006) Detection of an unusually large hydrothermal event plume above the slow-spreading Carlsberg Ridge: NW Indian Ocean. Geophysical Research Letters, doi: 10.1029/2006GL026048.

Mutter, C. Z. and Mutter, J. C. (1993). Variation in thickness of layer 3 dominates oceanic crustal structure. *Earth and Planetary Science Letters*, 117, 295–317.

Mutter, J. C. and Carton, H. D. (2013). The Mohorovičić discontinuity in ocean basins: Some observations from seismic data. *Tectonophysics*, 609, 314–330.

Mutter, J. C. and Karson, J. A. (1992). Structural processes at slow-spreading ridges. *Science*, 257, 627–634.

Mutter, J. C., Carbotte, S. M., Nedimović, M., Canales, J. P. and Carton, H. (2009). Seismic imaging in three dimensions on the East Pacific Rise. *Eos, Transactions of the American Geophysical Union*, 90(42), 374–375.

Naidoo, D. D. (1998). *Accretion of the Upper Oceanic Crust.* Ph.D. dissertation, University of Washington.

Nakamura, K., Morishita, T., Bach, W. *et al.* (2009). Serpentinized troctolites exposed near the Kairei hydrothermal field, Central Indian Ridge: Insights into the origin of the Kairei hydrothermal fluid supporting a unique microbial ecosystem. *Earth and Planetary Science Letters*, 280, 128–136.

National Research Council (1999). From monsoons to microbes: Understanding the ocean's role in human health. In *A Review of the Ocean Research Priorities Plan and Implementation Strategy*, Division of Earth and Life Studies, National Research Council, Ocean Studies Board, National Academies Press.

Natland, J. H. (1989). Partial melting of a lithologically heterogeneous mantle: Inferences from crystallization histories of magnesian abyssal tholeiites from the Siqueiros Fracture Zone. In Saunders, A. D. and Norry, M. J. (eds.), *Magmatism in the Ocean Basins*. Special Publication 42, London: Geological Society of London, pp. 41–70.

Natland, J. H. and Dick, H. J. B. (1996). Melt migration through high-level gabbro cumulates of the East Pacific Rise at Hess Deep: The origin of magma lenses and the deep crustal structure of fast-spreading ridges. In Mével, C., Gillis, K. M. and Allen, J. F. (eds.), *Proceedings of the Ocean Drilling Program, Scientific Results, Leg 148*. College Station, TX: Ocean Drilling Program, pp. 21–58.

Natland, J. H. and Dick, H. J. B. (2001). Formation of the lower ocean crust and the crystallization of gabbroic cumulates at a very slowly spreading ridge. *Journal of Volcanology and Geothermal Research*, 110(3–4), 191–233.

Natland, J. H. and Dick, H. J. B. (2009). Paired melt lenses at the East Pacific Rise and the pattern of melt flow through the gabbroic layer at a fast-spreading ridge. *Lithos*, 112(1), 73–86.

Neal, C. and Stranger, G. (1983). Hydrogen generation from mantle source rocks in Oman. *Earth and Planetary Science Letters*, 66, 315–320.

Nehlig, P. and Juteau, T. (1988). Flow porosities, permeabilities and preliminary data on fluid inclusions and fossil thermal gradients in the crustal sequence of the Sumail ophiolite (Oman). *Tectonophysics*, 151, 199–221.

Nehlig, P., Juteau, T., Bendel, T. and Cotton, J. (1994). The root zones of oceanic hydrothermal systems: Constraints from the Samail ophiolite (Oman). *Journal of Geophysical Research*, 99, 4703–4713.

Nelson, D., Haymon, R. M. and Lilley, M. (1991). Rapid growth of unusual hydrothermal bacteria observed at new vents during the ADVENTURE dive program to the EPR at 9°45′ –52′ N. *Eos, Transactions of the American Geophysical Union*, 72, 481.

Neo, N., Yamazaki, S. and Miyashita, S. (2009). Data report: Bulk rock compositions of samples from the IODP Expedition 309/312 sample pool, ODP Hole 1256D. In Teagle, D.A.H., Alt, J.C., Umino, S. *et al.* (eds.), *Proceedings of the Integrated Drilling Program, Scientific Results, Leg 312.* College Station, TX: Integrated Ocean Drilling Program, pp. 309–312.

Nicholson, R. and Pollard, D. D. (1985). Dilation and linkage of echelon cracks. *Journal of Structural Geology*, 20, 53–72.

Nicolas, A. (1989). *Structure of Ophiolites and Dynamics of Oceanic Lithosphere*. Dordrecht: Kluwer Academic Press.

Nicolas, A. and Boudier, F. (2000). Large mantle upwellings and related variations in crustal thickness in the Oman ophiolite. In Dilek, Y., Moores, E., Elthon, D. and Nicolas, A. (eds.), *Ophiolites and Oceanic Crust: New Insights from Field Studies and the Ocean Drilling Program*. Special Paper 349. Boulder, CO: Geological Society of America, pp. 67–73.

Nicolas, A., Boudier, F. and Ceuleneer, G. (1988). Mantle flow patterns and magma chambers at ocean ridges: Evidence from Oman ophiolite. *Marine Geophysical Researches*, 9, 293–310.

Nicolas, A., Boudier, F. and France, L. (2009). Subsidence in magma chamber and development of magmatic foliation in Oman ophiolite gabbros. *Earth and Planetary Science Letters*, 284, 76–87.

Nicolas, A., Boudier, F. and Meshi, A. (1999). Slow spreading accretion and mantle denudation in the Mirdita ophiolite (Albania). *Journal of Geophysical Research*, 104, 15,155–15,167.

Niu, Y. (2004). Bulk-rock major and trace element compositions of abyssal peridotites: Implications for mantle melting, melt extraction and post-melting processes beneath mid-ocean ridges. *Journal of Petrology*, 45(12), doi:10.1093/petrology/egh068.

Niu, Y. and Hékinian, R. (1997). Basaltic liquids and harzburgitic residues in the Garrett Transform: a case study at fast-spreading ridges. *Earth and Planetary Science Letters*, 146(1–2), 243–258.

Nooner, S. L. and Chadwick, W. W., Jr. (2009). Volcanic inflation measured in the caldera of Axial Seamount: Implications for magma supply and future eruptions. *Geochemistry, Geophysics, Geosystems*, 10(2), Q02002, doi:1029/2008GC002315.

Norton, D. (1984). Theory of hydrothermal systems. *Annual Review of Earth and Planetary Sciences*, 12, 155–157.

Nygaard, T. E., Bjerkgaard, T., Kelley, D., Thorseth, I. and Pedersen, R. B. (2003). Hydrothermal chimneys and sulphide mineralized breccias from the Kolbeinsey and Mohns Ridge. *Geophysical Research Abstracts*, 5, 11,863.

O'Hanley, D. S. (1996). *Serpentinites: Records of Tectonic and Petrological History*. New York: Oxford University Press.

Okal, E. A. and Stewart, L. M. (1982). Slow earthquakes along oceanic fracture zones: Evidence for asthenospheric flow away from hotspots? *Earth and Planetary Science Letters*, 57, 75–87.

Olafsson, J. S., Thors, K., Dtefansson, U. *et al.* (1990). Geochemical observations of a boiling hydrothermal site on the Kolbeinsey Ridge. *Eos, Transactions of the American Geophysical Union*, 71, 1650.

Ondréas, H., Cannat, M., Fouguet, Y. *et al.* (2009). Recent volcanic events and the distribution of hydrothermal venting at the Lucky Strike hydrothermal field, Mid-Atlantic Ridge. *Geochemistry, Geophysics, Geosystems*, 10(2), Q02006, doi:10.1029/2008GC002171.

Ondréas, H., Fouquet, Y., Voisset, M. and Radford-Knoery, J. (1997). Detailed study of three contiguous segments of the Mid-Atlantic Ridge, south of the Azores (37°N to 38°30′ N), using acoustic imaging coupled with submersible observations. *Marine Geophysical Researches*, 19, 231–255.

Oosting, S. E. and Von Damm, K. L. (1996). Bromide/chloride fractionation in seafloor hydrothermal fluids from 9–10° N East Pacific Rise. *Earth and Planetary Science Letters*, 144, 133–145.

Opatkiewicz, A. D., Butterfield, D. A. and Baross, J. A. (2009). Individual hydrothermal vents at Axial Seamount harbor distinct subseafloor microbial communities. *FEMS Microbial Ecology*, 70, 413–424.

Oreskes, N. (ed.) 2001. *Plate Tectonics: An Insider's History of the Modern Theory of the Earth*. Cambridge, MA: Westview Press.

OTTER (1984). The geology of the Oceanographer transform: The ridge-transform intersection. *Marine Geophysical Researches*, 6, 109–141.

OTTER (1985). The geology of the Oceanographer Transform: The transform domain. *Marine Geophysical Researches*, 7, 329–358.

Pace, N. R. (1997). A molecular view of microbial diversity and the biosphere. *Science*, 276, 734–740.

Pallister, J. S. and Hopson, C. A. (1981). Samail ophiolite plutonic suite: Field relations, phase variation and layering and a model of a spreading ridge magma chamber. *Journal of Geophysical Research*, 86, 2593–2644.

Paonita, A. and Martelli, M. (2007). A new view of the He–Ar–CO_2 degassing at mid-ocean ridges: Homogeneous composition of magmas from the upper mantle. *Geochimica et Cosmochimica Acta*, 7(7), 1747–1763.

Paonita, A. and Martelli, M. (2013). Comment on "CO_2 variability in mid-ocean ridge basalts from syn-emplacement degassing: Constraints on eruption dynamics" by Soule *et al.* *Earth and Planetary Science Letters*, 374, 251–253.

Paquet, F., Dauteuil, O., Hallot, E. and Moreau, F. (2007). Tectonics and magma dynamics coupling in a dyke swarm of Iceland. *Journal of Structural Geology*, 29, 1477–1493.

Pariso, J. E. and Johnson, H. P. (1989). Magnetic properties and oxide petrography of the sheeted dike complex in Hole 504B. In *Proceedings of the Ocean Drilling Program, Scientific Results Leg 148*. College Station TX: Ocean Drilling Program, pp. 159–166.

Paulick, H., Bach, W., Godard, M. *et al.* (2006). Geochemistry of abyssal peridotites (Mid-Atlantic Ridge, 15°20′ N, ODP Leg 209): Implications for fluid/rock interaction in slow spreading environments. *Chemical Geology*, 234, 179–210.

Pearce, J. A. (2003). Suprasubduction zone ophiolites: The search for modern analogs. In Dilek, Y. and Newcomb,

S. (eds.), *Ophiolite Concept and the Evolution of Geologic Thought*. Special Paper 373, Boulder, CO: Geological Society of America, pp. 269–293.

Pearce, J. A. and Cann, J. R. (1971). Ophiolite origin investigated by discriminant analysis using Ti, Zr, and Y. *Earth and Planetary Science Letters*, 12, 339–349.

Pearce, J. A. and Cann, J. R. (1973). Tectonic setting of basic volcanic rocks determined using trace element analyses. *Earth and Planetary Science Letters*, 19, 290–300.

Pearce, J. A., Ernewein, M., Bloomer, S. H. *et al.* (1994). Geochemistry of Lau Basin volcanic rocks: Influence of ridge segmentation and arc proximity. In Smellie, J. L. (ed.), *Volcanism Associated with Extension at Consuming Plate Margins*. Special Publication 81. London: Geological Society of London, pp. 53–75.

Pearce, J. A., Lippard, S. J. and Roberts, S. (1984). Characteristics and tectonic significance of supra-subduction zone ophiolites. In Kokelaar, B. P. and Howells M. F. (eds.), *Marginal Basin Geology: Volcanic and Associated Sedimentary and Tectonic Processes in Modern and Ancient Marginal Basins*. Special Publication 16. London: Geological Society of London, 77–94.

Pedersen, R. B., Malpas, J. and Falloon, T. (1996). Petrology and geochemistry of gabbroic and related rocks from Site 894 Hess Deep. In Mével, C., Gillis, K. M. and Allen, J. F. (eds.), *Proceedings of the Ocean Drilling Program, Scientific Results, Leg 147*. College Station, TX: Ocean Drilling Program, pp. 3–20, doi:10.2973/odp.proc.sr.147.001.1996.

Pedersen, R. B., Rapp, H. T., Thorseth, I. H. *et al.* (2010a). Discovery of a black smoker vent field and vent fauna at the Arctic Mid-Ocean Ridge. *Nature*, doi:10.1038/ncomms1124.

Pedersen, R. B., Thorseth, I. H., Hellevang, B. *et al.* (2005). Two vent fields discovered at the ultraslow spreading Arctic Ridge System. *Eos, Transactions of the American Geophysical Union*, 86(52), OS21C-01.

Pedersen, R. B., Thorseth, I. H., Lilley, M. D. *et al.* (2009). Discovery of the Loki's Castle vent field at the ultraslow spreading Arctic mid-ocean ridge. *Geochimica et Cosmochimica Acta*, 73, Supplement 1, A1008.

Pedersen, R., Thorseth, I. H., Nygaard, T.-E., Lilley, M. D. and Kelley, D. S. (2010b). Hydrothermal activity at the Arctic Mid-Ocean Ridge. In Rona, P., Devey, C. and Murton, B. (eds.), *Diversity of Hydrothermal Systems on Slow Spreading Ocean Ridges. Geophysical Monograph 188*. Washington, DC: American Geophysical Union, pp. 67–89.

Perfit, M. R., Langmuir, C. H., Baekisapa, M. *et al.* (1987). Geochemistry and petrology of volcanic rocks from the Woodlark basin: Addressing questions of ridge subduction. In *Marine Geology, Geophysics, and Geochemistry of the Woodlark Basin-Solomon Islands*. Houston, TX: Circum-Pacific Council for Energy and Mineral Resources Earth Science, 7, pp. 113–154.

Perfit, M. R. (2001). Mid-ocean ridge geochemistry and petrology. In Steel, J., Thorpe, S. and Turekian, K. (eds.), *Encylopedia of Ocean Sciences*. San Diego: Academic Press, pp. 1778–1788.

Perfit, M. R. and Chadwick, W. W., Jr. (1998). Magmatism at mid-ocean ridges: Constraints from volcanological and geochemical investigations. In Buck, W. R., Delaney, P. T., Karson, J. A. and Lagabrielle, Y. (eds.), *Faulting and Magmatism at Mid-Ocean Ridges*. Washington, DC: American Geophysical Union, pp. 59–115.

Perfit, M. R., Cann, J. R., Fornari, D. J. *et al.* (2003). Interaction of seawater and lava during submarine eruptions at mid-ocean ridges. *Nature*, 426, 62–64.

Perfit, M. R., Fornari, D. J., Malahoff, A. and Embley, R. W. (1983). Geochemical studies of abyssal lavas recovered by DSRV *Alvin* from Eastern Galápagos Rift, Inca Transform, and Ecuador Rift 3: Trace element abundances and petrogenesis. *Journal of Geophysical Research*, 88, 10,551–10,572.

Perfit, M. R., Fornari, D. J., Ridley, W. I. *et al.* (1996). Recent volcanism in the Siqueiros transform fault: Picritic basalts and implications for MORB magma genesis. *Earth and Planetary Science Letters*, 141, 91–108.

Perfit, M. R., Fornari, D. J., Smith, M. C. *et al.* (1994). Small-scale spatial and temporal variations in mid-ocean ridge crest magmatic processes. *Geology*, 22, 375–379.

Perfit, M. R. (1977). Petrology and geochemistry of mafic rocks from the Cayman Trench: Evidence for spreading. *Geology*, 5(2), 105–110.

Perfit, M. R., Ridley, W. I. and Jonasson, I. (1999). Geologic, petrologic and geochemical relationships between magmatism and massive sulfide mineralization along the eastern Galápagos Spreading Center. *Reviews in Economic Geology*, 8, 75–100.

Perfit, M. R., Wanless, V. D., Ridley, W. I. *et al.* (2012). Lava geochemistry as a probe into crustal formation at the East Pacific Rise. *Oceanography*, 25(1), 89–93.

Perk, N. W., Coogan, L. A., Karson, J. A. and Klein, E. M. (2007). Primitive gabbroic rocks from a tectonic window at Pito Deep: Implications for the accretion of plutonic rocks beneath the East Pacific Rise. *Contributions to Mineralogy and Petrology*, doi:10.1007/s00410-007-0210-z.

Perner, M., Kuever, J., Seifert, R. *et al.* (2007). The influence of ultramafic rocks on microbial communities at the Logatchev hydrothermal field, located 15°N on the Mid-Atlantic Ridge. *FEMS Microbial Ecology*, doi: 10.1111/j.1547–6941.2007.00325.x.

Petersen, S., Kuhn, K., Kuhn, T. *et al.* (2009). The geological setting of the ultramafic-hosted Logatchev hydrothermal field (14°45′ N, Mid-Atlantic Ridge) and its influence on massive sulfide formation. *Lithos*, 112, 40–56.

Peterson, D. W. and Swanson, D. A. (1974). Observed formation of lava tubes during 1970–71 at Kilauea Volcano, Hawaii. *Studies in Speleology*, 2, 209–222.

Peterson, D. W. and Tilling, R. I. (1980). Transition of basaltic lava from pahoehoe to 'a'a, Kilauea volcano

Hawaii: Field observations and key factors. *Journal of Volcanology and Geothermal Research*, 7, 271–293.

Peterson, J. J., Fox, P. J. and Schreiber, E. (1974). Newfoundland ophiolites and the geology of the oceanic layer. *Nature*, 247, 194–196.

Petrecca, R. F. and Grassle, J. F. (1990). Notes on fauna from several deep-sea hydrothermal vent and cold seep soft sediment communities. In McMurray, G. R. (ed.), *Gorda Ridge: A Seafloor Spreading Center in the United States Exclusive Economic Zone*. New York: Springer-Verlag, pp. 279–283.

Phipps Morgan, J. A. (1987). Melt migration beneath mid-ocean spreading centers. *Geophysical Research Letters*, 14, 1238–1241.

Pinkerton, H. and Wilson, L. (1994). Factors controlling the lengths of channel-fed lava flows. *Bulletin of Volcanology*, 56, 108–120.

Plüger, W. L. P., Herzig, M., Becker, K. P. *et al.* (1990). Discovery of hydrothermal fields at the Central Indian Ridge. *Marine Mining*, 9, 73–86.

Pockalny, R. A. and Larson, R. L. (2003). Implications for crustal accretion at fast-spreading ridges from observations in Jurassic oceanic crust in the western Pacific. *Geochemistry, Geophysics, Geosystems*, 4(1), 8903, doi:10.1029/2001GC000274.

Pockalny, R. A., Detrick, R. S. and Fox, P. J. (1988). Morphology and tectonics of the Kane Transform from Sea Beam bathymetry data. *Journal of Geophysical Research*, 93, 3179–3194.

Pockalny, R. A., Fox, P. J., Fornari, D. J., Macdonald, K. C. and Perfit, M. R. (1997). Tectonic reconstruction of the Clipperton and Siqueiros fracture zones: Evidence and consequences of plate motion change for the last 3 million years. *Journal of Geophysical Research*, 102, 3167–3181.

Podowski, E. L., Moore, T. S., Zelnio, K. A., Luther, G. W. I. and Fisher, C. R. (2009). Distribution of diffuse flow megafauna in two sites on the Eastern Lau Spreading Center, Tonga. *Deep-Sea Research I, Oceanographic Research Papers*, 56, 2041–2056.

Pollard, D. D. and Aydin, A. (1988). Progress in understanding jointing over the past century. *Geological Society of America Bulletin*, 100, 1181–1204.

Pollock, M. A., Klein, E. M., Karson, J. A. and Coleman, D. S. (2009). Compositions of dikes and lavas from the Pito Deep Rift: Implications for accretion at superfast spreading centers. *Journal of Geophysical Research*, 114 (B3), B03207, doi:10.1029/2007JB005436.

Pollock, M. A., Klein, E. M., Karson, J. A. and Tivey, M. A. (2005). Temporal and spatial variability in the composition of lavas exposed along the Western Blanco Transform Fault. *Geochemistry, Geophysics, Geosystems*, 6(11), Q11009, doi:10.1029/2005GC001026.

Pontbriand, C. W. and Sohn, R. A. (2014). Microearthquake evidence for reaction-driven cracking within the Trans-Atlantic Geotraverse active hydrothermal deposit. *Journal of Geophysical Research*, 119, 822–839, doi:10.1002/2013JB010110.

Pontbriand, C. W., Soule, S. A., Sohn, R. A. *et al.* (2012). Effusive and explosive volcanism on the ultraslow-spreading Gakkel Ridge, 85° E. *Geochemistry, Geophysics, Geosystems*, 13(10), doi:10.1029/2012GC004187.

Proskurowski, G., Kelley, D. S., Delaney, J. R. *et al.* (2012). Post-eruptive time series hydrothermal contribution to the water column above Axial Seamount. *EOS Proceedings of the American Geophysical Union*, OS22A–06.

Proskurowski, G., Lilley, M. D. and Brown, T. A. (2004). Isotopic evidence of magmatism and seawater bicarbonate removal at the Endeavour hydrothermal system. *Earth and Planetary Science Letters*, 225, 53–61.

Proskurowski, G., Lilley, M. D., Kelley, D. S. and Olson, E. J. (2006). Low temperature volatile production at the Lost City Hydrothermal Field: Evidence from a hydrogen stable isotope geothermometer. *Chemical Geology*, 229, 331–343.

Proskurowski, G., Lilley, M. D., Olson, E. J. *et al.* (2008). Abiogenic hydrocarbon production at Lost City Hydrothermal Field. *Science*, 319, 604–607.

Purdy, G. M. and Rabinowitz, P. D. (1978). The Kane fracture zone at the Mid-Atlantic Ridge: IPOD Sites 395 and 396. In Melson, W. G., Rabinowitz, P. *et al.* (eds.), *Initial Reports of the Deep Sea Drilling Project, Leg 45*. Washington, DC: US Govenment Printing Office, map supplement.

Putirka, K. D., Perfit, M. R., Ryerson, F. J. and Jackson, M. G. (2007). Ambient and excess mantle temperatures, olivine thermometry, and active vs. passive upwelling. *Chemical Geology*, 241(3), 177–206.

Putirka, K., Ryerson, F. J., Perfit, M. and Ridley, W. I. (2011). Mineralogy and composition of the oceanic mantle. *Journal of Petrology*, 52(2), 279–313.

Quick, J. E. and Denlinger, R. P. (1993). Ductile deformation and the origin of layered gabbro in ophiolites. *Journal of Geophysical Research*, 98, 14,015–14,027.

Radford-Knoery, J., Charlou, J.-L. Donval, J.-P. *et al.* (1998). Distribution of dissolved sulfide, methane, and manganese near the seafloor at the Lucky Strike (37°17′ N) and Menez Gwen (37°50′ N) hydrothermal vent sites on the mid-Atlantic Ridge *Deep-Sea Research Part I*, 45, 367–386.

Ranero, C. and Reston, T. J. (1999). Detachment faulting at oceanic core complexes. *Geology*, 27, 983–986.

Rautenschlein, M., Jenner, G. A., Hertogen, J. *et al.* (1985). Isotopic and trace element composition of volcanic glasses from the Akaki Canyon, Cyprus: Implications for the origin of the Troodos ophiolite. *Earth and Planetary Science Letters*, 75, 369–383.

Renninger, G. H., Kass, L., Gleeson, R. A. *et al.* (1995). Sulfide as a chemical stimulus for deep-sea hydrothermal vent shrimp. *Biological Bulletin*, 189, 69–76.

Resing, J. A., Rubin, K. H., Embley, R. W. *et al.* (2011). Active submarine eruption of boninite in the northeastern Lau Basin. *Nature Geoscience*, 4, 799–806, doi:10.1038/NGEO1275.

Reston, T. J. and Ranero, C. R. (2011). The 3-D geometry of detachment faulting at mid-ocean ridges. *Geochemistry, Geophysics, Geosystems*, 12(7), Q0AG05, doi:10.1029/2011GC003666.

Reston, T. J., Ranero, C. and Belykh, I. (1999). The structure of Cretaceous oceanic crust of the NW Pacific: Constraints on processes at fast spreading centers. *Journal of Geophysical Research*, 104(B1), 629–644.

Reynolds, T. J. and Beane, R. E. (1985). Evolution of hydrothermal fluid characteristics at the Santa Rita, New Mexico, porphyry copper deposit. *Economic Geology*, 80, 1328–1347.

Richards, J. G., Cann, J. R. and Jensenius, J. (1989). Mineralogical zonation and metasomatism of the alteration pipes of Cyprus sulfide deposits. *Economic Geology and the Bulletin of the Society of Economic Geologists*, 84(1), 91–115.

Riddihough, R. (1984). Recent movements of the Juan de Fuca Plate system. *Journal of Geophysical Research*, 89, 6980–6994.

Ridley, W. I., Perfit, M. R., Fornari, D. J. and Smith, M. C. (2006). Magmatic processes in developing oceanic crust revealed in a cumulate xenolith collected at the East Pacific Rise, 9°50′ N. *Geochemistry, Geophysics, and Geosystems*, 7, Q12O04, doi:10.1029/2006GC001316.

Riedel, C., Schmidt, M., Botz, R. and Theilen, F. (2001). The Grimsey hydrothermal field offshore North Iceland: Crustal structure, faulting and related gas venting. *Earth and Planetary Science Letters*, 193, 409–421.

Riou, V., Duperron, S., Halary, S. *et al.* (2010). Variation in physiological indicators in *Bathymodiolus azoricus* (Bivalvia: Mytilidae) at the Menez Gwen Mid-Atlantic Ridge deep-sea hydrothermal vent site within a year. *Marine Environmental Research*, 70, 264–271.

Robigou, V. R., Delaney, J. R. and Stakes, D. S. (1993). Large massive sulfide deposits in a newly discovered active hydrothermal system, the High Rise Field, Endeavour Segment, Juan de Fuca Ridge. *Geophysical Research Letters*, 20, 1887–1890.

Robinson, P. T., Malpas, J., Dilek, Y. and Zhou, M.-F. (2008). The significance of sheeted dike complexes in ophiolites. *GSA Today*, 18(11), doi:10.1130/GSATG22A.1, 4–10.

Robinson, P. T., Melson, W. G., O'Hearn, T. and Schmincke, H.-U. (1983). Volcanic glass compositions of the Troodos ophiolite, Cyprus. *Geology*, 11, 400–404.

Rogers, A. D., Tyler, P. A., Connelly, D. P. *et al.* (2012). The discovery of new deep-sea hydrothermal vent communities in the Southern Ocean and implications for biogeography. *PLoS Biology*, 10(1), doi:10.1371/journal.pbio.1001234.

Rona, P. A. (1980). TAG hydrothermal field: Mid-Atlantic Ridge crest at latitude 26°N. *Journal of the Geological Society London*, 137(4), 365–402.

Rona, P. A., Bougault, H., Charlou, J. L. *et al.* (1992). Hydrothermal circulation, serpentinization, and degassing at a rift valley–fracture zone intersection: Mid-Atlantic Ridge near 15°N, 45°W. *Geology*, 20, 783–786.

Rona, P. A., Hannington, M. D., Raman, C. V. *et al.* (1993). Active and relict sea-floor hydrothermal mineralization at the TAG hydrothermal field, Mid-Atlantic Ridge. *Economic Geology*, 88, 1989–2017.

Rona, P. A., Klinkhammer, G., Nelsen, T. A., Trefry, J. H. and Elderfield, H. (1986). Black smokers, massive sulfides, and vent biota at the Mid-Atlantic Ridge. *Nature*, 321, 33–37.

Rona, P. A., Thomson, G., Mottl, M. J. *et al.* (1984). Hydrothermal activity at the TAG hydrothermal field, Mid-Atlantic Ridge crest at 26°N. *Journal of Geophysical Research*, 89, 11,365–11,377.

Rona, P. A., Widenfalk, L. and Bostrom, K. (1987). Serpentinized ultramafic and hydrothermal activity at the Mid-Atlantic Ridge crest near 15°N. *Journal of Geophysical Research*, 92, 1417–1427.

Rouxel, O., Fouquet, Y. and Ludden, J. N. (2004). Subsurface processes at the Lucky Strike hydrothermal field, Mid-Atlantic Ridge: Evidence from sulfur, selenium, and iron isotopes. *Geochimica et Cosmochimica Acta*, 68, 2295–2311.

Rowland, S. K. and Walker, G. P. L. (1990). Pahoehoe and a'a in Hawaii: Volumetric flow rate controls the lava structure. *Bulletin of Volcanology*, 52(8), 615–628, doi:10.1007/BF00301212.

Rubin, A. M. (1992). Dike-induced faulting and graben subsidence in volcanic rift zones. *Journal of Geophysical Research*, 97, 1839–1858.

Rubin, A. M. (1995). Propagation of magma-filled cracks. *Annual Review of Earth and Planetary Sciences*, 23, 287–336.

Rubin, A. M. and Pollard, D. D. (1988). Dike-induced faulting in rift zones of Iceland and Afar. *Geology*, 16, 413–417.

Rubin, K. H. and Sinton, J. M. (2007). Inferences on mid-ocean ridge thermal and magmatic structure from MORB compositions. *Earth and Planetary Science Letters*, 260, 257–276.

Rubin, K. H., Embley, R. W., Clague, D. A. *et al.* (2009a). Lavas from active boninite and very recent basalt eruptions at two submarine NE Lau Basin sites. *Eos, Transactions of the American Geophysical Union*, 90(52), V43I–05.

Rubin, K. H., Macdougall, J. D. and Perfit, M. R. (1994). $^{210}Po - ^{210}Pb$ dating of recent volcanic eruptions on the seafloor. *Nature*, 369, 841–844.

Rubin, K. H., Sinton, J. M., Maclennan, J. and Hellebrand, E. (2009b). Magmatic filtering of mantle compositions at mid-ocean-ridge volcanoes. *Nature Geoscience*, 2(5), 321–328.

Rubin, K. H., Smith, M. C., Bergmanis, E. C. *et al.* (2001a). Geochemical heterogeneity within mid-ocean ridge lava flows: Insights into eruption, emplacement and global variations in magma generation. *Earth and Planetary Science Letters*, 188, 349–367.

Rubin, K. H., Smith, M. C., Bergmanis, E. C. *et al.* (2001b). Magmatic history and volcanological insights from individual lava flows erupted on the seafloor. *Earth and Planetary Science Letters*, 188, 349–367.

Rubin, K. H., Soule, S. A., Chadwick, W. W., Jr. *et al.* (2012). Volcanic eruptions in the deep sea. *Oceanography*, 25(1), 142–157, doi:org/10.5670/oceanog.2012.12.

Rudnicki, M. D., and German, C. R. (2002). Temporal variability of the hydrothermal plume above the Kairei vent field, 25°S, Central Indian Ridge. *Geochemistry, Geophysics, Geosystems*, 3(2), doi:10.1029/2001GC000240.

Ruelas-Inzunza, J., Soto, L. A. and Páez-Osuna, F. (2003). Heavy-metal accumulation in the hydrothermal vent clam *Vesicomya gigas* from Guaymas basin, Gulf of California. *Deep-Sea Research Part I*, 50(3), 757–761.

Russell, M. J., Hall, A. J. and Martin, W. (2010). Serpentinization as a source of energy at the origin of life. *Geobiology*, 8, 355–371.

Ryan, M. P. (1993). Neutral buoyancy and the structure of mid-ocean ridge magma reservoirs. *Journal of Geophysical Research*, 98, 22,321–22,338.

Rzhnanov, Y., Mayer, L., Beaulieu, S. *et al.* (2006). Deep-sea geo-referenced video mosaics. *IEEE Oceans 2006*, doi:10.1109/OCEANS.2006.307018.

Saal, A. E. and Van Orman, J. A. (2004). The ^{226}Ra enrichment in oceanic basalts: Evidence for melt-cumulate diffusive interaction processes within the oceanic lithosphere. *Geochemistry, Geophysics, Geosystems*, 5, Q02008, doi:10.1029/2003GC000620.

Sager, W. W., Zhang, J., Korenaga, J. *et al.* (2013). An immense shield volcano within the Shatsky Rise oceanic plateau, northwest Pacific Ocean. *Nature Geoscience*, 6, 976–981, doi:10.1038/ngeo1934.

Sakai, H., Des Marias, D. J., Ueda, A. and Moore, J. G. (1984). Concentrations and isotope ratios of carbon, nitrogen and sulfur in ocean-floor basalts. *Geochimica et Cosmochimica Acta*, 48, 2433–2441.

Salters, V. J. and Stracke, A. (2004). Composition of the depleted mantle. *Geochemistry, Geophysics, Geosystems*, 5(5), Q05B07, doi:10.1029/2003GC000597.

Sano, T., Sakuyama, T., Ingle, S., Rodriguez, S. and Yamasaki, T. (2011). Petrological relationships among lavas, dikes, and gabbros from Integrated Ocean Drilling Program Hole 1256D: Insight into the magma plumbing system beneath the East Pacific Rise. *Geochemistry, Geophysics, Geosystems*, 12(6), Q06013, doi:10.1029/2011GC003548.

Sansone, F. J., Resing, J. A., Tribble, G. W. *et al.* (1991). Lava–seawater interactions at shallow-water submarine lava flows. *Geophysical Research Letters*, 18, 1731–1734.

Sarrazin, J. and Juniper, S. K. (1999). Biological characteristics of a hydrothermal edifice mosaic community. *Marine Ecology Progress Series*, 185, 1–19, doi:10.3354/meps185001.

Sarrazin, J., Robigou, V., Juniper, S. K. and Delaney, J. R. (1997). Biological and geological dynamics over four years on a high temperature sulfide structure at the Juan de Fuca Ridge hydrothermal observatory. *Marine Ecology Program Series*, 152, 5–24.

Sarrazin J., Sarradin, P. M. and the MoMARETO cruise participants (2006). MoMARETO: A cruise dedicated to the spatio-temporal dynamics and the adaptations of hydrothermal vent fauna on the Mid-Atlantic Ridge. *InterRidge News*, 15, 24–33.

Sarrazin, J., Walter, C., Sarradin, P.-M. *et al.* (2006). Community structure and temperature dynamics within a mussel assemblage on the southern East Pacific Rise. In Felbeck, H. *et al.* (eds.), *Third International Symposium on Hydrothermal Vent and Seep Biology*, La Jolla, USA, September 12–16, 2005. *Cahiers de Biologie Marine*, 47(4), pp. 483–490.

Satake, K. and Atwater, B. F. (2007). Long-term perspectives on giant earthquakes and tsunamis at subduction zones. *Annual Review of Earth and Planetary Sciences*, 35, 349–374, doi:10.1146/annurev.earth.35.031306.140302.

Schander, C., Rapp, H. T., Kongsrud, J. A. *et al.* (2010). The fauna of hydrothermal vents on the Mohns Ridge (North Atlantic). *Marine Biology Research*, 6, 155–171.

Scheirer, D. S., Fornari, D. J., Humphris, S. E. and Lerner, S. (2000). High-resolution seafloor mapping using the DSL-120 sonar system: Quantitative assessment of side-scan and phase-bathymetry data from the Lucky Strike segment of the Mid-Atlantic Ridge. *Marine Geophysical Researches*, 21, 121–142.

Scheirer, D. S., Shank, T. M. and Fornari, D. J. (2006). Temperature variations at diffuse and focused flow hydrothermal vent sites along the northern East Pacific Rise. *Geochemistry, Geophysics, Geosystems*, 7(3), Q03002, doi:10.1029/2005GC001094

Schilling, J.-G., Zajac, M., Evans, R. *et al.* (1983). Petrologic and geochemical variations along the Mid-Atlantic Ridge from 29°N to 73°N. *American Journal of Science*, 283, 510–586.

Schmidt, A. K., Perfit, M. R., Rubin, K. H. *et al.* (2011). Rapid cooling rates at an active mid-ocean ridge from zircon thermochronology. *Earth and Planetary Science Letters*, 302, 349–358, doi:10.1016/j.epsl.2010.12.022.

Schmidt, C., Vuillemin, R., Le Gall, C., Gaill, F. and Le Bris, N. (2008). Geochemical energy sources from microbial primary production in the environment of hydrothermal vent shrimps. *Marine Chemistry*, 108, 18–31.

Schmidt, K., Garbe-Schönberg, D., Koschinsky, A. *et al.* (2011b). Fluid elemental and stable isotope composition of the Nibelungen hydrothermal field (8°18′ S, Mid-Atlantic Ridge): Constraints on fluid–rock interaction in heterogeneous lithosphere. *Chemical Geology*, 280, 1–18.

Schmidt, K., Koschinsky, A., Garbe-Schonberg, D., de Carvalho, L. M. and Seifert, R. (2007). Geochemistry of hydrothermal fluids from the ultramafic-hosted Logatchev hydrothermal field, 15°N on the Mid-Atlantic Ridge: Temporal and spatial investigation. *Chemical Geology*, 242, 1–21.

Schouten, H. and Denham, C. R. (2000). Comparison of volcanic construction in the Troodos ophiolite and oceanic crust using paleomagnetic inclinations from Cyprus Crustal Study Project (CCSP) CY-1 and CY-1A and Ocean Drilling Program (ODP) 504B drill cores. In Dilek, Y., Moores, E. M., Elthon, D. and Nicolas, A. (eds.), *Ophiolites and Oceanic Crust: New Insights from Field Studies and the Ocean Drilling Program*. Boulder, CO: Geological Society of America, pp. 181–194.

Schouten, H., Klitgord, K. and Gallo, D. (1993). Edge-driven microplate kinematics. *Journal of Geophysical Research*, 98, 6689–6702.

Schrenk, M. O., Kelley, D. S., Bolton, S. and Baross, J. D. (2004). Low archaeal diversity linked to sub-seafloor geochemical processes at the Lost City Hydrothermal Field, Mid-Atlantic Ridge. *Environmental Microbiology*, 6 (10), 1086–1095.

Schrenk, M. O., Kelley, D. S., Delaney, J. R. and Baross, J. A. (2003). Incidence and diversity of microorganisms within the walls of an active deep-sea sulfide chimney. *Applied Environmental Microbiology*, 69, 3580–3592.

Schroeder, T. and John, B. E. (2004). Strain localization on an oceanic detachment fault system, Atlantis Massif, 30°N, Mid-Atlantic Ridge. *Geochemistry, Geophysics, Geosystems*, 5, Q11007, doi:10.1029/2004GC000728.

Scott, R. B., Rona, P. A., McGregor, B. A. and Scott, M. R. (1974). The TAG hydrothermal field. *Nature*, 251, 301–302.

Searle, R. C. and Francheteau, J. (1986). Morphology and tectonics of the Galápagos triple junction. *Marine Geophysical Researches*, 8, 95–129.

Searle, R. C., Murton, B. J., Achenbach, K. *et al.* (2010). Structure and development of an axial volcanic ridge: Mid-Atlantic Ridge, 45°N. *Earth and Planetary Science Letters*, 299, 228–241.

Searle, R. C., Rusby, R. I., Engeln, J. *et al.* (1989). Comprehensive sonar imaging of the Easter microplate. *Nature*, 341, 701–705.

Segonzac, M., de Saint Laurent, M. and Casanova, B. (1993). L'énigme du comportement trophique des crevettes Alvinocarididae des sites hydrothermaux de la dorsale médio-atlantique. *Cahiers de Biologie Marine*, 34, 535–571.

Seher, T., Crawford, W. C., Singh, S. C. and Cannat, M. (2010). Seismic layer 2A variations in the Lucky Strike segment at the Mid-Atlantic Ridge from reflection measurements. *Journal of Geophysical Research*, 115, B07107, doi:10.1029/2009JB006783.

Seyfried, W. E., Jr. (1987). Experimental and theoretical constraints on hydrothermal alteration processes at mid-ocean ridges. *Annual Review of Earth and Planetary Science Letters*, 15, 317–335.

Seyfried, W. E., Jr., Berndt, M. E. and Seewald, J. S. (1988). Hydrothermal alteration processes at mid-ocean ridges: Constraints from diabase alteration experiments, hot spring fluids and composition of the oceanic crust. *Canadian Mineralogist*, 26, 787–804.

Seyfried, W. E., Jr., Foustoukos, D. I. and Allen, D. E. (2004). Ultramafic-hosted hydrothermal systems at mid-ocean ridges: Chemical and physical controls on pH, redox, and carbon reduction reactions. In German, C. R., Lin, J. and Parsons, M. (eds.), *Mid-Ocean Ridges: Hydrothermal Interactions between the Lithosphere and Oceans*. Geophysical Monograph 148, Washington, DC: American Geophysical Union, pp. 267–284.

Seyfried, W. E., Jr., Pester, N. J., Ding, K. and Rough, M. (2011). Vent fluid chemistry of the Rainbow hydrothermal system (39°N): Phase equilibria and in situ pH controls on subseafloor alteration processes. *Geochimica et Cosmochimica Acta*, 75, 1574–1593.

Seyfried, W. E., Jr., Pester, N. and Fu, Q. (2010). Phase equilibria controls on the chemistry of vent fluids from hydrothermal systems on slow spreading ridges: Reactivity of plagioclase and olivine solid solutions and the pH-silica connection. In Rona, P., Devey, C. and Murton, B. (eds.), *Diversity of Hydrothermal Systems on Slow Spreading Ocean Ridges*. Geophysical Monograph 188, Washington, DC: American Geophysical Union, pp. 297–320.

Shand, S. J. (1949). Rocks of the Mid-Atlantic Ridge. *Journal of Geology*, 57, 89–92.

Shank, T. M. (2004). The evolutionary puzzle of seafloor life. *Oceanus*, 42(2), 1–8.

Shank, T. M. (2006). Preliminary biological characterization of vent sites and the evolutionary relationships of vent fauna on the southern Atlantic Ridge. *Exploration and biogeography of deep-water chemosynthetic ecosystems on the Atlantic Equatorial Belt region. First Results and Planning for Future Research Workshop*, Barcelona, Spain.

Shank, T. M., Fornari, D. J., Von Damm, K. L. *et al.* (1998). Temporal and spatial patterns of biological community development at nascent deep-sea hydrothermal vents (9°50′ N, East Pacific Rise). *Deep-Sea Research Part II*, 45, 465–515.

Shank, T. M., Fornari, D. and Yoerger, D. *et al.* (2003). Deep submergence synergy: *Alvin* and *ABE* explore the Galápagos Rift at 86°W. *Eos, Transactions of the American Geophysical Union*, 84, 425–433, doi:10.1029/2003EO410001.

Shanks, W. C. III., Böhlke, J. K. and Seal, R. R. II (1995) Stable isotopes in mid-ocean ridge hydrothermal systems: Interactions between fluids, minerals, and organisms. In Humphris, S., Zierenberg, R., Mullineau, L. and Thomson, R. (eds.), *Seafloor Hydrothermal Systems: Physical, Chemical, Biological, and Geological Interactions*. Geophysical Monograph 91, Washington, DC: American Geophysical Union, pp. 194–221.

Shervais, J. W. (2001). Birth, death, and resurrection: The life cycle of suprasubduction zone ophiolites. *Geochemistry, Geophysics, Geosystems*, 2, 1010, doi:10.1029/2000GC 000080.

Shih, J. S.-F. (1980). *The Nature and Origin of Fine-Scale Sea-Floor Relief*. Ph.D. dissertation, Massachusetts Institute of Technology.

Shipboard Scientific Party (2005). *Expedition 305: Oceanic Core Complex Formation, Atlantis Massif 2*. USIO Preliminary Reports. College Station, TX: Integrated Ocean Drilling Program.

Siler, D. L. and Karson, J. A. (2009). Three-dimensional structure of inclined sheet swarms: Implications for crustal thickening and subsidence in the volcanic rift zones of Iceland. *Journal of Volcanology and Geothermal Research*, 188, 333–346.

Sims, K. W. W., Blichert-Toft, J., Perfit, M. R. *et al.* (2003). Aberrant youth: Chemical and isotopic constraints on the origin of off-axis lavas from the East Pacific Rise, 9°–10° N. *Geochemistry, Geophysics, Geosystems*, 4, doi:101029/2002GC000443.

Singh, S. C., Crawford, W. C., Carton, H. *et al.* (2006). Discovery of a magma chamber and faults beneath a Mid-Atlantic Ridge hydrothermal field. *Nature*, 442, 1029–1032, doi:10:1038/nature05105.

Sinha, M. C., Navin, D. A., Macgregor, L. M. *et al.* (1997). Evidence for accumulated melt beneath the slow spreading Mid-Atlantic Ridge. In Cann, J. R., Elderfield, H. and Laughton, A. (eds.), *Mid-Ocean Ridges: Dynamics of Processes Associated with Creation of New Ocean Crust*. Cambridge: Cambridge University Press, pp. 233–253.

Sinton, J. M. and Detrick, R. S. (1992). Mid-ocean ridge magma chambers. *Journal of Geophysical Research*, 97, 197–216.

Sinton, J. M., Bergmanis, E., Rubin, K. *et al.* (2002). Volcanic eruptions on mid-ocean ridges: New evidence from the superfast-spreading East Pacific Rise, 17°–19°S. *Journal of Geophysical Research*, 107, 301–314.

Sinton, J. M., White, S. M., Colman, A., Rubin, K. H. and Bowles, J. A. (2010). Volcanic eruptions on the western Galápagos Spreading Center: Connecting magma supply at depth to eruption rate on the surface. *Eos, Transactions of the American Geophysical Union*, Fall Meeting Supplement, V52A–06.

Sinton, J. M., Wilson, D. S., Christie, D. M., Hey, R. N. and Delaney, J. R. (1983). Petrologic consequences of rift propagation on oceanic spreading ridges. *Earth and Planetary Science Letters*, 62, 193–207.

Smith, D.K., (2004). Ears in the Ocean, Oceanus, 42 (2), ISSN 1559-1263.

Smith, D. K. and Cann, J. R. (1990). Hundreds of small volcanoes on the median valley floor of the Mid-Atlantic Ridge at 24–30°N. *Nature*, 348, 152–155, doi:10.1038/348152a0.

Smith, D. K. and Cann, J. R. (1993). Building the crust at the Mid-Atlantic Ridge. *Nature*, 365, 707–715.

Smith, D. K., Cann, J. R. and Escartín, J. (2006). Widespread active detachment faulting and core complex formation near 13°N on the Mid-Atlantic Ridge. *Nature*, doi: 10.1038/nature04950.

Smith, D. K., Escartín, J., Cannat, M. *et al.* (2003). Spatial and temporal distribution of seismicity along the northern Mid-Atlantic Ridge (15°–35°N). *Journal of Geophysical Research*, 108(B3), 2167, doi:10.1029/2002JB001964.

Smith, D. K., Escartín, J., Schouten, H. and Cann, J. R. (2008). Fault rotation and core complex formation: Significant processes in seafloor formation at slow-spreading mid-ocean ridges (Mid-Atlantic Ridge, 13–15° N). *Geochemistry, Geophysics, Geosystems*, 9(3), Q03003, doi:10.1029/2007GC001699.

Smith, D. K., Escartín, J., Schouten, H. and Cann, J. R. (2012). Active long-lived faults emerging along slow-spreading mid-ocean ridges. *Oceanography*, 25(1), 94–99, doi.org/10.5670/oceanog.2012.07.

Smith, D. K., Exon, N., Barriga, F., and Tatsumi,Y. (2010). Forty years of successful international collaboration in scientific ocean drilling, *Eos, Transactions of the American Geophysical Union*, 91, 393–394.

Smith, M. C., Perfit, M. R. and Jonasson, I. R. (1994). Petrology and geochemistry of basalts from the southern Juan de Fuca Ridge: Controls on the spatial and temporal evolution of mid-ocean ridge basalt. *Journal of Geophysical Research*, 99(B3), 4787–4812, doi:10.1029/93JB02158.

Smith, M. C., Perfit, M. R., Fornari, D. J. et al. (2001). Magmatic processes and segmentation at a fast spreading mid-ocean ridge: Detailed investigation of an axial discontinuity on the East Pacific Rise crest at 9°37′ N. *Geochemistry, Geophysics, Geosystems*, 2, 1–32, doi: 10.1029/2000GC000134.

Smith, W. H. and Sandwell, D. T. (1997). Global sea floor topography from satellite altimetry and ship depth soundings. *Science Magazine*, 277, issue 5334.

Snow, J., Jokat, W., Hellebrand, E., Mühe, R. and Shipboard Scientific Party (2001). Magmatic and hydrothermal activity in Lena Trough, Arctic Ocean. *Eos, Transactions of the American Geophysical Union*, 85(47), T11G-01.

Sogin, M., Morrison, H. G., Huber, J. A. *et al.* (2006). Microbial diversity in the deep sea and the unexplored 'rare biosphere'. *Proceedings of the National Academy of Sciences*, 103, 12,115–12,120.

Sohn, R. A., Willis, C., Humphris, S. E. *et al.* (2008). Explosive volcanism on the ultraslow-spreading Gakkel ridge, Arctic Ocean. *Nature*, (453), 1236–1238.

Solomon, S. C., Huang, P. Y. and Meinke, L. (1988). The seismic moment budget of slowly spreading ridges. *Nature*, 334, 58–60.

Sontag, S. and Drew, C. (1998). Blind Man's Bluff: The Untold Story of American Submarine Espionage. New York: Harper Collins, ISBN 0-06-103004-X.

Soule, S. A. and Cashman, K. V. (2005). Shear rate dependence of the pahoehoe-to-'a'a transition: Analog experiments. *Geology*, 33, 361–364.

Soule, S. A., Fornari, D. J., Perfit, M. R. *et al.* (2005). Channelized lava flows at the East Pacific Rise crest 9°–10°N: The importance of off-axis lava transport in developing the architecture of young oceanic crust. *Geochemistry, Geophysics, Geosystems*, 6, Q08005, doi:10.1029/2005GC000912.

Soule, S. A., Fornari, D. J., Perfit, M. R. *et al.* (2006). Incorporation of seawater into mid-ocean ridge lavas during emplacement. *Earth and Planetary Science Letters*, 252, 289–307.

Soule, S. A., Fornari, D. J., Perfit, M. P. and Rubin, K. (2007a). New insights into mid-ocean ridge volcanic processes from the 2005–06 eruption of the East Pacific Rise, 9°46′ –56′ N. *Geology*, 35, 1079–1082, doi:10.1130/G23924A.1.

Soule, S. A., Nakata, D. S., Fornari, D. J. *et al.* (2012). CO_2 variability in mid-ocean ridge basalts from syn-emplacement degassing: Constraints on eruption dynamics. *Earth and Planetary Science Letters*, 327, 39–49, doi:10.1016/j.epsl.2012.01.034.

Soule, S. A., White, S. M., Fornari, D. J., Ferrini, V. and Tivey, M. A. (2007b). Comparison of pre- and post- eruption surveys of the 2005–2006 East Pacific Rise Volcanic Event: Implications for fast-spreading mid-ocean ridge eruption processes. *Eos, Transactions of the American Geophysical Union*, 88(52), V21A–0383.

Sparks, D. W. and Parmentier, E. M. (1991). Melt extraction from the mantle beneath spreading centers. *Earth and Planetary Science Letters*, 105, 368–377.

Sparks, J. W. (1995). Geochemistry of the lower sheeted dike complex, Hole 504B, Leg 140. In Erzinger, J., Becker, K., Dick, H. J. B. and Stokking, L. B. (eds.), *Proceedings of the Ocean Drilling Program, Scientific Results, Leg 140*. College Station, TX: Ocean Drilling Program, pp. 81–97.

Sparks, R. S. J., Meyer, P. and Sigurdsson, H. (1980). Density variation amongst mid-ocean ridge basalts: Implications for magma mixing and the scarcity of primitive lavas. *Earth and Planetary Science Letters*, 46, 419–430.

Spiegelman, M. (1993). Physics of melt extraction: Theory, implications and applications. *Philosophical Transactions of the Royal Society of London, Series A*, 342, 23–41.

Spiegelman, M. (1996). Geochemical consequences of melt transport in 2D: The sensitivity of trace elements to mantle dynamics. *Earth and Planetary Science Letters*, 139, 115–132.

Spiegelman, M. and Kelemen, P. B. (2003). Extreme chemical variability as a consequence of channelized melt transport. *Geochemistry, Geophysics, Geosystems*, 4(7), 1055, doi:10.1029/2002GC000336.

Spiegelman, M. and McKenzie, D. (1987). Simple 2-D models for melt extraction at mid-ocean ridges and island arcs. *Earth and Planetary Science Letters*, 83, 137–152.

Spiegelman, M. and Reynolds, J. R. (1999). Combined dynamic and geochemical evidence for convergent melt flow beneath the East Pacific Rise. *Nature*, 402, 282–285.

Spiess, F. N. and Tyce, R. C. (1973). Marine Physical Laboratory Deep Tow Instrumentation System. La Jolla, CA: Scripps Institute of Oceanography Reference 73–4

Spiess, F. N., Macdonald, K. C., Atwater, T. *et al.* (1980). East Pacific Rise: Hot springs and geophysical experiments. *Science*, 207, 1421–1433.

Spooner, E. T. C. and Fyfe, W. S. (1973). Sub-seafloor metamorphism, heat and mass transfer. *Contributions to Mineralogy and Petrology*, 42, 287–304.

Squyres, S. W., Arvidson, R. E., Blaney, D. L. *et al.* (2006). Rocks of the Columbia Hills. *Journal of Geophysical Research*, 111, E02S11, doi:10.1029/2005JE002562.

Stakes, D. S., Perfit, M. R., Tivey, M. A. *et al.* (2006). The Cleft revealed: Geologic, magnetic and morphologic evidence for construction of upper oceanic crust along the southern Juan de Fuca Ridge. *Geochemistry, Geophysics, Geosystems*, 7, Q04003, doi:10.1029/2005GC001038.

Stakes, D. S., Shervais, J. W. and Hopson, C. A. (1984). The volcanic-tectonic cycle of the FAMOUS and AMAR valleys, Mid-Atlantic Ridge (36°47′ N): Evidence from basalt glass and phenocryst compositional variations for a steady state magma chamber beneath the valley midsections. *Journal of Geophysical Research*, 89, 6995–7028.

Staudigel, H., Gee, J., Tauxe, L. and Varga, R. J. (1992). Shallow intrusive directions of sheeted dikes in the Troodos ophiolite: Anisotropy of magnetic susceptibility and structural data. *Geology*, 20, 841–844.

Staudigel, H., Koppers, A. A. P., Lavelle, J. W., Pitcher, T. J. and Shank, T. M. (2010). From the guest editors 2010. *Oceanography*, 23(1), 18–19, doi:org/10.5670/oceanog.2010.59.

Steinmann, G. (1905). Geologische Beobachtungen in den Alpen, II. Die Schardtsche Ueberfaltungstheorie und die geologische Bedeutung der Tiefseeabsätze und der ophiolithischen Massengesteine. *Berichte der Naturforschenden Gesellschaft zu Freiburg*, 1(16), 44–67.

Stewart, M. A., Karson, J. A. and Klein, E. M. (2005). Four-dimensional upper crustal construction at fast-spreading mid-ocean ridges: A perspective from an upper crustal cross-section a the Hess Deep Rift. *Journal of Volcanology and Geothermal Research*, 144, 287–309.

Stewart, M. S., Klein, E. M. and Karson, J. A. (2002). The geochemistry of dikes and lavas from the north wall of the Hess Deep Rift: Insights into the four-dimensional character of crustal construction at fast-spreading mid-ocean ridges. *Journal of Geophysical Research*, 107(B9), 2181, doi:10.1029/2000JB000051.

Stewart, M. A., Klein, E. M., Karson, J. A. and Brophy, J. G. (2003). Geochemical relationships between dikes and lavas at the Hess Deep Rift: Implications for magma eruptibility. *Journal of Geophysical Research*, 108(B4), 2184, doi:10.1029/2001JB001622.

Stracke, A. and Bourdon, B. (2009). The importance of melt extraction for tracing mantle heterogeneity. *Geochimica et Cosmochimica Acta*, 73, 218–238.

Streams, M. E., Fisher, C. R. and Fiala-Médioni, A. (1997). Methanotrophic symbiont location and fate of carbon incorporated from methane in a hydrocarbon seep mussel. *Marine Biology*, 129, 465–476.

Streckeisen, A. (1976). To each plutonic rock its proper name. *Earth-Science Reviews*, 12(1), 1–33.

Stroup, J. B. and Fox, P. J. (1981). Geologic investigations in the Cayman Trough: Evidence for thin oceanic crust along the Mid-Cayman Rise. *Journal of Geology*, 89, 395–420.

Sturm, M., Goldstein, S., Klein, E. M., Karson, J. A. and Murrell, M. T. (2000). Uranium-series age constraints on lavas from the axial valley of the Mid-Atlantic Ridge, MARK Area. *Earth and Planetary Science Letters*, 181, 61–71.

Sturm, M., Klein, E. M., Karsten, J., Graham, D. and Karson, J. (2000). Evidence for subduction-related contamination of the mantle beneath the southern Chile Ridge: Implications for ambiguous ophiolite compositions. In Dilek, Y., Moores, E. M., Elthon, D. and Nicolas, A. (eds.), *Ophiolites and Oceanic Crust: New Insights from Field Studies and Oceanic Drilling*. Boulder, CO: Geological Society of America, pp. 13–20.

Summit, M. (2000). *Ecology, Physiology, and Phylogeny of Subseafloor Thermophiles from Mid-Ocean Ridge Environments*. Ph.D. thesis, University of Washington.

Summit, M. and Baross, J. A. (1998). Thermophilic subseafloor microorganisms from the 1996 North Gorda Ridge eruption. *Deep-Sea Research II*, 45, 2751–2766.

Sun, S. S. and McDonough, W. F. (1989). Chemical and isotopic systematics of oceanic basalts: Implications for mantle composition and processes. In Saunders, A. D. and Norry, M. J. (eds), *Magmatism in the Ocean Basins*. Special Publication 42, London: Geological Society of London, pp. 313–345.

Sun, S. S., Nesbitt, R. W. and Sharaskin, A. Y. (1979). Geochemical characteristics of mid-ocean ridge basalts. *Earth and Planetary Science Letters*, 44(1), 119–138.

Swanson, D. A. (1973). Pahoehoe flows from the 1969–1971 Mauna Ulu eruption, Kilauea volcano, Hawaii. *Geological Society of America Bulletin*, 84, 615–626.

Sykes, L. R. (1967). Mechanism of earthquakes and nature of faulting on the Mid-Oceanic Ridges. *Journal of Geophysical Research*, 72(8), 2131–2153.

Sykes, L. R. (1969). Seismicity of the mid-oceanic ridge system. In Hart, P. J. (ed.), *The Earth's Crust and Upper Mantle*. Geophysical Monograph 13. Washington, DC: American Geophysical Union, pp. 148–153.

Takai, K., Nakamura, K., Toki, T. *et al.* (2008). Cell proliferation at 122 °C and isotopically heavy CH_4 production by a hyperthermophilic methanogen under high pressure cultivation. *Proceedings of the National Academy of Sciences*, 105, 10,949–10,954.

Tandberg, A. H., Rapp, H. T., Schander, C. *et al.* (2012) *Exitomelita sigynae* gen. et sp. nov.: A new amphipod from the Arctic Loki Castle vent field with potential gill ectosymbionts. *Polar Biology*, 35, 705–716.

Tartarotti, P., Hayman, N. W., Anma, R. *et al.* (2006). Structure of Hole 1256D: The role of mechanical deformation in superfast-spread crust. *Eos, Transactions of the American Geophysical Union*, 87(52), B31B-1091.

Tauxe, L., Gee, J. S. and Staudigel, H. (1998). Flow directions in dikes from anisotropy of magnetic susceptibility data: The bootstrap way. *Journal of Geophysical Research*, 103, 17,775–17,790.

Taylor, C. D. and Wirsen, C. O. (1997). Microbiology and ecology of filamentous sulfur formation. *Science*, 277, 1483–1485.

Teixera, S., Cambon-Bonavita, M.-A., Serrao, E. A., Desbruyères, D. and Arnaud-Haond, S. (2011). Recent population expansion and connectivity in the hydrothermal shrimp *Rimicaris exoculata* along the Mid-Atlantic Ridge. *Journal of Biogeography*, 38, 563–574.

Teske, A., Hinrichs, K.-U., Edgcomb, V. *et al.* (2007). Microbial diversity of hydrothermal sediments in the Guaymas Basin: Evidence for anerobic methanotrophic communities. *Applied and Environmental Microbiology*, 68(4), 1994–2007.

Thompson, G., Bryan, W. B. and Humphris, S. E. (1989). Axial volcanism on the East Pacific Rise, 10–12°N. In Saunders, A. D. and Norry, M. J. (eds.), *Magmatism in the Ocean Basins*. Special Publication 42, London: Geological Society of London, pp. 181–200.

Thompson, G., Mottl, M. J. and Rona, P. A. (1985). Morphology, mineralogy, and chemistry of hydrothermal deposits from the TAG area, 26°N Mid-Atlantic Ridge. *Chemical Geology*, 49, 243–257.

Thordarson, T. and Self, S. (2003). Atmospheric and environmental effects of the 1783–1784 Laki eruption: A review and reassessment. *Journal of Geophysical Research*, 108(D1), 4011, doi:10.1029/2001JD002042.

Thordarson, T., Self, S., Miller, D. J., Larsen, G. and Vilmundardóttir, E. G. (2003). Sulphur release from flood lava eruptions in the Veidivötn, Grímsvötn and Katla volcanic systems, Iceland. In Oppenheimer, C., Pyle, D. M. and Barclay, J. (eds.), *Volcanic Degassing*. Special Publication 213, London: Geological Society of London, 103–121.

Thorseth, I. H., Pedersen, R. B., Kruber, C. and Kosler, J. (2007). Low-temperature hydrothermal deposits at the 71°N vent fields at the Arctic Mid-Ocean Ridge: Architecture, microstructures, and geochemistry. *Eos, Transactions of the American Geophysical Union*, 88(52), Fall Meeting Supplement, Abstract OS43A-0996.

Tiezzi, L. J. and Scott, R. B. (1980). Crystal fractionation in a cumulate gabbro, Mid-Atlantic Ridge, 26°N. *Journal of Geophysical Research*, 85, 5438–5454.

Titus, T. N., Kieffer, H. H. and Christensen, P. R. (2003). Exposed water ice discovered near the south pole of Mars. *Science*, 299, 1048–1051.

Tivey, M. A. and Johnson, H. P. (2002). Crustal magnetization reveals subsurface structure of Juan de Fuca Ridge hydrothermal vent fields. *Geology*, 30(11), 979–982.

Tivey, M. A., Johnson, H. P., Fleutelot, C. *et al.* (1998a). Direct measurement of magnetic reversal polarity boundaries in a cross-section of oceanic crust. *Geophysical Research Letters*, 25, 3631–3634.

Tivey, M. A., Rona, P. A. and Kleinrock, M. C. (1996). Reduced crustal magnetization beneath relict hydrothermal mounds: TAG hydrothermal field, Mid-Atlantic Ridge, 26°N. *Geophysical Research Letters*, 23, 3511–3514.

Tivey, M. A., Schouten, H. and Kleinrock, M. C. (2003). A near-bottom magnetic survey of the Mid-Atlantic Ridge axis at 26°N: Implications for the tectonic evolution of the TAG segment. *Journal of Geophysical Research*, 108(B5), 2277, doi:10.1029/2002JB001967.

Tivey, M. K. (1995). Modeling chimney growth and associated fluid flow at seafloor hydrothermal vent sites. In Humphris, S., Zierenberg, R., Mullineaux, L. and Thomson, R. (eds.), *Seafloor Hydrothermal Systems: Physical, Chemical, Biological, and Geological Interactions*. Geophysical Monograph 91, Washington, DC: American Geophysical Union, pp. 158–177.

Tivey, M. K. (2004). Environmental conditions within active seafloor vent structures: Sensitivity to vent fluid composition and fluid flow. In Wilcock, W., Cary, C., DeLong, E., Kelley, D. and Baross, J. (eds.), *Subseafloor Biosphere at Mid-Ocean Ridges*. Geophysical Monograph 144, Washington, DC: American Geophysical Union, pp. 137–152.

Tivey, M. K. (2007). Generation of seafloor hydrothermal vent fluids and associated mineral deposits. *Oceanography*, 20, 50–65.

Tivey, M. K. and Delaney, J. R. (1986). Growth of large sulfide structures on the Endeavour Segment of the Juan de Fuca Ridge. *Earth and Planetary Science Letters*, 77, 303–317.

Tivey, M. K., Humphris, S. E., Thompson, G., Hannington, M. D. and Rona, P. A. (1995). Deducing patterns of fluid flow and mixing within the TAG active hydrothermal mound using mineralogical and geochemical data. *Journal of Geophysical Research*, 100, 12,527–12,555.

Tivey, M. K., Mills, R. A. and Teagle, D. A. H. (1998b). Temperature and salinity of fluid inclusions in anhydrite as indicators of seawater entrainment and heating in the TAG active mound. In Herzig, P. M., Humphris, S. E. *et al.* (eds.), *Proceedings of the Ocean Drilling Program, Scientific Results, Leg 158*. College Station TX: Ocean Drilling Program, pp. 179–190.

Tivey, M. K., Stakes, D. S., Cook, T. L., Hannington, M. D. and Petersen, S. (1999). A model for growth of steep-sided vent structures on the Endeavour Segment of the Juan de Fuca Ridge: Results of a petrologic and geochemical study. *Journal of Geophysical Research*, 104, 22,859–22,883.

Tolstoy, M., Cowan, J. P., Baker, E. T. *et al.* (2006). A seafloor spreading event captured by seismometers. *Science*, 314, 1920–1922, doi:10.1126science.1133950.

Tolstoy, M., Waldhauser, F., Bohnenstiehl, D. R., Weekly, R. T. and Kim, W.-Y. (2008). Seismic identification of along-axis hydrothermal flow on the East Pacific Rise. *Nature*, 451, 181–184.

Tominaga, M. and Sager, W. W. (2010). Revised Pacific M-anomaly geomagnetic polarity timescale. *Geophysical Journal International*, 182, 203–232, doi:10.1111/j.1365-246X.2010.04619.x.

Tominaga, M. and Umino, S. (2010). Lava deposition history in ODP Hole 1256D: Insights from log-based volcanostratigraphy. *Geochemistry, Geophysics, Geosystems*, 11, Q05003, doi:10.1029/2009GC002933.

Tominaga, M., Teagle, D. A., Alt, J. C. and Umino, S. (2009). Determination of the volcanostratigraphy of oceanic crust formed at superfast spreading ridge: Electrofacies analyses of ODP/IODP Hole 1256D. *Geochemistry, Geophysics, Geosystems*, 10(1), doi:10.1029/2008GC002143.

Toner, B. M., Fakra, S. C., Manganini, S. J. *et al.* (2009). Preservation of iron (II) by carbon-rich matrices in a hydrothermal plume. *Nature Geosciences*, 2, 197–201.

Toomey, D. R. and Hooft, E. E. E. (2008). Mantle upwelling, magmatic differentiation, and the meaning of axial depth at fast-spreading ridges. *Geology*, 36, 679–682.

Toomey, D. R., Jousselin, D., Dunn, R. A., Wilcock, W. S. D. and Detrick, R. S. (2007). Skew of mantle upwelling beneath the East Pacific Rise governs segmentation. *Nature*, 446, 409–414.

Toomey, D. R., Purdy, G. M., Solomon, S. and Wilcox, W. (1990). The three dimensional seismic velocity structure of the East Pacific Rise near latitude 9°30′ N. *Nature*, 347, 639–644.

Toomey, D. R., Solomon, S. C. and Purdy, G. M. (1988). Microearthquakes beneath the median valley of the Mid-Atlantic Ridge near 23° N: Tomography and tectonics. *Journal of Geophysical Research*, 93, 9093–9112.

Toomey, D. R., Solomon, S. C., Purdy, G. M. and Murray, M. H. (1985). Microearthquakes beneath the median valley of the Mid-Atlantic Ridge near 23°N: Hypocenters and focal mechanisms. *Journal of Geophysical Research*, 90, 5443–5458.

Toomey, D. R., Solomon, S. C. and Purdy, G. M. (1994). Tomographic imaging of the shallow crustal structure of the East Pacific Rise at 9°30′ N. *Journal of Geophysical Research*, 99, 24,135–24,157.

Trask, J. L. and Van Dover, C. L. (1999). Site-specific and ontogenetic variations in nutrition of mussels (*Bathymodiolus* sp.) from the Lucky Strike hydrothermal vent field, Mid-Atlantic Ridge. *Limnology and Oceanography*, 44, 334–343.

Treves, B. and Harper, G. D. (1994). Exposure of serpentinites on the ocean floor: Sequence of faulting and hydrofracturing in the northern Apennine ophicalcites. *Ofioliti*, 19b, 435–466.

Tucholke, B. E. and Schouten, H. (1988). Kane fracture zone. *Marine Geophysical Researches*, 10, 1–39.

Tucholke, B. E. and Vogt, P. R. (1979). Western North Atlantic: Sedimentary evolution and aspects of tectonic history In Tucholke, B. E., Vogt, P. R. *et al.*, *Initial Reports Deep Sea Drilling Project, Leg 63*, Washington, DC: US Government Printing Office, pp. 791–825.

Tucholke, B. E., Behn, M. D., Buck, W. R. and Lin, J. (2008). Role of melt supply in oceanic detachment faulting and formation of megamullions. *Geology*, 36(6), 455–458, doi:10.1130/G24639A.

Tucholke, B. E., Lin, J. and Kleinrock, M. C. (1998). Megamullions and mullion structure defining oceanic metamorphic core complexes on the Mid-Atlantic Ridge. *Journal of Geophysical Research*, 103, 9857–9866.

Tucholke, B. E., Lin, J., Kleinrock, M. C. *et al.* (1997). Segmentation and crustal structure of the western Mid-Atlantic Ridge flank, 25°25′ –27°10′ N and 0–29 m. y. *Journal of Geophysical Research*, 102, 10,203–10,224.

Tunnicliffe, V. and Fowler, M. R. (1996). Influence of sea-floor spreading on the global hydrothermal vent fauna. *Nature*, 379, 531–533.

Tunnicliffe, V. A., McArthur, G. and McHugh, D. (1998). A biogeographical perspective of the deep-sea hydrothermal vent fauna. *Advances in Marine Biology*, 34, 353–442.

Twiss, R. J. and Moores, E. M. (1992). *Structural Geology*. New York: W. H. Freeman.

Umbgrove, J. H. F. (1947). *The Pulse of the Earth*. The Hague: Martinus Nijhoff.

Umino, S., Crispini, L., Tartarotti, P. *et al.* (2008). Origin of the sheeted dike complex at superfast spread East Pacific Rise revealed by deep ocean crust drilling at Ocean Drilling Program Hole 1256D. *Geochemistry, Geophysics, Geosystems*, 9, Q06O08, doi:10.1029/2007GC001760.

Umino, S., Nealson, K. and Wood, B. (2013). Drilling to Earth's mantle. *Physics Today*, 66(8), 36–41, doi:10.1063/PT.3.2082.

University National Laboratory System (UNOLS) (1994). *The Global Abyss: An Assessment of Deep Submergence Science in the United States*. Narragansett, RI: UNOLS Office, University of Rhode Island.

Van Ark, E. M., Detrick, R. S., Canales, J. P. *et al.* (2007). Seismic structure of the Endeavour Segment, Juan de Fuca Ridge: Correlations with seismicity and hydrothermal activity. *Journal of Geophysical Research*, 112, B02401, doi.org/10.1029/2005JB004210.

Van Dover, C. L. (1995). Ecology of Mid-Atlantic Ridge hydrothermal vents. In Parson, L. M., Walker, C. L. and Dixon, D. R. (eds.), *Hydrothermal Vents and Processes*, Special Publication 87, London: Geological Society of London, pp. 257–294.

Van Dover, C. L. (2000). *The Ecology of Deep-Sea Hydrothermal Vents*. Princeton, NJ: Princeton University Press.

Van Dover, C. L. (2003). Variation in community structure within hydrothermal vent mussel beds of the East Pacific Rise. *Marine Ecology-Progress Series*, 253, 55–66.

Van Dover, C. L. and Doerries, M. B. (2005). Community structure in mussel beds at Logatchev hydrothermal vents and a comparison of macrofaunal species richness on slow- and fast-spreading mid-ocean ridges. *Marine Ecology*, 26, 110–120.

Van Dover, C. L. and Trask, J. L. (2001). Biodiversity in mussel beds at a deep-sea hydrothermal vent and a shallow-water intertidal site. *Marine Ecology Program Series*, 195, 169–178.

Van Dover, C. L., Desbruyères, D., Segonzac, M. *et al.* (1996). Biology of the Lucky Strike hydrothermal field. *Deep-Sea Research Part I*, 43(9), 1509–1529.

Van Dover, C. L., German, C. R., Speer, K. G., Parson, L. M. and Vrijenhoek, R. C. (2002). Evolution and biogeography of deep-sea vent and seep invertebrates. *Science*, 95, 1253–1257.

Van Dover, C. L., Humphris, S. E., Fornari, D. *et al.* (2001). Biogeography and ecological setting of Indian Ocean hydrothermal vents. *Science*, 294, 818–823.

Van Dover, C. L., Szuts, E. A., Chamberlain, S. C and Cann, J. R. (1989). A novel eye in "eyeless" shrimp from hydrothermal vents of the Mid-Atlantic Ridge. *Nature*, 337, 458–460.

Vanko, D. A. (1988). Temperature, pressure, and composition of hydrothermal fluids, with their bearing on the magnitude of tectonic uplift at mid-ocean ridges, inferred from fluid inclusions in oceanic layer 3 rocks. *Journal of Geophysical Research*, 93, 4595–4611.

Vanko, D. A. and Batiza, R. (1982). Gabbroic rocks from the Mathematician Ridge failed rift. *Nature*, 300, 742–744.

Varga, R. J. (1991). Modes of extension at mid-ocean ridge spreading centers: Evidence from the Solea graben, Troodos ophiolite Cyprus. *Journal of Structural Geology*, 13, 517–538.

Varga, R. J., Gee, J. S., Staudigel, H. and Tauxe, L. (1998). Dike surface lineations as magma flow indicators within the sheeted dike complex of the Troodos Ophiolite, Cyprus. *Journal of Geophysical Research*, 103, 5241–5256.

Varga, R. J., Gee, J., Bettison-Varga, L., Anderson, R. S. and Johnson, C. L. (1999). Early establishment of sea-floor hydrothermal systems during structural extension: Paleomagnetic evidence from the Troodos ophiolite, Cyprus. *Earth and Planetary Science Letters*, 1999, 221–235.

Varga, R. J., Horst, A., Gee, J. S. and Karson, J. A. (2008). Direct evidence from anisotropy of magnetic susceptibility for lateral melt migration at superfast spreading centers. *Geochemistry, Geophysics, Geosystems*, 9(8), Q08008, doi:10.1029/2008GC002075.

Varga, R. J., Karson, J. A. and Gee, J. S. (2004). Paleomagnetic constraints on tilt and fault models for oceanic crust from oriented samples from the Hess Deep Rift, Equatorial Pacific Ocean. *Journal of Geophysical Research*, 109(2), 1–22, doi:10.1029/2003JB002486.

Veloso, E. E., Hayman, N. W., Anna, R. *et al.* (2014). Magma flow directions in the sheeted dike complex at superfast spreading mid-ocean ridges: Insights from IODP Hole 1256D, eastern Pacific. *Geochemistry, Geophysics, Geosystems*, 15, 1283–1295. doi:10.1002/2013GC004957.

Ver Eecke, H. C., Akerman, N. H., Huber, J. A. *et al.* (2013) Growth kinetics and energetics of a deep-sea hyperthermophilic methanogen under varying environmental conditions. *Environmental Microbial Reports*, 5(5), 665–671.

Vera, E. E. and Diebold, J. B. (1994). Seismic imaging of oceanic layer 2A between 9°30′ N and 10°N on the East Pacific Rise from two-ship wide-aperture profiles. *Journal of Geophysical Research*, 99, 3031–3041.

Vetriani, C., Chew, Y. S., Miller, S. M. *et al.* (2004b). Mercury adaptation among bacteria from a deep-sea hydrothermal vent. *Applied Environmental Microbiology*, 71, 220–226.

Vetriani, C., Speck, M. D., Ellor, S. V., Lutz, R. A. and Starovoytov, V. (2004a). *Thermovibrio ammonificans* sp. nov.: A thermophilic, chemolithotrophic, nitrate ammonifying bacterium from deep-sea hydrothermal vents. *International Journal of Systematic and Evolutionary Biology*, 54, 175–181.

Vine, F. J. and Matthews, D. H. (1963). Magnetic anomalies over oceanic ridges. *Nature*, 199, 947–949.

Von Damm, K. L. (1995). Controls on the chemistry and temporal variability of seafloor hydrothermal fluids. In Humphris, S., Zierenberg, R., Mullineau, L. and Thomson, R. (eds.), *Seafloor Hydrothermal Systems: Physical, Chemical, Biological, and Geological Interactions*. Geophysical Monograph 91, Washington, DC: American Geophysical Union, pp. 222–247

Von Damm, K. L. (2000). Chemistry of hydrothermal vent fluids from 9°–10°N, East Pacific Rise: "Time zero," the immediate post eruptive period. *Journal of Geophysical Research*, 105(B5), 11,203–11,222.

Von Damm, K. L. and Lilley, M. D. (2004). Diffuse flow hydrothermal fluids from 9°50′ N East Pacific Rise: Origin, evolution and biologeochemical controls. In Wilcock, W. D., Kelley, D. S., Baross, J. A., DeLong, E. and Cary, C. (eds.), *The Subseafloor Biosphere at Mid-Ocean Ridges*. Geophysical Monograph 144, Washington, DC: American Geophysical Union, pp. 245–268.

Von Damm, K. L., Bray, A. M., Buttermore, L. G. and Oosting, S. E. (1998). The geochemical controls on vent fluids from the Lucky Strike vent field, Mid-Atlantic Ridge. *Earth and Planetary Science Letters*, 160, 521–536.

Von Damm, K. L., Buttermore, L. G., Oosting, S. E. *et al.* (1997). Direct observation of the evolution of a seafloor 'black smoker' from vapor to brine. *Earth and Planetary Science Letters*, 149, 101–111.

Von Damm, K. L., Edmond, J. M., Measures, C. I. and Grant, B. (1985). Chemistry of submarine hydrothermal solutions at Guaymas Basin, Gulf of California. *Geochimica et Cosmochimica Acta*, 49, 2221–2237.

Von Damm, K. L., Lilley, M. D., Shanks, W. C. III *et al.* (2003). Extraordinary phase separation and segregation in vent fluids from the southern East Pacific Rise. *Earth and Planetary Science Letters*, 206, 365–378.

Von Damm, K. L., Oosting, S. E., Kozlowski, R. *et al.* (1995). Evolution of East Pacific Rise hydrothermal vent fluids following a volcanic eruption. *Nature*, 375, 47–50.

Von Damm, K. L., Parker, C. M., Zierenberg, R. A. *et al.* (2005). The Escanaba Trough, Gorda Ridge hydrothermal system: Temporal stability and subseafloor complexity. *Geochimica et Cosmochimica Acta*, 69, 4971–4989.

Vrijenhoek, R. C. (2010). Genetic diversity and connectivity of deep-sea hydrothermal vent metapopulations. *Molecular Ecology*, 19, 4391–4411.

Wager, L. R. and Brown, G. M. (1968). *Layered Igneous Rocks*. Edinburgh: Oliver and Boyd.

Wager, L. R., Brown, G. M. and Wadsworth, W. J. (1960). Types of igneous cumulates. *Journal of Petrology*, 1(1), 73–85.

Walker, G. P. L. (1986). Koolau dike complex, Oahu: Intensity and origin of a sheeted dike complex high in a Hawaiian volcanic edifice. *Geology*, 14, 310–313.

Walker, G. P. L. (1993). Basaltic-volcano systems. In Prichard, H. M., Alabaster, T., Harris, N. B. W. and Neary, C. R. (eds.), *Magmatic Processes and Plate Tectonics*. London: Blackwell Scientific, pp. 3–38.

Waller, R. G., Scanlon, K. M. and Robinson, L. F. (2011). Cold-water coral distributions from towed camera observations: Initial interpretations. *PLoS ONE*, 6(1), doi:10.1371/journal.pone.0016153.

Wang. T., Tucholke, B. E. and Lin, J. (2014). Spatial and temporal variations in crustal production at the Mid-Atlantic Ridge, 25°N–27°30′ N and 0–28 Ma. *Geochemistry, Geophysics, Geosystems* (submitted).

Wanless, V. D. and Shaw, A. M. (2012). Lower crustal crystallization and melt evolution at mid-ocean ridges. *Nature Geoscience*, 5(9), 651–655.

Wanless, V. D., Perfit, M. R., Klein, E. M., White, S. and Ridley, W. I. (2012). Reconciling geochemical and geophysical observations of magma supply and melt distribution at the 9°N overlapping spreading center, East Pacific Rise. *Geochemistry, Geophysics, Geosystems*, 13, Q11005, doi:10.1029/2012GC004168.

Wanless, V. D., Perfit, M. R., Ridley, W. I. and Klein, E. M. (2010). Dacite petrogenesis on mid-ocean ridges: Evidence for crustal melting and assimilation. *Journal of Petrology*, 51, 2377–2410.

Waters, C. L., Sims, K. W., Perfit, M. R., Blichert-Toft, J. and Blusztajn, J. (2011). Perspective on the genesis of E-MORB from chemical and isotopic heterogeneity at 9–10° N East Pacific Rise. *Journal of Petrology*, 52, 565–602.

Weekly, R. T., Wilcock, W. S. D., Hooft, E. E. E. *et al.* (2013). Termination of a 6 year ridge-spreading event observed using a seafloor seismic network on the Endeavour Segment, Juan de Fuca Ridge. *Geochemistry,*

Geophysics, Geosystems, 14, 1375–1398, doi:10.1002/ggge.20105.

Wegener, A. (1912). Die Entstehung der Kontinente. *Geologisches Rundschau*, 3, 276–292.

Welhan, J. K., Simoneit, B. R. T. and Tarafa, M. (1988). C_1–C_8 hydrocarbons in sediments from Guaymas Basin, Gulf of California: Comparison to Peru Margin, Japan Trench and California Borderlands. *Organic Geochemistry*, 12, 171–194.

Welhan, J. A. (1988). Origins of methane in hydrothermal systems. *Chemical Geology*, 71, 183–198.

Wendt, J. I., Regelous, M., Niu, Y., Hékinian, R. and Collerson, K. D. (1999). Geochemistry of lavas from the Garrett Transform Fault: Insights into mantle heterogeneity beneath the eastern Pacific. *Earth and Planetary Science Letters*, 173, 271–284

Wentworth, C. K. and Macdonald, G. A. (1953). Structures and forms of basaltic rocks in Hawai'i. *U.S. Geological Survey Bulletin*, 994, 1–98.

West, M., Menke, W., Tolstoy, M., Webb, S. and Sohn, R. (2001). Magma storage beneath Axial Volcano on the Juan de Fuca mid-ocean ridge. *Nature*, 413, 833–836.

Wetzel, L. R., Weins, D. A. and Kleinrock, M. C. (1993). Evidence from earthquakes for bookshelf faulting at large non-transform ridge offsets. *Nature*, 362, 235–237.

Wheat, C. G. and Fisher, A. T. (2007). Seawater recharge along an eastern bounding fault in Middle Valley, northern Juan de Fuca Ridge. *Geophysical Research Letters*, 34, L20602, doi:10.1029/2007GL031347.

White, R. S. and McKenzie, D. P. (1989). Magmatism in rift zones: The generation of volcanic continental margins and flood basalts. *Journal of Geophysical Research*, 94, 7685–7729.

White, R. S., Bown, J. W. and Smallwood, J. R. (1995). The temperature of the Iceland plume and the origin of outward-propagating V-shaped ridges. *Journal of the Geological Society, London*, 152, 1039–1045.

White, R. S., McKenzie, D. P. and O'Nions, R. K. (1992). Oceanic crustal thickness from seismic measurements and rare earth element inversions. *Journal of Geophysical Research*, 97, 19,683–19,715.

White, R. S., Minshull, T. A., Bickle, M. J. and Robinson, C. J. (2001). Melt generation at very slow-spreading ocean ridges: Constraints from geochemical and geophysical data. *Journal of Petrology*, 42, 1171–1196.

White, S. M., Macdonald, K. C. and Haymon, R. M. (2000). Basaltic lava domes, lava lakes, and volcanic segmentation on the southern East Pacific Rise. *Journal of Geophysical Research*, 105, 23,591–23,536.

White, S. M., Mason, J. L., Macdonald, K. C. *et al.* (2009). Significance of widespread low effusion rate eruptions over the past two million years for delivery of magma to the overlapping spreading centers at 9°N East Pacific Rise. *Earth and Planetary Science Letters*, 280, 175–184, doi:10.1016/j.epsl.2009.01.030.

White, S. M., Meyer, J. D., Haymond, R. M. *et al.* (2008). High-resolution surveys along the hotspot-affected Galápagos Spreading Center: 2. Influence of magma supply on volcanic morphology. *Geochemistry, Geophysics, Geosystems*, 9, Q09004, doi:10.1029/2008GC002036.

White, S. N., Chave, A. D. and Reynolds, G. T. (2002). Investigations of ambient light emission at deep-sea hydrothermal vents. *Journal of Geophysical Research*, 107, doi:10.1029/2000JB000015.

White, W. M. and Klein, E. M. (2014). Composition of the oceanic crust. In Holland, H. D. and Turekian, K. K. (eds.), *Treatise on Geochemistry (Second Edition)*. Oxford: Elsevier, pp. 457–496, doi:org/10.1016/B978-0-08-095975-7.00315-6.

Wilcock, W. S. D. (2004). Physical response of mid-ocean ridge hydrothermal systems to local earthquakes. *Geochemistry, Geophysics, Geosystems*, 5(11), Q11009, doi.org/10.1029/2004GC000701.

Wilcock, W. S. D. and Delaney, J. R. (1996). Mid-ocean ridge sulfide deposits: Evidence for heat extraction from magma chambers or cracking fronts? *Earth and Planetary Science Letters*, 145, 49–64.

Wilcock, W. S. D. and Fisher, A. T. (2004). Geophysical constraints on the subseafloor environment near mid-ocean ridges. In Wilcock, W. D., Kelley, D. S., Baross, J. A., DeLong, E. and Cary, C. (eds.), *The Subseafloor Biosphere at Mid-Ocean Ridges*. Geophysical Monograph 144, Washington, DC: American Geophysical Union, pp. 51–74.

Wilcock, W. S. D., Archer, S. D. and Purdy, G. M. (2002). Microearthquakes on the Endeavour segment of the Juan de Fuca Ridge. *Journal of Geophysical Research*, 107 (B12), 2336, doi:org/10.1029/2001JB000505.

Wilcock, W. S. D., DeLong, E. D., Kelley, D. S., Baross, J. A. and Cary, S. C. (2004). *The Subseafloor Biosphere at Mid-Ocean Ridges*. Geophysical Monograph 144, Washington DC: American Geophysical Union.

Wilcock, W. S. D., Hooft, E. E., McGill, P. R. *et al.* (2007). Microearthquakes beneath the Endeavour hydrothermal vent fields: Insights into reaction zone processes. *Eos, Transactions of the American Geophysical Union*, 88 (52), Fall Meeting Supplement, Abstract S13F-02.

Wilcock, W. S. D., Hooft, E. E. E., Toomey D. R. *et al.* (2009). The role of magma injection in localizing black smoker activity. *Nature Geoscience*, 2, 509–513.

Williams, D. L., Becker, K., Lawver, L. A. and Von Herzen, R. P. (1979). Heat flow at the spreading centers of the Guaymas Basin, Gulf of California. *Journal of Geophysical Research*, 84, 6757–6796.

Wilson, D. S., Teagle, D. A. H., Alt, J. A. *et al.* (2006). Drilling to gabbro in intact oceanic crust. *Science*, 312, 1016–1020.

Wilson, J. T. (1965). A new class of faults and their bearing on continental drift. *Nature*, 207, 343–347.

Wilson, L. and Head, J. W. (1981). Ascent and eruption of basaltic magma on the Earth and Moon. *Journal of Geophysical Research*, 86(B4), 2971–3001.

Wilson, L. and Head, J. W. (2007). Explosive volcanic eruptions on Mars: Tephra and accretionary lapilli formation, dispersal and recognition in the geologic record. *Journal of Volcanology and Geothermal Research*, 163, 83–97.

Wilson, L. and Parfitt, E. A. (1993). The formation of perched lava ponds on basaltic volcanoes: The influence of flow geometry on cooling-limited lava flow lengths. *Journal of Volcanology and Geothermal Research*, 56(1), 113–123.

Winn, C. D., Karl, D. M. and Massoth, G. J. (1986). Microorganisms in deep-sea hydrothermal plumes. *Nature*, 320, 644–746.

Wirsen, C. O., Tuttle, J. H. and Jannasch. H. W. (1986). Activities of sulfur-oxidizing bacteria at the 21°N East Pacific Rise vent site. *Marine Biology*, 92, 449–456.

Wolfe, C. J. and Solomon, S. C. (1998). Shear-wave splitting and implications for mantle flow beneath the MELT region of the East Pacific Rise. *Science*, 280, 1230–1232.

Wolfe, C. J., Purdy, G. M., Toomey, D. R. and Solomon, S. C. (1995). Microearthquake characteristics and crustal velocity structure at 29°N on the Mid-Atlantic Ridge: The architecture of a slow spreading segment. *Journal of Geophysical Research*, 100, 24,449–24,472.

Workman, R. and Hart, S. R. (2005). Major and trace element composition of the depleted MORB mantle (DMM). *Earth and Planetary Science Letters*, 321, 53–72.

Wright, T. J., Ebinger, C. J., Biggs, J. *et al.* (2006). Magma-maintained rift segmentation at continental rupture in the 2005 Afar dyking episode. *Nature*, 442, 291–294, doi:10.1038/nature04978.

Wyllie, P. J. (ed.) (1967). *Ultramafic and Related Rocks*. New York: John Wiley and Sons.

Yang, H. J., Kinzler, R. J. and Grove, T. L. (1996). Experiments and models of anhydrous, basaltic olivine-plagioclase-augite saturated melts from 0.001 to 10 kbar. *Contributions to Mineralogy and Petrology*, 124(1), 1–18.

Yaouancq, G. and MacLeod, C. J. (2000). Petrofabric investigation of gabbros from the Oman Ophiolite: Comparison between AMS and rock fabric. *Marine Geophysical Researches*, 21, 289–305.

Yeo, I. A. and Searle, R. C. (2013). High resolution ROV mapping of a slow-spreading ridge: Mid-Atlantic Ridge 45°N. *Geochemistry, Geophysics, Geosystems*, 14, doi:10.1002/ggge.20082.

Yeo, I., Searle, R. C., Achenbach, K., Lebas, T. and Murton, B. J. (2012). Eruptive hummocks: Building blocks of the upper ocean crust. *Geology*, 40, 91–94, doi:10.1130/G31892.

Yoerger, D. R., Jakuba, M., Bradley, A. M. and Bingham, B. (2007). Techniques for deep sea near bottom survey using an autonomous underwater vehicle. *International Journal of Robotics Research*, 26(1), 41–54.

Yoerger, D. R., Kelley, D. S. and Delaney, J. R. (1999). Fine-scale three dimensional mapping and imaging of the Mothra hydrothermal field, Juan de Fuca Ridge, *Eos, Transactions of the American Geophysical Union*, 80, 85.

Young, K. D., Jancin, M., Voight, B. and Orkan, N. I. (1985). Transform deformation of Tertiary rocks along the Tjörnes Fracture Zone, north central Iceland. *Journal of Geophysical Research*, 90, 9986–10,010.

Zhao, M., Canales, J. P., Sohn, R. A. (2012). Three-dimensional seismic structure of a Mid-Atlantic Ridge segment characterized by active detachment faulting (Trans-Atlantic Geotraverse, 25°55′ N-26°20′ N. *Geochemistry, Geophysics, Geosystems*, 13, Q0AG13, doi:10.1029/2012GC004454.

Zhu, W., Tivey, M. K., Gittings, H. and Craddock, P. R. (2007). Permeability–porosity relationships in seafloor vent deposits: Dependence on pore evolution processes. *Journal of Geophysical Research Letters*, 112, B05208, doi:10.1029/2006JB004716.

Zierenberg, R. A., Koski, R. A., Morton, J. L. and Bouse, R. M. (1993). Genesis of massive sulfide deposits on a sediment-covered spreading center, Escanaba Trough, southern Gorda Ridge. *Economic Geology*, 88(8), 2069–2098, doi:10.2113/gsecongeo.88.8.2069.

Zierenberg, R. A., Shanks, W. C. III, Seyfried, W. E., Jr., Koski, R. A. and Strickler, M. D. (1988). Mineralization, alteration, and hydrothermal metamorphism of the ophiolite-hosted Turner-Albright sulfide deposit, southwestern Oregon. *Journal of Geophysical Research*, 93(B5), 4657–4674, doi:10.1029/JB093iB05p04657.

Zonenshain, L. P., Kuzmin, M. I., Lisitsin, A. P., Bogdanov, Y. A. and Baranov, B. V. (1989). Tectonics of the Mid-Atlantic Rift Valley between the TAG and MARK Areas (26°–24°N): Evidence for vertical tectonism. *Tectonophysics*, 159, 1–23.

Index

INDEX